MW00763984

Synergism and Antagonism

in Chemotherapy

Synergism and Antagonism in Chemotherapy

Edited by

TING-CHAO CHOU

Memorial Sloan-Kettering Cancer Center
New York, New York

DARRYL C. RIDEOUT

The Research Institute of Scripps Clinic
La Jolla, California

ACADEMIC PRESS, INC.
Harcourt Brace Jovanovich, Publishers
San Diego   New York   Boston   London
Sydney   Tokyo   Toronto

This book is printed on acid-free paper. ∞

Copyright © 1991 by Academic Press, Inc.
All Rights Reserved.
No part of this publication may be reproduced or transmitted in any
form or by any means, electronic or mechanical, including photo-
copy, recording, or any information storage and retrieval system,
without permission in writing from the publisher.

Academic Press, Inc.
San Diego, California 92101

United Kingdom Edition published by
Academic Press Limited
24–28 Oval Road, London NW1 7DX

Library of Congress Cataloging-in-Publication Data

Synergism and antagonism in chemotherapy / edited by Ting-Chao
Chou, Darryl C. Rideout.
        p.   cm.
    Includes bibliographical references.
    Includes index.
    ISBN 0-12-174090-0   (alk. paper)
    1. Drug synergism. 2. Drug antagonism. 3. Chemotherapy,
  Combination.   I. Chou, Ting-Chao. II. Rideout, Darryl C., Date.
    [DNLM: 1. Drug Antagonism. 2. Drug Synergism. 3. Drug
Therapy, Combination.   QV 38 S998]
    RM302.3.S96   1991
    615.5'8—dc20
    DNLM/DLC
    for Library of Congress   90-898
                                                            CIP

Printed in the United States of America
91 92 93 94   9 8 7 6 5 4 3 2 1

Contents

PART I
Reviews and Methods of Quantitation

PART II

Mechanisms of Interaction

PART III

Condition-Selective Synergism and Antagonism

Contributors

Numbers in parenthesis indicate the pages on which the authors' contributions begin.

Avramis, V. I. (585) Division of Hematology–Oncology, Children's Hospital of Los Angeles; and Departments of Pediatrics, Pharmacy, and Biochemistry; University of Southern California, P.O. Box 54700; Los Angeles, CA 90054-0070

Berman, E. (715) Department of Medicine; Leukemia/Lymphoma Service; Memorial Sloan-Kettering Cancer Center; New York, NY 10021

Bertino, J. R. (449) Program of Molecular Pharmacology and Therapeutics; Memorial Sloan-Kettering Cancer Center and Cornell University Graduate School of Medicine; New York, NY 10021

Calogeropoulou, T. (507) Proteos 32 and Sirinon; N. Kifissia; Athens, GR-14564; Greece

Chang, B. K. (623) Department of Medicine; Hematology/Oncology Section; Medical College of Georgia; Augusta, GA 30912

Chou, J. H. (223) [1]Departmentof Biochemistry and Molecular Biology; Harvard University; Cambridge, MA 02138

Chou, T.-C. (3, 61) Laboratory of Biochemical Pharmacology; Memorial Sloan-Kettering Cancer Center and Cornell University Graduate School of Medical Sciences; 1275 York Ave. New York, NY 10021

Chang, T.-T. (715) Department of Pediatrics; Kaohsiung Medical College; Kaohsiung, Taiwan; Republic of China

Durand, R. E. (659) Medical Biophysics Unit, B. C. Cancer Research Centre; 601 W. 10th Avenue; Vancouver, B. C., Canada V5Z1L3

Eaton, J. W. (183) Department of Laboratory Medicine and Pathology; University of Minnesota; Minneapolis, Minnesota 55455

Eder, J. P. (541) Beth Israel Hospital; 330 Brookline Avenue; Boston, MA 02215

Ellis, W. Y. (183) Division of Experimental Therapeutics; Walter Reed Army Institute of Research; Washington, D. C. 20307-5100

Frei, E. III. Dana-Farber Cancer Institute and Harvard Medical School; 44 Binney St. Boston, MA 02115

[1]Present Address: M.D.–Ph.D. Program; University of California at San Francisco; San Francisco, CA 94143

Friedman, H. S. (245) Department of Pediatrics; Duke University Medical Center; Durham, North Carolina 27710

Fyfe, J. A. (285) Division of Experimental Therapy; Wellcome Research Laboratories; Research Triangle Park, North Carolina 27709

Galivan, J. (339) Wadsworth Center for Laboratories and Research; New York State Department of Health; Albany, New York 12201-0509

Greenberger, L. M. (311) Departments of Molecular Pharmacology and Cell Biology; Albert Einstein College of Medicine; Bronx, N.Y. 10461

Griffith, O.W. (245) Department of Biochemistry; Cornell University Medical Center; New York, New York 10021

Herman, T. S. (541) Dana-Faber Cancer Institute; 44 Binney Street; Boston, MA 02115

Holcenberg J. S. (585) Division of Hematology–Oncology; Children Hospital of Los Angeles; Departments of Pediatrics, Pharmacy and Biochemistry; University of Southern California, P.O. Box 54700, Los Angeles, CA 90054-0700

Horwitz, S. B. (311) Departments of Molecular Pharmacology and Cell Biology; Albert Einstein College of Medicine; Bronx, N.Y. 10461

Huang, S.-H. (585) Division of Hematology–Oncology; Children Hospital of Los Angeles; Departments of Pediatrics, Pharmacy and Biochemistry; University of Southern California; P. O. Box 54700, Los Angeles, CA 90054-0700

Jackson, R. C. (363) [2]Parke-Davis Pharmaceutical Research Division; Warner-Lambert Company; 2800 Plymouth Road; Ann Arbor, Michigan 48105

Keng, P. C. (689) Experimental Therapeutics Division; University of Rochester Cancer Center, 601 Elmwood Avenue, Box 704 Rochester, N.Y. 14642

Milhous, W. K. (183) Division of Experimental Therapeutics; Walter Reed Army Institute of Research; Washington, D.C. 20307-5100

Mini, E. (449) Dipartimento di Farmacologia; Preclinica e Clinica; Università degli Studi; Firenze, 50134 Italy

Rideout, D. C. (3, 507) Department of Molecular Biology; The Research Institute of Scripps Clinic; 10666 N. Torrey Pines Rd., La Jolla, CA 92037

Schinazi, R. F. (109) Emory University School of Medicine; Department of Pediatrics; Laboratory of Biochemical Pharmacology and Veterans Affairs Medical Center; 1670 Clairmont Road; Decatur, GA 30033

Siemann, D. W. (689) Experimental Therapeutics Division; University of Rochester Cancer Center; 601 Elmwood Avenue, Box 704; Rochester, NY 14642

Spector, T. (285) Division of Experimental Therapy; Wellcome Research Laboratories; Research Triangle Park; North Carolina 27709

Strekowski, L. (409) Department of Chemistry; Georgia State University; Atlanta, GA 30303

Teicher, B. A. (541) Dana-Farber Cancer Institute; 44 Binney Street; Boston, MA 02115

Vennerstrom, J. L. (183) College of Pharmacy; University of Nebraska Medical Center; Omaha, Nebraska 68198

Wilson, W. D. (409) Department of Chemistry; Georgia State University; Atlanta, GA 30303

Yang, C.-P. H. (311) Departments of Molecular Phamacology and Cel Biology; Albert Einstein College of Medicine; Bronx, N.Y. 10461

[2]Present Address: Dupont Medical Products Dept., 500 S. Ridgeway Ave., Glenolden, PA 19036

Preface

Synergism and Antagonism in Chemotherapy provides a practical, mechanism-oriented overview of chemotherapeutic drug combinations that have either achieved prominence in the treatment of human diseases, are currently being evaluated in clinical trials, or are under investigation in experimental model systems. Although the main emphasis is on cancer chemotherapy, the chemotherapy of antiviral and antimicrobial diseases is also presented in detail. The book should find a place on the reading lists of medical school curricula and graduate programs in pharmacology, chemistry, and biochemistry. It should be particularly useful for the research specialist involved in the design of new multi-agent strategies for the treatment of cancer and infectious diseases.

An introductory chapter provides an integrated overview of the many diverse aspects of drug interactions, including quantification of synergism and antagonism, mechanisms of interaction, and temperospatial aspects of synergism and antagonism. The many varieties of interactions are categorized in a way that will provide guidance to researchers who have discovered new examples of synergism or antagonism and wish to elucidate the mechanisms involved. These subjects are then dealt with in greater detail in chapters that address the physicochemical basis of chemotherapeutic interactions at molecular, cellular, and whole-animal levels. Contributions are included that deal with experimental design of drug combination studies and computer-aided data analysis. There are reviews of antimalarial, antineoplastic, and antiviral combinations, with additional chapters covering the topics of multiple drug resistance, fluorouracil combinations, inhibition of DNA repair, inhibition of multiple metabolic pathways, antifolate combinations, polymerase-targeted antiviral therapy, and inhibition of glutathione synthesis. Here, the reader will also find unique and timely treatments of drug interactions involving genetic changes, direct bond formation between two agents, and multiple drug binding to a single target. Four chapters are devoted to condition-dependent interactions: the influence of sequence and scheduling, tumor cell microenvironment, and cell phenotype on the degree of synergism between antineoplastic drugs.

Ting-Chao Chou
Darryl Rideout

PART I

Reviews and Methods
of Quantitation

CHAPTER 1

Synergism, Antagonism, and Potentiation in Chemotherapy: An Overview

Darryl C. Rideout
Department of Molecular Biology
The Research Institute of Scripps Clinic
La Jolla, California 92037

Ting-Chao Chou
Laboratory of Biochemical
Pharmacology
Memorial Sloan-Kettering
Cancer Center
and Cornell University
Graduate School of
Medical Sciences
New York, New York 10021

SYNERGISM AND ANTAGONISM IN CHEMOTHERAPY
Copyright © 1991 by Academic Press, Inc.
All rights of reproduction in any form reserved.

Chemotherapeutic agents are very often administered in combination with other drugs in the treatment of cancer, microbial infections and viral infections. The purpose of use of drugs in combination is to achieve therapeutic effects that are more favorable than using a single drug alone. The benefits may include increased therapeutic efficacy, decreased toxicity toward the host or nontarget tissues, increased selectivity of effect or therapeutic index, and minimization of or delay in development of drug resistance. By virtue of drug combination, the multiple-target approach is an effective way to achieve a therapeutic end point in heterogeneous cell populations or disease types.

An understanding of mechanisms of interaction between agents and an ability to measure and quantify such interactions in a meaningful way is important because of the influence such interactions can have on the success of clinical chemotherapy. The study of drug interactions can lead to a better understanding of mechanisms of drug action and drug resistance, in addition to providing insight into basic biological questions. An understanding of drug interactions can be useful in the design of novel individual agents as well as novel combinations of agents with potential clinical utility.

We wish to address the following questions in this chapter. How do we define and quantitatively measure synergism, antagonism, additivity, and potentiation? What mechanisms of synergism and antagonism have been proposed? How can one design experiments to support or rule out various possible mechanisms of interaction? How are these mechanisms relevant to clinical drug interactions? How can they best be exploited in the design of combination chemotherapy protocols and new drug combinations? The study of drug combinations is still effectively in its infancy, and these questions are often difficult to answer.

There have been a number of reviews that deal with synergism, potentiation, and antagonism. A selection from the 1983–1989 period includes those by Allan (1987), Brook (1987), Brown *et al.* (1988), Marget and Seeliger (1988), Eliopoulos (1987), Bonnem (1987), Hait (1987), Tyring (1987), Holm (1986), Giamarellou (1986), Calandra and Glauser (1986), Klastersky (1986), Neu (1989), Gill *et al.* (1989), Mitchell (1989), Rosenthal and Hait (1988), Coune (1988), Allan and Moellering (1985), Mayer and DeTorres (1985), Simpson (1985), Berenbaum (1989), Siemann (1985), Rahal (1983), Moellering (1983), Ling *et al.* (1983, 1984), Schabel *et al.* (1983), Kamataki *et al.* (1989), Fox and Roberts (1987), Peterson *et al.* (1985), Goldie (1983), Tsuruo (1983), Helson (1984), Ozols and Cowan (1986), Fojo *et al.* (1987), Klubes and Leyland-Jones (1989), Zaniboni (1989), Poon *et al.* (1989), Peters (1988), Kanamura *et al.* (1988), Borch and Markman (1989), O'Dwyer *et al.* (1987), Leyland-Jones and O'Dwyer (1986), Martin *et al.* (1985), and Kahl (1984). Although this list is not comprehensive, it should provide the reader with a reasonably complete picture of the current state of drug interaction studies when combined with the contents of this book.

Knowledge of alternative terminology is useful when searching for articles concerning synergism and antagonism using key words or title words. The distinctions between these terms are not trivial in terms of literature searches. For example, the number of references to the key word "synergism" differs dramatically from the number of references to "synergy". Synergism and potentiation are closely related to synergy, supraadditivity, therapeutic synergy, chemoenhancement, and enhancement. Collateral sensitivity and cocarcinogenesis are related to synergism and potentiation, whereas rescue, chemoprotection, and cross resistance are related to antagonism. Other useful terms include biochemical modulation, sequential blocking, two-route chemotherapy, drug resistance, combination chemotherapy, drug interactions, multidrug therapy, and pharmacological modulation. Chemical Abstracts Service has recently begun publishing biweekly selected abstracts (CA Selects®) on the topic of drug interactions (#SVC 075).

Many reviews on synergism/antagonism in drug combinations published prior to the mid-1980s have used the empirical approach (Ashford and Cobby, 1974; Berenbaum, 1977, 1980, 1981; Bliss, 1939; Goldin and Mantel, 1957; Goldin *et al.*, 1968; Loewe, 1928, 1953; Rothman, 1976; Scott, 1983). Due to lack of consensus on definition of terminology (for example, what is "synergy"?), an unambiguous discussion of this subject has been difficult. As shown in a literature survey conducted by Schinazi (Chapter 4) on antiviral drug combination studies, many different laboratories have used different interpretations or definitions of synergism or antagonism. The basic reason for this diversified interpretation of data is the lack of definition of the "additive effect" since, if the expected additive effect is not defined, synergism or antagonism cannot be evaluated quantitatively. The empirical approach often leads to considerable confusion since no physicochemical model is used and there is no actual derivation of equations in the treatment of the problem. Obviously, different approaches, different assumptions, and different methods for dose effect analysis used may lead to different interpretation or different conclusions on synergism/antagonism (Poch, 1981; Chou and Talalay, 1983, 1984, 1987). If the dose–effect relationship of a single drug alone is treated empirically, the combination of multiple drugs compounds the complexity, resulting in uncertainty. For example, although the formulas of probit (Bliss, 1939, 1967; Finney, 1952), logit (Reed and Berkson, 1939), or simple power function adequately linearize dose–effect relationships for the single drugs or their mixtures, the parameters of their plots have no physicochemical bearings. Using a fundamental natural principle as a guide, mass-action law has been selected as a model for dose effect analysis in order to obtain a generalized methodology that has sound theoretical basis and is easy to use (Chou, 1974, 1977; Chou and Talalay, 1977, 1981). Although many researchers have used this principle for various purposes in biochemistry and pharmacology (Cleland, 1963; Segel, 1975; Webb, 1963; Yagi and Ozawa, 1960; Yonetani and Theorell, 1964), the conclusions obtained often have been for specific mechanisms. A systematic analysis conducted by Chou (1974, 1976, 1977) and Chou and Talalay (1977, 1981) led to the deduction of only a few relatively general equations for dose–effect relationship from hundreds of equations derived. These equations contain no kinetic constants, are mechanism nonspecific, and have been shown to have a broad spectrum of applicability (see Chapter 2). Many researchers who have used empirical formulae to fit the experimental data in the past have begun to adopt mass-action law derived equations as a more meaningful and simple approach. It becomes more and more clear that, for the empirical approach, due to its lack of an underlying physical model, the actual derivation of equations is not possible.

In view of the complexity of biological systems, no method for dose effect analysis is considered perfect and universal. For example, the Michaelis–Menten equation is not valid for allosteric enzymes which exhibit higher order kinetics, and the Hill equation, although useful, does not take into account the minor intermediate species in the interactions. However, we cannot discredit their value just because of these deficiencies. The median-effect equation and the multiple-drug-effect equation, which are also based on the mass-action law but which involve inhibitors instead of substrates, will obviously face similar limitations. Therefore, in dose effect analysis the conformity test is routinely carried out prior to its application. The diagnostic test and the subsequent analysis can be carried out manually; however, it is more easily done with the aid of a microcomputer. So far numerous examples of enzyme, receptor, cellular, and animal studies support the general applicability of the mass-action approach using the median-effect principle (see Chapter 2).

I. Drug Combinations in Chemotherapy

A. OUTCOMES OF DRUG COMBINATIONS

The uses of the drug combination terminology and methods of dose effect analysis are discussed here based on modes of action (e.g., inhibitory, stimulatory, and modulative).

1. Combination of Inhibitory Drugs

The most common strategy used in chemotherapy involves the combination of two or more inhibitory drugs. For example, clinical protocols in cancer chemotherapy rarely use a single drug but use multiple drugs that affect macromolecule synthesis, interfere with metabolic pathways, and/or form covalent bonds with biotargets. The outcomes of combining inhibitory drugs are

1. synergism (supraadditivity) and potentiation
2. additivism (summation)
3. antagonism (subadditivity)

When each drug alone exhibits effects, the combined effect is called synergism, additivism, or antagonism. However, if one of the drugs has no effect by itself but increases the effects of other drugs, the result is called potentiation. Synergism, additivism, and antagonism are determined by the combination index or isobologram, whereas potentiation is usually presented as the percentage or fold increase in effect as a result of administration of the potentiator.

The prerequisite for meaningful determination of synergism or antagonism is the clear (unambiguous) definition for additivism. As is well known, the additive effect cannot be the simple arithmatic addition of effects. For example, for two drugs which inhibit 55% and 65%, respectively, the combined additive effect cannot be 120%. Therefore, rigorous treatment of additivism with a mathematical formulation is warranted. (For potentiation, one of the two agents has no activity by itself, but enhances the activity of the second agent. Interpretation is thus more straightforward.) In the field of anticancer and antimicrobial chemotherapy, many biomedical researchers are pursuing synergism. According to the Permutation Index of Science Citation Index (Institute for Scientific Information Service, Philadelphia), the terms synergism and synergy are cited over 1500 times annually. Unfortunately, the definition of additivism is rarely clarified. In one review alone (Goldin and Mantel, 1957), seven different definitions of synergism are given but none is supported by the others. The confusion in this area has reached crisis proportions since the same set of data subjected to different methods for dose effect analysis may reach different conclusions in terms of synergism or antagonism. Given the widespread use of this terminology, the confusion generated by the indiscriminate use of methods should have far reaching consequences in medical research and practice. The issues concerning the nature of dose–effect relationships, the parameters involved, and the definition of additivism will be discussed in detail in Chapter 2, taking into account the mass-action law principle, enzyme kinetic models, and pharmacological receptor theory.

2. Combination of Stimulatory and Inhibitory Drugs

Stimulatory drugs inducing pharmacological response are called agonists, whereas inhibitory drugs suppressing pharmacological response are called antagonists. Because agonists and antagonists exert opposite effects there is no obvious reason for their use in combination for general therapeutic purposes. However, rescues of overdoses caused by intentional or unintentional intoxication or side effects involve the use of agonists and antagonists. For example, a high dose of methotrexate which kills off all or most cancer cells can be followed by a dose of leucovorin which will rescue some intoxicated normal cells; overdose of insulin is treated with the antagonist glucagon; and trimetrexate and leucovorin exhibit host-selective antagonism in the treatment of *Pneumocystis carinii* in AIDS patients. In some common cold medications, an analgesic, an antihistamine, and a stimulant such as caffeine or an ephedrine analog are combined. While the antihistamine relieves the symptoms of the cold, it also induces undesirable drowsiness that can be partially counteracted by stimulation of the central nervous system. More frequently, agonists and antagonists are used in mechanistic investigations in which the stimulatory compound serves as primary ligand and the inhibitory compound serves as the reference ligand or probe. Reversal of the inhibitory effect induced by an antagonistic ligand (I) with a primary stimulatory or a substrate ligand (S) or vice versa provides a clue to the nature of interaction between the effector and the target receptor or the interaction between effectors.

When S concentration (or dose) is varied in the presence of different concentrations (or doses) of I, dose effect data (e.g., reaction velocity, induced response) may be evaluated with the steady-state mass-action law in different ways (Cleland, 1963; Segel, 1975).

If the dose–response curve of the primary ligand follows that of first-order relationship, i.e., the slope n of the Hill plot equals 1, the equation of Michaelis and Menten (1913)

$$v = \frac{SV}{S + K_m} \quad \text{or} \quad v = \frac{V}{1 + \dfrac{K_m}{S}} \quad \text{or} \quad \frac{v}{V} = \frac{1}{1 + \dfrac{K_m}{S}} \tag{1}$$

can be used, where v is the reaction velocity, V is the apparent maximal velocity or response, and K_m is the concentration or dose that produced one-half of the maximal response. Equation 1 has the following features.

1. A plot of $y = v$ vs $x = S$ yields a rectangular hyperbola.
2. A plot of $y = \log[v/(V - v)]$ vs $x = \log S$ (Hill plot) shows a slope (n) of 1.
3. A plot of $y = 1/v$ vs $x = 1/S$ (the double reciprocal plot or Lineweaver–Burk plot) gives a straight line with x intercept of $-1/K_m$, y intercept of $1/V$ and slope of K_m/V.
4. S is very much in excess over the available receptor binding site and a relatively small fraction of S is receptor bound.

When I and S are combined, they may show three types of inhibition.

Competitive Inhibition   This implies that I and S bind at the same site on the receptor. The dose response and the dose–effect relationship can be described by :

$$\frac{1}{v} = \frac{K_m}{V}\left(1 + \frac{I}{K_i}\right)\frac{1}{S} + \frac{1}{V} \tag{2}$$

For the double reciprocal plot, ($y = 1/v$ vs $x = 1/S$), the slope is now $(K_m/V)[1 + (I/K_i)]$ rather than K_m/V, taking into account the effect of the inhibitor. The intercept, $1/V$, is unchanged by the presence of the inhibitor.

Here,

    I: inhibitor concentration
    K_i: inhibition constant or enzyme inhibitor dissociation constant.

By studying the effects of I and S in combination, K_i can be calculated as follows:

$$\frac{1}{\dfrac{(\text{slope})_i}{(\text{slope})_0} - 1} = \frac{I}{\dfrac{\dfrac{K_m}{V}\left(1 + \dfrac{I}{K_i}\right)}{\dfrac{K_m}{V}} - 1} = \frac{I}{1 + \dfrac{I}{K_i} - 1} = K_i \tag{3}$$

where $(\text{slope})_i$ and $(\text{slope})_0$ are the slopes of the double-reciprocal plot(s) in the presence and absence of inhibitor, respectively.

    Noncompetitive Inhibition   This implies that I and S bind at different sites on the receptor or that I binds to the receptor nonspecifically. The dose response and the dose–effect relationship as shown in classical enzyme kinetic treatments can be described by:

$$\frac{1}{v} = \frac{K_m}{V}\left(1 + \frac{I}{K_i}\right)\frac{1}{S} + \frac{1}{V}\left(1 + \frac{I}{K_i}\right) \tag{4}$$

For the double reciprocal plot, $(y = 1/v \ vs \ x = 1/S)$, the slope is now $(K_m/V)[1 + (I/K_i]$ rather than K_m/V and the intercept is now $(1/V)[1 + (I/K_i)]$ rather than $1/V$, taking into account the effect of the inhibitor.

    Here, the calculation of K_{is} is the same as above and the calculation of K_{ii} (the K_i calculated from the intercept changes) is as shown here:

$$\frac{I}{\dfrac{(\text{int})_i}{(\text{int})_0} - 1} = \frac{I}{\dfrac{\dfrac{1}{V}\left(1 + \dfrac{I}{K_i}\right)}{\dfrac{1}{V}} - 1} = \frac{I}{1 + \dfrac{I}{K_i} - 1} = K_i \tag{5}$$

where $(\text{int})_i$ and $(\text{int})_0$ are the y intercepts of the double-reciprocal plots in the presence and absence of inhibitor, respectively.

    When $K_i = K_{is} = K_{ii}$, we see pure noncompetitive inhibition. In this case, $K_i = I_{50}$ as shown by Chou (1974).
    When $K_{is} \neq K_{ii}$, we see partial or mixed noncompetitive inhibition. In this case, $K_{is} > I_{50} > K_{ii}$ or $K_{ii} > I_{50} > K_{is}$ and I_{50} is the harmonic mean of K_{is} and K_{ii}

$$I_{50} = \frac{2K_{is}K_{ii}}{K_{is} + K_{ii}} \tag{6}$$

as indicated by Chou (1974).

When $K_{is} > I_{50} > K_{ii}$, the intercept of two plots (in the presence and absence of I) occurs below the X axis of the double reciprocal plot.
When $K_{is} < I_{50} < K_{ii}$, the intercept of two plots (in the presence and absence of I) occurs above the x axis of the double reciprocal plot.

Uncompetitive Inhibition This is relatively rare because it implies that in a single-substrate reaction I would not bind to the receptor unless the substrate binds to the receptor first. This type of situation occurs more frequently in multiple-substrate–multiple-product enzyme reactions with "ping-pong" mechanisms where the reaction sequence of substrate entry and product release is interrupted by an irreversible step(s) as indicated by Cleland (1963). For a simple system, the uncompetitive inhibition is described by:

$$\frac{1}{v} = \frac{K_m}{V}\frac{1}{S} + \frac{1}{V}\left(1 + \frac{I}{K_i}\right)$$

(7)

For the double reciprocal plot, ($y = 1/v$ vs $x = 1/S$), the slope K_m/V does not change but the intercept $1/V$ is now $(1/V)[1 + (I/K_i)]$ rather than $1/V$. The calculation of K_{ii} is the same as above.

Although Michaelis–Menten kinetics is derived from enzyme systems, the same mass-action law principle has been extended to receptor, cellular, and animal systems (see Chapter 2). As has been shown earlier, the median-effect equation derived by Chou (1977, 1980) can be readily used to derive the Michaelis–Menten (1913), Hill (1913), Scatchard (1949), and Henderson–Hasselbalch (Goldstein *et al.* 1968) equations.

3. Combination of Stimulatory Drugs

The stimulatory drugs such as substrates usually have high specificity. Selection of a single stimulatory drug with high efficacy (i.e., a drug which elicits a high maximal response) would usually suffice, making drug combinations unnecessary. In contrast, combination of inhibitor drugs (usually with diverse mechanisms) is a more logical choice for combination therapy.

Dose–Response Relationship and Parameters The dose–response relationships of the stimulatory drugs can be analyzed by the Michaelis–Menten equation (for the first order) or by the Hill equation (for the higher order). In contrast, the dose–effect relationships of the inhibitory drugs can be analyzed by the median-effect equation. The parameters for the stimulatory drugs are

K_m: The dose for half-maximal stimulatory effect or the dose for half-maximal saturation or occupancy of the receptor by the agonist ligand. K_m is a measurement of affinity. The lower the K_m concentration, the higher the affinity that is indicated.

V_{max}: The maximal stimulatory effect elicited by an agonist ligand. It is a measurement of efficacy.

n: The Hill coefficient or kinetic order. It signifies the cooperativity of effect among stimulatory ligands in relation to the receptor and it is also a measurement of the shape of the dose–response curve.

Dose Effect Parameters Distinct from the stimulatory drug or agonist, parameters for the inhibitory drug or antagonist (Chou, 1976) are

D_m: The median-effect dose (e.g., IC_{50}, ED_{50}, LD_{50}) signifying the potency of the drug.
m: The slope of the median-effect plot signifying the shape (sigmoidicity) of the dose–effect curves.

Although there is a V_{max} value for the stimulatory drug, there is no such restriction for the inhibitory drug since the effect is measured relative to an uninhibited control. The determination of V_{max} requires extrapolation and approximation whereas the determination of uninhibited control values is straightforward.

It is also of interest to note that different approaches have been used in deriving the median-effect equation (Chou, 1976; Chou and Talalay, 1981) as compared with the Michaelis–Menten or Hill equation (1913).

Equations for the Combination of Stimulatory Drugs The Michaelis–Menten equation and the Hill equation can be used for a two-drug combination as follows:

1. For the first order stimulatory drugs that have a hyperbolic dose–response curve, the combined response can be described by

$$v_{1,2} = \frac{V_1 S_1}{S_1 + K_{m_1}} + \frac{V_2 S_2}{S_2 + K_{m_2}} \tag{8}$$

The subscripts 1 and 2 indicate parameters or concentrations pertinent to drug 1 and drug 2, respectively.
2. For the higher order stimulatory drugs that have a sigmoidal dose–response curve, the combined response can be described by

$$v_{1,2} = \frac{V_1 (S_1)^n}{(S_1)^n + (K_{m_1})^n} + \frac{V_2 (S_2)^n}{(S_2)^n + (K_{m_2})^n} \tag{9}$$

The equations for the combination of inhibitory drugs are described in Chapter 2.

4. Combination of Effectors and Modifiers

There are situations in which drug 1 by itself has an effect or a response, drug 2 by itself has no effect or response, but the combination of these two drugs may result in an altered effect or response.

A. Drug 1 is a stimulator (agonist) and drug 2 is a modifier (modulator).
   1. If drug 1 response is increased by drug 2, then the effect of drug 2 is "potentiation," and drug 2 is a potentiator, an enhancer, or an up-regulator.
   2. If drug 1 response is decreased by drug 2, then the drug 2 effect is

inhibition, and drug 2 is an inhibitor, a suppressor, or a down-regulator.

B. Drug 1 is an antagonist and drug 2 is a modifier (modulator).

   1. If the inhibitory effect of drug 1 is increased by drug 2, then the effect of drug 2 is negative potentiation and drug 2 is an inhibition potentiator or an inhibition enhancer.

   2. If the inhibitory effect of drug 1 is decreased by drug 2, then the effect of drug 2 is alleviation and drug 2 is an alleviator or a rescuer. For example, methotrexate inhibits cancer cell growth and leucovorin serves as a rescuing agent.

The data analysis for the above four cases is simple and straightforward since one of the two drugs by itself has no dose–effect relationship parameters. We only need to present the results of how much of an increase or decrease of the response or effect of drug 1 is due to the presence of drug 2.

It is important to note that in situations in which both drug 1 and drug 2 by themselves have effects or responses, the outcomes of their combinations require more complicated calculations and should be presented in terms of synergism, additivism, or antagonism.

B. METHODS FOR DOSE EFFECT ANALYSIS OF SINGLE DRUGS

1. Empirical Curve-Fitting and Read-Off

This method (Berenbaum, 1989) is not based on any theory or principle. Rather, it relies on visual fitting of an empirical curve to a set of dose effect data. The dose required for a given effect or the effect produced by a given dose is then read from the graph by extrapolation or interpolation. To draw a reliable empirical curve to fit data usually requires a relatively large data mass and the interpolation usually relies heavily on the neighboring data points.

2. Linearity Method

Certain regions of a dose–effect curve such as low dose–low effect ranges may look linear on a simple arithmetic scale. It should be pointed out, however, that the linearity assumption contradicts the basic principle of the mass-action law in terms of pharmacological receptor theory or enzyme kinetics. As indicated by the median-effect equation or the Michaelis–Menten equation, no part of the dose–effect curve or the dose–response curve is actually linear. Thus, the linearity method for dose effect analysis or for low dose risk assessment has no justification.

3. Empirical Formula Method

Empirically or statistically conceived mathematical formulae are frequently referred to as models. These formulae, although they appear to linearize dose–effect or dose–response relationships, usually have no physicochemical bearing. The mathematical functions, such as the simple power law, probit (Finney, 1952), or logic (Reed and Berkson, 1939), do not have physical meaning in terms of their parameters such as constants, slope, and intercept for the plots when applied to simple, defined systems such as enzyme systems.

4. Method of the Median-Effect Principle

The median-effect equation was derived in 1974–1976 from the principle of the mass-action law using an enzyme kinetic system as a model (Chou, 1974, 1976). The equation is simple and describes the relationships between dose and effect regardless of the shape of the dose–effect curve (hyperbolic or sigmoidal). As has been shown earlier (Chou, 1977, 1980), the rearrangement of the median-effect equation automatically gives the Michaelis–Menten (1913), Hill (1913), Scatchard (1949), and Henderson–Hasselbalch (Goldstein *et al.*, 1968) equations. The median-effect equation and plot have been used in dose effect analysis in enzyme, receptor, cellular, and animal systems as will be illustrated in Chapter 2.

C. METHODS FOR QUANTITATION OF SYNERGISM AND ANTAGONISM

By definition, synergism is more than expected additivism and antagonism is less than expected additivism. Therefore, if the expected additive effect is not clearly defined, it will be difficult to discuss synergism or antagonism without confusion. In one review alone, seven definitions of synergism are listed (Goldin and Mantel, 1957). No refined guidelines were provided to indicate the situations in which each method would be appropriate.

Some methods that have been used in biomedical literature are presented here. Some of these methods are obviously wrong and others have various degrees of usefulness and limitations. Simple arithmetic addition of effects of two or more drugs is an erroneous way of calculating expected additive effects or concluding synergism/antagonism. Suppose that drug 1 and 2 each inhibit a system by 30%, and that the two drugs act additively (i.e., no synergism or antagonism). What is the expected additive effect? One may be tempted to say that the combined additive effect should be 60%, since we are talking about addition of effects. However, when the question is extended to a situation in which each drug alone inhibits 60%, the inhibition for the additive effect of 120% is physically impossible. When only a single dose of each drug is used for combination studies, the use of the simple arithmetic addition method is the most common error.

1. Fractional Product Method of Webb

Although the detailed derivation was not shown, Webb (1963) indicated that the combined effect of two inhibitors of an enzyme system can be calculated by $1 - i_{1,2} = (1 - i_1) (1 - i_2)$ where i_1, i_2, and $i_{1,2}$ are the fractional inhibition, of inhibitor 1, inhibitor 2, and their combination, respectively.

Therefore, if drug 1 and 2 each inhibit a system by 50%, the two drugs in combination, if additive, will inhibit 75%. This is calculated by $(1 - 0.5) \times (1 - 0.5) = 0.25$ and $1 - 0.25 = 0.75$. Thus if f_a and f_u are the fraction affected and the fraction unaffected, respectively, then $[1 - (f_a)_{1,2}] = [1 - (f_a)_1] \times [1 - (f_a)_2]$ or $(f_u)_{1,2} = (f_u)_1 (f_u)_2$, and $(f_a)_{1,2}$ will never be greater than one. This method is simple to use and seems to require only three data points (two for each drug alone and one for the combination) to determine synergism or antagonism, but is misused widely because of lack of understanding of its limitations. The major problem is that it does not take into account the shapes of the dose–effect curves. For example, if the shapes of dose–effect curves for drug 1 and drug 2 are hyperbolic and sigmoidal, respectively, a single dose of each drug alone will never reflect these facts. Thus, arbitrary use of the fractional product method will most likely yield an erroneous conclusion about synergism or antagonism.

More recent analysis with enzyme kinetic systems by Chou and Talalay (1977, 1981, 1984) indicated that the fractional product method is correct for the Michaelis–Menten-type first-order system but incorrect for higher order systems. In addition, the fractional product method is correct for mutually nonexclusive inhibitors (e.g., two inhibitors that have totally different or independent modes of actions), but not correct for mutually exclusive inhibitors (e.g., two inhibitors that have identical or similar modes of action). Thus, the fractional product method is not valid in three out of four different models described above and, therefore, suffers severe limitations in its usefulness in drug combination studies (Chou, 1987; Chou et al., 1991).

2. Classical Isobologram Method of Loewe

An isobologram is a graph that shows equipotency (equi–effect) doses for a given degree of effect. For example, ED_{50}-isobolograms show the doses that are required to inhibit 50% by drug 1 alone, drug 2 alone, and the various combinations of drug 1 and drug 2. In the classical ED_{50} isobologram, $(ED_{50})_{drug\,1}$ and $(ED_{50})_{drug\,2}$ are placed on x and y coordinates, respectively. If we connect $(ED_{50})_{drug\,1}$, and $(ED_{50})_{drug\,2}$ with a diagonal straight line, then

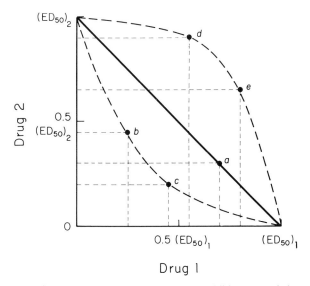

Drug 1

FIGURE 1   The classical ED_{50}-isobologram depicting the additive, synergistic, and antagonistic effects of two drugs. When an experimental combination data point falls on the hypotenuse, such as point *a*, the effects of two drugs are addictive. If experimental combination data points fall in the lower left, like *b* and *c*, then the effects of the two drugs are synergistic, if the data points fall in the upper right, like *d* and *e*, the effects of the two drugs are antagonistic. Points *b* and *c* require smaller doses of drug 1 and drug 2 to achieve 50% inhibition than point *a* but points *d* and *e* require larger doses of drug 1 and drug 2 to achieve the same effect.

this straight line depicts the isoeffective tracing (for inhibiting 50%) of inhibition by drug 1 and drug 2 at various combination mixtures as shown in Fig. 1.

The idea of isobolograms was conceived intuitively and has been in practice for about a century. Extensive studies of the isobologram were carried out by Loewe (1928, 1953), Finney (1952), Skipper (1974), Berenbaum (1977, 1981, 1985, 1988, 1989) and many others (Chou, 1987; Chou and Talalay, 1977, 1981, 1983, 1984, 1987; Elion *et al.*, 1954; Grindey and Nichol, 1972). Perhaps the most explicit applications of the classical isobologram to the experimental design of cytotoxic combinations are those described by Elion *et al.* (1954).

A logical inference for the classical isobologram that can be simply demonstrated is a sham test assuming that two drugs have the same identity. Imagine that one divides a solution of the test Compound A into two test tubes, labels them Drug 1 and Drug 2, and gives them double blinded to Individual 1 for bioassays. The experimental data are then given to Individual 2, who will determine whether Drug 1 and Drug 2 are synergistic, additive, or antagonistic. Naturally, Individual 1 would conduct an experiment by varying the concentrations of Drug 1, Drug 2, and their mixture (combination in constant or nonconstant ratios) and measure their effects. Individual 2 would use whatever means to construct an isobologram although an isobologram may not be the only way to analyze synergism, additivism, or antagonism. Suppose the measurements by Individual 1 are accurate and the construction of the isobologram by Individual 2 is done properly. We should expect that Drug 1 and Drug 2 are additive on the classical isobologram since they are, in fact, the same compound. They would also be mutually exclusive because they are, in fact, the same compound.

Because earlier studies did not use specific or well-defined models, the utility and limitations of isobolograms were not discussed to any appreciable extent. By systematic analysis of dose–effect relationships of single drugs by Chou (1974, 1976, 1977) and multiple drugs by Chou and Talalay (1977, 1981, 1984), general equations were obtained using enzyme kinetic models based on the mass-action law (see Chapter 2 for details).

In the multiple drug effect equation at a special case when $f_a = f_u = 0.5$, for the first time, the ED_{50}-isobologram equation was explicitly derived (Chou and Talalay, 1984). This classical isobologram equation was then extended to other effect levels (Chou and Talalay, 1984) as will be illustrated in the next subsection. Since the isobologram equations can be explicitly derived from the well-defined mass-action principle, their usefulness as well as limitations can be discussed clearly with more certainty than the empirical approach. As will be shown in Chapter 2, the classical isobologram equation correctly describes additive effects of two drugs when their effects are mutually exclusive (i.e., the same or similar modes of action). However, it is not applicable to two drugs when their effects are mutually nonexclusive (i.e., totally different or independent modes of action).

The classical isobol equation is the sum of two terms (Chou and Talalay, 1981, 1984):

$$\frac{(D)_1}{(D_x)_1} + \frac{(D)_2}{(D_x)_2} = 1 \tag{10}$$

where $(D_x)_1$ and $(D_x)_2$ are the doses of Drug 1 and Drug 2 that are required to inhibit a system $x\%$, whereas $(D)_1$ and $(D)_2$ in combination also inhibit $x\%$ (i.e., they are isoeffective). The sum of the two terms on the right side of the above equation is designated as the combination index or CI (Chou, 1987; Chou and Talalay, 1983, 1984, 1987) whereby $CI = 1$, $CI < 1$, and $CI > 1$ indicate additive, synergistic, and antagonistic interactions, respectively.

For two mutually nonexclusive drugs (such as two noncompetitive inhibitors in single-substrate Michaelis–Menten kinetics), a conservative isobologram equation was obtained which is the sum of three terms (Chou and Talalay, 1981, 1984):

$$\frac{(D)_1}{(D_x)_1} + \frac{(D)_2}{(D_x)_2} + \frac{(D)_1(D)_2}{(D_x)_1(D_x)_2} = 1 \tag{11}$$

The third term above is the product of the first two terms. The existence of the third term will invariably lead to higher CI values than in the mutually exclusive case (where CI is the sum of two terms). Therefore, the mutually nonexclusive assumption will always lead to a more conservative estimation of synergism (likewise, it will lead to conclusions of a somewhat higher degree of antagonism).

3. Combination Index Method of Chou and Talalay

The combination index method for drug combination studies proposed by Chou and Talalay (1983, 1984) is based on the median-effect principle (MEP) of the mass-action law. Two basic equations constitute the pillars of this methodology.

The Median-Effect Equation To relate dose and effect for a single drug in the simplest possible way, the median-effect equation derived by Chou in 1975 (1975, 1976) is given by:

$$f_a/f_u = (D/D_m)^m$$

or

$$D = D_m[f_a/(1-f_a)]^{1/m} \tag{12}$$

where the right side represents the dose and the left side represents the effect, in which f_a and f_u are the fractions affected and unaffected, respectively, D is dose, D_m is the median-effect dose signifying the potency, and m is a coefficient signifying the shape of the dose–effect curve.

The median-effect (Chou) plot with $x = \log(D)$ vs $Y = \log[f_a/(1-f_a)]$ determines two basic parameters: D_m (antilog of x-intercept) and m (the slope) as shown in Fig. 2. Once D_m and m are determined, one can calculate the dose for any given effect or the degree of effect for any given dose, thereby simulating the entire dose–effect curve (Chou and Chou, 1986; Chou and Talalay, 1984).

The Multiple-Drug-Effect Equation Based on the principle of the mass-action law, Chou and Talalay (1977, 1981, 1984) derived the general equation for two or more drugs. A condensed form of the equations for two drugs can be given by:

$$\left[\frac{(f_a)_{1,2}}{(f_u)_{1,2}}\right]^{1/m} = \left[\frac{(f_a)_1}{(f_u)_1}\right]^{1/m} + \left[\frac{(f_a)_2}{(f_u)_2}\right]^{1/m} + \alpha\left[\frac{(f_a)_1(f_a)_2}{(f_u)_1(f_u)_2}\right]^{1/m} \tag{13}$$

$$= \frac{(D)_1}{(D_m)_1} + \frac{(D)_2}{(D_m)_2} + \frac{\alpha(D)_1(D)_2}{(D_m)_1(D_m)_2}$$

where $m = 1$ is for first-order Michaelis–Menten-type kinetics and $m > 1$ (or $m < 1$) is for higher-order (or lower-order) Hill type kinetics. When $\alpha = 0$, the third term on the right side disappears and when $\alpha = 1$, the third term is conserved. $\alpha = 0$ is used for mutually exclusive drugs and $\alpha = 1$ is used for

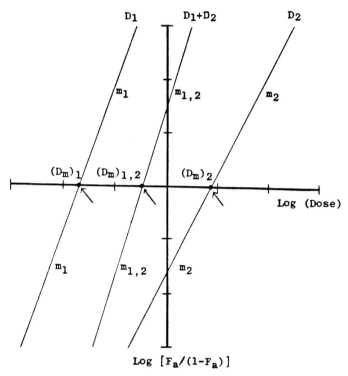

FIGURE 2 A typical median-effect plot of drug 1 (D_1), drug 2 (D_2), and the mixture of D_1 and D_2 (at a preassigned combination ratio). The antilog of x axis gives the median-effect concentration (D_m or IC_{50}, ED_{50}, LD_{50}, etc., depending on the end point of measurement). Thus D_m signifies the potency. The slope (m) signifies the shape of the dose–effect curve. When $m = 1, >1$, and <1, the curves are hyperbolic, sigmoidal, and negatively sigmoidal, respectively. Computer software can be used for obtaining plots and D_m and m parameters. The linear correlation coefficient (r) can also be obtained for diagnostic purposes.

mutually nonexclusive drugs. If the $(D)_1$ and $(D)_2$ combination ratio is P : Q, and this produces a median-effect with a dose $(D_m)_{1,2}$, then $(D)_1$ and $(D)_2$ in the numerators on the right can be calculated by:

$$(D)_1 = (D_m)_{1,2} \times \frac{P}{P + Q}, \quad (D)_2 = (D_m)_{1,2} \times \frac{Q}{P + Q} \qquad (14)$$

When $(f_a)_{1,2} = (f_u)_{1,2} = 0.5$, then Eq. 13 describes the ED_{50}-isobologram ($\alpha = 0$ for classical and $\alpha = 1$ for conservative isobolograms) for additive effect. In 1983 Chou and Talalay designated the right side of Eq. 13 as the combination index (CI); then CI $= 1, <1$, and >1 indicate summation, synergism, and antagonism, respectively. Similarly, the multiple drug effect equations for $x\%$ effect can be obtained. For more details, see Chapter 2.

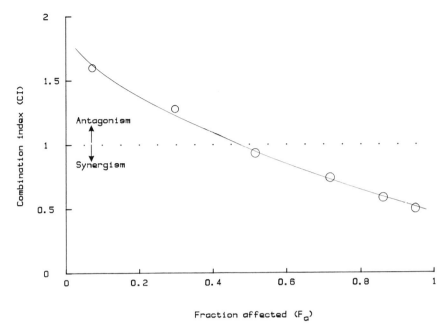

FIGURE 3 A typical F_a-CI plot indicating synergism or antagonism of two drugs at various effect levels in a mixture that is serially diluted. If several mixtures are made, it is possible to estimate the optimal combination ratio for maximal synergy. Different effect levels usually give different degrees of synergism/antagonism. $CI < 1$, $= 1$, and > 1 indicate synergism, additive effect, and antagonism, respectively. For anticancer or antimicrobial agents, synergism at high effect levels is clinically more relevant than synergism at low effect levels.

With the derived general equation available, it becomes relatively simple to develop computer software for automated dose effect analysis [e.g., *m* and D_m parameters, standard error (SE), and linear correlation coefficient of the median-effect plot (*r*)] and quantitation of synergism/antagonism (e.g., CI values at different dose or different effect levels) with the aid of a microcomputer (see Chou and Chou, 1986, and Chapter 6). A plot of fraction affected (F_a) vs combination index (CI) is called the F_a–CI plot as shown in Fig. 3.

By using the median-effect principle and computer simulation, considerable information about drug combination can now be readily obtained (Chou and Chou, 1986). Examples include the degrees of synergism/antagonism at different dose and different effect levels, the optimal combination ratio for maximal synergy, the schedule dependency of synergism/antagonism, and the degree of dose reduction allowed in a synergistic combination (compared with individual components) for a given degree of effect.

4. Dose-Reduction Index

The term "dose-reduction index" or DRI was first introduced in 1988 (Chou and Chou, 1988) and was used in 1989 for computerized automated analysis (Berman *et al.*, 1989) and for calculating DRI in a combination of 3'-azido-3'-deoxythymidine (AZT) and recombinant interferon Alpha A (rIFNαA) against human immunodeficiency virus type 1 (HIV-1) and normal bone marrow progenitor cells (see also Chapter 20 by Berman and Chang). DRI describes how many folds of dose reduction are allowed in a synergistic combination at a given degree of effect. The relationship between CI and DRI at a given degree of effect (e.g., $x\%$ inhibition) and at a given combination ratio ($D_1:D_2 = P:Q$) for mutually exclusive drugs can be described by:

$$CI = \frac{(D_x)_{1,2} \times P/(P+Q)}{(D_x)_1} + \frac{(D_x)_{1,2} \times Q/(P+Q)}{(D_x)_2} \tag{15}$$

$$= \frac{(D)_1}{(D_x)_1} + \frac{(D)_2}{(D_x)_2} = \frac{1}{(DRI)_1} + \frac{1}{(DRI)_2}$$

D_x can be calculated from the rearranged median-effect equation, $D_x = D_m[f_a/(1-f_a)]^{1/m}$ and D_m and m can be obtained from the median-effect plot as described earlier.

5. Empirical Methods

This method draws an empirical curve to fix experimental data without the guidance of the principle of the mass-action law or any designated models. The dose and the corresponding effect are then read from the graph by interpolation. Nearly all the early methods prior to 1980 for synergy analysis without a specified theoretical basis and the empirical method, still widely used at the present time, belong to this category. Since there is no fixed rule such as the mass-action law as a guide, the empirical curve can have an infinite number of shapes and, therefore, it is virtually impossible to assign a proper equation to describe an empirical curve. To ensure that a particular empirical dose–effect curve is the best fit over a wide range of doses, numerous data points are required. It can be costly in terms of time, effort, resources, and animal life, when animals are used in empirical, synergism/antagonism studies. For example, in an attempt to construct LD_{10}-isobologram for two drugs, using L1210 tumor bearing mice, over 600 animals were used in one study (Skipper, 1974). Using the median-effect method, the degrees of synergy were quantitatively determined for anti-simian varicella virus infection by human recombinant interferon β and antiviral nucleoside, 9-(1,3-dihydroxy-2-propoxymethyl)-guanine (DHPG), using only 30 monkeys (Soike *et al.*, 1987). No attempt has been made here to directly compare these two examples since different endpoints of measurements were used. However, there are basic differences in the approaches for data analyses and experimental designs. In contrast to using the empirical probit function to fit data, the median-effect method uses the data to fix a basic theory the mass-action law.

A new method claimed to be superior to the median-effect method has been proposed by Syracuse and Greco (1986) and reviewed by Gessner (1988). However, the review article incorrectly presented the Chou and Talalay equation (Eq. 22 in Chou and Talalay, 1981, shown as Eq. 4 in the Gessner review) by replacing f_u with f_a, thus misrepresenting 90% inhibition as 10% inhibition and 99% inhibition as 1% inhibition. In addition, the review article used an inappropriate equation (presented as Eq. 2 in the review). This equation as a general procedure to construct the isobologram is identical to the fractional product method of Webb (1963) except for the symbols used. Webb's method has severe limitations since it is not applicable in three out of four mass-action law models examined (see Chapter 2).

D. PREDICTION OF SYNERGISM/ANTAGONISM OR POTENTIATION

It is usually easier to predict potentiation than to predict synergism or antagonism since the potentiator by itself has no effect whereas each drug in combination for synergism/antagonism has its own effects. It would be a great advantage if synergism or antagonism could be predicted without conducting an experiment. However, this occurs only in relatively few straightforward and simple situations. Hypotheses of possible occurrence of synergism can be advanced but still need to be confirmed by experimental findings. More frequently, hypotheses are developed following the experimental findings or clinical trials and then additional experiments are carried out to prove or disprove the hypotheses.

The difficulty in the prediction of synergism/antagonism involves several factors. Each drug in combination has its own effects, hence, its own potency and a specific shape of dose-effect curve. These effects are also related to affinity (such as K_m and K_i) and efficacy (such as V_{max}), and the parameters can be further subclassified into individual rate constants and concentration factors. The biological complexity and intricacy usually do not allow easy access to all this information, especially when regulation factors (such as feedback inhibition) and spatial, temporal, and microenvironmental factors (such as pH, ionic strength, and temperature) are involved.

It is virtually impossible to take every conceivable factor (such as absorption, permeability, transport, metabolic activation, and inactivation.) into account in synergism/antagonism analysis. The usual way to conduct the analysis is to vary only a single variable, such as dose, at a time, and keep all other conditions constant. In addition, only a single endpoint—such as cell growth inhibition, cell kill, or inhibition of DNA synthesis—is measured at a time.

The determined synergism/antagonism is only valid in the experimental conditions selected and caution should be exercised to avoid overextrapolation of the conclusions. The experimental conditions are selected to reflect the physiological, pharmacological, and clinical situations.

Due to the complexity of biological systems and different drugs used in combination, a general basic principle such as the mass-action law is required to guide the analysis. Such a guideline should be mechanism nonspecific in order to be practical for broad applications. The assertion by Berenbaum (1989) that the detailed mechanism needs to be known in order to carry out synergism/antagonism analysis with the median-effect principle is not true. In fact, rarely, if ever is there a drug for which the detailed mechanism is known. After the degree of synergism or antagonism is determined with a mechanism-nonspecific general procedure, attempts are frequently made to understand how and why synergism/antagonism occurs by using biochemical and/or molecular biological techniques.

1. Frequent Sources of Confusion in Drug Combination Studies

The following are some of the issues frequently raised by investigators in drug combination studies.

Conclusions Based on Derived Equations

1. It is not possible to determine synergism/antagonism with a *single* dose of each drug alone and their combinations since the slopes of the dose–effect curves are unknown.
2. Both the isobologram equations and the fractional product equation can be derived from the median-effect principle of the mass-action law using enzyme kinetic models; thus the usefulness and limitations of these equations can be unambiguously discussed (Chou and Talalay, 1977, 1981, 1984).
3. The F_a–CI plot is effect oriented whereas an isobologram is a dose oriented description of synergism/antagonism or additive effect. Both graphs lead to identical conclusions since they are both based on the same multiple-drug-effect equation of Chou and Talalay (1977, 1981, 1984). Isobolograms tend to be messy and unreadable when more than four effect levels of isobolograms are presented on the same graph. This disadvantage can be avoided in the F_a–CI plot.
4. By using the median-effect principle of the mass-action law, dimensionless (i.e., no units) and mechanism-nonspecific (no kinetic constants) generalized equations can be developed to relate dose and effect in single-drug and drug-combination studies for synergism/antagonism determination (Chou and Talalay, 1981, 1984). The derived equations can be readily used for computerized analysis (Chou and Chou, 1986).
5. The quantitative determination of synergism/antagonism by itself does

not provide information about how and why synergism/antagonism has occurred. The latter is a separate issue and requires separate studies for its elucidation.

6. The mass-action law generated equations define what is the expected additive effect. Deviations of the experimentally observed results from the expected additive effect then define synergism or antagonism. There is no mass-action law generated "equation" that defines synergism or antagonism.

7. The classical isobologram (Chou, 1987; Chou and Talalay, 1984, 1987) is valid for mutually exclusive drugs but not valid for the mutually non-exclusive drugs. This is proven by mass-action law generated equations using a simple enzyme kinetic system (Chou and Talalay, 1984). The classical isobologram and its equation do not describe the additive effect of two noncompetitive inhibitors on a single substrate enzyme reaction and therefore a conservative isobologram method has been introduced for mutually nonexclusive drugs.

8. The fraction product equation of Webb (1913) is correct only for mutually nonexclusive drugs with a hyperbolic dose–effect curve (i.e., first order or $m = 1$). It is not valid for mutually exclusive drugs regardless of whether the dose–effect curve is hyperbolic or sigmoidal (Chou and Talalay, 1981, 1984).

9. No equation for dose response or dose effect analysis is perfect. This is true for the widely used Michaelis–Menten equation and the Hill equation. The median-effect equation and the multiple drug effect equation are also derived from the mass-action law, and thus have similar limitations. A diagnostic test for conformity is essential prior to their applications to dose effect analyses.

Questionable Claims Based on the Empirical Approach

1. An equation that is universally valid for synergism/antagonism analysis has been developed (Berenbaum, 1988). [There is no such equation.]
2. There is general validation for the classical isobologram (Berenbaum, 1989). [It is, in fact, invalid for nonexclusive inhibitors.]
3. The mass-action law *is* a mechanism (Berenbaum, 1989). [It is not.]
4. The median-effect equation *is* the Hill equation (Berenbaum, 1989). [They are different; see Chapter 2, Table 2.]

2. Oversimplification, Overextrapolation, and Overgeneralization

As indicated earlier, the complexity of biological systems does not permit every factor affecting the dose–effect relationship to be taken into account. One can only vary one factor at a time while keeping the rest of the factors constant to avoid overly complicated situations. Numerous studies *in vitro* and *in vivo* have shown that such an approach adequately conforms to the mass-action law as diagnosed by the linear correlation coefficient (r) of the

median-effect plot and the standard error of the median-effect plot parameters. Excellent consistency was also observed in interexperimental comparisons as well as in intraexperimental comparisons repeated two, three, and
four times. Despite this conformity, it is still recommended that routine
diagnostic tests be carried out. These requirements are intrinsic to a software
for dose effect analysis with microcomputers (Chou and Chou, 1986).

Experience with enzyme kinetic studies indicates that the simpler a general
equation is, the broader its applicability will be. The mechanism-specific
equations are usually very complicated as shown by Cleland (1963) and Segel
(1975), and are thus limited to special cases. The Michaelis–Menten equation
and the Hill equation for substrate-like ligands are very simple and have
proved their usefulness. It is well known, on the other hand, that both of
these equations are somewhat oversimplified since the former is not applicable to sigmoidal dose–response curves and the latter does not take into
account the minor intermediate species of interactions (Whitehead, 1970).
The median-effect equation (Chou and Talalay, 1977, 1981) for inhibitors are
also based on the mass-action law, and will obviously face similar limitations
as the Hill equation. However, the usefulness of the median-effect equation
and the multiple-drug-effect equation has been shown in numerous applications as indicated in recent reviews (Chou, 1987; Chou and Fanucchi, 1988;
Chou *et al.*, 1991; Chou and Talalay, 1983, 1984, 1987; see also Chapter 2).

With respect to extrapolation beyond the experimental dose range, the
median-effect equation provides a guideline. In contrast, the empirical
approach creates uncertainties especially at very low effect or very high effect
levels. As indicated earlier, the mass-action law generated equation for two
mutually nonexclusive inhibitors contradicts the classical isobologram and
conforms to the conservative isobologram (Chou and Talalay, 1984, 1987).
Earlier interpretation of this case by Chou and Talalay (1981), based on the
old definition at that time, was referred to as "necessarily synergistic." It was
later revised as an additive effect based on the conservative isobologram. This
change is based on two reasons.

1. The mass-action law generated equation describes only the additive
   effect, not the synergistic effect. The deviations or departures from the
   additive effect are then referred to as synergism or antagonism.
2. If two mutually nonexclusive drugs are necessarily producing synergism
   then this synergism is intrinsic synergism without any pharmacological
   interaction involved. Since it is zero-interaction, it is more appropriate
   to call it an additive effect.

Berenbaum's (1989) paper contradicts its own definition of zero-
interaction by referring to it as synergism in the case of mutually nonexclusive
drugs.

These are the different interpretations of the same equation that was
derived by Chou and Talalay (1981). There are numerous differences be-

tween the empirical approach and the mass-action law approach in dose effect analyses. It requires actual derivation of equations from physicochemical models before an assertion can be meaningfully made on such studies.

II. Mechanisms of Interaction

The purpose of this section is to present, in an organized manner, most if not all of the *proposed* mechanisms of chemotherapeutic drug interaction with one or more illustrative examples of each mechanism. Emphasis will be placed on work which includes well-designed experiments aimed at elucidating mechanisms of synergistic or antagonistic interactions and experiments aimed at ruling out possible mechanism of interaction. Because the field of drug interaction mechanisms is still at an early stage of development, *the mechanisms proposed in the outline are not meant to represent the last word on the subject, nor is the outline intended to provide a critical review of interaction mechanisms.* Nevertheless, this outline should help those who have encountered novel examples of chemotherapeutic synergism, potentiation, or antagonism to better decide which mechanisms to consider, which mechanisms to rule out, and which avenues of experimental design to pursue. The outline also serves as an introduction to the remainder of the book, since references will usually be made to other chapters whenever appropriate.

Many claims of synergism have not been determined with formally established procedures such as the isobologram technique or the combination index methodology. Sometimes qualitatively correct conclusions of synergism or antagonism may be obtained even when incorrect methodologies are used. This is particularly true when pronounced synergism or antagonism has occurred. In order to facilitate the discussion of various mechanisms that may lead to synergism or antagonism, the claims will be taken at face value here. The main criterion for the examples chosen in this section is simply their utility in explaining the proposed mechanisms of interaction.

The outline is organized in terms of mechanisms of *inter*action rather than mechanisms of action of drugs, drug potentiating agents, or drug inhibiting agents. The following examples illustrate the distinction between action and interaction mechanisms. Beta lactam/sulbactam and cyclophosphamide/4-dimethylaminobenzaldehyde combinations operate on different organisms through entirely different mechanisms of action (involving bacterial wall synthesis inhibition and mammalian cell DNA alkylation, respectively). However, they both involve the same subclass of mechanisms: synergism through inhibition of metabolic drug inactivation (by β-lactamase and aldehyde dehydrogenase, respectively; see Section IIA4). Although the mechanisms of chemotherapeutic action of benzylpenicillin/probenicid and amoxycillin/sulbactam combinations are identical (inhibition of bacterial cell was synthesis), the mechanisms of interaction differ: inhibition of excretion from the kidneys and inhibition of metabolic deactivation by beta-lactamase, respectively (see Sections IIA1 and IIA4).

For the purpose of explaining interaction mechanisms in Section II, the word "drug" is synonymous to its active form. Examples of drugs are benzyl-penicillin (which does not need to be metabolically activated) and acyclovir $5'$-triphosphate (the activated form of the prodrug, acyclovir; see Chapter 8 by Spector and Fyfe). *Target* is defined as a macromolecule or membrane which, through interaction with the active form of the drug, leads to the desired biological activity (e.g., an antiviral effect, inhibition of cell growth, or cell killing). DNA, reverse transcriptase, tubulin, and bacterial ribosomes are examples of targets. An *effector* is a macromolecule or membrane that does not produce the desired biological effect directly upon binding of an agent, but rather influences the biological activity of the *second* agent that acts on a target. Effectors include metabolic and catabolic enzymes, transport systems, biomembranes (in the sense of permeability barriers), serum albu-min, repair enzymes, and those macromolecules and membranes involved in the control of pH, oxygen concentration, and ion concentration.

Mechanisms of synergism, antagonism, potentiation, and collateral sensi-tivity can be divided into five distinct classes, depicted schematically in Fig. 4. The mechanisms of most examples of synergism and antagonism that are known to date involve the binding of each agent to a different effector or target. For example, sulfadiazine inhibits bacterial folate synthesis while trimethoprim inhibits dihydrofolate reductase, leading to synergism (Hitch-ings and Burchall, 1965). Vinblastine binds to tubulin and prevents tumor mitosis while verapamil binds to the export pump P-glycoprotein in multidrug resistant tumor cells and prevents the export of vinblastine, resulting in potentiation (see Chapter 9 by Yang *et al.*). The first three of the overall mechanistic classes of antagonism and synergism (Sections IIA, IIB, and IIC) all involve binding of the two agents to different macromolecular or mem-brane targets. These can be divided into three classes of interaction mechan-isms. In Class 1, drug A binds to an effector, thereby enhancing (potentiation, synergism) or diminishing (antagonism) the ability of drug B to reach the target and achieve a chemotherapeutic effect. In Class 2, drug A binds to an effector that alters the activity of drug B after drug B has bound to its target (this is *not* meant to imply that A must be administered or must begin acting after B). In Class 3, drugs A and B each bind to distinct targets, indepen-dently causing effects that potentiate, complement, or antagonize one another. The other two mechanistic classes (4 and 5, Sections IID and IIE) involve binding of drugs A and B to the same target and direct bond forma-tion between drugs A and B, respectively. Either, or both, drug A or B may be administered in the form of prodrugs. For each of these five classes, there are one or more chapters in this volume providing in-depth coverage of a particular aspect. These chapters are cross-referenced in the outline.

The outline headings and page numbers for this section are listed in Table 1

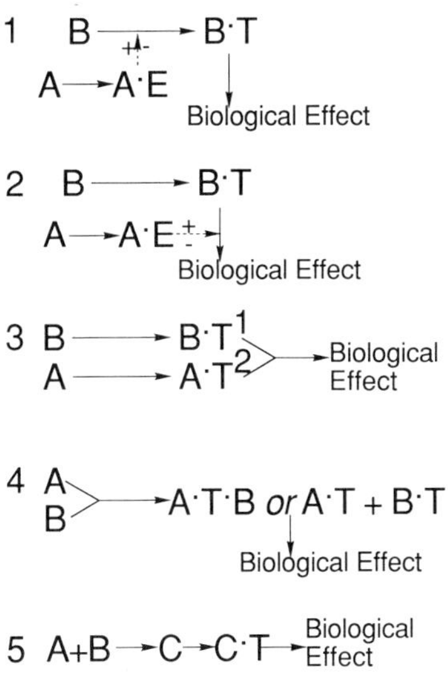

$$1 \quad \begin{array}{l} B \xrightarrow{\;+\!\!\!\!-\;} B{\cdot}T \\ A \longrightarrow A{\cdot}E \qquad \downarrow \\ \qquad\qquad \text{Biological Effect} \end{array}$$

$$2 \quad \begin{array}{l} B \longrightarrow B{\cdot}T \\ A \longrightarrow A{\cdot}E\!\!\stackrel{+}{}\!\!\rightarrow\!\!\mid \\ \qquad\qquad \text{Biological Effect} \end{array}$$

$$3 \quad \begin{array}{l} B \longrightarrow B{\cdot}T^1 \\ A \longrightarrow A{\cdot}T^2 \end{array}\!\!\!\searrow\!\!\!\nearrow \longrightarrow \begin{array}{l}\text{Biological}\\ \text{Effect}\end{array}$$

$$4 \quad \begin{array}{l} A \\ B \end{array}\!\!\!\searrow\!\!\!\nearrow \longrightarrow A{\cdot}T{\cdot}B \;\; or\; A{\cdot}T + B{\cdot}T \\ \qquad\qquad\qquad\qquad \downarrow \\ \qquad\qquad\qquad \text{Biological Effect}$$

$$5 \quad A{+}B \longrightarrow C \longrightarrow C{\cdot}T \longrightarrow \begin{array}{l}\text{Biological}\\\text{Effect}\end{array}$$

FIGURE 4   Schematic representation of interaction mechanism classes. A, B, and C represent drugs in their final active form (e.g., vinblastine, AZT triphosphate); E is an effector macromolecule or membrane (e.g., beta-lactamase, P-glycoprotein); T, T1, and T2 are target macromolecules or membranes (e.g., DNA, reverse transcriptase, tubulin); A·E and B·T, etc., are complexes (noncovalently or covalently bonded). The dotted arrow and ± symbol represent inhibition or enhancement of action. In mechanism class 1, A can also inhibit or accelerate the formation of B from a prodrug of B—not explicitly shown in this figure. In mechanism class 5, A and B are, strictly speaking, prodrugs of C rather than being the final active form of the drug.

for the convenience of the reader. The outline topics generally are *not* divided between synergism and antagonism, as these can so often result from identical mechanisms through the substitution of an agonist for an antagonist (or vice versa). Collateral sensitivity and the induction of cross resistance to one agent through mutagenesis induced by a second agent are treated as synergism and antagonism, respectively. It is not possible to distinguish collateral sensitivity from schedule-dependent "classical" synergism, nor induction of cross resistance from schedule-dependent "classical" antagonism, in terms of final effect (e.g., in terms of isobologram or combination index results). Similarly, phenomena involving selective inhibition of a different subpopulation by each agent (apparent synergism) and genetic selection by the first agent of a subpopulation exhibiting cross resistance to a second agent (apparent antagonism), both of which are of considerable clinical relevance, will be included alongside more "classical" examples of synergism and antagonism. The "nonclassical" mechanisms of synergism and antagonism can be of considerable clinical importance.

The organization according to interaction mechanisms is also intended to encourage the study of *comparative drug interactions*: the relationship between mechanisms of interaction in different systems (e.g., antiviral versus

TABLE 1
Proposed Mechanisms of Interaction

Class 1: Drug A binds to an effector, thereby altering the ability of drug B to reach its
   target   29
   Drug A alters the efflux of drug B   29
         Drug A interacts directly with an effector, thereby altering the efflux of drug B   30
         Drug A induces a genetic change that alters the efflux of drug B   30
         Drug A causes a transient change in gene expression, thereby altering the efflux of
         drug B   31
   Drug A alters the influx of drug B   31
         Drug A interacts directly with the effector, thereby altering the influx of drug B   31
         Drug A induces a genetic change that alters the influx of drug B   32
         Drug A alters the pH, thereby altering the influx of drug B   32
   Drug A alters the prodrug activation that is required to form drug B   32
         Drug A affects the biosynthesis of a metabolite that inhibits activation of the
         prodrug   32
         Drug A alters the pH, thereby altering the rate of drug B formation from the
         prodrug   33
         Drug A alters the oxygen concentration, thereby altering the rate of drug B
         formation from the prodrug   33
         Drug A induces genetic changes that affect the rate of formation of drug B from the
         prodrug   34
         Drug A alters metal ion concentrations that affect the rate of drug B formation from
         the prodrug   34
   Drug A alters the metabolic inactivation of drug B   35
         Drug A directly affects the enzyme that inactivates drug B   35
         Drug A induces genetic changes that affect the rate of inactivation of drug B   35
         Drug A induces transient changes in the expression of gene products that inactivate
         drug B   36
         Drug A alters the pH, thereby altering the rate of inactivation of drug B   36
         Drug A alters the metal ion concentration, thereby altering the rate of inactivation of
         drug B   36
         Drug A affects the synthesis of a metabolite that inactivates drug B through direct
         combination   36
   Drug A alters the incorporation of drug B into a macromolecule   37
   Drug A induces a genetic change that alters the concentration of the target of drug B   37
   Drug A alters the synthesis of a metabolite that competes with drug B for the target
         site   37
   Drug A alters the pH, thereby altering drug reactivity with the target   38
Class 2: Drug A binds to an effector, thereby altering the activity of drug B after it reaches its
   target   38
   Drug A affects the repair of DNA damage caused by drug B   38
         Drug A interacts directly with DNA repair-related proteins, thereby altering repair of
         drug B-induced damage   38
         Drug A induces genetic changes that alter the ability to repair DNA damage caused
         by drug B   39
         Drug A induces changes in oxygen concentration, thereby altering repair of drug
         B-induced damage   39
         Drug A alters the pH, thereby altering repair of drug B-induced damage   39
         Drug A alters the metal ion concentration, thereby altering repair of drug B-induced
         damage   40

TABLE 1 (*continued*)

Drug A affects cell cycle kinetics, thereby altering the biological effects of drug B after
  target binding   40
Drug A induces genetic changes that alter the effects of drug B after target binding without
  involving DNA repair   40

Class 3: Drugs A and B each bind to separate targets, causing effects that complement,
  potentiate, or antagonize one another   41
  Drugs A and B inhibit or potentiate sequential targets in a metabolic pathway   41
  Drugs A and B concurrently inhibit or potentiate parallel targets in a convergent metabolic
    pathway   42
      Drugs A and B both directly affect their parallel targets in convergent metabolic
      pathway   42
      Drug A induces a mutation that affects one metabolic pathway and drug B directly
      inhibits a parallel target in a convergent metabolic pathway   42
      Drug A and Drug B become incorporated into different sites on the same
      biopolymer   43
  Drugs A and B inhibit targets from distinct metabolic pathways within a single cell,
    microbe, or virally infected cell   43
  Drugs A and B each affect a different subpopulation of cells, microbes, or viruses   43

Class 4: Drugs A and B bind to the same macromolecule or membrane target   46
  Drugs A and B bind to identical sites on the same macromolecule or membrane target   46
  Drugs A and B bind to different sites on the same macromolecule or membrane target   46

Class 5: Drugs A and B combine with one another *in vivo*, forming a new drug with
  significantly altered biological effects   48
  Drugs A and B form an adduct, C, with a biological activity that is significantly different
    from either A or B   48
  Drug A is a catalyst or enzyme that activates a prodrug to form Drug B   49
  Drug A is a catalyst or enzyme that deactivates drug B   50
  Drug A and Drug B undergo an oxidation/reduction interaction, altering the activity of one
    or both agents   50

---

antineoplastic combinations or antiparasitic versus antibacterial combinations). Examples of comparative drug interactions can be found in subsequent sections (IIC1, IIA1, IIA4, and IID1). In addition, better understanding of individual drug action and drug-resistance mechanisms can result from combination studies. For examples, see Sections IIA1, IIA3, IIC1, IIC4.

A. CLASS 1: DRUG A BINDS TO AN EFFECTOR, THEREBY ALTERING THE ABILITY OF DRUG B TO REACH ITS TARGET

1. Drug A Alters the Efflux of Drug B

If drug A acts on an effector and thereby decreases the rate of efflux of B, the result is potentiation or synergism. If A causes an increase in the efflux of B, the result is antagonism.

Drug A Interacts Directly with an Effector, thereby Altering the Efflux of Drug B Synergism or potentiation occur through this mechanism when drug A acts on an effector, inhibiting efflux of B from the vicinity of the target. Drug A could antagonize the effects of B by direct interaction with the effector, leading to acceleration of the efflux of B. The inhibition of drug efflux from cells through inhibition of the effector P-glycoprotein (a plasma membrane-associated pump responsible for multidrug resistance) is reviewed by Yang *et al.* in Chapter 9 of this work. Inaba *et al.* (1988) showed that a number of polycyclic clinical drugs can reverse the resistance to vincristine efflux in a multiply resistant p388 line. The clinical agents tamoxifen and perhexiline maleate decrease the activity of P-glycoprotein and therefore enhance sensitivity to adriamycin in multiply drug resistant MCF7 breast carcinoma cells (Forster *et al.*, 1988). The development of nontoxic agents that reverse multiple drug resistance by blocking the export activity of P-glycoprotein directly has become a very active area of research. Recently, detailed structure/activity relationship studies of certain classes of such blockers have been carried out (e.g., Kiue *et al.*, 1990). An example of comparative potentiation is provided by the observation that verapamil is not only capable of reversing multiple drug resistance in mammalian tumor cells but is also capable of reversing the resistance of malarial parasites to chloroquine and other drugs through a similar mechanism (see Chapter 5 by Vennerstrom *et al.*).

Potentiation involving this mechanism can also occur at the pharmacological level. For example, probenicid enhances the concentration of benzylpenicillin in the human body by prohibiting the renal secretion of the latter. As a result, probenicid potentiates the antibacterial activity of benzylpenicillin in human beings through direct inhibition of efflux (Rowland and Tolzer, 1980).

If a doxorubicin-resistant tumor cell line is sensitive to combinations of doxorubicin and verapamil, but not sensitive to combinations of doxorubicin and buthionine sulfoximine, these observations suggest that active export of the drug (the multidrug resistant phenotype) plays a more significant role in doxorubicin resistance than intracellular detoxification of drug or of its metabolites due to high glutathione levels (See Chapter 7 by Griffith and Friedman and Chapter 9 by Yang *et al.*) In other words, drug combination studies can provide a useful tool for understanding mechanisms of drug resistance.

Drug A Induces a Genetic Change that Alters the Efflux of Drug B Drug A can also act indirectly, inducing genetic mutations that increase the efflux rate of B (causing "antagonism," i.e., induction of cross resistance) or decrease the efflux rate of B (causing "synergism" or "potentiation," i.e., collateral sensitivity). The induction of multiple drug resistance through exposure to doxorubicin most likely involves actual mutation rather than genetic selection given the mutagenicity of doxorubicin (De-yu *et al.*, 1988; Laidlaw *et al.*,

1989). In other words, the antagonism between doxorubicin and vinblastine observed when doxorubicin is administered first to mammalian cells over a long period probably involves a doxorubicin-induced genetic change. This change causes the heightened expression of the P-glycoprotein efflux pump capable of causing the efflux of both doxorubicin and vinblastine (see also Chapter 9 by Yang, *et al.*, concerning P-glycoprotein, and Chapter 16 by Avramis *et al.*, concerning synergism and antagonism at the genetic level). Antagonism that involves genetic selection can be difficult to distinguish from this genetic mutation mechanism (see also Section IIC4).

Drug A Causes a Transient Change in Gene Expression, thereby Altering the Efflux of Drug B If drug A enhances the expression of a gene responsible for the export of drug B, the result will be antagonism between A and B. If A decreases the expression of this gene, the result will be potentiation. Antagonism against HCT-15 human colon tumor cells was observed in combinations of vinblastine with differentiating agents such as sodium butyrate, dimethylsulfoxide, or dimethylformamide (Mickley *et al.*, 1989). The differentiating agents could cause a 5-fold increase in the expression of the P-glycoprotein gene product associated with multiple drug resistance, leading to a fall in vinblastine accumulation. Vinblastine accumulation could be restored in cells treated with differentiating agents by adding verapamil, providing further evidence that enhancement of P-glycoprotein-mediated efflux is responsible for antagonism between butyrate and vinblastine.

2. Drug A Alters the Influx of Drug B

Drug A can act on an effector to potentiate drug B by directly or indirectly enhancing the rate of influx of B into the vicinity of the target. Drug A can antagonize B by inhibiting the influx of B.

Drug A Interacts Directly with the Effector, thereby Altering the Influx of Drug B In this mechanism, A directly inhibits or enhances the effector responsible for the influx of B, resulting in antagonism or potentiation, respectively. In bacteria such as *Streptococcus mitus*, the beta lactam penicillin inhibits synthesis of cell wall components and thereby enhances the ability of drugs such as streptomycin to penetrate through the cell wall and reach their intracellular target (the ribosome). As a result, penicillin and streptomycin exhibit clinically useful antibacterial synergism (Yee *et al.*, 1986). (In addition to its ability to enhance streptomycin influx, penicillin has antibacterial activity of its own. Thus, synergism is observed as opposed to potentiation).

The antibacterial sulfa drugs sulfadimethoxine and sulfapyrazone both bind to a particular site on serum albumin. If combined with another drug which binds to this site, such as phenylbutazone or warfarin, the concentration of unbound (available) sulfa drug in serum can be increased considerably. This phenomenon, known as displacement, can lead to potentiated antibacterial effects and also to potentiated host toxicities (Rowland and

Tolzer, 1980). Displacement, a mechanism occurring at the pharmacological level, involves acceleration of drug influx into the target from serum. Serum albumin acts as the effector molecule, drug A is phenylbutazone or warfarin, and drug B is the sulfa drug.

Drug A Induces a Genetic Change That Alters the Influx of Drug B   In this mechanism, A causes a genetic mutation that enhances or inhibits the influx of B, resulting respectively in collateral sensitivity (similar to synergism and potentiation) or induction of cross resistance (similar to antagonism). Chronic exposure of F4 adenovirus-transformed rat brain cells to methylglyoxal bis-(guanylhydrazone) (MGBG) causes mutation to a variant expressing significant cross-resistance to doxorubicin, colchicine, and vincristine (Weber *et al.*, 1989). This cross resistance is *not* due to classical P-glycoprotein-mediated multiple drug resistance. MGBG-resistant F4 cells exhibit abnormally slow uptake of colchicine relative to the parent F4 line, whereas rates of colchicine efflux are identical for both lines. In summary, a combination of MGBG followed by colchicine exhibits antagonism involving a genetic change to a phenotype exhibiting decreased influx of colchicine. (See also section IIC4 and Chapter 16 by Avramis *et al.*).

Drug A Alters the pH, thereby Altering the Influx of Drug B   This is another indirect mechanism. Drug A alters the microenvironment (specifically, pH) in the vicinity of an effector. As a result of the pH change, the rate of influx of B is enhanced (causing potentiation) or depressed (causing antagonism). Glucose can induce a solid tumor-selective drop in extracellular pH when administered to humans and animals through enhancement of tumor glycolysis rates (Tannock and Rotin, 1989; Jähde *et al.*, 1989) Influx of chlorambucil into mammalian cells is enhanced as the pH drops (Jähde *et al.*, 1989). Glucose potentiates the antineoplastic activity of various alkylating agents in animals (Tannock and Rotin, 1989). Glucose may potentiate chlorambucil and other antineoplastic alkylating agents *in vivo* by decreasing extracellular pH and enhancing uptake. Enhancement of cancer chemotherapy through alterations in the cellular microenvironment has been reviewed by Brown *et al.* (1988).

3. Drug A Alters the Prodrug Activation That Is Required to Form Drug B

Drug A can potentiate or antagonize a prodrug by enhancing or inhibiting, respectively, the activation of this prodrug to form drug B.

Drug A Affects the Biosynthesis of a Metabolite That Inhibits Activation of the Prodrug   Two enzymes are involved in this mechanism. One synthesizes the nutrient metabolite and one activates the prodrug. If A enhances the synthesis of the nutrient metabolite, antagonism results through inhibition of prodrug activation of B by the nutrient metabolite. If A diminishes the generation of the metabolite, potentiation or synergism results. Allopurinol inhibits orotidine-5-monophosphate decarboxylase in gastrointestinal tissue and bone marrow, leading to accumulation of the nutrient metabolite, orotic acid. Orotic acid inhibits activation of the antineoplastic agent 5-fluorouridine by inhibiting fluorouridine monophosphate (FUMP) synthesis. The result is antagonism between allopurinol and FUra, useful for decreasing the toxicity of FUra (see Chapter 13 by Mini and Bertino).

The dihydrofolate reductase inhibitor metoprine, at noncytotoxic levels, decreases the level of cellular folates that inhibit the activation of the cytotoxic prodrug 10-propargyl-5, 8-dideazapteroylglutamic acid (PDDF) to active PDDF polyglutamates, which inhibit thymidylate synthase (see Chapter 10 by Galivan). Thus, metoprine and PDDF exhibit cytotoxic synergism through enhancement of prodrug activation (in addition to a second mechanism; see Section IIA7).

The antiretroviral agents AZT and ribavirin have both been tested clinically in the treatment of AIDS. AZT has unquestionable clinical utility, whereas the clinical efficacy of ribavirin is less certain. Unfortunately, a combination of these agents demonstrates antiviral antagonism, at least *in vitro*. Ribavirin inhibits the phosphorylation of AZT to its monophosphate, thereby decreasing the concentration of the active form of the drug, AZT triphosphate. Ribavirin appears to have this deleterious effect on conversion of prodrug AZT to the active AZT triphosphate by enhancing intracellular concentrations of deoxythymidinetriphosphate (dTTP). dTTP is an inhibitor of the phosphorylation of AZT (in addition to a second mechanism; see Section IIA5). This clinically relevant example of antagonism is detailed in Chapter 8 by Spector and Fyfe.

Drug A Alters the pH, thereby Altering the Rate of Drug B Formation from the Prodrug Alterations in pH can probably impinge on prodrug activation as well as on drug influx. A glycoside prodrug of hydroxycyclophosphamide is much more rapidly activated to the more cytotoxic drug (hydroxycyclophosphamide) at pH 6.0 than at pH 7.3 *in vitro* (Tietze *et al.*, 1989). It is possible to selectively decrease the extracellular pH in many solid tumors in humans and animals by administering glucose (Tannock and Rotin, 1989). Experiments are underway to determine whether glucose will potentiate the anticancer activity of the glycoside prodrug through tumor-selective alterations in extracellular microenvironment that accelerate prodrug activation.

Drug A Alters the Oxygen Concentration, thereby Altering the Rate of Drug B Formation from the Prodrug This mechanism also involves a drug A-induced microenvironment effect (involving oxygen concentration) on the rate of prodrug activation to form B. Flavone acetic acid potentiates the antitumor activity *in vivo* of the cytotoxic agent SR4233. Here, the potentiation is mediated through a decrease in oxygen concentration in the tumor environment induced by the flavone acetic acid, an agent that decreases tumor blood flow and *selectively* causes a decrease in available oxygen in the tumor as a result. SR4233 is a prodrug that must be bioreduced in order to exhibit its activity and is thus more active under low oxygen conditions (Sun and Brown, 1989).

Many antineoplastic drugs such as bleomycin and thiotepa require a high level of oxygen in order to achieve their chemotherapeutic effects. Strategy for enhancing cancer therapeutic activity of such drugs through alterations of microenvironment involves the use of agents that increase oxygen concentrations in tumors. An example is the fluorocarbon emulsion, fluosol DA, administered under high oxygen conditions (see Chapter 15 by Teicher *et al.*).

Drug A Induces Genetic Changes That Affect the Rate of Formation of Drug B from the Prodrug This is another mechanism involving a mutation induced by A. The expression of an enzyme required for the activation of a prodrug to B is enhanced or diminished by the mutation, resulting in collateral sensitivity ("synergism") or attainment of cross resistance ("antagonism"). Azacytidine is a clinical drug that can activate genes through demethylation of 5-methylcytosine residues in DNA, a mutagenic process involving incorporation of azacytidine into the DNA. Azacytidine can activate the cytidine kinase gene, which is inactivated through cytosine 5-methylation in cytarabine-resistant leukemia cells. These cytarabine-resistant cells revert to a sensitive state after azacytidine treatment, because the deoxycytidine gene is required for the phosphorylation and ultimate incorporation into the DNA of the cytotoxic prodrug araC. This work and other examples of synergism involving genetic changes are outlined in Chapter 16 of this book by Avramis *et al.*

The significance of the azacytidine/cytarabine studies for cancer chemotherapy becomes even more evident in light of the fact that increased DNA cytosine methylation can be induced in human carcinoma and sarcoma cells by exposure to 10 different clinical anticancer agents and the clinical antiviral agent azidodideoxythymidine (Nyce, 1989). In addition to azacytidine, the clinical antineoplastic agents cyclophosphamide and 6-thioguanine can decrease DNA cytosine methylation (Nyce, 1989).

This mechanism can be useful for clarifying drug resistance. If bacterial resistance to the antibiotic chloramphenicol can be permanently reversed by treatment with nonbactericidal doses of DNA intercalators, this implies that a gene or genes essential to chloramphenicol resistance are encoded on circular extrachromosomal DNA elements as opposed to the chromosomal DNA (see Chapter 16 by Avramis *et al.*).

Drug A Alters Metal Ion Concentrations That Affect the Rate of Drug B Formation from the Prodrug Here again, prodrug activation to form B is affected indirectly after A induces a microenvironment change: specifically, altered metal ion concentrations. The metal chelator O-phenanthroline antagonizes DNA damage mediated by the antineoplastic agent bleomycin in mammalian cells by sequestering iron (Larremendy *et al.*, 1989). Bleomycin must chelate iron in order to generate free radicals that damage DNA. Similarly, metal chelators antagonize the cytotoxicity and bactericidal activity of paraquat, an herbicide which has been responsible for the deaths of many thousands of human beings. The chelator binds divalent copper and iron ions which are required in combination with paraquat to complete the cyclic redox pathway that leads to the generation of toxic oxygen-derived free radicals (Korbashi *et al.*, 1989). In other words, chelators antagonize paraquat cytotoxicity through an alteration in the concentration of metal ions that occur naturally in biological systems.

4. Drug A Alters the Metabolic Inactivation of Drug B

The converse of prodrug activation is metabolic drug inactivation. If drug A acts on an effector, leading to enhancement or diminution of drug B inactivation, the result will be antagonism or potentiation, respectively.

Drug A Directly Affects the Enzyme That Inactivates Drug B Here, drug A directly inhibits or enhances the action of the enzyme that inactivates B, leading to potentiation or antagonism, respectively. An example of potentiation through inhibition of metabolic drug deactivation involves the enhancement of cyclophosphamide and other oxazaphosphorines by inhibitors of aldehyde dehydrogenase such as disulfuram. No potentiation is observed when aldehyde dehydrogenase inhibitors are combined with other types of nitrogen mustards, such as mechlorethamine. This selectivity is explained in terms of the unique metabolism of cyclophosphamide and other oxazaphosphorines. They become active through the formation of an aldehyde intermediate which can be inactivated by aldehyde dehydrogenase—mediated oxidation in drug resistant cells. Notably, disulfuram does not potentiate the activity of oxazaphosphorenes against tumor cells which lack high levels of aldehyde dehydrogenase (Sladek and Landkamer, 1985).

The activity of bleomycin against human A-253 squamous carcinoma cells is potentiated by leupeptin, an inhibitor of the amidase that degrades bleomycin to a less active form (Lazo *et al.*, 1989). Antibacterial potentiation involving inhibition of an amide—cleaving enzyme is exemplified by beta lactam/sulbactam combinations (English *et al.*, 1978). The utility of amidase inhibitors in potentiating chemotherapeutic drugs in two very different systems demonstrates the importance of comparative drug interaction studies.

Drug A Induces Genetic Changes That Affect the Rate of Inactivation of Drug B In this indirect mechanism, A causes a mutation that leads to an increase or decrease in the rate of degradation of B, causing induction of cross resistance or collateral sensitivity, respectively. Antineoplastic alkylating agents such as BCNU can cause mutations in tumor cells that lead to higher levels of glutathione and glutathione-S-transferase (see Chapter 7 by Griffith and Friedman). These mutations cause resistance to other alkylating agents, such as chlorambucil, due to enhanced metabolic inactivation of chlorambucil through glutathione conjugation. In short, BCNU exhibits schedule-dependent antineoplastic antagonism with chlorambucil due to induction of genetic changes that lead to accelerated metabolic deactivation of chlorambucil (see also Section IIC4).

Drug A Induces Transient Changes in the Expression of Gene Products that Inactivate Drug B   When A causes an increase or decrease in the expression of a gene required to inactivate B, this results in antagonism or potentiation, respectively. Enhancement of the expression of a gene product that metabolizes a drug can be useful if it leads to decreases in host toxicity without associated decreases in efficacy. Bismuth subnitrate is a nontoxic metal complex which can be used to enhance the synthesis of metallothionein. This protein, in turn, can inactivate cisplatinum and/or its toxic metabolites through thiol–metal bond formation. As a result, after bismuth subnitrate treatment it is possible to increase the doses of cisplatinum in animal models above those that are normally tolerated. Fortunately, the induced metallothionein does *not* significantly diminish the antitumor effect of cisplatinum (Satoh *et al.*, 1988).

Drug A Alters the pH, thereby Altering the Rate of Inactivation of Drug B   In this mechanism, drug A alters the pH at the location of drug B, thereby decreasing (potentiation, synergism) or increasing (antagonism) the rate of catabolism of B before B reaches its target. Lysosomotropic agents ammonium chloride, amantadine , and chloroquine, which raise the pH in mammalian cell lysosomes, decrease the rate of hydrolysis of cytotoxic ^{125}I-labeled monoclonal antibodies inside tumor cells. The lysosomal enzymes that degrade these labeled antibodies are less efficient at higher pH (Press *et al.*, 1990). Thus lysosomotropic agents potentiate the radioimmunotherapeutic activity of ^{125}I-labeled monoclonal antibodies.

Drug A Alters the Metal Ion Concentration, thereby Altering the Rate of Inactivation of Drug B   In this mechanism, the alterations in metal ion concentrations induced by A's action on an effector decrease (potentiation, synergism) or increase (antagonism) the rate of catabolism of B before B reaches the target site. Carboxylic ionophores monensin and nigericin, which facilitate transmembrane metal ion diffusion, decrease the rate of intracellular catabolism of ^{125}I-labeled monoclonal antibodies in neoplastic cells by metaloenzymes (Press *et al.*, 1990). Monensin and nigericin thereby potentiate the stability and cell-killing activity of ^{125}I-labeled monoclonal antibodies.

Drug A Affects the Synthesis of a Metabolite That Inactivates Drug B Through Direct Combination   This mechanism is similar to A affecting the enzyme that inactivates B, but less direct. If A inhibits synthesis of the metabolite, more B survives to reach the target, resulting in potentiation. Buthionine sulfoximine inhibits glutathione synthesis, thereby inhibiting the deactivation of alkylating agents that can result from thioether adduct formation with glutathione. Thus, buthionine sulfoximine potentiates the cytotoxicity of many alkylating agents against tumor cells (see Chapter 7 by Griffith and Friedman).

5. Drug A Alters the Incorporation of Drug B into a Macromolecule

In this mechanism, B is a drug that exhibits its chemotherapeutic effects after enzymatic incorporation into a macromolecule such as DNA. If A enhances this incorporation, the result is potentiation or synergism; if A inhibits the incorporation, A antagonizes B. The antiviral agent acyclovir, a nucleotide analog, is synergistic against the herpes simplex virus when combined with A1110U, an iron chelating ribonuclotide reductase inhibitor. Acyclovir must be activated to its 5′-triphosphate (AcTP) before it is incorporated into viral DNA. (After incorporation, the abnormal double-stranded oligonucleotide and the deoxynucleotidetriphosphate that would form part of the next base pair bind the viral DNA polymerase and cooperatively inhibit its action.) Deoxyguanosine triphosphate, the normal counterpart to acyclovir triphosphate, competes with AcTP for incorporation into the oligonucleotide, thereby decreasing the antiviral activity of AcTP. The iron chelator A1110U selectively inhibits the ribonucleotide reductase formed by the herpes virus, thereby inhibiting the synthesis of deoxyguanosine triphosphate. Thus, A1110U enhances incorporation of AcTP into the growing polydeoxynucleotide chain. This mechanism is discussed in detail in Chapter 8 by Spector and Fyfe.

6. Drug A Induces a Genetic Change That Alters the Concentration of the Target of Drug B

For this mechanism, drug A causes a mutation that alters the concentration of the target to which drug B binds. The outcome of this interaction (synergism or antagonism) depends on the mechanism of inhibition by B (competitive, noncompetitive, irreversible). Burkitt's lymphoma cells which have become resistant to nitrogen mustards become hypersensitive to novobiocin. This phenomenon is an example of collateral sensitivity, i.e., schedule-dependent synergism between nitrogen mustard and novobiocin. After mutagenesis through exposure to nitrogen mustard, the mustard-resistant lines expressed abnormally high levels of topoisomerase II, thus increasing the sensitivity to novobiocin (a topoisomerase II inhibitor) (Tan *et al.*, 1988). The hypothesis involving a genetic mechanism is reasonable, considering the mutagenicity of alkylating agents (Mariani *et al.*, 1988; Laidlaw *et al.*, 1989; van Zeeland, 1988). For other examples of synergism involving change at the genetic level, see Chapter 16 by Avramis *et al.*(see also Section IIC4).

7. Drug A Alters the Synthesis of a Metabolite That Competes with Drug B for the Target Site

In this mechanism, B is a metabolite analog that competes with the metabolite for the target site. If A causes an increase in the metabolite then

A + B will be antagonistic. If A induces a decrease in the metabolite then A + B will be synergistic. The synergism between PDDF and metoprine involves a second mechanism in addition to the mechanism outlined earlier. Metoprine enhances levels of dUMP by decreasing concentrates of folates that can react with dUMP. The dUMP and the PDDF polyglutamates bind to different sites on thymidylate synthase, causing synergistic inhibition of this vital enzyme (see Chapter 10 by Galivan). In other words, metoprine induces synthesis of a metabolite (dUMP) that binds to the same target as the second drug (PDDF polyglutamate).

8. Drug A Alters the pH, thereby Altering Drug Reactivity with the Target

This mechanism represents yet another way in which a pH change induced by A can impinge on B. Regardless of whether A causes an increase or decrease in pH, A can potentiate B if the pH change enhances target reactivity, whereas A can antagonize B if the pH change decreases target reactivity. The antineoplastic DNA cross-linking agent mechlorethamine (HN2) can be demethylated to bis(2-chloroethyl)amine (nor-HN2) in tumor cells. The DNA cross-linking ability of nor-HN2 rises as the pH drops (Jähde *et al.*, 1989). Glucose decreases the extracellular pH and, to a lesser extent, the intracellular pH in solid tumors (Jähde *et al.*, 1989; Tannock and Rotin, 1989). The potentiation of the antineoplastic activity of alkylating agent by glucose may be due in part to enhancement of nor-HN2 reactivity toward DNA due to glucose-induced decreases in the intracellular pH of solid tumors.

B. CLASS 2: DRUG A BINDS TO AN EFFECTOR, THEREBY ALTERING THE ACTIVITY OF DRUG B AFTER IT REACHES ITS TARGET

1. Drug A Affects the Repair of DNA Damage Caused by Drug B

If drug A inhibits repair of damage caused by a DNA-damaging drug B, the result is potentiation. If drug A accelerates repair, antagonism results.

Drug A Interacts Directly with DNA Repair-Related Proteins, thereby Altering Repair of Drug B-Induced Damage Cells and microbes have a number of mechanisms allowing them to recover from limited DNA damage caused by chemotherapeutic agents. If A inhibits repair, it can potentiate a DNA-damaging agent B; if A enhances repair, it will antagonize B. Nicotinamide, which inhibits the DNA repair-related enzyme poly(ADP-ribose) polymerase, potentiates the DNA-damaging agent, cisplatin. For a detailed discussion of chemotherapeutic potentiation and synergism involving DNA repair, see Chapter 15 by Teicher *et al.*

Drug A Induces Genetic Changes That Alter the Ability to Repair DNA Damage Caused by Drug B Here, drug A has a mutagenic effect on a gene involved in DNA repair. Chronic treatment of V79 Chinese hamster cells to the mutagenic alkylating agent MNNG leads to mutations which cause enhanced $O6$-methylguanine-DNA methyltransferase (MT) synthesis. MT is involved in the specific repair of alkylation damage to guanine oxygen atoms in DNA (Mariani *et al.*, 1988). As a result the MNNG-treated V79 cells become resistant to a number of other alkylating agents. In summary, MNNG exhibits antagonism in combination with other alkylating agents mediated through a genetic change leading to heightened expression of a speific DNA repair protein.

Raji/Burkitt lymphoma cells, when exposed to the cross-linking agent mechlorethamine, become cross resistant to hydroperoxycyclophosphamide and phenylalanine mustard (Frei *et al.*, 1985). It has been suggested that this cross resistance is due to the overexpression of the topoisomerase II gene product characteristic of the mechlorethamine–resistant lymphoma line. Quite possibly this resistance develops as a result of mutagenesis caused by the mechlorethamine. In summary, mechlorethamine and hydroperoxycyclophosphamide exhibit antagonism that is probably mediated through a genetic change induced by mechlorethamine (Mariani *et al.*, 1988; Laidlaw *et al.*, 1989; van Zeeland, 1988). This change causes heightened constitutive expression of topoisomerase II and superior DNA repair abilities (Tan *et al.*, 1988; see also Section IIC4 and Chapter 16 by Avramis *et al.*).

Drug A Induces Changes in Oxygen Concentration, thereby Altering Repair of Drug B-Induced Damage The ability of cells to repair certain kinds of DNA damage appears to be diminished at high concentrations of oxygen so that potentiation of DNA-damaging agents can be achieved with agents that enhance tumor oxygenation. When tumor-bearing animals are treated with cyclophosphamide (while breathing high concentrations of oxygen) the extent of DNA cross-linking in tumor cells and the antineoplastic activity are significantly higher when fluosol is also administered. Fluosol is a nontoxic fluorocarbon emulsion that can serve as an efficient carrier for oxygen *in vivo*, thereby presumably inhibiting enzymes which repair interstrand cross-links formed by metabolites of cyclophosphamide (see Chapter 15 by Teicher *et al.*).

Drug A Alters the pH, thereby Altering Repair of Drug B-Induced Damage This mechanism can be important when DNA repair efficiency depends on the pH. The hypotensive agent hydralazine increases the fraction of tumor cells that are hypoxic in mice bearing KAT and RAF-1 tumors. These agents enhance the anticancer activity of the alkylating agent melphalan. Although hydralazine decreases oxygen tension substantially in tumors, achieving close to 100% hypoxia, this is apparently *not* responsible for the enhanced activity of melphalan. The hypoxia-inducing agent BW–12C is incapable of enhancing melphalan activity (Adams *et al.*, 1989). Since melphalan is known to be more active against cells under acidic environments *in vitro* (Tannock and Rotin, 1989), the decreased pH induced by hydralazine is a reasonable candidate for the tumor-selective potentiation. It seems unlikely that decreased efflux of drug from the tumor site (due to vasoconstriction within the tumor) is an important factor, because hydralazine has an equal potentiating effect on melphalan whether given 1 hr before or 1 hr after the melphalan (Adams *et al.*, 1989). In summary, Adams and co-workers propose that hydralazine potentiates the antineoplastic activity and selectivity of melphalan in solid tumors by inducing a tumor-selective pH drop, thereby inhibiting recovery from melphalan damage.

Drug A Alters the Metal Ion Concentration, thereby Altering Repair of Drug B-Induced Damage   DNA repair efficiency can be sensitive to the concentration of certain metal ions. Etoposide cytotoxicity depends very much on the intracellular ionic environment. Ouabain, which blocks the sodium/potassium exchange pump in the outer membrane of the mammalian cells, antagonizes the action of etoposide. This antagonism is believed to involve a decrease in the effect of etoposide after it binds to its target, topoisomerase, due to alterations in the local intracellular microenvironmental (Lawrence *et al.*, 1989).

2. Drug A Affects Cell Cycle Kinetics, thereby Altering the Biological Effects of Drug B after Target Binding

In this mechanism, drug A blocks the cells at a certain point in the cell cycle, or at least alters the kinetics of the cell cycle (this applies primarily to cancer chemotherapy). Drug B acts selectively or specifically at a certain point X in the cell cycle. If A increases the proportion of cells at point X, it potentiates B, whereas A antagonizes B if A decreases the proportion of cells at point X. The activity of many anticancer drugs varies significantly throughout the cell cycle. Thus, it is possible to effect either potentiation or antagonism by adding a second agent which influences the kinetics of the cycle. For example, methotrexate is most active against cells in the S phase when DNA is being synthesized. However, methotrexate can cause arrest of cells in the G_1 phase through inhibition of RNA synthesis. Hypoxanthine potentiates methotrexate because hypoxanthine maintains RNA synthesis and allows the cells that have been arrested in G_1 to progress into the S phase where they are more sensitive to methotrexate (Fairchild *et al.*, 1988). See Chapter 17 by Chang for a more detailed discussion of the involvement of the cell cycle in synergism and antagonism.

3. Drug A Induces Genetic Changes That Alter the Effects of Drug B after Target Binding without Involving DNA Repair

Drug A has a mutagenic effect, altering a gene involved in the response (other than DNA repair) of the cell or pathogen to the interaction between B and its target. Mammalian tumor cells that have acquired multiple drug resistance through genetic change after exposure to drugs such as doxorubicin become hypersensitive to agents such as local anesthetics and hydrophobic Triton-X detergents. It has been proposed that this synergism is due to a heightened sensitivity to generalized membrane damage after doxorubicin-induced mutation to a phenotype expressing a membrane-spanning P-glyco-protein (Van Der Bliek and Borst, 1989). Here, the enhanced sensitivity to amphiphiles is presumably due to a change in the ability of the membrane to withstand damage *after* amphiphile binding. Postulation of genetic change as a mechanism for schedule-dependent synergism (i.e., collateral sensitivity) is reasonable in light of the mutagenicity of doxorubicin (De-yu *et al.*, 1988; Laidlaw *et al.*, 1989). See Section IIC4 concerning selective drug effects on different subpopulations and genetic selection.

C. CLASS 3: DRUGS A AND B EACH BIND TO SEPARATE TARGETS, CAUSING EFFECTS THAT COMPLEMENT, POTENTIATE, OR ANTAGONIZE ONE ANOTHER

1. Drugs A and B Inhibit or Potentiate Sequential Targets in a Metabolic Pathway

Although it may seem obvious that any combination of inhibitors that block different points along a metabolic pathway ("sequential blocking") or within a network will exhibit synergism, *this is in fact not always the case.* As Jackson shows in Chapter 11 of this book, combinations of antimetabolites may exhibit either synergism, antagonism, or mere additivity, depending on the nature of the inhibition (i.e., competitive or noncompetitive), the topology of the metabolic network (i.e., degree of branching, feedback inhibition, and/or enhancement), and other factors. Perhaps the most familiar example of sequential blocking is the synergistic antibacterial combination of trimethoprim, an inhibitor of dihydrofolate reductase, and sulfadiazine, which inhibits *p*-aminobenzoic acid incorporation into dihydrofolic acid (Hitchings and Burchall, 1965). "Maloprim" is a related combination of the sulfa drug dapsone and the antifolate pyrimethamine that exhibits antiprotozoal synergism. Maloprim is used clinically to treat chloroquine-resistant malaria. Dapsone was chosen over other sulfadrugs to match the slow excretion rate of pyrimethamine. For a discussion of Maloprim and related examples of sequential blocking, see Albert (1985). Antimalarial antifol/sulfa combinations are discussed in Chapter 5 by Vennerstrom *et al.*

A good example of the importance of comparative drug interaction studies involves combinations of antifolates that inhibit dideazafolate reductase and antifolates that inhibit thymidylate synthase. Kisliuk *et al.* (1985) demonstrated that the thymidylate synthase inhibitor $N10$-propargyl-5,8-dideaza-folate [PDDF] exhibits synergism in combination with the DHFR inhibitor trimethoprim in the inhibition of the bacterium *Lactobacillus casei*. Because of this observation, Galivan and co-workers (1987) examined the effects of cytotoxic DHFR inhibitors and thymidylate synthase inhibitors on hepatoma cell growth *in vitro*. They showed that PDDF exhibits cytotoxic synergism in combination with three different mammalian cell DHFR inhibitors: methotrexate, trimetrexate, and metoprine (see Chapter 10 by Galivan).

In an elegant set of experiments involving 3-way interaction, thymidine antagonizes the cytotoxic combination of trimetrexate and PDDF, indicating that the mechanism of synergism between trimetrexate and PDDF involves inhibition of thymidylate synthase. The *lack* of antagonism of trimetrexate and PPDF by inosine argues against the involvement of DHFR in the mechanism of trimetrexate/PDDF synergism (see Chapter 10 by Galivan). This is an example of the use of antagonism to elucidate the mechanism of action of a combination.

2. Drugs A and B Concurrently Inhibit or Potentiate Parallel Targets in a Convergent Metabolic Pathway

Inhibition of two different pathways that lead to the same product will lead to synergism in many instances (see Chapter 11 by Jackson).

Drugs A and B Both Directly Affect Their Parallel Targets in a Convergent Metabolic Pathway   The synthesis of ATP in mammalian cells involves a convergent metabolic pathway: it can occur either through cytoplasmic glycolysis or through oxidative phosphorylation in mitochondria. Rhodamine 123 selectively inhibits oxidative phosphorylation in carcinoma cells by inhibiting F_0F_1ATPase. The antineoplastic activity of rhodamine 123 can be potentiated both *in vitro* and *in vivo* through combination with 2-deoxyglucose, an inhibitor of glycolysis (Bernal *et al.*, 1983; Herr *et al.*, 1987).

Drug A Induces a Mutation that Affects One Metabolic Pathway and Drug B Directly Inhibits a Parallel Target in a Convergent Metabolic Pathway   A clinically useful example of synergism involves the combination of 6-mercaptopurine and methotrexate (see Chapter 3 by Frei). Tumors which become resistant to 6-mercaptopurine often to do so by losing the activity of the gene which is required to activate 6-mercaptopurine. This gene is also required for the synthesis of purines from hypoxanthine (the salvage pathway). The salvage pathway runs in parallel with the *de novo* pathway, which can be blocked by methotrexate. Once cells have lost the ability to synthesize purines from hypoxanthine due to mutation induced by 6-mercaptopurine, they become unusually susceptible to methotrexate because both salvage and *de novo* pathways are inoperative and nucleic acid synthesis stops altogether. Here 6-mercaptopurine, rather than directly inhibiting the salvage pathway, induces a mutation which eliminates this pathway completely. An alternative or simultaneous interaction mechanism would involve genetic selection of 6-

mercaptopurine resistant cells rather than mutation (see IIC4). Synergism involving genetic change is discussed in detail in Chapter 16 by Avramis, *et al.*

Drug A and Drug B Become Incorporated into Different Sites in the Same Biopolymer The synergism between AZT and ddI might be explained in terms of the ability of their active metabolites (5'-triphosphates AZTTP and ddITP or ddATP) to be incorporated into different sites (opposite A and U, respectively) on the same DNA strand formed by the HIV reverse transcriptase. In this way, neither drug competes for the incorporation of the other (see Chapter 4 by Schinazi).

Studies of synergism can occasionally be useful in the design of unique single agents. For example, a conjugate of AZT and ddI formed by linking 5' hydroxyl groups through a phosphodiester (AZTp-ddI) exhibits greater antiviral potency and selectivity *in vitro* than either AZT or ddI. AZTp-ddI is believed to hydrolyze intracellularly, forming AZT, ddI, and their 5'-monophosphates (see Chapter 4 by Schinazi).

3. Drugs A and B Inhibit Targets from Distinct Metabolic Pathways within a Single Cell, Microbe, or Virally Infected Cell

The antiretroviral reverse transcriptase inhibitors phosphonoformate, 3'-azidodeoxythymidine (AZT), and dideoxycytidine each show well-defined synergism against HIV-infected cells in combination with recombinant alpha-interferon. Alpha-A-interferon inhibits retroviruses through a mechanism other than the inhibition of reverse transcription. Thus, a combination of interferon and AZT exhibits synergism due to the inhibition of targets in two distinct vital metabolic networks of the retrovirus (Vogt *et al.*, 1988; Hartshorn *et al.*, 1987; see Chapter 20 by Berman and Chang).

Flucytosine is a cytosine analog that is selectively converted to the thymidylate synthase inhibitor fluororodeoxyuridine in fungi, but not in mammalian cells. Amphotericin B binds selectively to a steriod found only in fungal membranes and interferes with outer membrane function and integrity. In combination, these two agents were found to be superior to either agent alone in the clinical treatment of the life-threatening fungal infection cryptococcal meningitis (Bennett *et al.*, 1979).

4. Drugs A and B Each Affect a Different Subpopulation of Cells, Microbes, or Viruses

This mechanism of interaction is not responsible for synergism or antagonism in the classic sense, although it could not be distinguished from "classical" synergism using isobologram or combination index analysis (see Chapter 18 by Durand). Collateral sensitivity can involve genetic selection when exposure to one agent leads to hypersensitivity to the second agent because all of the remaining cells, microbes, and viruses are particularly sensitive to the second agent. For example, if a mixed population of tumor cells containing both sensitive and multiple drug-resistant cells is exposed to vinblastine, a nonmutagenic antineoplastic agent, the sensitive cells will be selectively killed and the multiply drug-resistant cells that remain will be unusually sensitive to local anesthetics (van der Bliek and Borst, 1989). In other words, vinblastine and certain local anesthetics should exhibit apparent synergism in their action on such mixed populations of cells with the vinblastine selectively killing off the subpopulation lacking the P-glycoprotein and the local anesthetics selectively killing off the subpopulation with the sensitive outer membrane due to the presence of the P-glycoprotein. Similarly, if such a mixed population of cells is first exposed to vinblastine and then to adriamycin, antagonism should result, since the vinblastine would increase the relative proportion of multiply drug-resistant cells in the mixed population through selection. For a detailed discussion of multidrug resistance, see Chapter 9 by Yang, *et al.*

An example in which exposure to a drug causes an increase in the activity of a second drug after reaching its target involves the collateral sensitivity of Chinese hamster ovary cells to the microtubular active agents taxol and colcemid (Schibler and Cabral, 1986). Taxol, which is currently undergoing clinical trials as an anticancer agent, inhibits cell growth by stabilizing assembled microtubules. In contrast, colcemid prevents the assembly of microtubules. The CHO cells which have been selected to resist the microtubular disruption action of colcemid become hypersensitive to the microtubular assembling activity of taxol, and vice versa. This collateral sensitivity (specifically, synergism mediated through genetic selection) may be of clinical relevance since taxol has demonstrated clinical utility and two clinically approved agents, vinblastine and vincristine, have mechanisms very similar to colcemid.

Siemann and Keng in Chapter 19 of this work have reviewed the synergistic interactions obtained against solid tumors by using combinations of agents targeted against two or more cell populations. The use of combinations of antitumor agents that can each destroy a different subpopulation of tumor cells in human subjects is the basis for many clinical anticancer regimens. The appearance of drug-resistant subpopulations of tumor cells, viruses, or microbes through selection or mutation is believed to be less likely after combination chemotherapy using two or more agents that do not always exhibit cross resistance (see Chapter 3 by Frei, Chapter 4 by Schinazi, and Chapter 5 by Vennerstrom *et al.*). Selective action by drug A against subpopulations resistant to drug B (and vice versa) has been proposed as a key mechanism suppressing development of drug resistance during combination chemotherapy.

It is important (but often difficult) to distinguish between selective effects on different subpopulations and mechanisms of synergism that involve mutagenesis induced by the first agent which impinges upon the effects of a second agent added at a later time. Examples of synergism involving such genetic changes have been discussed in earlier subsections. For each of these mechanisms involving genetic change there is a corresponding example of selective inhibitions of different subpopulations if both resistant *and* sensitive forms of the cell, microbe, or virus exist before the addition of either of the two agents. The mechanisms involving genetic change require mutation whereas the mechanisms involving the effects on different subpopulations involve genetic selection. In addition, both selection and mutation can readily occur together (a mixed mechanism).

If Chinese hamster ovary cells are treated with combinations of vinblastine and verapamil it is possible to isolate mutants that are resistant to verapamil because of abnormalities in the tubulin genes. This is true because the verapamil inhibits the action of the P-glycoprotein, thereby preventing the selection of mutants which resist vinblastine through a multidrug resistance mechanism. In contrast, verapamil does not interfere with the selection of tubulin mutants. This example demonstrates the utility of drug combinations in the elucidation of drug resistance mechanisms (Schibler *et al.*, 1989).

Mutual rescue in co-cultures of thymidine kinase negative (TK⁻) cells and hypoxanthine–guanine phosphoribosyltransferase negative [HGPRT⁻] cells represents a truly distinctive mechanism of subpopulation-specific antagonism. The HGPRT⁻ cells, in the presence of the antifolate aminopterin, can phosphorylate thymidine but fail to convert hypoxanthine to other purines. In contrast, TK⁻ cells can convert hypoxanthine to other purines but cannot phosphorylate thymidine. Neither of these cells individually can survive in the presence of aminopterin through salvage of thymine and hypoxanthine (i.e., they fail to grow in HAT medium). However, when the cells are in contact with one another and can communicate thorough gap junctions, both become resistant to aminopterin provided thymidine and hypoxanthine are both present. In other words, there is a selective antagonism of aminopterin activity by thymidine plus hypoxanthine, but only in HGPRT⁻ cells which are communicating with TK⁻ cells and TK⁻ cells which are communicating with HGPRT⁻ cells (Hooper and Subak-Sharpe, 1981). Antagonism of this type might be of clinical relevance in normal tissue resistance or tumor resistance to chemotherapy, considering the fact that gap junctions have been observed in both normal and tumor-derived human mammary epithelial cells (Eldridge *et al.*, 1989). The antagonism between thymidine, hypoxanthine, and aminopterin represents a mixed mechanism because it involves both enhancement of drug influx (via the gap junctions) and concurrent enhancement of reaction rates in a convergent metabolic pathway leading to DNA and RNA synthesis.

D. CLASS 4: DRUGS A AND B BIND TO THE SAME
MACROMOLECULE OR MEMBRANE TARGET

1. Drugs A and B Bind to Identical Sites on the Same Macromolecule or Membrane Target

If both A and B bind to identical sites on the same macromolecule, and both have the same kind of effect (i.e., both agonists or both antagonists), this will result in additive or antagonistic interaction because both drugs cannot bind the same site at the same time. The result will be antagonism if A is an antagonist and B is an antagonist, or if A binds noncovalently and B causes covalent modification of the target. Metabolic precursors, administered exogenously, can antagonize the action of antimetabolites. Leucovorin is a precursor for dihydrofolate, the substrate of dihydrofolate reductase (DHFR). Leucovorin thus inhibits the cytotoxic effects of DHFR inhibitors such as the antitumor antifolate methotrexate and the antimicrobial antifolate trimethoprim. Both dihydrofolate and antifolates bind to the same site on DHFR. Because the antagonism is specific for normal human tissues, leucovorin is clinically useful in anticancer chemotherapy (see Chapter 3 by Frei). Trimetrexate/leucovorin combinations are useful for the treatment of *Pneumocystis carinii*, an opportunistic pathogen that is the main cause of mortality in AIDS patients (D'Antonio *et al.*, 1986; Rosowsky *et al.*, 1989). Leucovorin can gain access to the cytoplasm of human cells, but fails to permeate and protect the pathogen. The success of "leucovorin rescue" for different antifolates in very different diseases underlines the importance of comparative interaction mechanism studies.

At high levels, intercalators such as ethidium bromide can antagonize bleomycin's cytotoxicity by blocking potential bleomycin binding sites on duplex DNA (see Chapter 12 by Strekowski and Wilson). The host toxicity of the antineoplastic fluorouracil can be antagonized by cytidine. Cytidine is partially converted to uracil and subsequently to deoxyuridine monophosphate (dUMP) in mammals. The dUMP exhibits antagonism with the FdUMP formed from fluorouracil through simple competitive inhibition of FdUMP binding to thymidylate synthase (Martin, 1987). Uridine probably antagonizes the cytotoxicity of azidodideoxythymidine (AZT) to bone marrow through a mechanism involving competition between a uridine metabolite (such as UTP) and an AZT metabolite (such as AZTTP) for the same mammalian enzyme site. See Chapter 4 by Schinazi for a discussion of the clinical potential of this combination.

2. Drugs A and B Bind to Different Sites on the Same Macromolecule or Membrane Target

When A and B bind different sites on the same target, this can lead to synergism, additivity or antagonism, depending on the nature of the allosteric interactions (if any) between these sites and the type of effect (agonistic, antagonistic, etc.) of A and B.

A variety of DNA binding molecules, especially those which tend to bind in the major groove of duplex DNA, have been shown to potentiate the DNA-cleaving activity of bleomycin *in vitro*. The observed antineoplastic synergism of these major groove-binding agents and bleomycin is thus due at least in part to simultaneous binding to the DNA molecule as opposed to being merely a matter of inhibition of DNA repair. This work and other examples of synergism involving simultaneous drug binding to the same target are detailed in Chapter 12 by Strekowski and Wilson. Some of the molecules that potentiate bleomycin when administered at low concentration can antagonize bleomycin when administered at higher concentrations by blocking potential bleomycin binding sites on the DNA. Thus, the potentiation is concentration dependent.

Another example is reflected in the synergism between tiazofurin and ribovirin against hepatoma cells. Tiazofurin is converted in mammals to its active drug form (TAD), as is ribovirin (to RMP). Both of these active drug metabolites, TAD and RMP, bind to inosine monophosphate dehydrogenase, a key enzyme in *de novo* purine biosynthesis. However, TAD and RMP do not inhibit one another because TAD binds to the NADH site while RMP binds to the inosine monophosphate/xanthine monophosphate binding site. (See Chapter 12 by Strekowski and Wilson).

An unusual example of synergism involving simultaneous binding of two agents to a membrane target was described by Hopfer *et al.* (1987). Amphotericin B demonstrates synergistic antifungal activity when combined with gramicidin S. Both of these antifungal agents disrupt fungal membranes, with some selectivity relative to mammalian cell membranes. It was shown that both the overall antifungal potency and the selectivity could be enhanced by using gramicidin S and amphotericin in combination and by incorporating both drugs in liposomes.

Perhaps the best known example of simultaneous drug binding to different sites on the same macromolecular target involves combinations of fluorouracil and leucovorin. Fluorouracil, after metabolic activation to fluorodeoxyuridine monophosphate (FdUMP), binds to thymidylate synthase. Leucovorin is converted to dihydrofolate, which binds to a different site on the same protein. The ternary complex dissociates very slowly. The resulting inhibition of synthesis of thymidylate by the enzyme blocks *de novo* DNA synthesis. Combinations of fluorouracil and leucovorin show a great deal of clinical promise in the treatment of gastrointestinal carcinomas because of this potentiation (Moran, 1989; see also Chapter 13 by Mini and Bertino).

The molecules tropolone methylether and *N*-acetyl mescaline (which resemble different segments of the antimitotic agent colchicine) are able to bind simultaneously to alpha tubulin without inhibiting one another (Andreu and Timasheff, 1986). Taxol (an experimental antimitotic agent that prevents tubulin depolymerization) and desacetamidocolchicine (an agent that prevents microtubular polymerization) can both bind to different sites on the same tubulin molecule (Bhattacharyya *et al.*, 1986).

E. CLASS 5: DRUGS A AND B REACT WITH ONE ANOTHER *IN VIVO*, FORMING A NEW DRUG WITH SIGNIFICANTLY ALTERED BIOLOGICAL EFFECTS

1. Drugs A and B Form an Adduct, C, with a Biological Activity That Is Significantly Different from Either A or B

Perhaps the most straightforward possible mechanism of synergism involves a combination of two reactive molecules which can combine with one another in or near the target under physiological conditions to form a product that is more bioactive than either of the two precursors. Although the mechanism of *action* of the product of this reaction may be quite complex, the mechanism of *interaction* involves a simple bimolecular reaction. No macromolecule or membrane is necessarily involved in the mechanism of synergism. An example is the *in vitro* cytotoxic and antibacterial synergism between decanal and *N*-amino *N'*-octylguanidine (AOG) (Rideout, 1986; Rideout *et al.*, 1988). Here, the synergism involves the formation of a cationic hydrazone detergent, decylideneiminooctylguanidine (DIOG) from decanal and AOG. DIOG is much more potent in terms of its cytotoxic activity under physiological conditions, causing complete oxblood erythrocyte lysis within 20 min at 14 μm whereas decanal (which is not cationic) and AOG (which exhibits little or no detergent action) show no activity after several hours at significantly higher concentrations. The mechanism of synergism was shown to involve direct covalent reaction through isolation of the product (DIOG) from cell suspensions that had been treated with decanal and AOG, and through demonstration of a lack of synergism in related, unreactive pairs of molecules. In another set of experiments, a combination of a triarylalkylphosphonium-derived aldehyde and a triarylalkylphosphonium-derived acylhydrazide exhibited a greater degree of synergism against Ehrlich ascites carcinoma cells than against untransformed CV-1 cells (Rideout *et. al.*, 1990). These and related examples are discussed in more detail in Chapter 14 by Rideout and Calogeropoulou.

Damage of human kidney tissue by the antineoplastic agent cisplatin only occurs after extracellular activation of cisplatin through replacement of a chloride ligand with water to form the aquated ion $(H_3N)_2(H_2O)PtCl^+$. This ion can combine with chloride anion to regenerate the less nephrotoxic cisplatin, again by displacement of the water ligand. Chloride anion (in the form of saline chlorouresis) is used clinically to prevent cisplatin-mediated nephrotoxicity (Conden, 1987; Ozols *et al.*, 1984). Here, antagonism of toxicity is caused by a chemical combination between chloride anion and $(H_3N)_2(H_2O)PtCl^+$, forming the less toxic cisplatin and water. Cisplatin does not need to be aquated *outside* of tumor cells to exhibit antineoplastic activity, and thus saline decreases cisplatin nephrotoxicity without compromising antineoplastic activity. Diethyl dithiocarbamate (DDTC) can combine with cisplatin/protein complexes, forming cisplatin/DDTC complexes and freeing the protein. DDTC antagonizes the toxic effects of cisplatin in animals presumably through direct covalent interaction (Gonias *et al.*, 1984). For further discussion of antagonism involving direct bond formation between agents, see Chapter 14 by Rideout and Calogeropoulou.

It is prudent to be cautious in postulating direct covalent combination between agents as an *in vivo* mechanism for antagonism of combinations such as thiols and alkylating agents or cisplatin. For example, Satoh *et al.* (1988) have shown that naturally occurring thiols such as metallothionein can be induced, and these can decrease the toxic side effects of drugs such as cisplatin. Similar mechanisms may operate in cyclophosphamide/thiosulfate and cisplatin/DITC combinations, if the thiosulfate and DITC can cause sufficient increases in glutathione concentrations or levels of enzymes that conjugate glutathione to alkylating agents.

Direct noncovalent interaction appears to explain the selective antagonism in combinations of sucrose monolaurate and amphotericin B. The amphiphilic sucrose monolaurate binds to the amphiphilic amphotericin B and prevents it from interacting with cholesterol that is found in mammalian cells. However, sucrose monolaurate fails to inhibit the binding of amphotericin to ergosterol, a steroid found in the membranes of the pathogenic fungus *Candida albicans* and other fungi, but not in mammalian cells. As a result, the nontoxic detergent sucrose monolaurate might be useful for enhancing the low therapeutic index of amphotericin B in the treatment of *Candida albicans* infections (Gruda *et al.*, 1988).

2. Drug A is a Catalyst or Enzyme That Activates a Prodrug to Form Drug B

A novel approach to drug potentiation involves the use of an enzyme or enzyme conjugate as one of the two components. Here, drug A itself (an enzyme or other catalyst) acts as the effector. The enzyme is generally covalently linked to a neoplastic antigen-specific antibody which targets it to the tumor. After the enzyme/antibody conjugate has achieved a suitably high tumor/normal tissue ratio, a prodrug is added in a second step. The prodrug is selectively converted to the active drug in the vicinity of the tumor, thereby exhibiting tumor-selective potentiation. The result is a greater antitumor selectivity than would be attained if the drug were administered in its active form (Kuefner *et al.*, 1988; Senter, 1990). Direct interaction between the two agents is involved in this mechanism.

3. Drug A Is a Catalyst or Enzyme That Deactivates Drug B

Here, again, drug A itself (the enzyme or other catalyst) acts as an effector, accelerating the breakdown of B by acting directly on B. Tannins, ellagic acid, and other phenols protect bacteria and mammalian cells from the mutagenic effects of the carcinogen benzpyrenediolepoxide (BPDE) by accelerating the rate of hydrolysis of the epoxide to a nonmutagenic tetraol. These phenols mediate epoxide hydrolysis by acting directly on the BPDE in aqueous solution in the absence of proteins, and presumably act through this direct mechanism in cultured cells as well (see Chapter 14 by Rideout and Calogeropoulou).

4. Drug A and Drug B Undergo an Oxidation/Reduction Interaction, Altering the Activity of One or Both Agents

The antimalarial agent artesunate is an endoperoxide that becomes re-duced as part of its mode of action. Doxorubicin, in its reduced form, can generate toxic free radicals by interacting with oxidants. The antimalarial synergism between these two is believed to involve redox chemistry. The antimalarial antagonism between vitamin E and artemisin is also believed to involve redox interactions (see Chapter 5 by Vennerstrom *et al.*).

F. MIXED MECHANISMS

Mechanisms of chemotherapeutic synergism, antagonism and potentiation can be mixed in nature and highly complex. A good illustrative example of this complexity is the proposed mechanism of synergism between lonidamine and doxorubicin. Lonidamine enhances membrane permeability, allowing higher concentrations of doxorubicin to enter the mitochondria. The doxoru-bicin inhibits oxidative phosphorylation in mitochondria, while the lonida-mine inhibits another form of ATP synthesis: glycolysis. The drop in ATP resulting from the action of the lonidamine thereby inhibits active export of doxorubicin (Floridi *et al.*, 1988). An additional complication in the mechan-ism of synergism is the fact that lonidamine is known to be capable of inhibiting repair of potential lethal damage induced by agents such as doxoru-bicin (see Chapter 15 by Teicher *et al.*). Another mixed mechanism, already mentioned (Sections IIA3 and A5) is the antiretroviral antagonism between AZT and ribavirin. Both inhibition of prodrug activation and inhibition of drug incorporation into DNA are involved. The antineoplastic combination PDDF + metoprine exhibits synergism through a combined mechanism (see Sections IIA3 and IIA7). Potentiation of the effects of doxorubicin on doxorubicin-resistant P388 lymphoma cells by the calcium channel blocker DMDP involved a mixed mechanism. DMDP, a lipophilic amine, blocks the

P-glycoprotein's ability to pump doxorubicin out of the cell. In addition, however, DMDP inhibits the repair of DNA single strand breaks induced by doxorubicin. DMDP selectively inhibits repair in resistant P388 cells as compared with sensitive P388 cells (Yin *et al.*, 1989).

III. Condition-Related Aspects of Synergism, Antagonism, and Potentiation

A. SPATIAL AND MICROENVIRONMENTAL EFFECTS (SUBPOPULATION-SELECTIVE DIFFERENCES IN SYNERGISM AND ANTAGONISM)

The extent of synergism or antagonism can vary for different cellular subpopulations in different locations. The degree of synergism can vary for different subpopulations of tumor cells in different environments. For example, the degree of synergism between etoposide and cisplatin (1:1 ratio) at a 99% kill level observed for V79 mammalian cell spheroids is substantially greater (i.e., the combination index is lower) for cells buried inside the spheroid than for cells on the surface (see Chapter 18 by Durand). An even stronger, similar trend (greater synergism for interior cells) is observed for misonidazole + CCNU. The spheroids represent models of solid tumors, with the interior cells representing tumor cells that are distant from capillaries. These interior cells are in a more hypoxic microenvironment in which molecules such as misonidazole can readily be reduced to the active form. Thus microenvironment differences may explain the extreme locational dependence of misonidazole/CCNU synergy. (See also IIC4 for an example of subpopulation-specific antagonism involving intercellular communication.) For a discussion of synergism involving combined drug effects on different tumor cell subpopulations, see Chapter 19 by Siemann and Keng. Chapter 20 by Berman and Chang describes synergism that is more pronounced for tumor cells than for bone marrow cells.

The microenvironment of the target can effect the extent of synergism or antagonism. The pH experienced by a pathogen in an infected animal can vary considerably. For example, acidosis can occur at the site of a severe bacterial infection. Some bacterial and protozoal pathogens, such as *Salmonella typhimurium* and *Leishmania donovanii*, can live in infected animals at both pH 7.4 (the cytoplasmic or extracellular environment) and pH between 4 and 5 (lysosomal environment). The degree of synergism between combinations of aldehydes and hydrazine derivatives that can combine *in situ* can be significantly greater at pH 5 than at pH 7.4, because the rate of reaction to form the more bactericidal hydrazones is faster at pH5 (Rideout *et al.*, 1988; Chapter 14 by Rideout and Calogeropoulou).

B. SCHEDULING EFFECTS

Scheduling effects on synergism in cancer chemotherapy are reviewed in detail in Chapter 17 by Chang, and discussed for fluorouracil combinations in Chapter 13 by Mini and Bertino. The order in which two agents are administered can dramatically affect the degree of synergism. For example, pretreatment of L1210 leukemia bearing mice with etoposide followed 6 hr later by high dose cytarabine produced a 100% cure rate. However, if an identical dose of cytarabine was administered 6 hr before etoposide, the cure rate drops significantly. This is an example of schedule-dependent therapeutic synergism. The following explanation was offered. When cytarabine is administered first, cells become blocked at the G_1/S border of the cell cycle. Etoposide, a topoisomerase II inhibitor, can only induce DNA strand breaks during the S phase, when DNA is being synthesized. Thus, pretreatment of cells with cytarabine antagonizes the action of VP16. In contrast, however, if cytarabine is administered after etoposide it may inhibit the repair of the DNA damage caused by the etoposide during S phase resulting in synergism and the dramatically high cure rate for this schedule (Ohkubo *et al.*, 1988). The mechanism of interaction differs depending on the order administration.

Tumor promotion by protein kinase C stimulators such as tetradecanoylphorbol acetate (TPA) represents a classic instance of schedule-dependent synergistic toxicity. For example, if mouse skin is treated with the DNA-damaging carcinogen BPDE, subsequent treatment with TPA will enhance the incidence of skin tumor formation considerably. However, pretreatment with TPA does not enhance BPDE carcinogenesis. This schedule dependence can be explained in terms of the reversibility of PKC stimulation by TPA in contrast with the irreversible nature of covalent DNA damage and subsequent mutations induced by BPDE. The effects of TPA stimulation reverse themselves after TPA is removed, and thus have no effect on subsequent BPDE treatment (Weinstein, 1988). Such adverse interactions need to be considered when designing combination chemotherapy regimens since several clinical chemotherapeutic agents, especially anticancer drugs, are mutagenic and carcinogenic (Pitot, 1989).

Combinations of trimetrexate and fluorouracil exhibit a schedule-dependent synergism in their cytotoxic activity that is dependent on a prodrug activation mechanism. If trimetrexate is given before fluorouracil, the trimetrexate induces an increase in the level of 5-phosphoribosyl 1-pyrophosphate (PRPP). PRPP and fluorouracil combine to form fluorodeoxyuridine monophosphate (FdUMP), the active form of fluorouracil, a process that is accelerated by higher PRPP levels. Cells are then killed or growth is inhibited through inhibition of thymidylate synthase. In contrast, if trimetrexate is given after fluorouracil, this enhancement of FdUMP foundation is not observed and the degree of synergism is much less pronounced. This schedule dependence was observed in cultured Chinese hamster ovary cells and also in mice infected with P388 leukemia (Elliott *et al.*, 1989).

The order in which two agents are administered to a system can actually *make the difference between synergism and antagonism*. It is clearly vital to take scheduling effects into account when determining the degree of synergism between two agents or designing combination chemotherapy protocols.

In summary, it is crucial to understand and quantify synergism and antagonism because of the growing significance of these phenomena in clinical combination chemotherapy and experimental pharmacology. The most useful and mathematically well-founded methods for quantifying synergism and antagonism are the interrelated combination index and isobologram methodologies. The mechanisms of interaction in most, if not all, combinations of two agents (represented as A and B) can be divided into five classes:

1. A binds an effector macromolecule (or membrane), thereby altering the ability of B to reach its target macromolecule (or membrane) and achieve a chemotherapeutic effect.
2. A binds an effector, thereby altering the activity of B after it reaches its target.
3. A and B each bind separate targets, causing effects that complement, potentiate, or antagonize one another.
4. A and B bind to the same target.
5. A and B react with one another, forming a new molecule with altered bioactivity.

References

Adams, G. E., Stratfor, I. J., Godden, J., and Howells, N. (1989). Enhancement of the antitumor effect of melphalan in experimental mice by some vasoactive agents. *Int. J. Rad. Onc.* **16**, 1137–1139.

Albert, A. (1985), "Selective Toxicity. The Physico-chemical Basis of Therapy," 7th Ed. Chapman & Hall, New York.

Allan, J. D., Jr. (1987). Antibiotic combinations. *Med Clin. North Am.* **71**, 1079–1091.

Allan, J. D., Jr., and Moellering, R. D., Jr. (1985). Management of infections caused by gram-negative bacilli: The role of antimicrobial combinations. *Rev. Infect. Dis.* **4**, S559–S71.

Andreu, J. M., and Timasheff, S. N. (1986). Tubulin–colchicine interactions and polymerization of the complex. *Ann. N.Y. Acad. Sci.* **466**, 676–689.

Ashford, J. R., and Cobby, J. M. (1974). A system of models for quantal response to the joint action of a mixture of drugs. *Biometrika* **37**, 457–474.

Bennett, J. E., Dismukes, W. E., Duma, R. J., Medoff, G., Sande, M. A., Gallis, H., Leonard, J., Fields, B. T., Bradshaw, M., Haywood, J., McGee, Z. A., Cate, T. R., Cobbs, C. G., Warner, J. F., and Alling, D. W. (1979). A comparison of amphotericin B alone and combined with flucytosine in the treatment of cryptococcal meningitis. *N. Engl. J. Med.* **301**, 126–131.

Berenbaum, M. C. (1977). Synergy, additivism and antagonism in immunosuppression: A critical review. *Clin. Exp. Immunol.* **28**, 1–18.

Berenbaum, M. C. (1980). Correlations between methods for measurement of synergy. *J. Infect. Dis.* **137**, 122–130.

Berenbaum, M. C. (1981). Criteria for analyzing interactions between biologically active agents. *Adv. Cancer Res.* **35**, 269–335.

Berenbaum, M. C. (1985). The expected effect of a combination of agents: The general solution. *J. Theor. Biol.* **114**, 413–431.

Berenbaum, M. C. (1988). Isobolographic, algebraic and search methods in the analysis of multi-agent synergy. *J. Am. Coll. Toxicol.*

Berenbaum, M. C (1989). What is synergy? *Pharmacol. Rev.* **41**, 93–141.

Berman, E., Duigou-Osterndorf, R., Krown, S. E., Fanucchi, M. P., Chou. J., Hirsch, M. S., Clarkson, B. D., and Chou, T.-C. (1989). Synergistic cytotoxic effect of azidothymidine and recombinant interferon alpha on normal human bone marrow progenitor cells. *Blood* **74**, 1281–1286.

Bernal, S. D., Lampidis, T. J., McIsaac, R. M., and Chen, L. B. (1983). Anticarcinoma activity *in vivo* of rhodamine 123, a mitochondrial-specific dye. *Science* **222**, 169–172.

Bhattacharyya, B., Ghoshchaudhuri, G., Maity, S., and Biswas, B. B. (1986). The role of the B-ring of colchicine on taxol-induced tubulin polymerization. *Ann. N.Y. Acad. Sci.* **466**, 791–793.

Bliss, C. I. (1939). The toxicity of poisons applied jointly. *Ann. Appl. Biol.* **26**, 585–615.

Bliss, C. I. (1967). "Statistics in Biology." McGraw-Hill, New York.

Bonnem, E. M. (1987). Alpha interferon: Combinations with other antineoplastic modalities. *Semin. Oncol.* **14**, 48–60.

Borch, R. F., and Markman, M. (1989). Biochemical modulation of cisplatin toxicity. *Pharmacol. Ther.* **41**, 371–380.

Brook, I. (1987). Synergistic combinations of antimicrobial agents against anaerobic bacteria. *Pediatr. Infect. Dis. J.* **6**, 332–335.

Brown, J. M., Hall, E. J., Hirst, D. J., Kinsella, T. J., Kligerman, M. M., Mitchell, J., and Travis, E. J. V. (1988). Chemical modification of radiation and chemotherapy. *Am. J. Clin. Oncol.* **11**, 288–303.

Calandra, T., and Glauser, M. P. (1986). Immunocompromised animal models for the study of antibiotic combinations, *Am. J. Med.* **30**, 45–52.

Chou, T.-C. (1974). Relationships between inhibition constants and fractional inhibitions in enzyme-catalyzed reactions with different numbers of reactants, different reaction mechanisms, and different types of mechanisms of inhibition. *Mol. Pharmacol.* **10**, 235–247.

Chou, T.-C. (1975). Derivation of Michaelis–Menten type and Hill type equations for reference ligands. *Proc. Int. Congr. Pharmacol., 6th* p. 619.

Chou, T.-C. (1976). Derivation and properties of Michaelis–Menten type and Hill type equations for reference ligands. *J. Theor. Biol.* **39**, 253–276.

Chou, T.-C. (1977). On the determination of availability of ligand binding sites in steady-state systems. *J. Theor. Biol.* **65**, 345–356.

Chou, T.-C. (1980). Comparison of dose–effect relationships of carcinogens following low-dose chronic exposure and high-dose single injection: An analysis by the median-effect principle. *Carcinogenesis* **1**, 203–213.

Chou, T.-C. (1987). Quantitative dose–effect analysis and algorithms: A theoretical study. *Asia Pac. J. Pharmacol.* **2**, 93–99.

Chou. J., and Chou, T.-C. (1986). "Dose–Effect Analysis with Microcomputers: Dose, Effect, Binding and Kinetics," computer software for IBM-PC series (2.1 DOS, 5.25" floppy disk). Elsevier–Biosoft, Cambridge, England.

Chou, J., and Chou, T.-C. (1988). Computerized simulation of dose reduction index (DRI) in synergistic drug combinations. *Pharmacologist* **30**, Abstract A231.

Chou, T.-C., and Fanucchi, M. P. (1988). Chemotherapy. *In* "Encyclopedia of Medical Devices and Instrumentation" (J. G. Webster, ed.), Vol 2, pp. 660–670. Wiley, New York.

Chou, T.-C., and Talalay, P. (1977). A simple generalized equation for the analysis of multiple inhibitions of Michaelis–Menten kinetic systems. *J. Biol. Chem.* **252**, 6438–6442.

Chou, T.-C., and Talalay, P. (1981). Generalized equations for the analysis of inhibitors of Michaelis–Menten and higher order kinetic systems with two or more mutually exclusive and nonexclusive inhibitors. *Eur. J. Biochem.* **115**, 207–216.

Chou, T.-C., and Talalay, P. (1983). Analysis of combined drug effects: A new look at a very old problem. *Trends Pharmacol. Sci.* **4**, 450–454.

Chou, T.-C., and Talalay, P. (1984). Quantitative analysis of dose–effect relationships: The combined effects of multiple drugs or enzyme inhibitors. *Adv. Enzyme Regul.* **22**, 27–55.

Chou, T.-C., and Talalay, P. (1987). Applications of the median-effect principle for the assessment of low-dose risk of carcinogens and for the quantitation of synergism and antagonism of chemotherapeutic agents. *In* "New Avenues in Developmental Cancer Chemotherapy" (K. R. Harrap and T. A. Connors, eds.), Bristol-Myers Symp. Ser. 8, pp. 37–64. Academic Press, Orlando, Florida.

Chou, T.-C., Rideout, D., Chou, J., and Bertino, R. J. (1991), Chemotherapeutic synergism, potentiation, and antagonism. *In* "Encyclopedia of Human Biology" (R. Dulbecco, ed.), Vol. 2, Academic Press, San Diego.

Cleland, W. W. (1963). The kinetics of enzyme catalyzed reactions with two or more substrates or products. *Biochim. Biophys. Acta* **67**, 173–196.

Conden, B. J. (1987), Reaction of platinum (II) antitumor agents with sulfhydryl compounds and the implications for nephrotoxicity. *Inorg. Clin. Acta* **137**, 125–130.

Coune, A. (1988). Amphotericin B as a potentiation agent to cytotoxic chemotherapy. *Eur. J. Cancer Clin. Oncol.* **24**, 117–121.

D'Antonio, R. G., Johnson, D. B., Winn. R. E., van Dellen, A. F., and Evans, M. E. (1986). Effect of folinic acid on the capacity of trimethoprim–sulfamethoxazole to prevent and treat *pneumocystis carinii* pneumonia in rats. *Antimicrob. Agents Chemother*, **29**, 327–329.

De-yu, L., Lahdetie, J., and Parvinen, M. (1988), Mutagenicity of gossypol analyzed by induction of meiotic micronuclei *in vitro. Mutat. Res.* **208**, 69–72.

Eldridge, S. R., Martens, T. W., Sattler C., A., and Gould, M. N. (1989). Association of decreased intercellular communication with the immortal but not the tumorigenic phenotype in human mammary epithelial cells. *Cancer Res.* **49**, 4326–4331.

Elion, G. B., Singer, S., and Hitchings, G. H. (1954). Antagonists of nucleic acid derivatives. VIII. Synergism in combinations of biochemically related antimetabolites. *J. Biol. Chem.* **208**, 477–488.

Eliopoulos, G. (1987). Antibiotic synergism and antagonism: Significance in clinical infections. *Chemioterapia* **6**, 188–190.

Elliott, W. L., Howard, C. T., Dykes, D. J., and Leopold, W. R. (1989). Sequence and schedule-dependent synergy of trimetrexate in combination with 5-fluorouracil *in vitro* and in mice. *Cancer Res.* **49**, 5586–5590.

English, A. R., Retsema, J. A., Girard, A. E., Lynch, J. E., and Barth, W. E. (1978). CP-45899, a beta-lactamase inhibitor extending the antibacterial spectrum of beta-lactams: Initial bacteriological characterization. *Antimicrob. Agents Chemother.* **14**, 414.

Fairchild, C. R., Maybaum, J., and Straw, J. A. (1988). Enhanced cytotoxicity with methotrexate in conjunction with hypoxanthine in L1210 cells in culture. *Cancer Chemother. Pharmacol.* **22**, 26–32.

Finney, D. J. (1952). "Probit Analysis," 2nd Ed., pp. 146–153. Cambridge Univ. Press, London.

Floridi, A., Gambacurta, A., Bagnato, A., Bianchi, C., Paggi, M. G., Silverstrini, B., and Caputo, A. (1988). Modulation of Adriamycin uptake by lonidamine in Ehrlich ascites tumor cells. *Exp. Mol. Pathol.* **49**, 421–431.

Fojo, A., Hamilton, T. C., Young, R. C., and Ozols, R. F. (1987). Multidrug resistance in ovarian cancer. *Cancer (Philadelphia)* **60**, 2075–2080.

Foster, B. J., Grotzinger, K. R., McKoy, W. M., Rubinstein, L. V., and Hamilton, T. C. (1988). Modulation of induced resistance to Adriamycin in two human breast cancer cell lines with tamoxifen or perhexiline maleate. *Cancer Chemother. Pharmacol.* **22**, 147–152.

Fox, M., and Roberts, J. J. (1987). Drug resistance and DNA repair. *Cancer metastasis Rev.* **6**, 261–281.

Frei, E., Cucchi, C. A., Rosowsky, A., Tantravahi, R., Bernal, S., Ervin, T. J., Ruprecht,

R. M., and Haseltine, W. A. (1985). Alkylating agent resistance: *In vitro* studies with human cell lines. *Proc. Natl. Acad. Sci. U.S.A.* **82**, 2158–2162.

Galivan, J., Nimec, Z., and Rhee, M. (1987). Synergistic growth inhibition of rat hepatoma cells exposed *in vitro* to N^{10}-Propargyl-5,8-dideazafolate with methotrexate or the lipophilic antifolates trimetrexate or metoprine. *Cancer Res.* **47**, 5256–5260.

Gessner, P. K. (1988). A straightforward method for the study of drug interactions: An isobolographic analysis primer. *J. Am. Coll. Toxicol.* **7**, 987–1012.

Giamarellou, H. (1986). Aminoglycosides plus beta-lactams against gram-negative organisms. Evaluation of *in vitro* synergy and chemical interactions. *Am. J. Med.* **80**, 126–137.

Gill, V. J., Witebsky, F. G., and MacLowry, J. D. (1989). Multicategory interpretive reporting of susceptibility testing with selected antimicrobial concentrations. Ten years of laboratory and clinical experience. *Clin. Lab. Med.* **9**, 221–238.

Goldie, J. H. (1983). Drug resistance and cancer chemotherapy strategy in breast cancer. *Breast Cancer Res. Treat.* **3**, 129–136.

Goldin, A., and Mantel, N. (1957). The employment of combinations of drugs in the chemotherapy of neoplasia: A review. *Cancer Res.* **17**, 635–654.

Goldin, A., Venditti, J. M., Mantel, N., Kline, I., and Gang, M. (1968). Evaluation of combination chemotherapy with three drugs. *Cancer Res.* **28**, 950–960.

Goldstein, A., Aronow, L., and Kalman, S. M. (1968). "Principles of Drug Action: The Basis of Pharmacology." Harper & Row, New York.

Gonias, S. L., Oakley, A. C., Walther, R. J., and Pizzo, S. V. (1984). Effects of diethyldithiocarbamate and nine other nucleophiles on the intersubunit cross-linking and inactivation of purified human α_2-macroglobulin by cis-diamininedichloroplatinum (II). *Cancer Res.* **44**, 5764–5570.

Grindey, G. B., and Nichol, C. A. (1972). Interaction of drugs inhibiting different steps in the synthesis of DNA. *Cancer Res.* **32**, 527–531.

Gruda, L., Gauthier, E., Elberg, S., Brajtburg, J., and Medoff, G., (1988). Effects of the detergent sucrose monolaurate on binding of amphotericin B to sterols and its toxicity for cells. *Biochem. Biophys. Res. Commun.* **154**, 954–958.

Hait, W. N. (1987). Targeting calmodulin for the development of novel cancer chemotherapeutic agents. *Anticancer Drug Des.* **2**, 139–149.

Hartshorn, K. L., Vogt, M. W., Chou, T.-C., Blumberg, R. S., Byinton, R., Schooley, R. T., and Hirsch, M. S. (1987). Synergistic inhibition of human immunodeficiency virus *in vitro* by azidothymidine and recombinant alpha A interferon. *Antimicrob. Agents Chemother.* **31**, 168–172.

Helson, L. (1984). Calcium channel blocker enhancement of anticancer drug cytotoxicity: A review. *Cancer Drug Deliv.* **1**, 353–361.

Herr, H. W., Huffman, J. L., Huryk, R., Heston, W. D. W., Melamed, M. R., and Whimore, W. F., Jr. (1987). Anticarcinoma activity of rhodamine 123 against a murine renal adenocarcinoma. *Cancer Res.* **48**, 2061–2063.

Hill, A. V. (1913). The combinations of hemoglobin with oxygen and carbon monoxide. *Biochem. J.* **7**, 471–480.

Hitchings, G., and Burchall, J. (1965). Inhibition of folate biosynthesis and function as a basis for chemotherapy. *Adv. Enzymol.* **27**, 415.

Holm, S. E. (1986). Interaction between beta-lactam and other antibiotics. *Rev. Infect. Dis.* **3**, S305–S314.

Hooper, M. L. S., and Subak-Sharpe, J. H. (1981). Metabolic cooperation between cells. *Int. Rev. Cytol.* **69**, 45–104.

Hopfer, R. L., Mehta, R., and Lopez-Berestein, G. (1987). Synergistic antifungal activity and reduced toxicity of liposomal amphotericin B combined with gramicidin S or NF. *Antimicrob. Agents Chemother.* **31**, 1978–1981.

Inaba, N., and Maruyama, E. (1988). Reversal of resistance to vincristine in P388 leukemia by various polycyclic clinical drugs, with a special emphasis on quinacrine. *Cancer Res.* **48**, 2064–2067.

Jähde, E., Glüsenkamp, K.-H., Klünder, I., Hulser, D. F., Tietze, L.-F., and Rajewsky, M. F. (1989), Hydrogen ion-mediated enhancement of cytotoxicity of bis-chloroethylating drugs in rat mammary carcinoma cells *in vitro. Cancer Res.* **49**, 2965–2972.

Kahl, R. (1984), Synthetic antioxidants: Biochemical actions and interference with radiation, toxic compounds, chemical mutagens and chemical carcinogens. *Toxicology* **33**, 185–228.

Kamataki, T., Kitada, M., Komori, M., and Inaba, N., (1989). Resistance to anticancer drugs in relation to cytochrome P-450. *Gan. To Kagaku Ryoho* **16**, 605–610.

Kanamaru, R., Ishioka, C., Konishi, Y., Ishikawa, A., Shibata, H., and Wakui, A. (1988). Biochemical modulation of anticancer drugs. *Sci. Rep. Res. Inst., Tohoku Univ., Ser. C* **35**, 18–28.

Kisliuk, R. L., Gaumont, Y., Kumar, P., Coutts, M., Nair, M. G., Nanavate, N. T., and Kalman, T. I. (1985). The effect of polyglutamylation on the inhibitory activity of folate analogs. *In* "Proceedings of the Second Workshop on Folyl and Antifolyl Polyglutamates" (I. D. Goldman, ed.), pp. 319–328. Praeger, New York.

Kiue, A., Sano, T., Sazuki, K., Inada, M., Okumana, M. Kikuchi, J., Sato, S., Kohno, K., and Kumano, M. (1990). Activities of newly synthesized dihydropyridines in overcoming of vincristine resistance, calcium antagonism, and inhibition of photoaffinity labeling of P-glycoprotein in rodents. *Cancer Res.* **50**, 310–317.

Klastersky, J. (1986). Concept of empiric therapy with antibiotic combinations. Indications and limits. *Am. J. Med.* **80**, 2–12.

Klubes, P., and Leyland-Jones, B. (1989). Enhancement of the antitumor activity of 5-fluorouracil by uridine rescue. *Pharmacol. Ther.* **41**, 289-302.

Korbashi, P., Katzhendler, J., Saltman, P., and Chevion, M. (1989). Zinc protects *Escherichia coli* against copper-mediated paraquat-induced damage. *J. Biol. Chem.* **264**, 8479–8482.

Kuefner, U., Lohrmann, U., Montejaro, Y., Vitols, K. S., and Huennekens, F. M. (1988). Chemotherapeutic potential of methotrexate peptides. *adv. Enzyme Regul.* **27**, 3–13.

Laidlaw, S. A., Dietrich, M. F., Lamtenzan, M. P.,Vargas, H. I., Block, J. B., and Kopple, J. D. (1989). Antimutagenic effects of taurine in a bacterial assay system. *Cancer Res.* **49**, 6600–6604.

Larremendy, M. L., Lopez-Larraza, D., Vital-Rioja, L., and Bianchi, N.O. (1989). Effect of the metal chelating agent *o*-phenanthroline on the DNA and chromosome damage induced by bleomycin in Chinese hamster ovary cells. *Cancer Res.* **49**, 6583–6586.

Lawrence, T. S., Canman, C. E., Maybaum, J., and Davis, M. A. (1989). Dependence of etoposide-induced cytotoxicity and topoisomerase II-mediated DNA strand breakage on the intracellular ionic environment. *Cancer Res.* **49**, 4775–4779.

Lazo, J.S., Mistry, J. S., and Morris, G. (1989). Synergistic cytotoxicity and enhanced cellular accumulation of bleomycin by cysteine proteinase inhibitors. *Proc. Am. Assoc. Cancer Res.* **30**, 572 (abstr. 2276).

Leyland-Jones, B., and O'Dwyer, P. J. (1986). Biochemical modulation: Application of laboratory models to the clinic. *Cancer Treat. Rep.* **70**, 219–229.

Ling, V., Kartner, N., Sudo, T., Siminovitch, L., and Riordon, J. R. (1983). Multi-drug resistance phenotype in Chinese hamster ovary cells. *Cancer Treat. Rep.* **67**, 869–874.

Ling, V., Gerlach J., and Karthner, N. (1984). Multidrug resistance. *Breast Cancer Res. Treat.* **4**, 89–94.

Loewe, S. (1928). Die Quantitationprobleme der Pharmakologie. *Ergeb. Physiol.* **27**, 47–187.

Loewe, S. (1953). The problem of synergism and antagonism of combined drugs. *Arzneim. Forsch.* **3**, 285–320.

Marget, W., and Seeliger, H. P. (1988). *Listeria monocytogenes* infections: Therapeutic possibilities and problems. *Infection (Munich)* **16**, S175–S177.

Mariani, L., Bertini, R., Fiorio, R., Gervasi, P., and Citti, L. (1988). The influence of chronic *N*-methyl-*N'*-nitro-*N*-nitrosoguanidine pre-treatments on mutagenic response and O_6-methylguanine–DNA methyltransferase activity in V79 Chinese hamster cells. *Mutat. Res.* **208**, 73–76.

Martin, D. S. (1987). Biochemical modulation: Perspectives and objectives. *In* "New Avenues in Developmental Cancer Chemotherapy" (K. R. Harrap and T. A. Connors, eds.), Bristol-Myers Symp. Ser. 8, pp. 113–152. Academic Press, Orlando, Florida.

Martin, D. S., Stolfi, R. L., Sawyer, R. D., and Young, C. W. (1985). Application of biochemical modulation with therapeutically inactive modulating agent in clinical trials of cancer chemotherapy, *Cancer Treat. Rep.* **69**, 421–423.

Mayer, K. H., and DeTorres, O. H. (1985). Current guidelines on the use of antibacterial drugs in patients with malignancies. *Drugs* **29**, 262–279.

Michaelis, L., and Menten, M. L. (1913). Die Kinetik der Invertinwirkung. *Biochem. Z.* **49**, 333–369.

Mickley, L. A., Bates, S. E., Richert, N. D., Currier, S., Tanaka, S., Foss, F., Rosen, N., and Fojo, A. T. (1989). Modulation of the expression of a multidrug resistance gene (mdr-1/P-glycoprotein) by differentiating agents. *J. Biol. Chem.* **264**, 18031–18040.

Mitchell, M. S. (1988). Combining chemotherapy with biological response modifiers in treatment of cancer. *JNCI, J.Natl. Cancer Inst.* **80**, 1445–1450.

Moellering, R. C., Jr. (1983) Rationale for use of antimicrobial combinations. *Am. J. Med.* **75**, 4–8.

Moran, R.G. (1989) Leucovorin enhancement of the effects of the fluoropyrimidines on thymidylate synthase. *Cancer* **63**, 1008–1012.

Neu, H. C. (1989) Synergy of fluoroquinolones with other antimicrobial agents. *Rev. Infect. Dis.* **5**, S1025–S1035.

Nyce, J. (1989). Drug-induced DNA hypermethylation and drug resistance in human tumors. *Cancer Res.* **49**, 5829–5836.

O'Dwyer, P. J., King, S. A., Hoth, D. F., and Leyland-Jones, B. (1987). Role of thymidine in biochemical modulation: A review. *Cancer Res.* **47**, 3911–3919.

Ohkubo, T., Higashigawa, M., Kawasaki, H., Kamiya, H., Sakural, M., Kagawa, Y., Kakito, E., Sumida, K., and Ooi, K. (1988). Sequence-dependent antitumor effect of VP-16 and 1-β-D-arabinofuranosylcytosine in L1210 ascites tumor. *Eur. J. Cancer Clin. Oncol.* **24**, 1823–1828.

Ozols, R. F., and Corden, K. (1986). New aspects of clinical drug resistance: The role of gene amplification and the reversal of resistance in drug refractory cancer. *Important Adv. Oncol.* pp. 129–157.

Ozols, R. F., Condey, B. J., Jacob, J., Wesley, M. W., Ostchega, Y., and Young, R. C. (1984). High-dose cisplatin in hypertonic saline. *Ann. Intern. Med.* **100**, 19–24.

Peters. G. J. (1988). Preclinical and clinical biochemical modulation of 5-fluorouracil by leucovorin and uridine. *Sci. Rep. Res. Inst., Tohoku Univ., Ser. C* **35**, 15–17.

Peterson, A. R., Danenberg, P. V., Ibric, L. L., and Peterson, H. (1985). Deoxyribonucleoside-induced selective modulation of cytotoxicity and mutagenesis. *Basic Life Sci.* **31**, 313–334.

Pitot, H. C. (1989). Principles of carcinogenesis: Chemical. *In* "Cancer: Principles and Practice of Oncology" (B. T. De Vita, Jr., S. Hellman, and S. A. Rosenberg, eds.), pp. 116–1350 Lippincott, Philadelphia, Pennsylvania.

Poch. G. (1981). The confusion about additive combinations. *Trends Pharmacol. Sci.* **2**, 256–257.

Poon, M. A., O'Connell, J. J., Moertel, C. G., Wieand, H. S., Cullinan, S. A., Everson, L. K., Krook, J. E., Mailliard, J. A., Laurie, J. A., Tschetter, L. K., and Wiesenfeld, M. (1989). Biochemical modulation of fluorouracil: Evidence of significant improvement of survival and quality of life in patients with advanced colorectal carcinoma. *J. Clin. Oncol.* **7**, 1407–1418.

Press, O. W., DeSantes, K., Andersan, S. K., and Geisler, F. (1990). Inhibition of catabolism of radiolabeled antibodies by tumor cells using lysosomotropic amines and carboxylic ionophones. *Cancer Res.* **50**, 1243–1250.

Rahal, J. J., Jr., (1983). Rationale for use of antimicrobial combinations in treatment of gram-negative infections. A review of recent reviews. *Am J. Med.* **75**, 68–71.

Reed, L. J., and Berkson, J. (1939). The application of the logistic function to experimental data. *J. Phys. Chem.* **33**, 760–779.

Rideout, D. (1986). Self-assembling cytotoxins. *Science* **233**, 561–563.

Rideout, D., Jaworski, J., and Dagnino, R., Jr. (1988). Environment-selective synergism using self-assembling cytotoxic and antimicrobial agents. *Biochem. Pharmacol.* **37**, 4505–4512.

Rideout, D., Calogeropoulou, T., Jaworski, J., and McCarthy, M. (1990). Synergism through direct covalent bonding between agents: A strategy for rational design of chemotherapeutic combinations. *Biopolymers* **29**, 247–262.

Rosenthal, S. A., and Hait, W. N. (1988). Potentiation of DNA damage and cytotoxicity by calmodulin antagonists. *Yale J. Biol. Med.* **61**, 39–49.

Rosowsky, A., Freisheim, J. H., Hynes, J. B., Queener, S. F., Bartlett, M., Smith, J. W., Lazarus, H., and Modest, E. J. (1989). Tricyclic 2, 4-diaminopyrimidines with broad antifolate activity and the ability to inhibit *pneumocystis carinii* growth in cultured human lung fibroblasts in the presence of leucovorin. *Biochem. Pharmacol.* **38**, 2677–2684.

Rothman, K. J. (1976). The estimation of synergy or antagonism. *Am. J. Epidemiol.* **103**, 506–511.

Rowland, M., and Tolzer, T. N. (1980). "Clinical Pharmacokinetics. Concepts and Applications." Lea & Febiger, Philadelphia, Pennsylavania.

Satoh, M., Naganuma, A., and Imura, N. (1988). Metallothionein induction prevents toxic side effects of cisplatin and Adriamycin used in combination. *Cancer Chemother. Pharmacol.* **21**, 176–178.

Scatchard, G. (1949). The attractions of proteins for small molecules and ion. *Ann. N.Y. Acad. Sci.* **51**, 660–672.

Schabel, F. M., Jr., Skipper, H. E., Trader, M. W., Laster, W. R. Jr., and Griswold, D. P., Jr. (1983). Establishment of cross-resistance profiles for new agents. *Cancer Treat. Rep.* **67**, 905– 22.

Schibler, M. J., and Cabral, F. (1986). Taxol-dependent mutants of Chinese hamster ovary cells with alterations in α- and β-tubulin. *J. Cell Biol.* **102**, 1522–1531.

Schibler, M. J., Barlow, S. B., and Cabral, F. (1989). Elimination of permeability mutants from selections for drug resistance in mammalian cells. *FASEB J.* **3**, 163–168.

Scott, B. R. (1983). Theoretical models for estimation of dose–effect relationships after combined exposure to cytotoxicants. *Bull. Math. Biol.* **45**, 323–345.

Segel, I. H. (1975). Multiple inhibition analysis. *In* "Enzyme Kinetics," pp. 465–503. Wiley, New York.

Senter, P. D. (1990). Activation of prodrug by antibody–enzyme conjugates: A new approach to cancer therapy. *FASEB J.* **4**, 188–193.

Siemann, D. W. (1984). Modification of chemotherapy by nitroimidazoles. *Int. J. Radiat. Oncol. Biol. Phys.* **10**, 1585–1594.

Simpson. W. G. (1985). The calcium channel blocker verapamil and cancer chemotherapy. *Cell Calcium* , 449–467.

Skipper, H. E. (1974). Combination therapy: Some concepts and results. *Cancer Chemother. Rep. (Suppl.)* **4**, 137–145.

Sladek, N. E., and Landkamer, G. J. (1985). Restoration of sensitivity to oxazaphosphorines by inhibitors of aldehyde dehydrogenase activity in cultured oxazaphosphorine-resistant L1210 and cross-linking agent resistant P388 cell lines. *Cancer Res.* **45**, 1549–1555.

Soike, K. F., Epstein, D. A., Gloff, C. A., Cantrell, C., Chou, T.-C., and Gerone, P. J. (1987). Effect of 9-(1,3-dihydroxy-2-propoxymethyl)-guanine and recombinant human β-interferon alone and in combination on simian varicella infection in monkeys. *J. Infect. Dis.* **156**, 607–614.

Sun, J. R., and Brown, J. M. (1989). Enhancement of the antitumor effect of flavone acetic acid by the bioreductive cytotoxic drug SR 4233 in a murine carcinoma. *Cancer Res.* **49**, 5664–5670.

Syracuse, K. C., and Greco, W. R. (1986). Comparison between the method of Chou and Talalay and a new method for the assessment of the combined effects of drugs: A Monte Carlo simulation study. *Proc. Am. Stat. Assoc., Biopharmacol. Sect.* pp. 127–132.

Tan, K. B., Mattern, M. R., Boyce, R. A., and Schein, P. S. (1988). Unique sensitivity of nitrogen mustard-resistance human Burkitt lymphoma cells to novobiocin. *Biochem. Pharmacol.* **37**, 4411–4413.

Tannock, I. F., and Rotin, D. (1989). Acid pH in tumors and its potential for therapeutic exploitation. *Cancer Res.* **49**, 4373–4384.

Tietze, L. F., Neumann, M., Möllers, T., Fischer, R., Glüsenkamp, K.-H., Rajewsky, M. F., and Jähde, E. (1989). Proton-mediated liberation of aldophosphamide from a nontoxic prodrug: A strategy for tumorselective activation of cytocidal drugs. *Cancer Res.* **49**, 4179–4184.

Tsuruo, T. (1983). Reversal of acquired resistance to vinca alkaloids and anthracycline antibiotics. *Cancer Treat. Rep.* **67**, 889–894.

Tyring, S. K. (1987). Antitumor actions of interferons. Direct, indirect, and synergy with other treatment modalities. *Int. J. Dermatol.* **26**, 549–556.

van der Bliek, A. M., and Borst, P. (1989). Multidrug resistance. *Adv. Cancer Res.* **52**, 165–203.

van Zeeland, A. A. (1988). Molecular dosimetry of alkylating agents: Quantitative comparison of genetic effects on the basis of DNA adduct formation. *Mutagenesis.* **3**, 179–191.

Vogt, M. W., Durno, A. G., Chou, T.-C., Coleman, L. A., Paradis, T. J., Shooley, R. T., Kaplan, J. C., and Hirsch, M. S. (1988). Synergistic interaction of 2',3'-dideoxycytidine and recombinant interferon-α-A on replication of human immunodeficiency virus type 1. *J. Infect. Dis.* **158**, 378–385.

Webb, J. L. (1963). "Enzyme and Metabolic Inhibitors," Vol. 1, pp. 66–79, 488–512. Academic Press, New York.

Weber, J. M., Sircar, S., Horvath, J., and Dion, P. (1989). Non-P-glycoprotein-mediated multidrug resistance in detransformed rat cells selected for resistance to methylglyoxal bis(guanylhydrazone). *Cancer Res.* **49**, 5779–5783.

Weinstein, I. B. (1988). The origins of human cancer: Molecular mechanisms of carcinogenesis and their implications for cancer prevention and treatment. *Cancer Res.* **48**, 4135–4143.

Whitehead, E. (1970). The regulation of enzyme activity and allosteric transition. *Prog. Biophys. Mol. Biol.* **21**, 321–397.

Yagi, K., and Ozawa. T. (1960). Complex formation of apo-enzyme, coenzyme and substrate of D-amino acid oxidase. *Biochim. Biophys. Acta* **42**, 381–387.

Yee, Y., Farber, B., and Mates, S. (1986). *J. Infect. Dis.* **154**, 531–534.

Yin, M.-B., Bankusli, I., and Rustum, Y. M. (1989). Mechanisms of the *in vivo* resistance to adriamycin and modulation by calcium channel blockers in mice. *Cancer Res.* **49**, 4729–4733.

Yonetani, T., and Theorell, H. (1964). Studies on liver alcohol dehydrogenase complexes. III. Multiple inhibition kinetics in the presence of two competitive inhibitors. *Arch. Biochem. Biophys.* **106**, 243–251.

Zaniboni, A. (1989). The emerging role of 5-fluorouracil and leucovorin in the treatment of advanced breast cancer. A review. *J.Chemother.* **1**, 330–337.

CHAPTER 2

The Median-Effect Principle and the Combination Index for Quantitation of Synergism and Antagonism

Ting-Chao Chou

Laboratory of Biochemical Pharmacology
Memorial Sloan-Kettering Cancer Center
and Cornell University Graduate School of Medical Sciences
New York, New York 10021

SYNERGISM AND ANTAGONISM IN CHEMOTHERAPY
Copyright © 1991 by Academic Press, Inc.
All rights of reproduction in any form reserved.

I. Introduction

The derivation of equations for dose effect analysis based on the median-effect principle of the mass-action law have been described in detail previously (Chou, 1974, 1975, 1976, 1977; Chou and Talalay, 1977, 1981) and have been followed by several reviews (Chou and Talalay, 1983, 1984, 1987; Chou, 1987; Chou and Fanucchi, 1988; Chou *et al.*, 1991). A step-by-step analysis of the experimental data is given in Section IVE2 of this chapter and the actual data analysis with a specific example is presented in Chapter 6. The purpose of this chapter is not to reiterate the derivations but rather to present the highlights of recent developments in and applications of dose effect analysis in single drugs or in drug combinations. Efforts have also been made to clarify the different interpretations by some investigators regarding the mass-action law derived equations as described by Chou and Talalay (1981). There are fundamental differences in conception and interpretation when different approaches are used for dose effect analysis. In this chapter, comparison of different conceptual bases are also discussed.

II. Dose Effect Analysis with Physicochemical Approach as Opposed to Empirical Approach

Dose–effect relationships in pharmacological systems have been subjected to various mathematical transformations in the past to linearize dose–effect plots based on empirical or statistical assumptions. For example, probit (Finney, 1952), logit (Reed and Berkson, 1929), and simple power law functions (Nordling, 1954) often provide adequate linearization of plots. However, the slopes and intercepts of such plots are usually devoid of any fundamental meaning (Table 1). When a simple biological system, such as an enzyme or receptor, is subjected to the probit, logit, or power law analysis, the parameters (such as slopes and intercepts) are without any physical meaning. How these formulations can be meaningful in complicated systems if they are without meaning in a simple system is an issue that remains unresolved.

A. THE POWER LAW

The power law in the form of $F_a = aD^k$ (Table 1), more frequently used in its logarithmic form, $\log(f_a) = \log(a) + k \log(D)$, is often used in the graphical representation of dose effect data. It is used for convenience of condensing the graphic scales rather than illustrating a theoretical principle. In fact,

TABLE 1

A Comparison of the Median-Effect Equation with the Power, Probit, and Logit Laws That Have Been Used in Dose-Effect Analysis for Single Drugs

---

The Median Effect Equation of the
MASS-ACTION LAW
(Chou, Mol. Pharmacol. <u>10</u>, 235, 1974) $f_a / f_u = (D/D_m)^m$
(Chou, J. Theor. Biol. <u>59</u>, 253, 1976)

POWER LAW
(Nordling, Brit. J. Cancer, <u>7</u>, 68, 1953)
(Armitage & Doll, Brit. J. Cancer, $f_a = a\, D^k$
<u>8</u>, 1, 1954)

PROBIT LAW $f_a = \dfrac{1}{\sigma\sqrt{2\pi}} \displaystyle\int_{-\infty}^{\log D} e^{-\frac{1}{2\sigma^2}(\log D - \log D_m)^2}\, d\,(\log D)$
(Bliss, Quart. J. Pharm. & Pharmacol.
<u>11</u>, 192, 1938)
(Finney, Probit Analysis, 1947)

where $Y = (\log D - \log D_m)/\sigma$

LOGIT LAW
(Berkson, J. Am. Stat. Assoc.
<u>41</u>, 40, 1946) $f_a = 1\Big/\left[1 + e^{-(\alpha + \beta \log D)}\right]$
(Thompson, Bact. Rev. <u>11</u>, 115, 1947)

D : Dose f_u : Fraction Unaffected
D_m : Median-Effect Dose σ^2 : Variance
m : Kinetic Order   Y : (Probit - 5) or Normal Equivalent Deviate
f_a : Fraction Affected   a,k,α and β: Constants

---

the simple power law for linearizing dose effect data contradicts the basic principle of the mass-action law, as exemplified in enzyme kinetics and pharmacological receptor theory. Simple Michaelis–Menten enzyme kinetic analysis indicates that $\log(f_a)$ vs $\log(D)$ (the power law plot) will not give a straight line but $\log[f_a/(1 - f_a)]$ vs $\log(D)$ (the median-effect plot) will. Although the power law plot gives a reasonably linear plot at low effect levels, the slope decreases when f_a reaches high levels and becomes nonlinear and levels off to reach a plateau (Chou, 1980).

B. THE PROBIT

The probit approach assumes a logarithmic normal distribution in dose–effect relationships (Bliss, 1939, 1967; Finney, 1947, 1952). However, there is no justification given for why this should be the case and no justifiable meaning of the slope and intercept of the probit plot has been provided by the proponents of this approach.

C. THE LOGIT

The logit approach (Reed and Berkson, 1929), on the other hand, has an underlying mathematical premise similar to the Hill equation and the median-effect equation of the mass-action law. However, the logit function is intuitively or empirically conceived rather than explicitly derived from a basic principle or model. The parameters (α, β, etc.) of the logit equation lack formal definitions when applied to simple biological systems.

Some investigators refer to an empirical/mathematical formula as a model without indicating what that model means in terms of physicochemical relevance. Because of the different approaches used, the data analysis and interpretation can be very different. A more stringent and rigorous definition of a model would involve a physically identifiable entity corresponding to a model or a basic principle with a physicochemical bearing.

D. OTHER EMPIRICAL APPROACHES

A statistical approach usually draws empirical curves to fit the experimental data. This would require a large number of data points in order to achieve an empirical correspondence between the data and the curves. Because the approach is empirical, the shapes of the possible number of fittings are unlimited and, therefore, can only give undefined meanings to the parameters. Consequently, the use of the parameters presents difficulties in further theoretical developments such as extending from a single drug to multiple drugs or from the first order to higher order kinetic systems. The main use of statistics involves studying probability, significance, variability, and errors of measurement and statistics alone does not describe the dose–effect relationships *per se* without an underlying theory concerning these relationships.

E. THE MASS-ACTION LAW APPROACH

Instead of drawing empirical curves to fit the data, the median-effect principle approach uses the data to fit the theory (i.e., the mass-action law). Because there is a fixed basic law, this approach requires relatively few data points to achieve the correspondence between the data and the theoretical function. It becomes a simple matter to determine how well the data fit the theoretical function by using linear correlation coefficient (r) and the standard error of the mean of parameters. Mass-action law parameters of this approach, such as m and D_m values of the median-effect plot (see Section III), can be readily used for further exploration of dose–effect relationships as in the case of multiple-drug-effect relationships (Chou and Talalay, 1977, 1981, 1984).

III. The Median-Effect Principle of the Mass-Action Law

The median-effect principle as described in Chapter 1 and in previously mentioned reviews (Chou and Talalay, 1984, 1987) was obtained from the derivation of enzyme reaction rate equations and dose–effect relationship equations.

A. THE MEDIAN-EFFECT EQUATION

The median-effect equation (Chou, 1976) states that:

$$f_a/f_u = (D/D_m)^m \tag{1}$$

where D is the dose; f_a and f_u are the fractions of the system affected and unaffected, respectively, by the dose D; D_m is the dose required to produce the median effect (analogous to the more familiar IC_{50}, ED_{50}, or LD_{50} values); and m is a coefficient signifying the sigmoidicity of the dose–effect curve, i.e., $m = 1$ for hyperbolic (first order or Michaelis–Menten) systems; $m > 1$ for sigmoidal, and $m < 1$ for negatively sigmoidal shapes. Since by definition, $f_a + f_u = 1$, several useful alternative forms of the equation can be derived as follows:

$$f_a/(1 - f_a) = [(f_a)^{-1} - 1]^{-1} = [(f_u)^{-1} - 1] = (D/D_m)^m \tag{2}$$

$$f_a = 1/[1 + (D_m/D)^m] \tag{3}$$

$$D = D_m[f_a/(1 = f_a)]^{1/m} \tag{4}$$

Therefore, when m and D_m values are determined, the entire dose–effect relationship is described. For a given dose (D), it is possible to calculate the effect (f_a), and for a given effect (f_a) it is possible to calculate the dose (D).

The median-effect equation describes the behaviour of many biological systems. It has been shown (Chou, 1977) to be a generalized form of the enzyme kinetic relationships of Michaelis and Menten (1913) and Hill (1913), the physical adsorption isotherm of Langmuir (1918), the pH-ionization equation of Henderson and Hasselbalch (Clark, 1933), the equilibrium binding equation of Scatchard (1949), and the pharmacological drug–receptor interaction (Tallarida and Jacob, 1979). For a more detailed comparison, see Section IIIC. Furthermore, the median-effect equation is directly applicable not only to primary ligands such as substrates, agonists, and activators, but also to secondary ligands such as inhibitors, antagonists, or environmental factors (Chou, 1976).

B. THE MEDIAN-EFFECT PLOT

The median-effect equation (Eq.1) describes the dose–effect curves linearized by taking the logarithms of both sides of the equation as follows:

$$\log(f_a/f_u) = m \log(D) - m \log (D_m)$$

or

$$\log[(f_a)^{-1} - 1]^{-1} = m \log(D) - m \log(D_m) \tag{5}$$

$$\log[(f_u)^{-1} - 1] = m \log(D) - m \log(D_m)$$

The median-effect plot (Fig.1) of $y = \log(f_a/f_u)$ or its equivalents with respect to $x = \log(D)$ is a general and simple method (Chou, 1976, 1980) for determining pharmacological median doses for lethality (LD_{50}), toxicity (TD_{50}), effect of agonist drugs (ED_{50}), and effect of antagonist drugs (IC_{50}). Thus, the median-effect principle of the mass-action law encompasses a wide range of applications (see Appendices I and II). The method has been used in scientific papers published in over seventy biomedical journals and monographs. The plot for dose effect analysis in single and multiple drugs gives the slope m, and the intercept of the dose–effect plot with the median-effect

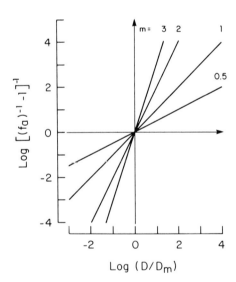

FIGURE 1   The median-effect plot at different slopes corresponding to m values of 0.5, 1, 2, and 3. The plot is based on the log-transformed median-effect equation (Eq. 5) in which the dose (D) has been normalized by dividing with the median-effect dose (D_m). Note that the ordinate $\log[(f_a)^{-1} - 1]^{-1}$ is identical to $\log[(f_u)^{-1} - 1]$, $\log[f_a/f_u]$, or $\log [f_a/(1 - f_a)]$. Reprinted with permission from *Advances in Enzyme Regulation* **22**, Chou and Talalay, Quantitative analysis of dose–effect relationships: The combined effects of multiple drugs or enzyme inhibitors. Copyright © 1984, Pergamon Press plc.)

axis (x axis) [i.e., when $f_a = f_u$, $f_a/f_u = 1$ and hence $y = \log(f_a/f_u) = 0$] gives $\log(D_m)$ and consequently the D_m value. Any cause–effect relationship that gives a straight line for this plot will provide the two basic parameters, m and D_m, and, thus, an apparent equation that describes such a system. The linearity of the median-effect plot (as determined from linear correlation coefficients) determines the applicability of the present method.

C. THE MEDIAN-EFFECT EQUATION AS A GENERALIZED EQUATION

The relationship between the median-effect equation and other major biomedical equations is shown in Scheme 1. Rearrangements of the median-effect equation and/or taking its logarithmic form gives the Michaelis–Menten (1913) equation for the first order primary ligand enzyme kinetics ($m = 1$), the Hill (1913) equation for the primary ligand occupancy of biological acceptors such as oxygen–hemoglobin interaction, the Scatchard (1949) equation for ligand–receptor binding and dissociation, or the Henderson–Hasselbalch equation (Goldstein *et al.*, 1968) for pH–ionization. Therefore, these equations are based on the same principle of the mass-action law but are derived for different purposes (Chou, 1977). When the end point of the measurement of the fractional inhibition in the median-effect equation is changed to

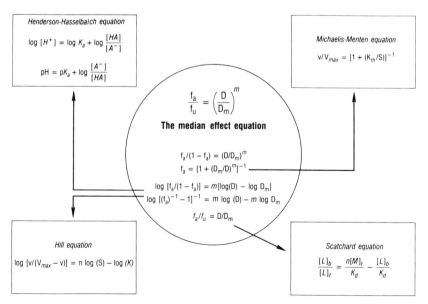

SCHEME 1 Relationship between the median effect equation (Chou, 1976) and other major biochemical equations.

fractional saturation, fractional occupancy, fractional binding, and fractional ionization, respectively, the correspondence among these equations becomes clear. By virtue of being fractional, in relativity ratios such as $f_a/f_u = (D/D_m)^m$, both sides of the median-effect equation are dimensionless quantities in equality, and thus represent the simplest possible form among these equations. The median-effect equation directly relates the dose (where D is the dose and D_m is the median-effect dose) on the right side and the effect (where f_a is the fraction affected and f_u is the fraction unaffected) on the left side. The m exponent on the right side signifies the shape of the dose–effect curve.

D. THE DIFFERENCES BETWEEN THE MEDIAN-EFFECT EQUATION AND THE HILL EQUATION

Some investigators refer to the median-effect equation as the Hill equation (e.g., Berenbaum, 1989). The median-effect equation is not the Hill equation nor is it the logit equation, just as the Scatchard and the Henderson–Hasselbalch equations are not the Michaelis–Menten equation. A comparison of the median-effect equation and the Hill equation (or the Michaelis–Menten equation) is given in Tables 2 and 3. They are different in many aspects.

TABLE 2
Comparison of the Median-Effect Equation and the Hill Equation

Median-effect equation	Hill equation and Michaelis–Menten equation
$\dfrac{f_a}{f_u} = \left(\dfrac{D}{D_m}\right)^m$	$\dfrac{v}{V_{max}} = \dfrac{S^n}{S^n + K}$
$\log[(f_a)^{-1} - 1]^{-1} = m \log D - m \log D_m$	$\log[v/(V_{max} - v)] = n \log S - \log K$
when $m = 1$	when $n = 1$
$f_a = 1/[1 + (D_m/D)]$	$v/V_{max} = 1/[1 + (K_m/S)]$

D: Inhibitory Drug Concentration	S: Substrate Concentration
f_a: Fraction Affected	v: Reaction Rate
f_u: Fraction Unaffected	V_{max}: Maximal Reaction Rate
D_m: Median-Effect Concentration	K: Constant
m: Slope of the Median-Effect Plot Signifying the Shape of Dose–Effect Curve	n: Hill Coefficient Signifying the Cooperativity of Binding

TABLE 3

Comparison of the Features of the Median-Effect Equation and the Hill Equation

Median-effect equation	Hill equation and Michaelis–Menten equation
For inhibitor ligands such as antagonists, reference ligands	For substrate-like ligands such as agonists or primary ligands
For dose–effect relationships derived by systematic analyses and mathematical induction[a]	For dose–response relationships; direct derivation
Any uninhibited effect as a control	Maximal response (V_{max}) as a control. The determination of V_{max} requires extrapolation
Involves median-effect dose	Involves equilibrium or aggregated dissociation constant
Inhibition mechanism and substrate mechanism independent	Substrate mechanism independent
A generalized mass-law equation[b]	A more specific mass-law equation

[a] See Chou (1974, 1976).

[b] The median-effect equation has been used for deriving the Hill equation, the Michaelis–Menten equation, the Scatchard equation, and the Henderson–Hasselbalch equation (Chou, 1977; also see Scheme 1).

IV. Dose–Effect Analysis of Combined Drug Effects

A. THE MULTIPLE-DRUG-EFFECT EQUATIONS AND F_a-CI PLOTS

The above analysis for single drugs using the median-effect principle can be extended to desribe the combined effects of two pharmacological agents, D_1 and D_2 (Chou and Talalay, 1977, 1981). In its simplest form (first-order mutually exclusive system), the median-effect equations describe the relationships between single drugs and their combinations as follows:

$$\frac{(f_a)_{1,2}}{(f_u)_{1,2}} = \frac{(f_a)_1}{(f_u)_1} + \frac{(f_a)_2}{(f_u)_2} = \frac{(D)_1}{(D_m)_1} + \frac{(D)_2}{(D_m)_2} \tag{6}$$

Similarly, if the slope (m) of the above plot is not 1 (e.g., higher-order mutually exclusive system), then

$$\left[\frac{(f_a)_{1,2}}{(f_u)_{1,2}}\right]^{1/m} = \left[\frac{(f_a)_1}{(f_u)_1}\right]^{1/m} + \left[\frac{(f_a)_2}{(f_u)_2}\right]^{1/m} \tag{7}$$

$$= \frac{(D)_1}{(D_m)_1} + \frac{(D)_2}{(D_m)_2}$$

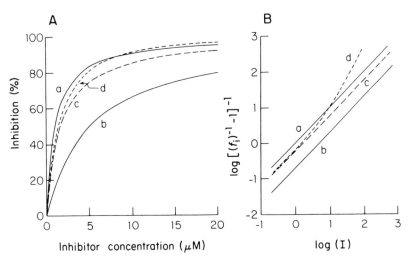

FIGURE 2 Theoretical plots for the combinations of effects of two inhibitory drugs in a first-order system ($m = 1$; Michaelis-Menten kinetics) assuming the inhibitory potency for D_1 is $(D_m)_1$ or $(ED_{50})_1 = 1 \mu M$ and for D_2 is $(D_m)_2 = 5 \mu M$. The fraction affected (f_a) is presented as a function of drug concentration for (a) D_1 alone; (b) D_2 alone; (c) a mixture of D_1 and D_2 (1:2) assuming the inhibitors are mutually exclusive in their effects; (d) a mixture of D_1 and D_2 (1:2) assuming they are mutually nonexclusive in their effects. In the case of mixtures of drugs, the abscissa represents the sum of $(D_1 + D_2)$, e.g., $D_1 = 3 \mu M$ plus $D_2 = 6 \mu M$ (total $9 \mu M$) when compared with $9 \mu M$ of D_1 or $9 \mu M$ of D_2. (A) Linear scale plot of f_a vs inhibitor concentration. (B) The median-effect plot of $\log [(f_a)^{-1} - 1]^{-1}$ vs log drug concentration. These graphs were constructed on the basis of Eq. 3 for (a) and (b) with $m = 1$; Eq. 7 with $m = 1$ for (c); and Eq. 8 with $m = 1$ for (d). Note that the curve of (d) clearly shows higher inhibitory effects at high concentrations when compared with its parent components on an equimolar basis. (Modified from Chou and Talalay, 1981.)

For two drugs that have the same or similar modes of action, the effects of both drugs are mutually exclusive (i.e., parallel median-effect plots for parent drugs and their mixtures, Figs. 2 and 3). For the situation in which the effects of both drugs are mutually nonexclusive (e.g., two drugs have different modes of action or act independently), the general equation for describing the summation of the effects has already been derived (Chou and Talalay, 1981, 1984). This general equation has an additional term when compared with Eq. 7 and is given by

$$
\left[\frac{(f_a)_{1,2}}{(f_u)_{1,2}}\right]^{1/m} = \left[\frac{(f_a)_1}{(f_u)_1}\right]^{1/m} + \left[\frac{(f_a)_2}{(f_u)_2}\right]^{1/m} + \left[\frac{(f_a)_1(f_a)_2}{(f_u)_1(f_u)_2}\right]^{1/m} \tag{8}
$$

$$
= \frac{(D)_1}{(D_m)_1} + \frac{(D)_2}{(D_m)_2} + \frac{(D)_1(D)_2}{(D_m)_1(D_m)_2}
$$

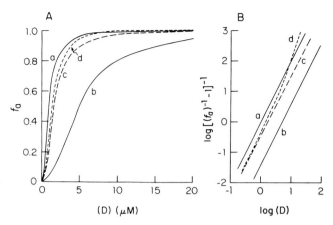

FIGURE 3 Theoretical plots for the combination of effects of two inhibitory drugs (each obeying second-order kinetics, i.e., $m = 2$), assuming inhibitory potency for D_1 is $(D_m)_1$ or $(ED_{50})_1 = 1 \mu M$ and for D_2 is $(D_m)_2$ or $(ED_{50})_2 = 5 \mu M$. Dose–effect relationships are given for (a) D_1 alone; (b) D_2 alone; (c) an equimolar mixture of I_1 and I_2, assuming that they are mutually exclusive in their effects; (d) an equimolar mixture of D_1 and D_2 assuming that they are mutually nonexclusive in their effects. (a) and (b) are based on Eq. 3 with $m = 2$; (c) is based on Eq. 7 with $m = 2$; and (d) is based on Eq. 8 with $m = 2$. (A) A plot of f_a vs drug concentration on a linear scale. (B) The median-effect plot of $\log[(f_a)^{-1} - 1]^{-1}$ vs log (D). (Modified from Chou and Talalay, 1981).

When a mixture of mutually nonexclusive drugs is used, the median-effect plot of $y = \log[(f_a)_{1,2}/(f_u)_{1,2}]$ with respect to $x = \log[(D)_1 + (D)_2]$ will give an upwardly concave curve which at high doses will intercept the plot of the more potent of the two drugs (Figs. 2 and 3). The theoretical plot for Eq. 8 and experimental examples have recently been described (see Chou and Talalay, 1984, 1987). In collaboration with Dr. P. Talalay, a general equation for the combination of more than two drugs was also derived (Chou and Talalay, 1977, 1981, 1984; Chou and Chou, 1989).

Based on Eqs. 6–8, the generalized multiple drug effect equations used for computer simulation have been developed recently (Chou and Chou, 1989). These equations have made it possible to calculate the combination index (CI) for quantifying synergism, summation, and antagonism as follows (Chou and Talalay, 1983, 1984):

$$CI = \frac{(D)_1}{(D_x)_1} + \frac{(D)_2}{(D_x)_2} \tag{9}$$

for mutually exclusive drugs and

$$CI = \frac{(D)_1}{(D_x)_1} + \frac{(D)_2}{(D_x)_2} + \frac{(D)_1(D)_2}{(D_x)_1(D_x)_2} \tag{10}$$

for mutually nonexclusive drugs.

For mutually exclusive or nonexclusive drugs, when

CI < 1, synergism is indicated.
CI = 1, summation is indicated.
CI > 1, antagonism is indicated.

To determine synergism, summation and antagonism at any effect level (i.e., for an f_a value), the general procedure involves three steps. (1) Generate the median-effect plot (Eq. 1) to determine m and D_m values for drug 1, drug 2, and their combinations. (2) Calculate the corresponding doses [i.e., $(D_x)_1$, $(D_x)_2$, and $(D_x)_{1,2}$] by using the alternative form of Eq. 1 for a given degree of effect (i.e., a given f_a value representing $x\%$ affected). (3) Calculate the combination index (CI) by using Eq. 9 or Eq. 10, where $(D_x)_{1,2}$ (from Step 2) is dissected into $(D)_1$ and $(D)_2$ by their known ratio. Thus CI values which are smaller then, equal to, or greater than one, represent synergism, summation, and antagonism, respectively. A plot consisting of a combination index as a function of effect levels is called the F_a–CI plot (Chou and Talalay, 1984). A typical example of this plot is given in Chapter 6. A detailed description of a step-by-step synergism/antagonism analysis is given in Section IVE2.

The general equation for isobolograms was recently derived, and its practical advantages and limitations have been defined (Chou and Talalay, 1984, 1987). The conventional (classical) isobologram has been shown to be valid for two drugs that are mutually exclusive (e.g., the same or similar modes of action) (Fig. 4). Another type of isobologram, the conservative isobologram, is used to accommodate two drugs that are mutually nonexclusive (e.g., totally independent or different modes of action) (Chou and Talalay, 1984; Chou, 1987). The exclusivity may be tested with the median-effect plot. The classical isobologram will always indicate slightly more synergism and less antagonism than the conservative isobologram because the CI value for the former is the sum of two terms and for the latter it is the sum of three terms (Eqs. 9 and 10). The isobolograms are dose oriented, and the F_a CI plots are effect oriented. The two graphic methods are based on the same general equation. Automated generation of both the F_a–CI plots and the isobolograms using microcomputers has been developed recently (Chou and Chou, 1989). Typical classical and extended isobolograms are given in Fig. 4 and Chapter 1, Fig. 1.

B. ISSUES AND INTERPRETATION OF EQUATIONS

1. The Explicit Derivation of the Classical and Conservative Isobologram Equations

The explicit derivation of the classical isobologram equation from the enzyme kinetic models was carried out by Chou and Talalay in 1977 and extended to its general form in 1981 and 1984.

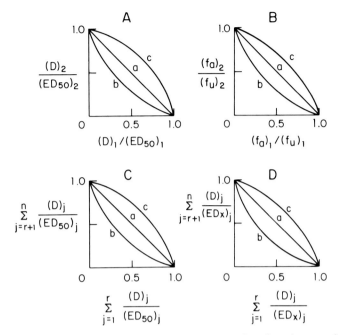

FIGURE 4 The classical and extended classical isobolograms based on the mutually exclusive assumptions. (A) The normalized median-effect dose isobologram (for ED_{50}, IC_{50}, LD_{50}, etc.) for two drugs. (B) The generalized isobologram for two drugs at different effect levels. (C) The generalized median-effect dose isobologram for n drugs. (D) The dose-normalized general isobologram for n drugs at different dose levels. D is the drug dose or concentration; ED_{50} and ED_x are the dose or concentration of drug for inhibiting 50 and X%, respectively; f_a is the fraction inhibited; f_u is the fraction uninhibited. The subscripts $1,2,\ldots n$, refer to drug 1, drug $2,\ldots$ drug n, respectively. Those combination data points that fall on the hypotenuse indicate additive effects (i.e., line a), those within the triangle indicate synergistic effects (e.g., curve b), and those within the upper right of the hypotenuse indicate antagonistic effects (e.g., curve c). The curves b and c can have irregular shapes. (Reprinted with permission from *Advances in Enzyme Regulation* **22**, Chou and Talalay, Quantitative analysis of dose–effect relationships: The combined effects of multiple drugs or enzyme inhibitors. Copyright © 1984, Pergamon Press plc.)

The classical ED_{50}-isobologram equation

$$\frac{(D)_1}{(D_m)_1} + \frac{(D)_2}{(D_m)_2} = 1 \tag{11}$$

can be explicity derived from the multiple-drug-effect equation as shown here.

For mutually exclusive drugs that have hyperbolic dose–effect curves (i.e., first order or $m = 1$), the combined dose–effect relationship is given by Eq. 6.

For mutually exclusive drugs that have sigmoidal dose–effect curves (i.e., higher order, or m^{th} order), the combined dose–effect relationship is given by Eq. 7.

At the median-effect doses, $(f_a)_{1,2} = 0.5$ and $(f_u)_{1,2} = 0.5$, we obtain Eq. 11.

Equation 11 exactly describes the classical isobologram at ED_{50}. It has been derived in the appendix of *Advances in Enzyme Regulation* (Chou and Talalay, 1984). The general form of the classical isobologram equation (i.e., at other effect levels) is

$$\frac{(D)_1}{(D_x)_1} + \frac{(D)_2}{(D_x)_2} = 1 \tag{12}$$

Thus, for mutually exclusive drugs, the classical isobologram correctly describes the additivity of the dose–effect relationships regardless of whether the two drugs have hyperbolic or sigmoidal dose–effect curves. Despite the fact that the isobologram has been used for decades, its explicit derivation and a clear description of its usefulness and limitations have not been given before (Chou and Talalay, 1984).

2. The Limitation of the Fractional Product Method and Classical Isobologram Method

Explicit Derivation of the Fractional Product Method of Webb (1963) For mutually nonexclusive drugs that have hyperbolic dose–effect curves (i.e., first-order kinetics or $m = 1$), the combined dose–effect relationship (Chou and Talalay, 1977, 1981) is

$$\frac{(f_a)_{1,2}}{(f_u)_{1,2}} = \frac{(f_a)_1}{(f_u)_1} + \frac{(f_a)_2}{(f_u)_2} + \frac{(f_a)_1(f_a)_2}{(f_u)_1(f_u)_2} \tag{13}$$

Since $f_a = 1 - f_u$,

$$\frac{1 - (f_u)_{1,2}}{(f_u)_{1,2}} = \frac{1 - (f_u)_1}{(f_u)_1} + \frac{1 - (f_u)_2}{(f_u)_2} + \frac{[1 - (f_u)_1][1 - (f_u)_2]}{(f_u)_1(f_u)_2}$$

$$= \frac{(f_u)_2[1 - (f_u)_1] + (f_u)_1[1 - (f_u)_2] + 1 + (f_u)_1(f_u)_2 - (f_u)_1 - (f_u)_2}{(f_u)_1(f_u)_2}$$

$$= \frac{(f_u)_2 - (f_u)_1(f_u)_2 + (f_u)_1 - (f_u)_1(f_u)_2 + 1 + (f_u)_1(f_u)_2 - (f_u)_1 - (f_u)_2}{(f_u)_1(f_u)_2}$$

$$= \frac{1 - (f_u)_1(f_u)_2}{(f_u)_1(f_u)_2}$$

Therefore,

$$(f_u)_1(f_u)_2 - (f_u)_{1,2}(f_u)_1(f_u)_2 = (f_u)_{1,2} - (f_u)_{1,2}(f_u)_1(f_u)_2$$

Hence,

$$(f_u)_{1,2} = (f_u)_1(f_u)_2 \tag{14}$$

Therefore, the fractional product concept of Webb (1963) correctly describes the additive effects of two mutually nonexclusive drugs (e.g., two noncompetitive inhibitors) in the Michaelis–Menten system (i.e., first order or $m = 1$).

The Classical Isobologram Method of Loewe (1957)   Even in the simplest (Michaelis–Menten) system, it can be shown that the classical isobologram concept describes the additive effect of mutually exclusive (e.g., competitive) inhibitors and that the fractional product concept describes the additive effect of mutually nonexclusive (e.g., noncompetitive) inhibitors (Chou and Talalay, 1981, 1984) (see Eq. 14). These two concepts are *not* compatible. Therefore both cannot be right at the same time and thus they cannot be used indiscriminately. The limitation (or the the lack of validity) of the isobologram method in certain situations (e.g., in the case of mutually nonexclusive inhibitors) can be seen clearly from equations derived from explicit models. If the classical isobologram is to be used as a universal procedure for drug combination studies as proposed by Berenbaum (1989), then it is necessary to assume that all drugs used in the biomedical fields have the same or similar modes of action, which is obviously not the case. However, the mechanisms or modes of action of drugs under study are frequently not known. Consequently, as a conservative measure both mutual exclusivity and nonexclusivity assumptions are made in synergism/antagonism calculations. However, if the assumption of mutual exclusivity or nonexclusivity is made, it should be explicitly stated so that the reader will not be misled.

   A comparison of the applicability of the Webb's fractional product method, the Loewe's classical isobologram method, and the method of median-effect plot and combination index is given in Table 4. Because the classical isobologram was intuitively conceived (Loewe, 1928, 1953, 1957) without a rigorous derivation or proof, its limitations, as indicated earlier, were not revealed until recently (Chou and Talalay, 1984). One simple way to check the classical isobologram equation is to use a "sham test," assuming two drugs (tested in double-blind fashion) have the same identity (Berenbaum, 1989). When two drugs have the same identity and thus act exactly the same way, they are necessarily mutually exclusive. What the classical isobologram fails to describe is the situation in which two drugs act totally differently and independently (i.e., when the two drugs are mutually nonexclusive). The simplest way to prove the limitation of the classical isobologram is to use the simplest enzyme kinetic system (the first order Michaelis–Menten equation for a single-substrate, single-product enzyme kinetic system in the presence of two mutually nonexclusive inhibitors, e.g., two noncompetitive inhibitors) as shown by Chou and Talalay in 1977. Since the classical isobologram is not valid in the first-order mutually nonexclusive systems, it is obvious that it is also not valid in the higher-order mutually nonexclusive systems.

   If the effects of two drugs are mutually nonexclusive but are analyzed by the classical isobologram method which intrinsically assumes that two drugs are mutually exclusive, the consequence is overestimation of synergism or underestimation of antagonism. This is because the CI equation for mutually nonexclusive drugs (even in simple first-order systems) contains a third term which is the product of the first two terms (Eq. 10). The higher CI value due

TABLE 4
Comparison of Applicability of Different Methods

	The nature of dose–effect curves and drug interactions			
	Mutually exclusive[a] (similar mode of action)		Mutually nonexclusive[b] (independent mode of action)	
	1st order[c]	Higher order[d]	1st order[c]	Higher order[d]
Webb's (1963) fractional product method[e]	No	No	Yes	No
Loewe's (1957) classical isobologram method[f]	Yes	Yes	No	No
Multiple drug effect equation (Chou and Talalay, 1984)	Yes	Yes	Yes	Yes

[a] Mutually exclusive drugs in mixture give a parallel median-effect plot with respect to the parent compounds.
[b] Mutually nonexclusive drugs in mixture give a upwardly concave dose–effect curve with respect to the parent compounds.
[c] Hyperbolic dose–effect curve.
[d] Sigmoidal dose–effect curve.
[e] $i_{1,2} = 1 - [(1 - i_1)(1 - i_2)]$ or $(f_u)_{1,2} = (f_u)_1(f_u)_2$ where i is the fractional inhibition and f_u is the fraction unaffected.
[f] The classical isobologram is for mutually exclusive drugs. The new conservative isobologram is for mutually nonexclusive drugs, and the $F_a - CI$ plots are for mutually exclusive and mutually nonexclusive drugs.

to the presence of the third term will predict less synergism or more antagonism. In this regard, the mutually nonexclusive assumption yields a more conservative estimate of synergism than the mutually exclusive assumption when the exclusivity of two drugs is unknown. Another limitation of the isobologram is the fact that only a few isoeffective levels can be presented on the same graph in order to avoid visual confusion caused by scattered data points. By contrast, the F_a–CI plot has no such limitation.

Another practical limitation of the isobologram is that, if the actual dose units for drug 1 and drug 2 are used on the coordinates of the isobologram without normalization with $(D_x)_1$ and $(D_x)_2$ (i.e., without normalization to unity), the combination data points are frequently jammed and are not readable from the graph. This limitation is partially evident when two drugs have very different potencies, when very high combination dose ratios are used, or when synergism is prominent.

3. Complexity in Higher Order Mutually Nonexclusive Drug Interactions

The dose–effect relationships for the first-order and higher-order mutually exclusive drugs and for the first-order mutually nonexclusive drugs are relatively straightforward as shown previously (Chou, 1974, 1976; Chou and Talalay, 1977, 1981). However, for higher-order mutually nonexclusive drugs, the dose–effect analysis requires a certain degree of approximation (Chou and Talalay, 1984). It is not appropriate to algebraically extend the first-order system to a higher-order system without taking into account the physicochemical (model) implications. Even for the single-drug higher-order systems, the equation derived by Hill (1913) from oxygen–hemoglobin interactions is not perfect since it assumes the existence of free and fully occupied complexes without taking into account the partially occupied intermediate complexes. Reexamining the model with modern enzyme kinetic analyses reveals that this simplification is necessary for practical considerations. Although the partially occupied (by oxygen) hemoglobin intermediates are minor species, their existence cannot be denied. However, if minor species are to be taken into account, the resulting equation is exceedingly complex because the contributions of each species to the overall system are not equal, and the detailed derivation requires complete understanding of the system (Segel, 1975). In biological systems, the microscopic rate constants are numerous and they can rarely be accurately determined. Thus demanding a complete equation for a biological system may be feasible theoretically but almost impossible practically at the present time (Whitehead, 1970). Therefore, the Hill plot and the median-effect plot should be considered practical approximations but not perfect solutions for higher-order systems.

Extension from a single drug to a two-drug combination for the first-order system is relatively simple, as shown by Chou and Talalay (1977), regardless of whether the drugs are mutually exclusive or nonexclusive. However, extension from a single drug to a two-drug combination for a higher-order system represents a considerable increase in complexity. Chou and Talalay (1984) proposed a first degree approximation by adding a third term to the multiple-drug dose–effect equation (Eq. 8) or CI equation (Eq. 10). This third term is the product of the first two terms and is fully justified in the first-order cases. The addition of the third term for the higher-order system is necessary but may not be sufficient (i.e., an additional compound term(s) may be added, if justified). In complex biological systems which show sigmoidal dose–effect curves in the presence of mutually nonexclusive drugs, the determination of exactly how many minor species exist and the relative contribution of each species is not a practical task. In a simple system such as hemoglobin, the first oxygen molecule that binds to one of the four subunits facilitates the binding of the second oxygen molecule to the second subunit, and so forth until four subunits are fully occupied. This "cooperative" rapid binding leads to a situation in which the free hemoglobin and the fully occupied hemoglobin are the predominant species. The intermediate minor species, although they may exist, cannot be counted equally with the initial or the final species (Segel, 1975). For allosteric enzymes, cellular systems, or animal systems, sigmoidal dose–effect curves which indicate higher-order systems are common (Ainslie *et al.*, 1972).

From the previous discussions and the derivation of the Chou and Talalay (1981) equations, the user of the Chou and Talalay method should be aware of the underlying assumptions. Those limitations for the Hill equation are also valid for the median-effect equation to some extent, although the inhibitor ligands are usually the dead-end inhibitors. Numerous applications in biomedical literature for enzyme, receptor, cellular, and animal systems indicate that the Hill (1913) equation, the median-effect equation derived by Chou (1976), and the multiple-drug-effect equation derived by Chou and Talalay (1981) are the practical approximations for dose effect analysis if the mass-action principle is to be observed.

4. Controversy and Comments

There are many aspects of dose effect analysis that can be the subject of controversy. Complexity in higher-order systems, especially those involving mutally nonexclusive drugs, is a main source of debate. It is clear from discussions presented earlier, that no one as yet has the perfect solution, Instead, a simple, practical, economical, and sensible method is to be sought. Another source of debate is the amount of statistics to be involved in dose effect analysis. When parameters of dose–effect relationships are to be analyzed, it is essential that multiple doses are used, just as it is essential to vary the substrate for K_m or to vary the inhibitor and/or substrate for K_i determinations. It is not customary nor economical to require a 95% confidence limit for K_m or K_i determination as can be seen in biomedical literature. Any set of biological data can be subjected to statistical analysis, including synergism/antagonism evaluation; hence, statistics *per se* is not the main issue. The main issue is the use of the proper method in the assessment of synergism or antagonism in situations where large sample size and accurate assays are possible. Another issue is the necessity of a rigorous definition of additivity of drug effects, as discussed earlier.

The median-effect principle with statistical diagnosis has been proven to be widely applicable in biological systems as evidenced in publications in numerous journals (Appendices I and II). These data can be subjected to side-by-side comparisons with other proposed methods for synergism/antagonism evaluations (Steel and Peckham, 1979; Berenbaum, 1977, 1980, 1981; Syracuse and Greco, 1986; Gessner, 1988; Sühnel, 1990) in terms of the thoroughness of theoretical basis, specific models used, equations derived, whether or not the equations were actually derived or "borrowed" or adopted, originality of theoretical theme, simplicity of usage, the minimum number of data points needed for analysis, availability of computer software, and acceptability by the scientific community. Many methods of synergism/antagonism evaluation are highly acclaimed, though in the final analysis it will be up to the scientific end users to make their own judgments. (The reader is referred to Chapter 1 for a more detailed discussion of different methods of dose effect analysis.)

C. COMPARISON OF DIFFERENT EXPERIMENTAL DESIGNS

1. Half-Dose Experimental Design

The experimental design given below for the combination of drug A and B is simple but not theoretically sound for synergism/antagonism determination.

A

B

A + B

A/2 + B/2

A + A

B + B

The following are some shortcomings of this design.

1. If A + B produce an effect greater than A or B alone, it proves nothing with respect to synergism because A + A or B + B also produce greater effects than A or B alone.
2. If A/2 + B/2 produce a greater effect than either A or B alone, this does not prove that there is synergism because the dose–effect curves of A and B may have totally different shapes. Single doses of single drugs or their combination do not indicate the shape of the dose–effect curve.

2. Nonconstant Combination Dose Ratios Design

A typical old-fashioned experimental design using nonconstant combination ratios.

Dose (μM)		Effect
Drug 1	Drug 2	Inhibition (%)
1		50
2		67
4		80
	2	5.7
	3	31.6
	5	85
1 +	3	70
2 +	3	85
4 +	3	92
1 +	2	52
2 +	2	71
4 +	2	82

While varying the concentrations of drug 1, the concentration of drug 2 is kept constant, or vice versa. The proportions of drug 1 and drug 2 in combination are changing most of the time, if not all the time, in this example. Since drug 1 and drug 2 alone have several doses, their effect can be measured and the parameters m and D_m can be determined (use of at least five different concentrations is recommended). Therefore, the CI values for each combination data point can be determined. However, the F_a–CI plot cannot be simulated for other effect levels because of the nonconstant ratio of combination design. This design, though acceptable, is not the most efficient one for data analysis.

D. COMPARISON OF DIFFERENT METHODS FOR DOSE-EFFECT ANALYSIS OF MULTIPLE DRUGS

If D_1 inhibits 40%, $(f_a)_1 = 0.4$. If D_2 inhibits 40%, $(f_a)_2 = 0.4$. Then, if additive, $D_1 + D_2$ inhibits ?%, i.e., $(f_a)_{1,2} = ?$

Methods	*Comments*
1. SIMPLE ARITHMETIC ADDITION $(f_a)_{1,2} = 0.4 + 0.4 = 0.8$	(i) Wrong method widely misused. For example, if each drug alone inhibits 60%, the combined effect of two drugs (if additive) is not expected to be 120%. (ii) Should not be used under any circumstances.
2. THERAPEUTIC SYNERGY Only the therapeutic end result relative to toxicity is of concern. Regardless of what combination dose of $D_1 + D_2$, when $(f_a)_{1,2} > 0.4$, therapeutic synergy is indicated whenever toxicity is tolerable.	(i) Pragmatic, usually nonquantitative, both therapeutic effect and toxic effect are taken into account simultaneously. (ii) Mainly used in clinical setting when dose effect analysis is not available or when the use of suboptimal doses is unethical. (iii) It is a selectivity question rather than a synergy question since synergy is not determined.

3. FRACTION PRODUCT
(Webb, 1963)

$$(f_a)_{1,2} = 1 - [1 - (f_a)_1][1 - (f_a)_2]$$

$$= 1 - [1 - 0.4][1 - 0.4]$$

$$= 1 - 0.36 = 0.64$$

(i) Does not take into account the shape of the dose–effect curve.

(ii) Not valid for sigmoidal dose–effect curve.

(iii) Not valid for mutually exclusive drugs.

(iv) Valid only for mutually nonexclusive drugs with hyperbolic dose–effect curves (i.e., follow the Michaelis–Menten type first-order kinetics, $m = 1$).

4. CLASSICAL ISOBOLOGRAM (Loewe, 1928)

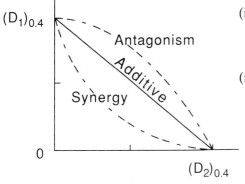

(i) Need dose–effect relationship for D_1, D_2, and $D_1 + D_2$

(ii) Valid for mutually exclusive drugs, but not for mutually nonexclusive drugs.

(iii) The old method requires visual interpolation, extrapolation, and large-scale experiment.

(iv) The classical isobologram equation has been explicitly derived from enzyme kinetic models by Chou and Talalay (1981). It is a special case of the multiple-drug-effect equation. Computer software allows the construction of an isobologram automatically and instantly.

(v) The isobologram and the F_a–CI plot are the two aspects of the same multiple-drug-effect equation. The former is dose oriented and the latter is effect oriented. Both give quantitatively identical conclusion about synergism/antagonism.

(vi) To avoid messy confusion only a few isobolograms can

be shown on a given graph. By contrast, the F_a–CI plot gives synergism/anatagonism for any number of effect levels in one graph.

5. MEDIAN-EFFECT PRINCIPLE (Chou and Talalay, 1981) Quantitation of combination index (CI)

CI = 1, additive
CI < 1, synergistic
CI > 1, antagonistic

$$CI = \frac{(D)_1}{(D_x)_1} + \frac{(D)_2}{(D_x)_2} +$$

$$\frac{\alpha(D)_1(D)_2}{(D_x)_1(D_x)_2}$$

$\alpha = 0$, mutually exclusive
$\alpha = 1$, mutually nonexclusive

(i) Synergism/antagonism at different dose or effect levels can be quantitively determined.

(ii) Need the dose–effect relationship for D_1 and D_2 alone, and for their combination(s).

(iii) Takes into account both relative potency (D_m) and the shape of dose–effect curve (m).

(iv) Does not require large-scale experiment (diagonal scheme for constant ratio drug combinations). Instead of drawing empirical curve to fit data, this method uses data to fit a theory (the mass-action law).

(v) Fully automated computer analysis available. Data from either constant or nonconstant combination ratio experiments can be used for calculating CI.

(vi) For additional comparisons, see Tables 3, 4 and 5.

E. LINKING THE THEORY, EQUATIONS, EXPERIMENTAL DESIGN, AND COMPUTERIZED SIMULATION

By using the mass-action law as a basic theoretical principle and using enzyme kinetics as a model, hundreds of equations can be derived (Chou, 1974, 1976, 1977; Chou and Talalay, 1977, 1981, 1984). These equations can be generalized to a few general equations by calculating the fractional effect to cancel out various kinetic constants such as K_m, and K_i, and V_{max}. A systematic derivation and simplification makes it possible to obtain

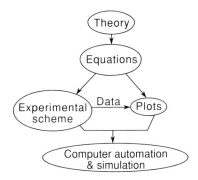

SCHEME 2 Flow chart of the development of a method for dose-effect analysis.

mathematically deduced generalized equations. With the availability of equations, it becomes a simple task to perform computerized analysis and to create graphics.

1. Theoretical Development and Experimental Design

A flow chart for conceptual development from theory to data analysis is given in Scheme 2. A more specific method for the development of dose effect analysis based on the median-effect principle for a single drug and for multiple drugs is shown in Schemes 3 and 4.

The experimental designs can be created so that dose range will be on, above (e.g., three data points), and below (e.g., two data points) the median-effect dose or concentration (e.g., IC_{50}, ED_{50}, or LD_{50}). The dose density for each drug, depending on the method and test system used usually has 5–8 data points.

2. A Step-by-Step Procedure for Synergism/Antagonism Analysis

1. Conduct a pilot study by arbitrarily varying drug concentration (usually 10^{-3}, 10^{-4}, 10^{-5}, 10^{-6}, 10^{-7}, 10^{-8} M) or by obtaining information from the scientific literature to estimate approximate potency (IC_{50} value) for each drug and for the approximate slope of the dose–effect curve.
2. Design the dose range (e.g., two- to fourfold serial dilution from the highest dose) for each drug of the experiment.
3. Make a mixture of D_1 and D_2 so that the two components will be equipotent [i.e., the dose ratio at $D_1/D_2 = (D_m)_1 (D_m)_2$]. For the most efficient data analysis, it is advisable that D_1 and D_2 are varied at a

The Median-Effect Principle

Effect	Dose	Effect	Derivations	Comment
$\dfrac{f_i}{f_v} =$	$\dfrac{I}{I_{50}}$		*Mol. Pharm.* **10** 235–247, 1974	1st order single
	$\dfrac{I}{I_{50}} = \left(\dfrac{f_i}{f_v}\right)^{1/m}$		*J. Theor. Biol.* **59** 235–276, 1976	Higher order single
	$\dfrac{I}{I_{50}} = \dfrac{I}{k_i} \times \dfrac{E_x}{E_f}$		*Mol. Pharm.* **10** 235–247, 1974	1st order single

$\longrightarrow$ DISTRIBUTION EQUATION

$$\frac{(f_i)_{1,2}}{(f_v)_{1,2}} \overset{x}{=} \frac{(I)_1}{(I_{50})_1} + \frac{(I)_2}{(I_{50})_2} = \frac{(f_i)_1}{(f_v)_1} + \frac{(f_i)_2}{(f_v)_2}$$
$$E_t = E + EI_1 + EI_2$$

J. Biol. Chem. **252** 6438–6442, 1977 — 1st order multiple mut. exclusive

$$\frac{(f_i)_{1,2}}{(f_v)_{1,2}} \overset{NX}{=} \frac{(I)_1}{(I_{50})_1} + \frac{(I)_2}{(I_{50})_2} + \frac{(I)_1(I)_2}{(I_{50})_1(I_{50})_2}$$
$$= \frac{(f_i)_1}{(f_v)_1} + \frac{(f_i)_2}{(f_v)_2} + \frac{(f_i)_1(f_i)_2}{(f_v)_1(f_v)_2}$$
$$E_t = E + EI_1 + EI_2 + EI_1I_2$$

J. Biol. Chem. Ibid — 1st order multiple mut. nonexclusive

$$\left[\frac{(f_i)_{1,2}}{(f_v)_{1,2}}\right]^{1/m} \overset{x}{=} \frac{(I)_1}{(I_{50})_1} + \frac{(I)_2}{(I_{50})_2} = \left[\frac{(f_i)_1}{(f_v)_1}\right]^{1/m} + \left[\frac{(f_i)_2}{(f_v)_2}\right]^{1/m}$$
$$E_t = E + E(I_1)_m + E(I_2)_m$$

Eur. J. Biochem. **115** 207–216, 1981 — Higher order multiple mut. exclusive

$$\left[\frac{(f_i)_{1,2}}{(f_v)_{1,2}}\right]^{1/m} \overset{NX}{=} \frac{(I)_1}{(I_{50})_1} + \frac{(I)_2}{(I_{50})_2} + \frac{(I)_1(I)_2}{(I_{50})_1(I_{50})_2}$$
$$= \left[\frac{(f_i)_1}{(f_v)_1}\right]^{1/m} + \left[\frac{(f_i)_2}{(f_v)_2}\right]^{1/m} + \left[\frac{(f_i)_1(f_i)_2}{(f_v)_1(f_v)_2}\right]^{1/m}$$
$$E_t = E + E(I_1)_m + E(I_2)_m + E(I_1)_m(I_2)_m$$

Eur. J. Biochem. Ibid — Higher order multiple mut. nonexclusive

SCHEME 3 The median-effect principle for dose-effect analysis in single drug and multiple drug conditions.

constant ratio. (This condition can be satisfied by serial dilution of a mixture. (See Table 5). For more detailed studies, make two or three more mixtures so that the potency ratio of D_1 and D_2 will be approximately 0.2:1 (emphasizing D_2 and de-emphasizing D_1), 5:1 (emphasizing D_1 and de-emphasizing D_2), etc. With several combination ratios, it is possible to estimate the optimal combination ratio for maximal synergy.

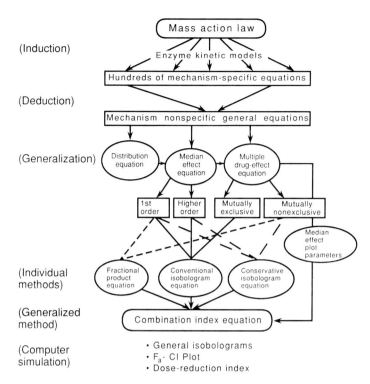

SCHEME 4   A flow chart for deriving generalized equations, method development, and computerized data analysis.

TABLE 5

An Idealized Experimental Design Showing the Outlay of Dose Range and Dose Density of Two Drugs for Drug Combination Analysis

		Drug 1					
		0	$0.25X$ $(ED_{50})_1$	$0.5X$ $(ED_{50})_1$	$(ED_{50})_1$	$2X$ $(ED_{50})_1$	$4X$ $(ED_{50})_1$
	0	Control $(f_a)_0$	$(f_a)_1$	$(f_a)_1$	$(f_a)_1$	$(f_a)_1$	$(f_a)_1$
	$0.25X$ $(ED_{50})_2$	$(f_a)_2$	$(f_a)_{1,2}$				
	$0.5X$ $(ED_{50})_2$	$(f_a)_2$		$(f_a)_{1,2}$			
Drug 2	$(ED_{50})_2$	$(f_a)_2$			$(f_a)_{1,2}$		
	$2X$ $(ED_{50})_2$	$(f_a)_2$				$(f_a)_{1,2}$	
	$4X$ $(ED_{50})_2$	$(f_a)_2$					$(f_a)_{1,2}$

4. Measure the dose–effect relationships for each drug alone and for the mixture(s).
5. Graph the data obtained in Step 4 with the median-effect plot and measure the parameters m_1, $(D_m)_1$, r_1, m_2, $(D_m)_2$, r_2, $(m)_{1,2}$, $(D_m)_{1,2}$, and $r_{1,2}$. The slope gives the m value (for the shape of the dose–effect curve, the x- intercept gives the D_m value for the potency of the drug, and r is the linear correlation coefficient of the median-effect plot.
6. Calculate the dose $(D_x)_1$, $(D_x)_2$, and $(D_x)_{1,2}$ at each given effect level $(f_a)_x$ with the use of Eq. 4, $D = D_m [f_a/(1-f_a)]^{1/m}$, and the corresponding m and D_m values obtained in Step 5. Note that $(D_x)_1$, $(D_x)_2$, and $(D_x)_{1,2}$ are doses of drug 1, drug 2, and their mixture which are required to affect a given system by $x\%$.
7. Calculate the contribution of D_1 and D_2 in the mixture $(D_x)_{1,2}$ from the known dose ratio of two drugs. For example, if $(D)_1/(D)_2 = P/Q$, then $(D)_1 = (D)_{1,2} \times P/(P + Q)$ and $(D)_2 = (D_x)_{1,2} \times Q/(P + Q)$, where $(D_x)_{1,2} = (D)_1 + (D)_2$.
8. Substitute $(D)_1$, $(D)_2$, $(D_x)_1$, and $(D_x)_2$, $(D_x)_2$ values in Eq. 9 or 10 in order to calculate the combination index (CI):

$$CI = \frac{(D)_1}{(D_x)_1} + \frac{(D)_2}{(D_x)_2} + \frac{\alpha(D)_1(D)_2}{(D_x)_1(D_x)_2} \tag{15}$$

If the effects of two drugs are mutually exclusive (i.e., the median-effect plots for D_1, D_2, and $D_{1,2}$ are parallel or if D_1 and D_2 have similar modes of action), then $\alpha = 0$ (i.e., the last term of Eq. 15 is dropped). If the effects of two drugs are mutually nonexclusive (i.e., the median-effect plots for D_1 and D_2 are parallel but $D_{1,2}$ is upwardly concave, or D_1 and D_2 have independent modes of action), then $\alpha = 1$. If the exclusivity of two drugs is not clear, it is suggested that the CI value be determined with both mutually exclusive $(\alpha = 0)$ and mutually nonexclusive $(\alpha = 1)$ assumptions. Note that a mutually nonexclusive assumption always predicts less synergism or more antagonisim than the mutually exclusive assumption. The former is called the conservative isobologram equation and the latter is called the classical isobologram equation.

9. When the CI value is equal to, less than, or greater than 1, then summation, synergism, or antagonism of effects is suggested, respectively.
10. Use the dose-reduction index equation to calculate the dose-reduction index (DRI).

A schematic presentation of the algorithms for computerized automation of quantitation of synergism and antagonism of two drugs is shown in Scheme 5 (for a sample analysis, see Chapter 6).

For f_a at x% affected by D_1, D_2, or their mixture (i.e., at isoeffective doses) and in the mixture where $(D_x)_{1,2} = (D)_1 + (D)_2$ and $(D)_1/(D)_2 = P/Q$, we get,

$$CI = \frac{(D)_1}{(D_x)_1} + \frac{(D)_2}{(D_x)_2} + \frac{\alpha(D)_1(D)_2}{(D_x)_1(D_x)_2}$$

Therefore,

$$CI = \frac{(D_x)_{1,2}[P/(P+Q)]}{(D_m)_1\{(f_{a_x})_1/[1-(f_{a_x})_1]\}^{1/m_1}} + \frac{(D_x)_{1,2}[Q/(P+Q)]}{(D_m)_2\{(f_{a_x})_2/[1-(f_{a_x})_2]\}^{1/m_2}}$$

$$+ \frac{\alpha[(D_x)_{1,2}]^2\,[P/(P+Q)][Q/(P+Q)]}{(D_m)_1\{(f_{a_x})_1/[1-(f_{a_x})_1]\}^{1/m_1}(D_m)_2\{(f_{a_x})_2/[1-(f_{a_x})_2]\}^{1/m_2}}$$

where $(D_x)_{1,2} = \{(f_{a_x})_{1,2}/[1-(f_{a_x})_{1,2}]\}^{1/m_{1,2}}[(D_m)_{1,2}]$

and $(f_{a_x})_1 = (f_{a_x})_2 = (f_{a_x})_{1,2}$ (i.e., isoeffective)

For *mutually exclusive* drugs, $\alpha = 0$
For *mutually nonexclusive* drugs, $\alpha = 1$
Diagnosis: CI = 1, Additive; CI < 1, Synergistic; and CI > 1, Antagonistic.

SCHEME 5   Algorithms for automated computerized quantitation of synergism and antagonism of two drugs.

Note that computer software for the Apple II series and the IBM-PC is available for semi-automated analysis and graphing from Steps 5 to 10. (Chou and Chou, 1989; see also Chapter 6). This analysis can also be done manually with a pocket calculator that has logarithmic and power functions.

Also note that the conservative isobologram or equation represents only "first degree conservation" since the intermediate complex species are expected to play only a minor role, especially for those inhibitor ligands which are the dead-end inhibitors rather than alternative-substrate-type inhibitors. The extreme conservatism used by Syracuse and Greco (1986) necessarily assumes that every intermediate species of the interaction is equally important, which is most unlikely. By contrast, first degree conservatism will definitely occur in all mutually nonexclusive cases although a further degree of conservatism may be needed in some cases. Also note that for three-drug combination studies, it is advisable that two-drug combinations also be carried out although this may result in a larger experiment (under the same experimental conditions). The reason for this suggestion is that D_1 and D_2 may be synergistic, D_2 and D_3 may be antagonistic, D_1 and D_3 may be synergistic, etc. Although the D_1, D_2, and D_3 combination would allow the CI value to be determined, it would not provide information of interactions among two component drugs. By conducting two- and three-drug combinations at the same time, some data points such as the single-drug dose-effect relationship and parameters can be shared and thus generate considerable savings in time, cost, and effort. It is also suggested that if the combination ratio for D_1, D_2, and D_3 is a:b:c, then it would be most efficient to conduct two drug combinations with ratios of $D_1:D_2 = a:b$, $D_2:D_3 = b:c$, and

$D_1 : D_3 = a : c$. Computer software for automated data analysis for a three-drug combination is available from Chou (1989; see also Chapter 6). If a three-drug combination study is to be carried out, the degree of conservatism for nonexclusivity is a complex issue. For simplicity, the classical isobologram equation assuming mutually exclusive interactions may be used especially for dead-end inhibitors. However, it is necessary to state this assumption so that it can be known that such a calculation may overestimate synergism. Alternatively, additional term(s) may be added to the calculation of CI value. For example, to product of terms corresponding to the mutually nonexclusive interactions (if known) need to be added in the CI calculation.

F. COMPUTERIZED AUTOMATION OF DATA ANALYSIS

For the automated quantitation of synergism/antagonism, initial computer software programs for the Apple II and IBM microcomputers have been developed and published (Chou and Chou, 1989; see also Chapter 6). A complete analysis for each set of multidrug combination studies takes less than 10 minutes. The printouts of the analyses consist of a dose–effect table, dose–effect curves, median-effect plot, F_a–CI tables, and F_a–CI plots based on both mutually exclusive and mutually nonexclusive assumptions, classical and conservative isobolograms, and summary tables. CI values for all drug combination data points regardless of constant dose ratio or nonconstant dose ratio experimental design can be calculated automatically. The diagnostic test of applicability of the experimental data to the median-effect principle is carried out in all data analyses. This diagnosis involves the linear correlation coefficient of the median-effect plot (r value), the standard error of the mean of the x intercept (the log of D_m value), and the slope (the m value). The statistical evaluation is routinely carried out at the beginning of the analysis rather than after the data have been subjected to various mathematical transformations. The statistical analysis may also be carried out at the final step on the CI values in different ways, for example, by assessing the standard error of the mean of CI value (Paquette *et al.*, 1988) or the distribution of combination data points in the F_a–CI plot (Chou and Chou, 1989).

V. Summary

There are demonstrated advantages of using drug combinations for chemotherapy of various conditions. The drug interactions in terms of synergism/antagonism can be assessed using various methods. Some of these methods are quite different and may yield quantitatively or qualitatively different conclusions.

The following are some of the major features of the method proposed by Chou and Talalay (1983, 1984, 1987) using the median-effect plot, the isobologram and combination index.

1. The mass-action law approach using enzyme kinetic models rather than using the empirical approach.
2. Derived equation for defining additivism and thus synergism and antagonism.
3. Linking equations, experimental design, and diagnostic plot; using experimental data to fit the mass-action law rather than using empirical curves to fit experimental data; routinely provide statistical diagnosis of applicability of data to the method prior to dose effect analysis.
4. Efficient data utilization.
   a. Minimum number of data points using diagonal experimental scheme.
   b. Conservation of laboratory animals, cost, time, and effort.
5. Either the constant combination dose ratio or the nonconstant combination dose ratio designs for multiple-drug-effect analysis may be used. The degrees of synergism/antagonism are expressed quantitatively with the combination indices at different dose levels and at different effect levels.
6. Allows the determination of the optimal combination dose ratio for maximal synergy.
7. Determines the dose-reduction index to provide a measure of how much (how many fold) the dose of each drug in a synergistic combination reduced at a given effect level, compared with the doses of each drug alone.
8. Has demonstrated applicability *in vitro* and *in vivo* including drug combination studies in enzyme, receptor, cellular, or animal studies; method including the median-effect principle has been used in scientific papers published in over 70 biomedical journals; also provides simple and rapid experimental assessments for prospective and retrospective clinical studies (see Appendices I and II).
9. Computer software for automated dose effect analysis for single-drug and multiple-drug combinations, graphics, statistics, and data storage and retrieval are available.

Appendix I

References for the theoretical background of the median-effect principle and multiple-drug-effect analysis and examples of their applications.

THEORY

Chou, T.-C.

- Mol. Pharmacol. **10**, 235–247 (1974)
- J. Theor. Biol. **59**, 253–276 (1976), **65**, 345–356 (1977)
- Carcinogenesis **1**, 203–213 (1980)

Chou, T.-C., and Talalay, P.

- J. Biol. Chem. **252**, 6438–6442 (1977)
- Eur. J. Biochem. **155**, 207–216 (1981)

REVIEWS

Chou, T.-C., and Talalay, P.

- Trends Pharmacol. Sci. **4**, 450–454 (1983)
- Adv. Enzyme Regul. **22**, 27–55 (1984)
- In "New Avenues in Developmental Chemotherapy", (K. R. Harrap and T. A. Connors, eds.), Bristol-Myer Symp. Ser. 8, pp. 37–64. Academic Press, Orlando, Florida, 1987

Chou, T.-C.

- Asia Pac. J. Pharmacol. **2**, 93–97 (1987)
- Chou, T.-C., et al., Chemotherapeutic synergism, potentiation , and antagonism. IN "Encyclopedia of Human Biology" (R. Dulbecco, ed.), Vol. 2. Academic Press, San Diego, California, 1991 (in press)

COMPUTER SOFTWARE

Chou, J., and Chou, T.-C.

- "Dose–Effect Analysis with Microcomputers," computer software and manual for IBM-PC and Apple II. Elsevier–Biosoft, Cambridge, England, 1987

EXAMPLES OF RECENT APPLICATION

Multiple Drug Combination Studies

Antitumor Agents

- Chou, T.-C., *et al.*, Cancer Res. **42**, 3957–3963 (1982)
- Steckel, J., *et al.*, Biochem. Pharmacol. **32**, 971–977 (1983)

- Chang, T. T., *et al.*, Cancer Res. **45**, 2434–2439 (1983)
- Chang, T. T., *et al.*, *Cancer Res.* **47**, 2247–2250 (1987)
- Benckhuijsen, C., *et al.*, *Cancer Res.* **47**, 4814–4820 (1987)
- Durand, R. E., and Chaplin, D. J., *Br. J. Cancer* **50**, 103–109 (1987)
- Durand, R. E., and Olive, P. L., *Cancer Res.* **47**, 5303–5309 (1987)
- Durand, R. E., and Goldie, J. H., *Cancer Treat. Rep.* **71**, 673–679 (1987)
- Naomoto, Y., *et al.*, *Gann* **78**, 87–92 (1987)
- Chou, T. C., and Fanucchi, M. F. *in* "Encyclopedia of Medical Devices and Instrumentation" (J. G. Webster, ed.) Vol. 2, pp. 660–670. Wiley, New York, 1988.
- Ackland, S. P., *et al.*, *Cancer Res.* **48**, 4244–4249 (1988)
- Kono, Y., *et al.*, *Cancer Res.* **48**, 351–356 (1988)
- Kong, X.-B., *et al.*, *Leuk. Res.* **12**, 853–859 (1988)
- Ratain, M. L., *et al.*, *JNCI, J. Natl. Cancer Inst.* **80**, 1412–1416 (1988)
- Rideout, D., *et al.*, *Biochem. Pharmacol.* **37**, 4505–4515 (1988)
- Hofmann, J., *et al.*, *Int. J. Cancer* **42**, 382–388 (1988)
- Kohno, N., *et al.*, *Br. J. Cancer* **58**, 330–334 (1988)
- Galivan, J., *et al.*, *Cancer Res.* **48**, 2421–2425 (1988)
- Chitambar, C. R., *et al.*, *Blood* **72**, 1930–1936 (1988)
- Miyake, M. *et al.*, *Exp. Cell Biol.* **56**, 297–302 (1988)
- Kohno, N., *et al.*, *Arch. Otolaryngol. Head Neck Surg.* **114**, 157–161 (1988)
- Howell, S. B., *et al.*, *Cancer Res.* **49**, 3178–3183 (1989)
- Hofmann, J., *et al.*, *Lipids* **24**, 312–317 (1989)
- Galivan, J., *et al.*, *Adv. Enzyme Regul.* **28**, 13–21 (1989)
- Romanini, A., *et al.*, *Adv. Enzyme Regul.* **28**, 323–338 (1989)
- Tsai, C. M., *et al.*, *Cancer Res.* **49**, 3390–3397 (1989)
- Triozzi P. L., *et al.*, *Leuk. Res.* **13**, 437–443 (1989)
- Hoffman, R., *et al.*, *Br. J. Cancer* **59**, 347–348 (1989)
- Peters, R. H., *et al.*, *Cancer Chemother. Pharmacol.* **23**, 129–134 (1989)
- Chang, B. K., *et al.*, *Anticancer Res.* **9**, 341–346 (1989)
- Howell, S. B., *et al.*, *Cancer Res.* **49**, 3178–3138 (1989)
- Romanini, A., *et al.*, *Cancer Res.* **49**, 6019–6023 (1989)
  See also *Proc. Am. Assoc. Cancer Res.* abstr. 1824, 2097, 2167, 2173, 2275, 2276, and 2307 (1989)
- Durand, R. E., *Intl. J. Cancer* **44**, 911–917 (1989)
- Zhou, J.-Y., *et al.*, *Cancer Res.* **50**, 2031–2035 (1990)
- Usui, N., *et al.*, *JNCI, J. Natl. Cancer Inst.* **81**, 1904–1909 (1989)
- Rideout, D., *et al.*, *Biopolymers* **29**, 247–262 (1990)
- Isonishi, S., *et al.*, *J. Biol. Chem.* **265**, 3623–3627 (1990)
- Fruehauf, J. P., *et al.*, *JNCI, J. Natl. Cancer Inst.* **82**, 1206–1208 (1990)
- Boike, G. M., *et al.*, *Gynecol. Oncol.* (in press)

Anti-HIV Agents

PFA + rIFN A, AZT + rIFN A, AZT + rsT4, AZT + Ribavirin, etc.

- Hartshorn, K. L., *et al.*, *Antimicrob. Agents Chemother.* **30**, 189–191 (1986)
- Sandstrom, E. G., *Drugs* **31**, 463–466 (1986)
- Sandstrom, E. G., and Kaplan, J. C., *Drugs* **34**, 372–390 (1989)
- Hartshorn, K. L., *et al.*, *Antimicrob. Agents Chemother.* **31**, 168–172 (1987)
- Vogt, M., *et al.*, *Science* **235**, 1376–1379 (1987)
- Vogt, M., *et al.*, *J. Infect. Dis.* **158**, 378–385 (1988)
- Schroder, H. C., *et al.*, *Biochem. Pharmacol.* **37**, 3947–3952 (1988)
- Montefiori, D. C., *et al.*, *AIDS Res. Human Retroviruses* **5**, 193–203 (1989)
- Ruprecht, R. M., *Intervirology* **30**, 2–11 (1989)
- Koshida, R., *et al.*, *Antimicrob. Agents Chemother.* **33**, 778–780 (1989)

- Johnson, V. A., *et al.*, *Antimicrob. Agents Chemother.* **33**, 53–57 (1989)
- Johnson, V. A., *et al.*, *J. Infect. Dis.* **159**, 837–844 (1989)
- Schinazi, R., *et al.*, *Antimicrob. Agents Chemother.* **33**, 115–117 (1989)
- Eriksson, B. F. H., and Schinazi, R. F., *Antimicrob. Agents Chemother.* **33**, 663–669 (1989)
- Szebeni, L., *et al.*, *Proc. Natl. Acad. Sci. U.S.A.* **86**, 3842–3846 (1989)
- Berman, E., *et al.*, *Blood* **74** 1281–1286 (1989)
- Hayashi, S., *et al.*, *Antimicrob. Agents Chemother.* **34**, 82–88 (1990)
- Johnson, V. A., *et al.*, *J. Infect. Dis.* **161**, 1059–1067 (1990)
  See also *Proc. Int. Conf. AIDS, 5th* abstr. MCP 55, 97, 110, 125, 128, and 131 (1989)
- Ruprecht, R. M., *et al.*, *J. AIDS* **3**, 551–560 (1990)
- Ruprecht, R. M., *et al.*, *Cancer Res.* (in press)

Anti-HSV Agents

- Schinazi, R., *et al.*, *Antimicrob. Agents Chemother.* **30**, 491–498 (1986)
- Soike, K. F., *et al.*, *J. Infect. Dis.* **156**, 607–614 (1987)
- Lin, J.-C., *et al.*, *J. Infect. Dis.* **159**, 248–254 (1989)
- Taylor, J. L., *et al.*, *Invest. Ophthalmol. Visual Sci.* **30**, 365–370 (1989)
- Soike, K. F., *et al.*, *Antiviral Res.* **13**, 165–174 (1990)

Receptor Analysis

- Lombardini, J. B., *et al.*, *Mol. Pharmacol.* **36**, 256–264 (1989)

Biological Response Modifiers

- Bregman, M., *et al.*, *Intl. J. Cancer* **37**, 101–107 (1986)
- Paquette, R. L., *et al.*, *Blood* **71**, 1596–1600 (1988)
- Fry, E. J. A., *et al.*, *Tromb. Haemostasis* **62**, 909–916 (1989)
- Vathsala, A., *et al.*, *Transplantation* **49**, 463–472 (1990)
- Secor, W. E., *et al.*, *J. Immunol.* **144**, 1484–1489 (1990)

Radiation Effect

- Potmesil, M., *et al.*, *Radiat. Res.* **105**, 147–157 (1986)

Insecticides, Carcinogens, Epidemiological Analysis and LD_{50}, ED_{50}, Therapeutic Index, Median-Effect Plot, F_a-CI Plot, Isobolograms
See *Reviews* and computer software manual given earlier.

- Sambasiva, K. R. S., *et al.*, *Ecotoxicol. Environ. Safety* **10**, 209–217 (1985)

Enzyme Inhibitors

- Steckel. J., *et al.*, *Biochem. Pharmacol.* **32**, 971–977 (1983)
- Nakamura, C. E., and Abeles, R. H., *Biochemistry* **24**, 1364–1376 (1985)
- Siravaraporn, W., and Yuthavong, Y., *Antimicrob. Agents Chemother.* **29**, 899–905 (1986)

Anesthetic Agents

- Norberg, L., and Wahlstrom, G., *Acta Pharmacol. Toxicol.* **58**, 96–104 (1986)

Cardiovascular Drugs

- Bennett, B. M., *et al.*, *Can. J. Physiol. Pharmacol.* **62**, 1194–1197 (1984)
- Collen, D., *Circulation* **77**, 731–735 (1988)
- Sharma, A. S., and Klein, G. J., *Am. J. Cardiol.* **61**, 330–335 (1988)

Single Agent Studies

Antiviral Agents

- Chou, T.-C., *et al.*, *Antimicrob. Agents Chemother.* **31**, 1355–1358 (1987)
- Chou, T.-C., *et al.*, *Mol. Pharmacol.* **26**, 587–593 (1984)
- Matulic-Adamic, J., *et al.*, *J. Med. Chem.* **31**, 1642–1647 (1988)
- Su, T. C., *et al.*, *J. Med. Chem.* **31**, 1209–1215 (1988)

Antitumor Agents

- Yeh, S. F. Y., *et al.*, *Planta Med.* **5**, 413–414 (1988)
- Muller, B., *et al.*, *Cancer Res.* **48**, 2454–2457 (1988)
- Doppler, *et al.*, *Cancer Res.* **48**, 2454–2457 (1988).
- Koyama, M. *et al.*, *J. Med. Chem.* **32**, 1594–1599 (1989)
- Chu, C. K., *et al.*, *Chem. Pharm. Bull.* **37**, 336–339 (1989)
- Chou, T.-C., *et al.*, *Phytother. Res.* **3**, 237–242 (1989)
- Tseng, C. C., *et al.*, *Kaohsiung J. Med. Sci.* **6**, 58–65 (1990)

Tumor Promoters

- Kopelovich, L., and Chou, T.-C., *Int. J. Cancer* **34**, 781–788 (1984)

Cell Differentiation Agents

- Kong, X B., *et al.*, *Leuk. Res.* **11**, 1031–1039 (1987)

Radioimmunoassay

- Smith, S. M., *J. Theor. Biol.* **98**, 475–499 (1982)

Enzyme–Ligand Interactions

- Kremer, A. B., *et al.*, *J. Biol. Chem.* **255**, 2405–2410 (1980)
- Bounias, M., *Experientia* **36**, 157–159 (1980)
- Viswanathan, T., and Alworth, W. F., *J. Med. Chem.* **24**, 822–830 (1981)
- Hata, K., *et al.*, *Int. J. Biochem.* **13**, 681–692 (1981)
- Mat Sukawa, R., *et al.*, *Biochwem. Biophys. Res. Commun.* **101**, 1305–1310 (1981)
- Kono, Y., *et al.*, *Comp. Biochem. Physiol.* **70**, 35–39 (1981)
- Colombo, R., and Pinelli, A. J., *Pharmacol. Methods.* **6**, 177–191 (1981)
- Soike, K., *et al.*, *Gen. Pharmacol.* **13**, 515–518 (1982)
- Takahashi, H. *et al.*, *Gen. Pharmacol.* **13**, 375–379 (1982)
- Rahier, A, *Biochem. Soc. Trans.* **11**, 537–543 (1983)
- Bounias, M., *Biochemistry* **31**, 2769–2775 (1983)
- Jenkins, W. T., *et al.*, *Biochem. Pharmacol.* **32**, 167–170 (1983)
- Rahier, A., *Biochem Pharmacol.* **24**, 1223–1232 (1985)
- Wiener, A., *Br. J. Biochem.* **157**, 351–363 (1986)
- Hasebe, K., *et al.*, *Enzyme* **37**, 109–114 (1987)

- Tsou, C. L., *in* "Advances in Enzymology," pp. 381–436. Cornell Univ., Ithaca, New York, 1988
- Tsujimoto, Y., *et al.*, *Cell Biol. Int. Rep.* **12,** 143–147 (1988)
- Rahier, A., *Eur. J. Biochem.* **181,** 615–6262 (1989)
- Barrie, S. E., *et al.*, *J. Steroid Biochem.* **33,** 1191–1195 (1989)

Receptor–Ligand Interactions

- Friedman, S. J., and Skehan, P., *Proc. Natl. Acad. Sci. U.S.A.* **77,** 1172–1176 (1980)
- Finotti, P., and Palantini, P., *J. Pharmacol. Exp. Ther.* **217,** 754–790 (1981)
- Rosenberg. R., *Biol. Biophys. Acta* **649,** 262–268 (1981)
- Murphy, K. M. M., and Snyder, S. M., *Mol. Pharmacol.* **22,** 250–257 (1982)
- Vauquelin, G., *et al.*, *Eur. J. Biochem.* **125,** 117–124 (1982)
- Andre, C., *et al.*, *EMBO J.* **2,** 499–504 (1983)
- Sharif, N. A., *Neurosci. Lett.* **41,** 301–306 (1983)
- Zahniser, N. R., and Molinoff, P. B., *Mol. Pharmacol.* **23,** 303–309 (1983)
- Vanderheyden, P., *et al.*, *Biochem. Pharmacol.* **33,** 2981–2987 (1984)
- Matsukawa, R. *et al.*, *Gen. Pharmacol.* **16,** 121–124 (1985)
- Baker, S. P., and Posner, P., *Mol. Pharmacol.* **30,** 411–418 (1986)
- Lakhdar-Ghazal and N., *et al.*, *Pharmacol. Biochem. Behav.* **25,** 903–911 (1986)
- Goldstein, A., and Barrett, R. W., *Mol. Pharmacol.* **31,** 603–609 (1987)
- Altman, H. J., *et al.*, *Behv. Neural Biol.* **48,** 49–62 (1987)
- Clark, J. A., *et al.*, *Mol. Pharmacol.* **34,** 308–317 (1988)
- Baker, S. P., *et al.*, *J. Neurochem.* **50,** 1044–1052 (1988)
- Christ, G. J., *et al.*, *Biochem. Pharmacol.* **37,** 1281–1286 (1988)
- Bylund, D., *et al.*, *J. Pharmacol. Exp. Ther.* **245,** 600–607 (1988)
- Andre, C., *et al.*, *Eur. J. Biochem.* **171,** 401–407 (1988)
- Price, M., *et al.*, *Mol. Pharmacol.* **35,** 67–74 (1989)
- Somers, W. J., *et al.*, *J. Urol.* **141,** 1230–1233 (1989)
- Standifer, K. M., *et al.*, *Arch. Pharmacol.* **339,** 129–137 (1989)
- Schoepp, D. D., *et al.*, *J. Neurochem.* **53,** 273–278 (1989)
- Schoepp, D. D., *et al.*, *J. Neural Transm.* **78,** 183–193 (1989)
- Clark, J. A., *et al.*, *J. Pharmacol Exp. Therap.* **251,** 461–468 (1989)
- Byland, D. B. and Ray-Prenger, C. J., *Pharmacol. Exp. Therap.* **251,** 640–644 (1989)

Cellular Studies

- Long, B. H., *et al.*, *Mol Pharmacol.* **22,** 152–157 (1982)
- Rosenberg, C., *Acta Physiol. Scand.* **116,** 321–330 (1982)
- Colombani, P. M., *et al.*, *Transplant. Proc.* **21,** 840–841 (1989)
- Kufe, D., *et al.*, *Cancer Res.* **44,** 69–73 (1984)
- Cavanaugh, P. F., *et al.*, *Cancer Res.* **45,** 4754–4759 (1985)
- Huang, P., and Plunkett, W., *Biochem. Pharmacol.* **36,** 2945–2950 (1987)
- Osty, J., *et al.*, *Endocrinology (Baltimore)* **123,** 2303–2311 (1988).
- Fykse, E. M., *et al.*, *J. Neurochem.* **52,** 946–951 (1989)
- Dewar, H. A., *et al.*, *Mol Pharmacol.* **35,** 787–794 (1989)
- Mori, H., *et al.*, *J. Clin. Immunol.* **10,** 45–51 (1990)
- Christansen, H., *et al.*, *J. Neurochem.* **54,** 1142–1147 (1990)

Tissue Studies

- Puig, M. M., *et al.*, *Anesthesiology* **68,** 559–562 (1988)
- Smith, J. B., *et al.*, *J. Biol. Chem.* **264,** 8723–8728 (1989)
- Smith, J. B., *et al.*, *J. Biol. Chem.* **264,** 7115–7118 (1989)

- Seo, J. K., *et al.*, *Pediatr. Res.* **25**, 225–227 (1989)
- Chu, S. H. W., *et al.*, *AM. J. Physiol.* **256**, G220–G225 (1989)
- Gonzales, R. A., *et al.*, *Mol. Pharmacol.* **35**, 787–794 (1989)
- Schoepp, D. D. and Johnson, B. G., *J. Neurochem.* **53**, 1865–1870 1989
- Woodward, J. J., and Gonzales, R. A., *J. Neurochem.* **54**, 712–715 (1990)

Carcinogen Studies

- Ramel, C., *Mutat. Res.* **168**, 327–342 (1986)

Radiation Studies

- Maisin, J. R., *et al.*, *Int. J. Radiat. Biol.* **51**, 1049–1057 (1987)

Cardiovascular Studies
- Herman, F., *et al.*, *Arch. Int. Pharmacol.* **300**, 281–291 (1989)

Appendix II

Examples of Applications of the Chou–Talalay Method in Anti-Viral Drug Combination Studies[a]

Drugs	Virus	Cell/Animal	Results[b]	References
ACV + Ara-A	HSV-2	Mice	Synergism (+++)	Schinazi et al. [Antimicrob. Agents Chemother. **30**, 491 (1986)]
FIAC + Ara-A	HSV-2	Mice	Synergism (++++)	Schinazi et al. [Antimicrob. Agents Chemother. **30**, 491 (1986)]
FIAC + ACV	HSV-2	Mice	Synergism (++++)	Schinazi et al. [Antimicrob. Agents Chemother. **30**, 491 (1986)]
FMAU + ACV	HSV-2	Mice	Synergism(++++)	Schinazi et al. [Antimicrob. Agents Chemother. **30**, 491 (1986)]
FMAU + Ara-A	HSV-2	Mice	Synergism (++++)	Schinazi et al. [Antimicrob. Agents Chemother. **30**, 491 (1986)]
PFA + rIFNαA	HTLV-3	PBL	Synergism(+++)	Hartshorn et al. [Antimicrob. Agents Chemother. **30**, 189 (1986)]
AZT + rIFNαA	HIV-1	PBL	Synergism(+++)	Hartshorn et al. [Antimicrob. Agents Chemother. **31**, 168 (1987)]
AZT + rIFNαA	HIV-1	BFUe, CFU-GM	Synergism (++)	Berman et al. [Blood **74**, 1281 (1989)]
AZT + Ribavirin	HIV-1	PBL, H9	Antagonism (– – –)	Vogt et al. [Science **235**, 1376 (1987)]

Combination	Virus	Cell	Result	Reference
DHPG + rHuIFNβ	Varicella	Monkey	Synergism (++++)	Soike et al. [*J. Infect. Dis.* **156**, 607 (1987)]
ddCyd + rIFNαA	HIV-1	PBL	Synergism (+++++)	Vogt et al. [*J. Infect. Dis.* **158**, 378 (1988)]
AZT + rsT4	HIV-1	PBL, H9, BT4	Synergism (++)	Johnson, et al. [*J. Infect Dis.* **159**, 837 (1989)]
CAS + rsT4	HIV-1	PBL, H9	Synergism (++++)	Johnson et al. [*Proc. Int. Conf. AIDS,* 4th abstr. (1988)]
AZT + PMEA	HIV-1	T-C3	Synergism (+++)	Smith et al. [*Proc. Int. Conf. AIDS,* 4th abstr. (1988)]
AZT + PFA	HIV-1	PBL	Synergism (+)	Erickson et al. [*Proc. Int. Conf. AIDS,* 4th abstr. (1988)]
ddI + sCD4	HIV-1	TM11	Synergism (+++)	Hayashi, et al. [*Antimicrob. Agents Chemother.* **34**, 82 (1990)]
ddI + SCD4	HIV-1	ATH8	Synergism (++)	Hayashi et al. [*Antimicrob. Agents Chemother.* **34**, 82 (1990)]
ddI + DS	HIV-1	ATH8	Synergism (+++)	Hayashi et al. [*Antimicrob. Agents Chemother.* **34**, 82 (1990)]
ddC + sCD4	HIV-1	ATH8	Synergism (++++)	Hayashi et al. [*Antimicrob. Agents Chemother.* **34**, 82 (1990)]

(continues)

Appendix II *(continued)*

Drugs	Virus	Cell/Animal	Results	References
ddC + DDS	HIV-1	ATH8	Synergism (++)	Hayashi et al. [*Antimicrob. Agents Chemother.* **34**, 82 (1990)]
AZT + sCD4	HIV-1	ATH8	Synergism (++)	Hayashi et al. [*Antimicrob. Agents Chemother.* **34**, 82 (1990)]
DS + sCD4	HIV-1	ATH8	Synergism (+)	Hayashi et al. [*Antimicrob. Agents Chemother.* **34**, 82 (1990)]
D4T + ddI	HIV-1	CEM	Synergism (++++)	Brankovan et al. [*Proc. Int. Conf. 5th AIDS*, abstr. (1989)]
AZT + Prad	HIV-1	CEM	Synergism (+++)	Brankovan et al. [*Proc. Int. Conf. AIDS, 5th* abstr. (1989)]
ddI + Prad	HIV-1	CEM	Synergism (++)	Brankovan et al. [*Proc. Int. Conf. AIDS, 5th* abstr. (1989)]
FEAU + IFN	Varicella	Monkey	Synergism (+++)	Soike et al. [*Antiviral Res.* **13**, 165 (1990)]
N-butyl-DNJ + AZT	HIV-1	H9	Synergism (++)	Johnson et al. [*Proc. ICAAC, 29th* (1989)]
AZT + IFNαA	MuRLV	Mouse	Synergism (++++)	Ruprecht et al. [*J. AIDS* **3**, 551 (1990)]
AZT + IFNα	EBV	HUCL	Synergism (++++)	Lin et al. [*J. Infect. Dis.* **159**, 248 (1989)]

AZT + IFNγ	EBV	HUCL	Synergism (++)	Lin et al. [J. Infect. Dis. 159, 248 (1989)]
IFNα + IFNγ	EBV	HUCL	Synergism (++++)	Lin et al. [J. Infect. Dis. 159, 248 (1989)]
AZT + CAS	HIV-1, HIV-2	PBL, H9	Synergism (+++)	Johnson et al. [Antimicrob. Agents Chemother. 33, 53 (1989)]
rsT4 + AZT + rIFNαA	HIV-1	H9	Synergism (+++++)	Johnson et al. [Proc. Int. Conf. AIDS, 5th abstr. (1989)]
S-dC28 + AZT	HIV-1	MT4	Antagonism (−)	Chou et al. [Proc. Int. Conf. AIDS, 6th abstr. (1990)]
S-dC28 + IFNαA	HIV-1	MT4	Synergistic (++)	Chou et al. [Proc. Int. Conf. AIDS, 6th abstr. (1990)]
S-dC28 + DS	HIV-1	MT4	Synergistic (+)	Chou et al. [Proc. Int. Conf. AIDS, 6th abstr. (1990)]
C164 + C102	GPCMV	GPE	Synergistic (+++)	Yang et al. [Proc. Int. Antiviral Symp., 4th (1990)]
C164 + DHPG	GPCMV	GPE	Synergistic (++)	Yang et al. [Proc. Int. Antiviral Symp., 4th (1990)]
HPMPC + DHPG	GPCMV	GPE	Synergistic (+)	Yang et al. [Proc. Int. Antiviral Symp., 4th (1990)]
C164 + ACV	GPCMV	GPE	Antagonistic (−)	Yang et al. [Proc. Int. Antiviral Symp., 4th (1990)]

(continues)

Appendix II (continued)

Drugs	Virus	Cell/Animal	Results	References
DHPG + AZT	GPCMV	GPE	Antagonistic (− −)	Yang et al. [Proc. Int. Antiviral Symp., 4th (1990)]
AZT + PFA	HIV-1	MT4	Synergism (+ +)	Zhu et al. [Proc. Int. Conf. AIDS, 6th abstr. (1990)]
AZT + DS	HIV-1	MT4	Synergism (+)	Zhu et al. [Proc. Int. Conf. AIDS, 6th abstr (1990)]
PFA + DS	HIV-1	MT4	Synergism (+)	Zhu et al. [Proc. Int. Conf. AIDS, 6th abstr. (1990)]
AZT + DS	HIV-1	MT4, H9	Synergism (+ + +)	Chou et al. (in preparation)
AZT + FLT	HIV-1	MT4, H9	Additive (+)	Chou et al. (in preparation)
AZT + DS + FLT	HIV-1	MT4, H9	Synergism (+ + +)	Chou et al. (in preparation)

[a] Abbreviations used are ACV, acyclovir; Ara-A, 9-β-arabinofuranosyl adenine or vidarabine; FIAC, 2′-fluoro-5-iodoarabinosylcytosine; FMAU, 2′-fluoro-5-methylarabinosyluracil; PFA, phosphonoformate; IFNαA, recombinant interferon alpha 2A; AZT, 3′-azido-3′-deoxythymidine; DHPG, 9-(1,3-dihydroxy-2-propoxymethyl) guanine or gancyclovir; rHUIFN-β, recombinant human interferon beta; ddC or ddCyd, 2′,3′-dideoxycytidine; ddI, 2′,3′-dideoxyinosine; FEAU, 2′-fluoro-5-ethylarabinosyluracil; FLT, 3′-fluoro-deoxythymidine; HPMPC, (S)-1-(3-hydroxy-2-phosphonylmethoxypropyl) cytosine; C164, compound 164 or [9-(2-hydroxy-1-3-2-dioxaphophorinan-5-yl) oxymethyl] guanine p-oxide; c-102, compound 102 or 4-amino-5-bromo-7-(2-hydroxylthoxy methyl-pyrrolo(2,3-d) pyrimidine; DS, dextran sulfate (molecular weight approximately 8000); N-butyl-DNJ or SC-48334, deoxynujirimycin; CAS, catanospermine; Prad, pradimicin; S-dC28, aphosphorothioate oligo-deoxycytidine 28 mer; HIV, human immunodeficiency virus; HSV, herpes simplex virus; EBV, Epstein–Barr virus; GPCMV, guinea pig cytomegalovirus; PBMC, peripheral blood mononuclear cell; MT4, human T cell leukemia cell line; PBL, phytohemagglutinin (PHA)-stimulated peripheral blood lymphocytes; HUCL, human umbilical cord lymphocytes; ATH8, T-cell clone; TM11, normal tetanus toxoid-specific helper-inducer T-cell clone; sCD4 or rsT4, recombinant soluble CD4 receptor protein; H9, CD4⁺ human lymphoblastoid T-cell clone; BT4, human monocytic cell line; GPE, guinea pig embryo cells; CFU-E, colony forming unit of normal bone marrow erythroid progenitor cells; CFU-GM, colony forming unit of normal bone marrow granulocyte-monocyte progenitor cells.

[b] Synergism, additivism, and antagonism are quantitatively determined by the combination index (CI) and correspond to CI < 1, =1, and >1, respectively. See original cited sources for actual CI values. Signs used and the approximated ranges of CI values (in parentheses) are + + + + +, (<0.01); + + + +, (0.01–0.2); + + +, (0.2–0.4); + +, (0.4–0.7); +, (0.7–0.9); −, (0.9–1.2); − −, (1.2–2); and − − −, (>2).

References

Ainslie, G. R., Jr., Shill, J. P., and Neet, K. E. (1972). Transients and cooperative kinetics of enzymes. *J. Biol. Chem.* **247,** 7088–7096.

Berenbaum, M.C. (1977). Synergy, additivism and antagonism in immunosuppression: A critical review. *Clin. Exp. Immunol.* **28,** 1–18.

Berenbaum, M. C. (1980). Correlations between methods for measurement of synergy. *J. Infect. Dis.* **137,** 122–130.

Berenbaum, M. C. (1981). Criteria for analyzing interactions between biologically active agents. *Adv. Cancer Res.* **35,** 269–335.

Berenbaum, M. C. (1989). What is synergy? *Pharmacol. Rev.* **41,** 93–141.

Berkson, J. (1946). Approximation of Chi-square by "probits" and by "logits." *J. Am. Stat. Assoc.* **41,** 70–74.

Bliss, C. I. (1938). The determination of the dosage–mortality curve from small numbers. *Q. J. Pharm. Pharmacol.* **11,** 192–216.

Bliss, C. I. (1939). The toxicity of poisons applied jointly. *Ann Appl. Biol.* **26,** 585–615.

Bliss, C. I. (1967). "Statistics in Biology." McGraw-Hill, New York.

Chou, T.-C. (1974). Relationships between inhibition constants and fractional inhibitions in enzyme-catalyzed reactions with different numbers of reactants, different reaction mechanisms, and different types of mechanisms of inhibition. *Mol. Pharmacol.* **10,** 235–247.

Chou, T.-C. (1975). Derivation of Michaelis–Menten type and Hill type equations for reference ligands. *Proc. Int. Congr. Pharmacol., 6th* p. 619.

Chou, T.-C. (1976). Derivation and properties of Michaelis–Menten type and Hill type equations for reference ligands. *J. Theor. Biol.* **59,** 253–276.

Chou, T.-C. (1977). On the determination of availability of ligand binding sites in steady-state systems. *J. Theor. Biol.* **65,** 345–356.

Chou, T.-C. (1980). Comparison of dose–effect relationships of carcinogens following low-dose chronic exposure and high-dose single injection: An analysis by the median-effect principle. *Carcinogenesis* **1,** 203–213.

Chou, T.-C. (1987). Quantitative dose–effect analysis and algorithms: A theoretical study. *Asia Pac. J. Pharmacol.* **2,** 93–99.

Chou, J., and Chou, T.-C. (1989). "Dose–Effect Analysis with Microcomputers: Dose, Effect, Binding and Kinetics," computer software for the IBM PC series (2.1 DOS, 5.25 "floppy disk). Elsevier–Biosoft, Cambridge, England.

Chou, T.-C., and Fanucchi, M. P. (1988). Chemotherapy. *In* "Encyclopedia of Medical Devices and Instrumentation" (J. G. Webster, ed.), Vol. 2, pp. 660–670. Wiley, New York.

Chou, T.-C., and Talalay, P. (1977). A simple generalized equation for the analysis of multiple inhibitions of Michaelis–Menten kinetic systems. *J. Biol. Chem.* **252,** 6438–6442.

Chou, T.-C., and Talalay, P. (1981). Generalized equations for the analysis of inhibitors of Michaelis–Menten and higher order kinetic systems with two or more mutually exclusive and nonexclusive inhibitors. *Eur. J. Biochem.* **115,** 207–216.

Chou, T.-C., and Talalay, P. (1983). Analysis of combined drug effects: A new look at a very old problem. *Trends Pharmacol. Sci.* **4,** 450–454.

Chou, T.-C., and Talalay, P. (1984). Quantitative analysis of dose–effect relationships: The combined effects of multiple drugs or enzyme inhibitors. *Adv. Enzyme Regul.* **22,** 27–55.

Chou, T.-C., and Talalay, P. (1987). Applications of the median-effect principle for the assessment of low-dose risk of carcinogens and for the quantitation of synergism and antagonism of chemotherapeutic agents. In "New Avenues in Developmental Cancer Chemotherapy" (K. R. Harrap and T. A. Connors, eds.), Bristol–Myers Symp. Ser. 8, pp. 37–64. Academic Press, Orlando, Florida.

Chou, T.-C., Rideout, D., Chou, J., and Bertino, R. J. (1991). Chemotherapeutic synergism, potentiation, and antagonism. *In* "Encyclopedia of Human Biology" (R. Dulbecco, ed.), Vol. 2. Academic Press, San Diego, California. In press.

Clark, A. J. (1933). "The mode of action of drugs on cells." Williams & Wilkins, Baltimore, Maryland.

Finney, D. J. (1947). "Probit Analysis." Cambridge Univ. Press, London.

Finney, D. J. (1952). "Probit Analysis," 2nd Ed., pp. 146–153. Cambridge Univ. Press, London.

Gessner, P. K. (1988). A straightforward method for the study of drug interactions: An isobolographic analysis primer. *J. Am. Coll. Toxicol.* **7,** 987–1012.

Goldstein, A., Aronow, L., and Kalman, S. M. (1968). "Principles of Drug Action: The Basis of Pharmacology." Harper & Row, New York.

Hill, A. V. (1913). The combinations of hemoglobin with oxygen and carbon monoxide. *Biochem. J.* **7,** 471–480.

Langmuir, I. (1918). The adsorption of gases on plane surfaces of glass, mica, and platinum. *J. Am. Chem. Soc.* **40,** 1361–1403.

Loewe, S. (1928). Die Quantitationprobleme der Pharmakologie. *Ergeb, Physiol.* **27,** 47–187.

Loewe, S. (1953). The problem of synergism and antagonism of combined drugs. *Arzneim.-Forsch.* **3,** 285–320.

Loewe, S. (1957). Antagonism and antagonists. *Pharmacol. Rev.* **9,** 237–242.

Michaelis, L., and Menten, M. L. (1913). Die Kinetik der Invertinwirkung. *Biochem. Z.* **49,** 333–369.

Nordling, C. O. (1954). Evidence regarding the multiple mutation theory of the cancer-inducing mechanisms. *Acta Genet. Stat. Med.* **5,** 94–104.

Paquette, R. L., Zhou, J.-Y., Yang Y. C., Clark, S. C., and Koeffler, H. P. (1988). Recombinant gibbon interleukin-3 acts synergistically with recombinant human G-CSF and GM,CSF in vitro. *Blood* **71,** 1596–1600.

Reed, L. J., and Berkson, J. (1929). The application of the logistic function to experimental data. *J. Phys. Chem.* **33,** 760–779.

Scatchard, G. (1949). The attractions of proteins for small molecules and ion. *Ann. N.Y. Acad. Sci.* **51,** 660–672.

Segel, I. H. (1975). Multiple inhibition analysis. *In* "Enzyme Kinetics," pp. 465–503. Wiley, New York.

Steel, G. G., and Peckham, M. J. (1979). Exploitable mechanism in combined radiotherapy-chemotherapy: The concept of additivity. *Int. J. Radiat. Oncol., Biol. Phys.* **5,** 85–91.

Sühnel, J. (1990). Evaluation of synergism or antagonism for the combined action of antiviral agents. *Antiviral Res.* **13,** 23–40.

Syracuse, K. C., and Greco, W. R. (1986). Comparison between the method of Chou and Talalay and a new method for the assessment of the combined effects of drugs: A Monte Carlo simulation study. *Proc. Am. Stat. Assoc., Biopharmacol. Sect.* pp. 127–132.

Tallarida, R. J., and Jacob, L. S. (1979). "Dose–Response Relation in Pharmacology." Springer-Verlag, New York.

Thompson, W. R. (1947). Use of moving average and interpolation to estimate median-effect dose. I. Fundamental formulas, estimation of error and relation to other methods. *Bacteriol. Rev.* **11,** 115–145.

Webb, J. L. (1963). Effect of more than one inhibitor. In "Enzyme and Metabolic Inhibitors," Vol. 1, pp. 66–79, 488–512. Academic Press, New York.

Whitehead, E. (1970). The regulation of enzyme activity and allosteric transition. *Prog. Biophys. Mol. Biol.* **21,** 321–397.

CHAPTER 3

Clinical Studies of Combination Chemotherapy for Cancer

Emil Frei III

Dana-Farber Cancer Institute
and Harvard Medical School
Boston, Massachusetts 02115

Starting in 1955, a series of quantitative treatment trials in patients with acute lymphocytic leukemia were undertaken. It became evident early that there were major advantages to combination chemotherapy. Among the thrust to conduct these studies were the following. (1) Combination chemotherapy for tuberculosis had been shown in Great Britain to be far superior to monochemotherapy. In the latter, drug resistance developed quickly. (2) Law, studying L1210 mouse leukemia, described the genetics of drug resistance and found that cross resistance between 6MP and methotrexate did not occur. This formed the basis for using these two agents in combination, which proved superior in mouse leukemia to single agents. (3) Dose limiting toxicity for some of the clinically available agents was qualitatively different, potentially allowing for no compromise in dose. (4) Beginning knowledge of tumor cell heterogeneity and the need, therefore, to address multiple targets became evident (Frei, 1972).

With this background, we found in acute lymphocytic leukemia that the complete remission induction rate for single agents ranged from 15 to 50%. When 6MP and methotrexate were employed in combination, the complete response rate was somewhat higher than for each agent used alone. However, to give these two agents together a 50% compromise in the dose of each agent was necessary. On the other hand, when active agents such as prednisone and 6MP and particularly prednisone and vincristine were employed in combination, full dose could be maintained and complete remission rates in excess of 90% were achieved. This was greater than the complete response rate calculated assuming independent drug action. Thus it was interpreted that this high complete remission rate resulted from an interaction between the drugs that was more than additive.

From this beginning, the criteria for the use of agents in combination were developed. Combination chemotherapy has swept the field. Essentially all forms of cancer that we can cure today, in either overt or micrometastatic

SYNERGISM AND ANTAGONISM IN CHEMOTHERAPY
Copyright © 1991 by Academic Press, Inc.
All rights of reproduction in any form reserved.

103

(adjuvant) form, require combination chemotherapy. The converse is that, with the possible exceptions of Burkitt's lymphoma and gestational chorio-carcinoma, single agents are never curative. Indeed, in the two examples cited, while single agents may be curative, combinations produce a higher cure rate (Frei, 1972).

The clinical criteria for combination chemotherapy are (1) that agents in the combination be individually active against the disease; (2) that they represent different mechanisms of action and, presumably, are not cross resistant; and (3) that a compromise in dose should not occur. Since both dose and combinations are highly important to optimizing therapy, the adding of an active agent to a combination may be neutralized if doses of the components are correspondingly reduced.

Treatment of the aforementioned example, acute lymphocytic leukemia, involved combinations of two agents. Using the criteria just listed, the MOPP program was developed for Hodgkin's disease. Each of the agents—mustargen, Oncovin, prednisone, and procarbazine—can produce a partial response rate of 40–50%, but a complete response rate of only approximately 10%. Also, remissions developed slowly. Because of differing dose-limiting toxicity, the MOPP components could be employed at full or nearly full doses in combination. There resulted an 80% complete response rate and a 50% cure rate. Important was the fact that tumor regression was dramatic—the median time to complete remission was 1–2 months. The same strategy with slightly different tactics was employed in non-Hodgkin's lymphoma where substantial cure rates were achieved with 4 or 5 agents used in combination (DeVita *et al.*, 1970, 1975).

Testis cancer is an important paradigm since it is not a hematologic malignancy. The cisplatin–bleomycin–vinblastine combination has proven highly effective and curative for this disease. Since these agents have differing dose-limiting toxicity, they can be employed at full dose in the combination (Einhorn and Donohue, 1977).

In the adjuvant situation—that is, the treatment of micrometastatic disease by chemotherapy—major examples include breast cancer and osteogenic sarcoma. In comparative studies, combination chemotherapy regimens such as CMF (cyclophosphamide, methotrexate, fluorouracil) of CAF (cyclophosphamide, Adriamycin, fluorouracil) have proven superior to single-agent studies with, for example, melphalan (Fisher *et al.*, 1984; Bonadonna and Valagussa, 1985). Similarly in osteogenic sarcoma, the combination of high dose methotrexate with leucovorin rescue and Adriamycin has been highly effective, producing a 60–70% cure rate, compared with 40% for single agents and 10–20% for the nonchemotherapy control (Eilber and Eckhardt, 1985).

Goldie and Coldman (1979), using mathematical modeling of mutation rates and drug resistance development, provide strong support for the use of agents in combination and, most particularly, for the use of alternating therapy, particularly alternating combination chemotherapy. This is based on the fact that the selection and development of drug-resistant cell lines which occurs with one combination will be suppressed by a second combination introduced at the appropriate time. It has been difficult to confirm this strategy in the clinic. Perhaps the best model is Hodgkin's disease, where ABVD (Adriamycin, bleomycin, vinblastine, dacarbazine) and MOPP have activity. Rotating these two combinations on a monthly basis has proven superior to MOPP only in comparative studies but, more recently, it has been found that ABVD only is at least as good as, and possibly better than, the rotating combination. One problem with the aforementioned rotating combination is that the dose intensity for any single agent in the rotation is reduced by 50%. Thus combination A may be given at monthly intervals, but rotation of A and B with each one given at bimonthly intervals amounts to a 50% reduction in dose intensity. In addition, there is no theoretical or actual evidence that, for example, a three-drug combination A, cycled with a three-drug combination B, would be better than the six-drug combination given at monthly intervals. This would be particularly true if dose reduction was not a major problem with the polychemotherapy.

Perhaps acute lymphocytic leukemia provides the most support for using combinations in sequence. Vincristine and prednisone are highly effective in inducing complete remission but relapse occurs quickly, even in the face of continuing the combination. On the other hand, if after achieving complete remission patients were treated with a combination of antimetabolites (i.e., 6MP and methotrexate), remission durations were long (Frei, 1972).

It generally has been accepted that the alkylating agents are cross resistant among themselves and that, therefore, the use of alkylating agents in sequence or in combination is not logical. Schabel was the first to challenge that doctrine when he found in L1210 mouse leukemia that when resistance was produced to a given alkylating agent, there was generally lack of cross resistance to the other alklating agents. (Schabel *et al.*, 1978). The production of alkylating-agent resistance in human tumor cell lines *in vitro* has also shown a general lack of cross resistance. Shabel proceeded to show synergism with combinations of alkylating agents experimentally. Since dose-limiting toxicity for the alkylating agents is primarily myelosuppression, their use in combination makes sense, most particularly in the autologous bone marrow transplantation arena where myelosuppression is controlled by the transplant. For most bone marrow transplantation regimens which include either cyclophosphamide plus TBI (total body irradiation) or cyclophosphamide plus busulfan, there is substantial evidence that such combinations are superior to single agents (Frei *et al.*, 1985).

In summary, while combination chemotherapy is ascendant in the clinic, there is no conclusive evidence that there is biochemical interaction between the components of combination chemotherapy or that true therapeutic synergism is observed. This is a soft statement, since synergism is difficult to define. Since the dose–respone curve of chemotherapeutic agents is steep, it is not surprising that two such agents for which dose is not compromised would be superior to either agent alone at the optimal dose. Thus the effects are independent and additive.

On the other hand, combinations of antitumor agents which include modulators do involve biochemical interactions and represent circumstances under which synergism based on biochemical interaction does occur. The classical modulation is leucovorin rescue following high dose methotrexate. One can administer much higher doses of methotrexate if leucovorin rescue is delivered in a timely fashion. This has produced superior results in osteogenic sarcoma and has provided some degree of control of microscopic and macroscopic meningeal leukemia. It has been most widely employed in non-Hodgkin's lymphoma. Thus methotrexate with rescue, if administered properly, is not myelosuppressive, and therefore can be added at full doses to combinations such as BACOP (bleomycin, Adriamycin, cyclophosphamide, vincristine, prednisone). In single-institution studies, M–BACOP (methotrexate with rescue + BACOP) would appear to be superior to BACOP. Comparative studies to date have not supported this assertion, but there are problems with experimental design and interpretation.

More compelling is the data with respect to modulation of fluorouracil with leucovorin. The inhibition of thymidylate synthase by FdUMP is more effective and prolonged to the extent that reduced folate in the form of 5,10-methylenetetrahydrofolate binds to the enzyme producing a ternary complex. Thus leucovorin has been found to modulate fluorouracil in the direction of improving antitumor activity experimentally against the appropriate enzyme system *in vitro* and experimentally *in vivo*. Thus its effect on the tumor is greater than that on the host. In the clinic there is now conclusive evidence on the basis of a number of comparative studies in metastatic colorectal cancer that fluorouracil plus leucovorin produces a higher initial response rate than fluorouracil only and that, in several of the studies, there is a concurrent improvement in survival. In noncomparative studies, fluorouracil plus leucovorin are providing a major increment in response rates in such diseases as head and neck cancer and metastatic breast cancer. In these clinical studies, fluorouracil toxicity is somewhat increased by leucovorin. There is some opinion that if fluorouracil were administered in increasing and more toxic doses, the same improvement in response would occur. The magnitude of the improvement in the therapeutic effect of fluorouracil after addition of leucovorin makes this interpretation unlikely, but the question remains to be answered rigorously.

The modulation approach has promise for other agents, such as alkylating agents. Alkylating agents must negotiate the cell membrane, and must deal with a high concentration of glutathione in the cytoplasm which can inactivate the alkylating agents through conjugation by glutathione transferases. They must then combine with DNA and produce cytotoxicity, a process which experimentally requires oxygen. This effect, therefore, is reduced in solid tumors which are commonly hypoxic. Finally, DNA repair may compromise the effect of the alkylating agent. All of these and other processes can be modulated by agents such as buthionine sulfoximine, ethacrynic acid, fluosol, topoisomerase inhibitors, and methylated xanthines (see also Chapter 15 by Teicher *et al.*). Thus there is a rich potential for clinical studies addressed to these observations, and such studies are ongoing. All of the aforementioned agents produce an improvement in the therapeutic index in preclinical *in vivo* systems. Thus, for reasons which generally are unknown, they modulate alkylating agents more in terms of antitumor effect than in terms of host toxicity. Perhaps the best evidence and background for selective antitumor effect resides with fluosol–oxygen which may correct the hypoxia, or with nitroimidazole derivatives which are oxygen mimics. Since tumors may be hypoxic and host tissues are not, such modulation may be specific for the tumor (see also Chapter 19 by Sieman and Keng).

Rapid advance in our knowledge concerning tumor cell heterogeneity strongly suggests that monotherapy—be it chemotherapy, hormonal therapy, or immunotherapy—while it may be effective, will not be curative. Some fraction of the cells within the tumor will be resistant initially and ascend. This is the primary experimental basis for combination therapy. From a pragmatic point of view, combination chemotherapy is highly important since in fact all curative chemotherapy in humans requires combinations of agents. It is difficult to show that the ascendancy of such combinations is more than the result of an independent additive effect of the agents or, conversely, that there is a biochemical interaction between the agents responsible for the improved effect. Such an interaction probably exists for at least some combinations, but has not been demonstrated rigorously.

On the other hand, the use of modulators selected largely for their known biochemical action, particularly with respect to modulating alkylating agents and antimetabolites, has a much stronger biochemical, pharmacological basis, and is proving effective not only experimentally but in some clinical studies.

References

Bonadonna, G., and Valagussa, P. (1985). Adjuvant systems therapy for resectable breast cancer. *J. Clin. Oncol.* **3**, 259.

DeVita, V. T., Serpick, A. A., and Carbone, P. (1970). Combination chemotherapy in the treatment of Hodgkin's disease. *Ann. Intern. Med.* **73**, 881–895.

DeVita, V. T., Canellos, G. P., Chabner, B. A., Schein, P., Hubbard, S. P., and Young, R. C. (1975). Histicytic lymphoma, a potentially curable disease. *Lancet* **1**, 248–250.

Eilber, F. R., and Eckhardt, J. (1985). Adjuvant therapy for osteosarcoma: A randomized prospective trial. *Proc. Am. Soc. Clin. Oncol.* **4**, 144.

Einhorn, L. H., and Donohue, J. (1977). Cisplatin, vinblastine and bleomycin combination chemotherapy in testicular cancer. *Ann. Intern. Med.* **87**, 293–298.

Fisher, B., Redmond, C., and Fisher, E. R. (1984). A summary of findings from NSABP trials of adjuvant therapy for breast cancer. *In* "Adjuvant Therapy of Cancer" (S. E. Jones and S. E. Salmon, eds.), Vol. IV. Grune & Stratton, Orlando, Florida.

Frei, E., III (1972). Combination cancer therapy: Presidential address. *Cancer Res.* **32,** 2593–2607.

Frei, E., III, Cucchi, C. A., Rosowski, A., Tantravahi, R., Bernal, S., Ervin, T. J., Ruprecht, R. M., and Haseltine, W. A. (1985). Alkylating agent resistance: In vitro studies with human cell lines. *Proc. Natl. Acad. Sci. U.S.A.* **82,** 2158–2162.

Goldie, J. H., and Coldman, A. J. (1979). A mathematic model for relating the drug sensitivity of tumors to their spontaneous mutation rate. *Cancer Treat. Rep.* **63,** 1727–1733.

Schabel, F. N., Jr., Trader, N. W., Laster, W. R., Jr., Wheeler, G. P., and Witt, M. H. (1978). Patterns of resistance and therapeutic synergism among alkylating agents. *Antibiot. Chemother. (Basel)* **23,** 200–215.

CHAPTER 4

Combined Chemotherapeutic Modalities for Viral Infections: Rationale and Clinical Potential

Raymond F. Schinazi

Department of Pediatrics
Emory University School of Medicine
and Veterans Affairs Medical Center
Decatur, Georgia 30033

SYNERGISM AND ANTAGONISM IN CHEMOTHERAPY
Copyright © 1991 by Academic Press, Inc.
All rights of reproduction in any form reserved.

I. Introduction

Viruses are among the most common infectious agents affecting humans. With the recent availability of new and essentially non toxic selective antiviral agents for the treatment of various viral infections, the debilitating and sometimes fatal results of these infections are being successfully prevented or treated. In the last decade, antiviral drugs have been introduced for the treatment of viral infections caused by herpes simplex viruses, influenza A virus, human immunodeficiency viruses (HIV), papillomavirus, and respiratory syncytial virus. However, there are increasing reports of patients, particularly with ocular herpes or those immunocompromised, who are clinically non responsive to antiviral therapy. In many instances, drug-resistant herpes simplex viruses (HSV) have been isolated from such patients. The potential utility of combination chemotherapy is now markedly increased because of the problem of drug toxicity and viral resistance to antiviral drugs (Crumpacker, 1983, 1989; Larder and Darby, 1984; Schinazi and Nahmias, 1984; Whitley, 1988).

The focus of this review is on combination chemotherapy of antiviral agents for HSV and HIV infections. There are more drugs approved or at an advanced preclinical stage for the treatment of infections caused by these viruses than for any others. The use of combinations is an attractive and logical extension of any therapeutic approach to enhance efficacy. The current emphasis on drug interactions is particularly timely in light of the clinical use of anti-AIDS drugs in combination with immunomodulators, anti-cytomegalovirus agents, and other drugs used to treat opportunistic infections in these individuals. The discussion will be essentially limited to those agents which are most likely to be of broad clinical usage. This review also presents historical and current perspective on the rationale for using various drugs in combinations as well as pharmacological and pharmacodynamic principles in using drug combinations. Combinations of drugs effective against viruses, including influenza and exotic viruses, have been reviewed previously (De Clercq, 1987; Hall, 1986; Hall and Duncan, 1988; Hayden, 1986).

II. Current Status of Research

With the advent of vidarabine (araA), acyclovir (ACV), and zidovudine (AZT), great strides have been made in the development of antiviral drugs. However, much remains to be done to improve delivery of these agents in order to increase their bioavailability to the target organ, and to find drugs with a different mechanism of action in order to combat drug-resistant variants. Most important, there is a need to uncover more potent and selective antivirals, or combined modalities, that can decrease or prevent the significant morbidity and mortality still resulting from severe infections despite current antiviral therapy.

An adequate rational basis for the design of combination of antivirals, based upon the structure and biological action of the drugs, is still lacking. Even with some understanding of the knowledge of the mechanism of drug action, a combined modality will be feasible only after knowing the critical viral enzymatic step(s) or viral receptors that are inhibited and which eventually lead to enhanced cell death. Some practical outlook is essential, and continued "hit-or-miss" screening of antiviral combinations is a necessary and potentially rewarding avenue. The increased amount of interest in the use of a combined modality is well evidenced by the mounting number of published reports (Fig 1). The earliest report on the use of combinations for viral infections is on the use of isatin thiosemicarbazone and certain phenoxypyrimidines in mice infected with vaccinia virus (Bauer, 1955). Since that time, the number of reports have gradually increased, expecially in the 1970s with the introduction of new selective antiviral drugs. Whereas the number of reports on herpesviruses and other viruses peaked in the mid-1980s the reports on combinations with HIV and other retroviruses is currently on an exponential rise. For this review, over 300 reports in the literature published from 1955 to 1990 were reviewed and tabulated (see Tables 1–3).[1] It is of interest that almost all of the combinations already examined produced at least enhanced (authors' conclusion) antiviral activity. Since combination studies are perfomed with a rationale for synergy, there might be a bias

---

[1] This article derives from results of experiments performed by a group composed of the author of this article and his collaborators; the use of "we, they, our" refers to that group.

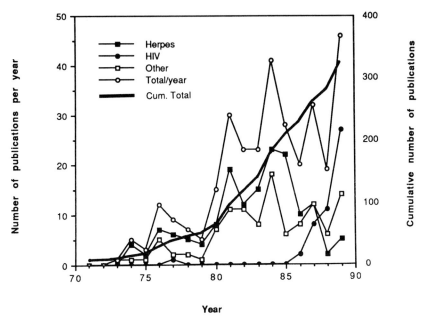

FIGURE 1   Two decades of combination chemotherapy for viruses.

toward finding synergistic interactions. In these published papers, articles, and abstracts, the authors have used different methods for performing the study and for analyzing their data.

With over 300 previous studies on multiple drug therapy, one may argue that there is no compelling reason for performing new studies. However, when most of these papers and abstracts are carefully reviewed, the following important points appear not to have been considered. (1) Since antagonism has been noted for some combinations, it is important to determine at the biochemical and pharmacological level or in an animal model that this will not occur, especially with a combination of antiviral drugs that is under consideration for the treatment of humans. (2) Few of the studies have been conducted in animal models—such studies are important since the levels of the two drugs in affected tissue, and the metabolism and half-life of the drugs may vary under *in vivo* conditions. (3) Detailed evaluation of the toxicity of the combination was not performed in the majority of the combinations studied. (4) The optimum ratio for the two or more drugs in combination was not established, (5) The minimum effective dose or concentration or median effective concentration was not determined. Lastly, (6) different methods for performing the study and for analyzing the data have often led to contradictory results. This last point is very important, since workers in this area have not attempted to standardize their techniques or at least subject their data to analysis by several of the methods reported.

TABLE 1

Combinations of Antivirals Used on Herpes Simplex Viruses (1963–1990)

Combination of compounds[a]	System/virus type[b]	Effect[c]	Basis[d]	Reference
Deaminase inhibitors				
Ara-A + DI	HSF and MK/HSV-1 and HSV-2	S	PR	Connor et al. (1974)
Ara-A + DI	AGMK/HSV-1	S	PR	Williams and Lerner (1975)
Ara-A + DI	KB/HSV-1	S	YR/ED_{50}	Schwartz et al. (1976), Drach and Shipman (1977)
Ara-A + DI	**Rabbit eye**/HSV	S	LS	Falcon and Jones (1977)
Ara-A + DI	HEL/HSV-1	S	PR	Bryson and Kronenberg (1977)
Ara-A + DI	HEp-2/HSV-1	S	PR	Sloan et al. (1977)
Ara-H + DI		I	PR	
Ara-A + DI	**Mouse enceph**/HSV-1	E	MDD; M	North and Cohen (1978, 1979)
Ara-A + AMP + DI		I	MDD; M	
Ara-A or Cordycepin + DI	HeLa/HSV-1	S	YR	Shannon et al. (1980)
Ara-A + DI	**Mouse enceph**/HSV-1	E	MDD; M	Bastow et al. (1983)
Ara-A + DI	AGMK/HSV-1	E	PR	Aduma et al. (1990)
BVDC + DI	Various cells/HSV-1	E	PR	
Interferon or interferon inducers				
Poly IC + IdUrd	**Rabbit eye**/HSV-1	E	LS	Oosterhuis et al. (1974)
Ara-A + IFN	AGMK/HSV-1	S	MT; MIC	Lerner and Bailey (1974)
Ara-A + IFN	AGMK/HSV-1 and HSV-2	S	MT; MIC	Lerner and Bailey (1976)
Ara-A + IFN or IFN inducers	**Mouse enceph**/HSV-1	A	M	Lefkowitz et al. (1976)

(continues)

TABLE 1 (continued)

Combination of compounds[a]	System/virus type[b]	Effect[c]	Basis[d]	Reference
PAA + Tilorone	Mouse enceph/HSV-1	I	M; VT	Fitzwilliams and Griffith (1976)
IFN; AraA (+ DI); AraH; AraAMP	HSF/HSV-2	E	PR	Bryson and Kronenberg (1977)
AraA + IFN	HSF; HEL; RK; MEF; AGMK; HE/HSV-1, HSV-2	V	MT; MIC	Lerner (1981)
AraH + IFN	Human cells/HSV-1 and HSV-2	S		Stanwick et al. (1981)
ACV + IFN$_\beta$	AGMK/HSV-1 and HSV-2	S	PR	Levin and Leary (1981)
ACV + IFN$_\alpha$	HSF/HSV-1	S	PR	Hammer et al. (1981, 1982)
ACV + IFN$_\alpha$	Lymphoid cells/HSV-1	E	YR	Langford and Stanton (1982)
IFN + Ab	Different cells/HSV	S	CPE	Sundmacher et al. (1978, 1981), Sundmacher (1982)
TFT + IFN$_\alpha$	Human eye/HSV	E	VT	
AraAMP, Ara-T, EtdUrd, FDT, PFA, PAA, IFN	HeLa: AGMK/HSV-1	E	YR; VR	Wigand and Janz (1981); Janz and Wigand (1982)
CMA + ACV or TFT	Rabbit eye/HSV-1	I	LS	O'Brien et al. (1981)
BVDU + IFN$_\alpha$	HEL/HSV-1	S	YR	Wigdahl et al. (1982)
IFN$_{\alpha,\beta}$ + IFN$_\gamma$	Murine brain tumor cell line/HSV-1	E/S	VR	Zerial et al. (1982)
ACV + IFN$_\alpha$	Human eye/HSV	E	CR	Collin et al. (1982)
TFT + IFN$_\alpha$	Human eye/HSV	E	HT	De Koning et al. (1981, 1982)
IFN$_\alpha$ + MC or Ab	Mice/HSV-1	S	MDD; M	Kohl et al. (1982a,b)
IFN + AraA + Dexamethasone	Human enceph/HSV-1	NA	M; Ab levels	Koskiniemi et al. (1982)

114

Drug combination	Host/Virus		Reference
IFN$_\alpha$ + ACV or MC	Mice/HSV-1	E/S	MDD; M	Kohl (1983)
ACV + IFN$_\alpha$	**Human eye/HSV**	E	HT	Colin et al. (1983)
ACV + IFN$_\alpha$	**Human eye/HSV**	E	DL; HT	De Koning et al. (1983)
IFN + Ab	Cell culture/HSV	S	CPE	Langford et al. (1983)
IFN$_\alpha$ + DHPG or ACV	HSF/HSV-1 and HSV-2	S	MIC	Moran et al. (1983)
TFT + IFN$_\alpha$	**Rabbit eye/HSV-1**	E	CR; VT	Trousdale and Nesburn (1984)
IFN$_\gamma$ + IFN α or β	**Human melanoma cells/HSV-1**	E	FIC	Czarniecki et al. (1984)
IFN + Ab + Anti-inflammatory drug	**Rabbit and human eye/HSV**	E	CS	Kolomiets et al. (1984)
AraA + IFN	FL cells/HSV-1	V	PR	Yoshida et al. (1984)
IFN$_\alpha$ + IFN$_\beta$		A/E		
DHPG + IFN$_\gamma$	NA/HSV-2	S	NA	Eppstein and Marsh (1984)
DHPG + IFN$_{\alpha,\beta,\gamma}$	**Mice/HSV-2**	S	FIC/ED$_{50}$	Fraser-Smith et al. (1984a–c)
Poly IC(LC) + ACV	HSF/HSV-1	S		Crane et al. (1984)
Poly IC(LC) + ACV + AraA		E		
Poly IC(LC) + AraA		E		
AraA + IFN + AraH or ACV		V		
AraA + IFN$_\alpha$ + IFN$_\beta$		S		
AraH + IFN				
AraH + ACV + IFN		I		
Prostaglandin D$_2$ + IFN$_\alpha$	FL cells/HSV	I/S	YR	Tanaka et al. (1984), Imanashi et al. (1984)
Antivirals + IFN	**Human eye/HSV**	NA	NA	Kaufman et al. (1984)

(*continues*)

115

TABLE 1 (continued)

Combination of compounds[a]	System/virus[b]	Effect[c]	Basic[d]	Reference
TFT + IFN$_{\alpha}$	Human eye/HSV	E	HS; VS	Sundmacher et al. (1984)
ACV + IFN$_{\alpha}$	Mice/HSV-1	S	M; FP	Connell et al. (1985)
ACV + IFN$_{\alpha}$	HSF/HSV-1 and HSV-2	S; E/S	MIC	Moran et al. (1985)
DHPG + IFN$_{\alpha}$	Mice/HSV-2	S	FIC/ED$_{50}$	Fraser-Smith et al. (1985)
DHPG + IFN$_{\gamma}$	HEL/HSV-2	S	YR	Rapp and Wrzos (1985)
Nonoxynol 9 + IFN$_{\alpha}$	Mice/HSV-2	S	M	Friedman-Kien et al. (1985)
Nonoxynol 9 + IFN$_{\alpha}$	Mice (cutaneous)/HSV-1	S	CS	Cerruti et al. (1985)
ACV + IFN$_{\alpha}$	Mice (weanling)/HSV-1	S	M; MDD; FP	Connell et al. (1985)
ACV + IFN$_{\alpha}$	Monkey eye/HSV	S	HT	Neuman-Haeflin et al. (1985)
IFN$_{\alpha}$ + IFN$_{\gamma}$	Human eye/HSV	E	CS, HT	Meurs and van Bijsterveld (1985a,b)
ACV + IFN$_{\alpha}$	Human eye/HSV	NA	NA	Sundmacher (1985)
ACV + IFN; IFN$_{\alpha}$ + IFN$_{\gamma}$	Eye/HSV	NA	NA	Sundmacher et al. (1987)
Nonoxynol 9 + IFN$_{\alpha}$	Human genital/HSV	E	CS; HT	Friedman-Kein et al. (1986)
AraA + IFN$_{\alpha}$, ACV + AraA, Ara-AMP + IFN$_{\alpha}$, or ACV + IFN$_{\alpha}$	Mice/HSV-2	S	M; MDD	Crane and Sunstrum (1988)
ACV or AraA + IFN$_{\beta}$	Human enceph/HSV	NA	NA	Prange and Henze (1988)
ACV + IFN$_{\alpha}$	HSF or MEF/HSV-1 or HSV-2	S	MDEA	Overall et al. (1988)
ACV + IFN$_{\alpha}$	Neural cells/HSV	NA	NA	Hanada et al. (1989)

Compound/combination	System/virus			Reference
Phosphonic acid derivatives + IFN	Mouse enceph/HSV and cells	NA	NA	L'vov et al. (1989), Samoilovich et al. (1989)
BVDU + IFN	Human eye/HSV	NA	NA	van Bijsterveld et al. (1989)
ACV + IFNα	AGMK/HSV	NA	NA	O'Brien et al. (1990)
Other combinations				
IdUrd + HC	Human and rabbit eye/HSV	E	LS	Kaufman et al. (1963)
IdUrd + AraC	Rabbit eye/HSV-1	E	LS	Kaufman (1965)
TFT + IdUrd or BrdUrd		E		
AraA + AraC or IdUrd	HEL/HSV-1	E	MT; MIC	Fiala et al. (1974)
AraC + IdUrd		A		
AraA + IdUrd	AGMK/HSV-1	I	MT; MIC	Lerner and Bailey (1974)
AraA + Ara-H		E		
MMdUrd + IdUrd or AraC or AraA	RK/HSV-1	E	PR	Babiuk et al. (1975)
IdUrd + AraC or AraA		E		
AraA + AraC		I		
IdUrd + 6AzUrd	Humans/genital herpes	S	CS	Bikbulatov and Grebenyuk (1977)
AraA + Levamasole	Rats/HSV-2	I	MDD; M	Fischer et al. (1976)
Immunomodulators	Newborn mice/HSV	NA	NA	Starr et al. (1976)
AraA or PAA + Ab	Mouse enceph/HSV-1	S	MDD; M	Cho et al. (1976), Cho and Feng (1977)
AraA + PFA	HEp-2/HSV-1	S	VR; CPE; YR	Shannon (1977)
AraA + AraH	RK/HSV-1	E	MT; MIC	Champney et al. (1978)

(continues)

TABLE 1 (*continued*)

Combination of compound[a]	System/Virus type[b]	Effect[c]	Basis[d]	Reference
AraA + (S)-DHPA	RK/HSV	S	NA	De Clercq et al. (1978)
EtdUrd + IdUrd	**Humans**/Labial and genital herpes	E	DL	Wassilew (1979)
AraAMP; AraA; TFT; EtdUrd; DDG IdUrd; PAA; dThd	**Rabbit eye**/HSV-1	V	LS	Gauri (1979)
ACV + AraA	AGMK/HSV-1 and HSV-2	E	PR; YR	Schinazi et al. (1981)
Ab + AraA	**Mouse eye**/HSV-1 and HSV-2	E	M	Davies et al. (1980)
MMdUrd + AraA or PAA	RK or AGMK/HSV-1 and HSV-2	S	PR (FIC)	Ayisi et al. (1980)
MMdUrd + EtdUrd or IdUrd		A		
MMdUrd + AraA	NA/HSV	S	NA	Meldrum et al. (1980)
AT + IdUrd	AGMK/HSV-1	S	PR	Fischer et al. (1979, 1980), Fischer (1981)
AraC, AraA, HU, 6MP, FdUrd, Amethopterin, Pactamycin	AGMK; HeLa/HSV-1	V	PD; YR; IQ	Wigand and Hassinger (1980)
AraA + TFT		S	YR	
ACV + Ab	**Mouse enceph**/HSV-1	S	MDD; M	Cho and Feng (1980)
AraA + AraH	HSF; HEL; RK; MEF; AGMK; HE	V	MT; MIC	Lerner (1981)
ACV + AraA or AraAMP	**Mouse enceph**/HSV-2	E/S	MDD; M; FP	Schinazi et al. (1981, 1982c)
AraA + Allopurinol	**Humans**/HSV	A	NA	Friedman and Grasela (1981)

118

Treatment	System			Reference
AraAMP, AraT, EtdUrd, FDT, PFA, PAA	HeLa; AGMK/HSV-1	E	YR, VR	Wigand and Janz (1981)
ACV + EtdUrd	**Rabbit eye/HSV**	E/A	HT	Gauri and Miestereck (1981)
TFT + Steroids		E		
AraA + Steroids		I		
Human leukocytes + Ab	**Mice/HSV-1**	E	MDD; M	Kohl et al. (1981)
ACV; AraA; BVDU; PFA; TFT	AGMK/HSV-1	E/SA/S	PR; YR; FP	Schinazi and Nahmias (1982)
ACV or AraA + HC or Indomethacin	Mouse enceph/HSV-2	E	MDD; M	Schinazi et al. (1982b)
ACV + AraA	Mice/HSV-2	S	MDD; M	Karim et al. (1982)
ACV + AraA + Ab		I		
Virazole + AraA or AraH	KB; BHK/HSV-1 and HSV-2	S	FIC	Allen et al. (1982)
Virazole + AraA (+DI)				
DHPG + PFA	HSF/HSV-1 and HSV-2	S	PR	Smith et al. (1982)
PAA + BVDU				
ACV + PFA or PAA		E/I		
BVDU + PFA		S		
Levamasole + Megasine	**Human genital/HSV**	NA	NA	Moschchik et al. (1982)
TFT + AT	AGMK/HSV-2	S	YR	Fischer et al. (1983)
ACV + HC	**Human eye/HSV**	E	CS	van Ganswijk et al. (1983)
ACV + Betamethasone	**Human eye/HSV**	E	CS; HT	Collum et al. (1983a,b)
MC + ACV	Mice/HSV-1	E	MDD; M	Kohl (1983)
MMdUrd + AraAMP	RK/HSV-1 and HSV-2	E	YR; FIC	Ayisi et al. (1983)
MMdUrdMP + AraAMP		A		
ACV + PFA	**Guinea pigs/HSV-1**	S	CS	Burkhardt and Wigand (1983)

(continues)

119

TABLE 1 (continued)

Combination of compounds[a]	System/Virus type[b]	Effect[c]	Basic[d]	References
FIAC, FMAU, ACV, AraA	**Mouse enceph**/HSV-2	E/S	MDD; M; FP	Schinazi et al. (1983a)
ACV + AraA	**Mouse enceph**/HSV-2 TK$^-$	I	MDD; M	Schinazi et al. (1983b)
ACV + AraA	AGMK/HSV-1	SA	YR; FP	Park et al. (1984)
	Mouse skin/HSV-1	S	CR; M	
	Viral DNA synthesis	E		
	Mice/latent virus	R		
ACV + AraA	**Mouse genital**/HSV-2	S	VT; M; FP	Crane et al. (1984)
	Mice/latent virus	R		
ACV + AraA	HSF/HSV-1	S	MIC; FIC	
ACV + AraH				
AraH + AraA				
AraA or AraH + IFN				
AraA + ACV + IFN				
AraA + AraH + ACV		I		
AraA + IFN + AraH or ACV				
Tunicamycin + DDG	**Rabbit eye**/HSV-1	I	CS	Raju et al. (1984)
AraT + AIU or AT	Cell culture/HSV-1 and HSV-2	E/SA	YR	Fischer and Ratschan (1984)
TFT + PFA	Cell culture/HSV-2 TK$^-$/DNA polymerase altered	E	NA	Grene et al. (1984)
ACV, PFA, TFT	**Rabbit eye**/HSV-2	I	CS	
TFT + HC	**Rabbit eye**/HSV-1	E	LS	Colin et al. (1984)

Drug combination	System/virus		Result	Reference
DHPG + PAA	AGMK/HSV-2	S	FIC	Crumpacker et al. (1984)
Rabavirin + AraA	Cell culture and **mice**/HSV-1 and HSV-2	S	FIC	Canonico et al. (1984a)
ACV + Dexamethasone	**Human eye**/HSV	E	LS	Lagoutte (1984)
ACV + PFA	**Guinea pigs**/HSV-1 resistant virus	E	CS	Poryo and Wigand (1984)
ACV + AraA ACV + BCG ACV + AraA + BCG AraA + BCG	**Mouse genital**/HSV-2	E	MDD	Karim et al. (1984)
MMdUrd + TFT or PFA MMdurd + ACV	RK/HSV-1 and HSV-2	S	FIC	Ayisi et al. (1985)
Aphidicolin, PFA, ACV-TP, and DHPG-TP	HSV DNA polymerase	E	K	Frank and Cheng (1985)
ACV + FIAC or FMAU	**Mice (newborn)**/HSV-2	S	M; MDD	Karim et al. (1985a)
ACV + AraA	**Mice (newborn)**/HSV-2	S	M; MDD	Karim et al. (1985b)
AraA + TFT	**Rabbit eye**/HSV-1	E	LS; VT	Pavan-Langston and Dunkel (1985)
AraA + IdUrd or TFT IdUrd + TFT	Cell culture/HSV-1	E	YR	
ACV, AraA, AraAMP, IFN	MEF and HSF/HSV-2	S	FIC	Crane and Milne (1985)
ACV + Ab	**Athymic mice**/HSV-1	E	M; MDD; LS; VT	Yamamoto et al. (1985)
ACV + A723U	AGMK/HSV-1	S	FIC/ED_{50}	Spector et al. (1985)
ACV or AraA + FIAC or FMAU	**Mouse enceph**/HSV-2	E/S	MDEA	Schinazi et al. (1986a)

(continues)

121

TABLE 1 (continued)

Combination of compound[a]	System/Virus type[b]	Effect[c]	Basis[d]	Reference
ACV + AraA (sequential)	**Human or mice/HSV**	E	M	Schinazi et al. (1986b)
ACV + AraA (sequential)	**Human enceph/HSV-2**	E	CR	Berger et al. (1986)
MMdUrd + AraA	RK/HSV-2 resistant virus	S	PR	Ayisi et al. (1986)
	Mice and guinea pigs/HSV-2	E		
AraA then ACV (sequential Rx)	**Human (pregnant)**	I	NA	Berger et al. (1986)
Hydroxynorvaline or 2-Deoxyglucose + TFT	Mouse eye/HSV-1	I	LS	Gordon et al. (1986)
ACV + Cotrimoxazole	**Human genital/HSV**	I/E	LS	Kinghorn et al. (1986)
ACV and AraA	**Mouse enceph/HSV-2 TK⁻**	I	M; MDD	Schinazi (1987)
ACV or BVDU + IgG	**Mouse (dermal)/HSV-1**	E/S	M	Hilfenhaus et al. (1987)
ACV + AraA	**Humans/HSV**	E	CR	Besser et al. (1987)
ACV or AraA + Betamethasone	**Human eye/HSV**	E	HT	Collum et al. (1983b, 1987)
AraA + ACV	**Human enceph/HSV**	E	CR	Besser et al. (1987)
PFA + PAA	Duck cells/Anatid herpesvirus	E	FIC	Johnson and Attanasio (1987)
ACV + PAA or PFA		S		
Ribavirin + Muramyl tripeptide (liposome)	**Mice/HSV**	E	M; MDD	Gangemi et al. (1987)
ACV + 2,2'-O-Cyclocytidine	**Mice**, latency/HSV-1	S	VT	Liu et al. (1986)
TFT, DHPG, AraA	**Rabbit eye/HSV-1**	NA	NA	Small et al. (1987)
ACV + Cyclosporine	**Human/HSV**	I/E	NA	Johnson et al. (1987)

Compound	Cell culture/Virus	Effect	Assay	Reference
BVDU + IgG	**Mice/HSV-1**	S		Hilfenhaus et al. (1987)
FEAU + ACV or AraA	**Mouse enceph** and Vero cells/HSV-2	E/S	M	Schinazi et al. (1987)
ACV + AraA	HSF/HSV-1	E/S	MDEA	Overall et al. (1988)
	HSF/HSV-2	A	MDEA	
	MEF/HSV-1 or HSV-2	A/I		
ACV + AraA	**Mouse enceph**/HSV-1	S	M; MDEA	Kern et al. (1988)
AraA + Various 2',3'-dideoxynucleosides	RK/HSV-1, HSV-2, VV	S	YR	Balzarini and De Clercq (1989)
ACV + A111OU	**Mice/HSV-1**	S	AUC	Ellis et al. (1987, 1989)
PFA + S-dC$_{28}$	HeLa/HSV-2	E	YR	Gao et al. (1990)
DHPG + S-dC$_{28}$		A/I		

[a] Compounds: 6AzCyd = 6-Azacytidine; 6AzUrd = 6-Azauridine; A723U = 2-Acetylpyridine-4-(2-morpholinoethyl)thiosemicarbazone; Ab = Antibody to virus (immune serum globulin, IgG); ACV = 9-(2-Hydroxyethoxymethyl)guanine; AIU = 5'-Amino-2',5'-dideoxy-5-iodouridine; AraA = 9-β-D-arabinofuranosyladenine; AraAMP = AraA-5'-monophosphate; AraC = 1-β-D-Arabinofuranosylcytosine; AraH = 9-β-D-Arabinofuranosylhypoxanthine; AraT = 1-β-D-Arabinofuranosylthymine; AT = 5'-Amino-5'-deoxythymidine; AZT = 3'-Azido-3'-deoxythymidine; BVDC = E-5-(2-Bromovinyl)deoxycytidine; BVDU = E-5-(2-Bromovinyl)deoxyuridine; CMA = Carboxymethylacridanone (IFN inducer); DDA = 2',3'-Dideoxyadenosine; DDC = 2',3'-Dideoxycytidine; DDG = 2-Deoxy-D-glucose; DDI = 2',3'-Dideoxyinosine; DHEA = Dehydroepiandrosterone; DHPG = 9-(1,3-Dihydroxy-2-propoxymethyl)guanine (also known as 2'NDG, BIOLF-62 or BW 759U); (S)-DHPA = (S)-9-(2,3-Dihydroxypropyl)adenine); DI = Deaminase inhibitor; dThd = Thymidine; EdUrd = 5-Ethyl-2'-deoxyuridine; FDT = 3'-Fluoro-3'-deoxythymidine (FLT, FddT); FIAC = 2'-Fluoro-5-iodoaracytosine; FMAU = 2'-Fluoro-5-methylarauracil; GMCSF = Granulocyte-macrophage colony-stimulating factor; HBB = 2-(α-Hydroxybenzyl)benzimidazole; HC = Hydrocortisone; IdUrd = 5-Iodo-2'-deoxyuridine; IFN = Interferon (alpha, beta or gamma) (some studies were done with recombinant IFN); MC = Mononuclear leukocytes; MMdUrd = 5-Methoxymethyl-2'-deoxyuridine; PAA = Phosphonoacetic acid; PFA = Phosphonoformate; PMEA = 9-(2-Phosphonylmethoxyethyl)adenine); PolyIC(LC) = Poly-L-lysine carboxymethylcellulose complex; Rec. 15/0209 = 4-Nitrophenylglyoxal-N-p-dichloroacetylanilineothoxy diacetal; SAM = S-Adenosylmethionine; S-dC$_{28}$ = Phosphorothioate homooligodeoxynucleoside; TFT = 5-Trifluoromethyl-2'-deoxyuridine; TNF = Tumor necrosis factor.

[b] Cell cultures: AGMK = African Green monkey kidney (Vero); BHK = baby hamster kidney; FTC = fetal tonsil; HE = horse epidermis; HEL = human embryonic lung; HSF = human skin fibroblast; KB = human epidermoid carcinoma; MEF = mouse embryo fibroblast; RK = rabbit kidney. Note that different strains of HSV were used by the authors. In vivo studies appear in bold type.

[c] Effect as described by authors (grouped arbitrarily): A = Antagonistic; E = Enhancing, additive, beneficial, or potentiating; I = Indifference (no effect greater than most effective agent alone); NA = Not available or no clear benefit; R = Reduced; S = Synergistic; SA = Subadditive; T = Toxic; V = Variable.

[d] Assay, end point, or basis for measured effect: AUC = Area under curve; CPE = Cytopathic effect; CR = Clinical response or signs; CS = Clinical score; DL = Duration of lesion; ED_{50} = Median effective (or inhibitory) dose; FIC = Fractional inhibitory concentration (isobologram; Loewe's method); FP = Fractional product method (Webb's method); HT = Healing time; IQ = Inhibitory quotient; LS = Lesion score; M = Mortality (or survival); MDD = Mean day of death; MDEA = Multiple-drug-effect analysis (combination index method); ME = Median effect; MIC = Minimum inhibitory concentration; MT = Microtitration; NA = Not available; PD = Plaque diffusion; PR = Plaque reduction; VR = Virus rating; VT = Virus titer; YR = Yield reduction.

TABLE 2

Combinations of Antivirals Used on Viruses Other Than Herpes Simplex Viruses (Arranged Chronologically from 1955–1990)

Combination[a]	Viruses[b]	Effect[c]	Reference
Isatin thiosemicarbazone + Phenoxypyridine	Vaccinia (**mice**)	S	Bauer (1955)
HBB + Guanidine HCl	Coxsackie (CC)	S	Eggers and Tamm (1963)
6AzUrd + N-formylbiuret	Vaccinia (CC)	S	Link et al. (1965–1966)
Amantadine + IFN	Influenza (CC)	E	Lavrov et al. (1968)
Amantadine + Rec. 15/0209	Influenza A (eggs and **mice**)	E/S	Altucci et al. (1969)
IFN + endogenous IFN induced with virus, poly IC or statolon	Encephalomyocarditis (**mice**)	I/E	Gresser et al. (1969)
Poly IC + Immune serum	Vaccinia (**mice**)	E	Worthington and Baron (1973)
AraA + DI	Varicella zoster or vaccinia (CC)	S	Connor et al. (1974)
AraA + DI	Vaccinia (CC)	S	Shannon (1975)
AraA + DI	Varicella zoster (CC)	S	Bryson and Connor (1976)
HBB + Guanidine	Coxsackie and Echo (**mice**)	E	Eggers (1976)
IFN + HPA-23	Encephalomyocarditis or VSV (**mice**)	E	Werner et al. (1976)
Rimantidine + Bonaphthone	Influenza A (**mice**)	E	Ilyenko et al. (1977)
Amantadine + Ab	Influenza (CC)	E	Tyrell (1976)
Rimantidine + Ribavirin	Influenza A and B (CC and **mice**)	E	Galegov et al. (1977)
IFN + Isoprinosine	Encephalomyocarditis (**mice**)	E	Chany and Cerutti (1977)
9 Purines/Pyrimidines	Adeno-2 (CC)	S	Wirjawan and Wigand (1978)

Combination	Virus system		Reference
AraA + (S)-DHPA	Vaccinia (CC)	S	De Clercq et al. (1978)
$IFN_{\alpha/\beta}$ + IFN_γ	Mengo (CC)	E	Fleischmann et al. (1979)
Amantadine + Ribavirin	Influenza A and B (**mice**)	E	Wilson et al. (1980)
Interferon + Antineoplastic agents (MTX; FUra; Adriamycin; Vincristine; 6-MP; HC; estradiol)	Newcastle disease (CC)	A	Cesario and Slater (1980)
AraA + PFA or DI	Murine leukemia (Gross)(CC)	S	Shannon and Schabel (1980)
IFN + TFT; ACV or PFA ACV/PFA: ACV/TFT; PFA/TFT	Cytomegalo (CC)	E S	Spector et al. (1980)
Rimatidine; Amantadine; Ribavirin	Influenza A (CC)	E	Hayden et al. (1980)
6AzCyd + halothane	Measles (CC)	E	Knight et al. (1980)
(S)-DHPA + 6AzaUrd	Vaccinia (CC)	S	Rada and Holy (1980)
AraA + Allopurinol	Varicella zoster (**human**)	A	Friedman and Grasela (1981)
AraA + IFN	Hepatitis B (**human**)	E	Scullard et al. (1981a,b)
ACV + IFN	Varicella zoster or Cytomegalo (CC)	S E	Levin and Leary (1981)
BVDU + ACV or AraA	Varicella zoster (CC)	E/S	Bryson et al. (1981)
ACV + AraA	Varicella zoster (CC)	E/S	Biron and Elion (1981)
ACV + TFT, IdUrd, PAA or BVDU		E	
6AzaUrd + Different nucleosides	Vaccinia (CC)	I	Surjono and Wigand (1981)

(*continues*)

TABLE 2 (*continued*)

Contribution[a]	Viruses[b]	Effect[c]	Reference
Amantidine + Ribavirin or rimantadine	Influenza A (CC)	S	Burlington et al. (1981, 1983)
Carbodine + DI	Influenza A (CC)	I	Shannon et al. (1981)
AraA + IFN	Hepatitis B (**human**)	NA	Sachs et al. (1981, 1982)
IFN$_{\alpha,\beta}$ or γ + Ab	AHCV, Adeno (CC)	S	Langford and Stanton (1982)
Ribavirin + IFN$_\beta$	Toga, Bunya, Arena (CC)	S	Luscri (1982)
Levamasole + Meganine	Genital herpes (human)	NA	Moschchik et al. (1982)
Immune + Virus type IFN	NA (CC)	E	Schwarz and Fleischmann (1982)
IFN$_{\alpha,\beta}$ + IFN$_\gamma$	Encephalomyocarditis (CC)	E/S	Zerial et al. (1982)
AraA + IFN$_\alpha$	Cytomegalo (**human**)	I	Meyers et al. (1982)
AraAMP + IFN$_\alpha$	Hepatitis B (**human**)	NA	C. I. Smith et al. (1982)
HBB + Guanidine	Coxsackie (CC)	S	Herrmann (1982)
LY 122771-72 + HBB + Guanidine	Coxsackie (**mice**)	E	Herrmann et al. (1982)
Guanidine + HBB or Enviroxime		S	
TFT + ACV or PAA	Cytomegalo (CC)	S	Spector et al. (1981, 1983)
TFT + IFN$_\beta$		E	
ACV + AraA	Cytomegalo (CC)	S	Spector and Kelley (1983, 1985)
ACV + IFN$_\beta$	Cytomegalo (**mice**)	I	Rose et al. (1982)
IFN + Ab	Coxsackie, Entero, Adeno (CC)	S	Langford et al. (1983)
IFN$_\alpha$ + ACV	Cytomegalo (CC)	S	Smith et al. (1983)

Drug/Treatment	Virus/Disease		Reference
IFN$_\alpha$ + BVDU, AraA or AraC		E	
ACV + IFN$_\alpha$	Cytomegalo (**human**)	NA	Wade et al. (1983)
Amantadine + Dyazide	Influenza A (**human**)	T	Wilson and Rajput (1983)
ACV + Amphotericin B	Pseudorabies (CC)	E	Malewicz et al. (1982, 1983)
ACV + Polyene Macrolide Antibiotics	Pseudorabies (CC)	E	Malewicz et al. (1984)
Rimantidine + Ribavirin IFN$_\alpha$ + Rimantidine + Ribavirin	Influenza A (CC)	E/S	Hayden et al. (1984)
ACV, AraA, BVDU, PFA, IFN$_\alpha$	Varicella zoster (CC)	E/S	Baba et al. (1984)
DHPG + IFN$_\alpha$ and $_\beta$	Cytomegalo (CC)	E/S	Rasmussen et al. (1984)
Selenazofurin + Ribavirin	Mumps, Measles, Parinfluenza type 3, Vaccinia, and HSV-2 (CC)	S	Kirsi et al. (1984)
Selenazofurin + 3-deazaguanosine		I/A	
Ribavirin + 3-deazaguanosine		E	
Selenazofurin + Ribavirin	Venezuelan equine encephalomyelitis, Japanese encephalitis, yellow fever, and Pichinde (CC)	S	Huggins et al. (1983, 1984)
Selenazofurin + Ribavirin	Korean hemorrhagic fever and Rift Valley fever (CC)	E	
Ribavirin + Tiazofurin	Japanese encephalitis and yellow fever (CC)	S	
Selenazofurin + Ribavirin	Venezuelan equine encephalomyelitis, yellow fever, and Japanese encephalitis (CC)	S	Canonico et al. (1984a,b)

(continues)

TABLE 2 (*continued*)

Contribution[a]	Viruses[b]	Effect[c]	Reference
Ribavirin + Tiazofurin	Japanese encephalitis and yellow fever (CC)	S	
IFN$_\alpha$ + IFN$_\gamma$	Yellow fever, Japanese encephalitis, Hantaan, Eastern equine encephalitis, Jamestown Canyon, Keystone, and Western equine encephalitis (CC)	S	
Ribavirin + Poly IC:LC	Rift Valley fever (**mice**)	E	Jahrling et al. (1984)
Ribavirin + Ab	Lassa fever (**monkey**)	E	Gorse et al. (1984)
IFN$_\alpha$ + Vincristine	Newcastle disease (**mice**)	I	Weigent et al. (1984)
IFN + IFN-induced transfer	Sindbis (CC)	S	Fleischmann and Fleischmann (1984)
IFN$_{\alpha/\beta}$ + IFN$_\gamma$	Newcastle disease (CC)	S	Langford et al. (1984)
IFN$_\beta$ + Ab	AHCV (CC)	S	Meyer and Horisberger (1984)
IFN$_\alpha$ + IFN$_\beta$	Influenza (Macrophages)	E	Goren et al. (1984)
IFN$_{\alpha,\beta,\gamma}$	NA (CC)	S	Watanabe et al. (1984)
Tunicamycin + IFN	Vesicular stomatitis (CC)	NA	Gil-Fernández et al. (1984)
IdUrd + Amantadine or Rifampicin	African swine fever (CC)	S	Nishimura and Ikeda (1984)
Muramyldipeptide + IFN Inducers	Vaccinia (**mice**)	S	Hayden et al. (1984)
IFN$_\alpha$ + Rimantadine or ribavirin	Influenza A or B	E	Langford et al. (1985)
Arildone + IFN$_\beta$	AHCV (CC)	E	Spector and Kelley (1985)
ACV + AraA	Cytomegalo (CC)	S	

ACV + IFN	Hepatitis B (**human**)	E	Schalm *et al.* (1985)
AraA + Prednisolone	Hepatitis B (**human**)	E	Yokosuka *et al.* (1985)
ACV + AraA, PAA or PFA	Channel catfish (CC)	E	Buck and Loh (1985)
ACV + Suramin	Duck hepatitis B (**ducks**)	E	Tsiquaye *et al.* (1986)
Amantadine + Ribavirin	Hepatitis A (CC)	I	Widell *et al.* (1986)
Levamisole, methisoprinol, adenine or adenosine + IFN	Encephalomyocarditis (CC)	E	Muñoz *et al.* (1986)
Rimantadine + Ribavirin	Influenza A (**mice**)	E	Hayden (1986)
Dichloroflavan, enviroxime chalcone Ro-09-0410, IFN$_{\alpha,\beta,\gamma}$	Rhino (CC)	S	Ahmad and Tyrell (1986)
AraA, IFN + Prednisolone (alternating)	Hepatitis B (**human**)	E	Omata and Uchiumi (1986)
AraAMP or IFN$_\alpha$ + Prednisolone (alternating)	Hepatitis B (**human**)	E	Perrillo (1986)
DHPG + HC	Cytomegalo (**human**)	I	Reed *et al.* (1986)
6-AzUrd, IdUrd, rifampicin, DL-p-fluorophenylalanine, ACV, DHPG, cyclohexamide, rimantadine, ribavirin	Fowl plague, VV (CC)	V	Rada and Hanusovskà (1987)
Ribavirin + muramyl tripeptide (liposome)	Influenza (**mice**)	E	Gangemi *et al.* (1987)
Quercetin + IFN$_{\alpha,\beta}$	Mengo (**mice**)	S	Veckenstedt *et al.* (1987)

(continues)

129

TABLE 2 (*continued*)

Combination[a]	Viruses[b]	Effect[c]	Reference
Glycyrrhizin + ACV, AraA BVDU, PFA or IFN$_\beta$	Varicella zoster (CC)	E/S	Baba and Shigeta (1987)
DHPG + IFN$_\beta$	Simian Varicella (**African Green Monkeys**)	S	Soike *et al.* (1987)
Ribavirin + Poly(ICLC)	Rift Valley fever (**mice**)	E	Kende *et al.* (1987)
ACV + isoprinosine or levamisole	Varicella zoster (**human**)	E	Tanphaichitra and Srimuang (1987)
DHPG + DMFO	Cytomegalo (CC)	S	Rush and Mills (1987)
Rimantadine + Aprotinin	Influenza A (**mice**)	S	Zhirnov (1987)
ACV + Prednisolone	Epstein–Barr (**human**)	E	Andersson *et al.* (1987)
Ribavirin + IgG	RSV (**rat**)	E	Gruber *et al.* (1987)
ACV + Cyclosporine	Varicella zoster (**human**)	I/E	Johnson *et al.* (1987)
DHPG + IgG	Cytomegalo (**human**)	E	Emanuel *et al.* (1988)
IFN$_\beta$ + IFN$_\gamma$	Hepatitis B (**human**)	I	Bissett *et al.* (1988)
Amantadine + Ribavirin	Influenza A (**pregnant women**)	E	Kirshon *et al.* (1988)
ACV + AraA	Varicella zoster (**human**)	E	Mooriyama *et al.* (1988)
DHPG + IgG	Cytomegalo (**human**)	E	Schmidt *et al.* (1988)
DHPG + IgG	Cytomegalo (**human**)	E	Reed *et al.* (1988)
Tiazofurin + Ribavirin	Hepatitis	E	Natsumeda *et al.* (1988)
DHPG + IgG	Cytomegalo (**human**)	NA	Verdonck *et al.* (1989)

ACV + IgG + HC	Epstein–Barr (**human**)	E	Hugo et al. (1989)
AZT + IFN$_{\alpha,\beta}$	Epstein–Barr (CC)	S	Lin et al. (1989)
IFN$_\alpha$ + IFN$_\beta$	Hepatitis B (**human**)	E	Caselmann et al. (1989)
DHPG + IgG	Cytomegalo (**mice**)	E	Rubin et al. (1989)
DHPG + GM-CSF	Cytomegalo (**human**)	E	Grossberg et al. (1989)
AZT + PFA	Cytomegalo (CC and CMV DNA polymerase)	E/S	Eriksson and Schinazi (1989)
IFN + CO$_2$ laser	Papilloma (**human**)	NA	Chiesa et al. (1989)
Antiviral Combinations	Hepatitis B (**human**)	E	de Man et al. (1989)
Homocysteine + SAM inhibitor	Vaccinia (CC)	E	Hasobe et al. (1989)
Glycyrrhizin + IFN	Hepatitis B (**human**)	NA	Hayashi et al. (1989)
AZT + IFN$_\alpha$	Hepatitis B (**human**)	NA	Hess et al. (1989)
Ribavirin + IFN$_\alpha$	Rabies (**human**)	A	Warrel et al. (1989)
ACV + IFN$_\alpha$	Feline herpes (CC)	S	Weiss (1989)
Ribavirin + IFN$_\alpha$	Feline herpes (CC)	NA	Weiss and Oostrom-Ram (1989)
Antisense Oligonucleotides	Influenza (CC)	NA	Kabanov et al. (1990)
DHPG + PFA	Cytomegalo (CC)	S	Manischewitz et al. (1990)
DHPG + IgG	Cytomegalo (**human**)	I	Jacobson et al. (1990)
FEAU + IFN$_\beta$	Simian varicella (**African Green monkeys**)	S	Soike et al. (1990)

[a] MTX = Methotrexate; FUra = 5-Fluorouracil; 6-MP = Mercaptopurine. For other abbreviations, see Table 1.
[b] Animal and human studies appear in bold type. CC = Cell culture; AHCV = Acute hemorrhagic conjunctivitis virus.
[c] For legend see Table 1.

TABLE 3

Combinations of Antivirals Used on HIV and Other Retroviruses (Arranged Chronologically from 1977–1990)

Combination	Viruses[a]	Effect[b]	Reference
AraA + DI	Murine leukemia (Rauscher)(CC)	E	Shannon (1977)
PFA + IFN$_\alpha$	HIV (CC)	S	Hartshorn et al. (1986)
ACV + Suramin	HIV (CC)	E	Resnick et al. (1986)
Glycyrrhizin + AZT	HIV (CC)	E	Ito et al. (1987)
AZT + Ribavirin	HIV (CC)	A	Vogt et al. (1987)
AZT + Dextran sulfate	HIV (CC)	S	Ueno and Kuno (1987)
AZT + GM-CSF	HIV (CC)	S	Hammer and Gillis (1987)
AZT + IFN$_\alpha$	HIV (CC)	S	Hartshorn et al. (1987)
DDA + 14-mer dC phosphothioate	HIV (CC)	S	Matsukura et al. (1987)
AZT, DDC, D4C, or D4T + Ribavirin	HIV (CC)	A	Baba et al. (1987)
DDA or DDG + Ribavirin		S	
DDpyrimydine nucleosides + DDpurine nucleosides		S	
DDC + dThd	HIV (CC)	S	Balzarini et al. (1987)
AZT + Ansamycin	HIV (CC)	I	Birch et al. (1988)
AZT + ACV	HIV (**human**)	NA	Surbone et al. (1988)
AZT + DDC	HIV (CC)	S	Vogt et al. (1988)
AZT + DDC alternating	HIV (**human**)	NA	Yarchoan et al. (1988)
AZT + GM-CSF	HIV (CC)	S	Perno et al. (1988)
AZT + Isoprinosine	HIV (CC and **mice**)	I	Schinazi et al. (1988)

Combination	System		Reference
AZT + Isoprinosine	HIV (**human**)	NA	De Simone et al. (1988)
AZT + Avarol	HIV (CC)	S	Schröder et al. (1988)
AZT + Uridine	HIV (CC)	I	Sommadossi et al. (1988)
AZT + DHPG	HIV, CMV (**human**)	T	Jacobson et al. (1988)
AZT + IFN$_\alpha$	Rauscher MLV (**mice**)	S	Ruprecht et al. (1988)
AZT + INF$_\beta$	HIV (CC)	S	Williams and Colby (1988, 1989)
DDC + IFN$_\alpha$($\pm$ TNF or DI)	Feline leukemia virus (CC and **cats**)	A	Zeidner et al. (1989)
AZT + IFN$_\alpha$	Human bone marrow cells	S	Berman et al. (1989)
AZT + Ampligen	HIV (CC)	S	Mitchell et al. (1989)
Ampligen + IFN$_{\alpha,\beta,\gamma}$, AZT, PFA, Amphotericin, or Castanospermine	HIV (CC)	S	Montefiori et al. (1989)
AZT + PMEA	HIV (CC)	E/S	Smith et al. (1989)
AZT + DDC (alternating)	HIV (CC)	E	Spector et al. (1989)
AZT + Dipyridamole	HIV (CC)	S	Szebeni et al. (1989)
AZT + ACV	HIV (**human**)	I	Brockmeyer et al. (1989)
AZT + PFA	HIV (CC and RT enzyme)	E/S	Eriksson and Schinazi (1989)
AZT + Castanospermine	HIV (CC)	S	Johnson et al. (1989c)
AZT + IFN$_\alpha$	HIV (**human**)	NA	Kovacs et al. (1989)
AZT + IFN$_\alpha$	HIV (**human**)	T/NA	Edlin et al. (1989)
AZT + sCD4	HIV (CC)	S	Johnson et al. (1989a)
AZT + Antimycobacterial	HIV (**human**)	I/E	Kavesh et al. (1989)
PFA + AME (Amphotericin B deriv.)	HIV (CC)	S	Plescia et al. (1989)

(continues)

TABLE 3 (*continued*)

Combination	Viruses[a]	Effect[b]	Reference
Adriamycin, Bleomycin, Vincristine	Epidemic Kaposi Sarcoma (**human**)	E	Gill *et al.* (1989)
AZT + 6-Thiodeazaguanine	HIV (CC)	E	Resnick and Mian (1989)
IFN + Actinomycin-D, Vinblastine, Bleomycin	HIV, KS (**human**)	I/T	Shepherd *et al.* (1989)
AZT + IFN$_\alpha$	HIV (CC)	S	Vogt *et al.* (1989)
AZT + Ribavirin		A	
PFA + IFN$_\alpha$		S	
AZT + IFN$_\alpha$	HIV (**human**)	E	Orholm *et al.* (1989)
AZT, IFN$_\alpha$	HIV latency (CC)	NA	Poli *et al.* (1989)
AZT + PFA	HIV (CC)	S	Koshida *et al.* (1989)
	HIV (RT enzyme)	A	
AZT + ACV (alternating)	HIV (**human**)	E	Lange *et al.* (1989)
AZT + PFA	HIV, CMV (**human**)	E	Lehoang *et al.* (1989)
AZT + N-Butyl 1-deoxynojirimycin	HIV (CC)	S	Johnson *et al.* (1989b)
Ribavirin + 3'-azido-2,6-diaminopurine-2',3'-dideoxyriboside	HIV (CC)	E	Balzarini *et al.* (1989)

Combination	Virus	Activity	Reference
2',3'-Dideoxynucleoside triphosphates + PFA	HIV (RT enzyme)	E	Starnes and Cheng (1989)
AZT + IFN$_\alpha$	HIV (**human**)	NA	Hess et al. (1989)
AZT, DDI, DDC + sCD4 or Dextran sulfate	HIV (CC)	S	Hayashi et al. (1990)
sCD4 + Dextran sulfate		A	
AZT + Vitamin E	HIV (CC)	S	Gogu et al. (1989)
AZT + Probenecid	HIV (**human**)	NA	Kornhauser et al. (1989)
AZT + ACV	HIV (**human**)	NA	Collier et al. (1989)
AZT + DDC (alternating)	HIV (**human**)	NA	Pizzo et al. (1989)
AZT + Dipyridamole	HIV (CC)	S	Szebeni et al. (1990)
AZT + DHEA	HIV (CC)	I	Schinazi et al. (1990a)
AZT + DDI	HIV (CC)	V	Schinazi et al. (1990b)
AZT + sCD4 and/or IFN$_\alpha$	HIV (CC)	S	Johnson et al. (1990)
AZT + Carbovir	HIV (CC)	S	Shannon (1990)
AZT + IFN$_\alpha$	Rauscher MLV (**mice**)	S	Ruprecht et al. (1990)
AZT + Acetaminophen	HIV (**human**)	NA	Steffe et al. (1990)
AZT + IFN$_\alpha$	Feline leukemia virus (CC and **cats**)	S	Zeidner et al. (1990)

[a] Animal and human studies appear in bold type. CC = Cell culture.
[b] For legend see Table 1.

III. Rationale for Combined Therapy with Antiviral Drugs

Combinations of antiviral drugs for HIV and HSV have generally focused on the most widely used and clinically relevant drugs (Figs. 2 and 3). There are several indications already reported of the potential advantages and problems of polychemotherapy of viral infections.

A. PREVENTION OR INHIBITION OF RESISTANT MUTANTS

Development of viruses that are resistant to antivirals has been reviewed (Crumpacker, 1983, 1989; Field, 1983, 1989; Larder and Darby, 1984). These workers have listed an impressive number of antiviral substances to which certain viruses developed resistance, mostly upon serial cultivation in cell culture. The hypothesis was entertained that development of viral resistance to a given drug is more likely the more selective the antiviral agent (Herrmann and Herrmann, 1977). Thus, in clinical situations, the rapid emergence of drug resistant variants or infection with an already resistant virus would impair the chances for cure. The addition of a second drug could then reduce the spread and amount of virus produced, or delay the emergence of a resistant strain. Drug-resistance development can be a problem in the treatment of herpes keratitis. Resistance to the newer drugs, such as acyclovir, is highly prevalent in immunocompromised patients, especially those affected with HIV (Crumpacker, 1989; Schinazi et al., 1986b). To compound the problem in these individuals, drug-resistance development of HIV has been demonstrated to occur clinically (Land et al., 1990; Larder et al., 1989, 1990; Larder and Kemp, 1989; Rooke et al., 1989; Zimmermann et al., 1988). Since a combination of drugs is expected to cause rapid healing, the duration of exposure of the virus to the drug can then be shortened reducing the likelihood of developing resistance (Collin et al., 1983). However, developing new strategies to overcome this increasing problem now will help us to cope this difficulty in the future.

B. REDUCTION OF TOXICITY

Even though some newer drugs are less toxic, there is still concern that acute toxicity can develop in certain types of patients, e.g., vidarabine or acyclovir in patients with chronic liver or renal disease (O'Brien and Campoli-Richards, 1989; Whitley, 1988; Whitley et al., 1985). The advantage of drug combinations is that lower levels of two effective antivirals can be used which can produce the same or higher effectiveness than one drug at high and possibly toxic dose. This approach is particularly useful when two compounds have overlapping toxicities, although combinations of drugs with nonoverlapping toxicities is preferred. Furthermore, the use of high doses of an antiviral to overcome drug resistance may be not only ineffective but also potentially dangerous (Schinazi, 1987). It has been shown that the combination of certain antivirals can inhibit the toxic effect of one or more of the agents, thus allowing the use of higher doses of the antiviral compounds (Fischer, 1981; Fischer *et al.*, 1979, 1980, 1983; Fischer and Murphy, 1983; Fischer and Phillips, 1984). In AIDS chemotherapy, certain compounds have been shown to reduce the toxicity of AZT *in vitro*. For example, uridine or dipyridamole in combination with AZT appear to reduce the toxicity of AZT in at least some cell types without reducing its antiviral activity (Sommadossi *et al.*, 1988; Szebeni *et al.*, 1989, 1990), i.e., these compounds show "host selective antagonism".

C. DECREASED INCIDENCE OF CLINICAL NONRESPONSIVENESS

Early effective treatment, particularly in severe HSV infections, is most important clinically (Whitley, 1988). With a combined modality, the likelihood of patients not responding to two or more antivirals is greately reduced.

D. ELIMINATION OF LATENT VIRUS OR PREVENTION OF ESTABLISHMENT OF LATENCY

One further potential value of combination chemotherapy is the use of a nontoxic drug which can reactivate latent virus and another which inhibits selectively the replication and establishment of latency of newly released virus in susceptible cells. However, this simplistic approach will not work unless the cells that are latently infected are eliminated.

The bulk of experimental evidence suggests that ACV treatment has no effect upon latent HSV infection *once established* (O'Brien and Campoli-Richards, 1989). In two independent studies (Park *et al.*, 1984; Crane *et al.*, 1984), the combination of ACV and araA significantly reduced the frequency of development of latent HSV-1 infection in the trigeminal ganglia of mice that had been treated 24 hr after infection, whereas ACV or araA alone did not. Preliminary data obtained in our laboratory suggest that latent virus can be prevented or eliminated in some animals with a combination of antivirals when treatment is delayed by 3 days after virus inoculation. Furthermore, when mice surviving primary intracerebral infection were treated with a combination of antivirals 1 year after infection (all the animals were HSV-seropositive by ELISA), the rate of recovery of HSV from the trigeminal ganglia was markedly reduced compared with untreated animals. The persistence of HSV in neural ganglia has been explained by two alternative hypotheses (Roizman and Sears, 1990). According to the *dynamic state hypothesis*, a small number of neural cells would constantly replicate virus. According to the *static state hypothesis*, the viral genome is maintained in a nonreplicating state, either in an extra chromosomal site or integrated into cellular DNA. One report suggests that the *predominant* form of the HSV-1 genome in latently infected cells is nonintegrated and linear (Wigdahl et al., 1984). One other report (Yamamoto et al., 1977) found evidence of thymidine kinase activity in sensory ganglia of mice with latent virus. Although the results with drug combinations in mice are controversial, they suggest that latent virus may be more accessible to antiviral attack. Perhaps the latent virus is metabolically active and is not in a "deep-frozen" state as is sometimes assumed. In such cases, if the antiviral drug can be transported in sufficient amount to the CNS and affect latent virus, then elimination of latent virus is a possibility. This would not be the case with integrated virus unless the genome and/or infected cell was eliminated. The pharmacokinetic characteristics of the drugs that are currently available clinically may not allow effective levels to accumulate in the CNS. However, by using a combination of drugs that are additive or synergistic, an effective therapeutic level may be achieved in the CNS. These findings also suggest that early treatment of humans with a combination of antiviral drugs may reduce the frequency of latent infection, probably by preventing the establishment of latency.

E. TREATMENT OF DIFFERENT VIRUSES CAUSING SIMILAR DISEASE

Ultimately, when antiviral drugs are developed that will affect different viruses, combinations of such antiviral agents will be advantageous for the treatment of a disease or a syndrome of multiple viral etiologies, e.g., pneumonitis caused by RSV or adenovirus.

F. OTHER CONSIDERATIONS

In life-threatening conditions, it may not be practical to give patients an unproven drug when one of limited effectiveness is already licensed for that treatment. A comparative study of the combination of a marginally effective licensed drug and an unlicensed new drug versus the licensed drug may be more acceptable.

The potential problems of combined therapy with antiviral drugs could well turn out to be similar to those already observed with combined antibacterial therapy and include the following. (1) Simultaneous administration of two or more antivirals may enhance adverse effects which are not serious with monotherapy; these effects include direct toxic effects and induction of hypersensitivity or allergic reaction. (2) Although unlikely, certain multiple therapies could actually favor the emergence of variants resistant to both drugs. Thus, care should be taken in selecting the two drugs to be used in combination, in order to decrease the probability of cross resistance. Another potential problem with combined treatment is that, by using lower levels of each drug, certain compartments such as the brain may receive suboptimal levels of the active drugs.

IV. Criteria for Usefulness of Combination Chemotherapy with Antiviral Agents

This author's research team was one of the first to emphasize the import-ance of combination antiviral therapy along the lines well demonstrated for antibacterial and anticancer therapy (DeVita *et al.*, 1975; Rahal, 1978). With the introduction of acyclovir, we were also among the first to study this drug in combination with araA and various interferons (Stanwick *et al.*, 1981). Our early experience with ACV and araA in cell culture and in mice (Schinazi *et al.*, 1981, 1982a, 1983a,b, 1985, 1986a–c; Schinazi and Chou, 1985; Schinazi and Del Bene, 1985) has led us to develop several criteria that should be applied for any new combination of drugs. To be considered useful, antivirals in combination should have one or more of three advantages. (1) They must interact to produce at least an additive effect, i.e., no interference or antago-nism should occur. (2) They should produce no increased toxicity relative to the agents alone (subadditive toxicity or nonoverlapping toxicity). Thus, the combination's therapeutic index (maximum tolerated dose divided by the minimum antiviral effective dose) would be substantially increased over the therapeutic index of either drug used alone. (3) They should inhibit emer-gence of variants resistant to either drug or to both drugs.

It has been repeatedly demonstrated that combinations of antivirals will most often produce an enhanced interaction compared with either of the single agents. Nevertheless, the use of antiviral drugs in combination can sometimes produce antagonistic interactions, additive toxicity, or may be ineffective against drug-resistant viruses. These problems can only be addres-sed in reliable and reproducible biochemical, cell culture, and animal sys-tems. For example, based on the accepted mechanism of action of drugs, we have been able to predict which combination of drugs was able to reduce the mortality of mice infected with a lethal dose of ACV-resistant clinical and laboratory strains of HSV (Schinazi, 1987; Schinazi *et al.*, 1983b, 1986b).

V. Method of Analysis of Drug Interactions

A problem in assessing the synergism/antagonism of drug combinations has been the lack of agreement among investigators on how to analyze the experimental results in a quantitative manner that is simple, theoretically sound, and does not require a large number of measurements. Theoretical models for drug interactions were proposed over a century ago (Fraser, 1872). The most widely used methods of analysis of such systems are the isobolo-gram introduced by Loewe, the fractional product method formalized by Webb, and the multiple-drug-effect analysis described by Chou and Talalay (1984). The merits and disadvantages of these methods are discussed in

Chapter 2 by Chou. It should be stressed that when using this method, a constant drug dose ratio is not always necessary, and that efficacy and toxicity can be analyzed simultaneously or sequentially.

Other methods of analysis are currently being developed by other groups (Sühnel, 1990). A method of analysis for drug interactions (COMBO analysis) which is based on logistic curves and extends the method of Chou and Talalay has been described recently (Weinstein *et al.*, 1990; Bunow and Weinstein, 1990). Compared with existing techniques for analysis of drug combinations, COMBO offers the following features. (1) The data can be used uncensored, i.e., data for low and high effects are not deleted from analyses. (2) Empirical parameters for potentiation, synergy, and antagonism are obtained. (3) A flexible choice of data models involving synergy, antagonism, and therapeutic index is provided. (4) A flexible choice of error structures is provided. (5) Confidence intervals on model parameters are obtained by Monte Carlo methods. (6) Statistical criteria for selection of outliers are generated. Therefore, in addition to the Chou and Talalay method, this extended statistical approach for drug interaction analysis is a welcome development. This method has been applied to interpret data on the interaction of AZT with dipyridamole against HIV *in vitro* (Szebeni *et al.*, 1989, 1990). The results obtained agreed well with the Chou and Talalay analysis. However, this modified method may provide more statistical evaluation. The determination of confidence intervals (as in the COMBO method) is particularly critical in animal and clinical evaluations where the number of groups are significantly smaller than in cell culture or enzyme combination studies.

Evaluation of biological response modifiers (BRMs) in combination with antiviral compounds, particularly *in vivo*, requires approaches somewhat different from those employed for antiviral combinations. Specifically, optimization of BRMs may be more readily achieved through the evaluation of immunological parameters that correlate with efficacy, in much the same way that surrogate markers are used to measure viral load. The subject of BRM combinations is beyond the scope of this review and is addressed in Chapter 20 by Berman and Chang.

VI. Mechanisms of Interaction

A. NUCLEOSIDE AND PYROPHOSPHATE ANALOGS

An understanding of the accepted mechanism of action and metabolism of the drug is essential in order to understand and develop rationally designed drug combinations. The mechanisms of action of various antiviral drugs have been reviewed previously (De Clercq, 1986, 1987; Hirsch and Kaplan, 1985;

O'Brien and Campoli-Richards, 1989; Richman, 1988; Schinazi and Prusoff, 1983; Schinazi, 1988).

Most antiviral nucleosides, including AZT and ACV, depend on cellular or virus-induced thymidine kinase for the initial phosphorylation. After further phosphorylation to the triphosphate form, they can interact with the viral polymerases and in most cases the compounds are incorporated into the viral nucleic acid. ACV and 9-(1,3-dihydroxy-2-propoxymethyl)guanine (DHPG) are examples which, as ACVTP and DHPGTP, compete with dGTP for incorporation into DNA (O'Brien and Campoli-Richards, 1989). AraATP also belongs to this class of compounds. AraATP is a competitive inhibitor of HSV-induced DNA polymerase with respect to dATP. The affinity of ara ATP for viral DNA polymerase is greater than for the host cell DNA polymerase; however, this specificity depends on the source of the viral enzyme. Recent studies also indicate that araATP is incorporated in internucleotide linkage of not only viral DNA, but also cellular DNA. In addition, it has been reported that araA interacts with the ribonucleotide reductase, the enzyme that converts ribonucleoside diphosphates to the corresponding 2'-deoxynucleotides. The role of arabinosyl hypoxanthine (araH), the main metabolite of araA, in the antiviral activity of this drug is not clear. ACV is also a competitive inhibitor and/or substrate for HSV-induced DNA polymerase. The 5'-triphosphate form of the drug, ACVTP, is incorporated into viral DNA as a chain terminator. The viral DNA polymerase becomes inactivated during the process; this reaction becomes irreversible. Therefore, ACVTP is considered a suicide inhibitor, i.e., once the ACVTP–enzyme complex is formed, the process is irreversible, dGTP can no longer compete, and the enzyme becomes inactive. This inactivation does not occur with the cellular DNA polymerase α (Furman *et al.*, 1984). More recently, Reardon and Spector (1989) have shown that in order for ACVTP to form a dead-end complex, the binding of an additional nucleotide, dCTP, is required. They state that ACVTP is not a suicide substrate in the absence of dCTP, and also that the ACVMP template–primer is not a potent inhibitor *per se*.

Another important class of antiviral agents are the pyrophosphate analogs. Phosphonoacetate (PAA) and its congener phosphonoformate (Foscarnet, PFA) inhibit pyrophosphorolysis competitively with pyrophosphate, but are noncompetitive with dNTPs and uncompetitive with DNA (Öberg, 1983).

B. INHIBITION OF DEGRADATIVE ENZYMES

Several antiviral drugs, such as araA, 5-ethyl-2'-deoxyuridine, and 5-(*E*)-(2-bromovinyl)-2'-deoxyuridine (BVDU) are susceptible to deaminases or phosphorylases. The use of an inhibitor of these enzymes in combination with the antiviral drug may increase the bioavailability of the drug, area-under-the-curve (AUC), and half-life, and hopefully increase its effectiveness without increased toxicity. A number of workers have studied the use of deaminase inhibitors in combination studies with vidarabine (see Table 1). However, in general, the toxic effects of the drug increased in the presence of the deaminase inhibitors and this combined modality has not been pursued clinically.

Another example is AZT which is extensively metabolized to the 5'-glucuronide derivative in humans by UDP-glucuronyl transferase (Richman, 1988; Schinazi, 1988). It has been suggested that inhibition of this liver enzyme may increase the bioavailability of AZT (De Miranda et al., 1989). However, prolonged use of inhibitors of microsomal enzymes in the liver, such as probenecid (Kornhauser et al., 1989), may produce additional side effects. Also, the intracellular metabolism of the antiviral drug in the presence of the inhibitor may be altered.

Recently it was shown that certain antivirals, such as BVDU, are rapidly metabolized by ubiquitous phosphorylases. This process is reversible provided a sugar donor nucleoside such as thymidine or 5-ethyl-2'-deoxyuridine is available. Such combinations have been suggested by Dr. E. De Clercq (Leuven University, Belgium).

C. ENHANCEMENT OF ANTIVIRAL ACTIVITY BY SUPPLEMENTING BLOCKADE OF A PATHWAY

Our knowledge of exploitable biochemical differences between virus-infected cells and noninfected cells is increasing. There are several viral or viral-induced enzymes, such as ribonucleotide diphosphate reductase, thymidine kinase, DNA polymerase, reverse transcriptase, RNase H, and proteases, which can be specifically inhibited by various antiviral drugs. Unless blockage of a given metabolic pathway is sufficient to reduce the end products to levels below which cell processes cannot continue, blockade of a single site may be ineffective in restricting viral production. Before the advent of kinetic analysis of competing inhibitors of enzymes, it was thought that the use of drugs acting on identical target sites in combination carried the risk of interference instead of additive or synergistic effects, since two competitive inhibitors acting on a single site would compete for that binding site. Analyses by Grindey and Cheng (1979) demonstrated that two inhibitors acting on the same site may provide more complete blockade than just one inhibitor. The probability of an enzyme escaping inhibition by two different compounds acting at the same site is markedly reduced (hit theory). Compounds inhibiting the same enzyme or an essential viral receptor may produce a more complete blockade resulting in additive to synergistic effects. In general inhibition of different viral targets is expected to produce synergy (see Chapter 11 by Jackson for an analysis of antimetabolite combinations).

D. PHARMACOLOGICAL APPROACH TO DRUG INTERACTIONS

Studies on the mechanism of action and metabolic disposition of the drugs alone and in combination in cells can provide information on the molecular basis for synergy. Such an approach, which has recently been used with nucleoside antimetabolites, is to determine the absolute levels of the nucleosides and nucleotides present in infected and noninfected cells for the drugs alone and in combination (Furman *et al.*, 1981; Spector *et al.*, 1985) in order to determine which nucleotide is accumulated in the cells and deduce from this information the rate-limiting step(s) in the biosynthesis of the viral products. For example, recently, Spector *et al.* (1985) demonstrated synergy for the combination of ACV and a ribonucleotide diphosphate reductase inhibitor known as A1110U. These workers noted a marked decrease in the level of dGTP and a concomitant increase in the level of ACVTP. The net result was an 80-fold increase in the ratio of ACVTP to dGTP. Since dGTP competes with ACVTP for the HSV DNA polymerase, they suggest that this imbalance in favor of ACVTP accounts for the mechanism of potentiation (see Chapter 8 by Spector and Fyfe). It should be stressed that some ribonucleotide reductase inhibitors did not produce a synergistic interaction with ACV (Spector *et al.*, 1985), suggesting that other mechanisms may also be involved. Of significance is that Ellis *et al.* (1989) also found that the combination of ACV and A1110U was synergistic in a cutaneous HSV mouse model. A1110U is a thiosemicarbazone; compounds of this class are known to produce hydrazine and thus could have tumor enhancing properties. Although the studies performed to date are with a topical preparation of A1110U, it is important to ascertain the tumor-enhancing potential of this compound prior to human studies with this combinations.

VII. Pharmacological Principles

A. DICHOTOMY BETWEEN *IN VITRO* AND *IN VIVO* STUDIES

One of the questions raised is "Will combination of agents synergistic *in vitro* behave similarly *in vivo*?" *In vitro*, concentrations of each agent probably remain relatively constant throughout the test period. *In vivo*, it is most likely that, with respect to concentration, events are in continuous flux (unless one uses an osmotic minipump or some other drug delivery system). In question are whether concentrations and concentration ratios approximate at any time those required for synergy *in vitro* and for how long such concentrations and ratios are sustained. In the clinical situation it is also not clear whether the levels of the drugs need to be sustained for a long time or whether peaks and troughs are more advantageous for some drugs. These are critical issues and, until they are put to test, there should be no surprise at the possible dichotomy between *in vitro* and *in vivo* appraisals. Interaction of antiviral substances in cell-free systems may be completely different from that observed in cell culture or in animal models. The science of equilibrium thermodynamics deals with closed systems, which do not exchange matter with the surroundings. However, living cells or whole organisms operate as open systems in which nutrients provide input of matter and energy, and waste products may be considered the output. This picture is even more complicated in viral-infected cells since host and viral products are being synthesized and catabolized at the same time. These complexities of drug metabolism and the existence of multiple feedback effects make it almost impossible to understand intuitively the perturbations in nucleotide pools that follow treatment of cells with physiological nucleosides or with antiviral drugs. In the presence of antiviral agents, alterations in the steady state can occur. In addition, differences in protein affinity of various antiviral agents in humans can result in an alteration of the metabolism of the drugs in combination compared with that found in other animals or in cell culture (Tozer, 1981). For these reasons, attempts have been made to use computer simulation of reactions involved in production and utilization of DNA precursors as a means of predicting drug interactions. Theoretical interactions for combinations of inhibitors of biochemical pathways have been proposed by Grindey and Cheng (1979; see also Schinazi and Nahmias, 1982). They predicted that, if the binding of two inhibitors with an enzyme is mutually exclusive additivity would be expected irrespective of the type of inhibition (competitive, uncompetitive, or noncompetitive). If, however, both agents bind simultaneously to the enzyme, partially interfere with each other, or bind in a nonexclusive way, the interaction will be synergistic. These expectations are not absolute and exceptions are possible (For further discussion, see Chapter 11 by Jackson). It is also not clear to us why two agents which primarily interfere with each other result in synergy.

Although studies on the interaction of two inhibitors of HSV DNA polymerase have been conducted (Frank and Cheng, 1985), only three studies of anti-HIV drugs with the purified HIV reverse transcriptase have been reported to date (Eriksson and Schinazi, 1989; Koshida *et al.*, 1989; Starnes and Cheng, 1989).

At present, and until more information becomes available, we would recommend evaluating combinations of antiviral drugs both *in vitro* and *in vivo* whenever possible. The finding of "antagonism" *in vitro*, in the absence of toxic effects of the drug combination, should suggest caution in the event that this combination is evaluated *in vivo*.

B. SCHEDULING OF DRUG COMBINATIONS

Another complex variable is that of the scheduling of drug combinations. Will combinations of clinically useful antivirals be more effective *in vivo* if administered simultaneously, sequentially, or by alternating the drug schedule? It has been shown repeatedly in cancer chemotherapeutic trials that optimal scheduling of combinations of drugs can make the difference between a good and poor response, or even success and failure. It is like that optimal schedules for different types of chemotherapeutic agents vary depending on their mechanism of action and pharmacokinetics. Information on optimum therapy with drugs can be obtained from animal model studies by comparing simultaneous, sequential, and alternating schedules of drug administration. Related to this objective is the need to determine the optimum ratio of these drugs in a combination that produces the greatest degree of synergy (with no increased toxicity) over a wide range of effect levels.

The importance of drug scheduling is exemplified by the pilot study of the Stanford group with the combination of araA and interferon in humans infected with hepatitis B virus (C. I. Smith *et al.*, 1982), which suggests that the toxicity problems associated with this combination could be reduced by using an alternating schedule. More recently, clinical studies in AIDS patients have been performed with AZT and ddC, compounds that do not have overlapping toxicities (Yarchoan *et al.*, 1988). In this case, the drugs were given in alternating sequence in order to allow the patient to recover from the reversible toxicity of the different agents, a practice commonly used in cancer chemotherapy (see Chapter 3 by Frei and Chapter 17 by Chang).

There are several parallels between the treatment of viral infections and cancer. However, some of the pharmacological principles of anticancer chemotherapy are distinctive from anti-infective therapy. Cell kill of cancer cells obeys first-order kinetics and therefore success of the therapy is dependent on the dose of the drug, the number of cells or tumor bulk, as well as the number of drugs, each drug having its first-order effect. If this principle is applied to viral infections, we should be aggressive when the number of infected cells is small and limited to non-CNS or accessible tissues. In cancer chemotherapy, the phase of the cell cycle is important and factors that regulate cell proliferation are important in the timing and selection of the drugs. This may be pertinent to the treatment of some viral infections. For example, certain cytokines *in vitro* can enhance virus replication, but they can also enhance drug potency (Perno *et al.*, 1988).

VIII. Treatment of HIV, Opportunistic Infections, and Malignancies

Various combinations of antiretroviral agents have been evaluated in cell culture, animal models, and humans infected with HIV; almost all currently available drugs have been studied in combination with AZT (Schinazi, 1988) Figs. 2 and 3, Table 3. A combination of recombinant alpha interferon and AZT is already being tested in humans, since in cell and in murine retrovirus models, this combination has been shown to be synergistic (Hartshorn *et al.*, 1986; Ruprecht *et al.*, 1988, 1990). Interferons may be important in controlling the development of drug-resistant viruses (virus selection or mutation is unlikely to occur in the presence of interferon). An additional advantage of using interferons in combination with AZT is that interferon-alpha was recently shown to suppress HIV expression in chronically infected cells (Poli *et al.*, 1989). However, interferons and other immunotherapies alone are unlikely to be useful in cases of overwhelming infections. Other combinations, such as ribavirin and AZT, have been found to be antagonistic in culture when tested at a ratio of 200:1 (Vogt *et al.*, 1987). The mechanism responsible for this antagonism appears to be inhibition of AZT phosphorylation. Ribavirin is a broad spectrum antiviral agent that is known to increase the intracellular levels of dTTP. This could result in feedback inhibition at the thymidine kinase level, the initial enzyme necessary for the phosphorylation of AZT. In addition, dTTP could compete effectively with AZTTP for the HIV reverse transcriptase, thus reducing the incorporation of AZT into viral DNA.

In terms of immunotherapy, there are many compounds that are being considered. Restoring immune function with one or more of these immunomodulators is unlikely to be successful unless it is combined with an effective antiviral drug (Hammer and Gillis, 1987).

Continued development of chemotherapies for opportunistic infection must be stressed. AIDS allows other infections to kill. Patients must be treated with antiviral drugs, not only to eliminate virus or prevent progression of disease, but also to eliminate the viral as well as nonviral opportunistic infections that frequently occur. Viral superinfections, such as those caused by cytomegalovirus, are responsible for considerable morbidity and mortality in AIDS patients. Fortunately, there are new investigational drugs such as DHPG and PFA that have been evaluated in controlled studies in various clinics in the United States and Europe for the treatment of CMV infections (Henry *et al.*, 1987; Rosecan *et al.*, 1986; Wamsley *et al.* 1987). It is important to determine how these drugs interact, not only at the basic level but also at the clinical level, with known antiretroviral agents such as AZT. Additionally, the interaction of antiretroviral agents with compounds used to treat malignancies or with immunomodulators needs to be evaluated.

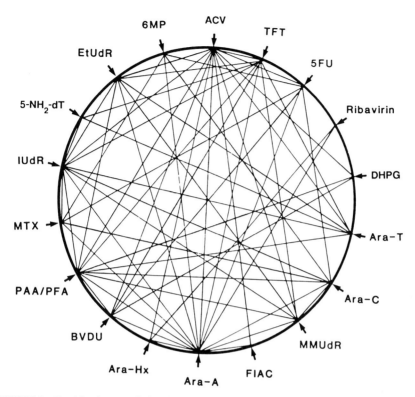

FIGURE 2 Combinations studied against HIV-1 and -2. See Table 1 for abbreviations. Not all compounds have specific antiherpetic activity.

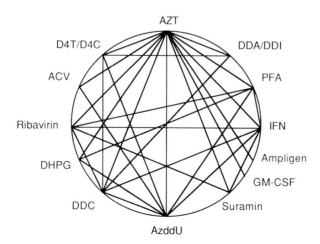

FIGURE 3 Combinations of antiviral drugs evaluated against HIV. Each line represents a particular dual combination that has been evaluated in culture. (Adapted from Schinazi, 1988.)

IX. Drug Combinations as a Challenge for Drug-Resistant Viruses

It is well appreciated that biological organisms will adapt genetically to almost any constant negative selective pressure as long as that pressure is neither lethal nor totally prohibitive of reproduction. Viruses are no exception. In fact, resistance is the rule with selective anti-infective agents. For example, the development of clinical drug-resistant HSV, the first treatable viral infection (herpes keratitis), is as old as antiviral chemotherapy. The widespread use of acyclovir (ACV) raises important concerns regarding the possible increased appearance *and transmission* of strains of ACV-resistant HSV. Basic information about the mechanism of drug resistance, the pathogenic consequences of infections caused by these variants, and their pattern of cross resistance to other antiviral drugs allows a rational selection of drug combinations.

Drug-resistance mutations have also proven valuable in identifying genes and assigning their functions. Once drug-resistance mutations identify a gene product, they can then be used to dissect it functionally. This approach has been used successfully to determine the molecular basis for the selective antiviral activity of ACV and other antiherpetic drugs and their interaction with the active center of the HSV thymidine kinase (Darby *et al.*, 1986; Larder and Darby, 1984) Hahn *et al.* (1986) determined that there was significant genetic variation in HIV over time in patients with AIDS or at risk for AIDS. The rate of evolution of HIV was estimated to be at least 10^{-3} nucleotide substitutions per site per year for the *env* gene and 10^{-4} for the *gag* gene, values a millionfold greater than for most DNA genomes and about tenfold greater than for some other RNA viruses, including certain retroviruses and influenza A virus. The influence of these genetic mutations on the biologic properties of these viruses has not been established. Drug resistance usually results from point mutations leading to alternations in proteins which ordinarily either activate the drug or are inhibited by it. For example, for HSV, a single point mutation in the thymidine kinase (TK) region of the viral genome can confer drug resistance (Larder and Darby, 1984). Nonlethal mutations in the DNA polymerase gene of HSV have been noted and suggest that mutations in the *pol* gene of HIV are probable, especially if the drug is administered for a prolonged period of time. The genomic heterogeneity of HIV should eventually lead to the selection of drug-resistant variants. In fact, resistance of ecotropic retroviruses to drugs like phosphonoformate in cell culture has been known for some time (Muratore *et al.*, 1984), and other types of resistant variants such as temperature-sensitive avian retrovirus have been isolated. These ts mutants are defective for replication at nonpermissive temperature, and it has been shown that the defect resides in the virus reverse transcriptase (Gerard and Grandgenett, 1980). More recently, several groups reported that AZT-resistant virus could be isolated from patients undergoing prolonged AZT therapy (Land *et al.*, 1990; Lader *et al.*, 1989; Larder and Kemp, 1989; Rooke *et al.*, 1989; Zimmermann *et al.*, 1988). It appears that multiple mutations in the reverse transcriptase portion of the genome are necessary to confer an AZT-resistant phenotype (Larder and Kemp, 1989).

To date there is no solid evidence that resistance is of clinical consequence in viral diseases. However, conclusive evidence is difficult to ascertain because there is currently little effort to define the primary cause of death in individuals that succumb to viral infections (Schinazi *et al.*, 1986b). In AIDS patients, the situation is further complicated by the presence of opportunistic infections that may mask the cause of death. Although a number of HIV strains with diminished sensitivity *in vitro* to inhibition by AZT and other antiviral drugs have been isolated from patients, the impact of these occurrences is still difficult to assess. It is likely that this problem will increase with the increased use of drugs such as AZT, and transmission of AZT-resistant variants is a distinct possibility. In question now is how we will assess the problem of clinical resistance to antiviral drugs. The molecular, virological, and clinical basis of drug resistance rests on answers to several key questions. How is resistance defined? What constitutes a drug-resistant strain? What standardized susceptibility assay should be used? Does clinical resistance to antiviral drugs always result in treatment failure? Does the isolation of resistant virus correlate with disease progression? Are mixed populations of resistant and sensitive strains of HIV (or HSV) in clinical isolates artifacts of the method of isolation or are they a real possibility? Is an HIV isolate representative of the HIV population in the patient from whom it was isolated? Are some of the drug-resistant variants really isolated from the patients or produced in cell culture in the presence of residual drug in the patients' organs or sera? Can drug resistant virus be generated in culture by the recombination of HIV with HTLV-1 (cells known to contain the HTLV-1 genome have been used for isolation of resistant virus)? Could the drug resistance seen with HIV also have a metabolic or cellular permeability basis in addition to a genetic basis? Can resistant variants increase the incidence of neurological problems in humans? Do resistant variants have altered pathological effects? Does the immune state of the patient influence the neurovirulence and pathogenicity of these resistant variants? We can speculate about and attempt to predict the answers to many of the questions just listed. However, it is important to answer them by a systematic practical approach and to develop methodologies to test the many hypotheses. In addition, the effectiveness of combinations of drugs against drug-resistant viruses needs to be addressed.

X. Combination Chemotherapy and Combined Modalities for Viruses: Results and Proposed Mechanisms of Interaction

Over the last decade, our group has been interested in combination chemotherapy for the treatment of viral infections. The summaries provided in this section are representative studies performed by our group using various antiviral agents, immunomodulators and other compounds which can produce synergy without increased toxicity. A rationale for these drug combinations is provided.

A. AZT–PHOSPHONOFORMATE (ERIKSSON AND SCHINAZI, 1989)

A unique property of PFA, an analog of pyrophosphate, is that it can inhibit both HIV and CMV *in vitro*. PFA is being investigated as a possible candidate for the treatment of HIV-1 and CMV infected patients. Since concomitant treatment of PFA is currently being considered in patients with CMV retinitis undergoing AZT therapy, it was important to determine the type of interaction produced by these drugs.

AZT and PFA produces a moderate synergistic inhibitory effect against HIV-1 *in vitro* at concentrations that are easily achieved in humans. The antiviral synergistic effect is more pronounced with increasing concentrations and is not due to suppression of virus due to toxicity. AZT neither inhibits the replication of human CMV in human embryonic lung fibroblasts nor interferes with the anti-CMV effect of PFA. By using partially purified reverse transcriptase of HIV and human CMV DNA polymerase, various combinations of AZT-5'-triphosphate and phosphonoformate produce strong indications of additive interactions. The synergistic interactions in infected cells and the additive effects observed at the reverse transcriptase level indicate that mechanisms other than the reverse transcriptase may be of importance for the inhibition of HIV replication by these two compounds. These basic studies suggest that concomitant treatment of CMV infections, such as CMV retinitis, with PFA in patients with AIDS receiving AZT may be appropriate, and this combination may also be useful in controlling HIV infection. Such studies are now ongoing in humans at the University of California, San Francisco.

B. AZT–URIDINE (SOMMADOSSI *et al.*, 1988)

The effects of natural purine and pyrimidine nucleosides on protection from or reversal of AZT cytotoxicity in human bone marrow progenitor cells was determined by using clonogenic assays. The selectivity of the "protection" or "rescue" agents was examined in evaluating the antiretroviral activity of AZT in combination with these modulating agents and of AZT alone. Following exposure of human granulocyte–macrophage progenitor cells for 2 hr to 5 μM AZT (70% inhibitory concentration), increasing concentrations of potential rescue agents were added. Cells were cultured, and colony formation was assessed after 14 days. At concentrations of up to 50 μM, no natural 2'-deoxynucleosides, including thymidine, were able to reverse the toxic effects of AZT. Dose-dependent reversal was observed with uridine and cytidine, and essentially complete reversal was achieved with 50 μM uridine. In the protection studies, 100 μM thymidine almost completely antagonized the inhibition of granulocyte–macrophage colony formation produced by 1 μM AZT (50% inhibitory concentration), and 50 μM uridine effected 60% protection against a toxic concentration of AZT (5 μM). The antiretroviral activity of AZT in human peripheral blood mononuclear (PBM) cells, assessed by reverse transcriptase assays, was substantially decreased in the presence of thymidine, whereas no impairment of suppression of viral replication was observed in the presence of uridine in combination with AZT at a molar ratio (uridine/AZT) as high as 10,000. This demonstration of the capacity of uridine to selectively rescue human bone marrow progenitor cells from the cytotoxicity of AZT suggests that use of a uridine rescue regimen with AZT may have potential therapeutic benefit in the treatment of AIDS. The mechanism for the reduced toxicity of AZT in the presence of uridine is currently under investigation. Since uridine has a short serum half-life, researchers have proposed to use uridine phosphorylase inhibitors in combination with AZT (J.-P. Sommadossi, personal communication). This alternative approach may prove to be clinically more practical.

C. ANTIVIRAL AGENT–ANTI-INFLAMMATORY AGENT

A new combined approach that we have pioneered (Schinazi *et al.*, 1982b) is the use of antiviral drugs with antiinflammatory agents for the treatment of experimental herpes encephalitis and cutaneous herpes. HSV causes acute inflammatory diseases of the mouth, genitals, skin, eyes, and central nervous system. With the new antivirals it is possible to effectively reduce the viral load; however, the inflammatory response that accompanies the infection persists after virus can no longer be detected in the lesion. Several reports of clinical trials of new topical anti-HSV drugs used for the treatment of genital and orolabial herpes have indicated that, whereas virus production may be reduced, the duration of lesions and pain is not usually affected (Corey *et al.*, 1982; Spruance *et al.*, 1984).

In theory, the antiviral drug should speed up the clearance of virus, while the anti-inflammatory agent should reduce the inflammation, thus decreasing the duration of clinical manifestations.

We have shown that, whereas topical ACV was effective in the guinea pig HSV genital model, it did not alter the clinical manifestations in the guinea pig HSV dorsal model; these results with topical ACV were analogous to the findings in humans undergoing therapy with topical ACV for the treatment of orolabial herpes (Corey *et al.*, 1982; Spruance *et al.*, 1984). Preliminary data indicate that the combinations of ACV and indomethacin, or phosphonoformate and hydrocortisone acetate (HCA), in the genital and dorsal models, respectively, appeared to be more effective than either agent alone. When used alone, the anti-inflammatory agents did not appear to increase the severity of disease, as has been reported with ocular herpes in humans and animal models (Kaufman *et al.*, 1963; Kolomiets *et al.*, 1984; Lagoutte, 1984). Our group has also shown that in mice infected intracerebrally with HSV, araA and indomethacin or HCA were more effective than single therapy with an antiviral drug (Schinazi *et al.*, 1982b).

D. AZT–ISOPRINOSINE (SCHINAZI *et al.*, 1988)

Since clinical trials are being planned with the immunomodulating drug isoprinosine combined with AZT in AIDS and ARC patients (De Simone *et al.*, 1988), it was important to determine the type of antiviral interaction produced by these drugs *in vitro*. Such a combined modality has the potential not only to produce enhanced antiviral effects, but also to have a valuable immunorestorative action. The interaction of several ratios of AZT and isoprinosine on the replication of HIV-1 in human PBM cells was determined by reverse transcriptase assay of disrupted virus obtained from supernatants of cells that were exposed to virus and the drugs separately and in combination, and by an HIV-1 p24 enzyme immunoassay (EIA) of the same supernatants. The correlation between the reverse transcriptase and EIA data was high. The antiviral activity of AZT alone was neither diminished nor augmented in combination with isoprinosine. Isoprinosine did not enhance virus yield when used alone or in combination with AZT in PBM cells, nor did it affect the growth of uninfected cells. The *in vitro* results indicate that this combination did not decrease the efficacy of AZT or exacerbate virus replication. However, it should be kept in mind that *in vitro* studies do not provide a test for immunomodulatory effects that could be observed in humans.

E. AZT–DEHYDROEPIANDROSTERONE (SCHINAZI *et al.*, 1990a)

Several beneficial biological attributes of dehydroepiandrosterone (DHEA) have been claimed. For example, it has been reported to have immunoregulatory, antidiabetic, anticancer, and antistress activity (Regelson *et al.*, 1988). DHEA has also been reported to prevent obesity without suppressing food consumption, to prolong life, and to depress the mutagenic effects of carcinogens. This compound is a potent noncompetitive inhibitor of mammalian glucose-6-phosphate dehydrogenase, a rate-limiting enzyme which leads to the production of NADPH for biosynthetic reductions in the cytosol and the production of ribose-5-phosphate as a precursor for nucleotide synthesis. DHEA also has immune up-regulatory effects in experimental animals. It was important to determine the effect of its interaction with AZT *in vitro*. First, the ability of DHEA and AZT to inhibit the replication of HIV was examined in human PBM cells. Using reverse transcriptase activity associated with virus to assess the antiviral activity, the median effective concentrations for DHEA and AZT were 17 μM and 0.0014 μM, respectively. Results obtained by an EIA were similar. Multiple-drug-effect analyses were used to quantitatively determine the interaction of AZT and DHEA in human PBM cells infected with HIV-1 at a ratio of 1:1000 and 1:4000. Analyses of the cell culture reverse transcriptase and EIA data indicated mostly antagonistic interactions. The combination index varied from 0.76 to 2.05 at the 90% inhibitory effect level. At therapeutic levels, no apparent toxicity to uninfected cells was observed for the drugs alone or in combination. The mechanism for the antiviral activity of DHEA and its interaction with AZT in infected human lymphocytes are unclear and may be related to regulation of *de novo* synthesis of nucleotides. Metabolic studies with this combination should provide information for the basis of the antagonism observed. These virological studies are especially timely since clinical trials with DHEA in individuals with AIDS are ongoing at the University of California, San Francisco, and the Community Research Initiative, New York.

F. AZT–DIPYRIDAMOLE (SZEBENI *et al.*, 1989, 1990; WEINSTEIN *et al.*, 1990)

Dipyridamole (DPM) is a potent inhibitor of nucleoside transport. It is commonly used as a coronary vasodilator and inhibitor of platelet function in the treatment of cardiovascular diseases and is being studied for combination chemotherapy of cancer. Recent studies indicate that DPM potentiates the effects of AZT against HIV-1 in cultured human monocytes and phytohemagglutinin-simulated T-lymphocytes, including PBM cells (Szebeni *et al.*, 1989, 1990). HPLC studies of nucleoside uptake suggest differential inhibition of the transport and phosphorylation of physiological nucleosides as mechanisms for the potentiation. At the same concentrations, DPM does not potentiate the toxic effects of AZT on these cells or on human bone marrow (granulocyte/monocyte) progenitors. Since the monocyte and T-cell lineages are important as hosts for HIV *in vivo*, these findings suggest the possibility of using DPM or its analogs in combination chemotherapy of HIV infection. However, DPM is complex in its activities, and *in vitro* studies such as these cannot be used to predict clinical efficacy or safety.

G. NUCLEOSIDE DIMERS (SCHINAZI *et al.*, 1990a)

Recently, Busso *et al.* (1988) and Hahn *et al.* (1989) showed that several nucleotide dimers of potentially useful antiretroviral agents inhibited HIV-1-induced cytopathic effects, reverse transcriptase production, and the expression of HIV-1 p24 antigens in the absence of toxic effects. It is not clear whether these novel dimers are functioning as dimers or monomers, nor how how the dimers dissociate intracellularly. Theoretically, if the molecule dissociates by cleavage of the phosphate bond linking the two known antiviral monomers, one nucleoside and one 5'-monophosphate form of the second nucleoside are released. Presumably the levels of cellular kinases and phosphorylases, as well as pharmacodynamic parameters, dictate the metabolic and anabolic fate of the components and their antiviral effectiveness as single or combined agents. The relative antiviral potency of AZT, ddI, and the dimer 3'-azido-3'-deoxythymidilyl-(5',5')-2',3'-dideoxy-5'-inosinic acid (AZT-P-ddI) was determined by our group in primary human PBM cells infected with HIV. The compounds listed in their order of potency were AZT-P-ddI > AZT > ddI. AZT-P-ddI exhibited the lowest toxicity in peripheral blood mononuclear cells, Vero cells, or CEM cells. Combination studies between AZT and ddI at nontoxic concentrations indicated a synergistic interaction at a drug ratio of 1 : 100 at all levels of antiviral effect. At higher ratios (1 : 500 and 1 : 1000), the interactions were synergistic only at concentrations that produced up to 75% virus inhibition. At higher levels of antiviral effects, this combination was additive or antagonistic as determined by the multiple drug effect analysis. AZT-P-ddI was about 10-fold less toxic than AZT to human granulocyte–macrophage and erythroid precursor cells. The greater antiviral activity and lower toxicity of this compound could not be attributed to the extracellular decomposition of the dimer in media at physiological temperature and pH. However, in acidic solutions, AZT-P-ddI decomposed in a pH-dependent manner. Of particular importance for this dimer is the finding that some AZT-resistant variants appear to be sensitive to ddI (Larder *et al.*, 1989), reducing the possibility for these resistant variants to develop in the presence of the dimer. These results suggest that advanced preclinical studies should be considered with this heterodimer of two clinically effective antiretroviral agents. They also demonstrate that studies of synergism can be used to rationally design novel single agents formed by chemically linking the two synergistic drugs to form a larger molecule.

H. ACYCLOVIR AND OTHER ANTIVIRAL NUCLEOSIDE ANALOGS

ACV and araA are approved for the treatment of herpes encephalitis. It is clear that, no matter which of the two drugs is used in humans, there is still significant mortality and morbidity (Whitley, 1988). Therefore, there is a need for improved treatment modalities and/or more effective drugs. Combination chemotherapy of araA and ACV, as we had suggested from our experiments in cell culture and in mice (Schinazi *et al.*, 1983a,b, 1986a–c; Schinazi and Nahmias, 1982), is currently under consideration and the preclinical studies performed by our group have been helpful for formulating improved protocols. We have provided further evidence substantiating the potential use of the combination of these drugs for life-threatening herpetic infections. Treatment of mice infected intracerebrally with HSV could be delayed by 3 days and still produce a synergistic effect over a wide range of doses of ACV and araA (Schinazi *et al.*, 1986a). This more stringent test for drug efficacy was selected, since it is more akin to the situation in humans. AraA and ACV were also evaluated in combination with 1-(2-deoxy-2-fluoro-β-D-arabinosyl)-5-iodocytosine (FIAC) and its 5-methyluracil analog (FMAU). Mice were inoculated intracerebrally with a lethal dose of HSV and then treated 3 days later intraperitoneally twice a day for 4 days with the drugs alone or in combination. Despite delayed treatment, most of the animals receiving FMAU alone or in combination with ACV or araA survived. *In vivo* synergy, as determined by multiple-drug-effect analysis, was noted for the combinations with no evidence of increased toxicity compared with the single treatments (Schinazi *et al.*, 1986a).

Despite these extensive preclinical studies there has been a reluctance to move into the clinic with these combinations, primarily because of toxicity concerns and because a number of alternative drugs are available. What should we study next in humans with HSV encephalitis? ACV in combination with araA? FMAU in combination with ACV or araA? The regimen of FMAU with araA may be more acceptable than using FMAU alone since in addition to nonoverlapping toxicities, araA is a TK independent compound and resistant viruses are less likely to occur with this combination. What about DHPG? In our encephalitis animal model, DHPG is clearly superior to araA or ACV, but is 50- to 100-fold less potent than FMAU when treatment is initiated early. However, if treatment is delayed by more than 24 hr, FMAU still appears to be more effective than DHPG. Because of the poor toxicological profile of DHPG in animals and humans, the widespread use of this drug may be limited to patients with severe herpetic infection (especially those with CMV infections). We have also performed studies of combinations of DHPG plus araA or ACV in mice infected intracerebrally with HSV. Additive to synergistic effects were noted, depending on the dose of drug used. No antiviral antagonism or synergistic toxicity was apparent.

I. COMBINATION OF INTERFERON AND OTHER ANTIVIRAL DRUGS

Since previous studies of the effect of interferons in combination with antivirals on HSV were conducted using relatively impure preparations (Levin and Leary, 1981; Stanwick *et al.*, 1981), the availability of cloned human interferon-alpha$_2$ (Hu IFNα_2) has permitted re-examination of its interaction with two licensed antivirals, ACV and araA, as well as two new selective nucleoside analogs discussed earlier FIAC and FMAU.

Vero cells were pretreated with increasing concentrations of IFN$_{\alpha2}$ for 20 hr prior to infection with HSV-1 (strain F) or HSV-2 (strain G). The cells were then treated with constant concentrations of the nucleosides. Additive or synergistic interactions were noted in such plaque reduction assays and no antagonism was observed (Schinazi *et al.*, 1982a). Studies with plaque-purified HSV variants resistant to high concentrations of araA, ACV, FIAC, or FMAU revealed no evidence of cross resistance to IFN$_{\alpha2}$. Combinations of high concentrations of drug and IFN$_{\alpha2}$ resulted in additive interactions with the resistant variants. These results suggest that IFN$_{\alpha2}$ could prove useful when combined with potent nontoxic antiviral drugs to prevent or reduce the emergence of resistant variants. (Synergy between IFN$_{\alpha2}$ and AZT against HIV is discussed in Chapter 20 by Berman and Chang.)

Pharmacokinetic studies indicated that no detectable interferon was present in the brain of mice 30 min after intramuscular treatment. In light of these findings, we do not believe that combinations of antivirals with interferon for the treatment of herpes encephalitis in humans can offer significant advantages over other combined modalities (Prange and Henze, 1988). However, for non-CNS systemic infections, such combinations may be extremely interesting. Fraser-Smith *et al.* (1984a,c) evaluated the combination of the acyclic nucleoside DHPG with murine IFN$_\beta$ in mice infected intraperitoneally with HSV-2 and showed a remarkable synergy (IFN$_\beta$ is essentially inactive when used alone).

More recently, we have evaluated the interaction of AZT or its uracil analog, AzddU (CS-87), in combination with IFN$_\beta$ in HIV-1 infected primary human lymphocytes. Of interest was the finding of synergy between AZT and this IFN, whereas with AzddU the interaction was mostly antagonistic over a wide range of drug ratios. The difference between these two related nucleosides is not known but suggests that the two drugs might have a different mode of action. Pharmacokinetic studies in humans with AZT and IFN$_\alpha$ indicated that the AUC (area under the curve) and half-life for AZT seemed to be unaffected in the presence of IFN (Kovacs *et al.*, 1989).

J. ACV-RESISTANT HSV VARIANTS AND DRUG COMBINATIONS

The purpose of these experiments was to determine if drug resistance in mice can be overcome by using high doses of antiviral agents alone or in combination. Various ACV-resistant HSV-2 TK$^-$ variants were prepared in our laboratory by single passage in the presence of 10 μM ACV. We have concentrated on studying the resistance of these HSV-2 clones, since the parent strain (G) is about 10,000 times more virulent than the KOS strain of HSV-1 and because earlier studies have focused on HSV-1 variants. An example of such a clone (G-ACV-C1) was found to be 40-fold more resistant to ACV than the parental clone, to be still susceptible to araA in plaque assays, and to be TK$^-$. G-ACV-C1 was found to be only 10-fold less virulent than the parental clone when inoculated intracerebrally in 6-week-old female ICR mice. When groups of 10 mice were inoculated intracerebrally with 40 plaque forming units (pfu) of the parental clone and treated 5 hr after infection with araA or ACV (100 mg/kg/day; bid × 4d), the mortality was significantly reduced. However, mice inoculated with 4000 pfu of G-ACV-C1 were essentially refractory to ACV treatment, but responded to araA or a combination of araA and ACV (Schinazi, 1987).

Even at doses of 300 mg/kg per day (bid × 4d), the mice were totally refractory to ACV treatment, but responded to araA (100–300 mg/kg) or a combination of araA/ACV (100/100 or 300/300 mg/kg). These results clearly demonstrate that in this model, high doses of ACV could not overcome the pathogenic effects of this variant and that a combination of ACV/araA is capable of preventing chronic disease and death.

For life-threatening HSV infections for which ACV may be shown to be clinically effective, a combination of ACV and another effective drug with a different mechanism should be considered rather than alternative single drug therapy (Schinazi *et al.*, 1986b). Although the combination of araA and ACV should be useful for patients with herpes encephalitis, araA is not effective in cutaneous HSV infections in humans. Therefore, it is unlikely that this combination can be used to treat cutaneous infections.

K. COMBINATION AND RESISTANCE STUDIES WITH ISOLATES FROM PATIENTS WHO RECEIVED TOPICAL AND SYSTEMIC ANTIVIRAL CHEMOTHERAPY

Studies on the mechanism which confers drug-resistance to HSV and investigations of the pathogenic potential of various resistant variants are of value in assessing the clinical importance of drug resistance. However, they only provide a background for careful evaluation on human isolates. We have been systematically interested in studying clinical isolates obtained from patients that do not respond to antiviral chemotherapy. Such studies are a good starting point to determine the impact of antiviral chemotherapy on the development of resistance. Our laboratory obtained clinical isolates from a patient with an acquired immunodeficiency who had received topical and systemic ACV (Schinazi *et al.*, 1986b). Briefly, studies with pre- and posttherapy isolates obtained from different body sites indicated that (1) TK$^-$ variants may be produced in humans after topical ACV therapy; (2) different ACV-

ACV-resistant or sensitive HSV-2 variants can establish latency at different body sites and be later reactivated; (3) TK⁻ HSV-2 variants can be virulent in adult mice. These results further suggest that, when drug-resistant viruses are isolated from patients with severe herpetic conditions who have multiple reactivations, combination therapy rather than alternating therapy should be considered. This is because the patients may also be shedding drug-sensitive virus at a different body site. These studies have implications not only for the treatment of HSV infections, but also other viruses including HIV.

XI. Prospects

Development of effective nontoxic combinations of selective antiviral agents for the treatment of various viral infections is an important goal in current antiviral research. The accelerated pace of development of drug combinations requires the establishment of the effective doses for the individual drugs. It should be emphasized that individualization of drug dose selection in humans will be necessary for combinations of antiviral agents. Different stages of infection may require higher doses of chemotherapy. At the very least, a drug combination should widen the therapeutic window of the single agent. Unfortunately, at the present time, the determination of the minimum effective dose cannot be accurately determined for a specific drug.

The development of drug-resistant viruses is not going to be limited to antiviral nucleosides. There are already reports of soluble CD4-resistant HIV that were selected *in vitro*, raising the possibility that these variants can be produced in humans. The paradox between the selectivity of the drug and the development of drug-resistant virus was mentioned earlier (Herrmann and Herrmann, 1977). Studies on the molecular genetics of viruses should suggest viral targets which lie in viral sequences that are essential for viral replication. Since mutations in these regions are less likely to occur, the selection of resistant viruses with that genotype should be rare. Thus, based on our understanding of the genetic mutation sequences of viruses it may be possible to synthesize antiviral agents that cannot produce drug resistant variants under selective pressures.

For an optimum combination to be developed, clinical studies will have to be performed using a great number of patients which could produce a major information problem. For example, with 70 available drugs, 2415 dual combinations are possible, irrespective of dose and dose regimen. The problem is compounded with triple combinations.[2] Therefore, general principles should

---

[2] For different triple combinations, the number would be 54,740. These values are calculated from the equation $n!/[k! \times (n-k)!]$, where n is the number of compounds, and k is the number of compounds in combination.

be developed for when a clinical combination study should be performed. Whenever possible, data should be available on drug interactions in cell-free systems, cell culture, and animal models. In addition, pertinent pharmacological and clinical information on the single drugs, such as their minimum effective doses and optimum schedules, could reduce the number of possible variables. Currently many basic principles are not being applied to combinations of anti-AIDS drugs. Combinations are being proposed without a comprehensive and integrated preclinical development program.

For severe life-threatening infections, when early treatment is essential, the use of combinations of two antiviral drugs should be advocated because the development of pathogenic or virulent drug-resistant viruses is possible. We should reserve compounds with different antiviral mechanisms in case resistant virus develops. With the development of more selective, potent, and nontoxic antiviral agents, we should reserve combined treatments for severe viral infections. Combination of two antiviral drugs should not be used for mild viral infections which are not life-threatening unless there is a specific rationale for using the compounds together. It is essential to develop rapid and accurate tests for determining if viral isolates are resistant to a particular drug and to define their pattern of cross resistance. The availability of such tests would allow physicians to make judicious choices when more than one antiviral agent is available. Combinations of drugs with compounds that enhance immune function or increase the therapeutic efficacy of the antiviral drugs have to be used cautiously. Enhancement of immune function could also reactivate latent virus causing severe infections. One also should consider how long to treat these patients with these combinations. The long-term toxicity of drug combinations may be difficult to assess. Modifying the ratio of the components in the combination and the schedule may reduce these side-effects. It is clear that an understanding of the pharmacokinetics of the components will dictate the efficacy and safety of these combinations.

Another consideration deals with the development of combined modalities by the pharmaceutical industry. Without the support of drug companies that manufacture antiviral drugs, it is unlikely that potentially useful combinations will be developed. Some pharmaceutical companies are reluctant to encourage the development of combinations of antivirals when they are not both manufactured by the same company. For example, combinations of two different compounds are being considered for use in humans, but may not be supported by the two different companies that produce them. With regard to Food and Drug Administration approval, combinations of drug are considered new entities; hence, preclinical studies as well as safety and efficacy studies are requested prior to New Drug Application. Until these studies are performed, physicians should be discouraged from using two or more approved drugs in combination.

Finally, only through well-implemented controlled trials that demonstrate clearly superior efficacy and safety of a combination over the single regimens can we make progress in this rapidly expanding area of research.

Acknowledgments

I thank my collaborators for allowing me to share unpublished data with the reader. My apologies if additional combination studies relevant to this review have escaped attention. I thank Lisa Friend, Anne Le Moine, David Feingold, and Eugenia Abbey for their assistance in preparing this manuscript. This review was completed in January 1990, and was supported by Public Health Service grants AI 26055, AI 25899, AI 05078, and AI 27196 from the National Institutes of Health, and the Department of Veterans Affairs.

References

Aduma, P. J., Gupta S. V., and De Clercq, E. (1990). Aniherpes virus activity and effect on deoxyribonucleoside triphosphate pools of (E)-5-(2-bromovinyl)-2'-deoxycytidine in combination with deaminase inhibitors. *Antiviral Res.* **13**, 111–126.

Ahmad, A. L. M., and Tyrell, D. A. J. (1986). Synergism between anti-rhinovirus antivirals: Various human interferons and a number of synthetic compounds. *Antiviral Res.* **6**, 241–252.

Allen, L. B., Vanderslice, L. K., Fingal, C. M., McCright, F. H., Harris, E. F., and Cook, P. D. (1982). Evaluation of anti-herpes virus drug combination: Virazole plus arabinofuranosylhypoxanthine and virazole plus arabinofuranosyladenine. *Antiviral Res.* **2**, 203–216.

Altucci, P. G., Varone, L., and Magrassi, F. (1969). Comparison of the effectiveness of two antiviral drugs and their combination. *Chemotherapy (Basel)* **14**, 140–150.

Andersson, J., Sköldenberg, B., Henle, W., Giesecke, J., Ötqvist, A., Julander, I., Gustavsson, E., Akerlund, B., Britton, S., and Ernberg, I. (1987). Acyclovir treatment in infectious mononucleosis: A clinical and virological study. *Infection (Munich)* (Suppl. 1), **15**, S14–S20.

Ayisi, N. K., Gupta, S. V., Meldrum, J. B., Taneja, A. K., and Babiuk, L. A. (1980). Combination chemotherapy: Interaction of 5-methoxymethyldeoxyuridine with adenine arabinoside, 5-ethyldeoxyuridine, 5-iododeoxyuridine, and phosphonoacetic acid against herpes simplex virus types 1 and 2. *Antimicrob. Agents Chemother.* **17**, 558–566.

Ayisi, N. K., Meldrum, J. B., Stuart, A. L., and Gupta, S. V. (1983). Comparison of the antiviral effects of 5-methoxymethyldeoxyuridine-5'-monophosphate with adenine arabinoside-5'-monophosphate. *Antiviral Res.* **3**, 161–174.

Ayisi, N. K., Gupta, S. V., and Babiuk, L. A. (1985). Combination chemotherapy: Interaction of 5-methoxymethyldeoxyuridine with trifluorothymidine, phosphonoformate and acycloguanosine against herpes simplex viruses. *Antiviral Res.* **5**, 13–27.

Ayisi, N. K., Gupta, S. V., and Babiuk, L. A. (1986). Efficacy of 5-methoxymethyl-2'-deoxyuridine in combination with arabinosyladenine for the treatment of primary herpes simplex genital infection of mice and guinea pigs. *Antiviral Res.* **6**, 33–47.

Baba, M., and Shigeta, S. (1987). Antiviral activity of glycyrrhizin against varicella–zoster virus in vitro. *Antiviral Res.* **7**, 99–107.

Baba, M., Ito, M, Shigeta, S., and De Clercq, E. (1984). Synergistic antiviral effect of antiherpes compounds and interferon on varicella–zoster in vitro. *Antimicrob. Agents Chemother.* **25**, 515–517.

Baba, M., Pauwels, R., Balzarini, J., Herdewijn, P., De Clercq, E., and Desmyter, J. (1987). Ribavirin antagonizes inhibitory effects of pyrimidine 2',3'-dideoxynucleosides but enhances inhibitory effects of purine 2',3'-dideoxynucleosides on replication of human immunodeficiency virus in vitro. *Antimicrob. Agents Chemother.* **31**, 1613–1617.

Babiuk, L. A., Meldrum, B., Gupta, S. V., and Rouse, B. T. (1975). Comparison of the antiviral effects of 5-methoxymethyl-deoxyuridine with 5-iodo-deoxyuridine, cytosine arabinoside, and adenine arabinoside. *Antimicrob. Agents Chemother.* **8**, 643–650.

Balzarini, J., and De Clercq, E. (1989). The antiviral activity of 9-β-D-arabinofuranosyladenine is enhanced by the 2',3'-dideoxyriboside, the 2',3'-didehydro-2',3'-dideoxyriboside and the 3'-azido-2',3'-dideoxyriboside of 2,6-diaminopurine. *Biochem. Biophys. Res. Commun.* **159,** 61–67.

Balzarini, J., Cooney, D. A., Dalal, M., Kang, G.-J., Cupp, J.E., De Clercq, E., Broder, S., and Johns, D. G. (1987). 2',3'-Dideoxycytidine: Regulation of its metabolism and anti-retroviral potency by natural pyrimidine nucleosides and by inhibitors of pyrimidine nucleotide synthesis. *Mol. Pharmacol.* **32,** 798–806.

Balzarini, J., Herdewijn, P., and De Clercq, E. (1989). Potentiating effect of ribavirin on the anti-retrovirus activity of 3'-azido-2,6-diaminopurine-2',3'-dideoxyriboside *in vitro* and in *vivo. Antiviral Res.* **11,** 161–171.

Bastow, K. F., Derse, D. D., and Cheng, Y.-C. (1983). Susceptibility of phosphonoformic acid-resistant herpes simplex virus variants to arabinosylnucleosides and aphidicolin. *Antimicrob. Agents Chemother.* **23,** 914–917.

Bauer, D. J. (1955). The antiviral and synergistic actions of isatin thiosemicarbazone and certain phenoxypyrimidines in vaccinia infection in mice. *Br. J. Exp. Pathol.* **36,** 105–114.

Berger, S. A., Weinberg, M., Treves, T., Sorkin, P., Geller, E., Yedwab, G., Tomer, A., Rabe, M., and Michaeli, D. (1986). Herpes encephalitis during pregnancy: Failure of acyclovir and adenine arabinoside to prevent neonatal herpes. *Isr. J. Med. Sci.* **22,** 41–44.

Berman, E., Duigou-Osterndorf, R., Krown, S. E., Fanucchi, M. P., Chou, J., Hirsch, M. S., Clarkson, B. D., and Chou, T.-C. (1989). Synergistic cytotoxic effect of azidothymidine and recombinant interferon alpha on normal human bone marrow progenitor cells. *Blood* **74,** 1281–1286.

Besser, R., Krämer, G., Rambow, A., and Hopf, H.C. (1987). Combined therapy with acyclovir and adenosine arabinoside in herpes simplex encephalitis. *Eur. Neurol* **27,** 197–200.

Bikbulatov, R. M., and Grebenyuk, V. N. (1977). Antiviral activity of experimental combination of 5-iodo-2'-deoxyuridine with 6-azauridine and its effectiveness in the treatment of herpetic infection with lesions of the skin and mucous membranes. *Vopr. Virusol.* **2,** 228–232.

Birch, C., Tachedjian, G., Lucas, C. R., and Gust, I. (1988). *In vitro* effectiveness of a combination of zidovudine and ansamycin against human immunodeficiency virus. *J. Infect. Dis.* **158,** 895.

Biron, K., K., and Elion, G. B. (1981). Effect of acyclovir combined with other antiherpetic agents on varicella zoster virus. *Am. J. Med.* **73,** 54–57.

Bissett, J., Eisenberg, M., Gregory, P., Robinson, W.S., and Merigan, T. C. (1988). Recombinant fibroblast interferon and immune interferon for treating chronic hepatitis B virus infection: Patients' tolerance and the effect on viral markers. *J. Infect. Dis.* **157,** 1076–1080.

Brockmeyer, N. H., Kreuzfelder, E., Mertins, L., Daecke, C., Goos, M., and Essen, U. (1989). Zidovudine therapy of asymptomatic HIV-1 infected patients and combined zidovudine-acyclovir therapy of HIV-1 infected patients with oral hairy leukoplakia. *J. Invest. Dermatol.* **92,** 647.

Bryson, Y., and Connor, J. D. (1976). *In vitro* susceptibility of varicella zoster virus to adenine arabinoside and hypoxanthine arabinoside. *Antimicrob. Agents Chemother.* **9,** 540–543.

Bryson, Y. J., and Kronenberg, L. H. (1977). Combined antiviral effects of interferon, adenine arabinoside, hypoxanthine arabinoside, and adenine arabinoside-5'-monophosphate in human fibroblast cultures. *Antimicrob. Agents Chemother.* **11,** 299–306.

Bryson, Y., Hebblewaite, D., and De Clercq, E. (1981). The *in vitro* sensitivity of clinical isolates of varicella zoster to BVDU alone and in combination with acyclovir and adenine arabinoside. *Proc. Int. Congr. Chemother. 12th* abstr. 131.

Buck, C., and Loh, P. C. (1985). *In Vitro* effect of acyclovir and other antiherpetic compounds on the replication of channel catfish virus. *Antiviral Res.* **5,** 269–280.

Bunow, B., and Weinstein, J. N. (1990). Combo: A new approach to the analysis of drug combinations *in vitro. Ann. N.Y. Acad. Sci.* (in press).

Burkhardt, U, and Wigand, R. (1983). Combined chemotherapy of cutaneous herpes simplex infection of the guinea pig. *J. Med. Virol.* **12**, 137–147.

Burlington, D. B., Meiklejohn, G., and Mostow, S. R. (1981). Synergistic anti-influenza A activity of amantadine and ribavirin in ferret tracheal ciliated epithelium. *Proc. intersci. Conf. Antimicrob. Agents Chemother. 21st* abstr. 875.

Burlington, D. B., Meiklejohn, G., and Mostow, S. R. (1983). Antiinfluenza A virus activity of combinations of amantadine and rimantadine in ferret tracheal ciliated epithelium. *J. Antimicrob. Chemother.* **11**, 7–14.

Busso, M., Mian, A. M., Hahn, E. F., and Resnick, L. (1988). Nucleotide dimers suppress HIV expression *in vitro*. *AIDS Res. Hum. Retroviruses* **4**, 449–455.

Canonico, P. G., Kende, M., Luscri, B. J., and Huggins, J.W. (1984a). Chemotherapy of toga- and bunyaviruses with liposome-mediated targeting and combinations of antiviral agents. *Proc. Int. Congr. Virol., 6th* abstr. W34-9.

Canonico, P. G., Kende, M., Luscri, B.J., and Huggins, J. W. (1984b). *In-vivo* activity of antivirals against exotic RNA viral infections. *J. Antimicrob. Chemother., Suppl. A* **14**, 27–41.

Caselmann, W. H., Eisenburg, J., Hofschneider, P. H., and Koshy, R. (1989). Beta- and gamma-interferon in chronic active hepatitis B. A pilot trail of short-term combination therapy. *Gastroenterology* **96**, 449–55.

Cerruti, R. L., Connell, E. V., Trown, P. W., and Sim, I. S. (1985). Synergistic interaction between interferon-α and acyclovir in the treatment of herpes simplex virus type 1 infection in mice. *Antiviral Res.* (Suppl. 1), 217–223.

Cesario, T. C., and Slater, L. M. (1980). Diminished antiviral effect of human interferon in the presence of therapeutic concentrations of antineoplastic agents. *Infect. Immun.* **27**, 842–845.

Champney, K. J., Lauter, C. B., Bailey, E. J., and Lerner, A. M. (1978). Antiherpesvirus activity in human sera and urines after administration of adenine arabinoside. *J. Clin. Invest.* **62**, 1142–1153.

Chany, C., and Cerutti, I. (1977). Enhancement of antiviral protection against encephalomyocarditis virus by a combination of isoprinosine and interferon. *Arch. Virol.* **55**, 225–231.

Chiesa, F., Donghi, R., Pilotti, S., Sala, L., and Stefanon, B. (1989). Human fibroblast interferon adjuvant to CO_2 laser in the treatment of recurrent juvenile laryngeal papillomatosis: Experience with 7 cases. *Tumor* **75**, 259–262.

Cho, C. T., and Feng, K. K. (1977). Synergistic effects of antiviral agents and humoral antibodies in experimental herpesvirus hominis encephalitis. *Ann. N.Y. Acad. Sci.* **284**, 321–333.

Cho, C. T., and Feng, K. K. (1980). Combined effects of acycloguanosine and humoral antibodies in experimental encephalitis due to herpesvirus hominis. *J. Infect. Dis.* **142**, 451.

Cho, C. T., Feng, K. K., and Brahmacupta, N. (1976). Synergistic antiviral effects of adenine arabinoside and humoral antibodies in experimental encephalitis due to herpesvirus hominis. *J. Infect. Dis.* **133**, 157–167.

Chou, T.-C., and Talalay, P. (1984). Quantitative analysis of dose–effect relationships: The combined effects of multiple drugs or enzyme inhibitors. *Adv. Enzyme Regul.* **22**, 27–55.

Colin, J., Chastel, C., Renard, G., and Cantell, K. (1982). Herpes oculaire. Activité antivirale synergique de l'acyclovir et de l'interféron leucocytaire humain. *Nouv. Presse Méd.* **11**, 2783.

Collin, J., Chastel, C., Renard, G., and Cantell, K. (1983). Combination therapy for dendritic keratitis with human leukocyte interferon and acyclovir. *Am. J. Ophthalmol.* **95**, 326–348.

Colin, J., Chastel, C., and Volant, A. (1984). Traitement par trifluorothymidine des kératites herpétiques cortisonées: Étude expérimentale chez le lapin. *J. Fr. Ophthalmol.* **7**, 603–607.

Collier, A., Coombs, R., Bozette, S., Spector, S., Pettinelli, C., Richman, D., Causey, D., Leedom, J., Davies, G., Gianola, F., and Corey, L. (1989). Virologic and clinical response to combination zidovudine and acyclovir in AIDS-related complex. *Proc. Intersci. Conf. Antimicrob. Agents Chemother., 29th*, Abstract 30, p. 105.

Collum, L. M. T., Logan, P., and Grant, D. M. (1987). A double-blind comparative trial of acyclovir and adenine arabinoside in combination with dilute betamethasone in the management of herpetic disciform keratitis. *Curr. Eye Res.* **6,** 221–224.

Collum, L. M. T., Logan, P., and Ravenscroft, T. (1983a). Acyclovir (Zovirax) in herpetic disciform keratitis. *Br. J. Ophthalmol.* **67,** 115–118.

Collum, L. M. T., O'Connor, M., and Logan, P. (1983b). Comparison of the efficacy and toxicity of acyclovir and of adenine arabinoside when combined with dilute betamethasone in herpetic disciform keratitis: Preliminary results of a double-blind trial. *Trans. Ophthalmol. Soc. U.K.* **103,** 597–599.

Connell, E. V., Cerruti, R. L., and Trown, P. W. (1985). Synergistic activity of combinations of recombinant human alpha interferon and acyclovir, administered concomitantly and in sequence, against a lethal herpes simplex virus type 1 infection in mice. *Antimicrob. Agents Chemother.* **28,** 1–4.

Connor, J. D., Sweetman, L., Carey, S., Stuckey, M. A., and Buchanan, R. (1974). Effect of adenosine deaminase upon the antiviral activity *in vitro* of adenine arabinoside for vaccinia virus. *Antimicrob. Agents Chemother.* **6,** 630–636.

Corey, L., Nahmias, A. J., Guinan, M. E., Benedetti, J. K., Critchlow, C. W., and Holmes, K. K. (1982). Trial of topical acyclovir in genital herpes simplex virus infections. *N. Engl. J. Med.* **306,** 1313–1319.

Crane, L. R., and Milne, D. A. (1985). Comparative activities of combinations of acyclovir, vidarabine or its 5′-monophosphate, and cloned human interferons against herpes simplex virus type 2 in human and mouse fibroblast cultures. *Antiviral Res.* **5,** 325–333.

Crane, L. R., and Sunstrum, J. C. (1988). Enhanced efficacy of nucleoside analogs and recombinant α interferon in weanling mice lethally infected with herpes simplex virus type 2. *Antiviral Res.* **9,** 1–10.

Crane, L. R., Milne, D. A., Sunstrum, J. C., and Lerner, A. M. (1984). Comparative activities of selected combinations of acyclovir, vidarabine, arabinosyl hypoxanthine, interferon, and polyriboinosinic acid–polyribocytidylic acid complex against herpes simplex virus type 2 in tissue culture and intravaginally inoculated mice. *Antimicrob. Agents Chemother.* **26,** 557–562.

Crumpacker, C. (1983). Resistance of herpes simplex virus to antiviral agents: Is it clinically important? *Drugs* **26,** 373–377.

Crumpacker, C. S. (1989). Molecular targets of antiviral therapy. *N. Engl. J. Med.* **321,** 163–172.

Crumpacker, C. S., Kowalsky, P. N., Oliver, S. A., Schnipper, L. E., and Field, A. K. (1984). Resistance of herpes simplex virus to 9-{[2-hydroxy-1-(hydroxymethyl)ethoxy]methyl} guanine: Physical mapping of drug synergism within the viral DNA polymerase locus. *Proc. Natl. Acad. Sci. U.S.A,* **81,** 1556–1560.

Czarniecki, C. W., Fennie, C. W., Powers, D. B., and Estell, D. A. (1984). Synergistic antiviral and antiproliferative activities of *Escherichia coli*-derived human α, β, and γ interferons. *J. Virol.* **49,** 490–496.

Darby, G., Larder, B. A., and Inglis, M. M. (1986). Evidence that the "active center" of the herpes simplex virus thymidine kinase involves an interaction between three distinct regions of the polypeptide. *J. Gen. Virol.* **67,** 753–758, and references cited therein.

Davies. W. B., Oakes, J. E., Weppner, W. A., and Taylor, J. A. (1980). Contribution of immune lymphocytes and antibody in chemotherapy of herpes simplex virus infected mice. *In* "Immunology of the Eye" (A. Suran, I. Gery, and R. B. Nussenblatt, eds.), pp. 467–477. IRL Press, Washington, D.C.

De Clercq, E. (1986). Chemotherapeutic approaches to the treatment of the acquired immune deficiency syndrome (AIDS). *J. Med. Chem.* **29,** 1561–1569.

De Clercq, E. (1987). Perspectives for the chemotherapy of AIDS. *Anticancer Res.* **7,** 1023–1038.

De Clercq, E., Descamps, J., and De Somer, P. (1978). (*S*)-9-(2,3-dihydroxypropyl)adenine: An aliphatic nucleoside analog with broad-spectrum antiviral activity. *Science* **200,** 563–565.

De Koning, E. W. J.,van Bijsterveld, O. P., and Cantell, K (1981). Human leukocyte interferon and trifluorothymidine in the treatment of dendritic keratitis. *In* "The Biology of the Interferon System" (E. De Maeyer, G. Galasso, and H. Schellekens, eds.), pp. 351–354. Elsevier/North-Holland, New York.

De Koning, E. W. J., van Bijsterveld, O. P., and Cantell, K. (1982). Combination therapy for dendritic keratitis with human leucocyte interferon trifluorothymidine. *Br. J. Ophthalmol.* **66,** 509–512.

De Koning, E. W. J., van Bijsterveld, O. P., and Cantell, K. (1983). Combination therapy for dendritic keratitis with acyclovir and α-interferon. *Arch. Ophthalmol. (Chicago)* **101,** 1866–1868.

de Man, R. A., Schalm, S. W., van der Heijden, A. J., ten Kate, F.W., Wolff, E. D., and Heijtink, R.A. (1989). Improvement of hepatitis B-associated glomerulonephritis after antiviral combination therapy. *J. Hepatol.* **8,** 367–372.

De Miranda, P., Good, S. S., Yarchoan, R., Thomas, R. V., Blum, R., Myers, C. E., and Broder, S. (1989). Alteration of zidovudine pharmacokinetics by probenecid in patients with AIDS or AIDS-related complex. *Clin. Pharmacol. Ther.* **46,** 494–500.

De Simone, C, Ferrazzi, M., Bitonti, F., Falciano, M., Tzantzoglou, S., Delia, S., and Sorice, F. (1988). Pharmacokinetics of zidovudine and concomitant inosine-pranobex in AIDS patients. *Immunopharmacol. Immunotoxicol.* **10,** 437–441.

DeVita, V. T., Jr., Young, R. C., and Canellos, G. P. (1975). Combination versus single agent chemotherapy: A review of the basis for selection of drug treatment of cancer. *Cancer (Philadelphia)* **35,** 98–110.

Drach, J. C., and Shipman, C., Jr. (1977). The selective inhibition of viral DNA synthesis by chemotherapeutic agents: An indicator of clinical usefulness? *Ann. N.Y. Acad. Sci.* **284,** 396–406.

Edlin, B. R., Falk, A., Weinstein, R. A., and Bitran, J. (1989). Interferon-α plus zidovudine in HIV infection. *Lancet* **1,** 156.

Eggers, H. J. (1976). Successful treatment of enterovirus-infected mice by 2-(α-hydroxybenzyl)-benzimidazole and guanidine. *J. Exp. Med.* **143,** 1367–1381.

Eggers, H. J., and Tamm, I. (1963). Synergistic effects of 2-(α-hydroxybenzyl)-benzimidazole and guanidine on picornavirus reproduction. *Nature (London)* **199,** 513–514.

Ellis, M. N., Lobe, D.C., and Spector, T. (1987). Combination antiviral chemotherapy for treatment of HSV-1 infection in athymic nude mice. *Am. Soc. Microbiol.* abstr. 625.

Ellis, M. N., Lobe, D. C., and Spector, T. (1989). Synergistic therapy by acyclovir and A1110U for mice orofacially infected with herpes simplex viruses. *Antmicrob. Agents Chemother.* **33,** 1691–1696.

Fmanuel, D., Cunningham, I., Jules-Elysée, K., Brochstein, J. A., Kernan, N. A., Laver, J., Stover, D., White, D. A., Fels, A., Polsky, B., Castro-Malaspina, H., Peppard, J. R., Bartus, P., Hammerling, U., and O'Reilly, R. J. (1988). Cytomegalovirus pneumonia after bone marrow transplantation successfully treated with the combination of ganciclovir and high-dose intravenous immune globulin. *Ann. Inter. Med.* **109,** 777–782.

Eppstein, D. A., and Marsh, Y. V. (1984). Potent synergistic inhibition of herpes simplex virus-2 by 9-[(1,3-dihydroxy-2-propoxy)methyl]guanine in combination with recombinant interferons. *Biochem. Biophys. Res. Commun.* **120,** 66–73.

Eriksson, B. F. H., and Schinazi, R. F. (1989). Combinations of 3'-azido-3'-deoxythymidine (zidovudine) and phosphonoformate (forcarnet) against human immunodeficiency virus type 1 and cytomegalovirus *in vitro*. *Antimicrob. Agents Chemother.* **33,** 663–669.

Falcon, M. G., and Jones, B. R. (1977). Antiviral activity in the rabbit cornea of adenine arabinoside, Ara-A 5' monophosphate, and hypoxanthine arabinoside, and interactions with adenosine deaminase inhibitor. *J. Gen. Virol.* **36,** 199–202.

Fiala, M., Chow, A. W., Miyasaki, K., and Guze, L. (1974). Susceptibility of herpes viruses to three nucleoside analogues and their combinations and enhancement of the antiviral effect at acid pH. *J. Infect. Dis.* **129,** 82–85.

Field, H. J. (1983). The problem of drug-induced resistance in viruses. *In* "Problems of Antiviral Therapy" (C. Stuart-Harris and J. S. Oxford, eds.), pp. 71–107. Academic Press, New York.

Field, H. J. (1989). The impact of drug resistance upon virus chemotherapy. *J. Antimicrob. Chemother.* **24**, 4–7.

Fischer, G. W., Balk, M. W., Crumrine, M. H., and Bass, J. W. (1976). Immunopotentiation and antiviral chemotherapy in a suckling rat model of herpes virus encephalitis. *J. Infect. Dis.* **133**, A217–A220.

Fischer, P. H. (1981). Preferential inhibition of iododeoxyuridine metabolism by 5'-aminothymidine. *Proc. Intersci. Conf. Antimicrob. Agents Chemother. 21st* abstr. 328.

Fischer, P. H., and Murphy, D. G. (1983) Preferential inhibition of trifluorothymidine phosphorylation by 5'-amino-5'-deoxythymidine in uninfected versus herpes simplex virus infected cells. *Fed Proc., Fed. Am. Soc. Exp. Biol.* **42**, 457 (abstr. 970).

Fischer, P. H., and Phillips, A. W. (1984) Antagonism of feedback inhibition. Stimulation of phosphorylation of thymidine and 5-iodo-2'-deoxyuridine by 5-iodo-5'-amino-2',5'-dideoxyuridine. *Mol. Pharmacol.* **25**, 446–451.

Fischer, P. H., and Ratschan, W. J. (1984). Interactions between thymidine (dThd) analogs in herpes simplex virus (HSV) types 1 (F) and 2 (G and MS) infected HeLa cells. *Fed. Proc., Fed. Am. Soc. Exp. Biol.* **43**, 563.

Fischer, P. H., Lee, J. J., Chen, M. S., Lin, T.-S., and Prusoff, W. H. (1979). Synergistic effect of 5'-amino-5'-deoxythymidine and 5-iodo-2'-deoxyuridine against herpes simplex virus infections *in vitro. Biochem. Pharmacol.* **28**, 3483–3486.

Fisher, P. H., Lee, J. J., Chen, M. S., Lin, T.-S., and Prusoff, W. H. (1980). Therapeutic synergism between 5'-amino-5'-deoxythymidine and 5-iodo-2'-deoxyuridine in the treatment of herpes simplex virus infections *in vitro. In* "Current Chemotherapy and Immunotherapy" (P. Periti and G. G. Grassi, eds.), Vol. II, pp.1366–1368. Am. Soc. Microbiol. Washington, D.C.

Fischer, P. H., Murphy, D. G., and Kawahara, D. R. (1983). Preferential inhibition of 5-trifluoromethyl-2'-deoxyuridine phosphorylation by 5'-amino-5'-deoxythymidine in uninfected versus herpes simplex virus-infected cells. *Mol. Pharmacol.* **24**, 90–96.

Fitzwilliams, J. F., and Griffith, J. F. (1976). Experimental encephalitis caused by herpes simplex virus: Comparison of treatment with tilorone hydrochloride and phosphonoacetic acid. *J. Infect. Dis.* (Suppl. 133), A221–A225.

Fleischmann, W. R., and Fleischmann, C. M. (1984). Potentiating effect of murine interferon-γ-containing lymphokine preparations on the antiviral and antiproliferative effects of murine interferon-α/β: Identification on the potentiation factor as murine interferon-γ itself. *Antiviral Res.* **4**, 221–228.

Fleischmann, W. R., Jr., Georgiades, J. C., Osborne, L. C., and Johnson, H. M. (1979). Potentiation of interferon activity by mixed preparations of fibroblast and immune interferon. *Infect. Immun.* **26**, 248–253.

Frank, K. B., and Cheng, Y.-C (1985). Mutually exclusive inhibition of herpes virus DNA polymerase by aphidicolin, phosphonoformate, and acyclic nucleoside triphosphates. *Antimicrob. Agents Chemother.***27**, 445–448.

Fraser, T. R. (1872). The antagonism between the actions of active substances. *Br. Med. J.* pp. 457–459 and 485–487.

Fraser-Smith, E. B., Eppstein, D. A., Marsh, Y. V., and Matthews, T. R. (1984a). Comparison of synergistic combinations of 9-(1,3-dihydroxy-2-propoxymethyl)guanine and α, β, or γ interferon against herpes simplex virus type 2 both *in vivo* and *in vitro. Proc. Intersci. Conf. Antimicrob. Agents Chemother.*, 24th abstr. 1182.

Fraser-Smith, E. B., Eppstein, D. A., Marsh, Y. V., and Matthews, T. R. (1984b).Enhanced efficacy of the acyclic nucleoside 9-(1,3-dihydroxy-2-propoxymethyl)guanine in combination with α-interferon against herpes simplex type 2 in mice. *Antimicrob. Agents Chemother.* **26**, 937–938.

Fraser-Smith, E. B., Eppstein, D. A., Marsh, Y. V., and Matthews, T. R. (1984c). Enhanced efficacy of the acyclic nucleoside 9-(1,3-dihydroxy-2-propoxymethyl)guanine in combination with β-interferon against herpes simplex type 2 in mice. *Antimicrob. Agents Chemother.* **25**, 563–565.

Fraser-Smith, E. B., Eppstein, D. A., Marsh, Y. V., and Matthews, T. R. (1985). Enhanced efficacy of the acyclic nucleoside 9-(1,3-dihydroxy-2-propoxymethyl)guanine in combination with γ interferon against herpes simplex virus type 2 in mice. *Antiviral Res.* **5**, 137-144.

Friedman, H. M., and Grasela, T. (1981). Adenine arabinoside and allopurinol: Possible adverse drug interaction. N. Engl. J. Med. **304**, 423.

Friedman-Kien, A. E., LaFleur, F., Glaser, R., Rosenberg, J., and Zang, E. (1985). Treatment of primary and recurrent genital herpes progenitalis with topical α interferon gel combined with nonoxynol-9 *Proc. Intersci. Conf. Antimicrob. Agents Chemother., 25th* abstr. 93.

Friedman-Kien, A. E., Klein, R. J., Glaser, R. D., and Czelusniak, S. M. (1986). Treatment of recurrent genital herpes with α interferon gel combined with nonoxynol 9. *J. Am. Acad. Dermatol.* **15**, 989–994.

Furman, P. A., De Miranda, P., St. Clair, M. H., and Elion, G. B (1981). Metabolism of acyclovir in virus-infected and uninfected cells. *Antimicrob. Agents Chemother.* **20**, 518–524.

Furman, P. A., St Clair, P. H., and Spector, T. (1984). Acyclovir triphosphate is a suicide inactivator of herpes simplex virus DNA polymerase. *J. Biol. Chem.* **259**, 9575–9579.

Galegov, G. A., Pushkarskaya, N. L., Obrosova-Serova, N. P., and Zhdanov, V. M. (1977). Combined action of ribavirin and rimantadine in experimental myxovirus infection. *Experientia* **33**, 905–906.

Gangemi, J. D., Nachtigal, M., Barnhart, D., Krech, L., and Jani, P. (1987). Therapeutic efficacy of liposome-encapsulated ribavirin and muramyl tripeptide in experimental infection with influenza or herpes simplex virus. *J. Infect. Dis.* **155**, 510–517.

Gao, W.-Y., Hanes R. N., Vazquez-Padua, M. A., Stein, C. A., Cohen, J. S., and Cheng, Y.-C. (1990). Inhibition of herpes simplex virus type 2 growth by phosphorothioate oligodeoxynucleotides. *Antimicrob. Agents Chemother.* **34**, 808–812.

Gauri, K. K. (1979). Anti-herpes virus polychemotherapy. *Adv. Ophthalmol.* **38**, 151–163.

Gauri, K. K., and Miestereck, H. (1981). Der Einfluss von Virustatika und Steroiden auf die stromale Wundheilung. *In* "Herpetische Augenerkrankungen" (R. Sundmacher, ed.), pp. 287–289. Bergmann Verlag, Munich.

Gerard. G. F., and Grandgenett, D.P. (1980). Retrovirus reverse transcriptase. *In* "Molecular Biology of RNA Tumor Viruses" (J. R. Stephenson, ed.), pp. 345–394. Academic Press, New York.

Gil-Fernández, C., Rodriquez, S., Sola, A., García, A., Gancedo, D., and Vilas, P. (1984). Combined antiviral effects of 5-iodo-2'-deoxyuridine, amantadine and rifampicin on the replication of African swine fever virus in Vero cells. *Proc. Int. Congr. Virol., 6th* abstr. P34-20.

Gill, P. S., Rarick, M. U., Krailo, M., Loureiro, C., Bernstein-Singer, M., and Levine, A. (1989). Combination chemotherapy (low dose Adriamycin, bleomycin, vincristine) for advanced epidemic Kaposi's sarcoma (EKS). Proc. Conf. AIDS, 5th abstr. THP144.

Gogu, B. S., Beckman, S. R. S., and Agrawal, K. C. (1989). Increased therapeutic efficacy of zidovudine in combination with vitamin E. *Biochem. Biophys. Res. Commun.* **165**, 407.

Gordon, Y., Cheng, D. P., Araullo-Cruz, T., Romanowsk, E., Johnson, B.J., and Blough, H. A. (1986). Efficacy of glycoprotein inhibitors alone and in combination with trifluridine in the treatment of murine herpetic keratitis. *Curr. Eye Res.* **5**, 93–99.

Goren, T., Orchansky, P., and Rubinstein, M. (1984). The mechanism of synergistic action among human interferon subtypes. *Proc. Int. Congr. Virol., 6th* abstr. W15-5.

Gorse, G. J., Slater, L. M., Kaplan, H. S., Tilles, J. G., and Cesario, T. C. (1984). Inhibition of interferon yield by vincristine. *Proc. Soc. Exp. Biol. Med.* **175**, 309–313.

Grene, R. B., Park, N.-H., and Pavan-Langston, D. (1984). *In vitro* and *in vivo* efficacy of single and combined antivirals utilizing acyclovir resistant HSV-2. *Invest. Ophthalmol. Visual. Sci.* (Suppl. 25), 23 (abstr. 24).

Gresser, I., Fontaine-Brouty-Boyé, D., Bourali, C., and Thomas, M.T. (1969). A comparison of the efficacy of endogenous, exogenous and combined endogenous–exogenous interferon in the treatment of mice infected with encephalomyocarditis virus (33529). *Proc. Soc. Exp. Biol. Med.* **130**, 236–242.

Grindey, G. B., and Cheng, Y.-C. (1979). Biochemical and kinetic approaches to inhibition of multiple pathways. *Pharmacol. Ther.* **4**, 307.

Grossberg, H. S., Bonnem, E. M., and Buhles, W. C., Jr. (1989). GM-CSF with ganciclovir for the treatment of CMV retinitis in AIDS. Correspondence. *N. Engl. J. Med.* **320**, 1560.

Gruber, W. C., Wilson, S. Z., Throop, B. J., and Wyde, P. R. (1987). Immunoglobulin administration and ribavirin therapy: Efficacy in respiratory syncytial virus infection of the cotton rat. *Pediatr. Res.* **21**, 270–274.

Hahn, B. H., Shaw, G. M., Taylor, M. E., Redfield, R. R., Markham, P. D., and Salahuddin, S. Z., Wong-Staal, F., Gallo, R. C., Parks, E. S., and Park, S. P. (1986). Genetic variation in HTLV-III/LAV over time in patients with AIDS or at risk for AIDS. *Science* **232**, 1548–1553.

Hahn, E. F., Busso, M., Mian, A. M., and Resnick, L. (1989). Nucleotide dimers as anti human immunodeficiency virus agents. *In* "Nucleotide Analogues as Antiviral Agents" (J. Martin, ed.), pp. 156–169. Am. Chem. Soc., Washington, D.C.

Hall, M. J., and Duncan, I. B. (1988). Antiviral drug and interferon combinations. *In* "Antiviral Chemother., Suppl. B* **18**, 165–176.

Hall, M. J., and Duncan, I. B. (1988). Antiviral drug and interferon combinations. *In* "Antiviral Agents: The Development and Assessment of Antiviral Chemotherapy" (H. J. Field, eds.), Vol.II, pp. 29–84. CRC Press, Boca Raton, Florida.

Hammer, S., and Gillis, J. M. (1987). Synergistic activity of granulocyte–macrophage colony-stimulating factor and 3'-azido-3'-deoxythymidine against human immunodeficiency virus *in vitro*. *Antimicrob. Agents Chemother.* **31**, 1046–1050.

Hammer, S. M., Kaplan, J. C., Lowe, B. R., and Hirsch, M. S. (1981). Interferon (IFN) and acyclovir (ACV) treatment of herpes simplex virus (HSV) in lymphoid cell cultures. *Proc. Intersci. Conf. Antimicrob. Agents Chemother., 21st* abstr. 24.

Hammer, S. M., Kaplan, J. C., Lowe, B. R., and Hirsch, M. S. (1982). Alpha interferon and acyclovir treatment of herpes simplex virus in lymphoid cell cultures. *Antimicrob. Agents Chemother.* **21**, 634–640.

Hanada, N., Kido, S., Kuzushima, K., Goto, Y., Rahman, M. M., and Morishima, T. (1989). Combined effects of acyclovir and human interferon-alpha on herpes simplex virus replication in cultured neural cells. *J. Med. Virol.* **29**, 345–351.

Hartshorn, K. L., Sandstrom, E. G., Neumeyer, D., Paradis, T. J., Chou, T.-C., Schooley, R. T., and Hirsch, M. S. (1986). Synergistic inhibition of human T-cell lymphotropic virus type III replication *in vitro* by phosphonoformate and recombinant α-A interferon. *Antimicrob. Agents Chemother.* **30**, 189–191.

Hartshorn, K. L., Vogt, M. W., Chou, T.-C, Blumberg, R. S., Byington, R., Schooley, R. T., and Hirsch, M. S. (1987). Synergistic inhibition of human immunodeficiency virus in vitro by azidothymidine and recombinant alpha A interferon. *Antimicrob. Agents Chemother.* **31**, 168–172.

Hasobe, M., McKee, J. G., Ishii, H., Cools, M., Borchardt, R. T., and De Clercq, E. (1989). Elucidation of the mechanism by which homocysteine potentiates the anti-vaccinia virus effects of the S-adenosylhomocysteine hydrolase inhibitor 9-(trans-2'-trans-3'-dihydroxy-cyclopent-4'-enyl)-adenine. *Mol. Pharmacol.* **36**, 490–496.

Hayashi, J., Kashiwagi, S., Noguchi, A., Ikematsu, H., Tsuda, H., Tsuji, Y., and Motomura, M. (1989). Combination therapy of glycyrrhizin withdrawal and human fibroblast interferon for chronic hepatitis B., Clin. Ther. **11**, 161–169.

Hayashi, S., Fine, R. L., Chou, T.-C., Currens, M. J., Broder, S., and Mitsuya, H. (1990). *In vitro* inhibition of the infectivity and replication of human immunodeficiency virus type 1 by combination of antiretroviral 2',3'-dideoxynucleosides and virus-binding inhibitors. *Antimicrob. Agents Chemother.* **34,** 82–88.

Hayden, F. G. (1986). Combinations of antiviral agents for treatment of influenza virus infections. *J. Antimicrob. Chemother., Suppl. B* **18,** 177–183.

Hayden, F. G., Schlepushkin, A. N., and Pushkarskaya, N. L. (1984). Combined interferon-α_2, rimantadine hydrochloride, and ribavirin inhibition of influenza virus replication *in vitro*. *Antimicrob. Agents Chemother.* **25,** 53–57.

Hayden, R. G., Douglas, R. G., Jr., and Simons, R. (1980). Enhancement of activity against influenza viruses by combinations of antiviral agents. *Antimicrob. Agents Chemother.* **18,** 536–541.

Henry, K., Cantrill, H., and Kish, M. A. (1987). Intravitreous ganciclovir for patients receiving zidovudine. *JAMA, J. Am. Med. Assoc.* **257,** 3066.

Herrmann, E. C. (1982). The advent of antiviral drugs and the search for drugs useful for human enteroviral diseases. *In* "Medical Virology" (L. M. de la Maza and E. M. Peterson, eds.), pp. 301–326. Elsevier, New York.

Herrmann, E. C., Jr., and Herrmann, J. A. (1977). A working hypothesis—virus resistance development as an indicator of specific antiviral activity. *Ann. N.Y. Acad. Sci.* **284,** 632–37.

Herrmann, E. C., Jr., Hermann, J. C., and Delong, D. C. (1982). Prevention of death of mice caused by coxsackie virus A16 using guanidine HCl mixed with enviroxime or 2-(α-hydroxybenzyl)-benzamidazole. *Proc. Annu. Meet. Am. Soc. Microbiol., 82nd* abstr. A31.

Hess, G., Rossol, S., Voth, R., Gerken, G., Ramadori, G., Drees, N., and Meyer zum Buschenfelde, K. H. (1989). Treatment of patients with chronic type B hepatitis and concurrent human immunodeficiency virus infection with a combination of interferon alpha and azidothymidine: A pilot study. *Digestion* **43,** 56–59.

Hilfenhaus, J., De Clercq, E., Koher, R., Geursen, R., and Seiler, F. (1987). Combined antiviral effects of acyclovir or bromovinyldeoxyuridine and human immunoglobin in herpes simplex virus infected mice. *Antiviral Res.* **7,** 227–235.

Hirsch, M. S., and Kaplan, J. C. (1985). Prospects of therapy for infections with human T-lymphotropic virus type III. *Ann. Intern. Med.* **103,** 750–755.

Huggins, J. W., Spears, C.T., Stiefel, J., Robins, R. K., and Canonico, P. G. (1983). Synergistic antiviral effects of ribavirin and C-nucleoside analogs tiazofurin and selenazofurin against togaviruses and arenaviruses. *Proc. Intersci. Conf. Antimicrob. Agents Chemother., 23rd* abstr. 155.

Huggins, J. W., Robins, R. K., and Canonico, P. G. (1984). Synergistic antiviral effects of ribavirin and C-nucleosides analog tiazofurin and selenazofurin against togaviruses, bunyaviruses, and arenaviruses. *Antimicrob. Agents Chemother.* **26,** 476–480.

Hugo, H., Linde, A., and Abom, P.-E. (1989). Epstein–Barr virus induced thrombocytopenia treated with intravenous acyclovir and immunoglobulin. *Scand. J. Infect. Dis.* **21,** 103–105.

Ilyenko, V. I., Platonov, V. G., and Khomenkova, I. K. (1977). Comparative study of the antiviral activity of rimantadine and bonafton. *Vopr. Virusol.* **2,** 223–228.

Imanishi, J., Tanaka, A., and Matsuoka, H. (1984). Inhibition de la multiplication du virus de l'herpès humaine I par la prostaglandine D_2 et effet synergique de l'interféron leucocytaire humain. *C. R. Seances Soc. Biol. Ses Fil.* **178,** 671–674.

Ito, M., Nakashima, H., Baba, M., Pauwels, R., De Clercq, E., Shigeta, S., and Yamamoto, N. (1987). Inhibitory effect of glycyrrhizin on the *in vitro* infectivity and cytopathic activity of human immunodeficiency virus [HIV (HTLV-III/LAV)]. *Antiviral Res.* **7,** 127–137.

Jacobson, M. A., De Miranda, P., Gordon, S. M., Blum, M. R., Volberding, P., and Mills, J. (1988). Prolonged pancytopenia due to combined ganciclovir and zidovudine therapy. *J. Infect. Dis.* **158,** 489–490.

Jacobson, M. A., O'Donnell, J. J., Rousell, R., Dionian, B., and Mills, J. (1990). Failure of adjunctive cytomegalovirus intravenous immune globulin to improve efficacy of ganciclovir in patients with acquired immunodeficiency syndrome and cytomegalovirus retinitis: A phase 1 study. *Antimicrob. Agents Chemother.* **34,** 176–178.

Jahrling, P. B., Peters, C. J., and Stephen, E. L. (1984). Enhanced treatment of Lassa fever by immune plasma combined with ribavirin in cynomolgus monkeys. *J. Infect Dis.* **149,** 420–427.

Janz, C., and Wigand, R. (1982). Combined interaction of antiherpes substances and interferon β on the multiplication of herpes simplex virus. *Arch. Virol.* **73,** 135–143.

Johnson, J. C., and Attanasio, R. (1987). Synergistic inhibition of anatid herpes virus replication by acyclovir and phosphonocompounds. *Intervirology* **28,** 89–99.

Johnson, P. C., Kumor, K., Welsh, M. S., Woo, J., and Kahan, B. D. (1987). Effects of coadministration of cyclosporine and acyclovir on renal function of renal allograft recipients. *Transplantation* **44,** 329–331.

Johnson, V. A., Barlow, M. A., Chou. T.-C., Fisher, R. A., Walker, B. D., Hirsch, M. S., and Schooley, R. T. (1989a). Synergistic inhibition of human immunodeficiency virus type 1 (HIV-1) replication *in vitro* by recombinant soluble CD4 and 3'-azido-3'-deoxythymidine. *J. Infect. Dis.* **159,** 837–844.

Johnson, V. A., Merrill, D. P., Chou, T.-C., and Hirsch, M. S. (1989b). Synergistic inhibition of HIV-1 replication by *N*-butyl deoxynojirimycin (*N*-butyl DNJ) and zidovudine (AZT). *Proc. Intersci. Conf. Antimicrob. Agents Chemother., 29th* p. 185 (abstr. 504).

Johnson, V. A., Walker, B. D., Barlow, M. A., Paradis, T. J., Chou, T.-C., and Hirsch, M. S. (1989c). Synergistic inhibition of human immunodeficiency virus type 1 and type 2 replication *in vitro* by castanospermine and 3'-azido-3'-deoxythymidine. *Antimicrob. Agents Chemother.* **33,** 53–57.

Johnson, V. A., Barlow, M. A., Merrill, D. P., Chou, T.-C., and Hirsch, M. S. (1990). Three drug synergistic inhibition of HIV-1 replication *in vitro* by zidovudine, recombinant soluble CD4, and recombinant interferon-alpha A. *J. Infect. Dis.* (in press).

Kabanov, A. V., Vinogradov, S. V., Ovcharenko, A. V., Krivonos, A. V., Melik-Nubarov, N. S., Kiselev, V. I., and Severin, E. S. (1990). A new class of antivirals: Antisense oligonucleotides combined with a hydrophobic substituent effectively inhibit influenza virus reproduction and synthesis of virus-specific proteins in MDCK cells. *FEBS Lett.* **259,** 327–330.

Karim, M. R., Benton, D. C., and Marks, M. I. (1982). Synergistic effect of acyclovir and vidarabine against *herpes virus hominis* type 2 infection in newborn mice. *Proc. Annu. Meet. Am. Soc. Microbiol., 82nd* abstr. A33.

Karim, M. R., Naylor, L., Kaup, L., Deuter, D., and Foster, D. (1984). Combination therapy against herpes simplex virus type 2 (HSV2) infection in female Sasco white mice. *Proc. Annu. Meet. Am. Soc. Microbiol., 84th* abstr. A22.

Karim, M R., Barney, S., Foster, D. E., Lopez, C., and Watanabe, K. A. (1985a). Comparative therapeutic activities of 2'-fluoro-5-iodoaracytosine, 2'-fluoro-5-methylarauracil, and acyclovir alone and in combination against herpes simplex virus (HSV-2) infection in newborn and adult mice. *Proc. Annu. Meet. Am. Soc. Microbiol., 85th* abstr. A39.

Karim, M. R., Marks, M. I., Benton, D. C., and Rollerson, W. (1985b). Synergistic antiviral effects of acyclovir and vidarabine on herpes simplex infection in newborn mice. *Chemotherapy (Basel)* **31,** 310–317.

Kaufman, H. E. (1965). *In vivo* studies with antiviral agents. *Ann. N.Y. Acad. Sci.* **130,** 168.

Kaufman, H. E., Martola, E. L., and Dohlman, C. H. (1963). Herpes simplex treatment with IDU and corticosteroids. *Arch. Ophthalmol. (Chicago)* **69,** 468–473.

Kaufman, H. E., Varnell, E. D., Centifanto-Fitzgerald, Y. M., and Sanitato, J. G. (1984). Virus chemotherapy: Antiviral drugs and interferon. *Antiviral Res.* **4,** 333–338.

Kavesh, N. G., Holzman, R. S., and Seidlin, M. (1989). The combined toxicity of azidothymidine and antimycobacterial agents. *Am. Rev. Respir. Dis.* **139,** 1094–1097.

Kende, M., Lupton, W. H., Rill, W. L., Levy, H. B., and Canonico, P. G. (1987). Enhanced therapeutic efficacy of poly(ICLC) and ribavirin combinations against Rift Valley fever virus infection in mice. *Antimicrob. Agents Chemother.* **31,** 986–990.

Kern, E. R., Vogt, P. E., and Overall, J. C. (1988). Effect of combination therapy with vidarabine (ara-A) and acyclovir (ACV) in a murine model of herpes simplex virus type 1 (HSV-1) encephalitis. *Antiviral Res.* **9**, 139 (abstr. II-24).

Kinghorn, G. R., Abeywickreme, I., Jeavons, M., Rowland, M., Barton, I., Potter, C. W., and Hickmott, C. (1986). Efficacy of oral treatment with acyclovir and cotrimoxazole in first episode genital herpes. *Genitourinary Med.* **62**, 33–37.

Kirshon, B., Faro, S., Zurawin, R. K., Samo, T. C., and Carpenter, R. J. (1988). Favorable outcome after treatment with amantadine and ribavirin in a pregnancy complicated by influenza pneumonia. A case report. *J. Reprod. Med.* **33**, 399–401.

Kirsi, J. J., McKernan, P. A., Burns, J. N., III, North, J. A., Murray, B. K., and Robins, R. K. (1984). Broad spectrum synergistic antiviral activity of selenazofurin and ribavirin. *Antimicrob. Agents Chemother.* **26**, 466–475.

Knight, P. R., Nahrwold, M. L., and Bedows, E. (1980). Anesthetic action and virus replication: Inhibition of measles virus replication in cells exposed to halothane. *Antimicrob. Agents Chemother.* **17**, 890–896.

Kohl, S. (1983). Additive effects of acyclovir and immune transfer in neonatal herpes simplex virus infection in mice. *Infect. Immun.* **39**, 480–482.

Kohl, S., Pickering, L. K., and Loo, L. S. (1981). Human leukocytes (leu) and immune serum globulin (ISG) protect newborn mice from herpes simplex virus (HSV) infection. *Proc. Intersci. Conf. Antimicrob. Agents Chemother., 21st* abstr. 235.

Kohl, S., Loo, L. S., and Greenberg, B. (1982a). Protection of newborn mice from lethal herpes simplex virus (HSV) infection by human mononuclear cells (MC) and human α interferon (IFN). *Fed. Proc., New Orleans, Louisiana, April* abstr. 1728, 566; see also *Pediatr. Res.* **16**, 226A (abstr. 882).

Kohl, S., Loo, L. S., and Greenberg, B. (1982b). Protection of newborn mice from a lethal herpes simplex virus infection by human interferon, antibody, and leukocytes. *J. Immunol.* **128**, 1107–1111.

Kolomiets, A., Chekina, A., Lolomiets, N., Boiko, V., Vichkanova, S., and Votyakov, V. (1984). Combined application of specific antiherpetic immunoglobulin and antiviral preparations in ophthalmoherpes. *Proc. Int. Congr. Virol. 6th* abstr. P34-17.

Kornhauser, D. M., Petty, B. G., Hendrix, B. W., Woods, A. S., Nerhood, L. J., Bartlett, J. G., and Lietman, P. S. (1989). Probenecid and zidovudine metabolism. *Lancet* **2**, 473–475.

Koshida, R. Vrang, L., Gilljam, G., Harmenberg, J., Öberg, B., and Wahren, B. (1989). Inhibition of human immunodeficiency virus *in vitro* by combinations of 3'-azido-3'-deoxythymidine and foscarnet. *Antimicrob. Agents Chemother.* **33**, 778–780.

Koskiniemi, M. L., Vaheri, A., Valtonen, S., Haltia, M., Kaste, M., Manninen, V., Salonen, E. M., Icen, A., Cantell, K., and Study Group (1982). Trial with human leucocyte interferon and vidarabine in herpes simplex virus encephalitis: Diagnostic and therapeutic problems. *Acta Med. Scand., Suppl.* **68**, 150–160.

Kovacs, J. A., Deyton, L., Davey, R., Falloon, J., Zunich, K., Lee, D., Metcalf, J. A., Bigley, J. W., Sawyer, L. A., Zoon, K. C., Masur, H., Fauci, A. S., and Lane, H. S. (1989). Combined zidovudine and interferon-α therapy in patients with Kaposi sarcoma and AIDS. *Ann. Intern. Med.* **111**, 280–287.

Lagoutte, F. (1984). Les kératites stromales herpétiques: Leur traitement par association acyclovir dexamethasone collyre—À propos de 32 cas. *Bull. Soc. Ophthalmol. Fr.* **134**, 989–991.

Land, S., Treolar, G., McPhee, D., Birch, C., Doherty, R., Cooper, D., and Gust, I. (1990). Decreased *in vitro* susceptibility to zidovudine of HIV isolates obtained from patients with AIDS. *J. Infect. Dis.* **161**, 326–329.

Lange, J. M. A., de Wolf, F., Mulder, J. W., Coutinho, R. A., van der Noordaa, J., and Goudsmit, J. (1989). Markers for progression to acquired immune deficiency syndrome and zidovudine treatment of asymptomatic patients. *J. Infect. Dis.* **18**, 85–91.

Langford, M. P. and Stanton, G. J. (1982). Synergistic antiviral effects of interferon plus antibody against acute hemorrhagic conjunctivitis viruses, herpes virus and adenovirus. **41,** 538–541.

Langford, M. P., Villarreal, A. L., and Stanton, G. J. (1983). Antibody and interferon act synergistically to inhibit enterovirus, adenovirus, and herpes simplex virus infection. *Infect. Immun.* **41,** 214–218.

Langford, M. P., Barber, J. C., and Stanton, G. J. (1984). Endogenous interferon and early appearing neutralizing activity act synergistically to inhibit acute hemorrhagic conjunctivitis (AHC) virus replication. *Invest. Ophthalmol. Visual Sci.* (Suppl. 25), 23 (abstr. 21).

Langford, M. P., Carr, D. J. J., and Yin-Murphy, M. (1985). Activity of arildone with or without interferon against acute hemorrhagic conjunctivitis viruses in cell culture. *Antimicrob. Agents Chemother.* **28,** 578–580.

Larder, B. A., and Darby, G. (1984). Virus drug-resistance: Mechanisms and consequences. *Antiviral Res.* **4,** 1–42.

Larder, B. A., and Kemp, S. D. (1989). Multiple mutations in HIV-1 reverse transcriptase confer high-level resistance to zidovudine (AZT). *Science* **246,** 1155–1158.

Larder, B. A., Darby, G., and Richman, D. D. (1989). HIV with reduced sensitivity to zidovudine (AZT) isolated during prolonged therapy. *Science* **243,** 1731–1734.

Larder, B. A., Cheseboro, B., and Richman, D. D. (1990). Susceptibilities of zidovudine-susceptible and -resistant human immunodeficiency virus isolates to antiviral agents determined using a quantitative plaque reduction assay. *Antimicrob. Agents Chemother.* **34,** 436–441.

Lavrov, S. V., Eremkina, E. I., Orlova, T. G., Galegov, V. D., Soloviev, V. D., and Zhdanov, V. M. (1968). Combined inhibition of influenza virus reproduction in cell culture using interferon and amantadine. *Nature (London)* **217,** 856–857.

Lefkowitz, E., Worthington, M., Conliffe, M. A., and Baron, S. (1976). Comparative effectiveness of six antiviral agents in herpes simplex type 1 infection of mice (39392). *Proc. Soc. Exp. Biol. Med.* **152,** 337–342.

Lehoang, P., Girard, B., Robinet, M., Marcel, P., Zazoun, L., Matheron, S., Rozenbaum, W., Katlama, C., Morer, I., Lernestedt, J. O., Saraux, H., Pouliquen, Y., Gentilini, M., and Rousselie, F. (1989). Foscarnet in the treatment of cytomegalovirus retinitis in acquired immune deficiency syndrome. *Ophthalmology (Rochester, Minn.)* **96,** 865–874.

Lerner, M. A. (1981). *In vitro* susceptibility studies of antiviral agents in combination versus HSV-1 and HSV-2: Adenine arabinoside (ara-A), arabinosyl-hypoxanthine (ara-Hx), and interferon (IF) in several tissue cultures. *In* "The Human Herpesviruses: An interdisciplinary Perspective" (A. J. Nahmias, W. E. Dowdle, and R. F. Schinazi, eds.), p. 670. Elsevier-North-Holland, New York.

Lerner, M. A., and Bailey, E. J. (1974). Synergy of 9-β-D-arabinofuranosyladenine and human interferon against herpes simplex virus, type 1. *J. Infect. Dis.* **130,** 549–552.

Lerner, M. A., and Bailey, E. J., (1976). Differential sensitivity of herpes simplex virus type 1 and 2 to human interferon: Antiviral effects of interferon plus 9-β-D-arabinofuranosyladenine. *J. Infect. Dis.* **134,** 400–404.

Levin, M. J., and Leary, P. L. (1981). Inhibition of human herpes viruses by combinations of acyclovir and human leukocyte interferon, *Infect. Immun.* **32,** 995–999.

Lin, J.-C., Zhang, Z.-X., Chou, T.-C., Sim, I., and Pagano, J. S. (1989). Synergistic inhibition of Epstein–Barr virus: Transformation of B lymphocytes by α and β interferon and by 3'-azido-3'-deoxythymidine. *J. Infect. Dis.* **159,** 248–254.

Link, F., Rada, B., and Blaskovic, D. (1965–1966). Problems of *in vitro* and *in vivo* testing of antiviral substances. *Ann N.Y. Acad. Sci.* **130,** 31–43.

Liu. W., Chen, A., Song, L., and Ma., Z. (1986). Effects of 2,2'-O-cyclocytidine and acyclovir on latent herpes simplex virus in trigeminal ganglia of mice. *Antimicrob. Agents Chemother.* **29,** 278–280.

Luscri, B. J. (1982). Synergy between human β interferon (IFN) and ribavirin (RVN). *Proc. Annu. Meeting Am. Soc. Microbiol. 82nd* p. 15 (abstr. A84).

Luscri, B. J., Canonico, P. G., and Huggins, J. W. (1984). Synergistic antiviral interactions of α and γ human interferons *in vitro*. *Proc. Intersci. Conf. Antimicrob. Agents Chemother., 24th* abstr. 1241, p. 311.

L'vov, N. D., Samoilovich, E. O., Tikhonchuk, I. A., Chepaikina, T. A., and Tsvetkov, E. N. (1989). Antiherpetic activity of Soviet-made phosphonic acid derivatives and their combinations with interferon inducers in a model of ophthalmic herpes. *Vopr. Virusol.* **34**, 186–192.

Malewicz, B., Momsen, M., and Jenkin, H. M. (1982). Amphotericin B potentiation of acyclovir antiviral activity. *Proc. Intersci. Conf. Antimicrob. Agents Chemother., 22nd* abstr. 417.

Malewicz, B., Momsen, M., and Jenkin, H. M. (1983). Combined effect of acyclovir and amphotericin B on the replication of pseudorabies virus in BHK-21 cells. *Antimicrob. Agents Chemother.* **23**, 119–124.

Malewicz, B., Momsen, M., Jenkin, H. M., and Borowski, E. (1984). Potentiation of antiviral activity of acyclovir by polyene macrolide antibiotics. *Antimicrob. Agents Chemother.* **25**, 772–774.

Manischewitz, J. F., Quinnan, G. V., Lane, H. C., and Wittek, A. E. (1990). Synergistic effect of ganciclovir and foscarnet on cytomegalovirus replication *in vitro*. *Antimicrob. Agents Chemother.* **34**, 373–375.

Matsukura, M., Shinozuka, K., Zon, G., Mitsuya, H., Reitz, M., Cohen, J. S., and Broder, S. (1987). Phosphorothioate analogs of oligodeoxynucleotides: Inhibitors of replication and cytopathic effects of human immunodeficiency virus. *Proc. Natl. Acad. Sci. U.S.A.* **84**, 7706–7710; see also *Proc. Inter. Conf. AIDS, 3rd, Abstr. Vol.* p. 54 (abstr. T.4.4.).

Matsukura, M., Zon, G., Shinozuka, K., Robert-Guroff, M., Shidada, T., Stein, C. A., Mitsuya, H., Wong-Staal, F., Cohen, J. S., and Broder, S. (1989). Regulation of viral expression of human immunodeficiency virus *in vitro* by an antisense phosphorothioate oligodeoxynucleotide against rev (art/trs) in chronically infected cells. *Proc. Natl. Acad. Sci. U.S.A.* **86**, 4244–4248.

Meldrum, J. B., Gupta, S. V., and Babiuk L. A. (1980). Comparative efficacy of 5-methoxymethyl-2′-deoxyuridine, 9-β-D-arabinofuranosyladenine and 5-iodo-2′-deoxyuridine in the treatment of experimental herpes keratitis. *Chemotherapy (Basel)* **26**, 54–63.

Meurs, P. J., and van Bijsterveld, O. P. (1985a). Combination therapy of recombinant human alpha 2 interferon and acyclovir in the treatment of herpes simplex keratitis. *Antiviral. Res.* (Suppl. 1), 225–228.

Meurs, P. J., and van Bijsterveld, O. P. (1985b). Berofor alpha 2 (r-Hu IFN-α 2 arg) und Aciclovir in der Behandlung der Herpes-Keratitis-Dendritica. *Klin. Monatsbl. Augenheilkd.* **187**, 40–42.

Meyer, T., and Horisberger, M. A. (1984). Combined action of mouse alpha and beta interferons in influenza virus-infected macrophages carrying the resistance gene *Mx*. *J. Virol.* **49**, 709–716.

Meyers, J. D., McGuffin, R. W., Bryson, Y. J., Cantell. K., and Thomas, E. D. (1982). Treatment of cytomegalovirus pneumonia after marrow transplant with combined vidarabine and human leukocyte interferon. *J. Infect. Dis.* **146**, 80–84.

Mitchell, W. M., Montefiori, D. C., Robinson, W. E., and Carter, W. A. (1989). Mismatched double-stranded RNA (ampligen) protects targets cells from HIV infection and reduces the concentration of 3′-azido-3′-deoxythymidine (AZT) required for virustatic activity. *Proc. Int. Conf. AIDS, 5th* abstr. MP5.

Montefiori, D. C., Robinson, W. E. and Mitchell, W. M. (1989). *In vitro* evaluation of mismatched double-stranded RNA (ampligen) for combination therapy in the treatment of acquired immunodeficiency syndrome. *AIDS Res. Hum. Retroviruses* **5**, 193–203.

Moran, D. M., Overall, J. C., Jr. and Kern, E. R. (1983). Combinations of interferon and acyclic nucleosides are synergistic against herpes simplex virus (HSV). *Proc. Intersci. Conf. Antimicrob. Agents Chemother. 23rd* p. 221 (abstr. 749).

Moran, D. M., Kern, E. R., and Overall, J. C., Jr. (1985). Synergism between recombinant human interferon and nucleoside antiviral agents against herpes simplex virus: Examination with an automated microtiter plate assay. *J. Infect. Dis.* **151**, 1116–1122.

Moriyama, K., Asano, Y., Fujimoto, K., Okamura, T., Shibuya, T., Harada, M., and Niho, Y. (1988). Successful combination therapy with acyclovir and vidarabine for disseminated varicella zoster virus infection with retinal involvement in a patient with B-cell lymphoma and adult T-cell leukemia. *Am. J. Med.* **85,** 885–886.

Moschchik, K. V., Grebeniuk, V. N., and Ershov, F. I. (1982). Treatment of genital herpes with levamisole in combination with megasine. *Vopr. Virusol.* **27,** 94–97.

Muñoz, A, Garcia, R. A., and Pérez-Aranda, A. (1986). Potentiation by levamisole, methisoprinol, and adenine or adenosine of the inhibitory activity of human interferon against encephalomyocarditis virus. *Antimicrob. Agents Chemother.* **30,** 192–195.

Muratore, O., Vanier, O. E., Raffanti, S. P., and Schito, G. C. (1984). *In vitro* recovery of resistant retrovirus isolates after exposure to phosphonoformate. *Eur. J. Clin. Microbiol.* **3,** 447–449.

Natsumeda, Y., Yamada, Y., Yamaji Y., and Weber, G. (1988). Synergistic cytotoxic effect of tiazofurin and ribavirin in hepatoma cells. *Biochem. Biophys. Res. Commun.* **53,** 321–327.

Neuman-Haeflin, D., Sundmacher, R., Frey, H., and Merk, W. (1985). Recombinant HuIFN-γ prevents herpes simplex keratitis in African green monkeys: Demonstration of synergism with recombinant HuIFN-α 2. *Med. Microbiol. Immunol.* **174,** 81–86.

Nishimura, C., and Ikeda, S. (1984). Synergistic action of MDP and its analogs on *in vivo* antiviral activity of interferon inducers. *Proc. Int. Congr. Virol. 6th* abstr. P34-33.

North, T. W., and Cohen, S. S. (1978). Erythro-9-(2-hydroxy-3-nonyl)-adenine as a specific inhibitor of herpes simplex virus replication in the presence and absence of adenosine analogues. *Proc. Natl. Acad. Sci. U.S.A.* **75,** 4684.

North, T. W., and Cohen, S. S. (1979). Aranucleosides and aranucleotides in viral chemotherapy. *Pharmacol. Ther.* **4,** 81–108.

Öberg, B. (1983). Antiviral effects of phosphonoformate (PFA, foscarnet sodium). *Pharmacol. Ther.* **19,** 387–415.

O'Brien J. J., and Campoli-Richards, D. M. (1989). Acyclovir: An update review of its antiviral activity, pharmacokinetics properties and therapeutic efficacy. *Drugs* **37,** 233–309.

O'Brien, W. J., Taylor, J. L., Grossberg, S. E., and Schultz, R. O. (1981). Interferon and carboxymethylacridanone therapy of herptic keratitis in rabbits. *Proc. Intersci. Conf. Antimicrob. Agents Chemother. 21st* abstr. 165.

O'Brien, W. J., Coe, E. C., and Taylor, J. L. (1990). Nucleoside metabolism in herpes simplex virus-infected cells following treatment with interferon and acyclovir, a possible mechanism of synergistic antiviral activity. *Antimicrob. Agents Chemother.* **34,** 1178–1182.

Omata, M., and Uchiumi, K. (1986). Combination of prednisolone withdrawal and antiviral agents (adenine arabinoside, interferon) in chronic hepatitis. *Br. J. Hepatol.* (Suppl. 2–3), S65–S69.

Oosterhuis, J. A., Nolen, W. A., Sie, S. H., and Versteeg, J. (1974). Poly I:C in herpetic keratitis in rabbits. *Exp. Eye Res.* **18,** 371–380.

Orholm, M., Pedersen, C., Mathiesen, L., Dowd, P., and Nielsen, J. O. (1989). Suppression of p24 antigen in sera from HIV-infected individuals with low-dose α-interferon and zidovudine: A pilot study. *AIDS* **3,** 97–100.

Overall, J. C., Moon, A., and Kern, E. R. (1988). Combination of acyclovir (ACV) plus vidarabine (ara-A) or ACV plus interferon (IFN) against herpes simplex virus type 1 (HSV-1) and HSV-2 in cell culture. *Antiviral Res.* **9,** 138 (abstr. II-23).

Park, N.-H., Callahan, J. G., and Pavan-Langston, D. (1984). Effect of combined acyclovir and vidarabine on infection with herpes simplex virus *in vitro* and *in vivo*. *J. Infect Dis.* **149,** 757–762.

Pavan-Langston, D., and Dunkel, E., C. (1985). Combination antiviral chemotherapy of HSV infection *in vitro* and *in vivo*. *Invest. Ophthalmol. Visual Sci.* **26,** 318 (abstr. 13).

Perno, C.-F., Yarchoan, R., Cooney, D. A., Hartman, N. R., Webb, D. S. A., Hao, Z., Mitsuya, H., Johns, D. G., and Broder, S. (1988). Replication of human immunodeficiency virus in monocytes. Granulocyte/macrophage colony-stimulating factor (GM-CSF) potentiates viral production yet enchances the antiviral effect mediated by 3'-azido-3'-deoxythymidine (AZT) and other dideoxynucleoside congeners of thymidine. *J. Exp. Med.* **168,** 933–951.

Perrillo, R. P. (1986). The use of corticosteroids in conjunction with antiviral therapy in chronic hepatitis B with ongoing viral replication. *J. Hepatol.* (Suppl. 2–3), S57–S64.

Pizzo, P. A., Einloth, M., Butler, K., Eddy, J., Jarosinski., P., Meer, J., Fallon, J., Moss, H., Wolters, P., Brouwers, P., Balis, F. M, and Poplack, D. G. (1989). A phase I-II study of dideoxycytidine (ddC) alone and in an alternating schedule with zidovudine in children with symptomatic human immunodeficiency virus infection. *Proc Intersci. Conf. Antimicrob. Agents Chemother., 29th.*

Plescia, O. J., Pontani, D., Schaffner, C., Sun, D., Sarin, P., and Shahied, S. D. (1989). Treatment of AIDS based on a combination of synergistic drugs. *Proc. Int. Conf. AIDS, 5th* abstr. MP226.

Poli, G., Orenstein, M. J., Kinter, A., Folks, T. M., and Fauci, A. S. (1989). Interferon-α but not AZT suppresses HIV expression in chronically infected cell lines. *Science* **244,** 575–577.

Poryo, A., and Wigand, R. (1984). Cutaneous herpes simplex virus infection of the guinea pig: Lack of resistance to acyclovir and phosphonoformic acid after topical treatment. *Med. Microbiol. Immunol.* **173,** 219–224.

Prange, H. W., and Henze, T. (1988). An antiviral combination treatment for virus encephalitis—Theoretical aspects and clinical experiences. *J. Neuroimmunol.* **20,** 165–167.

Rada, B., and Holy, A. (1980). Virus inhibitory effect of combination of 9-(S)-(2,3-dihydroxypropyl)adenine and 6-azauridine. *Chemotherapy (Basel)* **26,** 184–190.

Rada, B., and Hanusovskà, T. (1987). Rapid method for the detection of synergism in combinations of antiviral substances. *Acta Virol.* 31, 126–137.

Rahal, J. J. (1978). Antibiotic combinations: The clinical relevance of synergy and antagonism. *Medicine (Baltimore)* **57,** 179–195.

Raju, V. K., Varnell, E. D., and Kaufmann, H. E. (1984). The lack of effect of tunicamycin and 2-deoxy-D-glucose on corneal stromal herpes in rabbits. *Invest. Ophthalmol. Visual Sci.* **25,** 219–221.

Rapp, F., and Wrzos, H. (1985). Synergistic effect of human leukocyte interferon and nonoxynol-9 against herpes simplex virus type 2. *Antimicrob. Agents Chemother.* **28,** 449–451.

Rasmussen, L., Chen, P. T., Mullenax, J. G., and Merigan, T. C. (1984). Inhibition of human cytomegalovirus replication by 9-(1,3-dihydroxy-2-propoxymethyl)guanine alone and in combination with human interferons. *Antimicrob. Agents Chemother.* **26,** 441–445.

Reardon, J. E., and Spector, T. (1989). Herpes simplex virus type 1 DNA polymerase. Mechanism of inhibition by acyclovir triphosphate. *J. Biol. Chem.* **264,** 7405–7411.

Reed, E. C., Dandliker, P. S., and Meyers, J. D. (1986). Treatment of cytomegalovirus pneumonia with 9-[2-hydroxy-1-(hydroxymethyl)ethoxymethyl]guanine and high-dose corticosteroids. *Ann. Intern. Med.* **105,** 214–215.

Reed, E. C., Bowden, R. A., Dandliker, P. S., Lilleby, K. E., and Meyers, J. D. (1988). Treatment of cytomegalovirus pneumonia with ganciclovir and intravenous cytomegalovirus immunoglobulin in patients with bone marrow transplants. *Ann. Intern. Med.* **109,** 783–788.

Regelson, W., Loria, R., and Kalimi, M. (1988). Hormonal intervention.: "Buffer hormones" or "state dependency"—The role of dehydroepiandrosterone (DHEA), thyroid hormone, estrogen, and hypophysectomy in aging. *Ann. N.Y. Acad. Sci.* **521,** 260–273.

Resnick, L., and Mian, A. M. (1989). Enchanced *in vitro* suppression of HIV infectivity by a combination of nucleoside analogs. *Proc. Int. Conf. AIDS, 5th* abstr. WP1.

Resnick, L., Markham, J. D., Veren, K., Salahuddin, S. Z., and Gallo, R. C. (1986). *In vitro* suppression of HTLV-III/LAV infectivity by a combination of acyclovir and suramin. *J. Infect. Dis.* **154,** 1027–1030.

Richman, D. D. (1988). The treatment of HIV infection. *AIDS* **2,** S137–S142.

Roizman, B., and Sears, A. E. (1990). Herpes simplex viruses and their replication. *In* "Virology" (B. N. Fields and D. M. Knipe, eds.), 2nd Ed. pp. 1795–1841.

Rooke, R., Tremblay, M., Soudeyns, H., DeStephano, L., Yao, X.-J., Fanning, M., Montaner, J. S. G., O'Shaughnessy, M., Gelmon, K., Tsoukas, C., Gill, J., Ruedy, J., Wainberg, M. A., and the Canadian Zidovudine Multi-Center Study Group (1989). Isolation of drug-resistant variants of HIV-1 from patients on long term zidovudine therapy. *AIDS* **3**, 411–415.

Rose, R. M., Crumpacker, C., Warner, J. L., and Brain, J. D. (1982). Treatment of murine cytomegalovirus pneumonia with acyclovir and interferon. *Am. Rev. Respir. Dis.* **127**, 198–203.

Rosecan, L. R., Stahl-Bayliss, C. M., Kalman, C. M., and Laskin, O. L. (1986). Antiviral therapy for cytomegalovirus retinitis in AIDS with dihydroxy propoxymethyl guanine. *Am. J. Ophthalmol.* **101**, 405–418.

Rubin, R. H., Lynch, P., Pasternack, M. S., Schoenfeld, D., and Medearis, D. N., Jr. (1989). Combined antibody and ganciclovir treatment of murine cytomegalovirus-infected normal and immunosuppressed BALB/c mice. *Antimicrob. Agents Chemother.* **33**, 1975–1979.

Ruprecht, R. M., Gama-Sosa, A. M., and Rosas, H. D. (1988). Combination therapy after retroviral inoculation. *Lancet* **1**, 239–240.

Ruprecht, R. M., Chou, T.-C., Chipty, F, Gama-Sosa, M., Mullaney, S., O'Brien, L., and Rosas, D. (1990). Interferon-α and 3'-azido-3'-deoxythymidine are highly synergistic in mice and prevent viremia after acute retrovirus exposure. *J. Acquired Immune Deficiency Syndrome* **3**, 591–600.

Rush, J., and Mills, J. (1987). Effect of combinations of difluoromethylornithine (DFMO) and 9-[(1,3-dihydroxy-2-propoxy)methyl]guanine (DHPG) on human cytomegalovirus. *J. Med. Virol.* **21**, 269–276.

Sachs, S. L., Scullard, G. H., Pollard, R. B., Gregory, P. G., Robinson, W. S., and Merigan, T. C. (1981). Adenine arabinoside: Side effects and plasma levels in hepatitis B patients treated concomitantly with interferon. *Proc. Intersci. Conf. Antimicrob. Agents Chemother.* *21st* abstr. 870.

Sachs, S. L., Scullard, G. H., Pollard, R. B., Gregory, P. G., Robinson, W. S., and Merigan, T. C. (1982). Antiviral treatment of chronic hepatitis B virus infection: Pharmacokinetics and side effects of interferon and adenine arabinoside alone and in combination. *Antimicrob. Agents Chemother.* **21**, 93–100.

Samoilovich, E. O., L'vov, N. D., Chepaikina, T. A., Lidak, M. I., Tsvetkov, E. N., Bobkov, A. F., Garaev. M. M., and Barinski, I.F. (1989). Effectiveness of Soviet derivatives of phosphonic acid and their combination with interferon inducers in cell cultures and in a model of herpetic meningoencephalitis in mice. *Vopr. Virusol.* **34**, 466–474.

Schalm, S. W., van Buuren, H. R., Heytink, R. A., and de Man, R. A. (1985). Acyclovir enhances the antiviral effect of interferon in chronic hepatitis B. *Lancet* **2**, 358–360.

Schinazi, R. F. (1987). Drug combinations for treatment of mice infected with acyclovir-resistant herpes simplex virus. *Antimicrob. Agents Chemother.* **31**, 477–479.

Schinazi, R. F. (1988). Strategies and targets for anti-human immunodeficiency virus type 1 chemotherapy. *In* "AIDS in Children, Adolescents and Heterosexual Adults: An Interdisciplinary Approach to Prevention" (R. F. Schinazi and A. J. Nahmias, eds.), pp. 126–143. Elsevier, New York.

Schinazi, R. F., and Chou, T.-C. (1985). The median effect principle applied to studies on delayed treatment with combinations of antiviral drugs in mice inoculated intracerebrally with herpes simplex virus type 2. *Proc. Inter-Am. Soc. Chemother.*

Schinazi, R. F., and Del Bene, V. (1985). Characterization of acyclovir-resistant and sensitive herpes simplex viruses isolated from a patient with AIDS. Symposium on drug resistance in viruses, other micro-organisms and eukaryotic cells. *Br. Soc. Antimicrob. Chemother. Virol. Group Symp.*

Schinazi, R. F., and Nahmias, A. J. (1982). Different *in vitro* effects of dual combinations of anti-herpes simplex virus (HSV) compounds. *Am. J. Med.* **73**, 40–48.

Schinazi, R. F., and Nahmias, A. J. (1984). Herpes simplex virus infections. In "Current Therapy in Internal Medicine" (T. M. Bayless, M. C. Brain, and R. M. Cherniack, eds.), pp. 126–132. Decker, Trenton, New Jersey.

Schinazi, R. F., and Prusoff, W. H. (1983). Antiviral agents. *Pediatr. Clin. North Am.* **30**, 77–92.

Schinazi, R. F., Williams, C. C., and Nahmias, A. J. (1981). Additive antiviral effect of 9-β-D-arabinofuranosyladenine in combination with 9-(2-hydroxyethoxymethyl)guanine. In "The Human Herpesviruses: An Interdisciplinary Perspective" (A. J. Nahmias, W. R. Dowdle, and R. F. Schinazi, eds.), p. 671. Elsevier North-Holland, New York.

Schinazi, R. F., Niditch, A. S., and Nahmias, A. J. (1982a). Effect of combination of human interferon-alpha$_2$ and various nucleosides on herpes simplex viruses 1 and 2. *Meet. Southeast. Branch–Am. Soc. Microbiol.*

Schinazi, R. F., Peters, J., Chance, D., and Nahmias, A. J. (1982b). Studies *in vitro* and *in vivo* of combinations of antivirals and anti-inflammatory agents in relation to the treatment of herpes simplex viruses. In "Current Chemotherapy and Immunotherapy" (P. Periti and G. G. Grassi, eds.), Vol. II, pp. 1085–1087. Am. Soc. Microbiol, Washington, D.C.

Schinazi, R. F., Peters, J., Williams, C. C., Chance, D., and Nahmias, A. J. (1982c). Effect of combinations of acyclovir with vidarabine or its 5′-monophosphate on herpes simplex viruses in cell culture and in mice. *Antimicrob. Agents Chemother.* **22**, 499–507.

Schinazi, R. F., Peters, J., Sokol, M. K., and Nahmias, A. J. (1983a). Therapeutic activities of 1-(2-fluoro-2-deoxy-β-D-arabinofuranosyl)-5-iodocytosine and thymine alone and in combination with acyclovir and vidarabine in mice infected intracerebrally with herpes simplex virus. *Antimicrob. Agents Chemother.* **24**, 95–103.

Schinazi, R. F., Sokol, M. K., and Nahmias, A. J. (1983b). Drug combinations for the therapy of acyclovir (ACV)-resistant herpes simplex virus type 2 (HSV-2) variants in mice. *Proc. Intersci. Conf. Antimicrob. Agents. Chemother. 23rd*, Abstract 566, p. 187.

Schinazi, R. F., Chou, T.-C., Scott, R. T., Yao, J., and Nahmias, A. J. (1986a). Delayed treatment with combinations of antiviral drugs in mice infected with herpes simplex virus and application of the median-effect method of analysis. *Antimicrob. Agents Chemother.* **30**, 491–498.

Schinazi, R. F., Del. Bene, V., Scott. R. T., and Dudley-Thorpe, J. B. (1986b). Characterization of acyclovir-resistant and sensitive herpes simplex viruses isolated from a patient with an acquired immune deficiency. *J. Antimicrob. Chemother., Suppl. B* **18**, 127–134.

Schinazi, R. F., Fox, J. J., Watanabe, K., and Nahmias, A. J. (1986c). Activities of 1-(2-deoxy-2-fluoro-β-D-arabinofuranosyl)-5-iodocytosine and its metabolites against herpes simplex viruses type 1 and 2 in cell culture and in mice infected intracerebrally with herpes simplex virus type 2. *Antimicrob. Agents Chemother.* **29**, 77–84.

Schinazi, R. F., Scott, R. T., Watanabe, K. A., and Fox, J. J. (1987). Effect of 2′-fluoro-5-ethyl-arabinosyluracil (FEAU) in combination with acyclovir or vidarabine in BALB/c mice infected with herpes simplex virus. *Proc. Annu. Meet. Am. Soc. Microbiol., 87th* p. 311 (abstract S17).

Schinazi, R. F., Cannon, D. L., Arnold, B. H., and Martino-Saltzman, D. (1988). Combinations of isoprinosine and 3′-azido-3′-deoxythymidine in lymphocytes infected with human immunodeficiency virus type 1. *Antimicrob. Agents Chemother.* **32**, 1784–1787.

Schinazi, R. F., Eriksson, B. F. H., Arnold, B. H., Lekas, P., and McGrath, M. (1990a). Effect of dehydroepiandrosterone in lymphocytes and macrophages infected with human immunodeficiency viruses. In "The Biological Role of Dehydroepiandrosterone (DHEA)" (M. Kalimi and W. Regelson, eds.). de Gruyter, Berlin. In press.

Schinazi, R. F., Sommadossi, J. P., Saalmann, V., Cannon, D. L., Xie, M.-W., Hart, G. C., Smith, G. A., and Hahn, E. F. (1990b). Activity of 3′-azido-3′-deoxythymidine nucleotide dimers in primary lymphocytes infected with human immunodeficiency virus type 1. *Antimicrob. Agents Chemother.* **34**, 1061–1067.

Schmidt, G. M., Kovacs, A., Zaia, J. A., Horak, D. A., Blume, K. G., Nademanee, A. P., O'Donnell, M. D., Snyder, D. S., and Forman, S. J., (1988). Ganciclovir/immunoglobulin combination therapy for the treatment of human cytomegalovirus-associated interstitial pneumonia in bone marrow allograft recipients. *Transplantation* **46**, 905–907.

Schröder, H. C., Sarin, P. S., Rottmann, M., Wenger, R., Maidhof, A., Renneisen, M., and Müller, W. E. G. (1988). Differential modulation of host cell and HIV gene expression by combination of avarol and AZT *in vitro. Biochem. Pharmacol.* **37**, 3947–3952.

Schwartz, P. M., Shipman, C., Jr., and Drach, J. C. (1976). Antiviral activity of arabinosyladenine and arabinosylhypoxanthine in herpes simplex virus infected KB cells: Selective inhibition of viral deoxyribonucleic acid synthesis in the presence of an adenosine deaminase inhibitor. *Antimicrob. Agents Chemother.* **10**, 64–74.

Schwarz, L. A. and Fleischmann, W. R. (1982). Mechanism of action of the potentiated antiviral state induced by combined immune and virus-type interferons. *Proc. Annu. Meet. Am. Soc. Microbiol, 82nd* p. 253 (abstr. T24).

Scullard, G. H., Andres, L. L., Greenberg, H. B., Smith, J. L., Sawhney, V. K., Neal, A. E., Mahal, A. S., Proper, H., Merigan, T. C., Robinson, W, S., and Gregory, P. B. (1981a). Antiviral treatment of chronic hepatitis B virus infection: Improvement in liver disease with interferon and adenine arabinoside. *Hepatology* **1**, 228–232.

Scullard, G. H., Pollard, R. B., Smith, J. L., Sacks, S. L., Gregory, P. B., Robinson, W. S., and Merigan, T. C. (1981b). Antiviral treatment of chronic hepatitis B virus infection. I. Changes in viral markers with interferon combined with adenine arabinoside. *J. Infect. Dis.* **143**, 772–783.

Shannon, W. M. (1975). Adenine arabinoside: Cell culture studies. *In* "Adenine Arabinoside: An Antiviral Agent" (D. Pavan-Langston, R. A. Buchanan, and C. A. Alford, Jr., eds), p. 7. Raven, New York.

Shannon, W. M. (1977). Selective inhibition of RNA tumor virus replication *in vitro* and evaluation of candidate antiviral agents *in vivo. Ann. N.Y. Acad. Sci.* **284**, 472–507.

Shannon, W. M. (1990). Antiretroviral activity of carbocyclic nucleoside analogues. *In* "Advances in Chemotherapy of AIDS" (R. B. Diasio and J.-P. Sommadossi, eds.). Pergamon, New York.

Shannon, W. M., and Schabel, F. M., Jr (1980). Antiviral as adjuncts in cancer chemotherapy. *Pharmacol. Ther.* **11**, 263–390.

Shannon, W. M., Arnett, G., Schabel, F. M., Jr., North, T. W., and Cohen, S. S. (1980). Erthro-9-(2-hydroxy-3-nonyl)adenine alone and in combination with 9-β-D-arabinofuranosyladenine in treatment of systemic herpes virus infections in mice. *Antimicrob. Agents Chemother.* **18**, 598–603.

Shannon, W. M., Arnett, G., Westbrook, L., Shealy, Y.-M., O'Dell, C. A., and Brockman, R. M. (1981). Evaluation of carbodine, the carbocyclic analog of cytidine, and related carbocyclic analogs of pyrimidine nucleosides for antiviral activity against human influenza type A viruses. *Antimicrob. Agents Chemother.* **20**, 769–776.

Shepherd, F. A., Garvey, M. B., Evans, W. K., Fanning, M. M., Kline, M., and Read, S. E. (1989). Combination chemotherapy and interferon in Kaposi's sarcoma (KS) and AIDS. *Proc. Conf. AIDS, 5th.* abstr. TP224.

Sloan, B. J., Kielty, J. K., and Miller, F. A. (1977). Effect of a novel adenosine deaminase inhibitor (co-vidarabine, co-V) upon the antiviral activity *in vitro* and *in vivo* of vidarabine (Vira-A™) for DNA virus replication. *Ann. N.Y. Acad. Sci.* **284**, 60–80.

Small, G. H., Peyman, G. A., Srinivasan, A., Smith, R. T., and Fiscella, R. (1987). Retinal toxicity of combination antiviral drugs in an animal model. *Can J. Ophthalmol.* **22**, 300–303.

Smith, C. A., Wigdahl, B., and Rapp, F. (1983). Synergistic antiviral activity of acyclovir and interferon on human cytomegalovirus. *Antimicrob. Agents Chemother.* **24**, 325–332.

Smith, C. I., Kitchen, L. W., Scullard, G. H., Robinson, W. S., Gregory, P. B., and Merigan, T. C. (1982). Vidarabine monophosphate and human leukocyte interferon in chronic hepatitis B infection. *JAMA, J. Am. Med. Assoc.* **16**, 2261–2265.

Smith, K. O., Galloway, K. S., Ogilvie, K. K., and Cheriyan, U. O. (1982). Synergism among BIOLF-62, phosphonoformate, and other antiherpetic compounds. *Antimicrob. Agents Chemother.* **22,** 1026–1030.

Smith, M. S., Brian, E. L., De Clercq, E., and Pagano. J. S. (1989). Susceptibility of human immunodeficiency virus type 1 replication *in vitro* to acyclic adenosine analogs and synergy of the analogs with 3'-azido-3'-deoxythymidine. *Antimicrob. Agents Chemother.* **33,** 1482–1486.

Soike, K. F., Eppstein, D. A., Gloff, C. A., Cantrell, C., Chou, T.-C., and Gerone, P. J. (1987). Effect of 9-(1,3-dihydroxypropoxymethyl)guanine and recombinant human β interferon alone and in combination on simian varicella virus infection in monkeys. *J. Infect. Dis.* **156,** 607–614.

Soike, K. F., Chou, T.-C., Fox, J. J., Watanabe, K. A., and Gloff, C. A. (1990). Inhibition of simian varicella virus infection of monkeys by 1-(2-deoxy-2-fluoro-1-β-D-arabinosyl)-5-ethyluracil (FEAU) and synergistic effects of combination with human recombinant interferon β. *Antiviral Res.* **13,** 165–174.

Sommadossi, J.-P., Carlisle, R., Schinazi, R. F., and Zhou, Z. (1988). Uridine reverses the toxicity of 3'-azido-3'-deoxythymidine in normal human granulocyte–macrophage progenitor cells *in vitro* without impairment of antiretroviral activity. *Antimicrob. Agents Chemother.* **32,** 997–1001.

Spector, S. A., and Kelley, E. A. (1983). Inhibition of human cytomegalovirus by combined acyclovir and vidarabine. *Proc. Intersci. Conf. Antimicrob. Agents Chemother. 23rd* p. 253 (abstr. 920).

Spector, S. A., and Kelley, E. (1985). Inhibition of human cytomegalovirus by combined acyclovir and vidarabine. *Antimicrob. Agents Chemother.* **27,** 600–604.

Spector, S. A., Tyndall, M., and Connor, J. D. (1980). Additive and synergistic effects of acyclovir, interferon, phosphonoformic acid and trifluorothymidine against clinical isolate of cytomegalovirus. *Proc. Int. Conf. Antimicrob. Agents Chemother. 20th* abstr. 306.

Spector, S. A., Tyndall M., and Kelley, E. A. (1983). Inhibition of human cytomegalovirus by trifluorothymidine. *Antimicrob. Agents Chemother.* **23,** 113–118.

Spector, T., Averett, D. R., Nelson, D. J., Lambe, C. U., Morrison, R. W., Jr., St. Clair, M. H., and Furman, P. H. (1985). Potentiation of antiherpetic activity of acyclovir by ribonucleotide reductase inhibition. *Proc. Natl. Acad. Sci. U.S.A.* **82,** 4254–4257.

Spector, S. A., Ripley, D., and Hsia, K. (1989). Human immunodeficiency virus inhibition is prolonged by 3'-azido-3'-deoxythymidine alternating with 2',3'-dideoxycytidine compared with 3'-azido-3'-deoxythymidine alone. *Antimicrob. Agents Chemother.* **33,** 920–923.

Spruance, L. S., Crumpacker, C. S., Schnipper, L. E., Kern, E. R., Marlowe, S., Arndt, K. A., and Overall, J. C., Jr. (1984). Early, patient-initiated treatment of herpes labialis with topical 10% acyclovir. *Antimicrob. Agents Chemother.* **25,** 553–555, and references cited therein.

Stanwick, T. L., Schinazi, R. F., Campbell, D. E., and Nahmias, A. J. (1981). Combined antiviral effect of interferon and acyclovir on herpes simplex virus types 1 and 2. *Antimicrob. Agents Chemother.* **19,** 672–674.

Starnes, M. C., and Cheng, Y. C. (1989). Inhibition of human immunodeficiency virus reverse transcriptase by 2',3'-dideoxynucleoside triphosphates: Template dependance, and combination with phosphonoformate. *Virus Genes* **2,** 241–251.

Starr, S. E., Visintine, A. M., Tomeh, M. O., and Nahmias, A. J. (1976). Effects of immunostimulants on resistance of newborn mice to herpes simplex type 2 infection (39327). *Proc. Soc. Exp. Biol. Med.* **152,** 57–60.

Steffe, E. M., King, J. H., Inciardi, J. F., Flynn, N. F., Goldstein, E., Tonjes, T. S., and Benet, L. Z. (1990). The effect of acetaminophen on zidovudine metabolism in HIV-infected patients. *J. Acq. Immun. Def. Synd.* **3,** 691–694.

Sühnel, J. (1990). Evaluation of synergism or antagonism for the combined action of antiviral agents. *Antiviral Res.* **13,** 23–40.

Sundmacher, R. (1982). Interferon in ocular diseases. *In* "Interferon 4" (I. Gresser, ed.), pp. 177–200. Academic Press, New York.

Sundmacher, R. (1985). Therapie der herpetischen Erkrankungen des vorderen Augenabschnitts. *Klin. Monatsbl. Augenheilkd.* **186**, 421–423.

Sundmacher, R., Cantell, K., and Neumann-Haefelin, D. (1978). Combination therapy of dendritic keratitis with trifluorothymidine and interferon. *Lancet* **2**, 687.

Sundmacher, R., Neumann-Haefelin, D., and Cantell, K. (1981). Therapy and prophylaxis of dendritic keratitis with topical human interferon. *In* "Herpetische Augenerkrankugen" (R. Sundmacher, ed.), pp. 401–407. Bergman Verlag, Munich.

Sundmacher, R., Cantell, K., and Mattes, A. (1984). Combination therapy for dendritic keratitis. High-titer alpha-interferon and trifluridine. *Arch. Ophthalmol. (Chicago).* **102**, 554–555.

Sundmacher, R., Mattes, A., Neumann-Haefelin, D., Adolf, G., and Kruss, B. (1987). The potency of interferon α-2 and interferon γ in a combination therapy of dendritic keratitis: A controlled clinical study. *Curr. Eye Res.* **6**, 273–276; see also *Proc. Int. conf. Herpetic Eye Dis.* abstr. P26.

Surbone, A., Yarchoan, R., McAtee, N., Blum, M. R., Maha, M., Allain, J.-P., Thomas, R. V., Mitsuya, H., Lehrman, S. N., Leuther, M., Pluda, J. M., Jacobsen, H. A., Kessler, F. K., Myers, C. E., and Broder, S. (1988). Treatment for the acquired immunodeficiency syndrome (AIDS) and AIDS-related complex with a regimen of 3'-azido-2',3'-dideoxythymidine (azidothymidine or zidovudine) and acyclovir. *Ann. Intern. Med.* **108**, 534–540.

Surjono, I., and Wigand, R. (1981). Combined inhibition of vaccinia virus multiplication by inhibitors of DNA synthesis. *Chemotherapy (Basel)* **27**, 179–187.

Szebeni, J., Wahl, S. M., Popovic, M., Wahl, L. M., Gartner, S., Fine, R. L., Skaleric, U., Friedman, R. M., and Weinstein, J. N. (1989). Dipyridamole potentiates the inhibition by 3'-azido-3'-deoxythymidine and other dideoxynucleosides of human immunodeficiency virus replication in monocyte–macrophages. *Proc. Natl. Acad. Sci. U.S.A.* **86**, 3842–3846.

Szebeni, J., Wahl., S. M., Schinazi, R. F., Popovic, M., Gartner, S., Wahl, L. M., Weislow, O. S., Betageri, G., Fine, R. L., Dahlberg, J. E., Hunter, E., and Weinstein, J. N. (1990). Dipyridamole potentiates the activity of zidovudine and other dideoxynucleosides against HIV-1 in cultured cells. *Ann. N.Y. Acad. Sci.* (in press).

Tanaka, A., Matsuoka, H., and Imanishi, J. (1984). Antiherpetic actions of prostaglandin D_2 and the combined treatment with human leukocyte interferon. *Proc. Int. Congr. Virol., 6th* abstr. P34–49.

Tanphaichitra, A., and Srimuag, S. (1987). Efficacy of acyclovir combined with immuno-potentiating agents in the treatment of varicella–zoster. *J. Antimicrob. Chemother.* **19**, 255–262.

Tozer, T. N. (1981). Concepts basic to pharmacokinetics. *Pharmacol. Ther.* **12**, 109–131.

Trousdale, M. D., and Nesburn, A. B. (1984). Human interferon alpha A or alpha D and trifluridine treatment for herpetic keratitis in rabbits. *Invest. Ophthalmol. Visual. Sci.* **25**, 480–483.

Tsiquaye, K. N., Collins, P., and Zuckerman, A. J. (1986). Antiviral activity of the polybasic anion, suramin and acyclovir in hepadna virus infection. *J. Antimicrob. Chemother., Suppl. B* **18**, 223–228.

Tyrell, D. A. (1976). Changing ideas on antiviral agents for respiratory infection. *J. Infect. Dis.* (Suppl. 133), A79–A82.

Ueno, R., and Kuno, S. (1987). Anti-HIV synergism between dextran sulphate and zidovudine. *Lancet* **1**, 796–797.

van Bijsterveld, O. P., Meurs, P. J., De Clercq, E., and Maudgal, P. C. (1989). Bromovinyl-deoxyuridine and interferon treatment in ulcerative herpetic keratitis: A double masked study. *Br. J. Ophthalmol.* **73**, 604–607.

van Ganswijk, R., Oosterhuis, J. A., Sward-Vandenburg, M., and Versteeg, J. (1983). Acyclovir treatment in stromal herpetic keratitis. *Doc. Ophthalmol.* **55**, 57–61.

Veckenstedt, A., Guttner, J., and Beladi, I. (1987). Synergistic action of quercetin and murine alpha/beta interferon in the treatment of mengo virus infection in mice. *Antiviral Res.* **7**, 169–178.

Verdonck, L. F., de Gast, D. C., Dekker, A. W., de Weger, R. A., Schuurman, H.-J., and Rozenberg-Arska, M. (1989). Treatment of cytomegalovirus pneumonia after bone marrow transplantation with cytomegalovirus immunoglobulin combined with ganciclovir. *Bone Marrow Transplant.* **4**, 187–189.

Vogt, M., Hartshorn, K. L., Furman, P. A., Chou, T. C., Fyfe, J. A., Coleman, L. A., Crumpacker, C., Schooley, R. T., and Hirsch, M. S. (1987). Ribavirin antagonizes the effect of azidothymidine on HIV replication. *Science* **235**, 1376–1379.

Vogt, M. W., Durno, A. G., Chou, T.-C., Coleman, L. A., Paradis, T. J., Schooley, R. T., Kaplan, J. C., and Hirsch, M. S. (1988). Synergistic interaction of 2′,3′-dideoxycytidine (ddCyd) and recombinant interferon-alpha-A (r-IFN-alpha-A) on HIV-1 replication. *J. Infect. Dis.* **158**, 378–385.

Vogt, M. W., Chou, T.-C., Hartshorn, K. L., Colman, L. A., Neumeyer, D. A., and Hirsch, M. S. (1989). Synergism and antagonism *in vitro* among various antiviral drugs in the treatment of HIV infections. *Proc. Int. Conf. AIDS, 5th* abstr. TP30.

Wade, J. C., McGuffin, R. W., Springmeyer, S. C., Newton, B., Singer, J. W., and Meyers, J. D. (1983). Treatment of cytomegaloviral pneumonia with high-dose acyclovir and human leukocyte interferon. *J. Infect Dis.* **168**, 557–562.

Wamsley, S. L., Chew, E., Fanning, M. M., Read, S. E., Vellend, H., and Salit, I. E. (1987). Treatment of cytomegalovirus retinitis with trisodium phosphonoformate (foscarnet). *Proc. Int. Conf. AIDS, 3rd* abstr. TH.8.1.

Warrell, M. J., White, N. J., Looareesuwan, S., Phillips, R. E., Suntharasamai, P., Chanthavanich, P., Riganti, M., Fisher-Hock, S. P., Nicholson, K.G., Manatsathit S., *et al.* (1989). Failure of interferon alpha and ribavirin in rabies encephalitis. *Br. Med. J.* **2929**, 830–833.

Wassilew, S. W. (1979). Treatment of herpes simplex of the skin. Critical evaluation of antiherpetic drugs with reference to relative potency on the eye. *Adv. Ophthalmol.* **38**, 125–133.

Watanabe, M., Tarui, H., Shoji, M., Naruse, Y., Fujita, T., Kagamimori, S., and Honda, T. (1984). Combined effects of interferon and tunicamycin on the mouse sarcoma cell proliferation. *Proc. Int. Congr. Virol. 6th* abstr. P15–18.

Weigent, D. A., Lloyd, R. E., Blalock, J. E., and Stanton, G. J. (1984). Synergism of antiviral activity in cell cultures treated with low concentrations of interferon and interferon-treated lymphocytes (41940). *Proc. Soc. Exp. Biol. Med.* **177**, 257–261.

Weinstein, J. N., Bunow, B., Weislow, O. S., Schinazi, R. F., Wahl, S. M., Wahl, L. M., and Szebeni, J. (1990). Synergistic drug combinations in AIDS therapy: Dipyridamole–zidovudine in particular and principles of analysis in general. *Ann. N.Y. Acad. Sci.* (in press).

Weiss, R. C., (1989). Synergistic antiviral activities of acyclovir and recombinant human leukocyte (alpha) interferon on feline herpesvirus replication. *Am. J. Vet. Res.* **50**, 1672–1677.

Weiss, R. C., and Oostrom-Ram, T. (1989). Inhibitory effects of ribavirin alone or combined with human alpha interferon on feline infectious peritonitis virus replication *in vitro. Vet. Mircrobiol.* **20**, 255–265.

Werner, G. H., Jasmin, C., and Chermann, J. C. (1976). Effect of ammonium 5-tungsto-2-antimoniate on encephalomyocarditis and vesicular stomatitis infection in mice. *J. Gen. Virol.* **31**, 59–64.

Whitley, R. J. (1988). Antiviral treatment of a serious herpes simplex infection: Encephalitis. *J. Am. Acad. Dermatol.* **18**, 209–211.

Whitley, R. J., Soong, S. J., Alford, C. A., Hirsch, M. S., Schooley, R., Oxman, M. N., Connor, J. D., Betts, R., Dolin, R., and Reichman, R. C. (1985). Design of therapeutic studies in herpes simplex encephalitis. *Lancet* **1**, 284–285.

Widell, A., Hansson, B. G., Öberg, B., and Nordenfelt, E. (1986). Influence of twenty potentially antiviral substances on *in vitro* multiplication of hepatitis A virus. *Antiviral Res.* **6**, 103–112.

Wigand, R., and Hassinger, W. (1980). Combined antiviral effect of DNA inhibitors on herpes simplex virus multiplication. *Med. Microbiol. Immunol.* **168**, 179–190.

Wigand, R., and Janz, C. (1981). *Proc. Inter. Congr. Virol., 5th* abstr. P7/26.

Wigdahl, B. L., Scheck, A. C., De Clercq, E., and Rapp, F. (1982). High efficiency latency and activation of herpes simplex virus in human cells. *Science* **217**, 1145–1146.

Wigdahl, B., Scheck, A. C., Ziegler, R. J., De Clercq, E., and Rapp, F. (1984). Analysis of the herpes simplex virus genome during *in vitro* latency in human diploid fibroblasts and rat sensory neurons. *J. Virol.* **49**, 205–213.

Williams, B. B., and Lerner, A. M. (1975). Antiviral activity of an adenosine deaminase inhibitor: Decreased replication of herpes simplex virus. *J. Infect. Dis.* **131**, 673–677.

Williams, G. J., and Colby, C. B. (1988). Combined effects of r interferon-β and AZT on HIV replication in vitro. *Antiviral Res.* **9**, 138 (abstr. II-22).

Williams, G. J., and Colby, C. B. (1989). Recombinant human interferon beta supresses the replication of HIV and acts synergistically with AZT. *J. Interferon Res.* **9**, 709–718.

Wilson, T. W., and Rajput, A. H. (1983). Amantadine–dyazide interaction. *Can. Med. Assoc. J.* **129**, 974–975.

Wilson, S. Z., Knight, V., Wyde, P. R., Drake, S., and Cough, R. B. (1980). Amantadine and ribavirin aerosol treatment of influenza A and B infection in mice. *Antimicrob. Agents Chemother.* **17**, 642–648.

Wirjawan, E., and Wigand, J. R. (1978). Combined antiviral effect of DNA inhibitors on adenovirus multiplication. *Chemotherapy, (Basel)* **24**, 347–353.

Worthington, M., and Baron, S. (1973). Effect of polyriboinosinic–poly-ribocytidylic acid and antibody on infection of immunosuppressed mice with vaccinia virus. *J. Infect. Dis.* **128**, 308–311.

Yamamoto, J., Walz, M. A., and Notkins, A. L. (1977). Viral-specific thymidine kinase in sensory ganglia of mice infected with herpes simplex virus. *Virology* **76**, 866–869.

Yamamoto, M., Yoshinobu, H., Tang, L.-L., and Mori, R. (1985). Effects of combined use of acyclovir and antibody in athymic nude mice inoculated intracutaneously with herpes simplex virus. *Antiviral Res.* **5**, 83–91.

Yarchoan, R., Perno, C. F., Thomas, R. V., Allain, J.-P., Willis, R. J., McAtee, N., Fischl, M. A., Dubinsky, R., Mitsuya, H., Lawley, T. J., Safai, B., Myers. C. E., Perno, C. F., Klecker, R. W., Wills, R. J., Fischl, M. A., McNeely, M. C., Pluda, J. M., Leuther, M., Collins, J. M., and Broder, S. (1988). Phase I studies of 2′,3′-dideoxycytidine in severe human immunodeficiency virus infection as a single agent and alternating with zidovudine (AZT). *Lancet* **1**, 76–80.

Yokosuka, O., Omata, M., Imazeki, F., Hirota, K., Mori, J., Uchiumi, K., Ito, Y., and Okuda, K. (1985). Combination of short-term prednisolone and adenine arabinoside in the treatment of chronic hepatitis B. *Gastroenterology* **89**, 246–251.

Yoshida, I., Yoshida, T., Shinada, M., and Azuma, M. (1984). Effect of ara-A on the antiviral action of interferon. *Proc. Int. Congr. Virol., 6th* abstr. P15–20.

Zeidner, N. S., Strobel, J. D., Perigo, N. A., Hill, D. L., Mullins, J. I., and Hoover, E. A. (1989). Treatment of FeLV-induced immunodeficiency syndrome (FeLV-FAIDS) with controlled released capsular implantation of 2′,3′-dideoxycytidine. *Antiviral Res.* **11**, 147–160.

Zeidner, N. S., Rose, L. M., Mathiason-DuBard, C.K., Myles, M. H., Hill, D. L., Mullins, J. I., and Hoover, E. A. (1990). Zidovudine in combination with alpha interferon and interleukin-2 as prophylactic therapy for FeLV-induced immunodeficiency syndrome (FeLV-FAIDS). *J. Acg. Immun. Def. Syndr.* **3**, 787–796.

Zerial, A., Hovanessian, A. G., Stefanos, S., Huygen, K., Werner, G. H., and Falcoff, E. (1982). Synergistic activities of type I (α, β) and type II (γ) murine interferons, *Antiviral Res.* **2**, 227–239.

Zhirnov, O. P. (1987). High protection of animals lethally infected with influenza virus by aprotinin–rimantadine combination. *J. Med. Virol.* **21**, 161–167.

Zimmermann, F., Biesert, L., von Briesen, H., Wegner, R., Helm, E. B., Staszewski, S., and Rübsamen-Waigmann, H. (1988). Development of HIV variants with high resistance against AZT under treatment with AZT. *Proc. Int. Conf. AIDS, 4th* abstr. 3656.

CHAPTER 5

Antimalarial Synergism and Antagonism

Jonathan L. Vennerstrom
College of Pharmacy
University of Nebraska Medical Center
Omaha, Nebraska 68198

William Y. Ellis
Wilbur K. Milhous
Division of Experimental Therapeutics
Walter Reed Army Institute of Research
Washington, D.C. 20307

John W. Eaton
Department of Laboratory Medicine
and Pathology
University of Minnesota
Minneapolis, Minnesota 55455

I. Introduction

Plasmodium falciparum, an obligate intracellular protozoan parasite, is the etiological agent of the most serious and fatal form of human malaria, accounting for innumerable deaths and immeasurable mortality for over one

SYNERGISM AND ANTAGONISM IN CHEMOTHERAPY
Copyright © 1991 by Academic Press, Inc.
All rights of reproduction in any form reserved.

billion people at risk in the world. With the relatively recent emergence of drug-resistant strains of *P. falciparum*, considerable attention has been focused on the development of alternative chemotherapy, especially the use of drug combinations. The acquisition of drug resistance, however, appears to be independent of the chemical class to which the parasites are exposed (Milhous *et al.*, 1989).

This chapter describes synergistic, additive, and antagonistic antimalarial drug combinations including nutritional synergism and antagonism. Synergism describes an interaction of two or more antimicrobial agents in which the effect produced by the combination of drugs is greater than the sum of their individual effects when used alone (Moellering, 1979). Examples of drug antagonism are less common. Krogstad and Moellering (1980) have postulated that clinically important antagonism may actually occur more frequently than is documented because of the inability of investigators to recognize its effect or their reluctance to report an adverse result.

It is proposed (Peters, 1979) that drug combination therapy and prophylaxis will delay or prevent the emergence of resistance to newly introduced antimalarial drugs, especially if the mechanism for acquiring resistance is through a spontaneous mutation (Sande and Mandell, 1980). Indeed, the one, if not only, advantage of additive combinations is the prevention of drug resistance (Donno, 1989). Complementary drug combinations as defined by Peters (1987) will not be discussed; these include both combinations that act against different stages of a malaria parasite (e.g., chloroquine + primaquine), and the sequential use of a rapid schizontocide followed by a second slowly acting drug to destroy lingering blood stages (e.g. quinine + tetracycline). Finally, the recent discovery of drug-resistance modulators will be discussed.

II. Antifols

Drug potentiation is not necessarily limited to combinations of drugs acting on a single metabolic pathway, although the majority of synergistic combinations of antimalarial drugs which antagonize folate metabolism are *p*-aminobenzoic acid (PABA) analogs and dihydrofolate reductase (DHFR) inhibitors (Fig. 1). These drug combinations have played an important role in the chemotherapy and prophylaxis of malaria over the past three decades. Synergistic combinations of PABA competitors and DHFR inhibitors include sulfadiazine plus pyrimethamine, dapsone (DDS) plus trimethoprim, and sulfalene plus trimethoprim (McCormick and Canfield, 1972). Hurly (1959a) reported a potentiation of pyrimethamine with sulfadiazine in Gambians infected with *Plasmodium falciparum*, *P. malariae*, or *P. ovale*. Patients could be successfully treated with a combined dosage that was 10% the

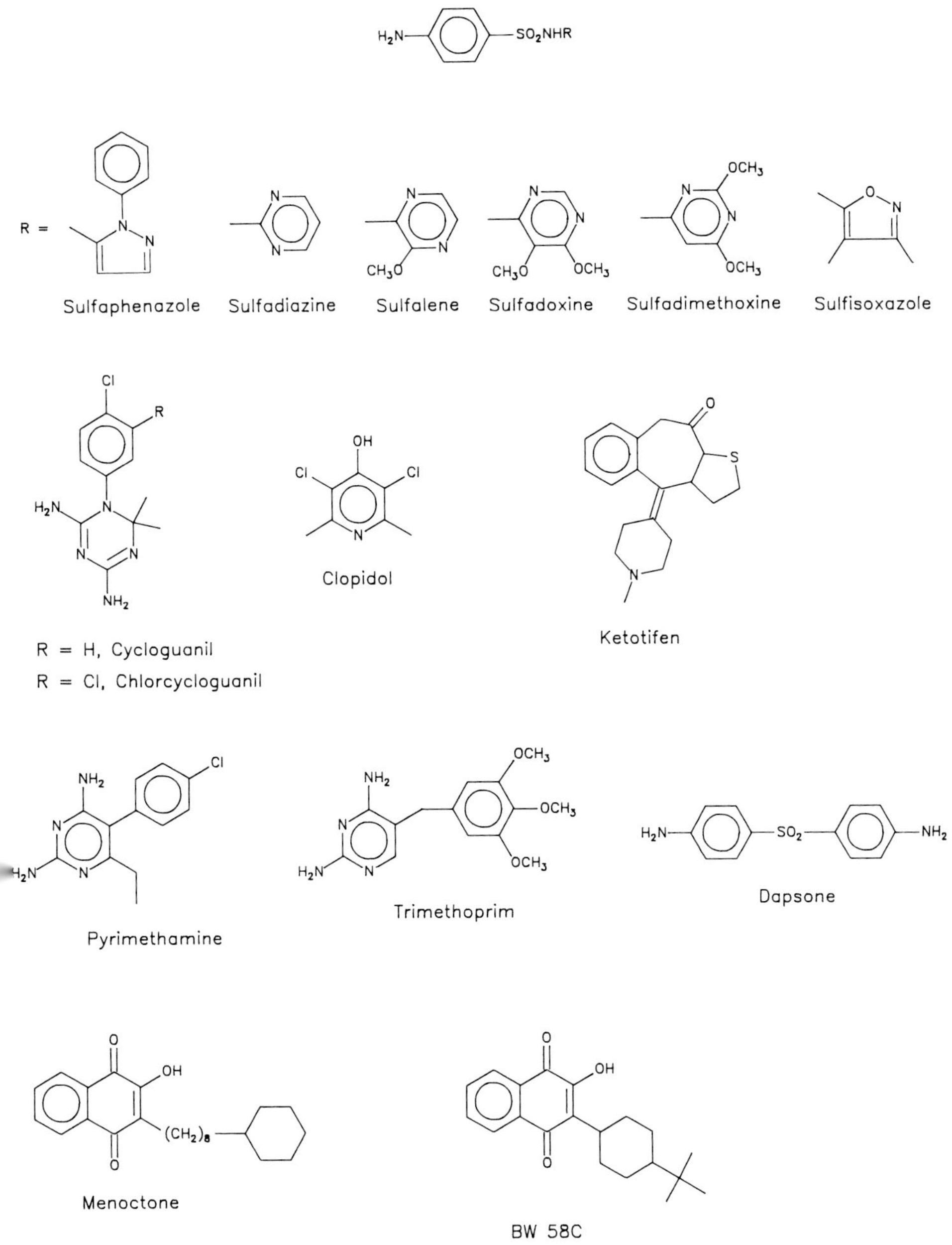
H2N SO2NHR
R =
Sulfaphenazole
Sulfadiazine
Sulfalene
Sulfadoxine
Sulfadimethoxine
Sulfisoxazole
R = H, Cycloguanil
R = Cl, Chlorcycloguanil
Clopidol
Ketotifen
Pyrimethamine
Trimethoprim
Dapsone
Menoctone
BW 58C

FIGURE 1

required dosage of pyrimethamine and 25% the required dosage of sulfadiazine when given alone. Lucas *et al.* (1969) reported similar findings in the use of pyrimethamine in combination with dapsone or sulfadoxine in a double-blind controlled study involving 280 children who were observed for one year in an endemic area of *P. falciparum*. (For a mathematical treatment of antifol/PABA analog synergism, see Chapter 11 by Jackson.)

Despite these well-documented capacities of sulfonamides to potentiate the activities of DHFR inhibitors, the ability of the malaria parasites to develop resistance to these drugs has greatly limited their clinical utility in many areas of the world. Recent clinical observations (Miller *et al.*, 1986) of severe allergic reactions (Stevens–Johnson syndrome and toxic epidermal neurolysis) associated with prophylactic use of sulfadoxine in combination with pyrimethamine (Fansidar®) raise serious questions concerning the safety of and tolerance for long-acting sulfonamides (such as sulfadoxine) in combination with pyrimethamine. New *in vitro* results, clinical reports, and field studies now suggest that parasites which are resistant to pyrimethamine may be more susceptible to proguanil. Accordingly, there is renewed interest in combining short-acting sulfonamides with proguanil for chemoprophylaxis in areas endemic for multidrug-resistant falciparum malaria (Shanks *et al.*, 1989). It is hoped that such combinations will not only improve prophylactic efficacy, but also reduce the risk for allergic reactions.

In vitro studies that demonstrate antagonism of antifolate antimalarials by folic acid have called into question current concepts of folate biochemistry in plasmodia in relation to the results from animal and human studies. In pursuit of quantitative methods to assess the *in vitro* efficacy of antifolate antimalarials, Milhous *et al.* (1985) find that physiologic levels of both PABA and folic acid result in a marked decrease in the intrinsic activity of both sulfadoxine and pyrimethamine. Currently there is little biochemical evidence to suggest that folic acid is utilized directly by *P. falciparum*. The very fact that antifolate antimalarials are such potent inhibitors of plasmodia disputes the hypothesis that plasmodia are able to use folate cofactors which are so abundant in the liver and host erythrocyte. The acquisition of this ability, though, would certainly provide a novel mechanism for development of drug resistance.

Table 1 provides a summary of findings conducted with plasmodia from both human and animal (rat, chicken, simian) hosts. While the antagonism of a PABA analog (dapsone) by PABA has been observed in humans (De-Gowin *et al.*, 1966), there is little experimental evidence to suggest that folic acid antagonizes the activity of any antifolate drugs in falciparum malaria. Although there are some inconsistencies in data from animals (probably attributable to differences in dose or route of administration), folic acid doses antagonize the activities of various antifolate antimalarials in animals. Frenkel and Hitchings (1957) also observed that both PABA and folic acid reversed the activities of pyrimethamine against toxoplasma in mice. After reviewing

TABLE 1
Examples of the Effects of PABA, Folic Acid Analogs, and Folinic Acid on the Activities of Antifolate Antimalarials in Animal and Human Studies

Species and references	Folate intermediate	Antifolate drug	Effect
Plasmodium berghei			
Thurston (1954)	Folic acid and PABA	Pyrimethamine	Antagonism
	Folic acid and PABA	Sulfadiazine	Antagonism
	Folic acid and PABA	Proguanil	Antagonism
Aviado and Ige (1968)	Folic acid and PABA	Metachloridine	Antagonism
Aviado et al. (1969)	Folic acid and PABA	Sulfadimethoxine	No effect
	Folic acid and PABA	Sulfadoxine	No effect
	Folic acid and PABA	Sulfalene	No effect
Plasmodium gallinaceum			
Bishop (1954)	Folic acid and PABA	Sulfadiazine	Antagonism
	PABA	Proguanil	No effect
Greenberg (1953)	PABA, PABG, folic and folinic acids	Chloroguanide	Antagonism
Greenberg (1954)	Folic acid	Sulfadiazine	Antagonism
	Folic acid analogs	Sulfadiazine	Antagonism
Rollo (1955)	Folic acid and PABA	Sulfadiazine	Antagonism
	Folic acid	Pyrimethamine	Antagonism
	PABA	Pyrimethamine	Slight antagonism

(continues)

TABLE 1 (*continued*)

Species and references	Folate intermediate	Antifolate drug	Effect
Bishop (1965)	Folic acid and PABA	Dapsone	Antagonism
Plasmodium knowlesi (in vitro)			
McCormick et al. (1971)	Folic acid, PABA, and folinic acid	Sulfalene	Antagonism
	Folic acid, PABA, and folinic acid	Pyrimethamine	No effect
Plasmodium falciparum			
Hurly (1959b)	Folic and folinic acids	Pyrimethamine	No effect
DeGowin et al. (1966)	PABA	Dapsone	Antagonism
Tong et al. (1970)	Folic and folinic acids	Chloroquine, pyrimethamine, and sulfisoxazole	No effect
Canfield et al. (1971)	Folinic acid	Quinine, pyrimethamine, and dapsone	No effect
Plasmodium falciparum (in vitro)			
Chulay et al. (1984)	Folic acid	Sulfadoxine	Antagonism
	Folic acid	Pyrimethamine	Antagonism
Watkins et al. (1985)	Folic acid and PABA	Sulfadoxine	Antagonism
	Folic acid and PABA	Pyrimethamine	Antagonism

these data, one could conclude that there are probably interspecies variations in the ability of plasmodia to utilize or break down exogenous folate. Peters (1987), Thompson and Werbel (1972), and Rollo (1964) have thoroughly reviewed the results of studies with antifolate antimalarials in both humans and animals.

As shown in Table 1, the first human experiment using a simultaneous administration of folic or folinic acid with pyrimethamine was conducted with Gambian infants and children (Hurly, 1959b). Using the minimal effective therapeutic dose of pyrimethamine and 30-fold higher doses of folinic and folic acid, there was no evidence of an adverse effect on these partially immune subjects.[1] Later, Strickland and Kostinas (1979) observed that American soldiers hospitalized for treatment of malaria suffered from megaloblastic anemia and reduced serum folic acid concentration. This deficiency was attributed to a combination of factors: (1) inadequate dietary folate; (2) reduced folate absorption by the intestine; (3) increased utilization of folic acid due to hemolysis and fever of malaria; and (4) drug inhibition by the use of pyrimethamine and sulfisoxazole. A subsequent study (Tong *et al.*, 1970) which is frequently cited concluded that coadministration of supplemental folates would not interfere with the activity of pyrimethamine. Although the presumed intent of the study was to examine the effects of folic and folinic acid supplements on the antimalarial treatment regimen (chloroquine + pyrimethamine + sulfisoxazole), one cannot overlook the obvious confounder in their study: all were treated with chloroquine in addition to pyrimethamine and sulfisoxazole. No mention was made of the prevalence of chloroquine-resistant malaria within the treatment group. Canfield *et al.* (1971) conducted a similar study in Vietnam with the combination of quinine, dapsone, and pyrimethamine and supplementation with folinic acid to treat anemia in patients with falciparum malaria. They concluded that the administration of folinic acid could be used to hasten the recovery from anemia without negating the effect of pyrimethamine and dapsone. Had there been any effect on drug activity, they would have observed a recrudescence of infection as is usually observed when quinine is given alone. Ethical considerations prevented the use of an appropriate control, i.e., folinic acid + dapsone + pyrimethamine without quinine.

The transport of folates *in vivo* has been reviewed by Herbert *et al.* (1980) and by Rowe (1978). The folate in human plasma appears to be distributed in three fractions: (1) free folate; (2) folate loosely bound to plasma proteins; and (3) folate bound to high-affinity protein binders. The low-affinity binding is apparently nonspecific and similar to the protein binding of various drugs.

Although the physiological and clinical properties of the specific folate-binding proteins (FABP) have not been fully characterized, their function or

---

[1] Gail and Herms (1969) have proposed that extremely high concentrations of PABA will not reverse the activity of pyrimethamine against *P. berghei* in immune mice.

dysfunction could have significant implications in the chemotherapeutics of antifolate drugs. In severe folate deficiency, the FABPs measure eight times greater than in normal serum. There is some evidence which suggests that malarial infection is closely linked to the nutritional status of the host. A recrudescence of malaria infection was reported by Murray *et al.* (1975, 1976) when starving patients were brought into food shelters or were hospitalized. This phenomenon, often referred to as "refeeding" malaria, is attributed to changing from famine to a near normal diet. Murray *et al.* (1976) has attributed this phenomenon to the availability of serum iron. One might speculate, though, that FABP serves to sequester serum folate as transferrin sequesters serum iron from invading pathogens.

Fernandes-Costa and Metz (1979) have demonstrated that levels of FABP in human umbilical cord serum were much higher than levels in maternal serum. Kamen and Caston (1975) have hypothesized that FABP may actually function by concentrating folate across the placenta and thus account for the preferential uptake of folate derivatives by the fetus. Gail and Herms (1969) investigated the role of exogenous folic acid in influencing the parasitemias of pregnant women with severe anemia in Nigeria. Two or four weeks after treatment with folic acid, a group of women who received the folic acid supplement had significantly higher reticulocyte counts, but equivalent median parasite densities compared with a control group. Since it is not contraindicated to give regular suppressive prophylaxis during pregnancy, it was necessary to determine if treatment of anemia with supplemental folic acid would exacerbate malaria infections. Brabin (1982) has hypothesized that folacin deficiency actually enhances the maternal immunosuppression that occurs with malarial infections in pregnant women and may diminish the passive immunity acquired by the fetus. The specific role of FABP in pregnancy and malaria infection remains to be investigated.

In summary, the precise mechanism for the antagonism of pyrimethamine and sulfadoxine by folic acid remains an enigma and will require elucidation of folate biosynthesis in falciparum malaria. The observations suggest that the biochemical or physiological factors which can modulate folate homeostasis in the host might very well modulate antagonism of antifolate antimalarials as well as the resulting clinical response in patients. (For a review of antineoplastic antifolate combinations, see Chapter 10 by Galivan.)

III. Combinations of Antifols with Other Antimalarial Drugs

Strong synergism is observed between cycloguanil and the hydroxynaphthoquinone menoctone (Peters 1970) and between a quinolone ester (ICI 56780) and sulfonamides or chlorcycloguanil (Ryley and Peters, 1970) against murine *Plasmodium berghei in vivo.* However, cycloguanil and menoctone produce only an additive effect against a multidrug-resistant Indochina III strain of

P. falciparum in vitro (Milhous, *et al.*, 1985). Another hydroxynaphthoquinone, BW58C, in combination with clopidol, is synergistic against *P. falciparum in vitro* (Latter *et al.*, 1984). Hydroxynaphthoquinones such as menoctone and BW58C (see Fig. 1) selectively inhibit the plasmodial respiratory chain-linked enzyme dihydroorotate dehydrogenase (Gutteridge and Coombs, 1977; Hammond *et al.*, 1985; Hudson, 1984; Hudson *et al.*, 1985). This inhibition produces a blockade of pyrimidine biosynthesis which would explain the synergism observed *in vivo.*

Exposure of mice infected with *P. berghei* NK 65 strain to a mixture of sulfaphenazole and chloroquine (Peters *et al.*, 1973) or sulfaphenazole and quinacrine (Peters, 1969) greatly decreases the rate at which the parasites become resistant to chloroquine and quinacrine, respectively. In fact, resistance to chloroquine develops rarely, if at all (Peters *et al.*, 1973). Similar results are obtained for combinations of mefloquine with pyrimethamine, sulfaphenazole, or primaquine (Peters *et al.*, 1977). A triple combination of pyronaridine, sulfadoxine, and pyrimethamine is additive against *P. berghei* ANKA (Shi *et al.*, 1990) but greatly inhibits the development of resistance to pyronaridine (Shao and Ye, 1986).

In both *P. berghei* and *P. yoelii*, drug resistance to both pyrimethamine and sulfadoxine develops only slowly and to a low level whereas no resistance develops to chloroquine when a triple combination of chloroquine–sulfadoxine–pyrimethamine is used (Peters, 1974). Therefore, this triple combination slows the rate at which *P. berghei* acquires resistance to any of the component drugs alone. Subsequently, chloroquine together with sulfadoxine–pyrimethamine came to be used in areas of drug-resistant *P. falciparum* (Peters, 1987). A regimen of quinine followed by sulfadoxine–pyrimethamine is also claimed to be effective against chloroquine-resistant falciparum malaria (Hall *et al.*, 1975a). In contrast, chloroquine in combination with pyrimethamine (Darachlor®) is not additive and does not appear to delay the onset of resistance to either agent (Rabinovitch, 1965; Peters *et al.*, 1973; Peters, 1987).

A similar result is obtained against *P. berghei* with the additive triple combination of mefloquine–sulfadoxine–pyrimethamine (Merkli *et al.*, 1980), where the resistance to mefloquine is greatly inhibited. Lines of *P. berghei* exposed to constant drug selection pressure with mefloquine alone developed resistance; for example, the dose that would still permit an 8–10% parasitemia was increased from 3 to 110 mg/kg after 43 serial passages in mice. Using mefloquine in combination with pyrimethamine and sulfadoxine, the largest dosage that would still permit an 8–10% parasitemia could only be increased, after 70 passages, to 14 mg/kg of mefloquine and 4.5 mg/kg of the pyrimethamine–sulfadoxine mixture (Merkli *et al.*, 1980). Resistance to mefloquine did not develop when *P. falciparum* was exposed to the preceding triple combination (Tan-Ariya *et al.*, 1984) which is slightly antagonistic (Milhous *et al.*, 1983) against this species. The mefloquine–sulfadoxine–pyrimethamine combination is effective for both prophylaxis and treatment of falciparum malaria as demonstrated in world-wide clinical trials (Desjardins et al., 1988; Navaratnam et al., 1989; Tin et al., 1985; Botero et al., 1985; Ekue et al., 1985; Harinasuta et al., 1985; Win et al., 1985; Anh et al., 1990) and is also more effective than is any component alone (Win et al., 1985). However, the potential for either sulfonamide allergy or dose-related reactions to mefloquine dictate caution in the use of this combination. As noted in Section II, the use of any long-acting sulfonamide for prophylaxis is contraindicated (Miller et al., 1986; Scholer et al., 1984).

Other antifol drug combinations of this category include that of ketotifen and sulfadoxine which is synergistic against P. cynomolgi-infected macaque monkeys (Huang et al., 1987). Combinations of dapsone with chloroquine and primaquine (Eppes et al., 1967) or with mefloquine (Puri and Dutta, 1989) are more effective against P. falciparum chloroquine-resistant and P. berghei mefloquine-resistant malaria, respectively, than is dapsone alone. Unfortunately, the combination of dapsone with either chloroquine or primaquine may also potentiate the methemoglobinemia observed in the use of dapsone alone for malaria prophylaxis; methemoglobinemia was greatest for the dapsone–primaquine combination (Clyde et al., 1970).

IV. Quinolines

In contrast to the antifols, the precise mechanism of action of the quinolines (Fig. 2) remains undefined. This makes it difficult to understand many of the synergistic and antagonistic interactions described next. Cinchonine is synergistic when combined with either quinine or quinidine or both against several strains of P. falciparum in vitro; in contrast, the combination of quinine and quinidine is additive (Druilhe et al., 1988). The triple combination of quinine–quinidine–cinchonine (Quinimax®) is effective for therapy of chloroquine-resistant P. falciparum infections in Africa (Deloron et al., 1989, 1990), whereas no combination of Cinchona alkaloids is more effective than quinine against clinical falciparum malaria in Asia (Bunnag et al., 1989). In combination with other antimalarial quinolines, quinine is generally antagonistic. For example, combinations of quinine and chloroquine (Stahel et al., 1988a; Rahman and Warhurst, 1985), quinine and mefloquine (Geary et al., 1986), and chloroquine and amodiaquine (Rahman and Warhurst, 1985) are each antagonistic against P. falciparum in vitro. In addition, the quinine–mefloquine combination can produce toxic synergism in the clinical setting (Göessi et al., 1987).

Other quinoline combinations that exhibit antagonism when tested against P. falciparum in vitro include those of chloroquine–mefloquine, chloroquine–amodiaquine, mefloquine–amodiaquine, mefloquine–piperaquine, and mefloquine–3-methylchloroquine (Geary et al., 1986; Fenchuan and Lin,

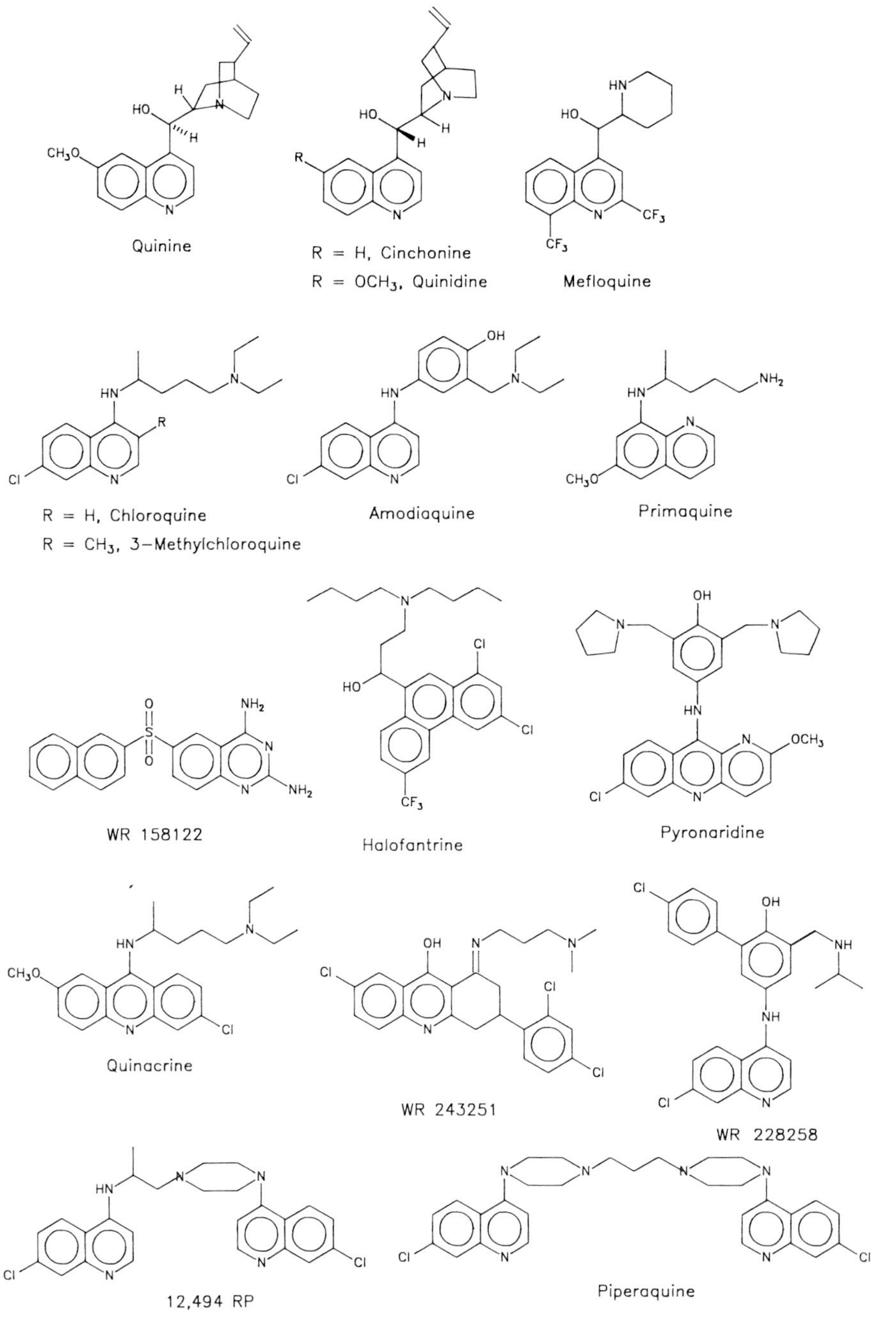

FIGURE 2 Quinolines.

1987; Stahel *et al.*, 1988a). Geary *et al.* (1986) suggest that the antagonism observed with mefloquine results from the opposing action of mefloquine and the other quinolines on membranes; mefloquine labilizes membranes whereas other quinolines stabilize membranes. For the other antimalarial quinolines, the results are not surprising when one considers that combinations of structurally similar drug molecules which act by the same mechanism often result in drug antagonism (Donno, 1989). An apparent exception to the antagonism that is observed among the quinoline antimalarial drugs is the reported synergy (Ye *et al.*, 1989) between chloroquine and tetrandrine (an isoquinoline alkaloid) against both chloroquine-sensitive and -resistant *P. falciparum in vitro*.

During the 1970s, a large number of 2,4-diamino-6-substituted quinolines were synthesized and tested in the U.S. Army antimalarial research program. One of the most active among these was WR 158122, which can effect radical cures in *Aotus trivirgatus* against the chloroquine-resistant Vietnam Oak Knoll strain of *P. falciparum* at low doses. WR 158122 also has shown excellent repository activity in rhesus monkeys but requires the coadministration of sulfonamides to prevent the rapid onset of resistance (Schmidt, 1979a,b). Balancing this, however, is a powerful synergy between sulfadiazine and WR 158122 in which activity is increased up to 75-fold (Schmidt, 1979b). A more intriguing form of synergy is that observed between WR 158122 and its tetrahydro analog WR 180872 (Kinnamon *et al.*, 1976). No explanation of this perplexing activity is available and its resolution presents an interesting challenge for the biochemist.

An equally curious behavior is exhibited by the floxacrine analog WR 243251. Against *P. berghei* in the Rane mouse model (Osdene *et al.*, 1967) this compound demonstrated an ED_{50} of 5 mg/kg. Surprisingly, neither the R enantiomer with an ED_{50} of 20 mg/kg nor the S enantiomer with an ED_{50} of 80 mg/kg was as active as the racemate. Subsequent studies (Mussallam *et al.*, 1987) have demonstrated a marked degree of synergism between the R and S stereoisomers against *P. falciparum in vitro* and against *P. berghei in vivo*.

A remarkably different and even selective mode of action is seen in the "autosynergy" of the amodiaquine analog WR 228258. Against *P. berghei* in the Rane mouse model (Osdene *et al.*, 1967) WR 228258 demonstrates a "normal" dose–response curve with maximum activity seen at 40 mg/kg and decreasing thereafter in a manner consonant with toxicity at higher doses. But the LD_{50} for this compound was approximately 900 mg/kg and toxicity was therefore wholly unexpected at such low dosing. Further studies revealed that at doses between 80 and 640 mg/kg, WR 228258 was less effective in clearing or suppressing plasmodia so that the cause of death in mice given these doses was apparently due more to high parasitemia than to chemical toxicity.

In vitro actvity against *P. falciparum* provides more challenge than light. When tested against a chloroquine-sensitive strain, WR 228258 exhibits an ED_{50} of 2 ng/ml, but against the chloroquine-resistant strain the activity of this drug is biphasic, showing an "initial" ED_{50} of 2 ng/ml, a no effect dose of 100 ng/ml, and a "final" ED_{50} of 300 ng/ml. No satisfactory explanation of this behavior has been determined.

V. Quinoline/Antibiotic Combinations

The most widely studied synergistic quinoline/antibiotic combination is that of chloroquine and erythromycin. The activity of chloroquine against two chloroquine-resistant lines (RC and NS) of *P. berghei* is considerably improved when coadministered with erythromycin, and the potentiation is greatest in the highly resistant RC strain; no potentiation is observed against the chloroquine-sensitive N strain (Warhurst *et al.*, 1976). Erythromycin itself is active only at high concentrations against *P. berghei*. Warhurst (1977) suggests that the activity of erythromycin on chloroquine-resistant *P. berghei* might be due to a chloroquine-induced increased permeability of the parasite mitochondria to erythromycin. The apparent lack of synergism against the sensitive (N) strain may be due to the fact that the direct antimicrobial action of chloroquine on the parasite is so rapid that it masks the relatively delayed effect of erythromycin on parasite mitochondria. Chloroquine–erythromycin synergism is also observed against various chloroquine-resistant strains of *P. falciparum* in culture (Gershon and Howells, 1984, 1986; Rahman and Warhurst, 1986). The chloroquine–erythromycin mixture fails, however, when used to treat patients infected with chloroquine-resistant *P. falciparum* (Phillips *et al.*, 1984).

Other macrolide antibiotics including spiramycin, oleandomycin, and tylosin are also synergistic with chloroquine against the drug-resistant K1 strain of *P. falciparum* in culture (Gershon and Howells, 1986). Against this same strain of *P. falciparum*, an antagonistic effect is observed between erythromycin and either quinine, amodiaquine, or mefloquine (Gershon and Howells, 1984; Rahman and Warhurst, 1985). The erythromycin–mefloquine combination is also not successful in preventing the development of mefloquine resistance in *P. berghei* (Puri and Dutta, 1989). Finally, spiramycin is synergistic with 12,494RP, a bis-4-aminoquinoline (Benazet and Godard, 1973), against chloroquine-resistant *P.y. nigeriensis* and with mefloquine and halofantrine against a drug-sensitive N strain of *P. berghei* (Peters, 1987).

Few quinine/antibiotic combinations have been studied aside from the oft-used complementary quinine–tetracycline combination. A synergistic effect of colchicine on quinine was reported (Reba and Sheehy, 1967) in the treatment of 22 U.S. soldiers with falciparum malaria in Vietnam. Quinine and clindamycin are synergistic against the chloroquine-resistant K1 strain of *P. falciparum* but additive against the chloroquine-sensitive NF54 strain (Rahman and Warhurst, 1985). This drug combination is also effective against multidrug-resistant falciparum malaria (Clyde *et al.*, 1975; Hall *et al.*, 1975b; Miller *et al.*, 1974; Kremsner *et al.*, 1988; Kremsner, 1990). Unfortunately, Hall *et al.* (1975b) and Clyde *et al.* (1975) report that the quinine–clindamycin combination is toxic to humans unless given sequentially with clindamycin following quinine. Hall *et al.* (1975b) suggest that this combination potentiates both antimalarial efficacy and toxicity. In any event, the sequential therapy is also useful (Hall *et al.*, 1975b; Miller *et al.*, 1974).

VI. Artemisinin Combinations

Mefloquine, primaquine, tetrandrine, and tetracycline are consistently synergistic in combination with artemisinin (Fig. 3), whereas chloroquine, amodiaquine, and antifols are usually antagonistic or at best additive. Mefloquine, tetrandrine, and tetracycline show synergism with artemisinin against both sensitive and resistant strains of *P. falciparum in vitro*, while primaquine is additive against the sensitive (NF54) and synergistic against the resistant (K1) strains (Chawira *et al.*, 1986b; Chawira and Warhust, 1987; Ye *et al.*, 1989). However, against the same strains of *P. falciparum*, both pyrimethamine and chloroquine are antagonistic to artemisinin. Against both drug-sensitive and -resistant strains of *P. berghei* in mice, mefloquine, primaquine, tetracycline, spiramycin, and clindamycin are synergistic, chloroquine and amodiaquine are additive, and all antifols tested are antagonistic (Chawira *et al.*, 1985, 1986a, 1987) in combination with artemisinin.

In *P. berghei*, cross resistance with artemisinin is uncommon. For example, only the RC strain of *P. berghei* shows cross resistance to an artemisinin–mefloquine combination (Chawira *et al.*, 1986a). Despite the fact that com-

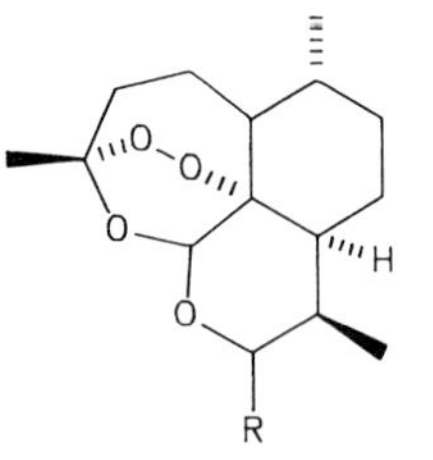

R = ═O, Artemisinin
R = OCH$_3$, Artemether
R = OCOCH$_2$CH$_2$COOH, Artesunate

FIGURE 3   Artemisinin derivatives.

binations of antifols with artemisinin are antagonistic, the triple combination of artemisinin–sulfadoxine–pyrimethamine markedly reduces the rate of development of resistance to the individual components in *P. berghei* (Zhou and Ning, 1988). It is suggested (Chawira and Warhurst, 1987; Chawira *et al.* 1985, 1987) that artemisinin may alter membranes or membrane receptors in such a way that it enhances the antiparasitic effect of mefloquine, primaquine, and tetracycline and conversely, in the same membrane-altering mechanism, interrupts chloroquine and antifol active concentration processes accounting for the observed antagonism.

In a randomized comparative clinical study, a combination of artemisinin with a regimen of mefloquine–sulfadoxine–pyrimethamine greatly increases the rate of parasite clearance in *P. falciparum* malaria compared with artemisinin alone (Li *et al.*, 1984). A combination of artemether (Fig. 3), an artemisinin analog, and mefloquine prevents the recrudescence associated with artemether alone (Shwe *et al.*, 1988). This combination is superior to quinine in the treatment of cerebral malaria due to *P. falciparum* (Shwe *et al.*, 1989).

The endoperoxide function in artemisinin is the one absolute essential for antimalarial activity (Gu *et al.*, 1980) suggesting an oxidant mode of action. Doxorubicin and miconazole, two drugs with oxidant effects (Pfaller and Krogstad, 1983; Powis, 1989), and two potential "catalytic" oxidant catechol flavones, casticin and artemitin, produce synergistic effects with artesunate (Fig. 3), a water-soluble artemisinin analog (Krungkrai and Yuthavong, 1987) and artemisinin (Elford *et al.*, 1987), respectively, against *P. falciparum* in culture. Artesunate induces lipid peroxidation to a greater extent (3-fold) in *P. falciparum*-infected versus uninfected red blood cells (Meshnick *et al.*, 1989), and dietary vitamin E deficiency in *P. yoelii*-infected mice is synergistic with artemisinin (Levander *et al.*, 1989a). A progressive increase in the potency of artemisinin with increasing oxygen tensions and a significant reduction in its potency by coadministration of dithiothreitol, vitamin E, catalase (Krungkrai and Yuthavong, 1987), ascorbate, or glutathione (Meshnick *et al.*, 1989) further support this hypothesis.

VII. Oxidants/Oxidant Drugs and Other Factors in Combination

In addition to artemisinin, numerous structurally diverse oxidants and oxidant drugs are antimalarial; their efficacy may depend in part on the well-known oxidant sensitivity of malaria parasites and their host erythrocytes (Brown and Green, 1987; Vennerstrom and Eaton, 1988; Clark *et al.*, 1989; Golenser and Chevion, 1989; Ames *et al.*, 1985). Hydrogen peroxide, either reagent or generated by glucose/glucose oxidase, is lethal for *P. yoelii* (Dockrell and Playfair, 1983, 1984) and *P. falciparum* (Wozencraft *et al.*, 1984; Kamchonwongpaisan *et al.*, 1989). In each case significant protection is afforded by the addition of catalase, demonstrating that the active principle is hydrogen peroxide. In contrast, antimalarial combinations of polyamine oxidase and various polyamines (Morgan *et al.*, 1981; Ferrante *et al.*, 1983; Morgan and Christensen, 1983; Rzepczyk *et al.*, 1984; Egan *et al.*, 1986) appear to involve a different mechanism of interaction. Catalase did not prevent this effect (Rzepczyk *et al.*, 1984) even though polyamine oxidase also produces hydrogen peroxide (Egan *et al.*, 1986).

Plasmodia species are microaerophilic and grow best under reduced oxygen pressures, i.e., 20 mm Hg or 1/7 atmospheric oxygen pressure (Scheibel *et al.*, 1979b). Increased oxygen tensions, particularly those greater than atmospheric, are poisonous to the parasite and act synergistically with some antimalarial drugs (Friedman, 1979; Divo *et al.*, 1985; Scheibel *et al.*, 1979b; Pfaller and Krogstad, 1983) (Fig. 4). Elevation of oxygen levels from 0.3–1% to 15–18% (near atmospheric) enhances the antimalarial activity of imidazoles such as ketoconazole (Pfaller and Krogstad, 1983), antibiotics such as oxytetracycline, chlortetracycline, minocycline, thiamphenicol, the mitochondrial-specific dye Janus Green, the riboflavin antagonist 8-methylamino-7-methyl-10-ribityl isoalloxazine (Divo *et al.*, 1985), and oxines such as 8-hydroxy-5-methyl quinoline (Scheibel and Adler, 1981) against *P. falciparum* in culture. This synergism suggests that the drugs themselves may exert oxidant stress on the infected erythrocyte. Indeed, minocycline, which contains a *p*-aminophenol function capable of redox cycling is the most potent of the tetracycline antibiotics against malaria (Willerson *et al.*, 1972; Puri and Dutta, 1981).

The effectiveness of many oxidant drugs against malaria may derive simply from the fact that infected erythrocytes are especially oxidant sensitive (Etkin and Eaton, 1975; Eckman and Eaton, 1979). However, it is likely that the selective effect of oxidant drugs on infected red cells may be partially attributable to the occurrence of reactive intracellular iron. The strongest, albeit indirect, evidence for this stems from the dramatic antagonistic effects of simultaneous administration of peroxides and the powerful iron chelator, desferrioxamine (Fig. 5). This chelator binds all six coordination positions of ferric iron (Keberle, 1964) thereby preventing any redox reactions on the part of the metal (Gutteridge *et al.*, 1979; Graf *et al.*, 1984). Both *t*-butyl hydroperoxide and hydrogen peroxide cause selective destruction of *P. vinckei* malaria-infected murine erythrocytes *in vivo*. When administered concurrently with desferrioxamine, however, the antimalarial and hemolytic effects of these agents are completely blocked (Clark *et al.*, 1983, 1984b). Thus, the *t*-butylalkoxy and *t*-butylperoxy radicals (O'Brien, 1969; Kalyanaraman *et al.*, 1983; Thornalley *et al.*, 1983; Davies, 1988) produced in reactions between *t*-butyl hydroperoxide and iron or "hemozoin" might be important contribu-

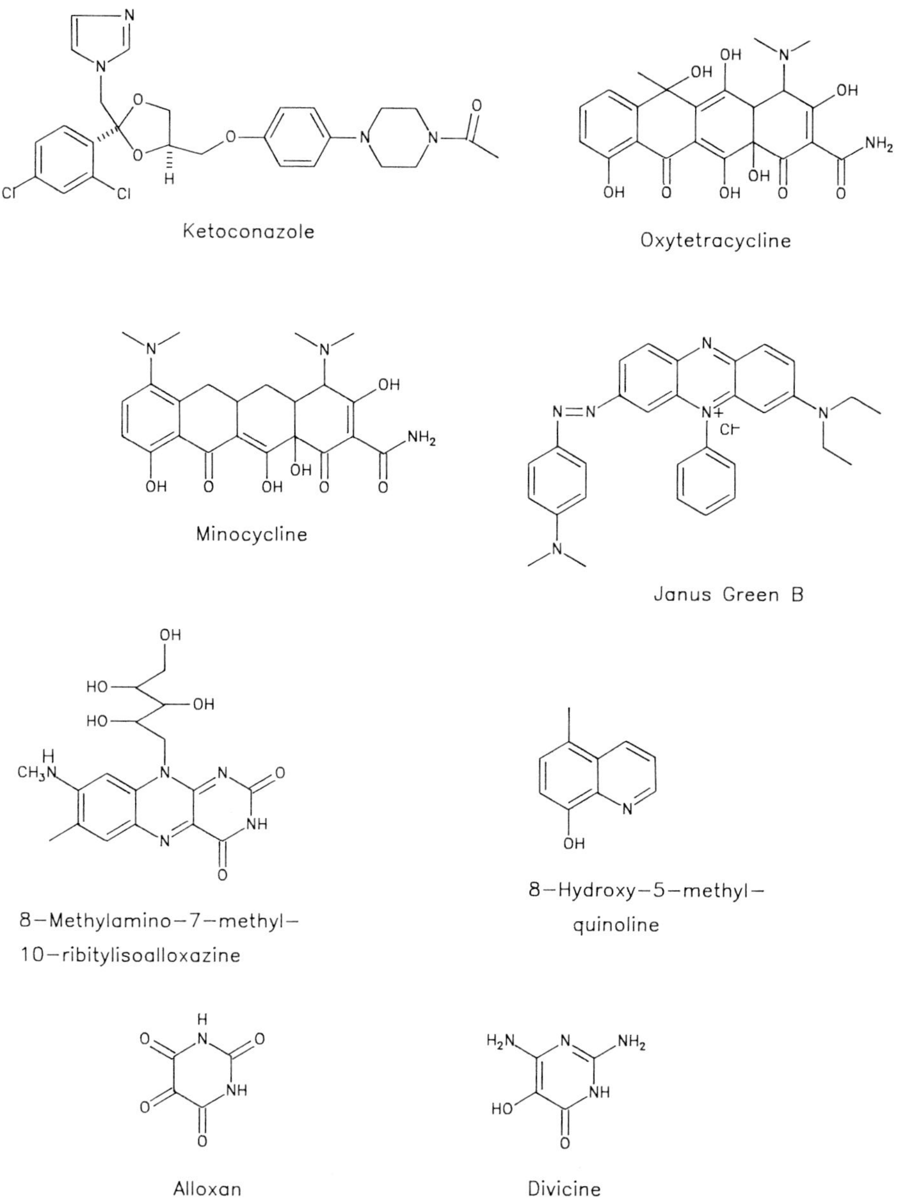

Ketoconazole

Oxytetracycline

Minocycline

Janus Green B

8−Methylamino−7−methyl−
10−ribitylisoalloxazine

8−Hydroxy−5−methyl−
quinoline

Alloxan

Divicine

FIGURE 4   Oxidant antimalarials.

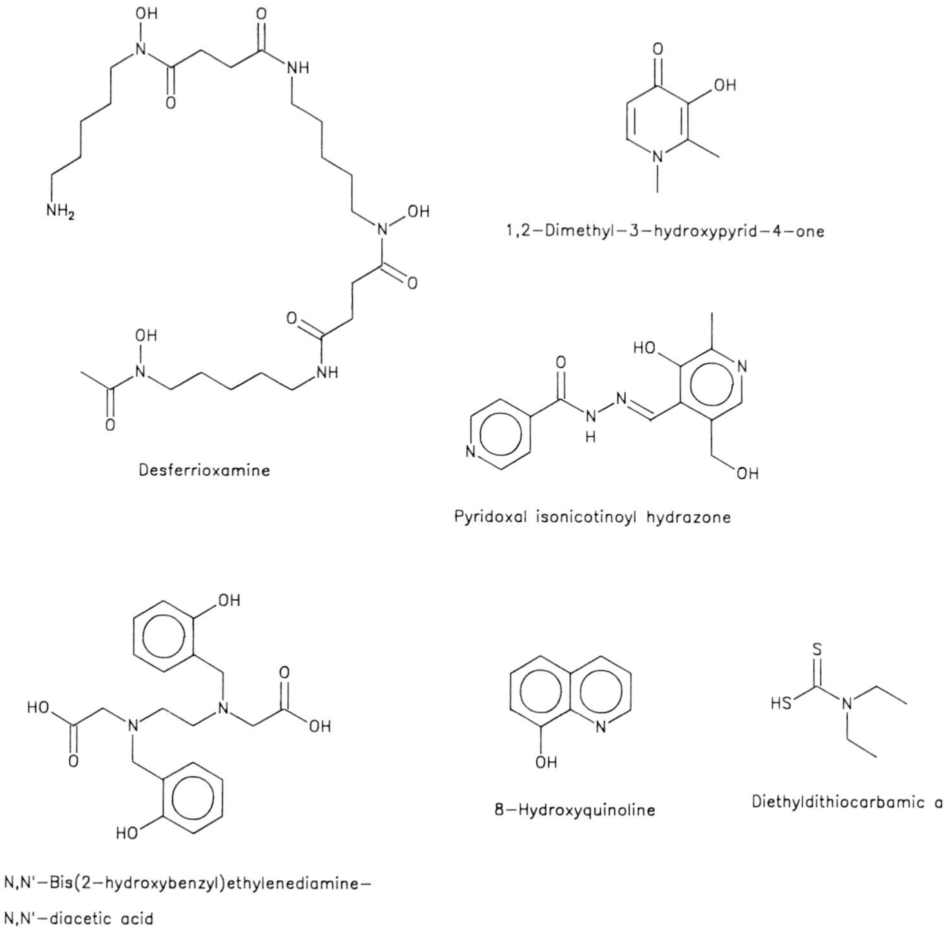

Desferrioxamine

1,2−Dimethyl−3−hydroxypyrid−4−one

Pyridoxal isonicotinoyl hydrazone

N,N'−Bis(2−hydroxybenzyl)ethylenediamine−
N,N'−diacetic acid

8−Hydroxyquinoline

Diethyldithiocarbamic a

FIGURE 5   Metal chelators.

tors to this antimalarial efficacy. Similarly, desferrioxamine completely abrogates the antimalarial and hemolytic effects of alloxan and divicine (Fig. 4), two redox cycling antimalarials (Clark and Hunt, 1983; Clark et al., 1984a). Diethyldithiocarbamate (Fig. 5), another metal chelator, also completely blocks the antimalarial activity of alloxan (Clark and Hunt, 1983). These observations suggest that free iron (and, perhaps in the case of diethyldithiocarbamate, copper) may ordinarily be present within infected cells or that it is generated in the oxidant-mediated degradation of hemoglobin and other metallic proteins (Trotta et al., 1981, 1982; Gutteridge, 1986) (see Section IX).

VIII. Nutritional Synergism and Antagonism

Several nutritional experiments indirectly support an oxidant mechanism of action for several antimalarial agents by enhancing or inhibiting their effects. For example, 2(3)-tert-butylated hydroxyanisole (BHA) prevents the antimalarial effect of both alloxan and divicine (Fig. 4) against *P. vinckei in vivo* (Clark and Cowden, 1985) while dietary (0.75%) BHA enhances parasitemias in *P. chabaudi* (Clark *et al.*, 1987). The combination of copper and ascorbate is synergistic against *P. falciparum in vivo*; this synergism is more pronounced in G6PD-deficient red cells (Marva *et al.*, 1989). These authors suggest that this copper/ascorbate synergism is mediated by active oxygen-derived species.

Godfrey (1957) demonstrated a direct antiparasitic action of dietary cod liver oil against *P. berghei* and showed that vitamin E completely reverses this inhibition. Recent experiments by Levander *et al.* (1989a–c, 1990) show that artemisinin and a variety of oils (cod liver, menhaden, anchovy, salmon, and linseed) containing elevated amounts of polyunsaturated fatty acids are synergistic with vitamin E deficiency against *P. yoelii* and *P. berghei in vivo*. In fact, no additional therapeutic effect of artemisinin was seen in *P. yoelii*-infected mice on a vitamin E deficient diet supplemented with cod liver oil (Levander *et al.*, 1989a). Mice deficient in vitamin E (a deficiency which causes red cells to be especially sensitive to lipid peroxidation and peroxide hemolysis) survive *P. berghei* infections much longer than do their normal counterparts (Eaton *et al.*, 1976). Not only do deficient animals have lower parasitemias, the parasites are also located preferentially in the younger and more oxidant-resistant reticulocytes. It is reasonable to suppose that the resistance to parasitemia conferred by vitamin E deficiency as well as the antimalarial synergy between vitamin E deficiency and both artemisinin and the prooxidant polyunsaturated oils is due to premature hemolysis of oxidant-sensitive infected erythrocytes.

IX. Chelators as Potential Antimalarials

A. INTRODUCTION

A number of metal chelators, principally drugs which bind either iron or copper, are variously effective against experimental animal malarias, cultured *P. falciparum*, and, in some cases, *P. falciparum* in primate hosts (Fig. 5). Unfortunately, none of these has proven therapeutic potential against human malaria infections. However, elucidation of the principles by which these drugs act may eventually lead to the development of effective metal-binding antimalarials.

In general, metal-binding drugs exert antimalarial effects in one of two ways. Desferrioxamine, α-ketohydroxypyridines, pyridoxal isonicotinoyl hydrazone and N,N'-bis(2-hydroxybenzyl)ethylenediamine-N,N'-diacetic acid evidently work through starvation of the parasite. Iron antagonizes the action of these drugs. By virtue of binding free iron within the infected erythrocyte, these chelators may prevent the synthesis of one or more critical iron-containing parasite proteins. In contrast, chelators such as diethyldithiocarbamate and 8-hydroxyquinoline probably act synergistically with trace metals present within infected red cells. The chelates which form within the infected cell are directly toxic to the parasite through presently unknown mechanisms.

B. CHELATORS WHICH "STARVE" THE MALARIA PARASITE OF METALS

As noted above, several iron-binding drugs are effective against malaria by virtue of forming tight, unreactive chelates. Desferrioxamine (DFO), long a mainstay in the therapy of iron overload disorders in humans, is one such agent that forms an unreactive DFO–Fe^{3+} complex (Gutteridge *et al.*, 1979). For example, the addition of DFO to systems containing polyunsaturated fatty acids and iron will block iron-mediated lipid peroxidation (see, e.g., Graf *et al.*, 1984). DFO will similarly prevent iron-catalyzed formation of reactive oxygen species such as hydroxyl radicals.

In this light, it is perhaps surprising that DFO is effective in arresting the growth of *P. falciparum in vitro* (IC_{50}, 4 μM) (Raventos-Suarez *et al.*, 1982). Similar effects have also been shown for two other bacterial siderophores, desferriferrithiocin and desferricrocin (Fritsch *et al.*, 1987). Furthermore, DFO slows the growth of *P. berghei* in rats and mice (Hershko and Peto, 1988; Singh *et al.*, 1985), *P. vinckei* in mice (Fritsch *et al.*, 1985), and *P. falciparum* infections in *Aotus* monkeys (Pollack *et al.*, 1987). Although most currently used antimalarial drugs have effects against only the erythrocytic or the hepatic stages of malaria infection, some iron chelators may not exhibit such stage-specific inhibition of parasite replication. For example, Stahel *et al.* (1988b) report that micromolar concentrations of DFO and desferriferrithiocin are effective in blocking hepatic schizogony by rodent (*P. yoelii*) and human (*P. falciparum*) malarias.

The probability that DFO works via iron deprivation of the parasite is supported by the observation that, against both cultured *P. falciparum* and experimental malaria infections, preformed DFO–Fe^{3+} has no antimalarial effect (Heppner *et al.*, 1988; Fritsch *et al.*, 1985; Pollack, 1983). It has been suggested that DFO may work through interception of intracellular (intraparasitic?) iron, a suggestion bolstered by the observation that *P. falciparum-*

infected human erythrocytes accumulate at least three times as much radiolabeled DFO as do uninfected red cells (Fritsch and Jung, 1986). Furthermore, recent work by Scott *et al.* (1990) indicates that high molecular weight polymers containing fully active DFO are incapable of arresting the growth of *P. falciparum* when added extracellularly or even when entrapped within the host erythrocyte. These findings are also in accord with the observation that the growth of *P. falciparum in vitro* is unaffected by removal of iron from the culture medium, implying that required iron is derived by the parasite from intracellular sources (Peto and Thompson, 1986; Heppner *et al.*, 1988). In summary, it now appears that DFO (and related chelators) may act by binding a labile pool of iron which exists within the parasite itself and that agents which deplete only extracellular iron or intraerythrocytic nonheme iron are ineffective.

Although DFO is modestly effective against experimental malarias *in vivo*, there are several drawbacks to the routine use of this drug including mandatory administration by iv infusion due to poor oral absorption and short plasma half-life, and toxicity (see, e.g., Mahoney *et al.*, 1989). DFO may also simply suppress parasite replication rather than being curative. Finally, DFO is quite expensive and ill-suited to routine treatment of a disease which largely afflicts impoverished people.

Therefore, orally active chelators of lower cost and, possibly, lesser toxicity may hold greater promise. At least two such agents recently have been tried. The α-ketohydroxypyridines (KHPs) are simple, low molecular weight derivatives of maltol and ethylmaltol (Kontoghiorghes and Sheppard, 1987). The KHPs form relatively high affinity 3:1 drug–metal chelates with iron. Tests against *P. falciparum in vitro* indicate that several KHPs (including 1,2-dimethyl-3-hydroxypyrid-4-one, which is now undergoing clinical trials for the therapy of iron overload) are at least as active as DFO (Heppner *et al.*, 1988). Like DFO, KHPs probably act through iron deprivation of the parasite because preformation of the chelate abrogates all antimalarial activity (Heppner *et al.*, 1988). KHPs have not yet been tested *in vivo* against experimental malarial infections.

At least one other orally active iron chelator—pyridoxal isonicotinoyl hydrazone (PIH)— is effective against *P. falciparum in vitro* (Clarke and Eaton, 1990). This drug, first described by Ponka and colleagues (Ponka *et al.*, 1979, 1982; Huang and Ponka, 1983) is a stable hydrazone formed between isoniazid (an antitubercular drug) and pyridoxal (vitamin B_6). PIH forms a lipophilic 2:1 PIH–Fe hexadentate chelate (Vitolo *et al.*, 1988; Webb and Vitolo, 1988). PIH is actually capable of substituting for transferrin in delivering iron to growing cells (Landshulz and Ekblom, 1985). Importantly, neither of the constituents of this hydrazone alone has antimalarial effect, but the simultaneous addition of both pyridoxal and isoniazid is as effective as preformed PIH (Clarke and Eaton, 1990; see also Chapter 14 by Rideout and Calogeropoulou for further discussion of synergism involving hydrazone formation). As is true of both DFO and KHP, saturation of PIH with iron causes antimalarial effects to disappear (Clarke and Eaton, 1990). Therefore, although PIH–Fe can evidently effect the delivery of iron to cultured cells, the iron-free drug may chelate and somehow remove iron from sensitive sites within cultured malaria parasites. In fact, PIH-mediated blockade of the incorporation of transferrin-bound iron into mammalian hepatocytes has been previously documented (Richardson *et al.*, 1988). Furthermore, PIH readily mobilizes nonheme ^{59}Fe from reticulocytes, demonstrating the ability of this drug to enter erythrocytes and support the efflux of iron across the intact erythrocyte membrane (Ponka *et al.*, 1984). Preliminary tests of the efficacy of PIH *in vivo* against murine *P. berghei* have shown a disappointing lack of effect (Clarke and Eaton, 1990), possibly due to a loss of the drug *in vivo* arising from the inherently unstable nature of this reversible hydrazone.

Recently, Yinnon *et al.* (1989) have reported that the polyanionic amine N,N'-bis(1-hydroxybenzyl)ethylenediamine-N,N'-diacetic acid (HBED) is effective against both *P. berghei in vivo* and *P. falciparum in vitro*. HBED is an orally active, lipophilic chelator with high affinity for iron. Overall, HBED is somewhat more effective than DFO both *in vitro* and *in vivo*, perhaps due to a greater ability of HBED to cross cell membranes and mobilize iron from intracellular iron stores (Hershko *et al.*, 1984).

Although reasonably strong evidence suggests that the above agents do act through iron deprivation of the developing parasite, the exact metabolic step(s) inhibited are not yet known. In view of the central role played by the iron-containing enzyme, ribonucleotide reductase, and its ease of inhibition via iron chelation (Thelander and Reichard, 1979), inhibition of this enzyme is a likely point of attack of these chelators. Indeed, DFO has previously been found to inhibit mammalian ribonucleotide reductase (Hoffbrand *et al.*, 1976; Lederman *et al.*, 1984).

C. CHELATORS WHICH FORM TOXIC DRUG–METAL COMPLEXES

Some of the earliest work on chelators of this sort was carried out by Scheibel and his co-workers. Among the more promising agents found by these investigators was 8-hydroxyquinoline (8HQ; also called oxine). Up to three molecules of 8HQ will bind one Fe^{3+} with high affinity to form a chelate which is neutral and highly lipophilic.

8HQ is an effective fungicide and bactericide with unusual properties including that of so-called "concentration quenching". In normal culture broth, bacteria are killed by, e.g., 10 μM 8HQ. However, the addition of much larger drug concentrations (e.g., 1 mM 8HQ) actually spares the bacteria (Albert *et al.*, 1953). As reviewed by Albert (1985), this is thought to be due to the much greater cytotoxicity of 1:1 8HQ–Fe complexes and the relative lack of toxicity of chelates having a higher drug–metal ratio.

Scheibel and Adler (1980, 1981, 1982) found 8HQ and the related agents 5-methyl-8-hydroxyquinoline and 2-mercaptopyridine-*N*-oxide, to be very effective against *P. falciparum in vitro*, with an ED_{50} of $< 0.1 \mu M$. It is likely that the antimalarial effects of 8HQ are due to the formation of a toxic drug–metal complex. This is supported by the observation that 5-hydroxyquinoline (incapable of forming charge-neutral iron chelates) is without effect in concentrations $> 10 \mu M$ (Scheibel and Adler, 1980). 8HQ may form an initial hydrophobic 2:1 8HQ–Fe complex which can cross cell membranes and dissociate to a more toxic 1:1 complex (Albert *et al.*, 1953; Albert 1985). Although in one report the antimalarial action of 8HQ did not show concentration quenching (Scheibel and Adler, 1980), the ratios of endogenous iron to added oxine were not known.

The primary target of this complex is unknown, but more recent results indicate that 8HQ–Fe may cause oxidant damage to cells and, particularly, cell membranes (Balla *et al.*, 1990). The addition of iron to malaria cultures simultaneously exposed to 8HQ does not diminish (and may enhance) the antimalarial effect (C. J. Clarke and J. W. Eaton, unpublished observations). As might be expected, 8HQ has the property of actually facilitating the uptake of iron by malaria-infected erythrocytes (Scheibel and Stanton, 1986) and by endothelial cells (Balla *et al.*, 1990). The oxine–iron chelate is clearly cytotoxic. For example, Balla *et al.* (1990) find that 8HQ–Fe is directly toxic to cultured endothelial cells. This toxicity is accompanied by a spontaneous (probably iron-mediated) peroxidation of cell membrane lipids. The combination of drug and iron is clearly necessary because neither agent alone exerts any toxicity. Furthermore, endothelial cells loaded with iron via exposure to 8HQ–Fe are markedly more susceptible to both endogenously and exogenously generated activated oxygen species such as those produced intracellularly by menadione and extracellularly by activated phagocytes (Balla *et al.*, 1989, 1990).

The antimalarial actions of 8HQ may derive from a combination of the drug with iron which is naturally present within malaria-infected erythrocytes. It has been found that *P. berghei*-infected red cells from mice and rats have a pool of nonhemoglobin and nonheme iron (Wood and Eaton, 1990). This pool, probably generated by parasite-mediated catabolism of host cell hemoglobin, may be available to interact with 8HQ. Thus, the formation of a toxic chelate may occur preferentially within the infected cell. Because significant amounts of free iron do not normally occur in the mammalian body, the actions of such chelating agents may be targeted selectively to the infected cell. Unfortunately, drugs of this class have not been well tested *in vivo* in experimental malaria infections.

A second antimalarial drug which may act synergistically with free metals is diethyldithiocarbamic acid (DDC) which has an exceptionally high affinity for copper (Albert, 1985). As was the case for 8HQ, the microbicidal effects of DDC exhibit the phenomenon of concentration quenching, whereby the addition of supraeffective concentrations of the drug actually spare the micro-organisms. The mechanism of the microbicidal action of DDC–Cu is unknown. (See Chapter 14 by Rideout and Calogeropoulou for a discussion of metal-chelation-mediated antagonism of cisplatin toxicity by DDC).

The antimalarial effectiveness of DDC was first noted by Scheibel *et al.*, (1979a). Against *P. falciparum in vitro*, DDC alone is effective, but at relatively high concentrations (Meshnick *et al.*, 1990). However, if small amounts of copper are added simultaneously, the antimalarial effects of DDC are greatly increased (Meshnick *et al.*, 1990). Interestingly, the endogenous erythrocyte enzyme, Cu/Zn superoxide dismutase, can act as an intracellular source for this synergistic copper. When human red cells are loaded (via osmotic lysis and resealing) with large amounts of Cu/Zn SOD and then infected with *P. falciparum*, the ED_{50} for DDC is only 3% that of the ED_{50} for control lysed and resealed cells (Meshnick *et al.*, 1990). DDC and copper are also synergistic in causing cell lysis. The combination of DDC and copper, but neither agent alone, causes lysis of normal human erythrocytes and of *E. coli* spheroplasts (Meshnick *et al.*, 1990; Agar *et al.*, 1990).

As was the case for 8HQ, DDC may exert preferentially toxic effects against malaria-infected erythrocytes by acting in concert with free metals (probably copper) which accumulate within the infected cell. Indeed, it has been found that *P. falciparum*—probably in the course of digesting host cell cytoplasm—accumulates copper-containing proteins such as Cu/Zn SOD (Fairfield *et al.*, 1983). The metallic residuum of proteolysis of this enzyme may then be available to form toxic chelates with DDC.

Although the precise mechanism of action of the DDC–Cu chelate against malaria is unknown, there are several clues which may eventually lead to its elucidation. First, as is true of the 8HQ–Fe chelate, the DDC–Cu chelate is very insoluble in water and is readily solvated in organic solvents such as chloroform or hexadecane. As might be expected, the complex accumulates in erythrocyte membranes (Agar *et al.*, 1990). This complex appears to be amphipathic, preferentially localizing at organic–aqueous interfaces. Like other amphipathic agents such as chlorpromazine (Lieber *et al.*, 1984), the DDC–Cu complex may then cause colloid osmotic lysis by enhancing the permeability of the lipid bilayer. The inclusion of large solutes such as dextran and sucrose will block red cell lysis by DDC–Cu. Secondary oxidative events appear unimportant in both the hemolytic and bactericidal actions of DDC–Cu because lysis of erythrocytes and killing of *E. coli* occur in the complete absence of oxygen (Agar *et al.*, 1990).

The principle of spontaneous assembly of a cytotoxic product from two relatively innocuous components is reminiscent of earlier work by Rideout and colleagues. In one instance, the hydrazone *N*-decylidenimino-*N'*-1-

octylguanidine can form through a biomolecular reaction involving N-amino-N'-1-octylguanidine and decanal. Although the two participants in this reaction are relatively ineffective in causing lysis of normal human erythrocytes, equimolar mixtures of the two, as well as the preformed hydrazone, are very potent hemolysins and bactericides (Rideout, 1986; Rideout et al., 1988, 1990).

Similarly, in the present instance, DDC, 8-HQ and, perhaps, isoniazid, may combine with metals already present within the malaria-infected erythrocyte to yield toxic or antimetabolic chelates. Chelating drugs which act synergistically with metals present in the malaria-infected erythrocyte may have great promise in the therapy of human malaria. This is particularly true of drugs which form toxic chelates because the malaria-infected red cell may be relatively unique in containing significant amounts of free metals, thereby targeting the actions of antimalarial agents which kill through the formation of active chelates.

X. Miscellaneous Combinations (Fig. 6)

An inhibitor of ornithine decarboxylase, DL-α-difluoromethylornithine, has antimalarial properties that are reversed by putrescine (1,4-diaminobutane), but not by the related homologues 1,3-diaminopropane and 1,5-diaminopentane (Assaraf et al., 1984). The authors suggest that this result indicates a specific requirement for putrescine by P. falciparum. DL-α-difluoromethylornithine produces an apparent potentiation of the activity of MDL 27695, a novel bis(benzyl)tetraamine against P. berghei in vivo; this same combination, however, was additive rather than synergistic against P. falciparum in vitro (Bitonti et al., 1989). These authors speculate that binding of MDL 27695 to DNA may explain the potentiation observed in vivo.

Weak synergism against P. falciparum in vitro is observed between tubercidin, a purine antimetabolite, and both pyrazofurin and menoctone, pyrimidine anti-metabolites (Scott et al., 1987). The observed synergism between tubercidin and pyrazofurin is attributed to depletion of the cellular pool of phosphoribosylpyrophosphate, which, in the malaria parasite, includes both inosine 5'-monophosphate produced by HPRTase and orotidine 5'-monophosphate produced in de novo pyrimidine biosynthesis (Scott et al., 1987). A combination of tubercidin, with either of two nucleoside transport inhibitors, nitrobenzylthioinosine or nitrobenzylthioguanine, is synergistic against P. falciparum in vitro (Gero et al., 1989; Gati et al., 1987). The rationale for the preceding combinations derives from the appearance of nitrobenzyl thioinosine- and nitrobenzyl thioguanine-insensitive nucleoside permeation sites in malaria-infected red cells. Coadministration of 5-fluoroorotic acid, a pyrimidine antimetabolite, and uridine cures P. yoelii-infected mice (Gómez

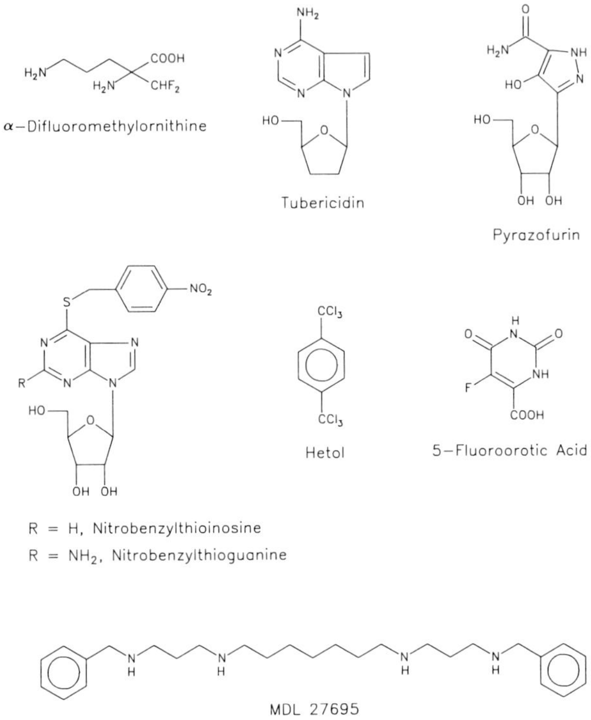

FIGURE 6   Miscellaneous antimalarials.

and Rathod, 1990). Mice treated with uridine before, during, and after 5-fluoroorotic acid treatment show no toxicity in contrast to mice treated with 5-fluoroorotic acid alone. The uridine seems to have little effect on the *in vivo* antimalarial efficacy of 5-fluoroorotic acid; similarly, for *P. falciparum* parasites *in vitro* (Rathod *et al.*, 1989) uridine offers no protection from 5-fluoroorotic acid cytotoxicity. Finally, significant antagonism is observed between several calmodulin antagonists (R24571, W-7, cyclosporin) and quinacrine, quinine, and chloroquine against both drug-resistant and -sensitive *P. falciparum* clones *in vitro*; greater antagonism was observed in the drug-resistant (W-2) clone. The authors (Scheibel *et al.*, 1987) interpret the results to suggest competition for the same receptor binding site.

The host of drug combinations showing additivity or synergy would appear to make the discovery of new combinations for malaria treatment a simple matter of pairing new leads with known active agents. That this is not the case is, however, well demonstrated by the U.S. Army antimalarial program's experience with hetol. Hetol was found to be more active that quinine in the *P. berghei*/mouse Rane test (Osdene *et al.*, 1967). Activity was also demonstrated in the *P. cynomolgi*/*Macaca mulatta* model and this assisted in the promotion of hetol to clinical status where it failed. Being unwilling to abandon so simple and unique a structure, it was decided that hetol could be more efficacious in combination with other active compounds. To that end hetol was combined, in divers ratios, with many active antimalarial compounds. This exercise failed, however, to discover any useful combinations.

XI. Resistance Modulators as Adjunct Antimalarial Chemotherapy

A new clinical rationale for drug combinations has recently emerged from studies which have helped elucidate the mechanism of multidrug-resistance in falciparum malaria. This novel method of adjunct chemotherapy has led to a new class of "resistance modulator" drugs (Fig. 7) targeted toward the falciparum malaria parasite's mechanism of drug resistance. *In vitro* studies demonstrate that the calcium channel blocking drug, verapamil, effectively reverses resistance to chloroquine in a chloroquine-resistant falciparum malaria clone but has no effect on the chloroquine-susceptible clone (Martin *et al.*, 1987). Similarly, verapamil reverses resistance to chloroquine in chloroquine-resistant clones and patient isolates obtained from many geographic regions (including Indochina, East and West Africa, and South and Central America) which suggests that the mechanism of resistance to chloroquine and to its active metabolite, desethylchloroquine, is similar in each of these isolates (Oduola *et al.*, 1988). Other calcium antagonists including diltiazem, nifedipine, chlorpromazine, and their analogs reverse resistance to chloroquine and enhance the intrinsic activities of quinine and quinidine (Kyle *et al.*, 1990). Peters *et al.* (1989) report that antihistamines, such as cyproheptadine and ketotifen (Fig. 1) also reverse chloroquine resistance *in vitro* against falciparum malaria and *in vivo* against rodent malaria.

The analogy between mammalian tumor cell lines and malaria parasites was first postulated by Chawira *et al.* (1986c) when explaining unrelated patterns of resistance in artemisinin (Fig. 3) resistant rodent malaria. Antimalarial drug resistance in *P. falciparum* is postulated to be analogous to multidrug resistance (MDR) in neoplastic cells where the induction of resistance with one drug confers resistance to other structurally and functionally unrelated drugs. The mechanism of MDR in tumor cells is thought to be mediated by enchanced outward transport of drug which reduces the intracellular accumulation to sublethal levels. With the development of an *in vitro* efflux model for falciparum malaria (Krogstad *et al.*, 1987), studies have also demonstrated that this drug transport process is a mechanism of MDR in malaria. If pharmacologically active drugs can be designed which will selectively inhibit drug efflux without significant toxicity to the host, this may prove to be a valuable new approach to the chemotherapy of multidrug-resistant malaria. Ye and Van Dyke (1988) have investigated the reversal properties of verapamil stereoisomers and found that the mechanism of

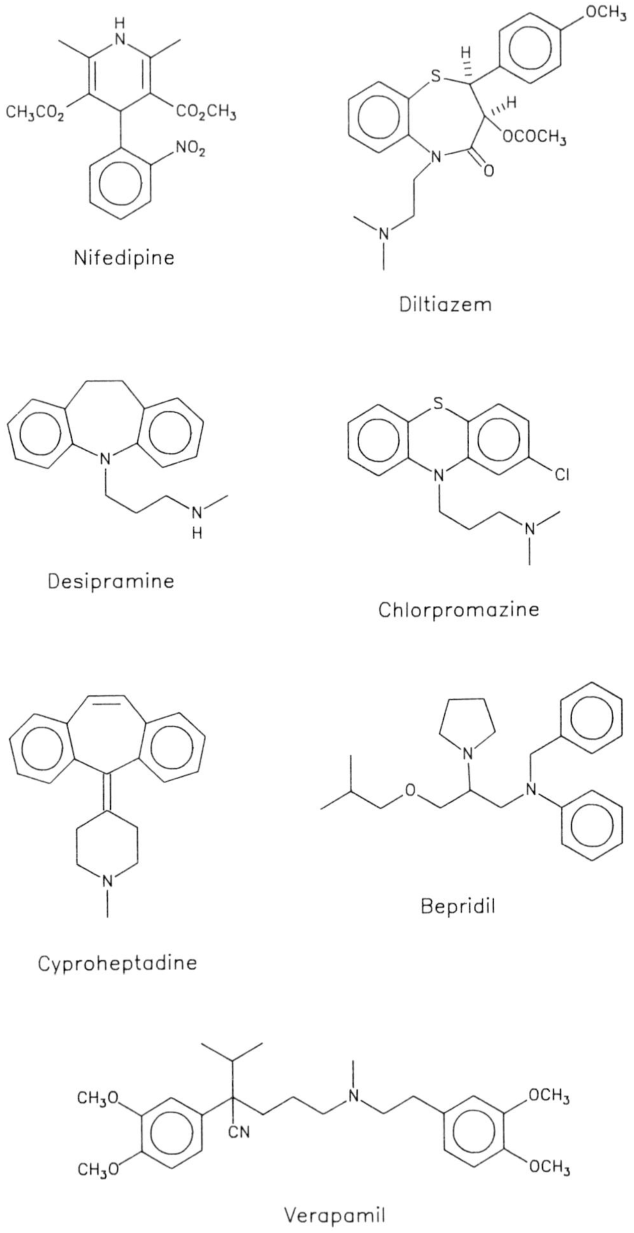

H
N
CH_3CO_2
CO_2CH_3
NO_2
Nifedipine
OCH_3
H
S
H
$OCOCH_3$
N
O
N
Diltiazem
N
N
H
Desipramine
S
N
Cl
N
Chlorpromazine
N
Cyproheptadine
N
O
N
Bepridil
CH_3O
CH_3O
CN
N
OCH_3
OCH_3
Verapamil

FIGURE 7 Resistance modulators.

reversal is independent of the stereospecific cardiovascular calcium channel binding properties.

Simian models for evaluation of these candidate drugs have been limited to the use of *Aotus* monkeys infected with a chloroquine-resistant strain (Vietnam Smith) of *P. falciparum* (Rossan *et al.*, 1985). Several candidate compounds including verapamil, nifedipine, diltiazem, bepridil, a tiapamil analog, a chlorpromazine analog and desipramine (Rossan, 1987; Bitonti *et al.*, 1988) have been evaluated in this model. Preliminary testing has involved the use of multiple dose regimens in combination with chloroquine. In combination with chloroquine, some of the compounds exhibit initial parasite suppression and clearance, but total cures have been achieved only in animals which were pretreated and which may have been partially immune as a result of previous infection. Toxicity is also noted with many of these drugs (Martin *et al.*, 1987; Watt *et al.*, 1990). The effect of these compounds on multiple-drug transporter proteins found in normal cells (Gottesman and Pastan, 1988) remains to be elucidated.

Fortunately, some information on the interactions of reversing and therapeutic agents may be gained from controlled studies of routine practice in the treatment of malaria where antinausea/antiemetic drugs known to be reversing agents are commonly given to malarial patients presenting side effects from antimalarial drug therapy. Such studies are currently in development (G. Watt, personal communication).

In summary, a better understanding of mechanisms of synergism and antagonism will be useful in the design of new, more effective antimalarial combinations. In this regard, significant progress is being made in the development of resistance modulators which may one day overcome the enormous problem of antimalarial drug resistance.

References

Agar, N. S., Mahoney, J. R., and Eaton, J. W. (1990). Hemolytic and microbicidal action of diethyldithiocarbamic acid. *Biochem. Pharmacol.*, (in press).

Albert, A., Gibson, M., and Rubbo, S. (1953). The influence of chemical constitution on anti-bacterial activity. Part VI: The bactericidal action of 8-hydroxyquinoline (oxine). *Br. J. Exp. Pathol.* **34**, 119.

Albert, A. (1985). "Selective Toxicity," 7th Ed. Chapman & Hall, London.

Ames, J. R., Ryan, M. D., Klayman, D. L., and Kovacic, P. (1985). Charge transfer and oxy radicals in antimalarial action. Quinones, dapsone metabolites, metal complexes, iminium ions, and peroxides. *J. Free Radicals Biol. Med.* **1**, 353–361.

Anh, T. K., Kim, N. V., Arnold, K., Chien, V. V., Bich, N. N., Thoa, K., and Ladinsky, J. (1990). Double-blind studies with mefloquine alone and in combination with sulfadoxine-pyrimethamine in 120 adults and 120 children with falciparum malaria in Vietnam. *Trans. R. Soc. Trop. Med. Hyg.* **84**, 50–53.

Assaraf, Y. G., Golenser, J., Spira, D. T., Bachrach, U., and Messer, G. (1984). *In vitro* arrest of *P. falciparum* growth by *DL*-α-difluoromethylornithine and its reversal by diamines. *J. Protozool.* **31**, 82A.

Aviado, D. M., and Ige, A. (1968). Pharmacology of metachloridine with special reference to its antimalarial activity. *Chemotherapy (Basel)* **13**, 289–302.

Aviado, D. M., Sing, G., and Berkley, R. (1969). Pharmacology of new anti-malarial drugs sulfonamides and trimethoprim. *Chemotherapy (Basel)* **13**, 37–53.

Balla, G., Vercellotti, G. M., Jacob, H. S., and Eaton, J. W. (1989). Pharmacologic elevation of cellular iron: Toxic consequences. *Clin. Res.* **37**, 901A.

Balla, G., Vercellotti, G. M., Eaton, J. W., and Jacob, H. S. (1990). Iron loading of endothelial cells augments damage by neutrophils and other oxidant systems: Protection by hydrophobic lazaroids and vitamin E. (in press).

Benazet, F., and Godard, C. (1973). Treatment of drug resistant malaria: Animal experiments. *WHO/SMR/73.15 Rep.* pp. 1–12.

Bishop, A. (1954). The action of 2,4-diamino-6,7-diisopropylpteridine upon *Plasmodium gallinaceum* and its relation to other compounds which are pteroylglutamic acid antagonists. *Parasitology* **44**, 450–464.

Bishop, A. (1965). Resistance to diamino-diphenylsulphone in *Plasmodium gallinaceum*. *Parasitology* **55**, 407–414.

Bitonti, A. J., Sjoerdsma, A., McCann, P. P., Kyle, D. E., Oduola, A. M. J., Rossan, R. N., Milhous, W. K., and Davidson, D. E. (1988). Reversal of chloroquine resistance in malaria parasite *Plasmodium falciparum* by desipramine. *Science* **242**, 1301–1303.

Bitonti, A. J., Dumont, J. A., Bush, T. L., Edwards, M. L., Stemerick, D. M., McCann, P. P., and Sjoerdsma, A. (1989). Bis(benzyl)polyamine analogs inhibit the growth of chloroquine-resistant human malaria parasites (*Plasmodium falciparum*) *in vitro* and in combination with α-difluoromethylornithine cure malaria. *Proc. Natl. Acad Sci. USA* **86**, 651–655.

Botero, D., Restrepo, M., and Montoya, A. (1985). Prospective double-blind trial of two different doses of mefloquine plus pyrimethamine-sulfadoxine compared with pyrimethamine-sulfadoxine alone in the treatment of falciparum malaria. *Bull. W.H.O.* **63**, 731–737.

Brabin, B. J. (1982). Hypothesis: The importance of folacin in influencing susceptibility to malarial infection in infants. *Am. J. Clin. Nutr.* **35**, 146–151.

Brown, O. R., and Green, T. J. (1987). Oxidant-stress: A possible approach to malaria chemotherapy. *Med. Sci. Res.* **15**, 563–565.

Bunnag, D., Harinasuta, T., Looareesuwan, S., Chittamas, S., Pannavut, W., Berthe, J., and Mondesir, J. M. (1989). A combination of quinine, quinidine, and cinchonine (LA40221) in the treatment of chloroquine-resistant falciparum malaria in Thailand: Two double-blind trials. *Trans. R. Soc. Trop. Med. Hyg.* **83**, 66.

Canfield, C. J., Keller, H. I., and Cirksena, W. J. (1971). Erythrokinetics during treatment of acute falciparum malaria. *Military Med.* **136**, 354–357.

Chawira, A. N., and Warhurst, D. C. (1987). The effect of artemisinin combined with standard antimalarials against chloroquine-sensitive and chloroquine-resistant strains of *Plasmodium falciparum in vitro*. *J. Trop. Med. Hyg.* **90**, 1–8.

Chawira, A. N., Robinson, B. L., Warhurst, D. C., and Peters, W. (1985). Drug combination with qinghaosu (artemisinine) against rodent malaria. *J. Protozool.* **31**, 47A.

Chawira, A. N., Warhurst, D. C., and Peters, W. (1986a). Drug combination studies with Qinghaosu (artemisinin) against sensitive and resistant strains of rodent malaria. *Trans. R. Soc. Trop. Med. Hyg.* **80**, 334–335.

Chawira, A. N., Warhurst, D. C., and Peters, W. (1986b). Artemisinin (qinghaosu) combinations against chloroquine-sensitive and resistant *Plasmodium falciparum in vitro*. *Trans. R. Soc. Trop. Med. Hyg.* **80**, 335.

Chawira, A. N., Warhurst, D. C., and Peters, W. (1986c). Qinghaosu resistance in rodent malaria. *Trans. R. Soc. Trop. Med. Hyg.* **80**, 477–480.

Chawira, A. N., Warhurst, D. C., Robinson, B. L. and Peters, W. (1987). The effect of combinations of qinghaosu (artemisinin) with standard antimalarial drugs in the suppressive treatment of malaria in mice. *Trans. R. Soc. Trop. Med. Hyg.* **81,** 554–558.

Chulay, J. D., Watkins, W. M., and Sixsmith, D. G. (1984). Synergistic antimalarial activity of pyrimethamine and sulfadoxine against *Plasmodium falciparum in vitro. Am. J. Trop. Med. Hyg.* **33,** 325–330.

Clark, I. A., and Cowden, W. B. (1985). Antimalarials. *In* "Oxidative Stress" (H. Sies, ed.), pp. 131–149. Academic Press, Orlando, Florida.

Clark, I. A., and Hunt, N. H. (1983). Evidence for reactive oxygen intermediates causing hemolysis and parasite death in malaria. *Infect. Immun.* **39,** 1–6.

Clark, I. A., Cowden, W. B., and Butcher, G. A. (1983). Free oxygen radical generators as antimalarial drugs. *Lancet* **1,** 234.

Clark, I. A., Cowden, W. B., Hunt, N. H., Maxwell, L. E., and Mackie, E. J. (1984a). Activity of divicine in *Plasmodium vinckei*-infected mice has implications for treatment of favism and epidemiology of G-6-PD deficiency. *Br. J. Haematol.* **57,** 479–487.

Clark, I. A., Hunt, N. H., Cowden, W. B., Maxwell, L. E., and Mackie, E. J. (1984b). Radical-mediated damage to parasites and erythrocytes in *Plasmodium vinckei* infected mice after injection of t-butyl hydroperoxide. *Clin. Exp. Immunol.* **56,** 524-530.

Clark, I. A., Hunt, N. H., Butcher, G. A., and Cowden, W. B. (1987). Inhibition of murine malaria (*Plasmodium chabaudi*) *in vitro* by recombinant interferon or tumor necrosis factor, and its enhancement by butylated hydroxyanisole. *J. Immunol.* **10,** 3493–3496.

Clark, I. A., Chaudhri, G., and Cowden, W. B. (1989). Some roles of free radicals in malaria. *Free Radical Biol. Med.* **6,** 315–321.

Clarke, C. J., and Eaton, J. W. (1990). Manuscript in preparation.

Clyde, D. F., Rebert, C. C., Du Pont, H. L., and Cucinell, S. A. (1970). Metabolic effects of the antimalarial diformyl diphenyl sulfone in man. *Fed. Proc., Fed. Am. Soc. Exp. Biol.* **29,** 808.

Clyde, D. F., Gilman, R. H., and McCarthy, V. C. (1975). Antimalarial effects of clindamycin in man. *Am. J. Trop. Med. Hyg.* **24,** 369–370.

Davies, M. J. (1988). Detection of peroxyl and alkoxyl radicals produced by reaction of hydroperoxides with heme-proteins by electron spin resonance spectroscopy. *Biochim. Biophys. Acta* **964,** 28–35.

DeGowin, R. L., Eppes, R. B., Carson, P. E., and Powell, R. D. (1966). The effects of diaphenylsulfone (DDS) against chloroquine-resistant *Plasmodium falciparum. Bull. W.H.O.* **34,** 671–681.

Deloron, P., Lepers, J. P., Verdier, F., Chougnet, C., Remanamirija, J. A., Andriamangatiana-Rason, M. D., Coulanges, P., and Jaureguiberry, G. (1989). Efficacy of a 3-day oral regimen of a quinidine–quinine–cinchonine association (Quinimax®) for treatment of falciparum malaria in Madagascar. *Trans. R. Soc. Trop. Med. Hyg.* **83,** 751–754.

Deloron, P., Lepers, J. P., Andriamangatiana-Rason, M. D., and Coulanges, P. (1990). Short-term oral cinchona alkaloids regimens for treatment of falciparum malaria in Madagascar. *Trans. R. Soc. Trop. Med. Hyg.* **84,** 54.

Desjardins, R. E., Doberstyn, E. B., and Wernsdorfer, W. H. (1988). The treatment and prophylaxis of malaria. *In* "Malaria—Principles and Practice of Malariology" (W. H. Wernsdorfer and I. McGregor, eds.), Vol. 1, pp. 827–864. Churchill Livingstone, Edinburgh, Scotland.

Divo, A. A., Geary, T. G., and Jensen, J. B. (1985). Oxygen- and time-dependent effects of antibiotics and selected mitochondrial inhibitors on *Plasmodium falciparum* in culture. *Antimicrob. Agents Chemother.* **27,** 21–27.

Dockrell, H. M., and Playfair, J. H. L. (1983). Killing of blood-stage murine malaria parasites by hydrogen peroxide. *Infect. Immun.* **39,** 456–459.

Dockrell, H. M., and Playfair, J. H. L. (1984). Killing of *Plasmodium yoelii* by enzyme-induced products of the oxidative burst. *Infect. Immun.* **43,** 451–456.

Donno, L. (1989). Drug combinations in the treatment of malaria. *J. Chemother.* **1**, 52–58.

Druilhe, P., Brandicourt, O., Chongsuphajaisiddhi, T., and Berthe, J. (1988). Activity of a combination of three cinchona bark alkaloids against *Plasmodium falciparum in vitro. Antimicrob. Agents Chemother.* **32**, 250–254.

Eaton, J. W., Eckman, J. R., Berger, E., and Jacob, H. S. (1976). Suppression of malaria infection by oxidant-sensitive host erythrocytes. *Nature (London)* **264**, 758–760.

Eckman, J. R., and Eaton, J. W. (1979). Dependence of plasmodial glutathione metabolism on the host cell. *Nature (London)* **278**, 754–756.

Egan, J. E., Haynes, J. D., Brown, N. D., and Eisemann, C. S. (1986). Polyamine oxidase in human retroplacental serum inhibits the growth of *Plasmodium falciparum. Am. J. Trop. Med. Hyg.* **35**, 890–897.

Ekue, J. M. K., Simooya, O. O., Sheth, U. K., Wernsdorfer, W. H., and Njelesani, E. K. (1985). A double-blind clinical trial of a combination of mefloquine, sulfadoxine and pyrimethamine in symptomatic falciparum malaria. *Bull. W. H. O.* **63**, 339–343.

Elford, B. C., Roberts, M. F., Phillipson, J. D., and Wilson, R. J. M. (1987). Potentiation of the antimalarial activity of qinghaosu by methoxylated flavones. *Trans. R. Soc. Trop. Med. Hyg.* **81**, 434–436.

Eppes, R. B., McNamara, J. V., DeGowin, R. L., Carson, P. E., and Powell, R. D. (1967). Chloroquine-resistant *Plasmodium falciparum*: Protective and hemolytic effects of 4,4'-diaminodiphenylsulfone (DDS) administered daily together with weekly chloroquine and primaquine. *Military Med.* **132**, 163–175.

Etkin, N. L., and Eaton, J. W. (1975). Malaria-induced erythrocyte oxidant sensitivity. *In* "Erythrocyte Structure and Function" (G. J. Brewer, ed.), pp. 219–232. Liss, New York.

Fairfield, A. J., Meshnick, S. R., and Eaton, J. W. (1983). Malaria parasites adopt host cell superoxide dismutase. *Science* **221**, 764–766.

Fenchuan, G., and Lin, C. (1987). Therapeutic effects of 13 combinations of antimalarials against *Plasmodium berghei* "NS" strain. *Chin. J. Parasitol. Parasitic Dis.* **5**, 253–255.

Fernandes-Costa, F., and Metz, J. (1979). Role of serum folate binders in the delivery of folate to the tissue and the fetus. *Br. J. Haematol.* **41**, 335–342.

Ferrante, A., Rzepczyk, C. M., and Allison, A. C. (1983). Polyamine oxidase mediates intra-erythrocytic death of *Plasmodium falciparum. Trans. R. Soc. Trop. Med. Hyg.* **77**, 789–791.

Frenkel, J. K., and Hitchings, G. H. (1957). Relative reversal of vitamins (p-aminobenzoic, folic, and folinic acids) on the effects of sulfadiazine and pyrimethamine on toxoplasma, mouse and man. *Antibiot. Chemother. (Basel)* **7**, 630–638.

Friedman, M. J. (1979). Oxidant damage mediates variant red cell resistance to malaria. *Nature (London)* **280**, 245–247.

Fritsch, G., and Jung, A. (1986). ^{14}C-Desferrioxamine B: Uptake into erythrocytes infected with *Plasmodium falciparum. Z. Parasitenkd.* **72**, 709–713.

Fritsch, G., Treumer, J., Spira, D. T., and Jung, A. (1985). *Plasmodium vinckei*: Suppression of mouse infections with desferrioxamine B. *Exp. Parasitol.* **60**, 171–174.

Fritsch, G., Sawatzki, G., Treumer, J., Jung, A., and Spira, D. (1987). *Plasmodium falciparum*: Inhibition *in vitro* with lactoferrin, desferriferrithiocin, and desferricrocin. *Exp. Parasitol.* **63**, 1–9.

Gail, K., and Herms, V. (1969). Der Einflud von pteroylglutaminsaure ("Folsaure") auf die Parasitendichte (*Plasmodium falciparum*) bei schwangeren Frauen in West Afrika. *Z. Tropenmed. Parasitol.* **20**, 440–450.

Gati, W. P., Stoyke, A. F.-W., Gero, A. M., and Paterson, A. R. P. (1987). Nucleoside permeation in mouse erythrocytes infected with *Plasmodium yoelii. Biochem. Biophys. Res. Commun.* **145**, 1134–1141.

Geary, T. G., Bonnani, L. C., Jensen, J. B., and Ginsburg, H. (1986). Effects of combinations of quinoline-containing antimalarials on *Plasmodium falciparum* in culture. *Ann. Trop. Med. Parasitol.* **80**, 285–291.

Gero, A. M., Scott, H. V., O'Sullivan, W. J., and Christopherson, R. I. (1989). Antimalarial action of nitrobenzylthioinosine in combination with purine nucleoside antimetabolites. *Mol. Biochem. Parasitol.* **34**, 87–98.

Gershon, P. D., and Howells, R. E. (1984). Combinations of the antibiotics erythromycin and tetracycline with three standard antimalarials against *Plasmodium falciparum in vitro. Ann. Trop. Med. Parasitol.* **78**, 1–11.

Gershon, P. D., and Howells, R. E. (1986). Synergy of four macrolide antibiotics with chloroquine against chloroquine-resistant *Plasmodium falciparum in vitro. Trans. R. Soc. Trop. Med. Hyg.* **80**, 753–757.

Godfrey, D. G. (1957). Antiparasitic action of dietary cod liver oil upon *Plasmodium berghei* and its reversal by vitamin E. *Exp. Parasitol.* **6**, 555–565.

Göessi, U., Hirsbrunner, R., and Truniger, B. (1987). Diffuse Lungeninfiltrate in Anschluss an eine Kombinationstherapie mit Mefloquine und Chininsulfat bei einer Malariamischinfektion. *Schweiz. Med. Wochenschr.* **117** (Suppl. 22), 31.

Golenser, J., and Chevion, M. (1989). Oxidant stress and malaria: Host–parasite interrelationships in normal and abnormal erythrocytes. *Semin. Hematol.* **26**, 313–325.

Gómez, Z. M., and Rathod, P. K. (1990). Antimalarial activity of a combination of 5-fluoroorotate and uridine in mice. *Antimicrob. Agents Chemother.* **34**, 1371–1375.

Gottesman, M. M., and Pastan, I. (1988). The multidrug transporter, a double-edged sword. *J. Biol. Chem.* **268**, 12163–12166.

Graf, E., Mahoney, J. R., Bryant, R. G., and Eaton, J. W. (1984). Iron-catalyzed hydroxyl radical formation—stringent requirement for free iron coordination site. *J. Biol. Chem.* **259**, 3620–3624.

Greenberg, J. (1953). Reversal of the activity of chloroguanide against *Plasmodium gallinaceum* by free or conjugated *p*-aminobenzoic acid. *Exp. Parasitol.* **2**, 271–279.

Greenberg, J. (1954). The effect of analogues of folic acid on the activity of sulfadiazine against *Plasmodium gallinaceum. Exp. Parasitol.* **3**, 351.

Gu, H. M., Lu, B. F., and Qu, Z. X. (1980). Activities of 25 derivatives of artemisinine against chloroquine-resistant *Plasmodium berghei. Acta Pharmacol. Sin.* **1**, 48–50.

Gutteridge, J. M. C. (1986). Iron promoters of the Fenton reaction and lipid peroxidation can be released from haemoglobin by peroxides. *FEBS Lett.* **201**, 291–295.

Gutteridge, J. M. C., Richmond, R., and Halliwell, B. (1979). Inhibition of the iron-catalyzed formation of hydroxyl radicals from superoxide and of lipid peroxidation by desferrioxamine. *Biochem. J.* **184**, 469–472.

Gutteridge, W. E., and Coombs, G. H. (1977). "Biochemistry of Parasitic Protozoa." Macmillan, London.

Hall, A. P., Doberstyn, E. B., Mettaprakong, V., and Sonkom, P. (1975a). Falciparum malaria cured by quinine followed by sulfadoxine–pyrimethamine. *Br. Med. J.* **2**, 15–17.

Hall, A. P., Doberstyn, E. B., Nanakorn, A., and Sonkom, P. (1975b). Falciparum malaria semi-resistant to clindamycin. *Br. Med. J.* **2**, 12–14.

Hammond, D. J., Burchell, J. R., and Pudney, M. (1985). Inhibition of pyrimidine biosynthesis *de novo* in *Plasmodium falciparum* by 2-(4-t-butylcyclohexyl)-3-hydroxy-1,4-naphthoquinone *in vitro. Mol. Biochem. Parasitol.* **14**, 97–109.

Harinasuta, T., Bunnag, D., Lassere, R., Leimer, R., and Vinijanont, S. (1985). Trials of mefloquine in vivax and of mefloquine plus Fansidar in falciparum malaria. *Lancet* **1**, 885–888.

Heppner, D. G., Hallaway, P. E., Kontoghiorghes, G. J., and Eaton, J. W. (1988). Antimalarial properties of orally active iron chelators. *Blood* **72**, 358–361.

Herbert, V., Colman, N., and Jacob, E. (1980). Folic acid and vitamin B12. *In* "Modern Nutrition in Health and Disease" (R. S. Goodhart and M. E. Shils, eds.), pp. 229–258. Lea & Febiger, Philadelphia, Pennsylvania.

Hershko, C., and Peto, T. E. (1988). Deferoxamine inhibition of malaria is independent of host iron status. *J. Exp. Med.* **168**, 375–387.

Hershko, C., Grady, R. W., and Link, G. (1984). Phenolic ethylenediamine derivatives: A study of orally effective iron chelators. *J. Lab. Clin. Med.* **103**, 337–346.

Hoffbrand, A. V., Ganeshaguru, K., Hooton, J. W. L., and Tattersall, M. H. N. (1976). Effect of iron deficiency and desferrioxamine on DNA synthesis in human cells. *Br. J. Haematol.* **33**, 517–526.

Huang, A. R., and Ponka, P. (1983). A study of the mechanism of action of pyridoxal isonicotinoyl hydrazone at the cellular level using reticulocytes loaded with non-heme ^{59}Fe. *Biochim. Biophys. Acta* **757**, 306–315.

Huang, W.-Z., Luo, M.-Z., Zhou, M.-X., and Pan, X.-Q. (1987). Study on treatment of *Plasmodium cynomolgi* infections of macaque with ketotifen. *Acta Pharm. Sin.* **22**, 409–412.

Hudson, A. T. (1984). Lapinone, menoctone, hydroxyquinolinequinones and similar structures. *Handb. Exp. Pharmacol.* **68** (II), 343–361.

Hudson, A. T., Randall, A. W., Fry, M., Ginger, C. D., Hill, B., Latter, V. S., McHardy, N., and Williams, R. B. (1985). Novel anti-malarial hydroxynaphthoquinones with potent broad spectrum anti-protozoal activity. *Parasitology* **90**, 45–55.

Hurly, M. G. D. (1959a). Potentiation of pyrimethamine by sulfadiazine in human malaria. *Trans. R. Soc. Trop. Med. Hyg.* **53**, 412–413.

Hurly, M. G. D. (1959b). Administration of pyrimethamine with folic and folinic acids in human malaria. *Trans. R. Soc. Trop. Med. Hyg.* **53**, 410–411.

Kalyanaraman, B., Mottley, C., and Mason, R. P. (1983). A direct electron spin resonance and spin-trapping investigation of peroxyl free radical formation by hematin/hydroperoxide systems. *J. Biol. Chem* **258**, 3855–3858.

Kamen, B. A., and Caston, J. O. (1975). Purification of folate binding factor in normal umbilical cord serum. *Proc. Natl. Acad. Sci. U.S.A.* **72**, 4261–4264.

Kamchonwongpaisan, S., Bunyaratvej, A., Wanachiwanawin, W., and Yuthavong, Y. (1989). Susceptibility to hydrogen peroxide of *Plasmodium falciparum* infecting glucose-6-phosphate dehydrogenase-deficient erythrocytes. *Parasitology* **99**, 171–174.

Keberle, H. (1964). The biochemistry of desferrioxamine and its relation to iron metabolism. *Ann. N.Y. Acad. Sci* **119**, 758–768.

Kinnamon, K. E., Ager, A. L., and Orchard, R. W. (1976). *Plasmodium berghei*: Combining folic acid antagonists for potentiation against malaria infections in mice. *Exp. Parasitol* **40**, 95–102.

Kontoghiorghes, G. J., and Sheppard, L. (1987). Simple synthesis of the potent iron chelators 1-alkyl-3-hydroxy-2-methylpyrid-4-ones. *Inorg. Chim. Acta* **136**, L11–L12.

Kremsner, P. G. (1990). Clindamycin in malaria treatment. *J. Antimicrob. Chemother.* **25**, 9–14.

Kremsner, P. G., Zotter, G. M., Feldmeier, H., Graninger, W., Rocha, R. M. and Wiedermann, G. (1988). A comparative trial of three regimens for treating uncomplicated falciparum malaria in Acre, Brazil. *J. Infec. Dis.* **158**, 1368–1371.

Krogstad, D. J., and Moellering, R. C. (1980). Combinations of antibiotics, mechanisms of interaction against bacteria. *In* "Antibiotics in Laboratory Medicine" (J. Lorian. ed.), pp. 298–341. Williams & Wilkins, Baltimore, Maryland.

Krogstad, D. J., Gluzman, I. Y., Kyle, D. E., Oduola, A. M. J., Martin, S. K., Milhous, W. K., and Schlesinger, P. H. (1987). Efflux of chloroquine from *Plasmodium falciparum*: Mechanism of chloroquine resistance. *Science* **238**, 1283–1285.

Krungkrai, S. R., and Yuthavong, Y. (1987). The antimalarial action on *Plasmodium falciparum* of qinghaosu and artesunate in combination with agents which modulate oxidant stress. *Trans. R. Soc. Trop. Med. Hyg.* **81**, 710–714.

Kyle, D. E., Oduola, A. M. J., Martin, S. K., and Milhous, W. K. (1990). Modulation of *Plasmodium falciparum* resistance to chloroquine, desethylchloroquine, quinine and quinidine by calcium antagonists *in vitro*. *Trans. R. Soc. Trop. Med. Hyg.* (in press).

Landshulz, W., and Ekblom, P. (1985). Iron delivery during proliferation and differentiation of kidney tubules. *J. Biol. Chem.* **260**, 15580–15584.

Latter, V. S., Hudson, A. T., Richards, W. H. G., and Randall, A. W. (1984). Antiprotozoal agents containing naphthoquinones and 4-pyridinol derivatives. Eur. Patent Appl. EP 123,239 (CA 102:32304t).

Lederman, H. M., Cohen, A., Lee, J .W. W., Freedman, M. H., and Gelfand, E. W. (1984). Deferoxamine: A reversible S-phase inhibitor of human lymphocyte proliferation. *Blood* **64**, 748–753.

Levander, O. A., Ager, A. L., Jr., Morris, V. C., and May, R. G. (1989a). Qinghaosu, dietary vitamin E, selenium, and cod liver oil: Effect on the susceptibility of mice to the malarial parasite *Plasmodium yoelii. Am. J. Clin. Nutr.* **50**, 346–352.

Levander, O. A., Ager, A. L., Jr., Morris, V. C., and May, R. G. (1989b). Comparative antimalarial action of plant and fish oil sources of ω-3 fatty acids in vitamin E-deficient mice. *FASEB J.* **3**, A339.

Levander, O. A., Ager, A. L., Jr., Morris, V. C., and May, R. G. (1989c). Menhaden–fish oil in a vitamin E-deficient diet: Protection against chloroquine-resistant malaria in mice. *Am. J. Clin. Nutr.* **50**, 1237–1239.

Levander, O. A., Ager, A. L. Jr., Morris, V. C., and May, R. G. (1990). *Plasmodium yoelii*: Comparative antimalarial activities of dietary fish oils and fish oil concentrates in vitamin E-deficient mice. *Exp. Parasitol.* **70**, 323–329.

Li, G., Guo, X., Arnold, K., Jian, H., and Fu, L. (1984). Randomized comparative study of mefloquine, qinghaosu, and pyrimethamine–sulfadoxine in patients within falciparum malaria. *Lancet* **2**, 1360–1361.

Lucas, A. O., Hedrickse, R. G., Okubadejo, O. A., Richards, W. H. G., Neal, R. A., and Kofie, A. A. K. (1969). The suppression of malarial parasitaemia by pyrimethamine in combination with dapsone or sulphormethoxine. *Trans. R. Soc. Trop. Med. Hyg.* **63**, 216–229.

Mahoney, J. R., Hallaway, P. E., Hedlund, B. E., and Eaton, J. W. (1989). Acute iron poisoning. Rescue with macromolecular iron chelators. *J. Clin. Invest.* **84**, 1362–1366.

Martin, S. K., Oduola, A. M. J., and Milhous, W. K. (1987). Reversal of chloroquine resistance in *Plasmodium falciparum* by verapamil. *Science* **235**, 899–901.

Marva, E., Cohen, A., Saltman, P., Chevion, M., and Golenser, J. (1989). Deleterious synergistic effects of ascorbate and copper on the development of *Plasmodium falciparum*: An *in vitro* study in normal and in G6PD-deficient erythrocytes. *Int. J. Parasitol.* **19**, 779–785.

McCormick, G. J., and Canfield, C. J. (1972). *In vitro* evaluation of antimalarial drug combinations. *Proc. Helminth. Soc. Wash.* **30**, 292–297.

McCormick, G. J., Canfield, C. J., and Willet, G. P. (1971). *In vitro* evaluation of antimalarial activity of folic acid inhibitors. *Exp. Parasitol.* **30**, 88–93.

Merkli, G., Richle, R., and Peters, W. (1980). The inhibitory effect of a drug combination on the development of mefloquine resistance in *Plasmodium berghei. Ann. Trop. Med. Parasitol.* **74**, 1-9.

Meshnick, S. R., Tsang, T. W., Lin, F. B., Pan, H. Z., Chang, C. N., Kuypers, F., Chiu, D., and Lubin, B. (1989). Activated oxygen mediates the antimalarial activity of qinghaosu, *In* "Malaria and the Red Cell" (J. W. Eaton, S. R. Meshnick, and G. J. Brewer, eds.) Vol. 2, pp. 95–104. Liss, New York.

Meshnick, S. R., Scott, M. D., Lubin, B., Ranz,´ A., and Eaton, J. W. (1990). Antimalarial activity of diethyldithiocarbamate: Potentiation by copper. *Biochem. Pharmacol.* **40**, 213–216.

Milhous, W. K., Weatherly, N. F., Bowdre, J. H., and Desjardins, R. E. (1983). Interaction of mefloquine and a fixed combination of sulfadoxine and pyrimethamine (Fansidar) against *Plasmodium falciparum in vitro. Am. Soc. Trop. Med. Hyg.* (abstr.).

Milhous, W. K., Weatherly, N. F., Bowdre, J. H., and Desjardins, R. E. (1985). *In vitro* activities of and mechanisms of resistance to antifol antimalarials. *Antimicrob. Agents Chemother.* **27**, 525–530.

Milhous, W. K., Gerena, L., Kyle, D. E., and Oduola, A. M. J. (1989). *In vitro* strategies for circumventing antimalarial drug resistance. *In* "Malaria and the Red Cell" (J. W. Eaton, S. R. Meshnick, and G. J. Brewer, eds.), Vol. 2, pp. 61–72. Liss, New York.

Miller, K. D., Lobel, H. O., Satriale, R. F., Kuritsky, J. N., Stern, R., and Campbell, C. C. (1986). Severe cutaneous reactions among American travelers using pyrimethamine–sulfadoxine combinations in malaria prophylaxis. *Am. J. Trop. Med. Hyg.* **35,** 451–458.

Miller, L. H, Glew, R. H., Wyler, D. J., Howard, W. A., Collins, W. E., Contacos, P. G., and Neva, F. A. (1974). Evaluation of clindamycin in combination with quinine against multidrug-resistant strains of *Plasmodium falciparum. Am. J. Trop. Med. Hyg.* **23,** 565–569.

Moellering, R. C., Jr. (1979). Antimicrobial synergism—An elusive concept. *J. Infect. Dis.* **140,** 639–641.

Morgan, D. M. L., and Christensen, J. (1983). Polyamine oxidation and the killing of intracellular parasites. *Adv. Polyamine Res.* **4,** 169–174.

Morgan, D. M. L., Christensen, J. R., and Allison, A. C. (1981). Polyamine oxidase and the killing of intracellular parasites. *Biochem. Soc. Trans.* **9,** 563–564.

Murray, M. J., Murray, C. J., Murray, A. B., and Murray, M. B. (1975). Refeeding—Malaria and hyperferraemia. *Lancet* **1,** 653–654.

Murray, M. J., Murray, A. B., Murray, M. B., and Murray, C. J. (1976). Somali food shelters in the Ogaden famine and their impact on health. *Lancet* **1,** 1283–1285.

Mussallam, H. A., Davidson, D. E., Jr., Rossan, R. N., Ager, A. G., Oduola, A. M. J., Kyle, D. E., Werbel, L. M., and Milhous, W. K. (1987). Novel antimalarial "synergism" *in vivo* and *in vitro* between two stereoisomers of the same drug. *Abstr. Am. Soc. Trop. Med. Hyg.* p. 222.

Navaratnam, V., Mohamed, M., Hussain, S., Kumar, A., Jamaludin, A., Sulaiman, I., Mahsufi, S., and Selliah, K. (1989). Chemosuppression of malaria by the triple combination mefloquine/sulfadoxine/pyrimethamine: A field trial in an endemic area in Malaysia. *Trans. R. Soc. Trop. Med. Hyg.* **83,** 755–759.

O'Brien, P. J. (1969). Intracellular mechanisms for the decomposition of a lipid peroxide. 1. Decomposition of a lipid peroxide by metal ion, heme compounds, and nucleophiles. *Can. J. Biochem.* **47,** 485–492.

Oduola, A. M. J., Kyle, D. E., and Milhous, W. K. (1988). Reversal of chloroquine resistance in *Plasmodium falciparum* from various geographical regions. *Abstr. Am. Soc. Trop. Med. Hyg.* 125 (abstr. 114).

Osdene, T. S., Russell, P. B., and Rane, L. (1967). 2,4,7-Triamino-6-ortho-substituted arylpteridines. A new series of potent antimalarial agents. *J. Med. Chem.* **10,** 431–434.

Peters, W. (1969). Partial inhibition by mepacrine of the development of sulphonamide resistance in *Plasmodium berghei. Nature (London)* **223,** 858–859.

Peters, W. (1970). A new type of antimalarial drug potentiation. *Trans. R. Soc. Trop. Med. Hyg.* **64,** 462–464.

Peters, W. (1974). Prevention of drug resistance in rodent malaria by the use of drug mixtures. *Bull. W.H.O.* **51,** 379–383.

Peters, W. (1979). Chemotherapy of malaria. *In* "Malaria, Epidemiology, Chemotherapy, Morphology and Metabolism" (J. P. Krier, ed.), Vol. 1. Academic Press, New York.

Peters, W. (1987). "Chemotherapy and Drug Resistance in Malaria", Vol. 2. Academic Press, Orlando, Florida.

Peters, W., Portus, J., and Robinson, B. L. (1973). The chemotherapy of rodent malaria, XVII. Dynamics of drug resistance, part 3: Influence of drug combinations on the development of resistance to chloroquine in *P. berghei. Ann. Trop. Med. Parasitol.* **67,** 143–154.

Peters, W., Portus, J., and Robinson, B.L. (1977). The chemotherapy of rodent malaria. XXVIII. The development of resistance to mefloquine (WR 142,490). *Ann. Trop. Med. Parasitol.* **71,** 419–427.

Peters, W., Ekong, R., Robinson, B. L., Warhurst, D. C., and Pan, X.-Q. (1989). Antihistaminic drugs that reverse chloroquine resistance in *Plasmodium falciparum. Lancet* **2,** 334–335.

Peto, T. E., and Thompson, J. L. (1986). A reappraisal of the effects of iron and desferriox-amine on the growth of *Plasmodium falciparum 'in vitro'*: The unimportance of serum iron. *Br. J. Haematol.* **63,** 273–280.

Pfaller, M A., and Krogstad, D. J. (1983). Oxygen enhances the antimalarial activity of the imidazoles. *Am. J. Trop. Med. Hyg.* **32,** 660–665.

Phillips, R. E., Karbwang, J., White, N. J., Looareesuwan, S., Warrell, D. A., Kasemsarn, P., and Warhurst, D. C., (1984). Failure of chloroquine–erythromycin and chloroquine–tetracy-cline combinations in treatment of chloroquine resistant falciparum malaria in Eastern Thailand. *Lancet* **1,** 300–302.

Pollack, S. (1983). Malaria and iron. *Br. J. Haematol.* **53,** 181-183.

Pollack, S., Rossan, R. N., Davidson, D.E., and Escajadillo, A. (1987). Desferrioxamine sup-presses *Plasmodium falciparum* in Aotus monkeys. *Proc. Soc. Exp. Biol. Med.* **184,** 162–164.

Ponka, P., Borova, J., Neuwirt, J., and Fuchs, O. (1979). Mobilization of iron from reticulo-cytes. Identification of pyridoxal isonicotinoyl hydrazone as a new iron chelating agent. *FEBS Lett.* **97,** 317–321.

Ponka, P., Schulman, H. M., and Wilczynska, A. (1982). Ferric pyridoxal isonicotinoyl hydra-zone can provide iron for heme synthesis in reticulocytes. *Biochim. Biophys. Acta* **718,** 151–156.

Ponka, P., Grady, R. W., Wilczynska, A., and Schulman, H. M. (1984). The effect of various chelating agents on the mobilization of iron from reticulocytes in the presence and absence of pyridoxal isonicotinoyl hydrazone. *Biochim. Biophys. Acta* **802,** 477–489.

Powis, G. (1989). Free radical formation by antitumor quinones. *Free Radical Biol. Med.* **6,** 63–101.

Puri, S. K., and Dutta, G. P. (1981). Blood schizontocidal activity of antibiotics against *Plasmo-dium berghei* sensitive as well as resistant to chloroquine, pyrimethamine and primaquine. *Indian J. Med. Res.,* **73,** Suppl. 29–35.

Puri, S. K., and Dutta, G. P. (1989). Delay in emergence of mefloquine resistance in *Plasmo-dium berghei* by use of drug combinations. *Acta Trop.* **46,** 209–212.

Rabinovitch, S. A. (1965). Experimental investigations of the antimalarial drug Haloquine. III. Investigation of the possibility to restrain the development of chemoresistance to chloridine (Daraprim) by combined administration of chloridine with Haloquine. *Med. Parazitol. Para-zit. Bolezni* **34,** 434–439.

Rahman, N. A., and Warhurst, D. C. (1985). Drug synergism against *Plasmodium falciparum in vitro. Trans. R. Soc. Trop. Med. Hyg.* **79,** 279.

Rathod, P. K., Khatri, A., Hubbert, T., and Milhous, W. K. (1989). Selective activity of 5-fluoroorotic acid against *Plasmodium falciparum in vitro. Antimicrob. Agents Chemother.* **33,** 1090–1094.

Raventos-Suarez, C., Pollack, S., and Nagel, R. L. (1982). *Plasmodium falciparum*: Inhibition of *in vitro* growth by desferrioxamine. *Am. J. Trop. Med. Hyg.* **31,** 919–922.

Reba, R. C., and Sheehy, T. W. (1967). Colchicine–quinine therapy for acute falciparum malaria acquired in Vietnam. *JAMA, J. Am. Med. Assoc.* **201,** 553–554.

Richardson, D., Baker, E., Ponka, P., Wilairat, P., Vitolo, M. L., and Webb, J. (1988). Effect of pyridoxal isonicotinoyl hydrazone and analogs on iron metabolism in hepatocytes and macrophages in culture. *Birth Defects, Orig. Artic. Ser.* **23,** 81.

Rideout, D. (1986). Self-assembling cytotoxins. *Science* **233,** 561–563.

Rideout, D., Jaworski, J., and Dagnino, R., Jr. (1988). Environment-selective synergism using self-assembling cytotoxic and antimicrobial agents. *Biochem. Pharmacol.* **37,** 4505–4512.

Rideout, D., Calogeropoulou, T., Jaworski, J., and McCarthy, M. (1990). Synergism through direct covalent bonding between agents: A strategy for rational design of chemotherapeutic combinations. *Biopolymers* **29,** 247–262.

Rollo, I. M. (1955). The mode of action of sulfonamides, proguanil and pyrimethamine on *Plasmodium gallinaceum. Br. J. Pharmacol.* **10,** 208–214.

Rollo, I. M. (1964). The chemotherapy of malaria. *In* "Biochemistry and Physiology of Pro-tozoa" (S. H. Hutner, ed.). Academic Press, New York.

Rossan, R. N. (1987). *In vivo* evaluation of calcium channel blockers in the reversal of chloroquine resistance in *Plasmodium falciparum*. *Abstr. Am. Soc. Trop. Med. Hyg.* (abstr. 359).

Rossan, R. N., Harper, J. S., Davidson, D. E., Escajadillo, A., and Christensen, H. A. (1985). Comparison of *Plasmodium falciparum* in Panamanian and Colombian owl monkeys. *Am. J. Trop. Med. Hyg.* **34**, 1037–1047.

Rowe, P. B. (1978). Inherited disorders of folate metabolism. *In* "The Metabolic Basis of Inherited Disease" (J. B. Stanbury, J. B. Wyngaarden, and D. S. Fredrickson, eds.), pp. 420–457. McGraw-Hill, New York.

Ryley, J. F., and Peters, W. (1970). The antimalarial activity of some quinoline esters. *Ann. Trop. Med. Parasitol.* **64**, 209–222.

Rzepczyk, C. M., Saul, A. J., and Ferrante, A. (1984). Polyamine oxidase-mediated intraerythrocytic killing of *Plasmodium falciparum*: Evidence against the role of reactive oxygen metabolites. *Infect. Immun.* **43**, 238–244.

Sande, M. A., and Mandell, G. L. (1980). Antimicrobial agents. *In* "The Pharmacological Basis of Therapeutics" (A. Goodman and L. S. Gillman, eds.), pp. 1098–1100. Macmillan, New York.

Scheibel, L. W., and Adler, A. (1980). Antimalarial activity of selected aromatic chelators. *Mol. Pharmacol.* **18**, 320–325.

Scheibel, L. W., and Adler, A. (1981). Antimalarial activity of selected aromatic chelators. II. Substituted quinolines and quinoline-N-oxides. *Mol. Pharmacol.* **20**, 218–223.

Scheibel, L. W., and Adler, A. (1982). Antimalarial activity of selected aromatic chelators. III. 8-Hydroxyquinolines (oxines) substituted in positions 5 and 7, and oxines annelated in position 5,6 by an aromatic ring. *Mol. Pharmacol.* **22**, 140–144.

Scheibel, L. W., and Stanton, G. G. (1986). Antimalarial activity of selected aromatic chelators. IV. Cation uptake by *Plasmodium falciparum* in the presence of oxines and siderochromes. *Mol. Pharmacol.* **30**, 364–369.

Scheibel, L. W., Adler, A., and Trager, W. (1979a). Tetraethylthiuram disulfide (Antabuse) inhibits the human malaria parasite *Plasmodium falciparum*. *Proc. Natl. Acad. Sci. U.S.A.* **76**, 5303–5307.

Scheibel, L. W., Ashton, S. H., and Trager, W. (1979b). *Plasmodium falciparum*: Microaerophilic requirements in human red blood cells. *Exp. Parasitol.* **47**, 410–418.

Scheibel, L. W., Colombani, P. M., Hess, A. D., Aikawa, M., Atkinson, C. T., and Milhous, W. K. (1987). Calcium and calmodulin antagonists inhibit human malaria parasites (*Plasmodium falciparum*): Implications for drug design. *Proc. Natl. Acad. Sci. U.S.A.* **84**, 7310–7314.

Schmidt, L. H. (1979a). Studies on the 2,4-diamino-6-substituted quinazolines. II. Activities of selected derivatives against infections with various drug-susceptible and drug-resistant strains of *Plasmodium falciparum* and *Plasmodium vivax* in owl monkeys. *Am. J. Trop. Med. Hyg.* **28**, 793–807.

Schmidt, L. H. (1979b). Studies on the 2,4-diamino-6-substituted quinazolines. III. The capacity of sulfadiazine to enhance the activities of WR 158,122 and WR 159,412 against infections with various drug-susceptible and drug-resistant strains of *Plasmodium falciparum* and *Plasmodium vivax* in owl monkeys. *Am. J. Trop. Med. Hyg.* **28**, 808–818.

Scholer, J., Leimer, R., and Richie, R. (1984). Sulfonamides and sulfones. *Handb. Exp. Pharmacol.* **68(I)**, 123–206.

Scott, H. V., Rieckmann, K. H., and O'Sullivan, W. J. (1987). Synergistic antimalarial activity of dapsone/dihydrofolate reductase inhibitors and the interaction of antifol, antipyrimidine and antipurine combinations against *Plasmodium falciparum in vitro*. *Trans. R. Soc. Trop. Med. Hyg.* **81**, 715–721.

Scott, M. D., Ranz, A., Kuypers, F. A., Lubin, B. H., and Meshnick, S. R. (1990). Parasite uptake of deferoxamine: A prerequisite for antimalarial activity. *Br. J. Haematol.* (in press).

Shanks, D., Karwacki, J. J., and Singharaj, R. T. A. (1989). Malaria prophylaxis during military operations in Thailand. *Military Med.* **154**, 500–502.

Shao, B.-R., and Ye, X.-Y. (1986). Delay in emergence of resistance to pyronaridine phosphate in *Plasmodium berghei. Acta Pharmacol. Sin.* **7**, 463–467.

Shi, X.-H., Shao, B.-R., and Ye, X.-Y. (1990). Combined action of pyronaridine and sulfadoxine/pyrimethamine against *Plasmodium berghei* ANKA strain in mice. *Acta Pharmacol. Sin.* **11**, 66–69.

Shwe, T., Myint, P. T., Htut, Y., Myint, W., and Soe, L. (1988). The effect of mefloquine–artemether compared with quinine on patients with complicated falciparum malaria. *Trans. R. Soc. Trop. Med. Hyg.* **82**, 665–666.

Shwe, T., Myint, P. T., Myint, W., Htut, Y., Soe, L., and Thwe, M. (1989). Clinical studies on treatment of cerebral malaria with artemether and mefloquine. *Trans. R. Soc. Trop. Med. Hyg.* **83**, 489.

Singh, C., Arif, A. J., Mathur, P. D., Chandra, S., and Sen, A. B. (1985). Effect of a specific iron chelator, desferrioxamine, on the host biochemistry and parasitaemia in mice infected with *Plasmodium berghei. Indian. J. Malariol.* **22**, 35–44.

Stahel, E., Druilhe, P., and Gentilini, M. (1988a). Antagonism of chloroquine with other antimalarials. *Trans. R. Soc. Trop. Med. Hyg.* **82**, 221.

Stahel, E., Mazier, D., Guillouzo, A., Miltgen, F., Landau, I., Mellouk, S., Beaudoin, R. L., Langlois, P., and Gentilini, M. (1988b). Iron chelators: *In vitro* inhibitory effect on the liver stage of rodent and human malaria. *Am. J. Trop. Med. Hyg.* **39**, 236–240.

Strickland, G. T., and Kostinas, J. E. (1970). Folic acid deficiency complicating malaria. *Am. J. Trop. Med. Hyg.* **19**, 910–915.

Tan-Ariya, P., Brockelman, C. R., and Menabandhu, C. (1984). Mefloquine–pyrimethamine–sulfadoxine combination delays emergence of mefloquine resistant *Plasmodium falciparum* in continuous culture. *Southeast Asian J. Trop. Med. Public Health* **15**, 531–535.

Thelander, L., and Reichard, P. (1979). Reduction of ribonucleotides. *Annu. Rev. Biochem.* **48**, 133–158.

Thompson, P. E., and Werbel, L. M. (1972). "Antimalarial Agents." pp. 527–562. Academic Press, New York.

Thornalley, P. J., Trotta, R. J., and Stern, A. (1983). Free radical involvement in the oxidative phenomena induced by tert-butyl hydroperoxide in erythrocytes. *Biochim. Biophys. Acta* **759**, 16–22.

Thurston, J.P. (1954). The chemotherapy of *Plasmodium berghei.* II, Antagonism of the action of drugs. *Parasitology* **44**, 99–110.

Tin, F., Ulaing, N., Tun, T., Win, S., and Lasserre, R. (1985). Falciparum malaria treated with a fixed combination of mefloquine, sulfadoxine and pyrimethamine: A field study in adults in Burma. *Bull. W.H.O.* **63**, 727–730.

Tong, M. J., Strickland, G. T., Votteri, B. A., and Gunning, J. (1970). Supplemental folates in the therapy of *Plasmodium falciparum* malaria. *JAMA, J. Am. Med. Assoc.* **214**, 2330–2333.

Trotta, R. J., Sullivan, S.G., and Stern, A. (1981). Lipid peroxidation and hemoglobin degradation in red blood cells exposed to t-butyl hydroperoxide—Dependence on glucose metabolism and hemoglobin status. *Biochim. Biophys. Acta* **679**, 230–237.

Trotta, R. J., Sullivan, S. G., and Stern, A. (1982). Lipid peroxidation and haemoglobin degradation in red cells exposed to t-butyl hydroperoxide. *Biochem. J.* **204**, 405–415.

Vennerstrom, J. L., and Eaton, J. W. (1988). Oxidants, oxidant drugs, and malaria. *J. Med. Chem.* **31**, 1269–1277.

Vitolo, M. L., Clare, B. W., Hefter, G. T., and Webb, J. (1988). Chemical studies of pyridoxal isonicotinoyl hydrazone relevant to its clinical evaluation. *Birth Defects, Orig. Artic. Ser.* **23**, 71.

Warhurst, D. C. (1977). Chloroquine–erythromycin potentiation of *P. berghei. Ann. Trop. Med. Parasitol.* **71**, 383.

Warhurst, D. C., Robinson, B. L., and Peters, W. (1976). The chemotherapy of rodent malaria,

XXIV. The blood schizontocidal action of erythromycin upon *Plasmodium berghei. Ann. Trop. Med. Parasitol.* **80,** 253–258.

Watkins, W. M., Sixsmith, D. G., Chulay, J. D., and Spencer, H. C. (1985). Antagonism of sulfadoxine and pyrimethamine antimalarial activity *in vitro* by p-aminobenzoic acid, p-aminobenzoylglutamic acid and folic acid. *Mol. Biochem. Parasitol.* **14,** 55–61.

Watt, G., Long, G. W., Grogl, M., and Martin, S. K. (1990). Reversal of drug-resistant falciparum malaria by calcium antagonists: Potential for host cell toxicity. *WHO/ MAL/90-1056* pp. 1–10.

Webb, J., and Vitolo, M. L. (1988). Pyridoxal isonicotinoyl hydrazone (PIH): A promising new iron chelator. *Birth Defects, Orig. Artic. Ser.* **23,** 63.

Willerson, D., Jr., Rieckmann, K. H., Carson, P. E., and Frischer, H. (1972). Effects of minocycline against chloroquine-resistant falciparum malaria. *Am. J. Trop. Med. Hyg.* **21,** 857–862.

Win, K., Thwe, Y., Lwin, T. T., and Win, K. (1985). Combination of mefloquine with sulfadoxine–pyrimethamine compared with two sulfadoxine–pyrimethamine combinations in malaria chemoprophylaxis. *Lancet* **2,** 694–695.

Wood, P. A., and Eaton, J. W. (1990). Diminished parasite heme in chloroquine-resistant *Plasmodium berghei.* Submitted

Wozencraft, A. O., Dockrell, H. M., Taverne, J., Targett, G. A. T., and Playfair, J. H. L. (1984). Killing of human malarial parasites by macrophage secretory products. *Infect. Immun.* **43,** 664–669.

Ye, Z., and Van Dyke, K. (1988). Reversal of chloroquine resistance in falciparum malaria independent of calcium channels. *Biochem. Biophys. Res. Commun.* **155,** 476–481.

Ye, Z., Van Dyke, K., and Castranova, V. (1989). The potentiating action of tetrandrine in combination with chloroquine or qinghaosu against chloroquine-sensitive and resistant falciparum malaria. *Biochem. Biophys. Res. Commun.* **165,** 758–765.

Yinnon, A. M., Theanacho, E. N., Grady, R. W., Spira, D. T., and Hershko, C. (1989). Antimalarial effect of HBED and other phenolic and catecholic iron chelators. *Blood* **74,** 2166–2171.

Zhou, Y.-Q., and Ning, D.-X. (1988). Inhibitory effect of artemisinin sulfadoxine–pyrimethamine combination on the development of drug-resistance in *Plasmodium berghei. Acta Pharmacol. Sin.* **9,** 453–457.

CHAPTER 6

Quantitation of Synergism and Antagonism of Two or More Drugs by Computerized Analysis

Joseph H. Chou
Department of Biochemistry and Molecular Biology
Harvard University
Cambridge, Massachusetts 02138

The primary purpose of this chapter is to describe how automated analysis of synergism and antagonism in two-drug combinations can be undertaken using personal computers. The analysis of three-drug combinations and other applications are also discussed.

SYNERGISM AND ANTAGONISM IN CHEMOTHERAPY
Copyright © 1991 by Academic Press, Inc.
All rights of reproduction in any form reserved.

I. Theory and Equations

This section describes the equations used for combination analysis. Derivations and a more detailed discussion of the equations are provided in Chapter 2.

A. MEDIAN-EFFECT EQUATION

The median-effect equation (Eq. 1) (T.-C. Chou, 1976) describes the dose–effect relationship for a single drug (or multiple drugs used in combination at a constant ratio).

$$\frac{f_a}{f_u} = \left(\frac{D}{D_m}\right)^m \tag{1}$$

D: dose administered

f_a: fraction affected by a dose D, for example, fractional growth inhibition of cells or colonies; fractional inhibition of enzyme activity. Note that f_a is a fractional activity compared with an untreated control sample with $D = 0$.

f_u: fraction unaffected. By definition, $f_u = 1 - f_a$.

D_m: median-effect dose. D_m is the dose required to produce a fractional effect of 0.5 (or 50%), hence the name median-effect dose. The D_m can also be used as a measure of potency.

m: a coefficient which describes the shape (sigmoidicity) of the dose–effect curve.

B. COMBINATION INDEX EQUATION

The combination index (Eq. 2) (Chou and Talalay, 1983; Chou and Talalay, 1984), abbreviated CI, describes quantitatively the extent of synergism or antagonism at a given effect level present in a mixture of drugs.

$$\text{CI} = \frac{(D_{comb})_1}{(D_{alone})_1} + \frac{(D_{comb})_2}{(D_{alone})_2} + \alpha \frac{(D_{comb})_1 (D_{comb})_2}{(D_{alone})_1 (D_{alone})_2} \tag{2}$$

$\text{CI} = 1$ indicates summation

< 1 indicates synergism

> 1 indicates antagonism

The combination index, CI, can be presented as a function of the fractional effect, f_a, by using Eq. 5 (p. 227) to solve for the individual components of Eq. 2 as follows:

$(D_{alone})_1$: the dose of drug 1 *alone* required for a given effect (f_a).

$(D_{comb})_1$: the dose of drug 1 *in the combination* required for a given effect (f_a).

$(D_{alone})_2$: the dose of drug 2 *alone* required for a given effect (f_a).

$(D_{comb})_2$: the dose of drug 2 *in the combination* required for a given effect (f_a).

$\alpha = 0$ if the effects of the two drugs are *mutually exclusive.*

$ = 1$ if the effects of the two drugs are *mutually nonexclusive.*

C. DOSE-REDUCTION INDEX

When several drugs are used in combination, the dose-reduction index (Eq. 3) (J. Chou and T.-C. Chou, 1988; Berman *et al.*, 1989), abbreviated DRI, describes the fold reduction in dose for each drug at a given degree of effect (f_a) in the combination. It is a more intuitive measure of drug dose reduction due to synergism (or dose increase due to antagonism) and is particularly helpful when attempting to minimize toxic side effects of one of the agents. However, each dose-reduction index describes the dose reduction of only a single drug at a time; the indices of *all* drugs used in a combination must be taken into account to determine whether significant synergism or antagonism is occurring.

$$(DRI)_1 = \frac{(D_{alone})_1}{(D_{comb})_1} \tag{3}$$

$(DRI)_1$: the dose-reduction index for drug 1

$(D_{alone})_1$: the dose of drug 1 *alone* required for a given effect (f_a).

$(D_{comb})_1$: the dose of drug 1 *in the combination* required for a given effect (f_a).

Analogous equations are used for drug 2.

II. Computerized Simulation and Automation

This section describes some of the possibilities available for computerized analysis. It includes a list of necessary data and a description of the calculations and analysis the computer can automate. In addition, analysis already incorporated into commercially available software is listed. A later section will provide an actual example of computerized analysis.

A. EXPERIMENTAL DATA NECESSARY FOR COMBINATION ANALYSIS

Since this method of analysis utilizes data very efficiently, only a small amount of data is required: dose effect data for each drug separately and dose effect data for the two drugs in combination. The combination ratio of the two drugs must also be known.

1. Single Agents

For each drug separately, one simply measures how the fraction affected (f_a) varies while varying dose.

2. Combinations of Agents at a Constant Ratio and at Nonconstant Ratios

For two drugs in combination, one still varies the doses of the two drugs while monitoring f_a. However, for a complete analysis of synergism and antagonism simulation, the doses must be varied such that a constant ratio of drug 1 to drug 2 is maintained. The actual combination ratio can be chosen arbitrarily. One possibility is to use the potency ratio (the ratio of the median-effect doses, (D_m). The simplest way to insure a constant ratio is to use serial dilutions of a mixture of drug 1 and drug 2 at known concentrations.

If a constant ratio of drug 1 to drug 2 is not maintained, the combination index can still be used to determine synergism/antagonism, but only at discrete fractional effects rather than across the entire range of effect.

For each set of data, five dose effect points are usually sufficient for analysis. Best results are obtained when there is a broad variation in f_a, perhaps with f_a varying from 0.4 to 0.99, rather than from 0.95 to 0.99.

B. CALCULATIONS DONE BY COMPUTER

Taking the logarithm of both sides of the median-effect equation (Eq. 1) yields

$$\log\left(\frac{f_a}{1 - f_a}\right) = m\log(D) - m\log(D_m) \tag{4}$$

This shows that a plot of $\log(D)$ on the horizontal axis and $\log[(f_a)/(1 - f_a)]$ on the vertical axis will yield a *line* with

slope $= +m$
y intercept $= -m\log(D_m)$

This plot is called the *median-effect plot* (also the Chou Plot). Thus, the user merely enters several dose (D) and effect (f_a) points and the computer can use the least squares method of linear regression to calculate the coefficients m and D_m. These coefficients will commonly be referred to as the *median-effect parameters*.

Note that in the case of a combination of two drugs at constant ratio, the user enters the *sum* of the two doses for each point. Since the concentrations of the two drugs are varied at a constant ratio, this sort of analysis is akin to pretending that the combination of two drugs is an entirely independent agent. Thus, a median-effect plot can also be made with the dose effect points of the combination, but *only* if the drug concentrations are varied *at a constant ratio*. From the sum of the two doses, the computer can use the combination ratio to determine the concentrations of each drug in the mixture. Starting with the median-effect equation (Eq. 1) and solving algebraically for D yields

$$D = D_m \left(\frac{f_a}{1 - f_a} \right)^{1/m} \tag{5}$$

Starting with the median-effect equation (Eq. 1) and solving algebraically for f_a yields

$$f_a = \frac{1}{1 + \left(\dfrac{D_m}{D} \right)^m} \tag{6}$$

Clearly, once the parameters m and D_m are known, one can easily predict the dose required to bring about a given effect (Eq. 5), or the effect resulting from a given dose (Eq. 6). Theoretically, since two points are sufficient to generate a line, only two dose and effect points are needed to allow prediction of an entire dose–response curve. However, utilization of only two points is unwise due to possible experimental error.

The combination index equation (Eq. 2) is the sum of several ratios of doses. In general, each ratio consists of the dose in a combination divided by the dose alone required to generate a given effect. Once the parameters m and D_m have been obtained from the median-effect plot (Eq. 4), each of the doses can be calculated for a given fractional effect (Eq. 5). These doses can then be used to calculate the combination index (Eq. 2). Note that, for the doses in the mixture, the combination ratio is used to determine the concentrations of each agent alone from the sum of the two agents. In general, it is best to calculate the combination index using both the mutually exclusive assumption and the mutually non exclusive assumption unless the modes of effect of the two drugs are clearly understood.

A plot of f_a on the horizontal axis and CI on the vertical axis will yield a complex *curve* describing the extent of synergism at each effect level. Note that if a nonconstant ratio in the mixture is used, only specific *points* can be determined rather than a complete curve. The calculations involved in generating an f_a–CI plot are quite laborious and greatly facilitated by computerized analysis.

The dose-reduction index (Eq. 3) at any effect level is also easily calculated once the median-effect parameters are determined.

C. COMPUTERIZED ANALYSIS POSSIBLE

The following describes the output possible from computer analysis after entry of dose effect data and combination ratios. Note that software for the Apple II and IBM PC are commercially available. In addition, a new version also for the IBM PC has been written and copyrighted, but has not yet been published. In the discussion that follows, features available in the Apple II software are indicated by (Apple) (J. Chou and T.-C. Chou, 1985), in the older IBM version by (IBM) (J. Chou and T.-C. Chou, 1987), and in the newer IBM version by (IBM*). Software can be obtained from Biosoft.[1]

Tabular Output

Crude Data (Apple, IBM, IBM)*   These are the dose effect data entered by the user.

Median-Effect Parameters (Apple, IBM, IBM)*   These include the parameters m and D_m, as calculated by least squares fit linear regression of the median-effect plot. Some statistical analysis, including the regression coefficient and standard errors of measurement, are also given.

Dose-Effect Analysis (Apple, IBM, IBM)*   This includes calculation of dose for a given effect level or calculation of effect for a given dose.

f_a–CI Table (Apple, IBM, IBM)*   This is a table showing how the combination index varies when fractional effect ranges from 0.05 to 0.95. This table is available only if the dose effect data for the combination is acquired using a constant ratio between the two agents. If a nonconstant ratio is used, the combination index can still be calculated by the newer IBM version (IBM*).

DRI Table (IBM)*   This is a table showing how the dose-reduction index varies when fractional effect ranges from 0.05 to 0.95. This is also possible only for a constant ratio analysis. With a nonconstant ratio of agents in the combination, the dose reduction index can be calculated only for discrete points.

[1] Biosoft, 22 Hills Road, Cambridge, CB2, 1JP, UK (Tel 0-223-68622) or Biosoft, P.O. Box 580, Miltown, NJ 08850 USA (Tel 201-613-9013).

Summary (IBM*) This is a one-page summary of median-effect parameters, doses required at several levels of effect (typically 50, 75, and 95%), and combination indices at the same levels of effect.

Graphical Output

Dose–Effect Curve (IBM, IBM*) A plot of fractional effect versus dose results in a curve which can be hyperbolic, sigmoidal, or negatively sigmoidal, depending on the parameter *m*.

Median-Effect Plot (Apple, IBM, IBM*) A plot of log $[(f_a)/(1 - f_a)]$ versus log (D), as described previously, results in a line, determined by linear regression.

f_a–CI Plot (Apple, IBM, IBM*) A plot of combination index versus fractional effect results in a curve displaying the extent of synergism/antagonism with varying effect levels. An f_a–log (CI) plot is also available (IBM*) for instances when the combination index varies greatly.

DRI Plot A plot of dose-reduction index versus fractional effect results in a curve displaying fold reduction of dose with varying effect levels.

Isobologram (IBM, IBM*) Both classical isobolograms (for the mutually exclusive assumption) and conservative isobolograms (for the mutually nonexclusive assumption) are constructed at effect levels of the user's choosing. The isobologram is a dose-oriented, graphical presentation of synergism and antagonism.

The Apple version requires appropriate graphics dump software to allow printing the graphs. The IBM versions require an installed graphics card as well as a monitor capable of displaying graphics. The older IBM version also allows some graphs to be plotted on a Hewlett–Packard[TM] plotter. Work is in progress to allow the newer IBM version to allow plotting as well.

III. Illustration of Analysis with Examples

This section shows an actual example of analysis using the newer version of the IBM software. The example shows inhibition of HIV reverse transcriptase activity in data previously described by Hartshorn *et al.*, 1987 and Berman *et al.*, 1989.

The computer printout shown here used the IBM-PC XT with a Hercules graphics card, an Epson FX 1050 printer, IBM version of software DOS 2.1 (Ref. 7) with added plotting feature (using a Hewlett–Packard 7470 A plotter). For automated printout of overall analysis, a dot matrix printer is usually used (e.g., Okidata U84, Epson FX 100, or FX 1050). In the printout, some comments are manually typed in the text using different character fonts in order to distinguish these from the actual computer output.

```
                    MULTIPLE DRUG EFFECT ANALYSIS

Experiment:    EFFECT OF rIFN-ALPHA-A  AND  AZT
               EXPERIMENT 2A

Date:

In filename:  B:\AZT

Drug A: AZT  (uM)
Drug B: rIFN-ALPHA-A   (U/ML)
a mixture of A and B at a 1 : 800 ratio


------------------------------------------------------------------------


Data for Drug A: AZT  (uM)

Dose                 Fraction Affected
----                 -----------------
 .01                  .2244
 .02                  .4634
 .04                  .6537
 .08                  .8488
 .16                  .9659

 5 data points entered.

X-int  : -1.63339
Y-int  :  2.60152  +/-  .17409
```

$\textcircled{M}$: 1.59271 +/- .11913 ⟶ Shape of dose-effect curve

$\textcircled{Dm}$: .02326 (μM)   ⟶ Potency (IC_{50})

$\textcircled{r}$: .99171   ⟶ Comformity

Data for Drug B: rIFN-ALPHA-A   (U/ML)

Dose	Fraction Affected
32	.0585
64	.1756
128	.2634
256	.4914

4 data points entered.

X-int : 2.42007
Y-int : -3.05507 +/- .264

M   : 1.26239 +/- .13297

Dm   : 263.0686 (U/ml)
r   : .98909

Data for a mixture of A and B at a 1 : 800 ratio

Dose	Fraction Affected
8.01	.5854
16.02	.7951
32.04	.9512
64.08	.9805
128.16	.9951

5 data points entered.

X-int : .83604
Y-int : -1.50756 +/- .11971

M   : 1.80322 +/- .07651

Dm   : 6.85552
r   : .99731

Data for a mixture of A and B at a 1 : 800 ratio

Mutually Exclusive Case

Fa	CI Values	Dose A	+	Dose B
.05	.50898	.00167		1.33766
.1	.47613	.00253		2.02446
.15	.45715	.00327		2.61655
.2	.4435	.00397		3.17407
.25	.43261	.00465		3.72309
.3	.42337	.00535		4.27987
.35	.41519	.00607		4.85742
.4	.40772	.00684		5.46819
.45	.40071	.00766		6.12586
.5	.39399	.00856		6.84696
.55	.3874	.00957		7.65294
.6	.38082	.01072		8.57337
.65	.37411	.01206		9.65139
.7	.36709	.01369		10.9538
.75	.35954	.01574		12.59193
.8	.35113	.01846		14.76995
.85	.34127	.0224		17.91706
.9	.32867	.02895		23.15726
.95	.30946	.04381		35.04693

CI values for actual experimental points of
a mixture of A and B at a 1 : 800 ratio

Mutually Exclusive Case

Dose A	Dose B	Fa	CI value
.01	8	.5854	.36934
.02	16	.7951	.3878
.04	32	.9512	.27801
.08	64	.9805	.30485
.16	128	.9951	.25192

Data for a mixture of A and B at a 1 : 800 ratio

Mutually Non-exclusive Case

Fa	CI Values	Dose A	+	Dose B
.05	.5329	.00167		1.33766
.1	.49509	.00253		2.02446
.15	.47357	.00327		2.61655
.2	.45823	.00397		3.17407
.25	.44608	.00465		3.72309
.3	.43583	.00535		4.27987
.35	.4268	.00607		4.85742
.4	.41858	.00684		5.46819
.45	.4109	.00766		6.12586
.5	.40357	.00856		6.84696
.55	.3964	.00957		7.65294
.6	.38927	.01072		8.57337
.65	.38201	.01206		9.65139
.7	.37444	.01369		10.9538
.75	.36635	.01574		12.59193
.8	.35736	.01846		14.76995
.85	.34685	.0224		17.91706
.9	.33351	.02895		23.15726
.95	.3133	.04381		35.04693

```
CI values for actual experimental points of
a mixture of A and B at a 1 : 800 ratio
```

Mutually Non-exclusive Case

Dose A	Dose B	Fa	CI value
.01	8	.5854	.37735
.02	16	.7951	.39542
.04	32	.9512	.2811
.08	64	.9805	.30806
.16	128	.9951	.25369

Data for a mixture of A and B at a 1 : 800 ratio

Fa	Dose A alone	Dose B alone	Dose reduction index A	B
.05	.00366	25.5333	2.19013	19.088
.1	.00585	46.14957	2.31343	22.79601
.15	.00783	66.57667	2.39322	25.44445
.2	.00974	87.73006	2.4551	27.6396
.25	.01167	110.184	2.50742	29.59478
.3	.01366	134.455	2.55403	31.41565
.35	.01577	161.1031	2.59712	33.16638
.4	.01803	190.8002	2.6381	34.89273
.45	.02051	224.4054	2.678	36.63244
.5	.02326	263.0687	2.71768	38.42124
.55	.02638	308.3933	2.75795	40.29738
.6	.03	362.71	2.79966	42.30656
.65	.03431	429.5706	2.84383	44.50866
.7	.0396	514.7085	2.89181	46.98904
.75	.04636	628.0872	2.94557	49.88012
.8	.05554	788.842	3.00834	53.40857
.85	.06912	1039.481	3.08613	58.01625
.9	.09241	1499.585	3.19258	64.75656
.95	.14774	2710.394	3.3723	77.33614

```
Dose reduction index values for actual experimental points of
a mixture of A and B at a 1 : 800 ratio
```

Fa	Dose A alone	Dose B alone	Dose reduction index A	B
.5854	.02889	345.7414	2.88851	43.21767
.7951	.05449	770.1033	2.72465	48.13146
.9512	.15013	2765.814	3.75314	86.43168
.9805	.27218	5859.22	3.40226	91.55031
.9951	.65389	17704.65	4.08681	138.3176

Isobolograms at the following effect levels will be generated.

	ED50	ED70	ED90
Drug A	.02326	.0396	.09241
Drug B	263.0686	514.7083	1499.582
Mixture #1	6.85552	10.96749	23.18618

Summary Table

Drug A: AZT   (uM)
Drug B: rIFN-ALPHA-A   (U/ML)

Drug	Combination Index Values at				Parameters		
	ED50	ED75	ED90	ED95	Dm	m	r
A					.02326	1.59271	.99171
B					263.0686	1.26239	.98909
Mix 1	.39399	.35954	.32867	.30946	6.85552	1.80322	.99731
	.40357	.36635	.33351	.3133	(at a 1 : 800 ratio)		

The first lines are under the mutually exclusive assumption.
The second lines are under the mutually non-exclusive assumption.

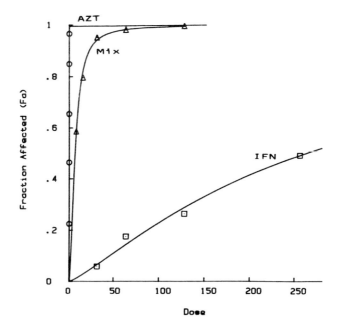

MEDIAN-EFFECT PLOT

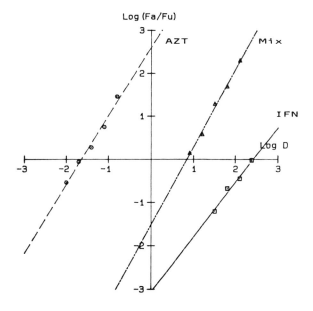

Fa-CI PLOT
Mutually Exclusive

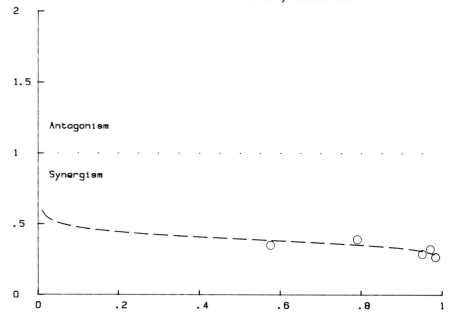

Fa-CI PLOT
Mutually Nonexclusive

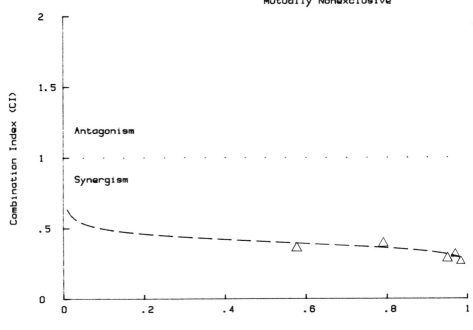

Conventional Isobolograms (Mutually Exclusive)
O ED 50, □ ED 70, △ ED 90

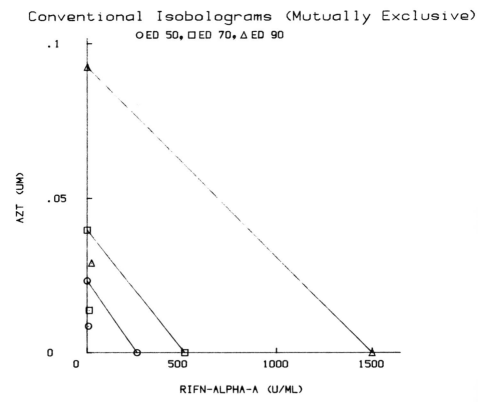

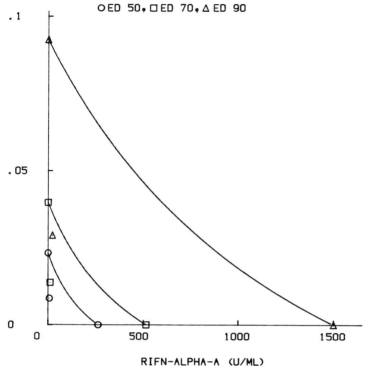

Conservative Isobolograms (Mutually Nonexclusive)

O ED 50, □ ED 70, Δ ED 90

RIFN-ALPHA-A (U/ML)

IV. Other Applications and Future Developments

This section describes various applications for median-effect based analysis. Future possibilities and inherent limitations are also discussed. Not all the topics relate only to computerized automation, but all can benefit greatly from the increased computational speed and accuracy.

A. SCHEDULE DEPENDENCY

The most common way to study drug combinations is to use both agents simultaneously. However, different results can be obtained if different drug administration schedules are employed. For example, if the agents are known to affect different stages of a cell's life cycle, it may be advantageous to administer one agent before the other. Parallel studies using agents either simultaneously or sequentially can be compared to determine whether greater synergy or antagonism is exhibited with either schedule. See Chapter 17 by Chang for a detailed treatment of schedule-dependent synergism.

B. OPTIMAL COMBINATION RATIO

Since the combination ratio in a drug mixture can be chosen arbitrarily, parallel studies can be conducted to determine the optimal ratio for maximal synergy or antagonism. If many ratios are studied, it would be possible to construct a three-dimensional plot, perhaps with axes of dose of drug 1, dose of drug 2, and combination index. Note that at present it is not possible to predict, using theoretical models, how varying the combination ratio will affect drug synergism and antagonism. An example of a combination that is very ratio dependent is the antibacterial combination of Fe^{3+} and 8-hydroxyquinoline, which can be either synergistic or antagonistic (see Chapter 5 by Vennerstrom *et al.*).

C. SELECTIVITY OF SYNERGISM: ANTITARGET VERSUS ANTIHOST

In some applications, notably cancer and AIDS therapy, effectiveness against the target is limited by toxicity towards the host. Ideally, drug combinations would be selected such that high synergy would be exhibited in antitarget effects, while strong antagonism would be found in the antihost effects. Computerized analysis can be very helpful in differentiating antihost effects from antitarget effects (Berman *et al.*, 1989). Note that even if synergism against target and antagonism against host cannot be achieved, the relative potencies of the agents against host and target can be exploited. For example, *in vitro* studies have shown that although the agents azidothymidine and interferon-alpa work synergistically both in anti-HIV effects and also against normal bone marrow progenitor cells, the concentrations necessary to inhibit HIV replication are many times higher than the concentrations necessary to cause toxicity against bone marrow cells (Berman *et al.*, 1989).

D. MODALITY OF COMBINATIONS

It is interesting to note that since *ratios* of doses are used in both the median-effect equation (Eq. 1) and the combination index equation (Eq. 2), the quantities being manipulated are *unitless*. Completely different modes of treatment can be combined and still be amenable to the analysis discussed in this chapter. Drug therapy, radiation treatment, oxygen tension, and other modes of therapy all can be combined for median-effect and combination index analysis without regard to mixing units because only ratios are used in the equations.

E. COMBINATIONS OF THREE OR MORE AGENTS: MUTUALLY EXCLUSIVE CASE ONLY

The materials and methods described in this section have not been previously reported. However, implementation of three-drug analysis is available on the newer IBM version of the software (IBM*).

Median-effect analysis can be conducted on three or more agents if the agents are used in constant ratio, analogous to the case with two drugs. However, the combination index equation becomes unworkable unless the effects of the agents are mutually exclusive. Note that in the two-drug case, an extra term is added only in the mutually nonexclusive case (Eq. 2). For more than two drugs, the extra terms to be added have been the subject of some controversy.

Without computerized automation, combination analysis of more than two agents becomes quite involved and impractical. The algorithm is best described as a "grouping" method. For example, with three drugs, A, B, C, one would start by deciding on a constant ratio for the three-drug combination. An arbitrary A:B:C ratio of 2:3:5 will be used.

One would first study the dose–effect relationship of each drug individualy. Then the effects of combination of A + B, A + C, and B + C at constant ratios of 2:3, 2:5, and 3:5, respectively, would be assessed. Finally, the effects of the A + B + C combination at a 2:3:5 ratio would be determined. Note carefully how each of the ratios was chosen, starting from the arbitrary 2:3:5 ratio.

Using standard two-drug combination analysis, one would determine the combination indices for the A + B, A + C, B + C combinations (Eq. 2). Each of these will be called a primary combination index.

Pretending that, for example, B + C was a completely independent agent, one could treat agent A and agent (B + C) as single drugs, and A + (B + C) as a two-drug combination. Standard two-drug combination analysis would yield a secondary combination index (Eq. 2).

The product of each primary and secondary combination index would yield a cumulative combination index describing the total synergism or antagonism in the drug combination.

Note that A + (B + C), B + (A + C), and C + (A + B) will each yield its own secondary combination index. Multiplying each secondary combination index with the corresponding primary combination index (for B + C, A + C, and A + B respectively) would yield three cumulative combination indices. These three cumulative indices should be comparable.

Also note that, up to this point, only standard two-drug combination analysis has been employed. Presumably, since only two-drug combination equations are being used, both mutually exclusive and nonexclusive agents can be compared. However, once again there is some controversy over this point, and only the purely mutually exclusive case is theoretically rigorously true. A combination index equation for three drugs under the mutually exclusive assumption would appear as

$$CI = \frac{(D_{comb})_1}{(D_{alone})_1} + \frac{(D_{comb})_2}{(D_{alone})_2} + \frac{(D_{comb})_3}{(D_{alone})_3} \tag{7}$$

This equation would yield yet a fourth combination index (Eq. 7) to describe the amount of synergism/antagonism in a three-drug combination. This entire set of analyses is available on the newer IBM version (IBM*).

One must shudder to think of the possibility of quantitatively analyzing four or more drugs used in combination. However, clinical cancer chemotherapy routinely employs as many as six agents in combination (see Chapter 3 by Frei).

F. STATISTICAL ANALYSIS

At present, statistical analysis available on published computer software include only regression coefficients and standard errors of slope and y intercept of the median-effect plots. It is also feasible to calculate 95% confidence intervals for many of the plots and parameters.

G. THREE-DIMENSIONAL ANALYSIS

In vitro studies of tumor spheroids have shown that oxygen tension, pH, fractional kill by added agents, and many other factors are all a function of distance into the cell mass. By using computer analysis and an automated cell sorter, Durand and Goldie, 1987 constructed f_a–CI plots with a third dimension of distance into the tumor spheroid. Automated cell sorting allowed a very large amount of data to be obtained for analysis by the median-effect principle. Similar studies can be imagined using a third dimension of oxygen tension or pH for the f_a–CI plot. This type of study can provide very valuable information about drug interactions in solid tumor models (see also Chapter 18 by Durand).

References

Berman, E., Duigou-Osterndorf, R., Krown, S. E., Fanucchi, M. P., Chou, J., Hirsch, M. S., Clarkson, B. D., and Chou, T.-C. (1989). Synergistic cytotoxic effect of azidothymidine and recombinant interferon alpha on normal bone marrow progenitor cells. *Blood* **74,** 1281–1286.

Chou, T.-C. (1976). Derivation and properties of Michaelis–Menten type and Hill type equations for reference ligands. *J. Theor. Biol.* **39,** 253–276.

Chou, J., and Chou, T.-C. (1985). "Dose–Effect Analysis with Microcomputers: Quantitation of ED_{50}, ID_{50}, Synergism, Antagonism, Low-Dose Risk, Receptor Ligand Binding, and Enzyme Kinetics," software for Apple II microcomputers. Elsevier–Biosoft, Cambridge, England.

Chou, J., and Chou, T.-C. (1987). Dose–Effect Analysis with Microcomputers: Quantitation of ED_{50}, ID_{50}, Synergism, Antagonism, Low-Dose Risk, Receptor Ligand Binding, and Enzyme Kinetics," software for IBM-PC microcomputers. Elsevier–Biosoft, Cambridge, England.

Chou, J., and Chou, T.-C. (1988). Computerized simulation of dose reduction index (DRI) in synergistic drug combinations. *Pharmacologist* **30,** A231.

Chou, T.-C., and Talalay, P. (1983). Analysis of combined drug effects: A new look at a very old problem. *Trends Pharmacol. Sci.* **4,** 450–454.

Chou, T.-C., and Talalay, P. (1984). Quantitative analysis of dose–effect relationships: The combined effects of multiple drugs or enzyme inhibitors. *Adv. Enzyme Regul.* **22,** 27–55.

Durand, R. W., and Goldie, J. H. (1987). Interaction of etoposide and cisplatin in an in vitro tumor model. *Cancer Treat. Rep.* **71,** 673–679.

Hartshorn, K. L., Vogt, M. W., Chou, T.-C., Blumberg, R. S., Byington, R., Schooley, R. T., and Hirsch M. S. (1987). Synergistic inhibition of human immunodeficiency virus in vitro by azidothymidine and recombinant interferon alpha-A. *Antimicrob. Agents Chemother.* **31,** 168–172.

PART II

Mechanisms of Interaction

CHAPTER 7

Inhibition of Metabolic Drug Inactivation: Modulation of Drug Activity and Toxicity by Perturbation of Glutathione Metabolism

Owen W. Griffith
Department of Biochemistry
Cornell University Medical Center
New York, New York 10021

Henry S. Friedman
Department of Pediatrics
Duke University Medical Center
Durham, North Carolina 27710

SYNERGISM AND ANTAGONISM IN CHEMOTHERAPY
Copyright © 1991 by Academic Press, Inc.
All rights of reproduction in any form reserved.

I. Introduction

A. PROTECTIVE ROLE OF GLUTATHIONE

Glutathione (L-γ-glutamyl-L-cysteinyl glycine, GSH) is the major intracellular nonprotein thiol in mammalian tissues. Glutathione is a required cofactor in several enzyme-catalyzed reactions and participates in a variety of metabolic processes including intraorgan and interorgan cysteine storage and transport, amino acid transport, and cellular protection against endogenous and exogenous free radicals, peroxides, and reactive electrophiles. The protective functions of GSH are dependent on both oxidation–reduction and nucleophilic reactions of the cysteinyl thiol (Arrick and Nathan, 1984); many of the nucleophilic thiol reactions lead ultimately to the formation of mercapturic acids and play a crucial role in cellular protection from toxic agents. Recognition of the role of GSH in modulating the activity and toxicity of electrophilic antineoplastic agents has elicited a number of studies examining the possibility that some antineoplastic therapies may be enhanced by pharmacological perturbations of GSH metabolism.

B. MANIPULATION OF GSH METABOLISM IN THE CHEMOTHERAPY OF CANCER

Recent studies showing enhanced chemotherapy of neoplasms by modifications of GSH metabolism extend early observations demonstrating the amelioration of nitrogen mustard toxicity in animals treated with cysteamine (Peczenik, 1953) or with cysteine, the amino acid which is generally limiting for GSH synthesis, (Brandt and Griffin, 1951). In other early studies Calcutt and Connors (1963) showed a correlation between murine tumor sensitivity to merophan and tumor sulfhydryl levels. Suzukake et al. (1982) later reported both increased GSH levels in an L1210 leukemia cell line resistant to melphalan and enhanced cytotoxicity by this alkylator following GSH depletion mediated by nutritional restriction. In 1979, it was suggested that recently developed inhibitors of GSH biosynthesis and metabolism could be used to enhance cancer therapies based on either ionizing radiation or alkylating agents (Meister and Griffith, 1979).

This chapter reviews the synthesis and turnover of GSH and the metabolic pathways involved in its several protective functions. We also review the treatment of trypanosomiasis by GSH depletion and summarize recent efforts by ourselves and others to enhance antineoplastic activity and minimize alkylator toxicity through the pharmacological control of GSH metabolism.

II. Glutathione Metabolism and Turnover

A. GLUTATHIONE BIOSYNTHESIS

Glutathione is synthesized in two ATP-dependent steps by the sequential action of γ-glutamylcysteine synthetase and glutathione synthetase (Reactions 1 and 2, respectively; see also Fig. 1). Both synthetases have been purified to homogeneity from several sources and have been extensively characterized with respect to physical properties, substrate specificity, mechanism of action, and metabolic control. The discussion here focuses on aspects of their mechanism and specificity relating directly to the pharmacological control of GSH metabolism or to the *in vivo* synthesis of GSH analogs of possible therapeutic interest; additional information is summarized in a review by Meister (1989).

(1)   L-Glutamate + L-Cysteine + ATP $\xrightarrow{Mg^{2+}}$ L-γ-Glutamyl-L-Cysteine + ADP + P_i

(2)   L-γ-Glutamyl-L-Cysteine + Glycine + ATP $\xrightarrow{Mg^{2+}}$ Glutathione + ADP + P_i

γ-Glutamylcysteine synthetase has been isolated and substantially purified from the erythrocytes, liver, and kidneys of several mammalian species (Meister, 1989). The rat kidney enzyme, which was recently cloned and sequenced (Yan and Meister, 1990), is best characterized in terms of kinetic and chemical mechanisms and physical properties; available data suggest it is similar in all important respects to the enzyme isolated from other mammalian tissues and species. Rat kidney γ-glutamylcysteine synthetase is a nonglycosylated, cytoplasmic dimer of nonidentical subunits (M_r 73,000 and 27,700) with the active site located exclusively on the larger subunit (Seelig *et al.*, 1984). Enzyme-catalyzed synthesis of γ-glutamylcysteine proceeds in two steps (Reactions 1a and 1b). Both ADP and γ-glutamylphosphate, the products of Reaction 1a, are tightly bound to the active site, and no products are released until the overall reaction is complete. The kinetic mechanism is of the Random BC type with ATP bound first and ADP released last; the other substrates and products bind and dissociate in random order. Amide bond formation thus requires the preliminary formation of a quarternary complex of enzyme, γ-glutamylphosphate, ADP, and cysteine (Yip and Rudolph, 1976; Schandle and Rudolph, 1981).

(1a)   L-Glutamate + ATP $\xrightarrow{Mg^{2+}}$ L-γ-Glutamylphosphate + ADP

(1b)   L-γ-Glutamylphosphate + L-Cysteine $\xrightarrow{Mg^{2+}}$ L-γ-Glutamyl-L-Cysteine + P_i

Under physiological conditions of ionic composition and substrate availability γ-glutamylcysteine synthetase is absolutely specific for L-glutamate (K_m = 1.6 mM) and is highly specific for L-cysteine (K_m = 0.3 mM) (Sekura

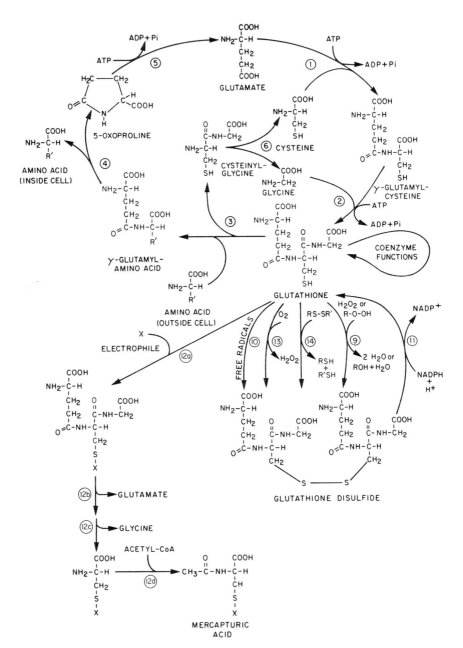

FIGURE 1 Reactions of GSH synthesis, turnover, and metabolism. The circled numbers correspond to the following enzymes or metabolic processes (reactions 7 and 8 mentioned in the text are not shown): 1, γ-glutamylcysteine synthetase; 2, glutathione synthetase; 3, γ-glutamyltranspeptidase; 4, γ-glutamylcyclotransferase; 5, 5-oxoprolinase; 6, cysteinylglycinase; 9, glutathione peroxidase; 10, nonenzymatic free radical oxidation of GSH; 11, glutathione reductase; 12a, glutathione transferase; 12b, γ-glutamyltranspeptidase; 12c, cysteinylglycinase; 12d, cysteine conjugate N-acetyltransferase; 13, glutathione oxidase; 14, enzymatic and nonenzymatic GSH-dependent transhydrogenations.

and Meister, 1977; O.W. Griffith and A. Meister, unpublished observations). The binding site for L-glutamate does not readily accommodate either L-aspartate or L-α-aminoadipate (the lower and higher homologs of L-glutamate, respectively) and neither amino acid has been shown to react *in vivo*. D-glutamate is a substrate for Reaction 1a, but the resulting D-γ-glutamylphosphate is unable to react with L-cysteine, dissociates from the active site, and cyclizes to 5-oxo-D-proline. Other unnatural glutamate analogs including L-α-methylglutamate, β-glutamate, L-*threo*-β-methylglutamate, and both enantiomers of α-aminomethylglutarate are moderately good substrates (Sekura and Meister, 1977; Sekura *et al.*, 1976; Bridges *et al.*, 1980). Since the products formed from these analogs are substrates of glutathione synthetase (see subsequent sections), it may be possible to exploit the broad glutamate analog specificity of γ-glutamylcysteine synthetase to allow *in vivo* formation of biochemically and pharmacologically interesting GSH analogs in animals administered suitable glutamate derivatives. Glutathione analogs have already been prepared *in vitro* using an immobilized *E. coli* enzyme (Moore and Meister, 1987).

Although L-cysteine is by far the best physiological substrate, γ-glutamylcysteine synthetase is also highly active with L-β-chloroalanine and L-β-cyanoalanine, amino acids that are not normally present *in vivo*; the enzyme is moderately active with L-α-aminobutyrate, a minor metabolite of L-methionine (Orlowski and Meister, 1971; Moore and Meister, 1987). Reaction with L-α-aminobutyrate to form L-γ-glutamyl-L-α-aminobutyrate occurs to a limited extent *in vivo* and, since the product is a substrate of glutathione synthetase, accounts for the presence of small amounts of ophthalmic acid (L-γ-glutamyl-L-α-aminobutyrylglycine) in many mammalian tissues. Because cysteine slowly inactivates the purified enzyme, L-α-aminobutyrate is also the preferred substrate for *in vitro* assays of γ-glutamylcysteine synthetase. L-Alanine and L-norvaline (the lower and higher homologs of L-α-aminobutyrate, respectively) have weak affinity for the L-cysteine binding site and react poorly. The extent to which L-cysteine is replaced *in vivo* by L-alanine or other structurally similar protein amino acids (e.g., L-serine) has not been directly determined but is thought to be small.

Glutathione synthetase has been obtained in highly purified form from yeast, *E. coli*, and rat kidney and erythrocytes (Meister, 1989). Rat kidney glutathione synthetase ($M_r = 118,000$) is a cytoplasmic dimer of apparently identical subunits; the highly purified enzyme is reported to contain 1.7 to 2.04% carbohydrate (Oppenheimer *et al.*, 1979). Although γ-glutamylcysteine synthetase and glutathione synthetase are located in the same cellular compartment and normally function in a highly integrated manner, there is presently no evidence that the enzymes are physically associated; direct studies of possible association of the synthetases are warranted. Even in the absence of physical association, efficient metabolic coupling between the synthetases is likely because glutathione synthetase exhibits a very high affinity for L-γ-glutamyl-L-cysteine and related L-γ-glutamylamino acids. Although accurate K_m values for the dipeptide substrates are technically difficult to determine, studies with the rat kidney enzyme indicate that the K_m for L-γ-glutamyl-L-α-aminobutyrate is about 20 μM when all substrates are present at concentrations comparable to those expected *in vivo*; the K_m for L-γ-glutamyl-L-cysteine is in the same range (Oppenheimer *et al.*, 1979).

The glutathione synthetase reaction is mechanistically similar to that catalyzed by γ-glutamylcysteine synthetase and also proceeds in two steps with an acyl phosphate intermediate (Reactions 2a and 2b). As noted, the enzyme is relatively nonspecific for the γ-glutamyl portion of its substrates and will, for example, utilize D-γ-glutamyl-L-α-aminobutyrate, L-γ-(α-methyl)-glutamyl-L-α-aminobutyrate, L-γ-(N-methyl)-glutamyl-L-α-aminobutyrate, β-aminoglutaryl-L-α-aminobutyrate, glutaryl-L-α-aminobutyrate, and some β-methylglutamyl and γ-methylglutamyl analogs at rates 15 to 100% of that observed with L-γ-glutamyl-L-cysteine and L-γ-glutamyl-L-α-aminobutyrate, which react at approximately equal rates (Oppenheimer *et al.*, 1979). Specificity for L-cysteinyl portion of L-γ-glutamyl-L-cysteine is somewhat higher. In addition to L-α-aminobutyrate, the enzyme is active with γ-glutamyl derivatives of L-(S-methyl)cysteine, L-alanine, and L-serine but is inactive or nearly inactive with γ-glutamyl derivatives of glycine, L-norvaline, and all D- and β-amino acids (Oppenheimer *et al.*, 1979). The relatively low specificity toward glutamate-modified L-γ-glutamyl-L-cysteine analogs supports the previously stated view that several interesting GSH analogs will be efficiently formed *in vivo* following administration of suitable glutamate analogs. In contrast to the wide range of γ-glutamylcysteine analogs accepted as substrates, glutathione synthetase is highly specific for glycine. L-Alanine, D-alanine, β-alanine, and sarcosine are not substrates. Tripeptide-like products are formed with aminooxyacetate and N-hydroxyglycine (Oppenheimer *et al.*, 1979), but it is possible that these highly nucleophilic glycine analogs react "nonenzymatically" with enzyme-bound L-γ-glutamyl-l-cysteinylphosphate.

(2a)   L-γ-Glutamyl-L-Cysteine + ATP → L-γ-Glutamyl-L-Cysteinylphosphate + ADP

(2b)   L-γ-Glutamyl-L-Cysteinylphosphate + Glycine → Glutathione + P_i

Glutathione synthesis is controlled *in vivo* by L-cysteine availability (Tateishi *et al.*, 1974, 1977) and by feedback inhibition of GSH on γ-glutamylcysteine synthetase (Richman and Meister, 1975). The relative importance of these factors depends on the nutritional state of the animal. Although cysteine is a nonessential amino acid, its synthesis via the transsulfuration pathway is dependent on sulfur from methionine, an essential amino acid. Since mammalian diets are often relatively poor in the sulfur amino acids, it is likely that GSH synthesis will be at times limited by cysteine availability. Consistent with this view, studies in rodents indicate that the GSH levels of major organs such as liver and kidney are significantly increased following oral or parenteral administration of L-cysteine or L-cysteine precursors. Similarly, in both rats and mice, the GSH concentration in liver fluctuates on a diurnal cycle; levels are high in the morning following nocturnal feeding and are low in the late evening after daytime fasting (Beck *et al.*, 1958). Glutathione levels in other tissues appear to be less sensitive to nutritional variations. As originally suggested by Tateishi *et al.* (1974, 1977), it is probable that GSH released by the liver is degraded peripherally to provide cysteine to extrahepatic tissues and permit those tissues to maintain relatively constant intracellular GSH levels (Tateishi *et al.*, 1977; Griffith and Meister, 1979a; Anderson *et al.*, 1980). The important role of hepatic GSH synthesis in the intraorgan delivery of cysteine suggests that pharmacological interventions decreasing GSH levels in liver may cause secondary GSH depletion in extrahepatic tissues.

In the presence of moderate to high levels cysteine, GSH synthesis is limited by feedback inhibition. Richman and Meister (1975) reported that rat kidney γ-glutamylcysteine synthetase is strongly inhibited by GSH and γ-glutamylcysteine; γ-glutamyl-α-aminobutyrate and ophthalmic acid are much weaker inhibitors. Inhibition by GSH is nonallosteric since part of the GSH binding site overlaps with or its identical to the substrate binding site for L-glutamate. The physiological significance of feedback inhibition by GSH is supported by the finding that the apparent K_i of GSH (about 2.3 mM) is comparable to the concentration of GSH maintained in rat kidney (3 to 4 mM) (Richman and Meister, 1975). Since there is no evidence for tissue specific isozymes of γ-glutamylcysteine synthetase, it is generally believed that the synthetase will be comparably inhibited in all tissues as the GSH concentration approaches and, in some tissues, exceeds 2 to 3 mM. Consistent with this view, in both liver and erythrocytes (tissues in which the GSH concentration is ≥ 2 mM) the normal *in vivo* rate of GSH synthesis represents only a small fraction of the total γ-glutamylcysteine synthetase activity present; full activity of the synthetase is presumably suppressed by GSH-mediated feedback inhibition (Griffith, 1981a; Meister, 1989). The critical importance of this control mechanism was dramatically illustrated with the discovery that the inherited disorder 5-oxoprolinuria represented a deficiency of glutathione synthetase. In affected patients, γ-glutamylcysteine is not effectively converted to GSH, and γ-glutamylcysteine synthetase is therefore not inhibited. γ-Glutamylcysteine is consequently synthesized in much larger than normal amounts and is converted intracellularly to 5-oxoproline and cysteine by γ-glutamylcyclotransferase (see Reaction 4). When the amount of 5-oxoproline produced exceeds the capacity of 5-oxoprolinase (see Reaction 5), 5-oxoproline is released into the blood producing an acidosis and is filtered into the urine producing 5-oxoprolinuria (Meister and Larsson, 1989).

Although feedback inhibition by GSH is an important mechanism limiting GSH synthesis, tissue GSH levels vary over a wide range, e.g., 0.5 to 8 mM in the mouse (Griffith and Meister, 1979a); it is apparent that the fixed K_i of

2.3 mM characterizing GSH-mediated inhibition of γ-glutamylcysteine synthetase cannot by itself account for the range of concentrations observed. Among other possible control mechanisms, cysteine availability (discussed earlier), the abundance of γ-glutamylcysteine synthetase in the tissue, and the rate of GSH utilization by the tissue (mainly transport out of the tissue, see subsequent discussion) are thought to be most important. Increased cysteine availability or decreased GSH transport are likely to account in part for the increased GSH levels exhibited by many alkylator-resistant tumors including some of those with the multidrug-resistant (MDR) phenotype. Understanding the mechanisms controlling GSH levels in specific tumors and surrounding normal tissues can in favorable circumstances suggest the design of pharmacological strategies for tumor-selective depletion of GSH.

B. TRANSPORT OF GSH ACROSS CELL MEMBRANES

Whereas intracellular GSH levels are typically in the millimolar range, extracellular GSH levels are typically less than 50 μM (Griffith and Meister, 1979b). This gradient of GSH concentration implies that the plasma membrane is generally impermeable to GSH and that facilitated diffusion of GSH does not occur. Consistent with that view, there is no convincing evidence that intact GSH enters cells at a significant rate even if the extracellular concentration is artificially made to exceed the intracellular GSH level. Intracellular GSH is, however, actively transported out of all mammalian cells examined. Studies by Inoue and Morino (1985) using isolated rat kidney cortical brush border membrane vesicles indicate that GSH transport is dependent on membrane potential and involves the transfer of negative charge. Transport was inhibited by S-benzylglutathione suggesting that the GSH transporter plays a role in the transport of glutathione S-transferase products out of the cell (see subsequent text). The specificity of transport with respect to other GSH analogs has not been characterized, but studies with human erythrocytes suggest that transport of glutathione disulfide (GSSG) is mediated by a distinct transport system (Kondo et al., 1981, 1989). Beutler and coworkers have suggested that specific transport of GSSG out of cells is a protective mechanism designed to minimize the effects of oxidative stress on the intracellular environment (Kondo et al., 1981).

As suggested earlier, regulation of the rate of GSH transport out of cells is a major factor in the control of tissue GSH levels. Although direct measurements of GSH transport are not available for most tissues, transport rates were estimated indirectly in mice administered buthionine sulfoximine, a potent inhibitor of γ-glutamylcysteine synthetase (see Section IV). In those studies, the rate of GSH depletion following administration of inhibitor was taken as an indication of the normal rate of GSH transport out of the cell. The results indicate that tissues vary widely in their rate of GSH transport. Liver and kidney GSH levels declined with $t_{1/2}$ of about 60 min and 30 min, respectively, whereas in other tissues such as heart, spleen, and colon GSH levels changed little in 2 hrs (Griffith and Meister, 1979a) (complete data are given in Table I, Section IV). Erythrocyte GSH turnover is too slow to measure by depletion, but isotope incorporation studies indicate that the turnover time of GSH in human erythrocytes is about 6 days (Griffith, 1981a).

C. EXTRACELLULAR METABOLISM OF GSH

The γ-glutamyl bond of GSH renders the tripeptide resistant to attack by most proteases and peptidases. Under physiological conditions, GSH appears to be degraded by only a single enzyme, γ-glutamyltranspeptidase, which catalyzes both the transpeptidation and hydrolysis of GSH as shown in Reactions 3a and 3b, respectively (Reaction 3a is shown in Fig. 1). γ-Glutamyltranspeptidase is a membrane-bound, extracellular glycoprotein that is found in particularly high amounts on the membranes of certain epithelial cells including those of the brush border of the kidney proximal tubule, the microvilli of the intestinal lumen, the ciliary body, and the choroid plexus. Erythrocytes, on the other hand, have negligible amounts of transpeptidase, and normal liver has only a small amount of enzyme, which is located exclusively on the bile canaliculi (Tate and Meister, 1985). Tumor cells vary widely in the amount of transpeptidase expressed on their plasma membrane surfaces, and there is preliminary evidence that an increase in the amount of transpeptidase may mediate one form of drug resistance (Ahmad *et al.*, 1987). γ-Glutamyltranspeptidase has been purified from a number of tissues and shows significant tissue-specific variability in its physical properties. The rat kidney enzyme is, however, catalytically typical and has been extensively characterized with respect to amino acid sequence, physical properties, and catalytic mechanism (Tate and Meister, 1985).

(3a)   Glutathione + L-Amino acid → L-γ-Glutamyl-L-Amino acid + Cysteinylglycine

(3b)   Glutathione + H_2O → L-Glutamate + Cysteinylglycine

As discussed previously, GSH is transported out of but not into cells. Extracellular metabolism of GSH by the sequential actions of γ-glutamyltranspeptidase and extracellular dipeptidases (cysteinylglycinase and aminopeptidase M) (Fig. 1, Reaction 6) allows the amino acids of GSH, including the nutritionally important L-cysteine residue, to be recovered. Cysteinylglycine can also be taken up intact and degraded intracellularly. The importance of γ-glutamyltranspeptidase in GSH turnover and cysteine recovery was demonstrated in studies where D- or L-γ-glutamyl-(*o*-carboxy)phenylhydrazide, potent transpeptidase inhibitors (see Section IV), were administered to mice; the treated mice exhibited markedly increased plasma GSH levels (73 μM vs 24 μM in controls) and excreted substantial amounts of GSH in their urine (200 to 800 nmol/hr vs 0.4 to 1.3 nmol/hr in controls). Additional studies established that the urinary GSH was derived both from the plasma by glomerular filtration and from the kidney itself by secretion into the tubule (Griffith and Meister, 1979b); the latter source normally accounts for about 80% of the total intratubular pool (Griffith, 1981b). Interestingly, the kidney GSH level of the treated mice decreased from 3.1 $\mu mol/gm$ to 2.2 $\mu mol/gm$ in 2 hr (Griffith and Meister, 1979b). The observed decrease suggests that renal GSH synthesis is dependent on the continuing recovery of the cysteinyl moiety of GSH through intratubular degradation of the tripeptide. Two patients having a congenital deficiency of γ-glutamyltranspeptidase have been identified by screening urine samples for thiol content; both patients exhibit markedly increased GSH levels in their plasma and urine (Griffith and Meister, 1980; Meister and Larsson, 1989).

The studies and observations just summarized indicate that the amino acids of GSH are not effectively reutilized in the absence of γ-glutamyltranspeptidase; the studies do not directly elucidate the extent to which transpeptidase activity and amino acid recovery are metabolically linked. It is not yet known, for example, if the γ-glutamylamino acid, glutamate, cysteine, and glycine formed extracellularly by the combined activities of transpeptidase and dipeptidases are taken up exclusively (or mainly) by the cells elaborating those enzymes. Studies of arteriovenous differences across specific organs suggest that the products released during GSH degradation are taken up within the organ where degradation was initiated (Abbott et al., 1984), but such studies do not establish the extent to which the amino acid and dipeptide products diffuse at the cellular level. The issue is of some importance in defining the extent to which overexpression of transpeptidase confers a selective advantage in terms of cysteine recovery and GSH synthesis specifically on the cells exhibiting extra transpeptidase. As noted, neoplastic cells are among those found to overexpress transpeptidase (Ahmad et al., 1987; Vistica and Ahmad, 1989).

D. INTEGRATION OF GLUTATHIONE METABOLISM: THE γ-GLUTAMYL CYCLE

The extent to which γ-glutamyltranspeptidase catalyzes the synthesis of a γ-glutamylamino acid (Reaction 3a) or glutamate (Reaction 3b) depends on the concentration and identity of the amino acids available to act as acceptors of the γ-glutamyl group in Reaction 3a. In vitro studies indicate that GSH hydrolysis and transpeptidation proceed at comparable rates when the γ-glutamyltranspeptidase reaction is carried out at physiological pH in the presence an amino acid acid mixture comparable to that in plasma (Allison and Meister, 1981). Preferred amino acid acceptors include cysteine, glutamine, methionine, alanine, and serine; branched chain amino acids and amino acids with ionic side chains are much less active. It is probable that L-cysteine and L-glutamine account for most transpeptidation *in vivo* (Meister, 1989). γ-Glutamylamino acids produced extracellularly are taken up by a specific transport system that is most active in transpeptidase-rich tissues. Intracellularly, the L-γ-glutamyl-L-amino acids are converted to L-glutamate and L-amino acid by the sequential reactions of γ-glutamylcyclotransferase (Reaction 4) and (Reaction 5) (see also Fig. 1).

(4)   L-γ-Glutamyl-L-Amino acid ⟶ 5-Oxo-L-Proline + L-Amino acid

(5)   5-Oxo-L-Proline + ATP $\xrightarrow{Mg^{2+}}$ L-Glutamate + ADP + P_i

As pointed out by Orlowski and Meister (1970), the reactions of GSH synthesis and degradation can be viewed as a metabolic cycle (the γ-glutamyl cycle, shown in the top portion of Fig. 1) in which intracellular GSH synthesis (Reactions 1 and 2) is followed by extracellular transpeptidation (Reaction 3a) and cysteinylglycine hydrolysis (Reaction 6); uptake of cysteine, glycine, and γ-glutamylamino acid is followed by the intracellular conversion of the γ-glutamylamino acid to glutamate and amino acid (Reactions 4 and 5) to complete the cycle. Overall the cycle results in the net transport of one molecule of amino acid (the transpeptidase acceptor amino acid) from the extracellular space to the cytosol for each molecule of GSH used (Orlowski and Meister, 1970). Subsequent studies using isotope tracer techniques and enzyme specific substrates and inhibitors clearly establish that the γ-glutamyl cycle operates *in vivo* and that amino acids are transported in the course of its operation (Van der Werf *et al.*, 1974; Griffith *et al.*, 1978). The cycle is probably not of major quantitative importance in relation to the total amino acid transport of any tissue, but it may play a significant role in the transport of certain amino acids such as cysteine. The γ-glutamyl cycle, defined to include both Reaction 3a and 3b, does fully account for GSH turnover.

The γ-glutamyl cycle can operate within a specific organ as in the case of the kidney proximal tubule cells which make, release, and degrade GSH, and then recover and reuse the amino acid products. The cycle can operate on an interorgan basis as is the case when the liver synthesizes and secretes GSH which is then degraded by extrahepatic tissues that ultimately recover and use the constituent amino acids (Griffith and Meister, 1979a; Anderson *et al.*, 1980).

III. Glutathione as a Cellular Protectant

A. REACTIONS LEADING TO GLUTATHIONE DISULFIDE

Although aerobic organisms derive an enormous energy benefit in being able to oxidize fatty acids, dicarboxylic acids, and carbohydrates to carbon dioxide, the high energy yield is obtained at the price of exposing cellular constituents to a continuing flux of reactive oxygen species. Studies of mitochondrial electron transport, for example, indicate that up to 5% of the electron flow may be diverted into the reduction of oxygen to superoxide (O_2^-) and hydrogen peroxide (H_2O_2) rather than water (Boveris et al., 1972; Naqui et al., 1986). Additional quantities of superoxide are generated by the spontaneous conversion of oxyhemoglobin to methemoglobin, by the respiratory burst of certain immune cells, and by the redox cycling of various cellular constituents or xenobiotics (e.g., menadione and certain drugs used in cancer chemotherapy) (Byczkowski and Gessner, 1988). Superoxide spontaneously dismutates to oxygen and hydrogen peroxide, a reaction also catalyzed by superoxide dismutase. Additional hydrogen peroxide is formed by direct reduction of oxygen, a reaction catalyzed by a variety of oxidases [e.g., monoamine oxidase, D-amino acid oxidase, and xanthine oxidase (but not xanthine dehydrogenase)]. Whereas superoxide and hydrogen peroxide exhibit only moderate direct reactivity with important cellular structures, they can together form hydroxyl radical ($HO^\cdot$), an extremely reactive and damaging species. As shown in Reactions 7a and 7b, formation of hydroxyl radical from superoxide and hydrogen peroxide is catalyzed by metal ions, especially iron (Fridovich, 1989). The fact that superoxide can apparently reduce the normally sequestered Fe^{3+} of ferritin and thereby release Fe^{2+} into the cytosol contributes to the danger of intracellular superoxide. It should be noted that hydroxyl radicals are also formed by the homolytic cleavage of hydrogen peroxide (Reaction 8); although such cleavage is normally very slow, it is catalyzed by certain drugs and has been exploited in the treatment of trypanosomiasis as discussed in Section V.

(7a) $$O_2^- + Fe^{3+} \rightarrow O_2 + Fe^{2+}$$

(7b) $$H_2O_2 + Fe^{2+} \rightarrow HO^\cdot + OH^- + Fe^{3+}$$

(8) $$H_2O_2 \rightarrow 2\,HO^\cdot$$

Glutathione plays an important role in both preventing and repairing oxidative damage to cellular structures. Damage prevention requires the rapid conversion of superoxide and hydrogen peroxide to nontoxic species; to this end, superoxide dismutase converts superoxide to oxygen and hydrogen peroxide. Within peroxisomes, hydrogen peroxide is itself dismutated to oxygen and water by catalase. Cytoplasmic and mitochondrial hydrogen peroxide, on the other hand, are reduced to water by glutathione peroxidase, a selenium-containing enzyme that catalyzes Reactions 9a and 9b. Glutathione peroxidase is widely distributed in mammalian tissues although levels vary considerably with species and organ. Levels are generally high in liver, kidney, pancreas, adrenals, and lung, but are lower in skeletal muscle and nerve tissue (Flohé, 1989). The relative importance of catalase and glutathione peroxidase in the detoxification of hydrogen peroxide remains somewhat controversial, particularly in erythrocytes where catalase is not sequestered in peroxisomes. However , since glutathione peroxidase has a higher affinity for hydrogen peroxide than does catalase, it is generally believed that under normal conditions most hydrogen peroxide is reduced by glutathione peroxidase. Since total catalase activity is often greater than glutathione peroxidase activity, catalase may play a quantitatively more important role under conditions of very high oxidative stress; under those circumstances, cytoplasmic hydrogen peroxide may diffuse into peroxisomes and react with catalase.

(9a) $\qquad$ $H_2O_2 + 2\ GSH \rightarrow 2\ H_2O + GSSG$

(9b) $\qquad$ $ROOH + 2\ GSH \rightarrow ROH + H_2O + GSSG$

Reduction and dismutation of hydrogen peroxide are reactions designed to prevent oxidative damage to cells. As indicated in Reaction 9b, glutathione peroxidase also has the ability to repair oxidative damage by reducing organic hydroperoxides such as those formed by reaction of lipid and nucleic acid free radicals with molecular oxygen. In contrast to hydrogen peroxide, organic peroxides are not attacked by catalase; they are, however, reduced by some glutathione *S*-transferases (see subsequent text) with an overall stoichiometry identical to that shown in Reaction 9*b*. Since the glutathione *S*-transferases do not contain selenium, the activities of the "authentic" glutathione peroxidase and that of glutathione *S*-transferases are distinguished by the terms Se-dependent and Se-independent glutathione peroxidase. The relative importance of these distinct activities varies with species, tissue, and substrate (Ketterer *et al.*, 1988). Since the reactions catalyzed are the same, the distinction is significant mainly in emphasizing the potential redundancy in cellular peroxide detoxification. In the context of the present discussion, it is important that both activities are GSH-dependent and oxidize two moles of GSH to GSSG for every mole of peroxide reduced.

As noted, hydroxyl radical is considerably more reactive and thus more cytotoxic than either superoxide or hydrogen peroxide. Hydroxyl radical is capable of extracting a hydrogen atom from unsaturated lipids to form water and a carbon-centered free radical; the latter rearranges, reacts with oxygen, and ultimately propagates the free radical chain reaction to additional cellular constituents. Ionizing radiation also produces oxygen- and carbon-centered free radicals that can react to form DNA cross-links and nucleic acid hydroperoxides such as 5-hydroperoxymethyluracil. Free radicals thus have the capacity to seriously damage both the membranes and genetic complement of cells; such damage is, of course, the intended outcome in the therapy of tumors using ionizing radiation. Whether adventitious or intended therapeutically, free radical damage can be prevented by intervention of GSH as shown in Reactions 10a and 10b. It is Reaction 10a that actually effects the detoxification since sulfur-centered free radicals are significantly less reactive than oxygen- or carbon-centered radicals; thus GS˙ does not continue the free radical chain reaction but rather reacts as shown in Reaction 10b to produce GSSG.

(10a) $R˙ + GSH → RH + GS˙$

(10b) $2 GS˙ → GSSG$

Although GSH-mediated reduction of peroxides and detoxification of free radicals results in the formation of GSSG, the intracellular GSH/GSSG ratio is normally 100 to 300 (Gilbert, 1989). The enzyme responsible for maintaining this ratio is glutathione reductase, an NADPH-dependent flavoprotein present in both the cytosol and mitochondrial matrix (Reaction 11). As indicated, reduction of GSSG results in the oxidation of NADPH; in most cells, NADP$^+$ is reduced to NADPH by the oxidative branch of the pentose phosphate pathway (i.e., by glucose-6-phosphate dehydrogenase and 6-phosphogluconate dehydrogenase). Although NADPH regeneration can be limiting under conditions of high oxidative stress or in inherited glucose-6-phosphate dehydrogenase deficiency, under most conditions the equilibrium of the glutathione reductase reaction is maintained well to the right.

(11) $GSSG + NADPH + H^+ → 2 GSH + NADP^+$

B. REACTIONS LEADING TO MERCAPTURIC ACIDS

In addition to the thiol-dependent oxidation-reduction reactions outlined earlier, the protective reactions of GSH encompass a series of reactions resulting in formation of mercapturic acids [S-(N-acetyl)-L-cysteine conjugates] (Stevens and Jones, 1989). Mercapturic acid synthesis is initiated by formation of a thioether between a reactive electrophile and GSH thiolate anion (Reaction 12a) and results ultimately in the formation of an N-acetyl-S-substituted cysteine derivative following reactions 12b–12d (see also Fig. 1 for structures). These reactions are catalyzed by γ-glutamyltranspeptidase, cysteinylglycinase, and cysteine conjugate N-acetyltransferase, respectively. It should be noted that some electrophiles are detoxified by reaction with the GSH thiolate anion without further metabolism through the mercapturate pathway.

(12a) $GSH + RX → GS\text{-}R + HX$

(12b) $GS\text{-}R → Glutamate + \underset{\overset{|}{R}}{CyS}\text{-}Gly$

(12c) $\underset{\overset{|}{R}}{CyS}\text{-}Gly → CyS\text{-}R + Glycine$

(12d) $CyS\text{-}R + Acetyl\text{-}CoA → N\text{-}Acetyl\text{-}CyS\text{-}R + CoASH$

Although the attack of GSH on an electrophile is catalyzed by intracellular glutathione *S*-transferases, nonenzymatic formation of GSH *S*-conjugates also occurs and may in some cases exceed the rate of enzyme-catalyzed reaction. Conjugates of GSH with the alkylating agents melphalan (Dulik *et al.*, 1986), chlorambucil (Dulik *et al.*, 1990), and cyclophosphamide (Yuan *et al.*, 1990) have been demonstrated. For most alkylating agents the relative contributions of enzymatic and nonenzymatic GSH conjugation remains poorly defined; in general, the importance of the enzyme-catalyzed reaction would be expected to increase with decreasing GSH concentration, decreasing electrophile reactivity, and increasing glutathione *S*-transferase activity. Quantitative studies of the relative importance of these factors is of considerable importance in the rational use of GSH modulation strategies in conjunction with alkylating agent therapy. Thus, although elevated levels of glutathione *S*-transferase (with normal levels of GSH) have been demonstrated in tumor cells resistant to nitrogen mustard (Wang and Tew, 1985) therapeutic benefits associated with modulation of glutathione *S*-transferase may be limited if enzymatic GSH conjugation is greatly exceeded by nonenzymatic conjugation or if the activity of glutathione *S*-transferase is not rate-limiting for detoxification.

IV. Pharmacological Control of Glutathione Levels and Metabolism

A. DEPLETION OF INTRACELLULAR GLUTATHIONE LEVELS

The demonstration that GSH is a major protectant against potentially cytotoxic peroxides, free radicals, and electrophiles suggests that tissues may be sensitized to such species by pharmacological interventions that decrease intracellular GSH levels. Under favorable circumstances, such sensitization may be of therapeutic benefit (see Section V). Several approaches to GSH depletion have been developed. Early investigations often relied on the reactions outlined in Section III, and GSH depletion was induced by deliberately stressing either the glutathione *S*-transferase or the glutathione peroxidase/glutathione reductase pathways. Exposure of cells, tissues, or intact animals to diethyl maleate (DEM), for example, results in a rapid and marked decrease in cellular GSH levels with formation of the GSH-diethyl maleate adduct (Plummer *et al.*, 1981). Wheras DEM is inherently electrophilic, other GSH depleting agents require initial activation by the cytochrome P-450 system (Peterson and Guengerich, 1988). Glutathione *S*-transferases are found in both the cytosol and mitochondrial matrix, and the GSH pool in both compartments can be depleted by suitable substrates

(Wahlländer *et al.*, 1979; Romero *et al.*, 1984). Since the cysteine moiety of GSH is irreversibly lost by adduct formation, resynthesis of GSH is often significantly delayed. Although GSH depletion is reliably obtained with DEM, phorone, and other electrophiles, thiols other than GSH are also derivatized. Nonenzymatic reaction of both protein and low molecular weight thiols with DEM seriously confounds interpretation of the results since the effects observed cannot be unambiguously attributed to the loss of GSH. Diethyl maleate is actuely toxic to both isolated cells and intact animals at doses only slightly higher than those required to effect substantial ($>95\%$) depletions of GSH.

Glutathione depletion has also been effected using agents able to oxidize GSH to GSSG. Two general approaches have been developed. In the more direct approach, a thiol specific oxidant is administered directly; of the several such reagents developed by Kosower and Kosower (1987), the most commonly used is diamide [diazenedicarboxylic acid bis (N,N-dimethylamide)]. In carefully done experiments it is possible to provide diamide in amounts just sufficient to fully oxidize the GSH pool to GSSG and to then investigate the metabolic or toxicologic effects of GSH depletion. Several studies indicate that diamide is highly but not absolutely specific for thiols, and, as a practical matter, reaction with cellular constituents other than thiols does not occur to a significant degree. However, although GSH reacts very rapidly with diamide, the reagent does oxidize other thiols, and studies in which diamide is used in amounts approaching or exceeding that needed to fully oxidize GSH are therefore difficult to interpret.

Beutler, Sies, and co-workers have popularized an alternative, indirect approach in which glutathione peroxidase substrates are administered in doses sufficient to cause oxidative depletion of GSH (Plummer *et al.*, 1981, and references therein). Among the more commonly used reagents are cumene hydroperoxide and *tert*-butyl hydroperoxide; although both are excellent glutathione peroxidase substrates, absolute specificity for GSH or even for thiol oxidation cannot be assured. It should be noted that with either the direct or indirect approach to oxidative GSH depletion, the effects are inherently short-lived since glutathione reductase remains active. Reduction of GSSG to GSH thus proceeds simultaneously with the oxidation of GSH to GSSG, and the extent of GSH depletion obtained depends largely on the relative rates of these reactions. In longer term studies, moderate depletion of total glutathione (i.e., GSH + GSSG) is also often observed because GSSG is preferentially transported out of cells (Sies and Summer, 1975). Although GSH oxidation can generally be made as rapid as necessary to achieve substantially complete depletion of GSH, the highly skewed GSH/GSSG ratio is obtained at the expense of running the glutathione reductase reaction at its maximum rate, often with attendant massive conversion of NADPH to $NADP^+$. The activities of enzymes sensitive to the thiol/disulfide ratio are also likely to be affected in such studies (Gilbert, 1989, 1990).

Although administration of glutathione S-transferase substrates or GSH oxidants causes GSH depletion in a wide range of tissues, the lack of specificity inherent in the use of such reagents seriously complicates interpretation of the results obtained. In 1975 Palekar *et al.* examined an alternative approach to GSH depletion and showed that liver and kidney GSH concentrations decreased 40 and 60%, respectively, in rats administered L-methionine-SR-sulfoximine (L-SR-MSO), an established inhibitor of both glutamine synthetase (Pace and McDermott, 1952) and γ-glutamylcysteine synthetase (Richman, *et al.*, 1973). Unfortunately, MSO causes convulsions, coma, and death when administered in doses causing significant GSH depletion; the toxic effects were thought to be attributable to inhibition of brain glutamine synthetase. Griffith and Meister subsequently explored the possibility of designing MSO analogs that selectively inhibited either glutamine or γ-glutamylcysteine synthetase, and in 1978 reported that α-ethyl-DL-methionine-SR-sulfoximine (α-Et-MSO) was a specific inhibitor of glutamine synthetase *in vitro* and *in vivo*. When administered to mice, α-Et-MSO caused glutamine depletion and convulsions but did not deplete tissue GSH (Griffith and Meister, 1978). Since α-Et-MSO, in contrast to MSO, is not metabolized to vinylglycine and other nonspecific cytotoxins, these studies convincingly demonstrated that inhibition of glutamine synthetase can fully account for the toxicity and convulsions seen following MSO administration. The alternative goal of selectively inhibiting γ-glutamylcysteine synthetase was achieved through the synthesis of higher S-alkyl homologs of MSO; analogs with S-n-alkyl groups of three to seven carbons are selective inhibitors of γ-glutamylcysteine synthetase (Griffith *et al.*, 1979a; Griffith and Meister, 1979c; Griffith, 1982). As shown in Figs. 2a and 2b, buthionine sulfoximine (S-(N-butyl)homocysteine sulfoximine, BSO) is an extremely effective inhibitor of isolated rat kidney γ-glutamylcysteine synthetase and is, for example, approximately 200-fold more effective than MSO.

The mechanism of inhibition by MSO and its analogs is well understood. As discussed previously, the γ-glutamylcysteine synthetase reaction proceeds with the initial ATP-dependent formation of γ-glutamylphosphate, an enzyme-bound intermediate that subsequently reacts with the amino group of cysteine (Reactions 1a and 1b). The glutamine synthetase reaction occurs by a similar mechanism in which γ-glutamylphosphate is attack by ammonia; the top panels of Fig. 3 illustrate this aspect of the overall mechanisms for both synthetases. A variety of studies have established that MSO, BSO, and related sulfoximines are initially bound to one or both synthetases as transition state analogs in which the tetrahedral sulfoximine structure mimics the tetrahedral intermediate formed when either ammonia or the amino group of cysteine attacks the activated carbonyl of γ-glutamylphosphate. Following initial binding, the sulfoximines are phosphorylated on the sulfoximine nitrogen by ATP to yield products that are very tightly, but noncovalently, bound in the active sites (Fig. 3, lower panels) (Griffith, 1982).

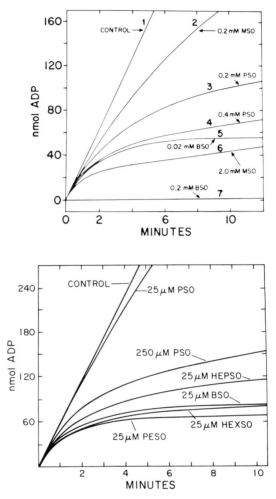

FIGURE 2   Time course of the γ-glutamylcysteine synthetase reaction in the presence and absence of analogs of methionine sulfoximine. *Top*, The effect of L-methionine-*SR*-sulfoximine (MSO), DL-prothionine-*SR*-sulfoximine (PSO), and DL-buthionine-*SR*-sulfoximine (BSO) on the reaction of purified rat kidney γ-glutamylcysteine synthetase carried out in a reaction mixture containing 5 mM L-glutamate, 10 mM L-α-aminobutyrate, 10 mM ATP, and a phospho-enolpyruvate/pyruvate kinase ATP-regenerating system. Product formation, indicated as nmol ADP, was determined from the flux through the ATP regenerating system. (Reproduced from Griffith and Meister, 1979c, with permission.) *Bottom*, A similar experiment comparing inhibition by PSO, hepathionine-*SR*-sulfoximine (HEPSO), BSO, DL-hexathione-*SR*-sulfoximine (HEXSO), and DL-pentathionine-*SR*-sulfoximine (PESO). (Reproduced from Griffith, 1982, with permission.)

Specificity of inhibition is controlled by the initial binding of the sulfoximines. Since both synthetases are inhibited only by the L, *S*-diastereomer of MSO, selective inhibition using isolated MSO isomers is not possible. However, as illustrated in Fig. 3, selective inhibition of glutamine synthetase is possible based on that enzyme's ability to accommodate α-hydrogen replacements at least as large as ethyl; γ-glutamylcysteine synthetase accommodates an α-methyl but not an α-ethyl substituent. Selective inhibition of γ-glutamylcysteine synthetase is possible based on the fact that *S*-alkyl substituents of at least four carbons are easily accommodated in the cysteine binding site, whereas the ammonia binding site of glutamine synthetase accommodates only the *S*-methyl group of MSO and, with very poor binding, the *S*-ethyl group of ethionine sulfoximine.

The observation that sulfoximines with *S*-alkyl groups of up to seven carbons are effective inhibitors of γ-glutamylcysteine synthetase suggests that the cysteine binding site is not sharply defined in the region normally occupied by sulfur. Although higher homologs of L-α-aminobutyrate are poor replacements for L-cysteine in γ-glutamylcysteine synthetase reaction, it was recently shown that L-γ-glutamyl derivatives of L-norvaline and L-norleucine

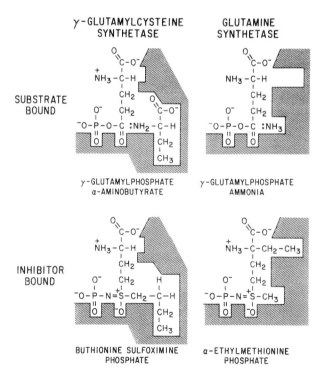

FIGURE 3   Binding of γ-glutamylphosphate and inhibitory sulfoximines to γ-glutamylcysteine synthetase and glutamine synthetase. (Modified from Griffith and Meister, 1978.)

are good inhibitors of the overall reaction (O.W. Griffith, unpublished observations). The product inhibition and sulfoximine specificity studies together suggest that energetically favorable binding interactions with the glutamate binding site are sufficient to overcome any negative interactions resulting from the requirement that the L-cysteine binding site accommodate groups significantly larger than cysteine. Since L-cysteinylglycine does not substitute for L-cysteine in the normal γ-glutamylcysteine synthetase reaction, but GSH (i.e., γ-glutamylcysteinylglycine) is nonetheless bound as a nonallosteric regulator, similar considerations may apply to the binding of γ-glutamyl derivatives that seemingly exceed the bounds of the carboxylate portion of the cysteine binding site.

Administration of BSO to experimental animals inhibits γ-glutamylcysteine synthetase and prevents further synthesis of GSH. In tissues exhibiting significant rates of GSH turnover, GSH is depleted by the continuing transport of GSH out of cells and by other reactions of GSH utilization (Griffith and Meister, 1979a,c; Griffith, 1982); L-buthionine-SR-sulfoximine is thus a potent GSH depleting agent. As illustrated in Fig. 4, liver and kidney GSH levels decrease to about 20% of control within 1 hr in mice given a single dose of BSO (4 mmol/kg, ip); comparable effects are observed when BSO is given subcutaneously (Table 1, Experiment A). Longer term studies are possible when BSO is included in the drinking water; the effects of including 20 mM BSO in the drinking water of mice for 15 days are summa-

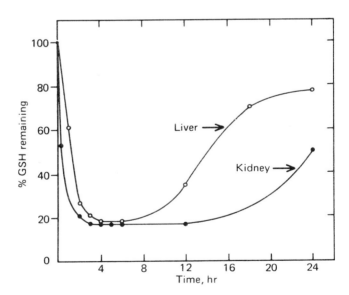

FIGURE 4 Levels of GSH in kidney and liver after administration of BSO (4 mmol/kg) to mice (starved for 12 hr) by intraperitoneal injection of a 100 mM solution. (Reproduced from Griffith and Meister, 1979a, with permission.)

rized in Table 1, Experiment B. Comparison of Experiments A and B indicates that for some tissues the extent of GSH depletion depends on the dose regimen; skeletal muscle, heart, and pancreas show markedly greater depletions in Experiment B, for example, whereas liver shows somewhat greater GSH depletions in Experiment A. Kidney is markedly depleted of GSH in both experiments. Maximum GSH depletions in a wide range of tissues are obtained by both injecting BSO (4 to 8 mmol/kg every 8 to 12 hr, ip or sc) and including it in the drinking water (10 to 30 mM; animals drink less water if it contains higher concentrations of BSO). Specific applications in the depletion of tumor GSH are discussed in Section V.

It is important to note that BSO shows negligible acute toxicity in normal mice and rats. Mice have been administered 400 mM BSO at a dose of 32 mmol/kg as a single intraperitoneal injection without overt toxicity. Mice do die following repeated administrations of BSO at this concentration, but death appears to be due to osmotic disturbances rather than to a specific effect on GSH metabolism. In this respect BSO is no more toxic than many of the

TABLE 1
Tissue GSH Levels after Administration of BSO[a]

Tissue	Tissue GSH [μmol/gm (%)]		
	Control	Experiment A	Experiment B
Brain	2.08 ± 0.15	1.93 ± 0.11 (93)	1.74 ± 0.13 (84)
Heart	1.35 ± 0.10	1.19 ± 0.14 (88)	0.20 ± 0.05 (15)
Lung	1.52 ± 0.13	1.42 ± 0.09 (94)	0.79 ± 0.07 (52)
Spleen	3.43 ± 0.35	3.46 ± 0.07 (101)	1.73 ± 0.18 (51)
Liver	7.68 ± 1.22	2.67 ± 1.15 (35)	4.26 ± 1.11 (56)
Pancreas	1.78 ± 0.31	0.81 ± 0.30 (46)	0.14 ± 0.06 (8)
Kidney	4.13 ± 0.15	0.75 ± 0.16 (18)	0.19 ± 0.05 (4)
Small intestine mucosa	2.94 ± 0.16	2.40 ± 0.36 (82)	1.64 ± 0.41 (56)
Colon mucosa	2.11 ± 0.19	1.83 ± 0.16 (87)	0.29 ± 0.14 (14)
Skeletal muscle	0.78 ± 0.05	0.52 ± 0.11 (67)	0.016 ± 0.009 (2)
Plasma, μM	28.4 ± 6.2	9.3 ± 3.8 (33)	9.1 ± 2.7 (32)

[a]Mice (28–35 gm) were fed *ad lib*. BSO was given subcutaneously on the back and neck by injection of a 100 mM solution (Experiment A) or was supplied (20 mM) in the drinking water. The BSO dose in Experiment A was 4 mmol/kg; the tissue GSH content was determined 2 hr after BSO administration. The BSO dose in Experiment B was approximately 5 mmol/kg per day; the tissue GSH content was determined 15 days after BSO treatment was begun. Values shown are averages ± S.D. of measurements on 4 to 10 mice; values in parentheses are percentages of the control value. (Reproduced from Griffith and Meister, 1979a, with permission)

protein amino acids. Very long term administration of L-SR-BSO (e.g., several weeks) causes cataracts in young mice and rats (Calvin et al., 1986; Martensson et al., 1989) and mitochondrial degeneration in the skeletal muscle of rats (Martensson and Meister, 1989). Toxicity appears to be caused by the inability of GSH-depleted tissues to adequately respond to normally occurring oxidative insults; in this respect, depletion of mitochondrial GSH is particularly serious since, as noted, the mitochondrial electron transport chain is the source of a continuing flux of reactive oxygen species. Fortunately, mitochondria appear to have a high affinity for GSH and mitochondrial GSH is thus depleted much more slowly than cytoplasmic GSH in BSO-treated mice and rats (Meredith and Reed, 1982; Griffith and Meister, 1985). For reasons that are not yet clear, mitochondrial GSH is extensively depleted in guinea pigs administered BSO; guinea pigs therefore die when administered BSO (4 mmol/kg every 8 hr, ip) for two or three days (O.W. Griffith, unpublished observations). To date, guinea pigs and neonatal mice or rats are the only animals in which acute nonosmotic toxicity of BSO has been demonstrated (O.W. Griffith, unpublished observations; Martensson et al., 1989).

In contrast to glutathione S-transferase substrates (e.g., DEM) or GSH oxidants (e.g., diamide), BSO is chemically unreactive and therefore cannot interact nonspecifically with cell structures. As noted, BSO does not significantly inhibit glutamine synthetase and thus does not cause glutamine depletion or convulsions. Possible metabolism of BSO was examined in adult mice. Although 15 to 20% of L-$[1 - {}^{14}C]$ methionine-SR-sulfoximine was metabolized to ${}^{14}CO_2$ in 6 hr, less than 3% of L-$[1 - {}^{14}C]$ buthionine-SR-sulfoximine was converted to ${}^{14}CO_2$ in similar experiments. The results suggest that metabolic degradation of BSO to vinylglycine and other cytotoxic metabolites is negligible (Griffith, 1982). Mice administered L-$[{}^{35}S]$buthionine-SR-sulfoximine excreted the radiolabel almost quantitatively in the urine; $>60\%$ of the material was unchanged BSO and $>90\%$ of the remainder was the N^{α}-acetyl derivative of BSO (Griffith, 1982). L-Buthionine-SR-sulfoximine is an analog of L-γ-glutamylamino acids and as such has been found to be a moderate inhibitor of γ-glutamylamino acid transport in the kidney tubule (Griffith et al., 1979b; Anderson and Meister, 1983). Brodie and Reed (1985) reported that BSO inhibits cysteine uptake by L-1210 cells. In an effort to elucidate and possibly minimize transport inhibition, we have recently separated the disastereomers of L-buthionine-SR-sulfoximine and examined their effects in vivo (Griffith, 1989). We find that the diastereomers are comparably taken up by various tissues of the mouse, but that only the L-S-diastereomer inhibits γ-glutamylcysteine synthetase in vitro and in vivo. As expected, administration of L-buthionine-S-sulfoximine at one half the dose previously used with L-SR-BSO causes GSH depletion of equal extent and duration; it is anticipated that the use of appropriate doses of L-buthionine-S-sulfoximine will allow investigators to reduce non-specific transport inhibition and deleterious osmotic effects by approximately 50%. L-Buthionine-R-sulfoximine, which does not inhibit γ-glutamylcysteine synthetase or effectively deplete tissue GSH, may find utility in control studies designed to investigate possible effects of BSO not attributable to specific inhibition of the synthetase.

B. AGENTS INCREASING INTRACELLULAR GLUTATHIONE LEVELS

Whereas GSH depletion is expected to sensitize tissues to damage by electrophiles, peroxides, and free radicals, pharmacological interventions that increase GSH levels may protect cells in cases where the GSH concentration is limiting. Thus, the rate of nonenzymatic processes such as the GSH-mediated detoxification of free radicals depends directly on the concentration of GSH and is proportionately accelerated by increases in GSH concentration. The effect of increasing GSH concentration on enzyme-mediated processes is less predictable but depends, to a first approximation, on the ratio of GSH concentration to the K_m for GSH of the enzyme under consideration. At a GSH concentration of $4 \times K_m$, for example, the reaction is already proceeding at 80% of V_{max}, and further increases in GSH concentration can increase the enzymatic rate by 25% at most. For several reasons the situation *in vivo* is considerably more complicated than is suggested by this simple analysis. Thus, the GSH-mediated detoxification of electrophiles including some of the cancer chemotherapeutic alkylators occurs by both nonenzymatic and glutathione S-transferase-mediated reactions. Changes in GSH concentration will proportionately affect the nonenzymatic reaction but will have a less than proportionate effect on S-transferase-catalyzed reactions in tissues where the GSH concentration is greater than the enzyme's K_m for GSH (typically 0.4 to 0.8 mM). Since the detoxification of peroxides depends on several enzymatic activities including glutathione peroxidase, glutathione reductase and the enzymes reducing $NADP^+$ to NADPH, the rate-determining reaction may or may not be sensitive to the GSH concentration depending on the tissue, the amount of peroxide present, and the availability of substrates for $NADP^+$ reduction. These and other complexities notwithstanding, it has been observed experimentally that interventions increasing tissue GSH levels are often protective. Not surprisingly, there is considerable interest in the development of multidrug protocols that increase GSH levels in nontarget tissues while decreasing GSH levels in target tissues; the synergistic effects of GSH depletion and cytotoxic drug administration are then confined to the target tissue.

Since GSH synthesis is often limited by cysteine availability, administration of L-cysteine or L-cysteine precursors increases GSH levels in many tissues. The effects are particularly dramatic in tissues previously challenged by electrophiles which react with GSH and cause net loss of cysteine through mercapturic acid formation. Although L-cysteine is inexpensive and very water soluble, it is fairly toxic when administered parenterally, is unpalatable when given orally, and is easily oxidized to the very insoluble disulfide, cystine; L-cysteine itself is thus not an attractive agent for increasing cysteine levels *in vivo*. Fortunately alternative agents are available. N-Acetyl-L-cysteine (Mucomist) is hydrolyzed to L-cysteine and acetate by intracellular amidases. Aldehyde adducts of L-cysteine are also effective. L-Thiazolidine-4-carboxylate (L-thioproline), formed from L-cysteine and formaldehyde, is converted by proline oxidase to N-formyl-L-cysteine, which is hydrolyzed to L-cysteine and formic acid. Larger aldehydes lead to 2-substituted thiazolidine-4-carboxylates; 2-RS-methylthiazolidine-4-R-carboxylate is taken up by cells and hydrolyzes spontaneously to liberate L-cysteine and acetaldehyde. L-2-Oxothiazolidine-4-carboxylate is a 5-oxo-L-proline analog that is hydrolyzed by 5-oxoprolinase to L-cysteine and carbon dioxide. In experimental animals it is an extremely efficient agent for increasing hepatic GSH levels following acetaminophen poisoning. Tissue specific increases in GSH levels may be possible in cases where the L-cysteine precursor is preferentially taken up or excluded by certain cell types or in cases where the enzymatic release of L-cysteine is limited to specific tissues. The possibility that some tumors may be deficient in 5-oxoprolinase suggests that L-2-oxothiazolidine-4-carboxylate may be of use in protecting nontumor tissues by selectively increasing their GSH content. Cysteine delivery systems have been recently reviewed (Anderson and Meister, 1987).

L-Cysteine and L-cysteine precursors increase GSH levels only in tissues containing active γ-glutamylcysteine synthetase. Such compounds are thus not generally effective in BSO-treated animals. Tissue GSH levels can be increased in BSO-treated animals, however, by administering compounds that yield GSH through γ-glutamylcysteine synthetase-independent pathways. γ-Glutamylamino acids and GSH esters are of particular use. L-γ-glutamyl-L-cysteine and L-γ-glutamyl-L-cystine are actively taken up by some transpeptidase-rich tissues, particularly kidney and pancreas (Griffith et al., 1978, 1979b, Anderson and Meister, 1983). Since L-γ-glutamyl-L-cystine is readily reduced to L-γ-glutamyl-L-cysteine intracellularly, both compounds are acted on by glutathione synthetase to yield GSH; since the γ-glutamylcysteine synthetase step of GSH biosynthesis is effectively bypassed, BSO is without effect.

A more direct approach to increasing GSH levels has been attempted through the oral or parenteral administration of GSH itself. Unfortunately, as discussed previously, intact GSH is not significantly taken up by cells. Since GSH is rapidly degraded extracellularly (and extracorporeally within the gut), exogenously supplied GSH is best viewed as an L-cysteine precursor; as such, it is moderately effective in increasing tissue GSH levels. Although plasma GSH levels are increased following administration of very large oral doses of GSH (Hagen and Jones, 1989), this effect is probably due to the leakage of GSH past the tight junctions of the intestinal epithelium and does not reflect uptake of intact GSH from the intestine. The problem of directly increasing intracellular GSH levels was solved in 1983 with the report by Puri and Meister that glutathione ethyl ester (Et-GSH, glycine carboxylate esterified), was taken up by mouse liver and kidney and hydrolyzed to GSH intracellularly, presumably by nonspecific esterases (Puri and Meister, 1983). A variety of subsequent studies have established that Et-GSH given orally or parenterally is effective in increasing the GSH content of a wide range of tissues. As expected, the increase occurs in a γ-glutamylcysteine synthetase and γ-glutamyltranspeptidase-independent manner. Convenient procedures for the synthesis of Et-GSH and related GSH esters have recently been published (Campbell and Griffith, 1989; Anderson and Meister, 1989). As discussed in Section V and in Chapter 15 by Teicher, use of GSH esters to selectively increase GSH levels and thus protect nontumor tissues may enhance several conventional cancer chemotherapies.

C. AGENTS INHIBITING ENZYMES OF GLUTATHIONE METABOLISM

As discussed earlier, the protective effects of GSH can be pharmacologically manipulated by protocols that directly alter the GSH level of tissues (e.g., administration of DEM, diamide, cysteine precursors, Et-GSH). Glutathione-mediated protective effects can also be controlled by use of agents that alter (generally inhibit) the activity of specific enzymes involved in GSH metabolism. The use of BSO to cause GSH depletion in tissues with moderate to high rates of GSH turnover is an example of this approach. Compounds and strategies for the inhibition of several other enzymes of GSH metabolism have also been developed. Of these, inhibitors of γ-glutamyltranspeptidase and of glutathione reductase have received considerable attention and are most likely to be of use in the development of synergistic drug regimens.

γ-Glutamyltranspeptidase is inhibited by serine plus borate and by a variety of glutamine antagonists including L-azaserine and 6-diazo-5-oxo-L-norleucine (L-DON). Although even limited inhibition of transpeptidase may complicate the interpretation of studies in which azaserine or DON is used as a glutamine antagonist, none of the agents mentioned is sufficiently potent to effect substantial transpeptidase inhibition *in vivo* at tolerated doses (Griffith and Meister, 1979c). γ-Glutamyltranspeptidase is potently and covalently inhibited by acivicin [(αS, 5S)-α-amino-3-chloro-4, 5-dihydro-5-isoxazole-acetic acid, AT-125], a glutamine antagonist developed by Upjohn (Allen *et al.*, 1980). Although acivicin effectively inhibits nucleic acid synthesis and showed initial promise as an anticancer agent, its clinical use is limited by adverse neurological changes. Since therapeutic (and ultimately toxic) doses of acivicin are lower than those normally used to inhibit transpeptidase, the observations with acivicin do not suggest that transpeptidase inhibition *per se* is necessarily associated with neurological changes. Since acivicin is the most effective irreversible γ-glutamyltranspeptidase inhibitor reported, we have synthesized acivicin diastereomers in an effort to obtain an effective transpeptidase inhibitor that does not affect glutamine metabolism or cause neurological disturbances. Because transpeptidase is known to be active with both D- and L-γ-glutamylamino acids, we anticipated that the acivicin diastereomer in which the α-amino group was in the D configuration would be an effective inhibitor; that expectation was verified with the finding that (αR, 5S)-α-amino-3-chloro-4,5-dihydro-5-isoxazoleacetic acid (D-threo-AT-125) is a strong inhibitor of purified rat kidney γ-glutamyltranspeptidase (O. W. Griffith, unpublished observations).

γ-Glutamyltranspeptidase is also strongly, but noncovalently inhibited both *in vitro* and *in vivo* by a variety of γ-glutamyl hydrazides of which the most effective are D- and L-γ-glutamyl-(*o*-carboxy) phenylhydrazide (Griffith and Meister, 1979b). If rat kidney transpeptidase is assayed using 0.1 to 1 mM D-γ-glutamyl-*p*-nitroanilide, a chromogenic substrate, as γ-glutamyl donor and 10 mM L-methionylglycine as γ-glutamyl acceptor, then L- and D-γ-glutamyl-(*o*-carboxy)phenylhydrazide inhibit with K_i values of 8.2 and 22.5 μM, respectively; under similar conditions GSH also inhibits the chromogenic reaction and exhibits a K_i of 175 μM. The phenylhydrazides are thus bound 8- to 21-fold more tightly than GSH by transpeptidase. Structural considerations and the observation that inhibition is competitive with respect to the γ-glutamyl donor substrate suggest that the inhibitors are bound in part to the glutamate region of the active site. The γ-glutamyl-(*o*-carboxy)-phenylhydrazides apparently also bind to the acceptor amino acid site of transpeptidase since the presence of the *o*-carboxy group, corresponding sterically to the carboxylate of an acceptor amino acid, greatly enchances inhibition. L-γ-Glutamyl-(*p*-carboxy)phenylhydrazide is bound with significantly reduced affinity (K_i = 800 μM) (Griffith and Meister, 1979b). As noted previously, both D- and L-γ-glutamyl-(*o*-carboxy)phenylhydrazide cause a substantial glutathionemia and glutathionuria when administered to mice (Griffith and Meister, 1979b). These compounds and other transpeptidase inhibitors may have application in blocking GSH degradation and L-cysteine recovery in transpeptidase-rich tumors.

Selective inhibitors of glutathione reductase would be of considerable interest since their use is expected to enhance the effects of both radiation therapy and therapies based on redox cycling drugs (e.g., Adriamycin). Although a completely selective inhibitor is not yet identified, Frischer and Ahmad (1976) reported that 1,3-*bis*-(2-chloroethyl)-1-nitrosourea (BCNU) caused severe generalized glutathione reductase deficiency when administered to patients or experimental animals at doses comparable to those used clinically. Although none of 19 other erythrocytic enzymes was inhibited (Frischer and Ahmad, 1976), concern regarding specificity for glutathione reductase persists on the basis of the known chemical reactivity of BCNU and of BCNU metabolites that are generated *in vivo*. Nevertheless, Nathan *et al.* (1981) have shown that BCNU significantly enhances the macrophage-mediated oxidative cytolysis of a variety of tumor cells in culture. These and a number of other studies suggest that much of the clinical activity of BCNU is directly related to the inhibition of glutathione reductase and, correspondingly, that the activity of the reductase may be manipulated by BCNU in the design of synergistic drug strategies.

V. Therapeutic Applications of Pharmacologic Manipulations of Glutathione Levels or Metabolism

A. GLUTATHIONE DEPLETION AND TRYPANOSOMIASIS

Trypanosoma brucei brucei is the causative agent of nagana, a disease affecting livestock in Africa. The trypanosome and its associated disease are biochemically and medically similar to African trypanosomiasis (African sleeping sickness). Studies by Cerami and co-workers demonstrated that *T. b. brucei* accumulate extraordinarily high levels of hydrogen peroxide (Meshnick *et al.*, 1977, 1978). The peroxide accumulation is due in part to the inability of the organism to synthesize cytochromes and the consequent inability to form active catalase. Trypanosomes also have little or no glutathione peroxidase. Studies in which the organisms were exposed *in vitro* or *in vivo* to agents that increase hydrogen peroxide levels further (e.g., menadione) or facilitate the homolytic cleavage of hydrogen peroxide to hydroxyl radicals (e.g., certain porphyrins) demonstrated that trypanosomes could be killed by augmentation of their endogenous peroxide-mediated oxidative stress. The importance of GSH in protecting trypanosomes from free radical damage was shown by treating trypanosome-infected mice with BSO (4 mmol/kg ip every 1.5 hr for up to 27 hr). The level of GSH in trypanosomes isolated from the blood stream of treated mice declined from 3 nmol/mg protein to about 1 nmol/mg protein in 8 hr. After 8 hr, intact trypanosomes could not be isolated from blood. Groups of 6 trypanosome-infected mice were then treated with BSO for 9, 18, or 27 hr and were observed for survival. A control group was not treated. All control mice and mice treated for 9 hr died after 4 or 5 days. Blood samples obtained from mice treated for 18 or 27 hr were parasite-free for several days; four of the mice treated for 27 hr showed significantly extended survival (11 to 16 days) and two mice were cured (Arrick *et al.*, 1981). These studies illustrate the importance of GSH (or its product, N^1, N^8-diglutathionylspermidine) in protection of trypanosomes from oxidative stress, and suggest a new therapeutic strategy for attacking the African trypanosomes producing disease in both animals and humans. Since *Trypanosoma cruzi*, the trypanosome causing Chagas disease in South America, is also catalase deficient, a similar therapeutic approach may be possible.

B. GLUTATHIONE DEPLETION AND ALKYLATOR THERAPY

Several studies have demonstrated that, in cell lines exhibiting elevated GSH levels, increases in classical alkylator cytotoxicity are often seen following BSO-mediated GSH depletion. L1210 leukemia cells resistant to melphalan were rendered melphalan-sensitive following BSO-mediated depletion of glutathione (Somfai-Relle *et al.*, 1984). Kramer *et al.* (1987) confirmed this *in vitro* observation, but were unable to increase the life span of mice injected intraperitoneally with these cells. Green *et al.* (1984) demonstrated enhanced melphalan activity against the melphalan-resistant human ovarian carcinoma cell line 1847ME following administration of BSO. Provocatively, Ozols *et al.* (1987) extended these observations to nude mice inoculated intraperitoneally with NIH: OVCAR-3 ovarian cancer cells and demonstrated prolongation of survival in mice treated with BSO plus melphalan compared with melphalan alone.

Treatment in our laboratory of the rhabdomyosarcoma xenograft TE-671 with BSO resulted in increased sensitivity to subsequently administered melphalan (Skapek *et al.*, 1988b). [Note that this cell line was initially reported by McAllister *et al.*, (1977) to be derived from a medulloblastoma but shown by Stratton *et al.* (1989) to be a subline of the rhabdomyosarcoma line RD.] Intraperitoneal administration to tumor-bearing mice of DL-BSO (2 doses at 12 hr intervals; 5 mmol/kg) decreased the GSH content of subcutaneous TE-671 xenografts to 25.7% of control. Administration of a 30 mM solution of L-BSO in drinking water for 96 hr decreased the GSH content to 17.4% of control. Depletion of GSH with these regimens resulted in a significant increase in the growth delay of the subcutaneous tumor over that produced by melphalan alone: 17.2 days vs 12.6 days for DL-BSO (ip) plus melphalan versus melphalan alone and 22.9 days vs 16.6 days for L-BSO (po) plus melphalan versus melphalan alone (Fig. 5). These studies demonstrate the increased cytotoxicity of melphalan resulting from BSO-mediated depletion of GSH in human subcutaneous xenografts and support further efforts to modulate cytotoxicity with this approach.

Similar studies conducted with the melphalan-resistant human rhabdomyosarcoma xenograft TE-671 MR suggested that resistance to melphalan in this human xenograft may be mediated in part by an elevated level of GSH (Rosenberg *et al.*, 1989). BSO pretreatment significantly increased TE-671 MR sensitivity to melphalan with growth delays increasing from 3.7/4.6 days (in duplicate trials) to 7.2/9.8 days. The fact that growth delays typical of the parent xenograft (12.2–16.6 days) were not obtained despite the reduction of

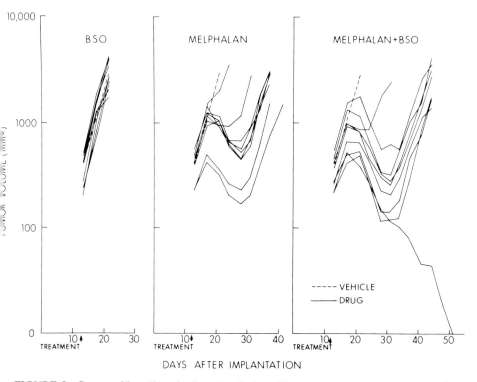

FIGURE 5   Groups of 9 to 10 randomly assigned mice with subcutaneous tumors were treated with L-BSO (p.o. administration at a concentration of 30 mM in the drinking water for 96 hr), melphalan (50% of the 10% lethal dose administered ip), melphalan plus BSO, or an equivalent volume of drug vehicle (administered ip). (Reproduced from Skapek *et al.*, 1988b, with permission.)

GSH to very low levels suggests that multiple mechanisms of resistance are operational. Frei *et al.* (1988) reached a similar conclusion regarding alkylator resistance in other neoplasms. It should be noted that the frequent observation that BSO-mediated GSH depletion increases alkylator sensitivity in both the parental and resistant cell lines (Hamilton *et al.*, 1985) raises the possibility that the operational mechanisms of resistance may in some cases be independent of the elevated levels of GSH. Although the increased GSH levels would then represent an epiphenomenon, GSH depletion would, of course, still confer a therapeutic advantage.

Application of GSH-modulating therapies in human central nervous system tumors such as medulloblastoma or glioma (which is the focus of our research program) required the demonstration that GSH metabolism within the central nervous system could be altered. Previous studies indicated that the central nervous system is relatively refractory to the effects of parenterally administered BSO (Griffith and Meister, 1979a), but the effect of BSO on tumors within the central nervous system had not been previously examined. Earlier studies did suggest that the blood–brain barrier is compromised by tumor growth (Groothuis and Blasberg, 1985) and that the rate of GSH turnover is often increased in neoplastic tissues relative to normal tissues (Minchinton et al., 1984). It was anticipated that either effect (or both) might allow selective BSO-mediated depletion of GSH in central nervous system neoplasms. This possibility was directly examined using appropriate xenografts.

Human medulloblastoma xenografts grow both as intraparenchymal masses and perivascular subarachnoid infiltrations in athymic mice (Friedman et al., 1983, 1985, 1988), a growth pattern that makes comparisons of tumor versus normal brain difficult. Since gliomas grow as discrete focal masses in

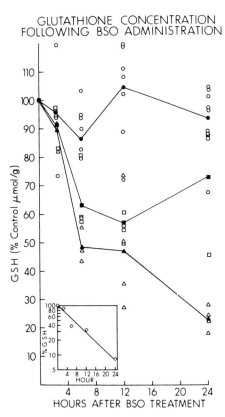

FIGURE 6   Glutathione concentration in subcutaneous and intracranial D-54 MG xenografts and contralateral, "normal brain" murine brain following intraperitoneal administration of 5 mmol/kg DL-BSO. Data represent individual samples (○, □, △) and mean values (●, ■, ▲) for subcutaneous and intracranial xenografts, and contralateral brain, respectively. *Insert*: Depletion of cytoplasmic GSH in intracranial xenografts, corrected for a mitochondrial pool of 16%. (Reproduced from Skapek et al., 1988a, with permission.)

athymic mice, our initial studies have focused on D-54 MG, a human glioma-derived continuous cell line which can be grown as subcutaneous or intracranial (parenchymal) xenografts (Skapek *et al.*, 1988b). Intraperitoneal administration of a single dose of BSO (5 mmol/kg) resulted in depletion of intracellular GSH at 12 hr to 57 and 47% of control in subcutaneous and intracranial xenografts, respectively. At 24 hr GSH was depleted to 73 and 23% of control in subcutaneous and intracranial xenografts, respectively. Concurrent measurement of GSH in the contralateral (tumor-free) cerebral hemisphere in mice bearing intracranial D-54 MG xenografts demonstrated insignificant GSH depletion (Fig. 6). Multiple doses of BSO, at 12 hr intervals, resulted in further depletion of GSH to 27% (sc) and 16.5% (ic) of control 12 hr following the final dose of BSO (Skapek *et al.*, 1988b). Quantitative analysis of BSO delivery to xenograft and contralateral brain tissue revealed transfer constants (K_1) of 5.3 to 6.3 μl/gm·min and 0.23 to 1.35

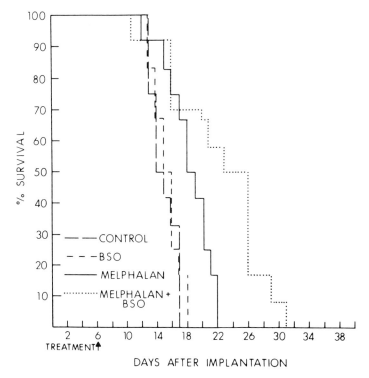

FIGURE 7 Groups of 10 randomly assigned mice bearing intracranial D-54 MG xenografts were treated with melphalan at 50% of the 10% lethal dose on day 6 after transplantation (control), L-BSO (2.5 mmol/kg every 12 hr for four doses) plus L-BSO available concomitantly as a 20 mM solution in the drinking water, drug vehicle (16.7% dimethyl sulfoxide), or melphalan plus L-BSO (melphalan given 4–6 hr after the last L-BSO injection). (Reproduced from Friedman *et al.*, 1988, with permission.)

μl/gm·min for xenograft and normal brain, respectively, indicating the enhanced delivery of BSO to the xenograft relative to normal brain (Fekete et al., 1990). These studies demonstrated that BSO-mediated inhibition of GSH synthesis allows an intracranial tumor to be depleted of GSH without significantly affecting normal brain GSH levels.

The therapeutic ramifications of this intervention for alkylator sensitive or resistant tumor cells were defined in further studies. Athymic mice bearing intracranial human glioma (D-54 MG) or rhabdomyosarcoma (TE-671) xenografts were treated with melphalan alone or BSO followed by melphalan. Administration of BSO depleted intracellular GSH to 7.5% of the control level. BSO plus melphalan resulted in a significant increase in median survival over that produced by melphalan alone: 45.3 vs 26.4% in TE-671 and 69 vs 27.6% in D-54 MG (Friedman et al., 1989) (Fig. 7). These studies justify further efforts to modulate chemotherapeutic interventions of primary malignant brain tumors by depletion of GSH.

C. GLUTATHIONE DEPLETION AND ALKYLATOR TOXICITY

The extensive depletion of GSH in many tissues following therapy with BSO raises the concern that significant toxicity may result from the subsequent administration of an alkylating agent (Griffith and Meister, 1979a). As noted in Section IV, the toxicity of BSO alone in normal mice is negligible with only minimal weight loss observed (Griffith and Meister, 1979c; Page et al., 1987; Soble and Dorr, 1987). There are, however, conflicting reports regarding systemic toxicity due to administration of an alkylating agent following BSO-mediated depletion of GSH. Smith et al., (1988) stated that BSO pretreatment did not "significantly" enhance or alter the toxic effects of melphalan in mice. Similarly, Russo et al. (1986) reported that the marrow toxicity seen with melphalan in C3H mice was not increased by pretreatment with BSO. However, Kramer et al. (1985, 1986) have noted an increase in the toxicity of melphalan in BDF mice when the alkylator was administered after BSO pretreatment; increased toxicity was demonstrated by a decrease in the LD_{50}, by hepatotoxicity and by enhanced nephrotoxicity following administration of BSO and MeCCNU to Fischer 344 rats. Additionally, Soble and Dorr (1987) have demonstrated that cyclophosphamide treatment of DC-1 mice following BSO-mediated depletion of GSH resulted in sudden death within hours of administration; interestingly, enhanced toxicity with BCNU and melphalan was not seen. In the cyclophosphamide studies, central nervous system and pulmonary congestion were observed in necropsied animals as well as evidence for acute tubular necrosis. However, the precise mechanism(s) of death was not identified.

Studies in our laboratory confirmed the enhanced toxicity of cyclophosphamide in athymic mice and Fischer 344 rats previously treated with BSO

(Friedman *et al.*, 1990). Combination therapy caused sudden cardiac deaths due to ventricular fibrillation within 3 hr. Reduction in the dose of cyclophosphamide prolonged survival to 24–72 hr, and death then resulted from renal tubular cell injury secondary to massive skeletal muscle damage. These studies have defined the crucial role of GSH in protecting cardiac and skeletal muscle from cyclophosphamide-induced toxicity. Cardiac toxicity is dose-limiting with this alkylator when bone marrow support (autologous or allogeneic) is used, and repletion of cardiac GSH levels (depleted by either alkylating agents or BSO) may potentially provide protection from alkylator induced toxicity without decreasing the antitumor activity of cyclophosphamide or BSO–cyclophosphamide combinations.

D. INCREASES IN GLUTATHIONE LEVELS AND ALKYLATOR THERAPY

The observation that depletion of GSH may sensitize normal or neoplastic cells to endogenous or exogenous electrophiles has led to attempts to protect against these cellular insults by increasing GSH levels. Although administration of cysteine or GSH are not particularly effective in augmenting GSH levels, compounds such as *N*-acetylcysteine or L-2-oxothiazolidine-4-carboxylate result in intracellular delivery of cysteine with subsequent increases in GSH concentration (see Section IV). Since cysteine delivery systems are limited by the negative feedback of GSH on γ-glutamylcysteine synthetase and are ineffective in BSO-treated animals, therapeutic attention has focused on Et-GSH and related glutathione monoesters. As noted, this agent effects significant increases in the GSH level of liver, kidney, spleen, pancreas, and heart (Anderson *et al.*, 1985). Protection from the hepatic toxicity of acetaminophen (Anderson *et al.*, 1985) and BCNU or cyclophosphamide (Teicher *et al.*, 1988) has been demonstrated with this compound. Interestingly, protection of a murine fibrosarcoma from the antineoplastic activity of these compounds was not seen (Teicher *et al.*, 1988).

VI. Glutathione Modulation: Perspective on Further Applications

Glutathione modulation may play an important role in future modifications of chemotherapeutic activity and toxicity. BSO-medited depletion of GSH has been clearly shown to sensitize parental or resistant cells to subsequently administered alkylating agents such as melphalan. However, depletion of GSH is not limited to tumor cells, and enhanced alkylator toxicity to normal organs may also result from this approach. Enhanced toxicity may represent an exacerbation of previously seen complications (e.g., cyclophosphamide-induced cardiotoxicity occurring at lower alkylator doses) or the appearance of novel complications (e.g., skeletal muscle damage from lower dose cyclophosphamide plus BSO).

Despite these concerns and potential difficulties, any of several approaches may allow the rational use of GSH depletion to increase drug activity without enhancing toxicity. Identification and quantitation of differential rates of GSH depletion or repletion between normal and tumor cells may allow therapeutic "windows" to be defined. Identification of such "windows" would allow drug treatment at nadirs of tumor GSH levels with relatively normal host organ GSH levels (Lee *et al.*, 1987). Alternatively, compartmental therapy, as is used in the treatment of leptomeningeal neoplasia by direct, subarachnoid instillation of a GSH modulator and an alkylating agent, may facilitate enhanced antineoplastic activity without enhanced systemic side effects (Fuchs *et al.*, 1990).

Although BSO-mediated GSH depletion can enhance alkylator activity in resistant tumor cells known to demonstrate elevated GSH levels, this observation does not in itself confirm a causal relationship between drug resistance and elevations of GSH. Enhanced alkylator activity may result from modulation of the specific mechanism mediating drug resistance (i.e., elevated GSH levels) or represent a nonspecific modulation enhancing alkylator activity. For example, if the MDR phenotype results in decreased intracellular drug levels, depletion of GSH may enhance the activity of the residual intracellular alkylator. Thus, modulation of GSH levels may provide significant therapeutic benefit even when reversal of the specific mechanism of resistance is not possible. Furthermore, the steep dose–response curves seen with alkylating agents suggests that even small decreases in drug inactivation are likely to be important.

Pharmacological interventions that augment GSH levels may also be therapeutically relevant. It may be possible to exploit differences in the rates by which normal and neoplastic cells replete GSH following depletion by BSO or other agents; both normal repletion mechanisms and glutathione-ester-mediated repletion should be examined. Such studies may demonstrate therapeutic windows. Similarly, differences in protection of previously unperturbed normal vs neoplastic cells (Teicher *et a.*, 1988) may afford protection of normal cells from drug-induced damage and potentially allow use of higher drug dosages. Future studies are likely to demonstrate that the therapeutic index of chemotherapeutic intervention can be increased by these and other perturbations of GSH metabolism.

Acknowledgments

Work from our laboratories was supported by NIH grants DK26912 (O.W.G.) and CA44640 (H.S.F.) and ACS grant CH-403 (H.S.F.)

References

Abbott, W. A., Bridges, R. J., and Meister, A. (1984). Extracellular metabolism of glutathione accounts for its disappearance from the basolateral circulation of the kidney. *J. Biol. Chem.* **259**, 15393–15400.

Ahmad, S., Okine, L., Wood, R., Aljian, J., and Vistica, D. T. (1987). γ-Glutamyl transpeptidase (γ-GT) and maintenance of thiol pools in tumor cells resistant to alkylating agents. *J. Cell. Physiol.* **131**, 240–246.

Allen, L., Meck, R., and Yunis, A. (1980). The inhibition of γ-glutamyl transpeptidase from human pancreatic carcinoma cells by (αS, 5S)-α-amino-3-chloro-4,5-dihydro-5-isoxazoleacetic acid (AT-125; NSC-163501). *Res. Commun. Chem. Pathol. Pharmacol.* **27**, 175–182.

Allison, R. D., and Meister, A. (1981). Evidence that transpeptidation is a significant function of γ-glutamyl transpeptidase. *J. Biol. Chem.* **256**, 2988–2992.

Anderson, M. E., and Meister, A. (1983). Transport and direct utilization of γ-glutamylcyst(e)ine for glutathione synthesis. *Proc. Natl. Acad. Sci. U. S. A.* **80**, 707–711.

Anderson, M. E., and Meister, A. (1987). Intracellular delivery of cysteine. In "Methods in Enzymology" W. B. Jakoby and O. W. Griffith, eds.), Vol. 143, pp. 313–325. Academic Press, Orlando, Florida.

Anderson, M. E., and Meister, A. (1989). Glutathione monoesters. *Anal. Biochem.* **183**, 13–15.

Anderson, M. E., Bridges, R. J., and Meister, A. (1980). Direct evidence for interorgan transport of glutathione and that the non-filtration renal mechanism for glutathione utilization involves γ-glutamyl transpeptidase. *Biochem. Biophys. Res. Commun.* **96**, 848–853.

Anderson, M. E., Powrie, F., Puri, R. N., and Meister, A. (1985). Glutathione monoethyl ester: Preparation, uptake by tissues, and conversion to glutathione. *Arch. Biochem. Biophys.* **239**, 538–548.

Arrick, B. A., and Nathan, C. F. (1984). Glutathione metabolism as a determinant of therapeutic efficacy: A review. *Cancer Res.* **44**, 4224–4232.

Arrick, B. A., Griffith, O. W., and Cerami, A. (1981). Inhibition of glutathione synthesis as a chemotherapeutic strategy for trypanosomiasis. *J. Exp. Med.* **153**, 720–725.

Beck, L. V., Rieck, V. D., and Duncan, B. (1958). Diurnal variation in mouse and rat liver sulfhydryl. *Proc. Soc. Exp. Biol. Med.* **97**, 229–213.

Boveris, A., Oshino, N., and Chance, B. (1972). The cellular production of hydrogen peroxide. *Biochem. J.* **128**, 617–630.

Brandt, E. L., and Griffin, A. C. (1951). Reduction of toxicity of nitrogen mustards by cysteine. *Cancer (Philadelphia)* **4**, 1030–1035.

Bridges, R. J., Griffith, O. W., and Meister, A. (1980). L-γ-(*threo-β*-methyl)Glutamyl-L-α-aminobutyrate, a selective substrate of γ-glutamyl cyclotransferase. *J. Biol. Chem* **225**, 10787–10792.

Brodie, A. E., and Reed, D. J. (1985). Buthionine sulfoximine inhibition of cystine uptake and glutathione biosynthesis in human lung carcinoma cells. *Toxicol. Appl. Pharmacol.* **77**, 381–387.

Byczkowski, J. Z., and Gessner, T. (1988). Biological role of superoxide ion-radical. *Int. J. Biochem.* **20**, 569–580.

Calcutt, G., and Connors, T. A. (1963). Tumour sulphydryl levels and sensitivity to the nitrogen mustard merophan. *Biochem. Pharmacol.* **12**, 839–645.

Calvin, H. I., Medvedovsky, C., and Worgul, B. V. (1986). Near-total glutathione depletion and age-specific cataracts induced by buthionine sulfoximine in mice. *Science* **233**, 553–555.

Campbell, E. B., and Griffith, O. W. (1989). Glutathione monoethyl ester: High-performance liquid chromatographic analysis and direct preparation of the free base form. *Anal. Biochem.* **183**, 21–25.

Dulik, D. M., Fenselau, C., and Hilton, J. (1986). Characterization of melphalan–glutathione adducts whose formation is catalyzed by glutathione-S-transferase. *Biochem. Parmacol.* **35,** 3405–3409.

Dulik, D. M., Colvin, O. M., and Fenselau, C. (1990). Characterization of glutathione conjugates of chlorambucil by fast atom bombardment and thermospray liquid chromatography/mass spectrometry. *Biomed. Environ. Mass Spectrom.* **19,** 248–252.

Fekete, I., Schlageter, K., Griffith, O. W., Bigner, S. H., Friedman, H. S., Bigner, D. D., and Groothuis, D.R. (1990). The rate of buthionine sulfoximine entry into brain and xenotransplantable human gliomas. *Cancer Res.* **50,** 1251–1256.

Flohé, L. (1989). The selenoprotein glutathione peroxidase. *In* "Glutathione: Chemical, Biochemical, and Medical Aspects" (D. Dolphin, O. Avaramović, and R. Poulson, eds.) Part B, pp. 643–731, Wiley, New York.

Frei, E., Teicher, B. A., Cucchi, C.A., Rosowsky, A., Flatow, J. L., Kelley, E., and Genereux, P. (1988). Resistance to alkylating agents: Basic studies and therapeutic implications. *In* "Mechanisms of Drug Resistance in Neoplastic Cells" (P. V. Woolley and K. D. Tew, eds.). Academic Press, San Diego, California.

Fridovich, I. (1989). Superoxide dismutases: An adaptation to a paramagnetic gas. *J. Biol. Chem.* **264,** 7761–7764.

Friedman, H. S., Bigner, S. H., McComb, R. D., Schold, S. C. Jr., Pasternak, J. F., Groothuis, D. R., and Bigner, D. D. (1983). A model for human medulloblastoma: Growth, morphology and chromosomal analysis *in vitro* and in athymic mice. *J. Neuropathol. Exp. Neurol.* **42,** 485–503.

Friedman, H S., Burger, P. C., Bigner, S. H., Trojanowski, J. O., Wikstrand, C. J., Halperin, E. C., and Bigner, D. D. (1985). Establishment and characterization of the human medulloblastoma cell line and transplantable xenograft D283. *J. Neuropathol. Exp. Neurol.* **44,** 592–605.

Friedman, H. S., Burger, P. C., Bigner, S. H., Trojanowski, J. O., He, X., Wikstrand, C. J., Kurtzberg, J., Berens, M. E., Halperin, E. C., and Bigner, D. D. (1988). Phenotypic and genotypic analysis of a human medulloblastoma cell line and transplantable xenograft (D341 MED) demonstrating amplification of c-*myc*. *Am. J. Pathol.* **130,** 472–484.

Friedman, H. S., Colvin, O. M., Griffith, O. W., Lippitz, B., Elion, G. B., Schold, S. C., Hilton, J., and Bigner, D. D. (1989). Increased melphalan activity in intracranial human medulloblastoma and glioma xenografts following buthionine-sulfoximine mediated glutathione depletion. JNCI, *J. Natl. Cancer Inst.* **81,** 524–527.

Friedman, H. S., Colvin, O. M., Aisaka, K., Popp, J., Bossen, E. H., Reimer, K. A., Powell, J. B., Hilton, J., Gross, S. S., Levi, R., Bigner, D. D., and Griffith, O. W., (1990). Glutathione protects cardiac and skeletal muscle from cyclophosphamide-induced toxicity. *Cancer Res.* **50,** 2455–2462.

Frischer, H., and Ahmad, T. (1976). Severe generalized glutathione reductase deficiency after antitumor chemotherapy with BCNU (1,3-*bis*-(chloroethyl)-1-nitrourea). *J. Lab. Clin. Med.* **89,** 1080–1091.

Fuchs, F. H., Archer, G., Colvin, O. M., Bigner, S. H., Schuster, J. M., Fuller, G. N., Schold, S. C., Muhlbaier, L. H., Friedman, H. S., and Bigner, D. D. (1990). A nude rat model of human neoplastic meningitis: Activity of intrathecal 4-hydroperoxycyclophosphamide. *Cancer Res.* **50,** 1954–1959.

Gilbert, H. F. (1989). Thermodynamic and kinetic constraints on thiol/disulfide exchange involving glutathione redox buffers. *In* "Glutathione Centennial" (N. Taniguchi, T. Higashi, Y. Sakamoto, and A. Meister, eds), pp. 73–87. Academic Press, San Diego, California.

Gilbert, H. F. (1990). Molecular and cellular aspects of thiol–disulfide exchange. *Adv. Enzymol.* **63,** 69–172.

Green, J. A., Vistica, D. T., Young, R. C., Hamilton, T. C., Rogan, A. M., and Ozols, R. F. (1984). Potentiation of melphalan cytotoxicity in human ovarian cancer cell lines by glutathione depletion. *Cancer Res.* **44,** 5427–5431.

Griffith, O. W. (1981a). Glutathione turnover in human erythrocytes. *J. Biol. Chem.* **256**, 4900–4904.

Griffith, O. W. (1981b). The role of glutathione turnover in the apparent renal secretion of cystine. *J. Biol. Chem.* **256**, 12263–12268.

Griffith, O. W. (1982). Mechanism of action , metabolism and toxicity of buthionine sulfoximine and its higher homologs, potent inhibitors of glutathione biosynthesis. *J. Biol. Chem.* **257**, 13704–13712.

Griffith, O. W. (1989). L-Buthionine-SR-sulfoximine: Mechanism of action, resolution of diastereomers and use as a chemotherapeutic agent. *in* "Glutathione Centennial" (N. Taniguchi, T. Higashi, Y. Sakamoto, and A. Meister, eds.), pp. 285–299. Academic Press, San Diego, California.

Griffith, O. W., and Meister, A. (1978). Differential inhibition of glutamine and γ-glutamylcysteine synthetases by α-alkyl analogs of methionine sulfoximine that induce convulsions. *J. Biol Chem.* **253**, 2333–2338.

Griffith, O. W., and Meister, A. (1979a). Glutathione: Interorgan translocation, turnover, and metabolism. *Proc. Natl. Acad. Sci. U.S.A.* **76**, 5606–5610.

Griffith, O. W., and Meister, A. (1979b). Translocation of intracellular glutathione to membrane-bound γ-glutamyl transpeptidase as a discrete step in the γ-glutamyl cycle: Glutathionuria after inhibition of transpeptidase. *Proc. Natl. Acad. Sci. U.S.A.* **76**, 268–272.

Griffith, O. W., and Meister, A. (1979c). Potent and specific inhibition of glutathione synthesis by buthionine sulfoximine (*S*-*n*-butyl homocysteine sulfoximine). *J. Biol. Chem.* **254**, 7558–7560.

Griffith, O. W., and Meister, A. (1989). Excretion of cysteine and γ-glutamylcysteine moieties in human and experimental animal γ-glutamyl transpeptidase deficiency. *Proc. Natl. Acad. Sci. U.S.A.* **77**, 3384–3387.

Griffith, O. W., and Meister, A. (1985). Origin and turnover of mitochondrial glutathione. *Natl. Acad. Sci. U.S.A.* **82**, 4668–4672.

Griffith, O. W., Bridges, R. J., and Meister, A. (1978). Evidence that the γ-glutamyl cycle functions *in vivo* using intracellular glutathione: Effects of amino acids and selective inhibition of enzymes. *Proc. Natl. Acad. Sci. U.S.A.* **75**, 5405–5408.

Griffith, O. W., Anderson, M.E., and Meister, A. (1979a). Inhibition of glutathione biosynthesis by prothionine sulfoximine (*S*-*n*-propylhomocysteine sulfoximine), a selective inhibitor of γ-glutamylcysteine synthetase. *J. Biol. Chem.* **254**, 1205–1210.

Griffith, O. W., Bridges, R. J., and Meister, A. (1979b). Transport of γ-glutamylamino acids: Role of glutathione and γ-glutamyl transpeptidase. *Proc. Natl. Acad. Sci. U.S.A.* **76**, 6322.

Groothuis, D. R., and Blasberg, R. G. (1985). Rational brain tumor chemotherapy. *Neurol. Clinl.* **3**, 801–816.

Hagen, T. M., and Jones, D. P. (1989). Role of glutathione transport in extrahepatic detoxication. *In* "Glutathione Centennial" (N. Taniguchi, T. Higashi, Y. Sakamoto, and A. Meister, eds.), pp. 423–433. Academic Press, San Diego, California.

Hamilton, T. C., Winker, M. A., Louis, K. G., Batist, G., Behrens, B. C., Tsuruo, T., Grotzinger, K. R., McKoy, W. M., Young, R. C., and Ozols, R. F. (1985). Augmentation of Adriamycin, melphalan, and cisplatin cytotoxicity in drug-resistant and -sensitive human ovarian carcinoma cell lines by buthionine sulfoximine mediated glutathione depletion. *Biochem. Pharmacol.* **34**, 2583–2586.

Inoue, M., and Morino, Y. (1985). Drect evidence for the role of the membrane potential in glutathione transport by renal brush-border membranes. *J. Biol. Chem.* **260**, 326–331.

Ketterer, B., Meyer, D. J., and Clark, A. G. (1988). Soluble glutathione transfers isozymes. *In* "Glutathione Conjugation: Mechanism and Biological Significance" (H. Sies and B. Ketterer, eds.), pp. 73–135. Academic Press, San Diego, California.

Kondo, T., Dale, G. L., and Beutler, E. (1981). Studies on glutathione transport utilizing inside-out vesicles prepared from human erythrocytes. *Biochim. Biophys. Acta* **645**, 132–136.

Kondo, T., Kawakami, Y., and Taniguchi, N (1989). Glutathione disulfide-stimulated Mg^{++}-ATPase of human erythrocytes. *In* "Glutathione Centennial" (N. Taniguchi, T. Higashi, Y. Sakamoto, and A. Meister, eds.) pp. 369–380. Academic Press, San Diego, California.

Kosower, N. S., and Kosower, E. M. (1987). Formation of disulfides with diamide. *In* "Methods in Enzymology" (W. B. Jakoby and O. W. Griffith, eds.), Vol. 143, pp. 264-270. Academic Press, Orlando, Florida.

Kramer, R. A., Schuller, H. M., Smith, A. C., and Boyd, M. R. (1985). Effects of buthionine sulfoximine on the nephrotoxicity of 1-(2-chloroethyl)-3-(trans-4-metholocyclohexyl)-1-nitrosourea (MeCCNU). *J. Pharmacol. Exp Ther.* **234**, 498–506.

Kramer, R. A., Ahmad, S., and Vistica, D. (1986). Toxicologic considerations in chemosensitization of melphalan (L-PAM) by buthionine sulfoximine (BSO) in mice. *Proc. Am. Assoc. Cancer Res.* **27**, 419.

Kramer, R. A., Greene, R., Ahmad, S., and Vistica, D. T. (1987). Chemosensitization of L-phenylalanine mustard by the thiol-modulating agent buthionine sulfoximine. *Cancer Res.* **47**, 1593–1597.

Lee, F. Y. F., Allanlunis-Turner, M. J., and Sieman, D. W. (1987). Depletion of tumor versus normal tissue glutathione by buthionine sulfoximine. *Br. J. Cancer* **56**, 33–38.

Martensson, J., and Meister, A. (1989). Mitochondrial damage in muscle occurs after marked depletion of glutathione and is prevented by giving glutathione monoester. *Proc. Natl. Acad. Sci. U.S.A* **86**, 471–475.

Martensson, J., Steinherz, R., Jain, A., and Meister, A. (1989). Glutathione ester prevents buthionine sulfoximine-induced cataracts and lens epithelial cell damage. *Proc. Natl. Acad. Sci. U.S.A.* **86**, 8727–8731.

McAllister, R. M., Isaacs, H., Rongey, R., Peer, M., Au, W., Soulap, S. W., and Gardner, M. B. (1977). Establishment of a medulloblastoma cell line. *Int. J. Cancer* **20**, 206–212.

Meister, A. (1989). Metabolism and function of glutathione. *In* "Glutathione: Chemical, Biochemical, and Medical Aspects" (D. Dolphin, O. Avaramović, and R. Poulson, eds.), Part B, pp. 367–474, Wiley, New York.

Meister, A., and Griffith, O. W. (1979). Effects of methionine sulfoximine analogs on the synthesis of glutamine and glutathione; possible chemotherapeutic implications. *Cancer Chemother. Rep.* **63**, 1115–1121.

Meister, A., and Larsson, A. (1989). Glutathione synthetase deficiency and other disorders of the γ-glutamyl cycle. *In* "The Metabolic Basis of Inherited Disease" (C. R. Scriver, A.L. Beaudet, W. S. Sly, and D. Valle, eds.), pp. 855–868, McGraw-Hill, New York.

Meredith, M. J., and Reed, D. J. (1982). Status of the mitochondrial pool glutathione in the isolated hepatocyte. *J. Biol. Chem.* **257**, 3747–3753.

Meshnick, S. R., Chang, K. P., and Cerami, A. (1977). Heme lysis of the bloodstream forms of *Trypanosoma brucei. Biochem. Pharmacol.* **26**, 1923–1928.

Meshnick, S. R., Grady, R. W., Blobstein, S. H., and Cerami, A. (1978). Porphyrin-induced lysis of *Trypanosoma brucei:* A role for zinc. *J. Pharmacol. Exp. Ther.* **207**, 1041–1050.

Minchinton, A. J., Rojas, A., Smith, A., Soranson, J. A., Shrieve, D. C., Jones, N. R., and Bremer, J. C. (1984). Glutathione depletion in tissues after administration of buthionine sulfoximine. *Int. J. Radiat. Oncol., Biol. Phys.* **10**, 1261–1164.

Moore, W. R., and Meister, A. (1987). Enzymatic synthesis of novel glutathione analogs. *Anal. Biochem.* **161**, 487–493.

Naqui, A., Chance, B., and Cadenas, F. (1986). Reactive oxygen intermediates in biochemistry *Annu. Rev. Biochem.* **55**, 137–166.

Nathan, C. F., Arrick, B. A., Murray, H. W., De Santis, N. M., and Cohn, Z.A. (1981). Tumor cell anti-oxidant defenses. *J. Exp. Med.* **153**, 766–782.

Oppenheimer, L., Wellner, V. P., Griffith, O. W., and Meister, A. (1979). Glutathione synthetase. *J. Biol. Chem.* **254**, 5184–5190.

Orlowski, M., and Meister, A. (1979). The γ-glutamyl cycle: A possible transport system for amino acids. *Proc. Natl. Acad. Sci. U.S.A.* **67**, 1248–1255.

Orlowski, M., and Meister, A. (1971). Isolation of highly purified γ-glutamylcysteine synthetase from rat kidney. *Biochemistry* **10**, 372–380.

Ozols, R. F., Louis, K G., Plowman, J., Behrens, B. C., Fine, R. L., Dykes, D., and Hamilton, T. C. (1987). Enhanced melphalan cytotoxicity in human ovarian cancer *in vitro* and in tumor-bearing nude mice by buthionine sulfoximine depletion of glutathione. *Biochem. Pharmacol.* **36**, 147–153.

Pace, J., and McDermott, E.E (1952). Methionine sulphoximine and some enzyme systems involving glutamine. *Nature (London)* **169**, 415–416.

Page, J. G., Carolton, B. D., Smith, A. C., Kastello, M. D., and Grieshaber, C. K. (1987). Preclinical toxicology and pharmacokinetic studies of buthionine sulfoximine (BSO, NSC-326231) in CD2F1 mice. *Proc. Am. Assoc. Cancer Res.* **28**, 440.

Palekar, A. G., Tate, S. S., and Meister, A. (1975). Decrease in glutathione levels in kidney and liver after injection of methionine sulfoximine into rats. *Biochem. Biophys. Res. Commun.* **62**, 651–657.

Peczenik, O. (1953). Influence of cysteamine, methylamine and cortisone on the toxicity and activity of nitrogen mustard. *Nature (London)*. **172**, 454–455.

Peterson, L. A., and Guengerich, F. P. (1988). Comparison of and relationships between glutathione *S*-transferase and cytochrome P-450 systems. *In* "Glutathione Conjugation: Mechanism and Biological Significance" (H. Sies and B. Ketterer, eds.), pp. 193–233. Academic Press, San Diego, California.

Plummer, J. L., Smith, B. R., Sies, H., and Bend, J. R. (1981). Chemical depletion of glutathione. *In* "Methods in Enzymology" (W. B. Jakoby, ed.), Vol. **77**, pp. 50–59. Academic Press New York.

Puri, R. N., and Meister, A. (1983). Transport of glutathione, as γ-glutamylcysteinylglycyl ester, into liver and kidney. *Proc. Natl. Acad. Sci. U.S.A.* **80**, 5258–5260.

Richman P. G., and Meister, A. (1975). Regulation of γ-glutamylcysteine synthetase by nonallosteric feedback inhibition by glutathione. *J. Biol. Chem.* **250**, 1422–1426.

Richman, P. G., Orlowski, M., and Meister, A. (1973). Inhibition of γ-glutamylcysteine synnthetase by methionine-*S*-sulfoximine. *J. Biol. Chem.* **248**, 6684–6690.

Romero, F. J., Soboll, S., and Sies, H. (1984). Mitochondrial and cytosolic glutathione after depletion by phorone in isolated hepatocytes. *Experientia* **40**, 365–367.

Rosenberg, M. C., Colvin, O. M., Griffith, O. W., Bigner, S. H., Elion, G. B., Horton, J. K., Lilley, E., Bigner, D. D., and Friedman, H. S. (1989). Establishment of a melphalan-resistant human rhabdomyosarcoma xenograft with cross-resistance to vincristine and enhanced depletion. *Cancer Res.* **49**, 6917–6922.

Russo, A., Tochner, Z., Phillips, T., Carmichael, J., DeGraff, W., Friedman, N., Fisher, J., and Mitchell, J. B. (1986). In vivo modulation of glutathione by buthionine sulfoximine effect on marrow response to melphalan. *Int. J. Radiat. Oncol., Biol. Phys.* **12**, 1187–1189.

Schandle, V. B., and Rudolph, F. B. (1981). Isotope exchange at equilibrium studies with rat kidney γ-glutamylcysteine synthetase. *J. Biol. Chem.* **256**, 7590–7594.

Seelig, G. F., Simondsen, R. P., and Meister, A. (1984). Reversible dissociation of γ-glutamylcysteine synthetase into two subunits. *J. Biol. Chem.* **259**, 9345–9347.

Sekura, R., and Meister, A. (1977). γ-Glutamylcysteine synthetase. *J. Biol. Chem.* **252**, 2599–2605.

Sekura, R., Hochreiter, M., and Meister, A. (1976). α-Aminomethylglutarate, a β-amino acid analog of glutamate that interacts with glutamine synthetase and the enzymes that catalyze glutathione synthesis. *J. Biol. Chem.* **251**, 2263–2270.

Sies, H., and Summer, K. H. (1975). Hydroperoxide-metabolizing systems in rat liver. *Eur. J. Biochem.* **57**, 503–512.

Skapek, S. X., Colvin, O. M., Griffith, O. W., Groothuis, D. R., Colapinto, E. V., Lee, Y. S., Hilton, J., Elion, G. B., Bigner, D. D., and Friedman, H. S., (1988a). Buthionine sulfoximine-mediated depletion of glutathione in intracranial human glioma-derived xenografts. *Biochem. Pharmacol.* **37**, 4313–4317.

Skapek, S. X., Colvin, O. M., Griffith, O. W., Elion, G. B., Bigner, D. D., and Friedman, H. S. (1988b). Enhanced cytotoxicity of melphalan secondary to BSO mediated depletion of glutathione in medulloblastoma xenografts in athymic nude mice . *Cancer Res.* **48**, 2764–2767.

Smith, A. C., Placke, M. E., Ryan, M. J., Kastella, M. D., and Grieshaber, C. K. (1988). The effect of buthionine sulfoximine (BSO, NSC 326231) on melphalan (NSC-8806) induced toxicity in CD2F1 mice. *Proc. AM. Assoc. Cancer Res.* **29**, 511.

Soble, M. J., and Dorr, R. T. (1987). Lack of enhanced myelotoxicity with buthionine sulfoximine and sulfhydryl-dependent anticancer agents in mice. *Res. Commun. Chem. Pathol. Pharmacol.* **55**, 161–180.

Somfai-Relle, S., Suzukake, K., Vistica, B. P., and Vistica, D. T. (1984). Reduction in cellular glutathione by buthionine sulfoximine and sensitization of murine tumor cells resistant to L-phenylalanine mustard. *Biochem. Pharmacol.* **33**, 485–490.

Stevens, J. L., and Jones, D. P., (1989). The mercapturic acid pathway: Biosynthesis, intermediary metabolism and physiological disposition. *In* "Glutathione: Chemical, Biochemical and Medical Aspects" (D. Dolphin, O. Avaramović, and R. Poulson, eds.), Part B, pp. 46–84 Wiley, New York.

Stratton, M. R., Darling, J., Pilkington, G. J., Lantos, P. L., Reevs, B. P., and Cooper, C. S. (1989). Characterization of the human cell line TE-671. *Carcinogenesis* **10**, 889–905.

Suzukake, K., Petro, B. J., and Vistica, D. T. (1982). Reduction in glutathione content of L-PAM resistant L1210 cells confers drug sensitivity. *Bio chem. Pharmacol.* **31**, 121–124.

Tate, S. S., and Meister, A. (1985). γ-Glutamyl transpeptidase from rat kidney. *In* "Methods in Enzymology" (A. Meister, ed.), Vol. 113, pp. 400–437, Academic Press Orlando, Florida.

Tateishi, N., Higashi, T., Shinya, S., Naruse, A., and Sakamoto, Y. (1974). Studies on the regulation of glutathione level in rat liver. *J. Biochem. (Tokyo)* **75**, 93–103.

Tateishi, N., Higashi, T., Naruse, A., Nakashima, K., Shiozaki, H., and Sakamoto, Y. (1977). Rat liver glutathione: Possible role as a reservoir of cysteine. *J. Nutr.* **103**, 51–60.

Teicher, B. A., Crawford, J. M., Houden, S. A., Lin, Y., Cathcart, K. N. S., Luchette, C. A., and Flatow, J. L. (1988). Glutathione monoethyl ester can selectively protect liver from high dose BCNU or cyclophosphamide. *Cancer (Philadelphia)* **62**, 1275–1281.

Van der Werf, P., Stephani, R. A. and Meister, A. (1974). Accumulation of 5-oxoproline in mouse tissues after inhibition of 5-oxoprolinase and administration of amino acids: Evidence for function of the γ-glutamyl cycle. *Proc. Natl. Acad. Sci. U.S.A.* **71**, 1026–1029.

Vistica, D. T., and Ahmad, S. (1989). Acquired resistance of tumor cells to L-phenylalanine mustard: Implications for the design of a clinical trial involving glutathione depletion. *In* "Glutathione Centennial" (N. Taniguchi, T. Higashi, Y. Sakamoto, and A. Meister, eds.), pp. 301–316. Academic Press, San Diego, California.

Wahlländer, A., Soboll, S., and Sies, H. (1979). Hepatic mitochondrial and cytosolic glutathione content and the subcellular distribution of GSH S-transferases. *FEBS Lett.* **97**, 138–140.

Wang, A. L., and Tew, K. D. (1985). Increased glutathione-S-transferase activity in a cell line with acquired resistance to nitrogen mustards. *Cancer Treat. Rep.* **69**, 677–682.

Yan, N., and Meister, A. (1990). Amino acid sequence of rat kidney γ-glutamylcysteine synthetase. *J. Biol Chem.* **265**, 1588–1593.

Yip, B., and Rudolph, F. B. (1976). The kinetic mechanism of rat kidney γ-glutamylcysteine synthetase. *J. Biol. Chem.* **251**, 3563–3568.

Yuan, Z., Fenselau, C., Dulik, D. M., Martin, W., Emany, W. B., Brundrett, R. B., Colvin, O. M., and Cotter, R. J. (1990). Laser desorption electron impact: Application to a study of the mechanism of conjugation of glutathione and cyclophosphamide. *Anal. Chem.* **62**, 868–870.

CHAPTER 8

Synergy and Antagonism in Polymerase-Targeted Antiviral Therapy: Effects of Deoxynucleoside Triphosphate Pool Modulation on Prodrug Activation

Thomas Spector
James A. Fyfe
Division of Experimental Therapy
Wellcome Research Laboratories
Research Triangle Park, North Carolina 27709

Several viruses of the herpes group and the human immunodeficiency virus (HIV) differ greatly with respect to their viral classification and the nature of the disease produced by their respective infections. Nevertheless, as objects of antiviral chemotherapy, they share several common features. First, they are both susceptible to therapy by nucleoside analogs. Second, in both cases, the therapeutic analogs serve as prodrugs since they require conversion to the

SYNERGISM AND ANTAGONISM IN CHEMOTHERAPY
Copyright © 1991 by Academic Press, Inc.
All rights of reproduction in any form reserved.

corresponding triphosphate form for activation. Third, the antiviral mechanism of the activated analogs involves highly specific interactions with the respective viral-encoded nucleic acid polymerase. Two examples of attempts to enhance the antiviral potency of the nucleoside analogs by combination therapy with additional agents have yielded very different results.

The activity of acyclovir, Zovirax® (ACV)[1], an antiherpetic agent, was markedly potentiated by A1110U, an inhibitor of the viral-encoded ribonucleotide reductase. Conversely, the anti-HIV activity of 3'-azido-3'-deoxythymidine, zidovudine, Retrovir® (AZT) was strongly antagonized by ribavirin (RBV). A1110U and RBV, by entirely different mechanisms, modulate pool sizes of the various nucleoside triphosphates. In each specific situation, these modulations resulted in either synergistic or antagonistic antiviral interactions. This chapter will review the molecular basis for these effects.

I. Synergistic Antiherpetic Chemotherapy by Acyclovir (ACV) and an Inhibitor of Herpes Virus Ribonucleotide Reductase

A. INTRODUCTION

The first section of this chapter discusses an agent which modifies deoxynucleotide pools to produce the desirable effect of potentiating the antiviral activity of ACV. It will begin with a review of the mechanism of action of this nucleoside analog to establish the background for understanding the basis of the synergistic interaction. ACV is used clinically to treat infections by herpes simplex virus type 1 (HSV-1), herpes simplex virus type 2 (HSV-2) and herpes varicella zoster virus (VZV). The growing acceptance of ACV therapy can be attributed to both its safety and its efficacy. These features reflect unusual interactions of ACV and its metabolites with two enzymes, thymidine kinase and DNA polymerase, that are specified by these herpes viruses. Although it may seem odd that these viruses encode two enzymes that are isofunctional with the cellular enzymes, the viral enzymes have several distinguishing features, among which are those that account for their selective and therapeutically beneficial interactions with ACV. The third viral-encoded enzyme that is apparently redundant with a cellular enzyme is ribonu-

---

[1] Abbreviations used are HSV-1, herpes simplex virus type 1; HSV-2, herpes simplex virus type 2; VZV, varicella zoster virus; A1110U, 2-acetylpyridine 5-[(dimethylamino)thiocarbonyl]-thiocarbonohydrazone; ACV, acyclovir (9-[(2-hydroxyethoxy)methyl]guanine); ACVTP, acyclovir triphosphate; AZT, 3'-azido-3'-deoxythymidine; AZTMP/AZTTP, 5'-mono- and triphosphate derivatives of AZT; RBV, ribavirin; RBVTP, 5'-triphosphate derivative of RBV; HIV, human immunodeficiency virus; 5'-AT, 5'-amino-5'-deoxythymidine.

cleotide reductase. The importance of the distinct features of herpes ribonucleotide reductases will be described later in this chapter. The first feature to be addressed here is the selective activation of ACV by viral thymidine kinase.

B. SELECTIVE ACTIVATION OF ACV

ACV (see Fig. 1) itself is inert as an antiviral agent. The usefulness of ACV is contingent on its conversion to a nucleoside triphosphate analog. The first phosphorylation step in this conversion represents the gateway to activation and is selectively catalyzed by herpes-encoded thymidine kinase (Fyfe *et al.*, 1978). The viral enzyme recognizes the deoxyguanosine analog, ACV, as a thymidine analog and accepts it as an alternative substrate. Fortunately, cellular kinases do not pander to the fraudulent nucleoside and very little activation occurs in uninfected cells (Elion *et al.*, 1977). The trace quantities of phosphorylated ACV detected in uninfected cells has been attributed to an unusual reaction catalyzed in the reverse direction by 5'-nucleotidase (Keller *et al.*, 1985). Therefore, significant concentrations of ACV monophosphate (ACVMP) are only produced in herpes-infected cells. Rapid conversion of ACVMP to ACV triphosphate (ACVTP) is then catalyzed by cellular enzymes (Miller and Miller, 1980, 1982). The fully activated drug now assumes the more reasonable identity of a dGTP analog and competes with its physiological counterpart for binding to herpes DNA polymerase.

C. SELECTIVE INHIBITION OF HERPES DNA POLYMERASE

Although totally devoid of a structure analogous to a 2',3'-carbon bridge, ACVTP is well tolerated as a substrate and is incorporated into the growing viral DNA chain (Furman *et al.*, 1979; Derse *et al.*, 1981). However, because ACV monophosphate (ACVMP) lacks a 3'-hydroxyl group, it does not

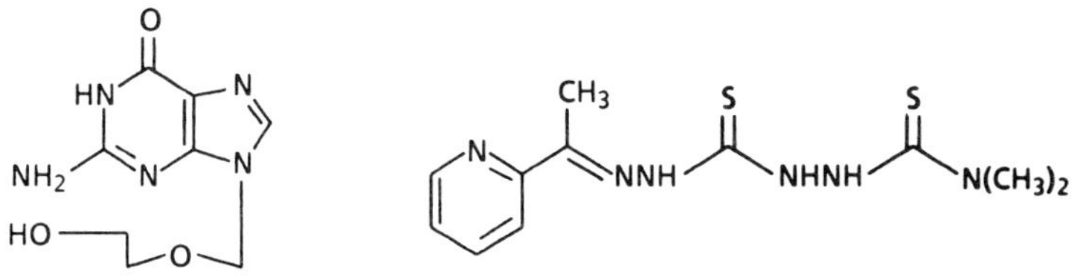

ACV **A1110U**

FIGURE 1   Structure of acyclovir (ACV) and 2-acetylpyridine 5-[(dimethylamino) thiocarbonyl]-thiocarbonohydrazone (A1110U).

support chain extension and termination occurs (Reid *et al.*, 1988). Moreover, the proofreading activity of the viral polymerase lacks the capacity to excise ACVMP (Derse *et al.*, 1981). The polymerase reaction has been studied in detail with HSV-1 DNA polymerase and has resulted in several interesting observations. The data of Fig. 2 show that the enzyme does not simply dissociate from the ACVMP-terminated primer. Instead, it loses enzymatic activity in a time-dependent manner. Kinetic analyses (Furman *et al.*, 1984) revealed that ACVTP initially competed very favorably with dGTP to form a Michaelis–Menten intermediate with the enzyme–primer–template complex (4 nM K_m for ACVTP vs 200 nM K_m for dGTP). Furthermore, once ACVMP is incorporated, the DNA polymerase appeared to become inactivated with a first-order rate constant of 0.24 min^{-1}. Thus, with the exception of plausible covalent enzyme modification, ACVTP had all the quantifications of an effective suicide substrate inactivator.

Additional studies using a synthetic primer–template in place of the customary calf thymus DNA further elucidated the fate of HSV-1 DNA polymerase (Reardon and Spector, 1989). Since the model primer–template was designed to accept dGTP as the next nucleotide in the sequence, incorporation of either dGTP or ACVTP could be monitored in the absence of the other deoxynucleoside triphosphate substrates. This simplified system revealed that, under these conditions, ACVTP was a conventional substrate

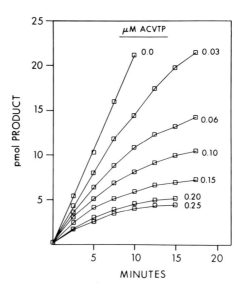

FIGURE 2   Progressive inhibition of HSV-1 DNA polymerase by ACVTP. The incorporation of [^{3}H]dGMP into activated calf thymus DNA from 5 μM [^{3}H]dGTP in the presence of 100 μM dATP, dCTP, and dTTP, and the indicated concentrations of ACVTP were determined at the indicated reaction times. (Reproduced from Furman *et al.*, 1984, by permission.)

that did not cause inactivation of the polymerase. However, when dCTP, the deoxynucleotide that follows ACVTP in the template sequence, was present, significant incorporation (greater than stoichiometric with enzyme) of ACVMP was not detected. If dCTP were added while the incorporation of ACVMP was already in progress, the reaction would immediately cease. The study revealed that the incorporation of ACVMP actually induced the next nucleotide of the template sequence to act as an inhibitor. The K_i for dCTP as the next nucleotide inhibitor was about 1/35 its K_m value as a substrate on this synthetic primer–template, indicating that the incorporation of ACVMP induces a 35-fold increase in the apparent (nonproductive) affinity of the next sequential nucleotide. Thus, the deoxynucleotides (other than dGTP), at concentrations commonly used in DNA polymerase assays, force the ACVMP-terminated primer and enzyme into an inhibited dead-end complex. Moreover, when both dGTP, ACVTP and the next sequential deoxynucleotide were present, the enzyme partitions between productive DNA synthesis and dead-end complex formation, but would eventually be siphoned off into the latter in a manner that mimicked suicide inactivation. The scheme for this mechanism is presented in Fig. 3.

The earlier observation that ACVMP-terminated calf thymus DNA was a potent inhibitor of HSV-1 DNA polymerase (Derse *et al.*, 1981) was also investigated using the just-described synthetic primer–template that had been completely terminated by previous incorporation of ACVMP (Reardon and Spector, 1989). This synthetic DNA was an unimpressive competitive

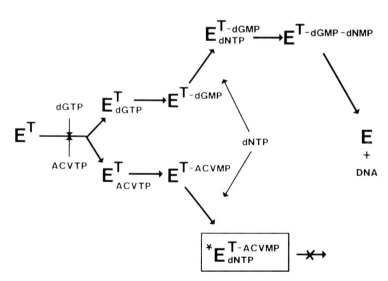

FIGURE 3   Scheme for the ACVTP induced inhibition by the next nucleotide of the template sequence. E, HSV-1 DNA polymerase; T, primer template;*, trapped enzyme complex. (Derived from Reardon and Spector, 1989, by permission.)

inhibitor vs. non terminated synthetic primer–template with a K_i value equal to the K_m value of the latter. However, addition of dCTP, the next nucleotide of the template sequence, forced the enzyme into the dead-end complex with the ACVMP-terminated primer–template and potent inhibition was observed. Thus potent inhibition by the next sequential nucleotide could be induced by ACVMP-terminated primer–template which was either added exogenously or generated from ACVTP as a substrate.

The interactions of ACVTP with human DNA polymerases sharply contrast with those with the viral enzyme. First, the initial binding of ACVTP to DNA polymerase α is less favorable than that of dGTP (Furman et al., 1979; Reardon, 1989). Second the incorporation of ACVMP into calf thymus DNA in the presence of the four deoxynucleotide substrates did not cause the inhibition of human DNA polymerase α to increase progressively with time[2] (Furman et al., 1984). Evidence for the weak ability of ACVTP to induce inhibition by the next nucleotide in the template sequence was only detected for the incorporation of dGMP into the small defined primer–template described earlier when high concentrations of ACVTP and 5 mM dCTP (the nucleotide following ACVMP incorporation in the template sequence) were present (Reardon, 1989). Moreover, ACVTP was neither a substrate nor an inhibitor of human DNA polymerase β (Derse et al., 1981; Reid et al., 1988; Reardon, 1989). Thus, the unique substrate specificity of herpes thymidine kinase and the magnified sensitivity of herpes DNA polymerase to inhibition created the basis for the safety and efficacy of ACV as an antiviral agent. These properties also presented a serious challenge to attempts to improve the usefulness of ACV without disturbing this balance of desirable features. However, the clinical need for a more potent topical therapy, together with the detection of a possible site of potentiation by a second agent provided impetus for the following studies.

D. SELECTIVE INHIBITION OF HERPES RIBONUCLEOTIDE REDUCTASE

An undesirable consequence of a competitive inhibitor is that it can promote the build-up of the physiological competing substrate which subsequently diminishes the effectiveness of that inhibitor. However, under most circumstances where a dGTP analog inhibits a DNA polymerase, the pool of dGTP would not be expected to increase significantly. The reason is that mammalian ribonucleotide reductase, the enzyme responsible for deoxynucleotide synthesis, is subject to an elaborate feedback mechanism

---

[2] Additional unpublished data of S.E. Hopkins and T. Spector confirmed the linear reactions by monitoring dGMP incorporation into calf thymus DNA at inhibitory concentrations of ACVTP using a stable preparation of HeLa cell DNA polymerase α.

which strictly limits overproduction of deoxynucleotides (Thelander and Reichard, 1979; Hunting and Henderson, 1982). On the other hand, herpes simplex and varicella zoster viruses circumvent this limitation by encoding for ribonucleotide reductases that are devoid of regulation by feedback-inhibition and are therefore capable of catalyzing unrestricted synthesis of dGTP (reviewed by Spector, 1985, 1989). Thus, treatment of HSV-1 infected cells with ACV and subsequent inhibition of herpes DNA polymerase inadvertently increases dGTP pool sizes (Furman *et al.*, 1982; Spector *et al.*, 1985, 1989). It therefore became apparent that inhibition of herpes ribonucleotide reductase might prevent the ACV-induced build-up of dGTP and its ability to compete with ACVTP for binding to herpes DNA polymerase.

Herpes ribonucleotide reductases are much less complicated than their cellular counterparts (reviewed by Spector, 1985, 1989). They are neither activated nor inhibited by the physiological triphosphate nucleosides which allosterically regulate mammalian ribonucleotide reductase. Reduction of all four natural substrates in the presence of a sulfhydryl reducing agent is catalyzed at a common site on the enzyme. No other exogenous cofactors have been found to activate the enzyme. Although the viral enzymes were insensitive to inhibition by nucleotides, the fact that they clearly differed from the cellular enzyme presented them as potential chemotherapeutic targets. Several series of 2-acetylpyridine thiosemicarbazone inhibitors of herpes ribonucleotide reductase emerged from two independent laboratories (Shipman *et al.*, 1981, 1986; Spector *et al.*, 1985, 1989). These compounds were considerably less cytotoxic variations of a group of inhibitors originally designed as anticancer agents (reviewed by Moore and Sartorelli, 1984). It appeared that their selective toxicity towards herpes viruses was expressed at the level of enzyme inhibition (Spector *et al.*, 1985, 1989; Turk *et al.*, 1986). A1110U, 2-acetylpyridine 5-[(dimethylamino)thiocarbonyl]-thiocarbonohydrazone (Fig. 1) is a particularly potent inhibitor that has been studied in depth (Spector *et al.*, 1989).

Sub-micromolar concentrations of A1110U produced rapid inactivation of partially purified HSV-1, HSV-2, and VZV ribonucleotide reductases. The compound also inhibited human ribonucleotide reductase, but not as effectively. At concentrations above $5 \mu M$ it inactivated the human enzyme but below 5 μM it produced only weak (noninactivating) inhibition. An explanation for the differential inhibitory effects was provided by a study of the role of the iron complex of A1110U in the inhibition process (Porter *et al.*, 1990).

A1110U forms a very stable complex with iron. Studies in iron-free assay media (Fig. 4) revealed that A1110U in either the free or iron-complexed form was capable of inactivating HSV-1 ribonucleotide reductase. In contrast, human ribonucleotide reductase was not inhibited by uncomplexed A1110U, and was only weakly inhibited (not inactivated) by iron-complexed A1110U. Futhermore, the combination of uncomplexed and iron-complexed A1110U produced considerably greater than additive (synergistic)

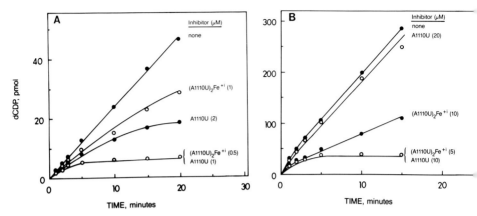

FIGURE 4   Inhibition of herpes and human ribonucleotide reductase. Effects of A1110U, A1110U₂Fe complex, and the combination of A1110U and A1110U₂Fe complex on the reduction of [¹⁴C]CDP by HSV-1 (*left*) and human (*right*) ribonucleotide reductase. Reaction temperature was lowered from 37°C (human enzyme) to 30°C for the viral enzyme to observe the very fast rates of inactivation. (Reproduced from Porter *et al.*, 1990, by permission.)

rates of inactivation of either enzyme. Here again, herpes ribonucleotide reductase is more susceptible to inactivation because the second-order rate constant for inactivation by iron-complexed A1110U in the presence of a small amount of uncomplexed A1110U was approximately 10-fold larger for the viral than for the human enzyme. A summary of these differences is presented in Table 1. HSV-1 ribonucleotide reductase that had been very rapidly inactivated by the combination of inhibitor species could be completely reactivated by removing the inhibitors and replenishing the iron. Thus, it was postulated that A1110U removes iron from the enzyme's active site, and that iron-complexed A1110U binds to the enzyme to greatly facilitate this process. Interestingly, manganese could competitively retard the reactivation by iron, indicating that it can bind to the iron site, but cannot support catalysis. Moreover, the finding that the ATP, used as an essential activator of human ribonucleotide reductase in the original experiments, was contamin-

TABLE 1
Effects of A1110U Species on Ribonucleotide Reductases

	Herpes	Human
Free A1110U	Inactivation	No inhibition
A1110U₂Fe	Inactivation	Weak inhibition
Free A1110U plus A1110U₂Fe	Very fast inactivation[a]	Fast inactivation[b]

[a] $K_{inactivation} = (7\ hr^{-1}) + (2.5 \times 10^7\ M^{-1}hr^{-1})[A1110U_2Fe]$ at 30°C.
[b] $K_{inactivation} = (0.38 \times 10^7\ M^{-1}hr^{-1})[A1110U_2Fe]$ at 37°C.

ated with traces of iron provided an explanation for the just-described results with this enzyme. At low concentrations, all the A1110U was converted to the iron complex by the contaminating iron in the assay media for human ribonucleotide reductase and only weak inhibition occurred. Higher concentrations of A1110U provided uncomplexed A1110U as well iron-complexed A1110U, and this combination of species produced inactivation. These studies point to the importance of controlling for exogenous iron. If A1110U were assayed in iron-free media, the selective inhibition of herpes ribonucleotide reductase would have appeared to be absolute. Although real biosystems are not iron-free, the actual selectivity of A1110U was adequate to permit *in vitro* studies of its antiviral effects in the presence and absence of ACV.

E. SYNERGISTIC INHIBITION OF VIRUS REPLICATION *IN VITRO*

A1110U produced independent inhibition of HSV-1, HSV-2 and VZV replication in cell culture at concentrations (IC_{50} = about 1.5 μM) that were similar to those required to inactivate the respective viral ribonucleotide reductases (Spector *et al.*, 1989). A1110U also produced a marked potentiation of the activity of ACV by lowering its IC_{50} by one or two orders of magnitude. The terms potentiation or synergy are used in this text to denote the condition where the inhibition observed by two inhibitors in concert is significantly greater than the calculated sum of their independent effects. In some assays, potentiation was evident at subinhibitory concentrations of A1110U. In other assays, synergy did not occur until concentrations of A1110U that were independently inhibitory were present. Since inhibition by A1110U was typically characterized by a very steep dose–response curve, the latter experiments displayed abruptly appearing synergistic inhibition followed by total inhibition and required careful titration to fully illustrate. Marked synergy between A1110U and ACV has been clearly demonstrated against the three herpes viruses, HSV-1, HSV-2, and VZV, which are independently inhibited by A1110U. A double-drug isobologram depicting the synergistic activities of A1110U and ACV against HSV-2 replication is presented in Fig. 5.

Additional evidence supporting the hypothesis that inhibition of herpes ribonucleotide reductase can result in the potentiation of ACV was recently provided by Coen *et al.*, (1989). They found that two mutated HSV strains that did not express active ribonucleotide reductase were hypersensitive to ACV.

Although A1110U was well tolerated by the confluent host cells used to support viral replication, it was not without toxicity toward rapidly dividing mammalian cells. Fortunately, and in contrast to its antiviral activity, its

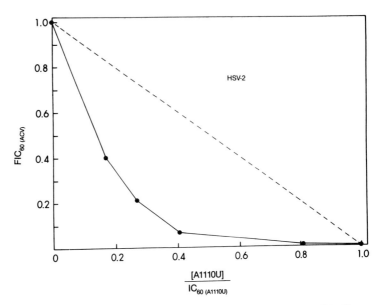

FIGURE 5   Synergistic inhibition of HSV-2 replication by A1110U and ACV. The concentration of each inhibitor was covaried and the inhibition of plaque formation assessed. The $FIC_{60(ACV)}$ is the ratio of the concentration of ACV required to inhibit plaque formation by 60% in the presence of a fixed concentration of A1110U to the concentration required in the absence of A1110U. The *x* axis is the ratio of the fixed concentration of A1110U to the concentration of A1110U that produced 60% inhibition of plaque reduction in the absence of ACV. IC_{60} values were used because > 50% inhibition was observed at some of the combination doses. The dashed line shows the theoretical plot for independent inhibitors. (Reproduced from Spector *et al.*, 1989, by permission.)

cellular toxicity was not augmented by ACV. These selective and potent activities *in vitro* prompted the following studies of the effects of the two agents on herpes infections in animals.

F. SYNERGISTIC THERAPY FOR
HERPES-INFECTED ANIMALS

Most studies of the efficacy of the combination therapy were conducted on nude (athymic) mice which are immunodeficient. This model has the advantage of supporting infections by the less pathogenic mutant virus strains that are resistant to ACV therapy. Mice were either infected on the snout (orofacial lesions) or on the back (zosteriform lesions). Both areas of infection resulted in the formation of severe lesions and fatality. Hairless, immunocompetent mice were also studied. However, the infections in these mice were less severe and self limiting. Finally, infected guinea pigs were also used, but resulted in labor-intensive studies. The data from these experiments are reported by Ellis *et al.* (1986, 1989) and by Lobe *et al.* (1987) or are unpublished data of M. N. Ellis, D. C. Lobe, and T. Spector. Studies with the nude mice are reviewed here.

Combination therapy was tested against wild type strains of HSV-1 and HSV-2 and several HSV-1 ACV-resistant mutant strains. A1110U was applied topically and ACV was either applied topically or administered in the drinking water. Statiscally significant synergistic therapy was demonstrated against both wild-type viruses and two resistant viruses. Graphic display of an experiment with nude mice is shown in Fig. 6. The partial effectiveness of the individual drugs was typically abrogated shortly after termination of treatment, and lesions developed in a delayed but parallel time-frame to the untreated controls. In contrast, the effectiveness of the combination therapy in some experiments was complete with neither detectable lesions nor virus during the 2 weeks posttreatment. In other experiments where synergy was demonstrated, reversal of the effectiveness of the combination therapy only occurred in a few of the mice or was significantly delayed.

Synergy was not demonstrable against the third ACV-resistant virus which expressed a DNA polymerase with diminished sensitivity to inhibition by ACVTP. A1110U by itself produced unusually potent efficacy against this strain, which left inadequate room for synergistic improvement. Although

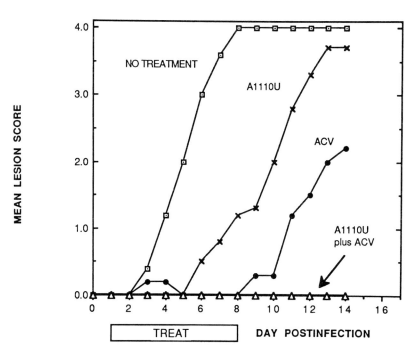

FIGURE 6   Synergistic therapy for HSV-1 infected mice. Snouts of athymic mice were infected with HSV-1 and treated 3 or 4 times per day with topical preparations of ACV, A1110U, or the combination of ACV and A1110U as indicated. (Constructed from data of Ellis *et al.*, 1989, by permission.)

combination therapy was very effective against this ACV-resistant strain, it was not statistically better than therapy with A1110U alone. Since A1110U was equally potent against every wild-type and mutant virus studied *in vitro* (IC_{50} = about 1.5 μM), it is not known why it was particularly effective against the polymerase-mutated virus *in vivo*.

The two ACV-resistant strains against which synergistic therapy was demonstrable were a strain partially deficient in thymidine kinase and a strain expressing a thymidine kinase with a diminished capacity to accept ACV as a substrate. The thymidine kinase deficient mutant is the most common resistant virus isolated from humans (reviewed by Larder and Darby, 1986).

Studies were performed to confirm that A1110U was actually producing the effects on dGTP pools that were expected for a ribonucleotide reductase inhibitor, and to ascertain whether the effects were adequate to account for the extensive synergy observed between ACV and A1110U.

G. METABOLIC PERTURBATIONS

The data in Table 2 show the effects of either ACV, A1110U, or the combination of the two antiviral agents on the deoxynucleoside triphosphate and ACVTP pools in HSV-1 infected cells. ACV, via inhibition of HSV DNA polymerase by ACVTP, caused the pool of dGTP to increase. Furthermore, as expected from inhibition of herpes ribonucleotide reductase, A1110U prevented this increase and even depressed the dGTP pool to levels below those of untreated infected cells. Moreover, A1110U had the bonus effect of promoting a large increase in the level of ACVTP. This double faceted effect

TABLE 2
Effects of A1110U on Deoxynucleoside Triphosphates and ACVTP Levels[a]

Treatment		Intracellular pools (pmol/10^6 cells)					Ratio of ACVTP/dGTP
A1110U	ACV	dCTP	dTTP	dATP	dGTP	ACVTP	
None	None	20	225	20	16	—	—
None	10 μM	21	240	82	68	2.6	0.038
2 μM	None	16	260	20	8.0	—	—
2 μM	10 μM	16	260	33	6.4	23	3.6

[a] Confluent cultures of Vero cells were infected with 10 PFU/cell of HSV-1 (Patton strain). After 1 hr incubation at 37°C, the media was replaced with fresh media and A1110U and [2-^{3}H] ACV where indicated. After an additional incubation of 8 hr, the cells were collected by centrifugation and extracted with 3.5% PCA. Supernatants were neutralized with KOH and treated with periodate and methylamine to remove ribonucleotides. Treated extracts were chromatographed on a Whatman Partisil PXS 10 column with a linear gradient of 0.3 M to 0.8 M KH$_2$PO$_4$, pH 3.5. Pools of deoxynucleotides were quantitated by their uv absorbances and ACVTP by its specific activity.

resulted in a > 90-fold increase in the ACVTP:dGTP ratio, which should greatly facilitate the binding of ACVTP to herpes DNA polymerase.

Although it is not definitively known why inhibition of ribonucleotide reductase causes ACVTP pools to increase, Karlsson and Harmenberg (1988) have reported pertinent data. A723U, another inhibitor of herpes ribo-nucleotide reductase and potentiator of ACV (Spector *et al.*, 1985, 1987), caused a decrease of thymidine excretion (in the presence or absence of ACV) into the media of HSV-infected cells. If this reflects a decrease in the intracellular pools of thymidine, then the ability of thymidine to competitively inhibit phosphorylation of ACV by HSV thymidine kinase would also be reduced. It can be further postulated that inhibition of ribonucleotide re-ductase should result in a decrease in the pool of deoxyuridine, which is another efficient substrate of HSV thymidine kinase (Cheng, 1976; Dobersen and Greer, 1978) that could also block the phosphorylation of ACV.

Examination of Table 2 reveals the third favorable aspect of the combined therapy. Since pyrimidine deoxynucleotides are more readily replenished by salvage pathways, inhibition of ribonucleotide reductase selectively depletes purine deoxynucleotides (reviewed by Reichard, 1988). The net effect is to decrease the amount of competing substrate (dGTP) while maintaining two of the other deoxynucleotides at levels sufficient to trap the polymerase in a dead-end complex with ACVMP-terminated primer and the next nucleotide (see Fig. 3).

H. CONCLUSIONS

A1110U, an inhibitor of HSV ribonucleotide reductase and potentiator of ACV, caused dGTP pools to decrease, the pool of ACVTP to increase, and had no net effect on the pyrimidine deoxynucleotide triphosphate pools. These three aspects of metabolic modulation, which ensure the binding of ACVTP to, and the subsequent trapping of, herpes DNA polymerase, prob-ably account for the synergy between ACV and A1110U. It is hoped that this type of combination therapy will be useful treatment for human disease.

II. Antagonism of 3′-Azido-3′-deoxythymidine (AZT) Antiviral Action by Ribavirin (RBV)

A. INTRODUCTION

This section describes the enzymatic basis by which RBV antagonizes the inhibition of HIV replication by AZT. Both of these compounds (Fig. 7) can act as analogs of common nucleosides. Otherwise, they have little structural

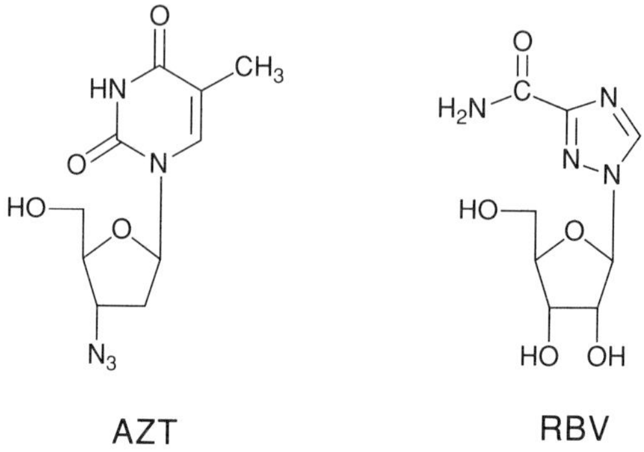

AZT   RBV

FIGURE 7   The structural formulas for the antiviral compounds 3'-azido-3'-deoxythymidine (AZT) and ribavirin (RBV).

similarity. AZT is metabolized as an analog of the pyrimidine, dThd, and RBV as one of several purine derivatives. The crystal structure of RBV resembles that of guanosine (Prusiner and Sundaralingam, 1973), but RBV may also mimic an adenine or inosine derivative as it interacts with purine metabolizing enzymes.

It is these interactions of each individual drug and any consequent effects on nucleotide levels that need to be understood before the results from the combination of the two drugs can be predicted. The following two subsections describe what is known about how each drug works separately.

B. MECHANISM OF ACTION OF AZT

The synthesis of AZT was first reported in 1964 (Horwitz et al., 1964) and was later used as an intermediate to synthesize 3'-amino nucleosides as potential antineoplastic agents (Lin and Prusoff, 1978). The first experiments involving AZT with a retrovirus were reported in 1978 (Krieg et al., 1978; Ostertag et al., 1978). AZT appeared to indirectly reduce the number of C-type particles but increased the number of A-type particles released in an endogenous Friend virus system. In the first reported (DeClerq et al., 1980) direct tests of antiviral activity, AZT was reported to be inactive against vaccinia, HSV-1, and vesicular stomatitis viruses. Shortly thereafter, it was discovered that AZT inhibited gram negative bacteria, so studies to delineate its metabolism and mode of inhibition were initiated. This activity was dependent on the conversion of AZT to its 5'-monophosphate derivative catalyzed by the bacterial dThd kinase. Subsequent to its conversion to AZTTP, it inhibited DNA replication (Elwell et al., 1987).

In 1984, AZT was studied as a potential anti-HIV agent by testing it against other murine retroviruses (Furman and Barry, 1988). It had potent activity, and thus was tested directly against HIV replication. Against this causative agent of AIDS, it was highly active (Mitsuya *et al.*, 1985). It has since been demonstrated to be a useful antiviral agent in patients infected with HIV (Fischl *et al.*, Creagh-Kirk *et al.*, 1988).

AZT enters cells by passive diffusion (Zimmerman *et al.*, 1987). The lipophilicity of the azido group appears to facilitate the passage of the drug across the blood–brain barrier where the concentration of AZT reached 75% of the plasma level (Yarchoan *et al.*, 1986). Since its entry into cells is independent of transporters, other nucleosides or nucleoside anaogs should not interfere with the uptake of AZT. Intracellular AZT is efficiently anabolized to form the monophosphate (AZTMP) by a reaction catalyzed by cytosolic dThd kinase (Furman *et al.*, 1986). AZT had an apparent K_m value similar to that for dThd (3.0 vs 2.9 μM) and a relative maximal velocity value only 40% lower. In human (H9) cells, anabolism of AZTMP to form the diphosphate, catalyzed by cytosolic dTMP kinase, was not nearly as efficient. The analog monophosphate appeared to compete well for the enzyme with an apparent K_m value of 9 μM vs 4 μM for thymidylate, but was a poor phosphate acceptor with a maximal velocity value of only 0.3% of that for thymidylate. Thus, the compound was an effective alternative-substrate inhibitor of dTMP kinase. The net metabolic effect was that AZT was readily anabolized to its monophosphate, which accumulated in cells; the AZT di- and triphosphate levels were much lower. Formation of all three of the AZT phosphate derivatives was dose dependent. The triphosphate of AZT (AZTTP) competed with dTTP to inhibit DNA synthesis catalyzed by HIV reverse transcriptase. Inhibition of this enzyme was 100- to 300-fold more potent that the inhibition of cellular polymerase α. If AZTMP were incorporated into DNA, the 3'-azido group of AZT would be expected to block chain elongation. Incorporation of AZTMP and termination of DNA elongation has been demonstrated with purified HIV reverse transcriptase (St. Clair *et al.*, 1987). In studies with an avian retrovirus (Olsen *et al.*, 1987), chain termination during viral DNA reverse transcription with purified reverse transcriptase was also observed. However, inhibition of virus replication in infected cells showed an absence of premature termination products. Inhibition of DNA synthesis (full length DNA product) correlated well with the expected effects from competitive inhibition of reverse transcription by the AZTTP and with the level of inhibition of virus production. Whether inhibition of reverse transcriptase or termination of DNA elongation is the predominant effect on HIV replication is not yet known.

The mechanism described here (Fig. 8) indicates that the enzymes involved in AZT anabolism and in reverse transription are crucial to the mechanism of antiviral activity of AZT. Metabolic changes that might affect the antiviral activity would be likely to involve these enzymes. It has been shown, for instance, that high levels of exogenously supplied dThd will reverse the antiviral effects of AZT (Mitsuya *et al.*, 1985). It is not likely that this reversal is caused by competition for entry into the cell since AZT and dThd enter

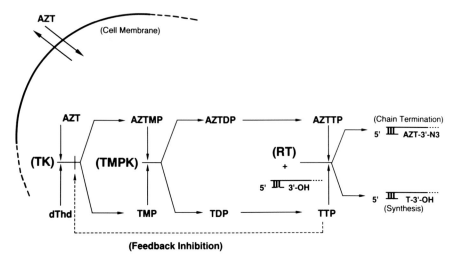

FIGURE 8   AZT anabolism and interference with reverse transcription. TK and TMPK are the abbreviations for the cellular dThd and thymidylate kinases and RT for the HIV reverse transcriptase.

cells by different pathways. However, competition between AZT and dThd for dThd kinase or between AZTMP and dTMP for dTMP kinase could slow the anabolism of AZT. In addition, accumulated dTTP, a known feedback inhibitor of dThd kinase (Cheng and Prusoff, 1974), could slow anabolism of AZT and also diminish the ability of AZTTP to compete with dTTP for binding to the HIV reverse transcriptase. Thus, inhibition of reverse transcription and possible incorporation into, and termination of, the reverse transcribed DNA would be expected to decrease. On the other hand, in cells treated only with AZT (no exogenous dThd) the concentration of dTTP was decreased (Furman et al., 1986; Frick et al., 1988), presumably by the inhibitory effect of AZTMP on the TMP kinase. This could serve to self-potentiate the action of AZT by decreasing the concentration of the competing substrate, dTTP, and thus provide a binding advantage at the reverse transcriptase catalytic site. This effect on the concentration of dTTP seems to be transitory and varies with cell type, however, and its importance is difficult to assess.

An understanding of the antiviral mechanism of the second drug under consideration, RBV, is less complete.

C. MECHANISM OF RBV ACTION

Ribavirin is a broad-spectrum antiviral agent that has shown clinical efficacy against influenza virus, respiratory syncytial virus, parainfluenza virus and Lasa fever (Gilbert and Knight, 1986). In 1984, McCormick *et al.* (1984) reported that RBV suppressed HIV virus replication in lymphocyte cultures. The lowest effective concentration of the drug, about 100–200 μM, was substantially higher than that required to inhibit the replication of the clinically susceptible RNA viruses mentioned earlier (Sidwell, 1980; Browne, 1981; Smee and Matthews, 1986). No mechanism of action has been delineated for RBV against HIV but studies have indicated at least three possible mechanisms for the inhibition of replication of the other viruses.

Experiments with the nucleoside transport inhibitor nitrobenzylthioinosine (NBMPR) indicate that RBV can enter cells by facilitated diffusion (Paterson *et al.*, 1979). Early studies with RBV have shown that in most cells RBV is readily converted to its mono-, di-, and triphosphate derivatives. The first step of the phosphorylation is catalyzed by adenosine kinase (Zimmerman and Deeprose, 1978; Miller *et al.*, 1979; Willis *et al.*, 1978). The monophosphate of RBV masquerades as an inosine analog and inhibits inosinate dehydrogenase (Streeter *et al.*, 1973), which catalyzes the first committed step in the transition from *de novo* purine ribonucleotide synthesis to guanylate synthesis. Inhibition of this enzyme appears to significantly affect concentrations of both ribonucleotide and deoxyribonucleoside triphosphates. The level of GTP decreases, but that of UTP and CTP increase (Zimmerman and Deeprose, 1978; Lowe *et al.*, 1977; Drach *et al.*, 1980). Of the deoxyribonucleotide concentrations, dGTP levels drop and dTTP levels increase substantially. These altered concentrations of nucleotides can be restored to normal by addition of guanosine or GMP (Lowe *et al.*, 1977).

Since influenza is an RNA virus, the drop in GTP levels was suspected to play a role in the mechanism of antiviral activity against this virus. However, although low concentrations of RBV produced a partial drop in GTP levels which correlated with repression of virus production, higher concentrations of RBV caused further reductions of virus production without further decreasing GTP levels (Wray *et al.*, 1985a). The implication of a second mechanism was substantiated (Eriksson *et al.*, 1977) by direct demonstration of inhibition of virus RNA polymerase by RBV triphosphate (RBVTP) and inhibition of primer generation and RNA elongation catalyzed with influenza virus capsids (Wray *et al.*, 1985b). Recently, a thorough study on a related RNA virus, reovirus, supported and extended these conclusions (Rankin *et al.*, 1989). It showed that synthesis of virus RNA and protein was inhibited to about the same extent as was virus yield. The primary target for this activity was the elongation step of plus-stranded messenger RNA synthesis. No effect on cap formation was detected. Interestingly, neither GTP nor the other ribonucleoside triphosphates reversed the inhibitory effect of RBVTP. These investigators proposed a model in which RBVTP inhibits the plus-strand formation (double-stranded to single-stranded synthesis) by binding to the viral RNA polymerase near the catalytic site and interfering with unwinding of the double-stranded RNA during the replicative process.

A third proposed mechanism of RBV action is based on an observation of its effect on capping of vaccinia virus RNA. Messenger RNA cap formation catalyzed by extracts of vaccinia virus cores was inhibited by RBVTP (Goswami *et al.*, 1979). This correlates with the inhibition of vaccinia virus replication by RBV (Sidwell *et al.*, 1972). Polio virus, on the other hand, lacks a 5'-cap on its genome and its replication is not significantly inhibited by RBV (Sidwell *et al.*, 1972). This tenuous correlation , however, breaks down in the case of reovirus (previous paragraph). Furthermore, whether host cell capping enzymes are subject to inhibition by RBVTP and, if they are, whether that inhibition can disrupt the replication of viruses such as HIV which rely on them, has not been demonstrated.

Data from studies (Frank *et al.*, 1987) with visna virus, which is of the same subfamily as HIV, may be somewhat more directly relevant to the considerations discussed here. Ribavirin inhibited the replication of visna virus, though only by about 50% at concentrations close to 1mM. The reverse transcriptase extracted from the virus particles was inhibited by RBVTP, but the inhibition was noncompetitive (vs dGTP) and only partial with 50% inhibition at about 250μM. Thus, marginal activity correlated roughly with marginal inhibition of reverse transcriptase. Similar studies with reverse transcriptase from HIV have not been reported.

Trying to predict possible effects of metabolites on the action of RBV, then, is problematic since the mechanism is ill defined. Perturbation of nucleotide pools, in and of itself, may be part of the mechanism of RBV inhibition of HIV. On the other hand, the effect of RBVTP on RNA elongation of influenza and reovirus would not seem to have a counterpart in the HIV-infected cell. In addition, RBVTP has no significant effect on eukaryotic RNA polymerase I or II, eukaryotic DNA polymerase α or β, or poly(A) polymerase (Muller *et al.*, 1977). If there is an effect of RBVTP on host cell capping of HIV RNA, perhaps levels of GTP would influence the extent of the effect. Finally, if RBVTP directly inhibits HIV reverse transcriptase, but in a noncompetitive manner (as with visna virus reverse transcriptase; see previous paragraph), perhaps the only significant metabolic perturbation would be one which interferes with anabolism of RBV to form its triphosphate derivative.

Experimentally, the major detectable interaction between AZT and RBV has been an effect of RBV on the activity of AZT. This is discussed in the next section.

D. EFFECT OF RBV ON AZT ANTIVIRAL ACTIVITY

Based on the general idea that if two drugs had independent modes of action their combined effect might be synergistic, it was anticipated that RBV might be synergistic with AZT. As indicated earlier, there was no evidence that the targets of the two drugs were similar. The observed effect of the combination, however, was that RBV *antagonized* the antiviral effect of AZT (Vogt *et al.*, 1987; Baba *et al.*, 1987).

The first metabolic effect noted was that in cells treated with RBV less AZTTP was formed (Vogt *et al.*, 1987). Thus it was reasonable to think that either RBV, one of its metabolites, or a nucleic acid precursor whose concentration was affected by RBV was interfering with the anabolism of AZT (Fig. 8). No direct effects were detected by RBV or its metabolites on either cytosolic dThd kinase or dTMP kinase, the anabolizing enzymes that catalyze the conversion of AZT to its mono- and diphosphate derivatives. This confirmed similar observations published previously (Drach *et al.*, 1980). However, it has been reported that RBV induced an increase in the concentration of dTTP in treated cells. Since dTTP is a feedback inhibitor of dThd kinase [competitive with ATP and noncompetitive with dThd (Cheng and Prusoff, 1974)], it was logical to compare the effects of RBV on dTTP levels and on AZT anabolism (Fyfe *et al.*, 1988). In cells (H9) that were treated with a subtoxic (30μM) concentration of RBV, a temporal inverse correlation between the intracellular concentration of dTTP the ability of the cells to phosphorylate AZT was observed (Fig. 9). This reciprocal relationship was also observed with a structurally unrelated compound, mycophenolic acid

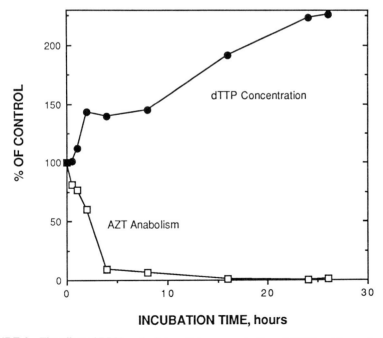

INCUBATION TIME, hours

FIGURE 9   The effect of RBV on the intracellular concentration of dTTP and the anabolism of AZT. RBV (30 μM) was added to a suspension of H9 cells at 0 hr. Samples were removed at the indicated times and assayed for the concentration of dTTP and the anabolism of AZT (1 hr accumulation spanning the indicated times).

(MPA), which also is known to induce an increase in the levels of dTTP (Lowe *et al.*, 1977). Furthermore, it was verified that GMP, which reverses the effects of RBV on dTTP pool sizes (Lowe *et al.*, 1977), also reversed the effect of RBV on AZT anabolism. To further establish a direct link between the proposed effect of dTTP on AZT phosphorylation, the dThd analog, 5'-amino-5'-deoxythymidine (5'-AT) was utilized. This compound directly interferes with dTTP feedback inhibition of dThd kinase (Fischer *et al.*, 1986). If RBV-induced increases of dTTP concentration were interfering with the phosphorylation of AZT, then the inhibition of AZT anabolism by RBV should be reversed by 5'-AT. In fact, this was observed. The 5'-AT completely reversed the RBV-induced inhibition of AZT anabolism and even stimulated it above that observed in the absence of RBV. This would be expected if dTTP levels in untreated cells are high enough to control the activity of the dThd kinase by feedback inhibition. Furthermore, since the dTTP levels remain elevated with 5'-AT treatment, it is evident that the predicted effect of 5'-AT is happening by the predicted mechanism.

To confirm the cellular metabolism experiments, both the feedback inhibition of dThd kinase, in this case with AZT as a substrate, and the reversal by 5'-AT were demonstrated with purified H9 cytosolic dThd kinase.

The adjustments of cellular metabolism in response to RBV and 5'-AT are depicted in Fig. 10. Since 5'-AT interacts specifically to block the feedback

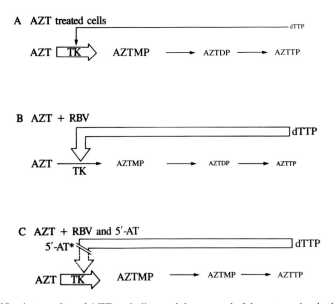

FIGURE 10   Antagonism of AZT anabolism and the reversal of the antagonism by 5'-amino-5'-deoxythymidine.

inhibition of dThd kinase by dTTP, the reversal of the RBV effect by 5'-AT supports the idea that the feedback inhibition by dTTP was the crucial cause of decreased anabolism.

The documented increase of dTTP concentration may also interfere with the AZT antiviral effect in a second way. In addition to the lower levels of AZTTP formed, the higher concentration of dTTP means that the AZTTP will be less successful at competing for binding to the reverse transcriptase, the target enzyme of the antiviral effect of AZT. Both the effect on anabolism of AZT and the effect on competition for the reverse transcriptase would be expected to play a role in the antagonism exhibited by RBV.

Recently, similar antagonism was also observed between RBV and another analog of dThd, 2',3'-dideoxythymidin-2'-ene (Baba *et al.*, 1987). This analog may also rely on dThd kinase for its anabolism (Balzarini *et al.*, 1989) and it is likely that its activity is antagonized by RBV by the same mechanism described for AZT. This additional example supports the expectation that any antiviral agents that rely on cytosolic dThd kinase for activation may well be antagonized by RBV.

E. CONCLUSIONS

These studies with RBV and AZT demonstrate the crucial role that perturbation of nucleotide levels can play in the antiviral activities of these drugs. This and other examples illustrate the point that information about associated metabolic effects may be as important as that about specific mechanisms of drug action in understanding and even predicting possible synergistic or antagonistic drug interactions.

III. Summary

This chapter describes attempts to potentiate the antiviral activities of ACV and AZT by combining each with a second therapeutic agent. Since the chemotherapeutic targets of the nucleoside analogs are viral nucleic acid polymerases, the analogs are prodrugs and require conversion to their respective triphosphate nucleoside. In both cases, the putative potentiating agent produced modulations of the cellular levels of physiological deoxynucleotides. However, synergy was observed in one situation, and antagonism in the other.

Combination therapy of ACV and A1110U for treatment of herpes infections was synergistic. A1110U, an inhibitor of herpes ribonucleotide reductase, had three positive effects: a decrease in the pool size of dGTP, the competing physiological polymerase substrate; an increase in the pool of

ACV triphosphate; and maintenance of the levels of noncompeting deoxynucleotides, which serve to trap the polymerase in an inhibited complex with ACVMP-terminated DNA.

In contrast, combination therapy of AZT and RBV for treatment of HIV infections was antagonistic. RBV caused the cellular pools of dTTP to increase, which had two deleterious effects. The phosphorylation of AZT catalyzed by thymidine kinase was diminished by enhanced feedback inhibition, and the ability of AZT triphosphate to effectively compete for binding to HIV reverse transcriptase was impaired.

It is apparent that each case of drug combination will, to some extent, be a special case. The combination of drugs with different modes of action does not assure that synergy will result. These results stress the need for understanding each agent's mechanism of action as well as its effects on the cellular metabolism to successfully predict the outcome of combination therapy.

References

Baba, M., Pauwels, R., Balzarini, J., Herdewijn, P., DeClercq, E., and Desmyter, J. (1987). Ribavirin antagonizes inhibitory effects of pyrimidine 2′,3′-dideoxynucleosides but enhances inhibitory effects of purine 2′,3′-dideoxynucleosides on replication of human immunodeficiency virus in vitro. *Antimicrob. Agents Chemother.* **31**, 1612–1617.

Balzarini, J., Herdewijn, P., and DeClercq, E. (1989). Differential patterns of intracellular metabolism of 2′,3′-didehydro-2′3′-dideoxythymidine and 3′-azido-2′,3′-dideoxythymidine, two potent anti-human immunodeficiency virus compounds. *J. Biol. Chem.* **264**, 6127–6133.

Browne, M. J. (1981). Comparative inhibition of influenza and parainfluenza virus replication by ribavirin in MDCK cells. *Antimicrob. Agents Chemother.* **19**, 712–715.

Cheng, Y.-C. (1976). Deoxythymidine kinase induced in HELA TK⁻ cells by herpes simplex virus type I and type II: Substrate specificity and kinetic behavior. *Biochim. Biophys. Acta* **452**, 370–381.

Cheng, Y.C., and Prusoff, W. H. (1974). Mouse ascites sarcoma 180 deoxythymidine kinase: General properties and inhibition studies. *Biochemistry* **13**, 1179–1185.

Coen, D. M., Goldstein, D. J., and Weller, S. K. (1989). Herpes simplex virus ribonucleotide reductase mutants are hypersensitive to acyclovir. *Antimicrob. Agents Chemother.* **33**, 1395–1399.

Creagh-Kirk, T., Doi, P., Andrews, E., Nusinoff-Lehrman, S., Tilson, H., Hoth, D., and Barry, D. W. (1988). Survival experience among patients with AIDS receiving zidovudine: Follow-up of patients in a compassionate plea program. *JAMA, J. Am. Med. Assoc.* **260**, 3009–3015.

DeClercq, E., Balzarini, J., Descamps, J., and Eckstein, F. (1980). Antiviral, antimetabolic and antineoplastic activities of 2′- or 3′-amino or -azido-substituted deoxyribonucleosides. *Biochem. Pharmacol.* **29**, 18490–1851.

Derse, D., Cheng, Y. C., Furman, P. A., St. Clair, M. H., and Elion, G. B. (1981). Inhibition of purified human and herpes simplex virus-induced DNA polymerases by 9-(2-hydroxyethoxymethyl)guanine triphosphosphate: Effects on primer–template function. *J. Biol. Chem.* **256**, 11447–11451.

Dobersen, M. J., and Greer, S. (1978). Herpes simplex virus type 2 induced prymidine nucleoside kinase: Enzymatic basis for the selective antiherpetic effect of 5-halogenated analogues of deoxycytidine. *Biochemistry* **17**, 920-928.

Drach, J. C., Barnett, J. W., Thomas, M. A., Smith, S. H., and Shipman, C., Jr. (1980). Inhibition of viral and cellular DNA synthesis by ribavirin. *In* "Ribavirin: A Broad-Spectrum Antiviral Agent" (R. A. Smith and W. Kirkpatrick, eds.), pp. 119–128. Academic Press, New York.

Elion, G. B., Furman, P. A., Fyfe, J. A., De Miranda, P., Beauchamp, L. M., and Schaeffer, H. J. (1977). Selectivity of the action of an antiherpetic agent, 9-(2-hydroxyethoxymethyl)-guanine. *Proc. Natl. Acad. Sci. U.S.A.* **74**, 5716–5720.

Ellis, M. N., Lobe, D. C., and Spector, T. (1986). Combination antiviral chemotherapy for treatment of HSV-1 infection in athymic mice. *Proc. ICAAC, 26th* p. 211 (abstr. 625).

Ellis, M. N., Lobe, D. C., and Spector, T. (1989). Synergistic therapy by acyclovir and A1110U for mice orofacially infected with herpes simplex virus. *Antimicrob. Agents Chemother.* **33**, 1691–1696.

Elwell, L. P., Ferone, R., Freeman, G. A., Fyfe, J. A., Hill, J. A., Ray, P. H., Richards, C. A., Singer, S. C., Knick, V. B., Rideout, J. L., and Zimmerman, T. P. (1987). Antibacterial activity and mechanism of action of 3′-azido-3′-deoxythymidine (BW A509U). *Antimicrob. Agents Chemother.* **31**, 274–280.

Eriksson, B., Helgstrand, E., Johansson, N. G., Larsson, A., Misiorny, A., Noren, J. O., Philipson, L., Stenberg, K., Stening, G., Stridh, S., and Oberg, B. (1977). Inhibition of influenza virus ribonucleic acid polymerase by ribavirin triphosphate. *Antimicrob. Agents Chemother.* **11**, 946–951.

Fischer, P. H., Vazquez-Padua, M. A., and Reznikoff, C. A. (1986). Perturbation of thymidine kinase regulation: A novel chemotherapeutic approach. *Adv. Enzyme Regul.* **25**, 21–34.

Fischl, M. A., Richman, D. D., Grieco, M. H., Gottlieb, M. S., Volberding, P. A., Laskin, O. L., Leedom, J. M., Groopman, J. E., Mildvan, D., Schooley, R. T., Jackson, G. C., Durack, D. T., King, D., and the AZT Collaborative Working Group (1987). The efficacy of azidothymidine (AZT) in the treatment of patients with AIDS and AIDS-related complex: A double-blind, placebo-controlled trial. *N. Engl. J. Med.* **317**, 185–191.

Frank, K. B., McKernan, P. A., Smith, R. A., and Smee, D. F. (1987). Visna virus as an in vitro model for human immunodeficiency virus and inhibition by ribavirin, phosphonoformate, and 2′,3′-dideoxynucleosides. *Antimicrob. Agents Chemother.* **31**, 1369–1374.

Frick, L. W., Nelson, D. J., St. Clair, M. H., Furman, P. A., and Krenitsky, T. A. (1988). Effects of 3′-azido-3′-deoxythymidine on the deoxynucleotide triphosphate pools of cultured human cells. *Biochem. Biophys. Res. Commun.* **154**, 124–129.

Furman, P. A., and Barry, D. W. (1988). Spectrum of antiviral activity and mechanism of action of zidovudine. *Am. J. Med.* **85** (Suppl. 2A), 176–181.

Furman, P. A., St. Clair, M. H., Fyfe, J. A., Rideout, J. L., Keller, P. M., and Elion, G. B. (1979). Inhibition of herpes simplex virus-induced DNA polymerase activity and viral DNA replication by 9-(2-hydroxyethoxymethyl)guanine and its triphosphate. *J. Virol.* **32**, 72–77.

Furman, P. A., Lambe, C. U., and Nelson, D. J. (1982). Effect of acyclovir on the deoxyribonucleoside triphosphate pool levels in Vero cells infected with HSV-1. *Am J. Med.* **73**, (Suppl. 1A), 14–17.

Furman, P. A., St. Clair, M. H., and Spector, T. (1984). Acyclovir triphosphate is a suicide inactivator of the herpes simplex virus DNA polymerase. *J. Biol. Chem.* **259**, 9575–9579.

Furman, P. A., Fyfe, J. A., St. Clair, M. H., Weinhold, K., Rideout, J. L., Freeman, G. A., Nusinoff-Lehrman, S., Bolognesi, D. P., Broder, S., Mitsuya, H., and Barry, D. W. (1986). Phosphorylation of 3′-azido-3′-deoxythymidine and selective interaction of the 5′-triphosphate with human immunodeficiency virus reverse transcriptase. *Proc. Natl. Acad. Sci. U.S.A.* **83**, 8333–8337.

Fyfe, J. A., Keller, P. M., Furman, P. A., Miller, R. L., and Elion, G. B. (1978). Thymidine kinase from herpes simplex virus phosphorylates the new antiviral compound, 9-(2-hydroxyethoxymethyl)guanine. *J. Biol. Chem.* **253**, 8721–8727.

Fyfe, J. A., Furman, P., Vogt, M., and Sherman, P. (1988). Mechanism of ribavirin antagonism of anti-HIV activity of azidothymidine. *Proc. Conf. AIDS, 4th* abstr. 3617.

Gibert, B. E., and Knight, V. (1986). Minireview: Biochemistry and clinical applications of ribavirin. *Antimicrob. Agents Chemother.* **30**, 201–205.

Goswami, B. B., Borek, E., Sharma, O. K., Fujitaki, J., and Smith, R. A. (1979). The broad spectrum antiviral agent ribavirin inhibits capping of mRNA. *Biochem. Biophys. Res. Commun.* **89**, 830–836.

Horwitz, J. P., Chua, J., and Noel, M. (1964). Nucleosides. V. The monomesylates of 1-(2'deoxy-β-D-lyxofuranosyl)thymine. *J. Org. Chem.* **29**, 2076–2078.

Hunting, D., and Henderson, J. F. (1982). Models of the regulation of ribonucleotide reductase and their evaluation in intact mammalian cells. *CRC Crit. Rev. Biochem.* **13**, 325–348.

Karlsson, A., and Harmenberg, J. (1988). Effects of ribonucleotide reductase inhibition on pyrimidine deoxynucleotide metabolism in acyclovir-treated cells infected with herpes simplex virus type 1. *Antimicrob. Agents Chemother.* **32**, 1100–1102.

Keller, P. M., McKee, S. A., and Fyfe, J. A. (1985). Cytoplasmic 5'-nucleotidase catalyzes acyclovir phosphorylation. *J. Biol. Chem.* **260**, 8664–8667.

Krieg, C. J., Ostertag, W., Clauss, U., Pragnell, I. B., Swetly, P., Roesler, G., and Weimann, B. J. (1978). Increase in intracisternal A-type particles in Friend cells during inhibition of Friend Virus (SFFV) release by interferon or azidothymidine. *Exp. Cell Res.* **116**, 21–29.

Larder, B. A., and Darby, G. (1986). Susceptibility to other antiherpes drugs of pathogenic variants of herpes simplex virus selected for resistance to acyclovir. *Antimicrob. Agents Chemother.* **29**, 894–898.

Lin, T.-S., and Prusoff, W. H. (1978). Synthesis and biological activity of several amino analogues of thymidine. *J. Med. Chem.* **21**, 109–112.

Lobe, D. C., Spector, T., and Ellis, M. N. (1987). Evaluation of combination antiviral chemotherapy using the "zosteriform rash" back model in HSV-2 infected athymic mice. *Proc. ICAAC, 27th* abstr. 1157.

Lowe, J. K., Brox, L., and Henderson, J. F. (1977). Consequences of inhibition of guanine nucleotide synthesis by mycophenolic acid and Virazole. *Cancer Res.* **37**, 736–743.

McCormick, J. B., Getchell, J. P., Mitchell, S. W., and Hicks, D. R. (1984). Ribavirin suppresses replication of lymphadenopathy-associated virus in cultures of human adult T lymphocytes. *Lancet* **2**, 1367–1369.

Miller, W. H., and Miller, R. L. (1980). Phosphorylation of acyclovir (acycloguanosine) monophosphate by GMP kinase. *J. Biol. Chem.* **255**, 7204–7207.

Miller, W. H., and Miller, R. L. (1982). Phosphorylation of acyclovir diphosphate by cellular enzymes. *Biochem. Pharmacol.* **31**, 3879–3884.

Miller, R. L., Adamczyk, D. L., Miller, W. H., Koszalka, G. W., Rideout, J. L., Beacham, L. M., Chao, E. Y., Haggerty, J. J., Krenitsky, T. A., and Elion, G. B. (1979). Adenosine kinase from rabbit liver: II. Substrate and inhibitor specificity. *J. Biol. Chem.* **254**, 2346–2352.

Mitsuya, H., Weinhold, K. J., Furman, P. A., St. Clair, M. H., Nusinoff-Lehrman S., Gallo, R. C., Bolognesi, D., Barry, D. W., and Broder, S. (1985). 3'-Azido-3'-deoxythymidine (BWA509U): An antiviral agent that inhibits the infectivity and cytopathic effect of human T-lymphotropic virus type III/lymphadenopathy-associated virus *in vitro*. *Proc. Natl. Acad. Sci. U.S.A.* **82**, 7096–7100.

Moore, E. C., and Sartorelli, A. C. (1984). Inhibition of ribonucleotide reductase by α-(N)-heterocyclic carboxaldehyde thiosemicarbazones. *Pharmacol. Ther.* **24**, 439–447.

Muller, W. E. G., Maidhof, A., Taschner, H., and Zahn, R. K. (1977). Virazole (1-β-D-Ribofuranosyl-1,2,4-triazole-3-carboxamide); a cytostatic agent. *Biochem. Pharmacol.* **26**, 1071–1075.

Olsen, J. C., Furman, P., Fyfe, J. A., and Swanstrom, R. (1987). 3'-Azido-3'-deoxythymidine inhibits the replication of avian leukosis virus. *J. Virol.* **61**, 2800–2806.

Ostertag, W., Krieg, C. J., Cole, T., Pragnell, I. B., Swetly, P., Weimann, B. J., and Dube, S. K. (1978). Induction of intracisternal A-type particles during Friend cell differentiation. *Exp. Cell Res.* **116**, 31–37.

Paterson, A. R. P., Yang, S.-E., Lau, E. Y., and Cass, C. E. (1979). Low specificity of the nucleoside transport mechanism of RPMI 6410 cells. *Mol. Pharmacol.* **16**, 900–908.

Porter, D. J. T., Harrington, J. A., and Spector, T. (1990). Herpes simplex type I ribonucleotide reductase: Selective inactivation by A1110U and its iron complex. *Biochem. Pharmacol.* **39**, 639–646.

Prusiner, P., and Sundaralingam, M. (1973). A new class of synthetic nucleoside analogues with broad-spectrum antiviral properties. *Nature (London)* **244**, 116–118.

Rankin, J. T., Jr., Eppes, S. B., Antczak, J. B., and Joklik, W. K. (1989). Studies on the mechanism of the antiviral activity of ribavirin against reovirus. *Virology* **168**, 147–158.

Reardon, J. E. (1989). Herpes simplex virus type I and human DNA polymerase interactions with 2′-deoxyguanosine 5′-triphosphate analogs: Kinetics of incorporation and induction of inhibition. *J. Biol. Chem.* **264**, 19039–19044.

Reardon, J. E., and Spector, T. (1989). Herpes simplex virus type I DNA polymerase: Mechanism of inhibition by acyclovir triphosphate. *J. Biol. Chem.* **264**, 7405–7411.

Reichard, P. (1988) Interactions between deoxyribonucleotide and DNA synthesis. *Annu. Rev. Biochem.* **57**, 349–374.

Reid, R., Mar, E.-C., Huang, E.-S., and Topal, M. D. (1988). Insertion and extension of acyclic, dideoxy, and Ara nucleotides by herpesviridae, human α and human β polymerases. *J. Biol. Chem.* **263**, 3898–3904.

Shipman, C., Jr., Smith, S. H., Drach, J. C., and Klayman, D. L. (1981). Antiviral activity of 2-acetylpyridine thiosemicarbazones against herpes simplex virus. *Antimicrob. Agents Chemother.* **19**, 682–685.

Shipman, C., Jr., Smith, S. H., Drach, J. C., and Klayman, D. L. (1986). Thiosemicarbazones of 2-acetylpyridine, 2-acetylquinoline, 1-acetylisoquinoline, and related compounds as inhibitors of herpes simplex virus *in vitro* and in a cutaneous herpes guinea pig model. *Antiviral Res.* **6**, 197–222.

Sidwell, R. W. (1980). Ribavirin: *In vitro* antiviral activity. *In* "Ribavirin: A Broad-Spectrum Antiviral Agent" (R. A. Smith and W. Kirkpatrick, eds.), pp. 23–42. Academic Press, New York.

Sidwell, R. W., Huffman, J. H., Khare, G. P., Allen, L. B., Witkowski, J. T., and Robins, R. K. (1972). Broad-spectrum antiviral activity of Virazole: 1-β-D-Ribofuranosyl-1,2,4-triazole-3-carboxamide. *Science* **177**, 705–706.

Smee, D. F., and Matthews, T. R. (1986). Metabolism of ribavirin in respiratory syncytial virus-infected and uninfected cells. *Antimicrob. Agents Chemother.* **30**, 117–121.

Spector, T. (1985). Inhibition of ribonucleotide reductases encoded by herpes simplex viruses. *Pharmacol. Ther.* **31**, 295–302.

Spector, T. (1989). Ribonucleotide reductases encoded by herpes viruses: Inhibitors and chemotherapeutic considerations. *In* "Inhibitors of Rinonucleotide Diphosphate Reductase Activity" (J. G. Cory and A. H. Cory, eds.). Sect. 128, pp. 235–243. Pergamon, New York.

Spector, T., Averett, D. R., Nelson, D. J., Lambe, C. U., Morrison, R. W., Jr., St. Clair, M. H., and Furman, P. A. (1985). Potentiation of antiherpetic activity of acyclovir by ribonucleotide reductase inhibition. *Proc. Natl. Acad. Sci. U.S.A.* **82**, 4254–4257.

Spector, T., Stonehuerner, J. G., Biron, K., Averett, D. R., and Furman, P. A. (1987). Ribonucleotide reductases induced by varicella zoster virus: Characterization, and potentiation of acyclovir by its inhibition. *Biochem. Pharmacol.* **36**, 4341–4346.

Spector, T., Harrington, J. A., Morrison, R. W., Jr., Lambe, C. U., Nelson, D. J., Averett, D. R., Biron, K., and Furman, P. A. (1989). 2-Acetylpyridine 5-[(dimethylamino)thiocarbonyl]thiocarbonohydrazone (A1110U), a potent inactivator of ribonucleotide reductases of herpes simplex and varicella zoster viruses and a potentiator of acyclovir. *Proc. Natl. Acad. Sci. U.S.A.* **86**, 1051–1055.

St. Clair, M. H., Richards, C. A., Spector, T., Weinhold, K. J., Miller, W. H., Langlois, A. J., and Furman, P. A. (1987). 3'-Azido-3'deoxythymidine triphosphate as an inhibitor and substrate of purified human immunodeficiency virus reverse transcriptase. *Antimicrob. Agents Chemother.* **31,** 1972–1977.

Streeter, D. G., Witkowski, J. T., Khare, G. P., Sidwell, R. W., Bauer, R. J., Robins, R. K., and Simon, L. N. (1973). Mechanism of action of 1-β-D-ribofuranosyl-1,2,4-triazole-3-carboxamide (Virazole), a new broad-spectrum antiviral agent. *Proc. Natl. Acad. Sci. U.S.A.* **70,** 1174–1178.

Thelander, L., and Reichard, P. (1979). Reduction of ribonucleotides. *Annu. Rev. Biochem.* **48,** 133–158.

Turk, S. R., Shipman, C., Jr., and Drach, J.-C. (1986). Selective inhibition of herpes simplex virus ribonucleoside diphosphate reductase by derivatives of 2-acetylpyridine thiosemicarbazone. *Biochem. Pharmacol.* **35,** 1539–1545.

Vogt, M. W., Hartshorn, K. L., Furman, P. A., Chou, T.-C., Fyfe, J A., Coleman, L. A., Crumpacker, C., Schooley, R. T., and Hirsch, M. S. (1987). Ribavirin antagonizes the effect of azidothymidine on HIV replication. *Science* **235,** 1376–1379.

Willis, R. C., Carson, D. A., and Seegmiller, J. E. (1978). Adenosine kinase initiates the major route of ribavirin activation in a cultured human cell line. *Proc. Natl. Acad. Sci. U.S.A.* **75,** 3042–3044.

Wray, S. K., Gilbert, B. E., Noall, M. W., and Knight, V. (1985a). Mode of action of ribavirin: Effect of nucleotide pool alterations on influenza virus ribonucleoprotein synthesis. *Antiviral Res.* **5,** 29–37.

Wray, S. K., Gilbert, B. E., and Knight, V. (1985b). Effect of ribavirin triphosphate on primer generation and elongation during influenza virus transcription *in vitro. Antiviral Res.* **5,** 39–48.

Yarchoan, R., Klecker, R. W., Weinhold, K. J., Markham, P. D., Lyerly, H. K., Durack, D. T., Gelmann, E., Lehrman, S. N., Blum, R. M., Barry, D. W., Shearer, G. M., Fischl, M. A., Mitsuya, H., Gallo, R. C., Collins, J. M., Bolognesi, D. P., Myers, C. E., and Broder, S. (1986). Administration of 3'-azido-3'-deoxythymidine, an inhibitor of HTLV-III/LAV replication, to patients with AIDS or AIDS-related complex. *Lancet* **1,** 575–580.

Zimmerman, T. P., and Deeprose, R. D. (1978). Metabolism of 5-amino-1-β-D-ribofuranosylimidazole-4-carboxamide and related five-membered heterocycles to 5'-triphosphates in human blood and L5178Y cells. *Biochem. Pharmacol.* **27,** 709–716.

Zimmerman, T. P., Mahony, W. B., and Prus, K. L. (1987). 3'-Azido-3'-deoxythymidine: An unusual nucleoside analogue that permeates the membrane of human erythrocytes and lymphocytes by nonfacilitated diffusion. *J. Biol. Chem.* **262,** 5748–5754.

CHAPTER 9

Reversal of Multidrug Resistance in Tumor Cells

Chia-Ping Huang Yang
Lee M. Greenberger
Susan Band Horwitz
*Departments of Molecular Pharmacology
and Cell Biology
Albert Einstein College of Medicine
Bronx, New York 10461*

I. Introduction

Effective cancer chemotherapy can be severely limited by the presence of drug resistant cells within a tumor population. Some malignant tumors are intrinsically resistant to standard antineoplastic agents, while other tumors respond initially to chemotherapy and then relapse after acquiring resistance

SYNERGISM AND ANTAGONISM IN CHEMOTHERAPY
Copyright © 1991 by Academic Press, Inc.
All rights of reproduction in any form reserved.

to a wide variety of drugs. To understand the mechanisms of intrinsic and acquired multidrug resistance (MDR) and to develop strategies for overcoming chemotherapeutic insensitivity, drug resistant mammalian cell cultures have been used as model systems. These multidrug-resistant cell lines have been established by growing cells in the presence of stepwise increments of various antitumor drugs, such as the *vinca* alkaloids or anthracyclines. Many of these cell lines display the MDR phenotype, which includes (1) cross resistance to structurally and functionally unrelated drugs, (2) a net decrease in drug accumulation, (3) the overproduction of a plasma membrane glycoprotein, known as P-glycoprotein, (4) amplification and/or transcriptional activation of the *mdr* gene, and (5) reversal of the MDR phenotype by a variety of different compounds. An example of such an agent is verapamil, which is being studied extensively as a reversal agent.

II. Review of the Multidrug Resistance Phenotype

A. CROSS RESISTANCE

Cell lines exhibiting the MDR phenotype have been selected for their ability to replicate in the presence of a single cytotoxic agent. However, they display cross resistance to a wide variety of unrelated compounds. Drugs associated with MDR include the *vinca* alkaloids, anthracyclines, taxol, colchicine, epipodophyllotoxins, and actinomycin D. Although these drugs have different cellular targets and inhibit diverse cellular activities such as microtubule assembly, DNA and RNA synthesis, protein synthesis, and topoisomerase II activity, they have a common characteristic, i.e., they are all hydrophobic and many are of natural origin.

In general, cells display their greatest resistance against the agent with which they were selected, although exceptions exist (Tsuruo *et al.*, 1986). In a typical example, various MDR cell lines, such as J7.V1-1, J7.C1-100, and J7.T1-50, derived from the mouse macrophage-like cell line J774.2 demonstrated their greatest resistance against vinblastine, colchicine, and taxol, respectively, the drugs against which each was selected. They also displayed cross resistance, although at a much reduced level, to many other drugs (Roy and Horwitz, 1985). It is not completely understood why MDR cells in general display more resistance to the agents against which they were selected. Various MDR cells derived from J774.2 also demonstrated a small amount (2- to 5-fold) of collateral sensitivity to bleomycin (Roy and Horwitz, 1985). The basis for the collateral sensitivity may be related to the hydrophilic nature of bleomycin. Collateral sensitivity has also been observed with local anesthetics, certain steroids, and detergents (Riehm and Biedler, 1972; Bech-Hansen *et al.*, 1976).

B. REDUCED DRUG ACCUMULATION

MDR cells avoid drug cytotoxicity by maintaining the intracellular drug concentration at an extremely low level (see Riordan and Ling, 1985; Endicott and Ling, 1989). The steady state level of drug associated with the cell is a combination of several factors: (1) influx of drug through the plasma membrane, (2) efflux of drug from cells, (3) binding of drug to membranes and intracellular targets, and (4) sequestration of drug in acidic cellular compartments such as lysosomes. Since hydrophobic compounds absorb easily to cell membranes, the initial rate of drug influx is difficult to measure. Reduced drug entry which has been reported for MDR cells (Dano, 1973; Beck *et al.*, 1983) probably represents an experimental artifact (see van der Bliek and Borst, 1989). However, drug binding to plasma membranes of MDR cells has been demonstrated in many cell lines (see Section IIIA3). There is also evidence that the concentration of drug is reduced at cellular targets in MDR cells compared with sensitive cells (Willingham *et al.*, 1986; Masurovsky and Horwitz, 1989). Although drug may be distributed to distinct cellular compartments when it is initially loaded into cells, it can be redistributed when cells are placed into a drug-free medium for efflux experiments. The overall efflux is enhanced in MDR cells compared with drug sensitive cells (see van der Bliek and Borst, 1989). In addition, the efflux of drugs in MDR cells is an energy-dependent process since enhanced efflux in MDR cells is abolished by glucose deprivation and by addition of metabolic inhibitors such as sodium azide and dinitrophenol (Dano, 1973; Skovsgaard, 1978; Inaba *et al.*, 1979). Therefore, it is generally agreed that drug efflux is the major determinant of the low steady-state level of drug in MDR cells.

C. OVERPRODUCTION OF P-GLYCOPROTEIN

A 130–180 kDa membrane protein known as P-glycoprotein is overproduced in MDR cells (see Endicott and Ling, 1989; Gottesman and Pastan, 1988; Horwitz *et al.*, 1988). Transfection of sensitive cells with the mouse or human *mdr* gene confers the multidrug resistance phenotype, strongly suggesting that P-glycoprotein plays an important role in MDR (Gros *et al.*, 1986a; Deuchars *et al.*, 1987; Ueda *et al.*, 1987). P-glycoprotein has been hypothesized to be an energy-dependent drug efflux dump with broad specificity for hydrophobic compounds, based on the following observations: (1) strong structural homology with bacterial transport proteins (Gros *et al.*,

1986b; Chen *et al.*, 1986; Gerlach *et al.*, 1986), (2) the presence of two nucleotide binding sites (Gros *et al.*, 1986b; Chen *et al.*, 1986), (3) the binding of photoactive analogs of *vinca* alkaloids, anthracycline, and ATP to P-glycoprotein (Safa *et al.*, 1986; Cornwell *et al.*, 1986b, 1987b; Busche *et al.*, 1989), (4) the ATP-dependence of the uptake of vinblastine by membrane vesicles isolated from MDR cells (Horio *et al.*, 1988; Naito *et al.*, 1988), and (5) the energy dependence of drug accumulation and efflux in MDR cells (see Endicott and Ling, 1989; van der Bliek and Borst, 1989). Based on hydropathy plot analysis, it was predicted that the P-glycoprotein molecule spans the membrane 12 times. The protein is actually composed of a single motif repeated twice; each motif is composed of six membrane-spanning domains followed by an ATP-binding consensus region. P-glycoprotein shares various amounts of homology with several other proteins likely to be involved in transport, such as the yeast STE 6 gene product (McGrath and Varshavsky, 1989), the cystic fibrosis transmembrane conductance regulator (CFTR) (Riordan *et al.*, 1989), and a putative gene product from an amplified gene (*pfmdr*) in chloroquine-resistant *Plasmodium falciparum* (Wilson *et al.*, 1989; Foote *et al.*, 1989). Adenylate cyclase also contains this motif (Krupinski *et al.*, 1989).

P-glycoprotein is encoded by a small gene family comprising three *mdr* genes in rodents and two *mdr* genes in humans (Hsu *et al.*, 1989, 1990; Ng *et al.*, 1989; Chin *et al.*, 1989). An examination of the P-glycoprotein precursors in several independently isolated murine J774.2 cell lines which were selected for resistance to either colchicine (J7.C1-100), vinblastine (J7.V1-1 or J7.V3-1), or taxol (J7.T1-50) indicated that these cell lines synthesized either a 120- or 125-kDa precursor, except for one cell line that was selected with taxol and produced both precursors simultaneously (Roy and Horwitz, 1985; Greenberger *et al.*, 1987). These two distinct P-glycoprotein precursors are the products of different members of the *mdr* gene family. Whereas J7.C1-100 and J7.V1-1 overproduced mainly the *mdr*1b gene products, J7.V3-1 overproduced mainly the *mdr*1a gene products, and J7.T1-50 overproduced both gene products. Overexpression of the *mdr*1a or the *mdr*1b gene was found to correlate with the differential overproduction of either the 120- or the 125-kDa P-glycoprotein precursor, respectively (Greenberger *et al.*, 1988; Hsu *et al.*, 1989).

P-glycoprotein is not limited to MDR tumor cells. Both *mdr* gene expression and P-glycoprotein have been found in normal tissues, such as human liver, kidney, colon, adrenal gland, and placenta (Fojo *et al.*, 1987; Sugawara *et al.*, 1988; Hitchins *et al.*, 1988). Except for the adrenals, where P-glycoprotein is dispersed throughout the organ, it is found predominantly at the apical border of cells lining the lumen (Thiebaut *et al.*, 1987). P-glycoprotein has also been detected in human endothelial cells at blood–brain barrier sites (Cordon-Cardo *et al.*, 1989). Interestingly, Arceci *et al.* (1988) have demonstrated that P-glycoprotein is overproduced during pregnancy at the luminal surface of the endometrium in the mouse. Further, the analysis of the *mdr* multigene family in mouse tissues indicates a distinct pattern of tissue-specific expression (Croop *et al.*, 1989). These findings suggest that P-glycoprotein has a physiological role in normal tissues, and that each *mdr* family member may have a specialized transport function. However, the endogenous substrates have not been identified. The location of P-glycoprotein in tissues involved in excretion suggests that, at least in part, P-glycoprotein may participate in the removal of metabolites or xenobiotics (Thiebaut *et al.*, 1987).

D. GENE AMPLIFICATION AND/OR TRANSCRIPTIONAL ACTIVATION OF *mdr* GENES

MDR in eukaryotic cells is often accompanied by both chromosomal and extrachromosomal changes. The presence of homogeneously staining regions (HSRs) and double-minute chromosomes (DMs) suggested that gene amplification occurred in MDR cells (see Horwitz *et al.*, 1988). In a colchicine-resistant MDR variant, J7.C1-20, derived from J774.2 cells, there was a reasonable correlation between the average number of DMs per cell, the quantity of P-glycoprotein, and the level of drug resistance (Lothstein and Horwitz, 1986). However, DNA amplification is not the only mechanism involved in overexpression of *mdr* genes. In some cases, particularly in low-level drug-resistant cell lines, overexpression is associated primarily with transcriptional activation of *mdr* genes (Shen *et al.*, 1986; Bradley *et al.*, 1989). Specific hybridization probes that have been derived from three classes of mouse P-glycoprotein cDNAs have revealed the differential amplification and/or transcriptional activation of three distinct but closely related *mdr* genes—*mdr*1a, *mdr*1b, and *mdr*2—in mouse (Hsu *et al.*, 1989). Recent evidence has indicated that the P-glycoproteins encoded by *mdr*1a and *mdr*1b are functionally distinct (Lothstein *et al.*, 1989).

III. Reversal of Multidrug Resistance Phenotype

The MDR phenotype can be reversed by diverse hydrophobic agents (see Fig. 1). The most effective reversal agents reported include verapamil (Tsuruo *et al.*, 1981, 1983a; Rogan *et al.*, 1984; Beck *et al.*, 1986; Willingham *et al.*, 1986), progesterone (Yang *et al.*, 1989), reserpine (Inaba *et al.*, 1981; Beck *et al.*, 1988; Akiyama *et al.*, 1988), and the cyclosporines (Slater *et al.*, 1986; Twentyman, 1988; Hait *et al.*, 1989). Many other agents, such as tamoxifen

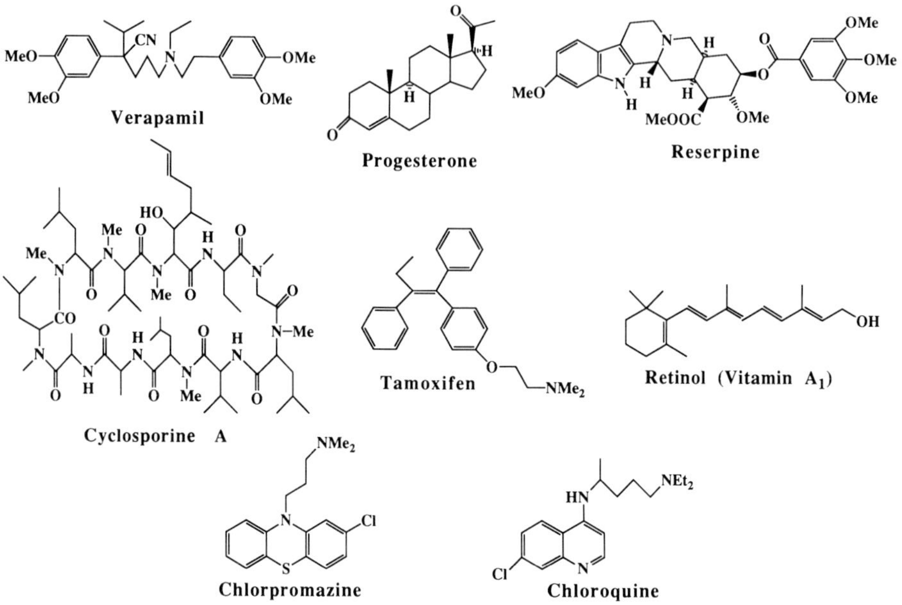

FIGURE 1   Structures of reversal agents.

(Ramu *et al.*, 1984b), vitamin A (Nogae *et al.*, 1987), derivatives of phenothiazines such as chlorpromazine (Ford *et al.*, 1989), and lysosomotropic agents such as chloroquine (Shiraishi *et al.*, 1986; Klohs and Steinkampf, 1988) have also been reported to reverse MDR.

A. METHODS FOR ANALYSIS

Several methods have been used to study the effect of reversing agents on MDR. These include (1) measurement of cell viability, (2) measurement of steady-state drug accumulation, and (3) photoaffinity labeling of P-glycoprotein.

1. Cell Viability

The ability of drug-sensitive and MDR cells to grow in the presence of cytotoxic drugs is usually assessed by measuring colony formation after drug exposure or by determining cell replication after growth in the presence of drug for 2–3 days. In this way, the ED_{50}, the concentration of drug that inhibits cell division by 50% after a specific time, is determined. A putative reversal agent is tested in the presence of a cytotoxic agent to determine its effect on the ED_{50} of the cytotoxic drug. A decrease in ED_{50} suggests that the compound may be a reversal agent. Since some reversal agents may be toxic, it is important to determine the effect of the reversal agent on the growth of both sensitive and resistant cells in the absence of the antitumor drug.

2. Steady-State Level of Drug Accumulation

The steady-state level of drug can be measured by incubating drug-sensitive or MDR cells with labeled drug, for example, [^{3}H]vinblastine, then washing the cells free of drug adhering to the outside of the cell, and measuring the amount of cell-associated drug using either a filtration or centrifugation method. In general, drug accumulation is much reduced in highly resistant MDR cells compared with drug-sensitive cells (see Section IIB). A reversal agent is most likely capable of enhancing drug sensitivity by maintaining a high intracellular drug concentration by inhibiting drug efflux. Therefore, measurement of drug accumulation in the presence and absence of a test compound can determine if the compound is a reversal agent. In addition, the extent of enhancement of drug accumulation usually reflects the effectiveness of the reversal agent.

In addition to radiolabeled drug, accumulation of fluorescent drugs, such as daunomycin, can be measured with a fluorometer or fluorescence microscope. The latter method, which involves the measurement of inherent fluorescence of daunomycin, allows the examination of drug accumulation and drug efflux in a single cell (Willingham *et al.*, 1986). In MDR cells derived from a human KB cell line, the single-cell measurements, performed in the presence and absence of glucose, indicated that diminished daunomycin accumulation in MDR cells resulted from accelerated energy-dependent efflux across the plasma membrane (Willingham *et al.*, 1986). This efflux was inhibited by verapamil and quinidine, compounds known to reverse MDR.

3. Photoaffinity Labeling of P-Glycoprotein

It was demonstrated that [^{3}H]vinblastine bound specifically to membrane vesicles prepared from MDR cells of human KB lines and that this binding was inhibited by reversal agents such as verapamil (Cornwell *et al.*, 1986a). However, this binding was reversible and did not demonstrate a direct interaction between P-glycoprotein and drug. Experiments utilizing photoactive analogs of [^{3}H]vinblastine, [^{3}H]N-(p-azidobenzoyl)-N'-(β-aminoethyl)-vindesine (NABV) and ^{125}I-labeled N-(p-azidosalicyl)-N'-(β-aminoethyl)-vindesine (NASV), to specifically photolabel P-glycoprotein presented the first evidence that antitumor drugs bound specifically to the P-glycoprotein molecule (Safa *et al.*, 1986; Cornwell *et al.*, 1986b).

Several other photoactive compounds have been synthesized and used to specifically photolabel P-glycoprotein. These include photoactive analogs of calcium channel blockers, such as azidopine (Safa *et al.*, 1987; Yang *et al.*, 1988) and verapamil derivatives (Safa, 1988; Yusa and Tsuruo, 1989); a doxorubicin derivative, iodomycin (Busche *et al.*, 1989); colchicine analogs [³H]*N*-(*p*-azidobenzoyl)aminohexanoyldeacetylcolchicine (NABC) and ¹²⁵I-labeled *N*-(*p*-azidosalicyl)aminohexanoyldeacetylcolchicine (NASC) (Safa *et al.*, 1989b); the α-adrenergic receptor photoaffinity probe [¹²⁵I]iodo aryl azidoprazosin (Greenberger *et al.*, 1989); [³H]cyclosporine diazirine analog (Foxwell *et al.*, 1989); and the synthetic isoprenoid ¹²⁵I-labeled *N*-solanesyl-*N*,*N*′-*bis*(3,4,-dimethoxybenzyl) ethylenediamine (SDB) photoanalog (Akiyama *et al.*, 1989). Each photoaffinity probe contains an arylazido group with the exception of iodomycin and the cyclosporine photoanalog. [³H]azidopine and [¹²⁵I]iodo aryl azidoprazosin (see Fig. 2) have the advantage of being commercially available, relatively stable in the absence of ultraviolet light, and capable of forming covalent bonds via nitrene intermediates upon ultraviolet light irradiation. Analysis of the tryptic peptides from the [³H]azidopine-labeled P-glycoprotein (encoded by the mouse *mdr*1b gene) using reverse phase HPLC indicated that this calcium antagonist has at least one specific binding site on the P-glycoprotein molecule (Yang *et al.*, 1988).

All of the photoactive probes reported are selectively inhibited by vinblastine > doxorubicin > colchicine, suggesting that these agents interact with P-glycoprotein in a similar manner. Consistent with this hypothesis, peptides derived from P-glycoprotein labeled with [³H]azidopine and [¹²⁵I]iodo aryl azidoprazosin are identical. The common binding domain is no larger than approximately 6 kDa (Greenberger *et al.*, 1990).

Immunological mapping studies with azidopine-labeled P-glycoprotein encoded by the human *mdr*1 gene indicated that azidopine bound to two

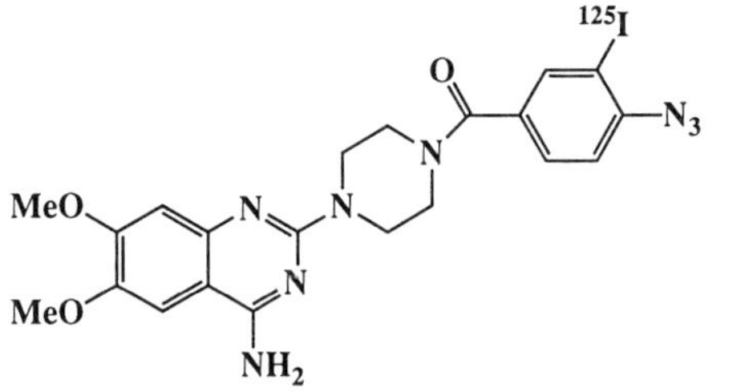

Iodoarylazidoprazosin

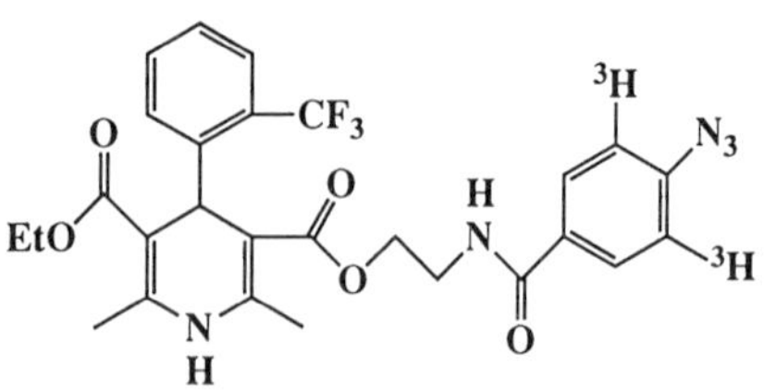

Azidopine

FIGURE 2 Structures of photoaffinity probes for P-glycoprotein.

different domains in P-glycoprotein. A major photolabeled domain was found in putative transmembrane regions 7 through 12, while a minor domain was located within putative transmembrane regions 1 through 6 (Bruggemann *et al.*, 1989; Yoshimura *et al.*, 1989). In the mouse MDR cell line J7.V3-1, which overproduces the *mdr*1a gene product, two classes of vinblastine binding sites have been suggested (Yang *et al.*, 1990) by kinetic analysis. It is not clear if these two classes of vinblastine binding sites are reflected in the two photoaffinity binding domains.

Verapamil, progesterone, reserpine, and several other compounds have been reported to interact with P-glycoprotein by demonstrating that photoaffinity labeling of P-glycoprotein was inhibited by the addition of these agents (Cornwell *et al.*, 1987a; Safa *et al.*, 1987; Akiyama *et al.*, 1988; Beck *et al.*, 1988; Yang *et al.*, 1988, 1989; Nogae *et al.*, 1989). Therefore [^{3}H]azidopine, for example, can be used as a first screen to select compounds that may reverse MDR (Yang *et al.*, 1989). However, it should be noted that good inhibitors of photolabeling need not necessarily be good reversal agents. For example, hydrocortisone has been shown to be a good inhibitor of [^{3}H]azidopine photoaffinity labeling of P-glycoprotein, but it does not increase the accumulation of vinblastine in MDR cells (Yang *et al.*, 1989).

B. CALCIUM CHANNEL BLOCKERS

Verapamil was the first calcium antagonist that was shown to reverse MDR by increasing intracellular drug levels (Tsuruo *et al.*, 1981, 1983a,b; Rogan *et al.*, 1984; Ramu *et al.*, 1984a; Kessel and Wilberding, 1985; Beck *et al.*, 1986; Willingham *et al.*, 1986) although its effects vary among different cell lines (see Endicott and Ling, 1989). However, calcium metabolism is not involved in the reversion of MDR (see Section IV). Verapamil specifically inhibits both the reversible binding of vinblastine to membrane vesicles isolated from MDR cells and the photoaffinity labeling of vinblastine analogs to P-glycoprotein, thereby suggesting that it interacts with P-glycoprotein (Cornwell *et al.*, 1986a, 1987a; Safa *et al.*, 1987; Yang *et al.*, 1988). Verapamil also reverses chloroquine resistance in *Plasmodium falciparum* (Martin *et al.*, 1987).

The World Health Organization has subdivided the calcium antagonists into six types (Vanhoutte and Paoletti, 1987): verapamil-like (I), nifedipine-like (II), diltiazem-like (III), flunarizine-like (IV), prenylamine-like (V), and others such as perhexiline (VI). The effects of this series of drugs on [^{3}H]azidopine photoaffinity labeling in drug-resistant cell lines derived from the murine cell line J774.2 have been examined. Types I, II, III, and IV inhibit the labeling of P-glycoprotein in three drug-resistant cell lines, J7.V1-1, J7.C1-100, and J7,T1-50. Types V, VI, and bepridil have stimulatory effects of [^{3}H]azidopine labeling of P-glycoprotein in J7.V1-1 and J7.C1-100 cells. In J7.T1-50 cells, an unusual MDR cell line that produces two P-glycoproteins that differ by approximately 10 kDa, these latter compounds have a stimulatory effect on the photoaffinity labeling of the upper P-glycoprotein band but not on the lower band as assessed by sodium dodecyl-sulfate-polyacrylamide gel electrophoresis (SDS-PAGE). In addition, the effects of prenylamine, nicardipine, and vinblastine on [3H]azidopine photo-affinity labeling of P-glycoprotein in J7.V1-1, J7.V3-1, and J7.T1-50 cells, which overproduce the *mdr*1b, *mdr*1a, and both *mdr*1a and *mdr*1b gene products, respectively, have been compared (Table 1). In J7.T1-50 cells the upper band is the product of *mdr*1b whereas the lower band is the product of *mdr*1a (Greenberger *et al.*, 1988; Hsu *et al.*, 1989). Interestingly, it was found that prenylamine enhanced [3H]azidopine photolabeling of P-glycoprotein from both J7.V1-1 cells and J7.T1-50 cells (upper P-glycoprotein band) by about 80%, while it had a slight inhibitory effect (20–30%) on the photolabel-ing of P-glycoprotein from J7.V3-1 cells and J7.T1-50 cells (lower P-glycoprotein band). However, both nicardipine and vinblastine inhibited the photolabeling of all P-glycoproteins. Therefore, prenylamine, a blocker for fast-open calcium channels, can be used to differentiate two *mdr* gene products.

Although the effects of nicardipine and prenylamine on [3H]azidopine photoaffinity labeling of P-glycoprotein encoded by *mdr*1b (e.g., J7.V1-1 cells) were opposite, both drugs enhanced the sensitivity of J7.V1-1 cells to vinblastine (Yang *et al.*, 1988). Among the six types of calcium channel blockers, Type I (verapamil, desmethoxyverapamil), Type II (diltiazem), Type III (nicardipine, nifedipine), Type V (prenylamine), Type VI (perhex-iline maleate), and bepridil have been reported to reverse MDR (Tsuruo

TABLE 1

Effects of Prenylamine, Nicardipine, and Vinblastine on [3H]Azidopine Photoaffinity Labeling of P-Glycoprotein[a]

| Cell line | Disintegrations/min/mg protein | | | |
	Control	Prenylamine	Nicardipine	Vinblastine
J774.2	252	—	—	—
J7.V1–1	3271 (100)	5757 (176)	133 (4)	547 (17)
J7.V3–1	1435 (100)	1031 (72)	119 (8)	215 (15)
J7.T1–50 (U)	1123 (100)	2070 (184)	33 (3)	180 (16)
J7.T1–50 (L)	878 (100)	710 (81)	30 (3)	141 (16)

[a] Microsomal membranes from J774.2, J7.V1–1, J7.V3–1, and J7.T1–50 were prepared and photoaffinity labeled with 50 nM [3H]azidopine. Proteins (60 μg) were analyzed by gel electrophoresis as described (Yang *et al.*, 1988), to quantitate the amount of [3H]azidopine labeling, P-glycoprotein was excised from the gels and counted. Labeling was done in the absence or presence of a 1000-fold molar excess (50 μM) of prenylamine, nicardipine, or vinblastine. U, upper band; L, lower band.

et al., 1981; Ramu *et al.*, 1984a; Cornwell *et al.*, 1987a; Sriram *et al.*, 1987; Yang *et al.*, 1988).

C. STEROIDS

During pregnancy in the mouse, *mdr* gene expression (*mdr*1b specifically) is induced significantly in the endometrium of the uterus (Arceci *et al.*, 1988). Due to this observation and an interest in endogenous substrates for P-glycoprotein, the [^{3}H]azidopine photoaffinity labeling assay was used to screen steroids, specifically those for which concentration is altered during pregnancy (Yang *et al.*, 1989a). Several steroids, notably progesterone, were found to inhibit [^{3}H]azidopine photoaffinity labeling of P-glycoprotein from both the mouse gravid uterus and MDR J774.2 cells. In MDR cells, particularly those that overproduced the *mdr*1b gene product, progesterone also (1) inhibited vinblastine binding to membrane vesicles, (2) enhanced vinblastine accumulation in cells, (3) blocked efflux of vinblastine from cells, and (4) increased the sensitivity of MDR cells to vinblastine. The inhibitory effect of steroids on [^{3}H]azidopine labeling of P-glycoprotein and on vinblastine binding to membrane vesicles prepared from MDR cells correlated with their effect on vinblastine accumulation in MDR cells. Moreover, the effect of steroids on vinblastine accumulation was proportional to the hydrophobicity of the steroids (Fig. 3). Progesterone was the most hydrophobic steroid that was tested and it had the greatest inhibitory effect on [^{3}H]azidopine photoaffinity labeling of P-glycoprotein (Yang *et al.*, 1989). The inhibitory effect of progesterone on drug binding was also observed in Adriamycin-resistant K562 cells (Naito *et al.*, 1989).

In J7.V1-1 cells, progesterone and verapamil (1–10 μM) were equally effective in increasing the sensitivity of cells to vinblastine. In this concentration range, both compounds were equally effective in inhibiting [^{3}H]azidopine photoaffinity labeling of P-glycoprotein as well as enhancing vinblastine accumulation in cells. However, progesterone, but not verapamil, has more potent effects on cell lines overproducing the *mdr*1b gene product than on those overproducing the *mdr*1a gene product (Yang *et al.*, 1990).

D. RESERPINE AND RELATED DRUGS

Several classes of compounds which include indole alkaloids (reserpine, vindoline, catharanthine, yohimbine), quinolines (quinine), acridines (quinacrine, acridine), and chloroquine (a lysosomotropic agent) are capable of enhancing drug cytotoxicity in MDR cells derived from the human T-cell leukemia line, CEM (Beck *et al.*, 1988). Quinacrine also reverses vincristine resistance in a murine P388 leukemia line (Inaba and Maruyama, 1988). The

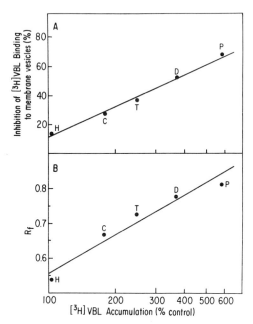

FIGURE 3   Correlation of the accumulation of [³H]vinblastine with percentage inhibition of [³H]vinblastine binding to membrane vesicles (A) and R_f values for steroids (B). Steady state level of [³H]vinblastine was measured in J7.V1-1 cells. Cells were incubated with 1 μM [³H]vinblastine for 1 hr at 37°C. Specific [³H]vinblastine binding to J7.V1-1 membrane vesicles was measured using a filtration method. R_f values for steroids were determined using thin layer chromatography. Steroids were dissolved in methanol and run vertically in a solvent system consisting of chloroform:methanol (1:5.7). H, hydrocortisone; C, corticosterone; T, testosterone; D, deoxycorticosterone; P, progesterone. From Yang et al., 1989a, with permission.

physical properties common to these agents are lipid solubility, cationic charge at physiological pH, and similar molar refractivities (Zamora et al., 1988). Among these compounds, reserpine, chloroquine, quinacrine, vindoline, and catharanthine, each of which enhanced cytotoxicity in CEM/VBL cells, also inhibited ¹²⁵I-labeled NASV labeling of P-glycoprotein. However, neither quinine nor yohimbine inhibited this labeling. In MDR human KB carcinoma cells, several agents that suppressed the multidrug resistance phenotype, including cepharanthine, a bisbenzylisoquinoline alkaloid (Shiraishi et al., 1987), quinidine, and reserpine, were found to inhibit photolabeling of P-glycoprotein by ¹²⁵I-labeled NASV at doses comparable to those that (reversed multidrug resistance (Akiyama et al., 1988). Zamora et al., (1988) suggested that a common "pharmacophore"—the minimum set of structural and functional features required for modulator-binding to P-glycoprotein—contains a basic nitrogen atom and two planar aromatic rings. Reserpine, a potent modulator of antineoplastic drug cytotoxicity in MDR cells and a strong inhibitor of binding of a photoactive vinblastine analog to P-glycoprotein, does contain these substructural domains. A series of reserpine and yohimbine analogs that varied in the spatial orientation of these domains were tested for their ability to reverse MDR and to compete with ^{125}I-labeled NASV for binding to P-glycoprotein. It was found that compounds that retained the pendant benzoyl function in an appropriate spatial orientation all modulated MDR. Therefore, it was proposed that the relative disposition of aromatic rings and basic nitrogen atoms is important for modulators of P-glycoprotein activity (Pearce *et al.*, 1989). Whether the "pharmacophore" hypothesis is valid in all classes of compounds that reverse MDR needs to be further examined.

E. PHENOTHIAZINES AND RELATED COMPOUNDS

Phenothiazines and structurally related antipsychotics inhibit the activity of a multifunctional calcium binding protein, calmodulin. Phenothiazines and other calmodulin antagonists possess antiproliferative and cytotoxic effects that are proportional to their anticalmodulin activity (see Ford *et al.*, 1989). Among calmodulin inhibitors, trifluoperazine and clomipramine (antidepressants) and prenylamine (a coronary vasodilator) were shown to increase accumulation of vincristine and Adriamycin in MDR P388 leukemia cells (Tsuruo *et al.*, 1982). In human cancer cells three phenothiazines—thioridazine, chlorpromazine, and trifluoperazine—reversed MDR (Akiyama *et al.*, 1988) but did not inhibit ^{125}I-labeled NASV labeling effectively at concentrations at which they reversed resistance. Therefore, these agents may act in different ways than verapamil, which is known to bind specifically to P-glycoprotein and to inhibit drug-binding to P-glycoprotein competitively (Naito and Tsuruo, 1989). These compounds may interact with lipid (Shiraishi *et al.*, 1987) and change the environment in which P-glycoprotein resides and functions. Alternatively, these compounds may influence the transport of drug by interacting with P-glycoprotein at a site different from the one to which antitumor agents bind. Ford *et al.*, (1989) studied the structure–activity relationship of 30 phenothiazines and related compounds on MDR reversion in human breast cancer cells. Compounds with an increased ability to reverse MDR had (1) substitutions on the phenothiazine ring that increased hydrophobicity, (2) a 4-carbon bridge and a piperazinyl amine, and (3) tertiary amines instead of secondary or primary amines. Structure–activity relationships suggest that an ideal phenothiazine structure for reversing MDR has a hydrophobic nucleus with a -CF3 ring substitution at position 1 connected by a 4-carbon alkyl bridge to *p*-methyl-substituted piperazinylamine. Further substitution studies indicate that (*trans*) flupenthixol is the most active of the compounds that enhances the potency of doxorubicin as a cytotoxic agent against MDR cells. This relatively nontoxic compound increases the accumulation of doxorubicin in MDR cells, but not in sensitive cells, suggesting that it modulates MDR by interacting with a specific cellular target in these resistant cells. Examination of whether photoaffinity labeling of P-glycoprotein is influenced by (*trans*)flupenthixol would help to clarify if the specific cellular target of this compound is P-glycoprotein.

F. CYCLOSPORINES

Cyclosporine A, a hydrophobic 11-amino acid cyclic polypeptide, is a unique immunosuppressant that acts by inhibiting an early stage of T-lymphocyte activation. Slater *et al.* (1986) demonstrated that MDR Ehrlich ascites carcinoma cells were sensitized to daunomycin in the presence of cyclosporine A. A series of non- or minimally-immunosuppressive analogs of cyclosporine A were examined in the MDR human small-cell lung cancer cell line NCI-H69, and several of these analogs, such as W8-032 and B3-243, were found to be highly effective modifiers of resistance to Adriamycin and vincristine (Twentyman, 1988). It was concluded that the immunosuppressive activity of cyclosporines and their ability to modulate drug resistance were unrelated. More recently, Hait *et al.* (1989) demonstrated that 11-methyl-leucine cyclosporine (11-met-leu CsA), a nonimmunosuppressive analog of cyclosporine, is as effective as the parent drug in increasing the level of antitumor drugs in MDR cells. This sensitization by cyclosporine was independent of its effects on cyclophilin (a cytosolic cyclosporine A binding protein), on calmodulin, or on protein kinase C (Hait *et al.*, 1989). Although it was suggested (Hait *et al.*, 1989) that P-glycoprotein was not the site of action of cyclosporine A, since there were no differences in its binding to MDR cells compared with sensitive cells, Safa *et al.* (1989a) demonstrated that P-glycoprotein labeling with ^{125}I-labeled NASV was inhibited in a dose-dependent manner by cyclosporine A and its nonimmunosuppressive analogs. Recently, Foxwell *et al.* (1989) demonstrated, using a [^{3}H]cyclosporine diazirine analog, that cyclosporine bound directly to P-glycoprotein and that this binding was not dependent on the immunosuppressive potential of the cyclosporine derivative. Furthermore, kinetic analysis indicated that cyclosporine A competitively inhibited high affinity ATP-dependent vincristine binding to the plasma membrane of MDR Chinese hamster lung cells, DC-3F/VCRd-5L (Safa *et al.*, 1989a). Therefore, it was suggested that cyclosporines modulated *vinca* alkaloid resistance through an interaction with P-glycoprotein. The effects of cyclosporine A on [^{3}H]azidopine photoaffinity labeling of P-glycoprotein in both J7.V1-1 and J7.V3-1 cells, which overproduced *mdr*1b and *mdr*1a gene products, respectively, were also examined (Fig. 4). Cyclosporine A at a 1000-fold molar excess (50 μM) inhibited photolabeling by 70–85% in membranes from both cell lines. This inhibitory effect was comparable to that caused by verapamil in both cell lines and by progesterone in J7.V1-1 cells (Yang *et al.*, 1988, 1989).

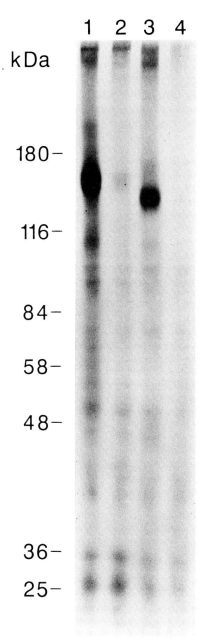

FIGURE 4 Effect of cyclosporine A on [³H]azidopine photoaffinity labeling of membrane proteins from J7.V1-1 and J7.V3-1 cells. Membranes prepared from J7.V1-1 (lanes 1 and 2) and J7.V3-1 cells (lanes 3 and 4) were photolabeled using 50 nM [³H]azidopine. Photolabeled membranes were analyzed by SDS-PAGE on 5–12% gradient gels. Lanes 1 and 3, [³H]azidopine labeling of membranes. Lanes 2 and 4, [³H]azidopine labeling of membranes in the presence of 1000-fold molar excess (50μM) of cyclosporine A.

G. OTHER COMPOUNDS

Many other classes of compounds have been reported to reverse MDR. These include drug analogs such as *N*-acetyl daunorubicin (Skovsgaard, 1980), antiestrogens such as tamoxifen (Ramu *et al.*, 1984b), other triphenylethylene compounds such as toremifene (DeGregorio *et al.*, 1989), vitamin A (Nogae *et al.*, 1987), synthetic isoprenoids such as SDB-ethylenediamine (Nakagawa *et al.*, 1986), antiarrhythmic agents such as amidarone (Chauffert *et al.*, 1987), secretory inhibitors such as monensin (Klohs and Steinkampf, 1988), and cytochalasin B (Tsuruo and Iida, 1986). In addition, compounds

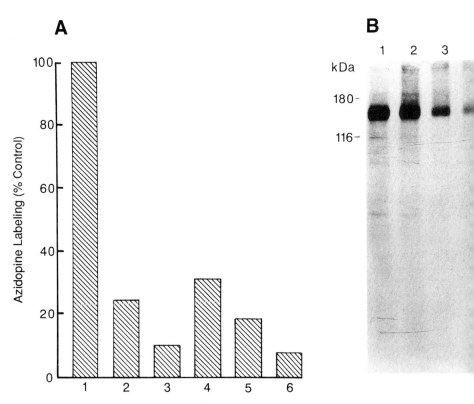

FIGURE 5   Effect of tamoxifen on [³H]azidopine photoaffinity labeling of membrane proteins from J7.V1-1 cells. (A) Membranes prepared from J7.V1-1 were photolabeled using 50 nM [³H]azidopine in the absence of drugs (lane 1) or presence of 1000-fold molar excess (50 μM) of tamoxifen (lane 2), vinblastine (lane 3), verapamil (lane 4), nifedipine (lane 5), and vitamin A (lane 6). Photolabeled membranes were analyzed by SDS-PAGE. To quantitate the amount of [³H]azidopine labeling, P-glycoprotein was excised from the gels, extracted with 90% Beckman tissue solubilizer, and counted. (B) Membranes were photolabeled using 50 nM [³H]azidopine in the absence of tamoxifen (lane 1) or the presence of 250-, 500-, and 1000-fold molar excess of tamoxifen (lanes 2, 3, and 4, respectively). Photolabeled membranes were analyzed by SDS-PAGE and fluorography.

such as dipyridamol (Ramu et al., 1984c), histidinol (Warrington and Fang, 1989), ACH-52 [a partial analog of nifedipine (Shinoda et al., 1989)], and erythromycin (Hofsli and Nissen-Meyer, 1989) have been demonstrated to increase sensitivity to antitumor drugs in MDR cells. The effects of tamoxifen and vitamin A on P-glycoprotein in MDR cells derived from J774.2 cells were examined (Fig. 5). It was found that 4-hydroxytamoxifen and retinyl acetate inhibited [³H]azidopine photoaffinity labeling by 76 and 88%, respectively, suggesting that both compounds interacted with P-glycoprotein. The effect of 4-hydroxytamoxifen on drug-sensitive and vinblastine-resistant J774.2 cells were also studied and it was found that tamoxifen (1) increased vinblastine accumulation in MDR J7.V1-1 cells (Fig. 6A); (2) decreased vinblastine accumulation in drug-sensitive J774.2 cells (Fig. 6B); and (3) enhanced sensitivity of J7.V1-1 cells to vinblastine. The mechanism by which tamoxifen decreases vinblastine in drug-sensitive cells is not understood.

It has been demonstrated that P-glycoprotein is present in rat liver canalicular membrane vesicles where it is likely to function as an ATP-driven efflux pump (Kamimoto et al., 1989). A possible function of P-glycoprotein in these membranes may be related to the transport of bile acids (Yang et al., 1989). Taurocheno-, glycocheno-, and taurolithocholic acid produced dose-dependent inhibition of daunomycin transport by canalicular membrane vesicles (Mazzanti et al., 1989). In vinblastine-resistant MDR J7.V1-1 cells, the inhibitory effect of several bile acids (dehydro-, litho-, and chenodeoxycholic

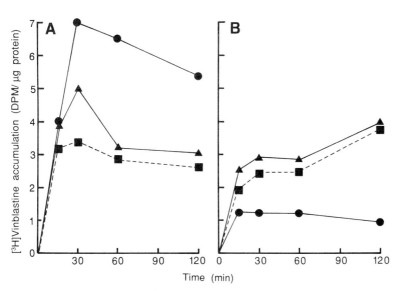

FIGURE 6   Effect of tamoxifen on [³H]vinblastine accumulation in drug sensitive and vinblastine-resistant cells. J774.2 (A) and J7.V1-1 (B) cells were incubated with 1 μM [³H]vinblastine at 37°C in the absence (●) or presence of 50 μM 4-hydroxytamoxifen (4-OH-T, ▲) or tamoxifen (T, ■) and the steady-state level of [³H]vinblastine measured. Data are expressed as dpm/μg protein and each value is the mean of 3 replicates.

acid and cholic acid methyl ester) on vinblastine binding to membrane vesicles correlated with their effect on the enhancement of vinblastine accumulation in whole cells (data not shown). It appeared that the hydrophobicity of bile acids was related to the inhibition of drug binding.

Furthermore, fluorescent dyes which stained various intracellular structures in sensitive cells, did not stain or only weakly stained MDR cells (Neyfakh, 1988). The efflux of the dyes was energy dependent (Lalande *et al.*, 1981), indicating that these fluorescent dyes were pumped out of MDR cells by a mechanism similar to the transport of antitumor drugs. It is reasonable to propose that such fluorescent dyes could increase the level of antitumor drugs in resistant cells.

IV. Mechanisms Involved in Reversion of Multidrug Resistance

Although P-glycoprotein is believed to transport structurally diverse hydrophobic molecules, the exact mechanism involved in this glycoprotein-mediated drug transport is not well understood. Earlier studies demonstrated that (1) calcium channel blockers and calmodulin inhibitors reversed MDR (see Sections IIC and IIE (2) there was a higher level of Ca^{2+} in MDR cells than in sensitive cells (Tsuruo *et al.*, 1984), and (3) a calcium binding protein (sorcin) was present in some MDR lines (Meyers and Biedler, 1981; Horwitz *et al.*, 1988). Therefore, it was suggested that the MDR phenotype was related to calcium metabolism. However, several findings indicated that calcium metabolism was not the mechanism involved in reversion of MDR. (1) There was no evidence for a voltage-gated calcium channel in an MDR variant of a human T-cell leukemia line CCRF-CEM (Lee *et al.*, 1988). (2) There was no correlation between calcium channel blocking activity and reversion of MDR among analogs of the dihydropyridine class of calcium channel blockers (Kamiwatari *et al.*, 1989). (3) Ca^{2+} was not required for vincristine binding (Naito and Tsuruo, 1989).

Although it is true that reversal agents interact with P-glycoprotein and inhibit the efflux of antitumor agents, the exact site(s) of interaction between antitumor drugs and reversal agents is not clear. Since some antitumor agents inhibited photolabeling of P-glycoprotein more strongly than others, it was proposed that different classes of compounds might bind to overlapping or distinct sites in the P-glycoprotein molecule. Alternatively, there may be a common binding site for all drugs, but the affinity of individual drugs for this site could differ. The results reported by Greenberger *et al.* (1990) support the latter possibility since [^{3}H]azidopine and [^{125}I]iodo aryl azidoprazosin bind to a common domain in P-glycoprotein. However, this finding does not exclude the possibility that (1) there is diversity in the drug binding site within this domain and (2) these drugs could interact with P-glycoprotein at distinct sites in a manner that does not allow photochemical linkage to occur. Clearly, an understanding of the conformation of the drug binding site, along with detailed mapping studies, will be needed to resolve these possibilities. If a single site is responsible for drug interaction, it would explain why (1) a variety of agents known to reverse MDR competitively inhibited *vinca* alkaloids from binding to resistant cell membranes (Naito and Tsuruo, 1989; Safa *et al.*, 1989a; Yang *et al.*, 1990) and (2) photoactive analogs of antitumor drugs or reversal agents used so far (azidopine, verapamil, vinblastine, colchicine, daunomycin, and iodo aryl azidoprazosin) were preferentially competed by vinblastine > doxorubicin > colchicine, and this same preferential order was observed when [^{3}H]vinblastine-binding to membrane vesicles was measured (Cornwell *et al.*, 1986a). Presumably the drug binding site(s) have hydrophobic pockets to which both antitumor drugs and reversing agents bind. The affinity of binding is strongly influenced by the hydrophobicity of the molecule. The more hydrophobic the reversing agent, the more strongly it competes for the drug binding site and prevents antitumor drugs from being pumped out of MDR cells. Since P-glycoprotein is embedded in the membrane, it would not be surprising to learn that the hydrophobic pocket contains a membrane spanning domain(s).

However, several reversing agents, such as verapamil, have been shown to modulate the MDR phenotype not only by inhibiting drug binding to P-glycoprotein, but also by affecting the function of P-glycoprotein or mechanisms unrelated to P-glycoprotein. For example, (1) verapamil increased ATP consumption in MDR cells to a greater extent than some drugs which induced the MDR phenotype, and the concentration required for maximum activity was much lower (1 μM) than that required for inhibiting drug efflux (Broxterman *et al.*, 1988); (2) verapamil stimulated ATPase activity associated with P-glycoprotein (Hamada and Tsuruo, 1988); (3) verapamil caused an increase in the phosphorylation of P-glycoprotein (Hamada *et al.*, 1987); (4) verapamil inhibited the activity of protein kinase C, whose stimulation may play a role in drug transport in MDR cells (Fine *et al.*, 1988); and (5) a number of MDR cells were collaterally sensitive to verapamil and yet maintained a lower intracellular verapamil concentration than the sensitive cells (Cano-Gauci and Riordan,1988). These effects caused by verapamil and related reversing agents may give insight into the normal physiological functions of P-glycoprotein.

V. Clinical Implications

Although increased expression of P-glycoprotein can be detected in some leukemias and solid tumors (Goldstein *et al.*, 1989), a strict correlation between the presence of P-glycoprotein in tumor cells and nonresponse to chemotherapy has not been established. If the presence of P-glycoprotein proves to have prognostic value, as in the case of children with soft tissue sarcoma (Chan *et al.*, 1990), then chemotherapeutic strategies, based on the characteristics and function of P-glycoprotein, could be designed. In addition, antibodies specific for P-glycoprotein are being evaluated for clinical and diagnostic use.

Verapamil is a potent reversing agent in MDR cells. However, it has not been widely applied in the treatment of cancer since the dose required to reverse resistance in tissue culture is either toxic in people or not clinically achievable (see Endicott and Ling, 1989). It has been reported that a mean verapamil concentration of 0.45 μM was achieved with continuous iv infusion (Benson *et al.*, 1985). However, micromolar concentrations of verapamil are needed to reverse MDR. DMDP [N(3,4-dimethoxyphenethyl)-N-methyl-2-(2-naphthyl)-m-dithiane-2-propylamine hydrochloride], a calcium channel blocker, is less toxic *in vivo* and more effective in restoring cellular drug concentration than verapamil. It has been reported that this tiapamil analog, when injected into mice harboring Adriamycin-resistant cells, can restore accumulation of doxorubicin in resistant cells at achievable nontoxic plasma concentrations (Yin *et al.*, 1989). There is evidence that administration of verapamil with chemotherapeutic drugs may partially circumvent drug resistance in B-cell neoplasms in humans (Dalton *et al.*, 1989).

An effort to synthesize analogs of dihydropyridine calcium channel blockers which do not have calcium channel blocking activity but have strong reversal activity may be beneficial (Kamiwatari *et al.*, 1989). Other reversal agents which cause fewer side effects and therefore may be better tolerated are being considered as alternatives. Examples are cyclosporine A analogs that are not immunosuppressive (see Section IIF) or progesterone analogs. Megastrol acetate, like progesterone, has a strong inhibitory effect on [^{3}H]azidopine photoaffinity labeling of P-glycoprotein (Fig. 7A). Both megastrol acetate and medroxyprogesterone acetate increased [^{3}H]vinblastine accumulation in cells 4-fold (Fig. 7B), an effect comparable to that of progesterone. Since megastrol acetate and medroxyprogesterone acetate have been used at high doses in the treatment of breast cancer, application of these two drugs to the problem of drug resistance should be considered.

Recently Tsuruo *et al.* reported that a new fluorine-containing anthracycline derivative, ME2303, showed excellent antitumor activity against various experimental tumor models. This compound alone was effective against drug resistant human and murine tumor cells. Other studies by Coley *et al.* (1989) used several analogs of Adriamycin with 9-alkyl substituents and/or specific changes in the sugar residue. The cells were less resistant to these analogs than to Adriamycin. When these analogs were used in combination with relatively low concentrations of verapamil or cyclosporine A, MDR was circumvented.

In general, several factors should be considered when designing a good reversal agent for chemotherapy: (1) The concentration of reversal agent required should be nontoxic and clinically achievable. (2) Since P-glyco-

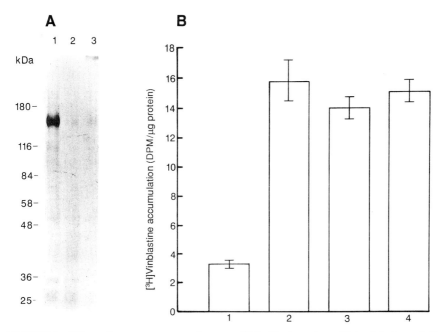

FIGURE 7 Effect of progesterone analogs on [^{3}H]azidopine photoaffinity labeling of P-glycoprotein (A) and on [^{3}H]vinblastine accumulation (B) in J7.V1-1 cells. (A) Membrane vesicles prepared from J7.V1-1 cells were photolabeled using 50 nM [^{3}H]azidopine in the absence (lane 1) or presence of 1000-fold molar excess (50 μM) of progesterone (lane 2) or megastrol acetate (lane 3). Photolabeled membranes were analyzed by SDS-PAGE and fluorography. (B) J7.V1-1 cells were incubated with 1 μM [^{3}H]vinblastine at 37°C for 1 hr and the amount of [^{3}H]vinblastine associated with the cells measured. Lane 1, no steroid; lanes 2–4, 50 μM steroid; lane 2, progesterone; lane 3, megastrol acetate; lane 4, medroxyprogesterone acetate. Data are expressed as percentage of control with standard deviations (n = 3).

protein is present in many normal tissues and the normal physiological functions of P-glycoprotein are not known, application of reversal agents may result in unexpected side effects related to affecting the normal functions of P-glycoprotein. (3) The long-term application of reversal agents may increase P-glycoprotein levels in normal and malignant cells or select cells that express high levels of P-glycoprotein.

VI. Summary

P-glycoprotein is overproduced in multidrug-resistant tumor cells. The protein mediates resistance by acting as an energy-dependent drug efflux pump with broad specificity for hydrophobic antitumor drugs. Reversal agents interact with P-glycoprotein in MDR cells and (1) inhibit photoaffinity labeling of P-glycoprotein, (2) increase antitumor drug accumulation, (3) decrease drug efflux, and (4) enhance drug sensitivity. Kinetic studies have indicated that most reversal agents competitively inhibit drug binding, although comparative domain mapping of the binding site(s) of antitumor drugs and reversal agents needs to be accomplished. Application of our knowledge concerning reversal agents for multidrug resistance to the clinical setting will require systematic study of the occurrence of P-glycoprotein overexpression in tumor cells as well as careful analysis of side effects produced by reversal agents. Since distinct *mdr* gene products are present in different normal mouse tissues and at least one reversal agent, progesterone, has differential effects on different *mdr* gene products, it is possible that certain reversal agents will selectively interfere with only specific types of P-glycoprotein. The information acquired concerning drug resistance in the past fifteen years holds great potential for improving the efficacy of cancer chemotherapy in the immediate future.

Acknowledgments

The authors are indebted to Ms. Twee Do for her contributions to this manuscript. Research that originated in the authors' laboratory was supported in part by USPHS grant CA39821.

References

Akiyama, S.-I., Cornwell, M. M., Kuwano, M., Pastan, I., and Gottesman, M. M. (1988). Most drugs that reverse multidrug resistance also inhibit photoaffinity labeling of P-glycoprotein by a vinblastine analog. *Mol. Pharmacol.* **33,** 144–147.

Akiyama, S.-I., Yoshimura, A., Kikuchi, H., Sumizawa, T., Kuwano, M., and Tahara, Y. (1989). Synthetic isoprenoid photoaffinity labeling of P-glycoprotein specific to multidrug-resistant cells. *Mol. Pharmacol.* **36,** 730–735.

Arceci, R. J., Croop, J. M., Horwitz, S. B., and Housman, D. (1988). The gene encoding multidrug resistance is induced and expressed at high levels during pregnancy in the secretory epithelium of the uterus. *Proc. Natl. Acad. Sci. U.S.A.* **85,** 4350–4354.

Bech-Hansen, N. T., Till, J. E., and Ling, V. (1976). Pleiotropic phenotype of colchicine-resistant CHO cells: Cross resistance and collateral sensitivity. *J. Cell. Physiol.* **88,** 23–32.

Beck, W. T., Cirtain, M. C., and Lefko, J. L. (1983). Energy-dependent reduced drug binding as a mechanism of vinca alkaloid resistance in human leukemic lymphoblasts. *Mol. Pharmacol.* **24,** 485–492.

Beck, W. T., Cirtain, M. C., Look, A. T., and Asmun, R. A. (1986). Reversal of *vinca* alkaloid resistance but not multiple drug resistance in human leukemic cells by verapamil. *Cancer Res.* **46,** 778–784.

Beck, W. T., Cirtain, M. C., Glover, C. J., Felsted, R. L., and Safa, A. R. (1988). Effects of indole alkaloids on multidrug resistance and labeling of P-glycoprotein by a photoaffinity analog of vinblastine. *Biochem. Biophys. Res. Commun.* **153,** 959–966.

Benson, A. B., III, Trump, D. L., Koeller, J. M., Egorin, M. I., Olman, E. A., Witte, R. S., Davis, T. E., and Tormey, D. C. (1985). Phase I study of vinblastine and verapamil given by concurrent iv infusion. *Cancer Treat. Rep.* **69,** 795–799.

Bradley, G., Naik, M., and Ling, V. (1989). P-glycoprotein expression in multidrug-resistant human ovarian carcinoma cell lines. *Cancer Res.* **49**, 2790–2796.

Broxterman, H. J., Pinedo, H. M., Kuiper, C. M., Kaptein, L.C.M., Schuurhuis, G. J., and Lankelma, J. (1988). Induction by verapamil of a rapid increase in ATP consumption in multidrug-resistant tumor cells. *FASEB J.* **2**, 2278–2282.

Bruggemann, E. P., Germann, U. A., Gottesman, M M., and Pastan, I. (1989). Two different regions of phosphoglycoprotein are photoaffinity-labeled by azidopine. *J. Biol. Chem.* **264**, 15483–15488.

Busche, R., Tummler, B., Riordan, J. R., and Cano-Gauci, D. F. (1989). Preparation and utility of a radioactivated analogue of daunomycin in the study of multidrug resistance. *Mol. Pharmacol.* **35**, 414–421.

Cano-Gauci, D. F., and Riordan, J. R. (1987). Action of calcium antagonists on multidrug resistant cells. *Biochem. Pharmacol.* **36**, 2115–2123.

Chan, H. S. L., Thorner, P. S., Haddad, G., and Ling, V. (1990). Immunohistochemical detection of P-glycoprotein: Prognostic correlation in soft tissue sarcoma of childhood. *J. Clin. Oncol.* **8**, 689–704.

Chauffert, B., Rey, D., Coudert, B., Dumas, M., and Martin, F. (1987). Amiodarone is more efficient than verapamil in reversing resistance to anthracyclines in tumour cells. *Br. J. Cancer* **56**, 119–12.

Chen, C. J., Chin, J. E., Ueda, K., Clark, D. P., Pastan, I., Gottesman, M. M., and Roninson, I. B. (1986). Internal duplication and homology with bacterial transport proteins in the mdr1 (P-glycoprotein) gene from multidrug-resistant human cells **47**, 381–389.

Chin, J. E., Soffir, R., Noonan, K. E., Choi, K., and Roninson, I. B. (1989). Structure and expression of the human MDR (P-glycoprotein) gene family. *Mol. Cell. Biol.* **9**, 3808–3820.

Coley, H. M., Twentyman, P. R., and Workman, P. (1989). Identification of anthracyclines and related agents that retain preferential activity over Adriamycin in multidrug-resistant cell lines, and further resistance modification by verapamil and cyclosporin A. *Cancer Chemother. Pharmacol.* **24**, 284–290.

Cordon-Cardo, C., O'Brien, J. P., Casals, D., Rittman-Graner, L., Biedler, J. L., Melamed, M. R., and Bertino, J. R., (1989). Multidrug-resistance gene (P-glycoprotein) is expressed by endothelial cells at blood–brain barrier sites. *Proc. Natl. Acad. Sci. U.S.A.* **86**, 695–698.

Cornwell, M M., Gottesman, M. M., and Pastan, I. H. (1986a). Increased vinblastine binding to membrane vesicles from multidrug-resistant KB cells. *J. Biol. Chem.* **261**, 7921–7928.

Cornwell, M. M., Safa, A. R., Felsted, R. L., Gottesman, M. M., and Pastan, I. (1986b). Membrane vesicles from multidrug-resistant human cancer cells contain a specific 150- to 170-KDa protein detected by photoaffinity labeling. *Proc. Natl. Acad. Sci. U.S.A.* **83**, 3847–3850.

Cornwell, M. M., Pastan, I., and Gottesman, M. M. (1987a). Certain calcium channel blockers bind specifically to multidrug-resistant human KB carcinoma membrane vesicles and inhibit drug binding to P-glycoprotein. *J. Biol. Chem.* **262**, 2166–2170.

Cornwell, M M., Tsuruo, T., Gottesman, M. M., and Pastan, I. (1987b). ATP-binding properties of P-glycoprotein from multidrug-resistant KB cells. *FASEB J.* **1**, 51–54.

Croop, J. M., Raymond, M., Haber, D., Devault, A., Arceci, R. J., Gross, P., and Housman, D. E. (1989). The three mouse multidrug resistance (mdr) genes are expressed in a tissue-specific manner in normal mouse tissues. *Mol. Cell. Biol.* **9**, 1346–1350.

Dalton, W. S., Grogan, T. M., Meltzer, P. S. Scheper, R. J., Durie, B. G. M., Taylor, C. W., Miller, T. P., and Salmon, S. E. (1989). Drug-resistance in multiple myeloma and non-Hodgkin's lymphoma: Detection of P-glycoprotein and potential circumvention by addition of verapamil to chemotherapy. *J. Clin. Oncol.* **7**, 415–424.

Dano, K. (1973). Active outward transport of daunomycin in resistant Ehrlich ascites tumor cells. *Biochim. Biophys Acta* **323**, 466–483.

DeGregorio,M. W., Ford, J. M., Benz, C. C., and Wiebe, V. J. (1989). Toremifene: Pharmacologic and pharmacokinetic basis of reversing multidrug resistance. *J. Clin. Oncol.* **7**, 1359–1364.

Deuchars, L. L., Du, R.-P., Naik, M., Evernden-Porelle, D., Kartner, N., van der Bliek, A., and Ling, V. (1987). Expression of hamster P-glycoprotein and multidrug resistance in DNA-mediated transformants of mouse LTA cells. *Mol. Cell. Biol.* **7,** 718–724.

Endicott, J. A., and Ling, V. (1989). The biochemistry of P-glycoprotein-mediated multidrug resistance. *Annu. Rev. Biochem.* **58,** 137–171.

Fine, R. L., Patel, J., and Chabner, B.A. (1988). Phorbol esters induce multidrug resistance in human breast cancer cells. *Proc. Natl. Acad. Sci. U.S.A.* **85,** 582–586.

Fojo, A. T., Ueda, K., Slamon D. J., Poplack, D. G., Gottesman, M. M., and Pastan, I. (1987). Expression of a multidrug-resistance gene in human tumors and tissues. *Proc. Natl. Acad. Sci. U.S.A.* **84,** 265–269.

Foote, S. J., Thompson, J. K., Cowman, A. F., and Kemp, D. J. (1989). Amplification of the multidrug resistance gene in some chloroquine-resistant isolates of P. falciparum. *Cell* **57,** 921–930.

Ford, J. M., Prozialeck, W. C., and Hait, W. N. (1989). Structural features determining activity of phenothiazines and related drugs for inhibition of cell growth and reversal of multidrug resistance. *Mol. Pharmacol.* **35,** 105–115.

Foxwell, B. M. J., Mackie, A., Ling, V., and Ryffel, B. (1989). Identification of the multidrug resistance-related P-glycoprotein as an cyclosporine binding protein. *Mol. Pharmacol.* **36,** 543–546.

Gerlach, J. H., Endicott, J. A., Juranka, P. F., Henderson, G., Sarangi, F., Deuchars, K. L., and Ling, V. (1986). Homology between P-glycoprotein and a bacterial haemolysin transport protein suggests a model for multidrug resistance. *Nature (London)* **324,** 485–489.

Goldstein, L. J., Galski, H., Fojo, A., Willingham, M., Lai, S.-L., Gazdar, L. A., Pirker, R., Green, A., Crist, W., Brodeur, G. M., Lierber, M., Cossman, J., Gottesman, M. M., and Pastan, I. (1989). Expression of a multidrug resistance gene in human cancers. *JNCI, J. Natl. Cancer Inst.* **81,** 116–124.

Gottesman, M. M., and Pastan, I. (1988). The multidrug transporter, a double-edged sword. *J. Biol. Chem.* **263,** 12163–12166.

Greenberger, L. M., Williams, S. S., and Horwitz, S. B. (1987). Biosynthesis of heterogeneous forms of multidrug resistance-associated glycoproteins. *J. Biol. Chem.* **262,** 13685–13689.

Greenberger, L. M., Lothstein, L., Williams, S. S., and Horwitz, S. B. (1988). Distinct P-glycoprotein precursors are overproduced in independently isolated drug-resistant cell lines. *Proc. Natl. Acad. Sci. U.S.A.* **85,** 3762–3766.

Greenberger, L. M., Yang, C.-P. H., Gindin, E., and Horwitz, S. B. (1990). Photoaffinity probes for the α1-adrenergic receptor and the calcium channel bind to a common domain in P-glycoprotein. *J. Biol. Chem.* **265,** 4394–4401.

Gros, P., Ben Neriah, Y., Croop, J. M., and Housman, D. E. (1986a). Isolation and expression of a complementary DNA that confers multidrug resistance. *Nature (London)* **323,** 728–731.

Gros, P., Croop, J., and Housman, D. (1986b). Mammalian multidrug resistance gene: Complete cDNA sequence indicates strong homology to bacterial transport proteins. *Cell* **47,** 371–380.

Hait, W. N., Stein, J. M., Koletsky, A. J., Harding, M. W., and Handschummacher, R. E. (1989). Activity of cyclosporin A and a non-immunosuppressive cyclosporin against multidrug resistant leukemic cell lines. *Cancer Commun.* **1,** 35–43.

Hamada, H., and Tsuruo, T. (1988). Characterization of the ATPase activity of the Mr 170,000 to 180,000 membrane glycoprotein (P-glycoprotein) associated with multidrug resistance in K562/ADM cells. *Cancer Res.* **48,** 4926–4932.

Hamada, H, Hagiwara, K.-I., Nakajima, T., and Tsuruo, T. (1987). Phosphorylation of the Mr 170,000 to 180,000 glycoprotein specific to multidrug-resistant tumor cells: Effects of verapamil, trifluoperazine and phorbol esters. *Cancer Res.* **47,** 2860–2865.

Hitchins, R, N., Harman, D. H., Davey, R. A., and Bell, D. R. (1988). Identification of a multidrug resistance associated antigen (P-glycoprotein) in normal human tissues. *Eur. J. Cancer Clin. Oncol.* **24,** 449–454.

Hofsli, E., and Nissen-Meyer, J. (1989). Reversal of drug resistance by erythromycin: Erythromycin increases the accumulation of actinomycin D and doxorubicin in multidrug-resistant cells. *Int. J. Cancer* **44**, 149–154.

Horio, M., Gottesman, M. M., and Pastan, I. (1988). ATP-dependent transport of vinblastine in vesicles from human multidrug-resistant cells. *Proc. Natl. Acad. Sci. U.S.A.* **85**, 3580–3584.

Horwitz, S. B., Goei, S., Greenberger, L., Lothstein, L., Mellado, W., Roy, S. N., Yang, C.-P. H., and Zeheb, R. (1988). Multidrug resistance in the mouse macrophage-like cell line J774.2. In "Mechanisms of Drug Resistance in Neoplastic Cells" (P. V. Woolley and K. D. Tew, eds.), pp. 223–242. Academic Press, San Diego, California.

Hsu, S. I.-H., Lothstein, L., and Horwitz, S. B. (1989). Differential overexpression of three *mdr* gene family members in multidrug-resistant J774.2 mouse cells. *J. Biol. Chem.* **264**, 12053–12062.

Hsu, S. I.-H., Cohen, D., Kirschner, L. S., Lothstein, L., Hartstein, M., and Horwitz, S. B. (1990). Structural analysis of the mouse *mdr*1a (P-glycoprotein) promotor reveals the basis for differential transcript heterogeneity in multidrug-resistant J774.2 Cells. *Mol. Cell. Biol.* **10**, 3596–3606.

Inaba, M., and Maruyama, E. (1988). Reversal of resistance to vincristine in P388 leukemia by various polycyclic clinical drugs, with a special emphasis on quinacrine. *Cancer Res.* **48**, 2064–2067.

Inaba, M., Kobayashi, H., Sakurai, Y., and Johnson, R. K. (1979). Active efflux of daunorubicin and Adriamycin in sensitive and resistant sublines of P388 leukemia. *Cancer Res.* **39**, 2200–2203.

Inaba, M., Fujikura, R., Tsukagoshi, S., and Sakurai, Y. (1981). Restored *in vitro* sensitivity of Adriamycin- and vincristine-resistant P388 leukemia with reserpine. *Biochem. Pharmacol.* **30**, 2191–2194.

Kamimoto, Y., Gatmaitan, Z., Hsu, J., and Arias, I. M. (1989). The function of Gp 170, the multidrug resistance gene product, in rat liver canalicular membrane vesicles. *J. Biol. Chem.* **264**, 11693–11698.

Kamiwatari, M., Nagata, Y., Kikuchi, H., Yoshimura, A., Sumizawa, T., Shudo, N., Sakoda, R., Seto, K., and Akiyama, S.-I. (1989). Correlation between reversing of multidrug resistance and inhibiting of [³H]azidopine photolabeling of P-glycoprotein by newly synthesized dihydropyridine analogues in a human cell line. *Cancer Res.* **49**, 3190–3195.

Kessel, D., and Wilberding, C. (1985). Anthracycline resistance in P388 murine leukemia and its circumvention by calcium antagonists. *Cancer Res.* **45**, 1687–1691.

Klohs, W. D., and Steinkampf, R. W. (1988). The effect of lysosomotropic agents and secretory inhibitors on anthracycline retention and activity in multiple drug-resistant cells. *Mol. Pharmacol.* **34**, 180–185.

Krupinski, J., Coussen, F., Bakalyar, H. A., Tang, W.-J., Feinstein, P.G., Orth, K., Slaughter, C., Reed, R. R., and Gilman, A.G. (1989). Adenylyl cyclase amino acid sequence: Possible channel- or transporter-like structure. *Science* **244**, 1558–1564.

Lalande, M. E., Ling, V., and Miller, R. G. (1981). Hoechst 33342 dye uptake as a probe of membrane permeability changes in mammalian cells. *Proc. Natl. Acad. Sci. U.S.A.* **78**, 363–367.

Lee, S. C., Deutsch, C., and Beck, W. T. (1988). Comparison of ion channels in multidrug-resistant and -sensitive human leukemic cells. *Proc. Natl. Acad. Sci. U.S.A.* **85**, 2019–2023.

Lothstein, L., and Horwitz, S. B. (1986). Expression of phenotypic traits following modulation of colchicine resistance in J774.2 cells. *J. Cell. Physiol.* **127**, 253–260.

Lothstein, L., Hsu, S. I.-H., Horwitz, S. B., and Greenberger, L. M. (1989). Alternative overexpression of two phosphoglycoprotein genes is associated with changes in multidrug resistance in a J774.2 cell line. *J. Biol. Chem.* **264**, 16054–16058.

Martin, S. K., Oduola, A. M. J., and Milhous, W. K. (1987). Reversal of chloroquine resistance in Plasmodium falciparum by verapamil. *Science* **235**, 899–901.

Masurovsky, E. B., and Horwitz, S. B. (1989). Ultrastructural effects of colchicine, vinblastine, and taxol in drug-sensitive and multidrug-resistant J774.2 cells. *Protoplasma* **148,** 138–149.

Mazzanti, R., Kamimoto, Y., Gatmaitan, Z., and Arias, I. M. (1989). Bile acid inhibition of Gp 170, an ATP-dependent efflux pump in bile canaliculi. *Proc. Am. Assoc. Stud. Liver Dis.* p. 64.

McGrath, J. P., and Varshavsky, A. (1989). The yeast STE6 gene encodes a homologue of the mammalian multidrug resistance P-glycoprotein. *Nature (London)* **340,** 400–404.

Meyers, M.B., and Biedler, J. L. (1981). Increased synthesis of a low molecular weight protein in vincristine resistant cells. *Biochem. Biophys. Res. Commun.* **99,** 228–235.

Naito, M., and Tsuruo, T. (1989). Competitive inhibition by verapamil of ATP-dependent high affinity vincristine binding to the plasma membrane of multidrug-resistant K562 cells without calcium ion involvement. *Cancer Res.* **49,** 1452–1455.

Naito, M., Hamada, H., and Tsuruo, T. (1988). ATP/Mg^{2+}-dependent binding of vincristine to the plasma membrane of multidrug-resistant K562 cells. *J. Biol. Chem.* **263,** 11887–11891.

Naito, M., Yusa, K., and Tsuruo, T. (1989). Steroid hormones inhibit binding of vinca alkaloid to multidrug resistance related P-glycoprotein. *Biochem. Biophys. Res. Commun.* **158,** 1066–1071.

Nakagawa, M., Akiyana, S.-I., Yamaguchi, T., Shiraishi, N., Ogata, J., and Kuwano, M. (1986). Reversal of multidrug resistance by synthetic isoprenoids in the KB human cancer cell line. *Cancer Res.* **46,** 4453–4457.

Neyfakh, A. A. (1988). Use of fluorescent dyes as molecular probes for the study of multidrug resistance. *Exp. Cell Res.* **174,** 168–176.

Ng, W. F., Sarangi, F., Zastawny, R. L., Veinot-Drebot, L., and Ling, V. (1989). Identification of members of the P-glycoprotein multigene family. *Mol. Cell. Biol.* **9,** 1224–1232.

Nogae, I., Kikuchi, J., Yamaguchi, T., Nakagawa, M., Shiraishi, N., and Kuwano, M. (1987). Potentiation of vincristine by vitamin A against drug-resistant mouse leukaemia cells. *Br. J. Cancer* **56,** 267–272.

Nogae, I., Kohno, K., Kikuchi, J., Kuwano, M., Akiyama, S.-I., Kiue, A., Suzuki, K.-I., Yoshida, Y., Cornwell, M. M., Pastan, I., and Gottesman, M.M. (1989). Analysis of structural features of dihydropyridine analogs needed to reverse multidrug resistance and to inhibit photoaffinity labeling of P-glycoprotein. *Biochem. Pharmacol.* **38,** 519–527.

Pearce, H. L., Safa, A. R., Bach, N. J., Winter, M. A., Cirtain, M. C., and Beck, W. T. (1989). Essential features of the P-glycoprotein pharmacophore as defined by a series of reserpine analogs that modulate multidrug resistance. *Proc. Natl. Acad. Sci. U.S.A.* **86,** 5128–5132.

Ramu, A., Fuks, Z., Gatt, S., and Glaubiger, D. (1984a). Reversal of acquired resistance to doxorubicin in P388 murine leukemia cells by perhexiline maleate. *Cancer Res.* **44,** 144–148.

Ramu, A., Glaubiger, D., and Fuks, Z. (1984b). Reversal of acquired resistance to doxorubicin in P388 murine leukemia cells by tamoxifen and other triparanol analogues. *Cancer Res.* **44,** 4392–4395.

Ramu, A., Spanier, R., Rahaminoff, H., and Fuks, Z. (1984c). Restoration of doxorubicin responsiveness in doxorubicin-resistant P388 murine leukaemia cells. *Br. J. Cancer* **50,** 501–507.

Riehm, H., and Biedler, J. L. (1972). Potentiation of drug effect by Tween 80 in Chinese hamster ovary cells resistant to actinomycin D and daunomycin. *Cancer Res.* **32,** 1195–1200.

Riordan, J. R., and Ling, V. (1985). Genetic and biochemical characterization of multidrug resistance. *Pharmacol. Ther.* **28,** 51–75.

Riordan, J. R., Rommens, J. M., Kerem, B., Alon, N., Rozmahel, R., Grzelczak, Z., Zielenski, J., Lok, S., Plavsic, N., Chou, J.-L., Drumm, M. L., Iannuzzi, M. C., Collins, F. S., and Tsui, L.-C. (1989). Identification of the cystic fibrosis gene: Cloning and characterization of complementary DNA. *Science* **245,** 1066–1073.

Rogan, A. M., Hamilton, T. C., Young, R. C., Klecker, R. W., Jr., and Ozols, R. F. (1984). Reversal of Adriamycin resistance by verapamil in human ovarian cancer. *Science* **224**, 994–996.

Roy, S. N., and Horwitz, S. B. (1985). A phosphoglycoprotein associated with taxol resistance in J774.2 cells. *Cancer Res.* **45**, 3856–3863.

Safa, A. R. (1988). Photoaffinity of labeling of the multidrug resistance related P-glycoprotein with photoactive analogs of verapamil. *Proc. Natl. Acad. Sci. U.S.A.* **85**, 7187–7191.

Safa, A. R., Glover, C. J., Meyers, M. B., Biedler, J. L., and Felsted, R. L. (1986). Vinblastine photoaffinity labeling of a high molecular weight surface membrane glycoprotein specific for multidrug-resistant cells. *J. Biol. Chem.* **261**, 6137–6140.

Safa, A. R., Glover, C. J., Sewell, J. L., Meyers, M. B., Biedler, J., and Felsted, R. L. (1987). Identification of the multidrug resistance-related membrane glycoprotein as an acceptor for calcium channel blockers. *J. Biol. Chem.* **262**, 7884–7888.

Safa, A. R., Choe, M. M., Morrow, M., and Manley, S. A. (1989a). Cyclosporin A and its nonimmunosuppressive analogs reverse *vinca* alkaloid resistance by interacting with P-glycoprotein. *Proc. Am. Assoc. Cancer Res.* **30**, 498.

Safa, A. R., Mehta, N. D., and Agresti, M. (1989b). Photoaffinity labeling of P-glycoprotein in multidrug resistant cells with photoactive analogs of colchicine. *Biochem. Biophys. Res. Commun.* **162**, 1402–1408.

Shen, D.-W., Fojo, A., Chin, J. E., Roninson, I. B., Richert, N., Pastan, I., and Gottesman, M. M. (1986). Human multidrug-resistant cell lines: Increased *mdr* expression can precede gene amplification. *Science* **232**, 643–645.

Shinoda, H., Inaba, M., and Tsuruo, T. (1989). *In vivo* circumvention of vincristine resistance in mice with P388 leukemia using a novel compound, AHC-52. *Cancer Res.* **49**, 1722–1726.

Shiraishi, N., Akiyama, S.-I., Kobayashi, M., and Kuwano, M. (1986). Lysosomotropic agents reverse multiple drug resistance in human cancer cells. *Cancer Lett.* **30**, 251–259.

Shiraishi, N., Akiyama, S.-I., Nakagawa, M., Kobayashi, M., and Kuwano, M. (1987). Effect of bisbenzylisoquinoline (biscoclaurine) alkaloids on multidrug resistance in KB human cancer cells. *Cancer Res.* **47**, 2413–2416.

Skovsgaard, T. (1978). Mechanisms of resistance to duanorubicin in Ehrlich ascites tumor cells. *Cancer Res.* **38**, 1785–1791.

Skovsgaard, T. (1980). Circumvention of resistance to daunorubicin by N-acetyldaunorubicin in Ehrlich ascites tumor. *Cancer Res.* **40**, 1077–1083.

Slater, L. M., Sweet, P., Stupecky, M., and Gupta, S. (1986). Cyclosporin A reverses vincristine and daunorubicin resistance in acute lymphatic leukemia in vitro. *J. Clin. Invest.* **77**, 1405–1408.

Sriram, K., Felsted, R. L., and Safa, A. R. (1987). Reversal of drug resistance by calcium blockers correlates with their ability to block vinblastine photoaffinity labeling of gp150–180. *Proc. Am. Assoc. Cancer Res.* **28**, 285.

Sugawara, I., Kataoka, I., Morishita, Y., Hamada, H., Tsuruo, T., Itoyama, S., and Mori, S. (1988). Tissue distribution of P-glycoprotein encoded by a multidrug-resistant gene as revealed by a monoclonal antibody, MRK16. *Cancer Res.* **48**, 1926–1929.

Thiebaut, F., Tsuruo, T., Hamada, H., Gottesman, M. M., Pastan, I., and Willingham, M. C. (1987). Cellular localization of the multidrug-resistance gene product P-glycoprotein in normal human tissues. *Proc. Natl. Acad. Sci. U.S.A.* **84**, 7735–7738.

Tsuruo, T., and Iida, H. (1986). Effects of cytochalasins and colchicine on the accumulation and retention of daunomycin and vincristine in drug resistant tumor cells. *Biochem. Pharmacol.* **35**, 1087–1090.

Tsuruo, T., Iida, H., Tsukagoshi, S., and Sakurai, Y. (1981). Overcoming of vincristine resistance in P388 leukemia in vivo and in vitro through enhanced cytotoxicity of vincristine and vinblastine by verapamil. *Cancer Res.* **41**, 1967–1972.

Tsuruo, T., Iida, H., Tsukagoshi, S., and Sakurai, Y. (1982). Increased accumulation of vincristine and Adriamycin in drug-resistant P388 tumor cells following incubation with calcium antagonists and calmodulin inhibitors. *Cancer Res.* **42**, 4730–4733.

Tsuruo, T., Iida, H., Naganuma, K., Tsukagoshi, S., and Sakurai, Y. (1983a). Promotion by verapamil of vincristine responsiveness in tumor cell lines inherently resistant to the drug. *Cancer Res.* **43**, 808–813.

Tsuruo, T., Iida, H., Nojiri, M., Tsukagoshi, S., and Sakurai, Y. (1983b). Circumvention of vincristine and Adriamycin resistance *in vitro* and *in vivo* by calcium influx blockers. *Cancer Res.* **43**, 2905–2910.

Tsuruo, T., Iida, H., Kawabata, H., Tsukagoshi, S., and Sakurai, Y. (1984). High calcium content of pleiotropic drug-resistant P388 and K562 leukemia and Chinese hamster ovary cells. *Cancer Res.* **44**, 5095–5099.

Tsuruo, T., Iida-Saito, H., Kawabata, H., Oh-hara, T., Hamada, H., and Utakoji, T. (1986). Characterization of resistance to Adriamycin in human myelogenous leukemia K562 resistant to Adriamycin and in isolated clones. *Gann* **77**, 1967–1972.

Tsuruo, T., Yusa, K., Sudo, Y., Takamori, R., and Sugimoto, Y. (1989). A fluorine-containing anthracycline (ME2303) as a new antitumor agent against murine and human tumors and their multidrug-resistant sublines. *Cancer Res.* **49**, 5537–5542.

Twentyman, P. R. (1988). Modification of cytotoxic drug resistance by non-immuno-suppressive cyclosporins. *Br. J. Cancer* **57**, 254–258.

Ueda, L., Cardarelli, C., Gottesman, M. M., and Pastan, I. (1987). Expression of a full-length cDNA for the human "MDR1" gene confers resistance to colchicine, doxorubicin and vinblastine. *Proc. Natl. Acad. Sci. U.S.A.* **84**, 3004–3008.

van der Bliek, A. M., and Borst, P. (1989). Multidrug resistance. *Adv. Cancer Res.* **52**, 165–203.

Vanhoutte, P. M., and Paoletti, R. (1987). The WHO classification of calcium antogonists. *Trends Pharmacol. Sci.* **8**, 4–5.

Warrington, R. C., and Fang, W. D. (1989). Reversal of the multidrug resistant phenotype of Chinese hamster ovary cells by L-histidinol. *JNCI, J. Natl. Cancer Inst.* **81**, 798–803.

Willingham, M. C., Cornwell, M. M., Cardarelli, C. O., Gottesman, M. M., and Pastan, I. (1986). Single cell analysis of daunomycin uptake and efflux in multidrug-resistant and -sensitive KB cells: Effects of verapamil and other drugs. *Cancer Res.* **46**, 5941–5946.

Wilson, C. M., Serrano, A. E., Wasley, A., Bogenschutz, M. P., Shankar, A. H., and Wirth, D. F. (1989). Amplification of a gene related to mammalian *mdr* genes in drug resistant *Plasmodium falciparum*. *Science* **244**, 1184–1186.

Yang, C.-P. H., Mellado, W., and Horwitz, S. B. (1988). Azidopine photoaffinity labeling of multidrug-resistance associated protein. *Biochem. Pharmacol.* **37**, 1417–1421.

Yang, C.-P. H., DePinho, S. G., Greenberger, L. M., Arceci, R. J., and Horwitz, S. B. (1989). Progesterone interacts with P-glycoprotein in multidrug-resistant cells and in the endometrium of gravid uterus. *J. Biol. Chem.* **264**, 782–788.

Yang, C. -P. H., Cohen, D., Greenberger, L. M., Hsu, S. I. -H., and Horwitz, S. B. (1990). Different transport properties of two *mdr* gene products are distinguished by progesterone. *J. Biol. Chem.* **265**, 10282–10288.

Yin, M.-B., Bankusli, I., and Rustum, Y. M. (1989). Mechanisms of the *in vivo* resistance to Adriamycin and modulation by calcium channel blockers in mice. *Cancer Res.* **49**, 4729–4733.

Yoshimura, A., Kuwazuru, Y., Sumizawa, T., Ichikawa, M., Ikeda, S.-I., Uda, T., and Akiyama, S.-I. (1989). Cytoplasmic orientation and two domain structure of the multidrug transporter, P-glycoprotein, demonstrated with sequence-specific antibodies. *J. Biol. Chem.* **264**, 16282–16291.

Yusa, K., and Tsuruo, T. (1989). Reversal mechanism of multidrug resistance by verapamil: Direct binding of verapamil to P-glycoprotein on specific sites and transport of verapamil outward across the plasma membrane of K562/ADM cells. *Cancer Res.* **49**, 5002–5006.

Zamora, J. M., Pearce, H. L., and Beck, W. T. (1988). Physical–chemical properties shared by compounds that modulate multidrug resistance in human leukemic cells. *Mol. Pharmacol.* **33**, 454–462.

CHAPTER 10

Biochemical Mechanisms of the Synergistic Interaction of Antifolates Acting on Different Enzymes of Folate Metabolic Pathways

John Galivan
Wadsworth Center for Laboratories and Research
New York State Department of Health
Albany, New York 12201

I. Introduction

Until recently the antifolates have consisted of a series of folate analogs which inhibit dihydrofolate reductase (DHFR)[1]. The combination of the use of two inhibitors of dihydrofolate reductase to gain synergistic interactions has had little theoretical or experimental support. In fact such attempts have generally

---

[1] Abbreviations used are MTX, methotrexate (4-NH$_2$-10-CH$_3$PteGlu); DHFR, dihydrofolate reductase: PDDF, 10-propargyl-5,8-dideazapteroylglutamic acid; DMPDDF, 2-desamino-2-methyl-10-propargyl-5,8-dideazapteroylgutamic acid; GAR, glycine-aminoribotide; DDATHF, 5,10-dideazatetrahydrofolate; H$_4$PteGlu, 5,6,7,8-tetrahdrofolate; H$_2$PteGlu, 7,8-dihydrofolate; 5,10-CH$_2$H$_4$PteGlu, 5,10-methylenetetrahydrofolate; 5CH$_3$H$_4$PteGlu; 5-methyltetrahydrofolate; 10-HCOH$_4$PteGlu, 10-formyltetrahydrofolate; when a subscript n = 2 to 7 follows Glu it indicates that these are cellular poly-γ-glutamate with 2 to 7 glutamate residues; dUMP, deoxyuridine monophosphate; FdUMP, 5-fluorodeoxyuridine monophosphate.

SYNERGISM AND ANTAGONISM IN CHEMOTHERAPY
Copyright © 1991 by Academic Press, Inc.
All rights of reproduction in any form reserved.

led to antagonistic interactions (Roos and Schimke, 1988). With the emergence of new agents which directly inhibit the 1-carbon transfer from the folate coenzymes the possibility of synergy has been re-evaluated. At least one theory of the activity of the antifolates is that the inhibition of dihydrofolate reductase leads to the accumulation of $H_2PteGlu_n$ and the depletion of $5,10\text{-}CH_2H_4PteGlu_n$, and $10\text{-}HCOH_4PteGlu_n$, which are coenzymes for thymidylate and purine biosynthesis (Jackson and Grindey, 1984) (Fig. 1). If such a depletion occurs it is entirely possible that treatment with DHFR inhibitors can make those pathways more sensitive to inhibition by analogs which directly inhibit 1-carbon transfer. Two examples of these latter compounds are 10-propargyldideazafolate (PDDF) and 5,10-dideazatetrahydrofolate (DDATHF) which block thymidylate synthase and glycinamide ribonucleotide formyltransferase, respectively. This chapter deals with the way in which DHFR inhibitors can interact synergistically with each of these compounds in tumor cells in culture.

The use of folate analogs for selective oncolytic activity has been a major component of experimental and clinical therapeutics for the past four decades (Frei *et al.*, 1975; Goldman, 1977; Bertino, 1979; Schornagel and McVie, 1983; Jolivet *et al.*, 1983; Jackson and Grindey, 1984; Sirotnak and DeGraw, 1984; Schimke, 1986). The initial compounds developed, represented by aminopterin and methotrexate (MTX), were folate analogs having a 4-amino group in place of the 4-hydroxy that exists in folic acid and, in some cases,

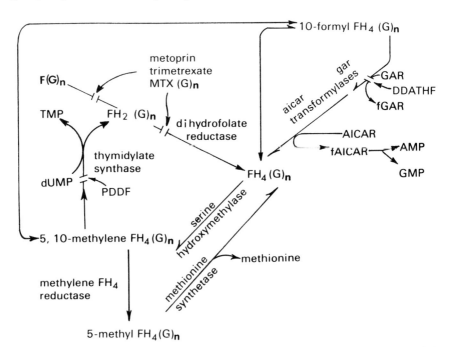

FIGURE 1   Pathways of folate metabolism and selected inhibitors. FH_4, $H_4PteGLU$; (G)n polyglutamate.

additional modifications (Fig. 2). These analogs had the common feature of being tight binding inhibitors ($K_D < 10^{-9}$ M) (Jackson et al., 1984) of dihydrofolate reductase. Certain modifications, in particular 10-ethyl-10-deazaaminopterin, appear to have generated a new subclass that has greater therapeutic activity (Sirotnak et al., 1984; Kris et al., 1988). Since the advent of these antimetabolites, three major changes in the structural features of the folate analogs have resulted in new classes of activities which are in the process of being evaluated experimentally and clinically.

The first of these new classes comprises the lipid-soluble antifolates (Werbel, 1984). Among the most widely studied of these are 2,4-diaminopyrimidines represented by metoprine and 2,4-diaminoquinazolines of which trimetrexate is an example (Fig. 2). These are similar to the folate analogs of highly conserved structure (called classical antifolates) in that they contain a 2,4-diaminopyrimidine that is requisite for binding to DHFR but differ in that the p-aminobenzoylglutamate species is replaced by large hydrophobic groups. These compounds are inhibitors of DHFR [although they generally have a lower affinity than MTX (Jackson et al., 1984] but enter cells independently of the methotrexate/reduced folate coenzyme transport system, cannot be converted to polyglutamate forms, and, unlike MTX, are not likely to inhibit folate-requiring enzymes other than DHFR (Bertino, 1979; Greco and Hakala, 1980; Diddens et al., 1983; Werbel, 1984; Jackson et al., 1984; Kruger-McDermott et al., 1987).

The two other structural modifications involve the pteridine ring and the 10 nitrogen, yielding derivatives that directly inhibit enzymes catalyzing 1-carbon transfer instead of DHFR (Fig. 1,2). The first of these, PDDF, has had the 5 and 8 nitrogens replaced by carbon and a propargyl group added at the 10 position. The resulting compound is a potent inhibitor of thymidylate synthase (Jones et al., 1981; Jackson et al., 1983); the polyglutamate derivatives are approximately two orders of magnitude more effective than the monoglutamates (Cheng et al., 1985; Nair et al., 1985; Pawelczak et al., 1988; Sikora et al., 1988). A recent more modified form of this compound has been generated which replaces the 2-amino group with a methyl substituent [2-desamino-2-methyl-10-propargyl-5,8-dideazatetrahydrofolate (DMPDDF)] that is 50–100 times more effective on a concentration basis against tumor cells in vitro (Jackman et al., 1988; Hughes et al., 1988; Patel et al., 1989). The third structural modification has resulted in a unique folate (DDATHF, Fig. 1) that inhibits the first folate-dependent step in purine biosynthesis, GAR formyltransferase (Fig. 1) (Taylor et al., 1985; Piper et al., 1988; Boschelli et al., 1988; Beardsley et al., 1989). This derivative replaces the 5 and 10 nitrogens with carbon and has a fully reduced pyrazine ring (Fig. 2). Although it has not yet been demonstrated, it is likely that the polyglutamate forms of this folate analog are probably the active agents, as is the case with PDDF (Cheng et al., 1985; Nair et al., 1985; Pawelczak et al., 1988; Sikora et al., 1988; Johnson et al., 1988; Galivan et al., 1988b, 1989b) and MTX (Galivan, 1979; Fry et al., 1982; Jolivet et al., 1983; Galivan et al., 1985).

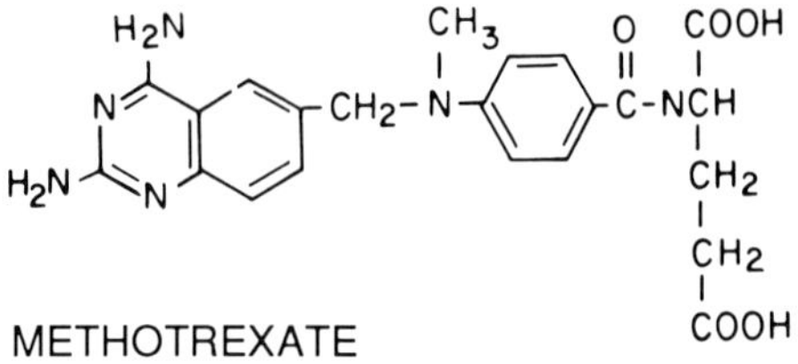

FIGURE 2   Structure of the antifolates.

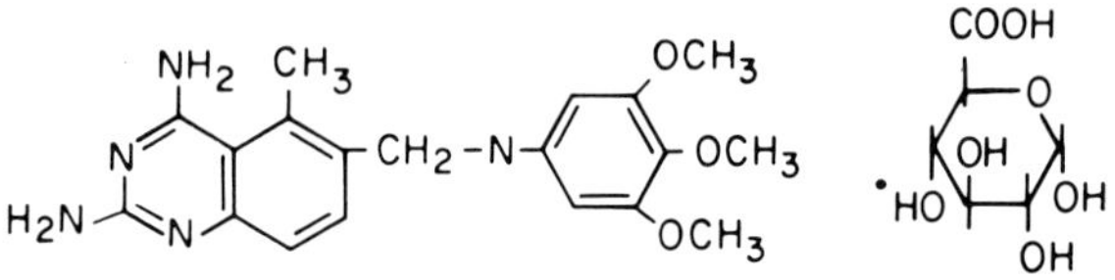

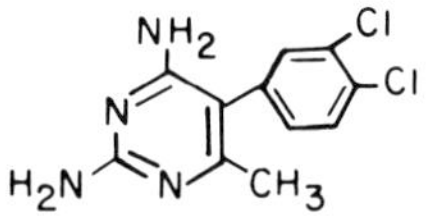

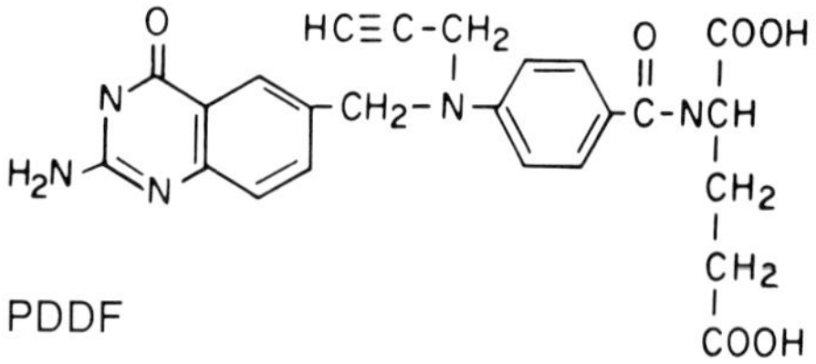

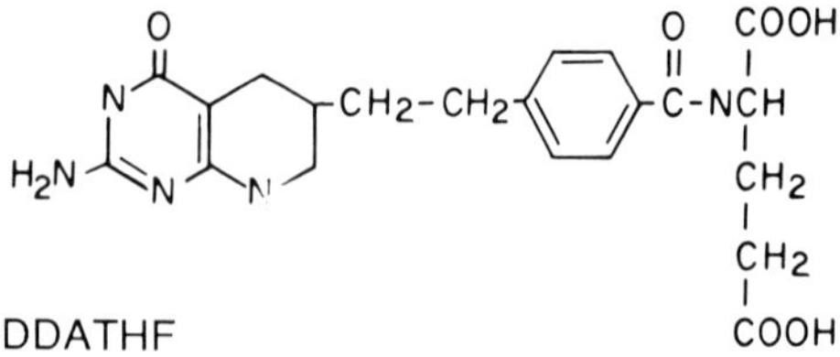

II. Rationale

A. BACKGROUND

Early theoretical analysis of the potential interactions of combinations of inhibitors of two enzymes in the folate pathway(s) were directed at thymidylate synthase and DHFR as targets (Jackson and Harrap, 1973; Harvey, 1978, 1982). The theoretical treatments concluded that positive or synergistic drug interactions were not feasible using drug combinations directed at those two sites (Harvey, 1982). As indicated in subsequent sections, this has since been shown not to be true, at least *in vitro*. Although reports of the drug interaction between MTX and fluorouracil (which inhibits thymidylate synthase after conversion to 5-fluorodeoxyuridylate) are variable, synergy can be observed (Bertino *et al.*, 1977; Kufe and Egan, 1981; Cadman *et al.*, 1981; Piper *et al.*, 1983). This topic has been the subject of two recent reviews (Cadman, 1984; Browman, 1984). In addition, recent studies (Kisliuk, 1986; Galivan *et al.*, 1987, 1988a) that are discussed here on the synergistic interaction of two folate analogs acting on two different targets do not support the previous concept. The reasons for these differences probably reside in the pervasive changes in cell metabolism caused by DHFR inhibitors that are not accounted for in theoretical treatments and the fact that steady-state conditions are assumed to be maintained in spite of inhibition (Harvey, 1982). This assumption does not appear to be applicable in all situations.

Several lines of reasoning have contributed to the attempts to use the combination of two folate analogs, one of which inhibits DHFR and a second which inhibits TS. Initially it should be remembered that MTX is an extensively used therapeutic agent, and attempts to enhance its activity by using it with other agents are a natural corollary of its widespread application. Moreover, a knowledge of two specific areas within the domain of folate and antifolate metabolism and enzymology (Sections IIB and IIC) lead to the tentative suggestion that a combination of DHFR and thymidylate synthase inhibitors could be synergistic.

B. INTERACTION OF PDDF AND THYMIDYLATE SYNTHASE

The first question to be addressed is the nature of the interaction between thymidylate synthase and the inhibitory folate analog, PDDF. It should be noted that the polyglutamate derivatives are probably the active species due to their higher affinity for the enzyme, and their greater cellular formation and retention in cultured cells (Cheng *et al.*, 1985; Nair *et al.*, 1985; Pawelczak *et al.*, 1988; Sikora *et al.*, 1988; Johnson *et al.*, 1988). Kinetic studies with

PDDF and thymidylate synthase have shown that the inhibition is competitive with the substrate $5,10\text{-}CH_2H_4PteGlu$ (Jackson et al., 1983; Pogolotti et al., 1986), although kinetics for the more inhibitory PDDF polyglutamates have not been conducted. The studies by Pogolotti et al., (1986) with Lactobacillus casei thymidylate synthase demonstrated a complex pattern of inhibition. PDDF and $5,10\text{-}CH_2H_4PteGlu$ compete for binding and PDDF initially forms dissociable complexes even in the presence of dUMP. PDDF–thymidylate synthase–dUMP isomerizes to a nondissociable complex in which the inhibition by PDDF becomes noncompetitive with respect to $5,10\text{-}CH_2H_4PteGlu$. This complex form slowly ($t_{1/2} = 0.88$ hr) and dissociates even more slowly ($t_{1/2} = 26$ hr). Qualitatively similar results were obtained in the same study using human thymidylate synthase. The nature of enhanced binding of the folate and pyrimidine nucleotides with thymidylate synthase in ternary complex compared with binary interactions has been established for a number of years (Galivan et al., 1976) and is consistent with the previously mentioned studies of PDDF and dUMP. Extrapolation of these results to cell culture systems suggests that a reduction in cellular $5,10\text{-}CH_2H_4PteGlu_n$ would enhance the capacity of PDDF to inhibit thymidylate synthase owing to the initial competitive interaction of the two compounds. In addition an increase in dUMP which can occur following inhibition of DHFR or thymidylate synthase (Jackson et al., 1983) could substantially favor the formation of the inhibited complex or prevent its dissociation.

C. EFFECT OF MTX ON FOLATE POOLS

The second area is related to the effects of antifolates on cellular folate metabolism. Although several possible ways exist to reduce the cellular levels of $5,10\text{-}CH_2H_4PteGlu_n$, using a DHFR inhibitor is one theoretical possibility. As seen in Fig. 1, the only folate-utilizing enzymatic reaction that catalyzes concomitant oxidation of the $H_4PteGlu_n$ to $H_2PteGlu_n$ during 1-carbon transfer is thymidylate synthase. For each molecule of dUMP synthesized, a molecule of $H_2PteGlu_n$ is formed. In order for this to again become a functional folate coenzyme it must first be reduced by NADPH in the presence of DHFR to $H_4PteGlu_n$. Thus inhibition by MTX can theoretically cause an accumulation of $H_2PteGlu$ and a diminution in some or all of the $H_4PteGlu_n$ species.

Data concerning the effects on pyrimidine and purine biosynthesis suggest that blockage of the thymidylate synthase cycle by DHFR inhibitors impairs not only thymidylate biosynthesis but also purine biosynthesis. The extent of inhibition of each of these pathways appears to vary among different cell lines (Tattersall et al., 1974). Thymidine alone restores growth of L-cell mouse fibroblasts after MTX exposure (Borsa and Whitmore, 1969) whereas MTX can exert a greater antipurine effect in L5178Y lymphoma cells (Hryniuk,

1972). Direct evidence for the capacity to inhibit *de novo* purine biosynthesis has been documented (Cadman *et al.*, 1981; Allegra *et al.*, 1986). The dual inhibitory effect of MTX led to the general concept that MTX (as its poly-glutamates) causes the accumulation of $H_2PteGlu_n$ at the expense of the N_{-5}, N_{-10}-substituted $H_4PteGlu_n$ to the extent that insufficient concentrations of substrates were available to support purine and/or thymidylate biosynthesis (Jackson and Harrap, 1973; Jackson and Grindey, 1984). The depletion of the reduced folates would be a consequence of continued thymidylate synthesis under conditions (MTX blockade) that disallowed the reduction of $H_2PteGlu_n$. It should be noted that thymidylate synthase activity is necessary to deplete tetrahydrofolate cofactors and generate elevated levels of $H_2PteGlu_n$ (Moran *et al.*, 1979). For this reason, tissues or cells with high levels of thymidylate synthase can deplete reduced folate more rapidly in the presence of a DHFR inhibitor and consequently will be more sensitive to growth inhibition (Washtien, 1982).

Experimental analysis of the effects of DHFR inhibitors on cellular folate pools has been somewhat limited until recently. The problems reside in the large number of possible structures due to the variation in poly-γ-glutamate chain length and N_5 and N_{10} substituents. In addition the reduced pteridine ring is quite susceptible to oxidation. Thus the presence, in theory, of over a hundred structures and their instability have made it impossible to have a comprehensive analysis of cellular folates.

To circumvent these problems a number of approaches have been used. A widely used procedure to determine the structure is the conversion of cellular folates to monoglutamate with γ-glutamyl hydrolase, followed by HPLC analysis under reductive conditions (Duch *et al.*, 1983). This gives reasonable recoveries for most folate species, although it has not yet been adapted to analyzing the substrate for thymidylate synthase, $5,10$-$CH_2H_4PteGlu$, which can be unstable on HPLC. This instability may be due in part to the dissociation of the methylene group to form formaldehyde and also the sensitivity to oxidation. $5,10$-$CH_2H_4PteGlu_n$ has been measured using purified thymidylate synthase and entrapment in ternary complex with FdUMP. The latter can be isolated by various filtration methods (Priest *et al.*, 1982; Bunni *et al.*, 1988). This assay has also been adapted to allow quantitation of $H_4PteGlu_n$, $H_2PteGlu_n$, 5-$CH_3H_4PteGlu_n$, and 10-$HCOH_4PteGlu$ by the cycling of cellu-lar folates with added folate-converting enzyme preparations (Bunni *et al.*, 1988; Galivan *et al.*, 1989b). Houghton *et al.*, (1988) have also investigated $5,10$-$CH_2H_4PteGlu$ by using it as a substrate for thymidylate synthase and evaluating the kinetic tritium release from $[5\text{-}^3H]dUMP$. The utility of these assays is not extensive with regard to quantitation of $5,10$-$CH_2H_4PteGlu$ after insult with DHFR inhibitors, but the available results are summarized here.

Early studies showed that treatment with cytotoxic concentrations of MTX resulted in the expansion of the $H_2PteGlu_n$ pool (Nixon *et al.*, 1973; Moran *et al.*, 1975; White and Goldman, 1976; Jackson *et al.*, 1983). More recent studies employing HPLC have concurred with the earlier reported increases in H_2-folate pools (Allegra et al., 1986, 1987; Baram et al., 1987, 1988; Matherly et al., 1987). The surprising result in several of these studies was the maintenance of the $10\text{-}HCOH_4PteGlu_n$ pool in the face of inhibition of DHFR (Allegra et al., 1986, 1987; Baram et al., 1987; Matherly et al., 1987). This has been rationalized by the ability of $H_2PteGlu_n$ to inhibit 5-amino-4-imidazolecarboxamide formyl transferase. However the possibility has recently been pointed out that this may be the consequence of an isolated mitochondrial pool of $10\text{-}HCOH_4PteGlu_n$ (Case and Steele, 1989; Horne et al., 1989). Since HPLC was used, the cellular amounts of $5,10\text{-}CH_2H_4PteGlu_n$ were not quantitated. A limited number of studies have been conducted on this point. In Krebs ascites cells in vivo, a modest reduction (40–60%) in $5,10\text{-}CH_2H_4PteGlu_n$ is observed after MTX exposure (Sur et al., 1983). Exposure of L1210 cells to MTX in vitro results in significant reduction in $5,10\text{-}CH_2H_4PteGlu_n$ only when the cultures are grown on high levels of folic acid (Bunni et al., 1988). Thus a precedent for the depletion of the $5,10\text{-}CH_2H_4PteGlu_n$ does exist, but this pertains to a limited number of experimental systems and conditions. Prior to the studies described here no evidence had been presented in a cell culture or in vivo indicating that a diminution in $5,10\text{-}CH_2H_4PteGlu_n$ could lead to more effective blockade of thymidylate biosynthesis by PDDF.

An interesting and opposite modulation of $5,10\text{-}CH_2H_4PteGlu_n$ to gain a therapeutic advantage in the case of fluorouracil (FU) therapy is worth comment. In order for the FU metabolite fluorodeoxyuridylate (FdUMP) to inhibit thymidylate synthase, an excess of both and FdUMP and $5,10\text{-}CH_2H_4PteGlu_n$ relative to the enzyme is needed to form an inhibited covalent ternary complex (Danenberg, 1977). Hence the use of FU (as FdUMP) as a cytotoxic agent requires adequate cellular levels of $5,10\text{-}CH_2H_4PteGlu_n$ to effectively block thymidylate synthase. Several in vitro and in vivo studies have shown that cells can be deficient in $5,10\text{-}CH_2H_4PteGlu_n$. This can be compensated for by exogenous folinic acid, resulting in enhanced activity of FU (Ullman et al., 1978; Houghton et al., 1981, 1988; Evans et al., 1981; Keyomarsi and Moran, 1988). In contrast, elevated cellular amounts of $5,10\text{-}CH_2H_4PteGlu_n$ could be expected to interfere with the interaction of a folate analog inhibitor of thymidylate synthase (e.g., PDDF) and the target enzyme. As a result, a reduction in the substrate would aid inhibition. In addition an increase in dUMP, which would impair the activity of FdUMP (Berger and Hakala, 1984), would be expected to augment the activity of PDDF by stabilizing the PDDF–thymidylate synthase complex (Pogolotti et al., 1986).

As will be seen in subsequent sections an apparently analogous interaction has been evaluated between inhibitors of DHFR and the GAR formyltransferase inhibitor DDATHF (Galivan et al., 1988a; Thorndike et al., 1988).

However that result rests on a less certain rationale since, in some systems, 10-HCOH$_4$PteGlu$_n$ is maintained in the face of MTX toxicity (Allegra *et al.*, 1986, 1987; Baram *et al.*, 1987; Matherly *et al.*, 1987) and the kinetic interaction between DDATHF and 10-HCOH$_4$PteGlu$_n$ with regard to GAR formyltransferase has not been elucidated.

D. CHOICE OF INHIBITORS

In deciding which inhibitors to use, the choice of analogs that block 1-carbon transport are restricted to a large extent by availability. The best understood folate-based inhibitor of thymidylate synthase is PDDF, which has been employed in most of the studies which will be described here. However preliminary studies with the more growth inhibitory analog DMPDDF (Jackman *et al.*, 1988; Patil *et al.*, 1989) have also been conducted. DDATHF is a recently developed folate analog which blocks the purine pathway by inhibiting GAR formyltransferase (Beardsley *et al.*, 1989).

The choice of a DHFR inhibitor is somewhat more complex due to the large number of compounds available and the existence of two major classes: the classical antifolates such as MTX and the lipid-soluble antifolates. MTX was shown initially to exhibit a positive drug interaction with PDDF under certain conditions (Galivan *et al.*, 1987) but it was not chosen for mechanistic studies for several reasons. Because of the structural similarities between MTX, PDDF, DMPDDF, and DDATHF it is possible that these all complete for transport. Recent studies have clarified that point by demonstrating that MTX, DMPDDF, and DDATHF enter H35 hepatoma cells via the reduced folate carrier, whereas PDDF appears to gain entry by an additional, undefined mechanism (Kruger-McDermott *et al.*, 1987; Galivan *et al.*, 1988a; Patil *et al.*, 1989). Thus, attempts to study the pharmacological interaction of antifolates acting at different enzyme targets could be obfuscated by an interaction at the cell membrane. Second, it is likely that all four of these analogs will complete for glutamylation by folylpolyglutamate synthetase (McGuire and Coward, 1984; Moran *et al.*, 1985; Cook *et al.*, 1987) resulting in a second site of interaction that will cause drug interactions complicating the analysis of target enzyme interactions. Lastly, MTX polyglutamates can directly inhibit thymidylate synthase and GAR formyltransferase (Kisliuk *et al.*, 1979; Allegra *et al.*, 1985, 1987). This interaction is extremely difficult to quantitate in intact cells and, as a result, when MTX is used in combination it can compromise an understanding of the effects of PDDF and DDATHF on their targets.

The experimental complications associated with MTX are obviated to a large extent by the use of lipid-soluble inhibitors of DHFR. In the studies reported from our laboratory metoprine and trimetrexate were used. As noted earlier these share none of the characteristics just described that present difficulties in the use of MTX in combination and are known only to inhibit DHFR when present at low concentrations. Presumably other lipid-soluble inhibitors of DHFR could serve the same purpose.

III. The Effects of Combined Antifolates

A. BACKGROUND

The earliest use of antifolate synergy was the combination of a sulfonamide with an inhibitor of dihydrofolate reductase, which was reviewed earlier by Hitchings and Baccanari (1984) and will be treated in Chapter 11 by Jackson. Briefly, this combination causes inhibition of dihydrofolate synthetase and dihydrofolate reductase resulting in an enhanced reduction in the tetrahydrofolate pool and drug synergy. Since mammals cannot synthesize folic acid from its components utilizing p-aminobenzoate (the site of sulfonamide inhibition), the sulfonamide cannot impair cellular folate metabolism as it does in bacteria and protozoa.

The initial observations of combined drug synergism with DHFR and thymidylate synthase inhibitors was observed in bacteria (Kisliuk et al., 1985a,b; Kisliuk, 1986). In the first of these studies it was shown that various concentrations of PDDF could exhibit a modest synergistic effect with the lipid-soluble DHFR inhibitor, trimethoprim, against the growth of *Lactobacillus casei* (Kisliuk et al., 1985a,b). An additional study (Kisliuk, 1986) demonstrated that similar results occurred when the diglutamate derivative of PDDF was combined with trimethoprim and the inhibition of thymidylate biosynthesis was measured in a cell-free system. Later studies in the *L. casei* system were extended to the use of DDATHF (Thorndike et al., 1988). In that study DDATHF was used with trimethoprim and the combined activity was greater than the fractional product of each of the agents alone, suggesting synergy.

B. THE EFFECTS OF ANTIFOLATES IN COMBINATION AGAINST MAMMALIAN CELLS

The studies just described with *Lactobacillus casei* presented by Kisliuk (1986) and Kisliuk et al. (1985a,b, 1989) along with an understanding of the properties of PDDF (Section IIB) prompted the assessment of the drug interaction of PDDF with a DHFR inhibitor in mammalian cells. For the reasons described earlier (Section IID) we chose the lipid-soluble DHFR inhibitors metoprine and trimetrexate. All the studies performed in this laboratory which are described here utilized the chemically induced rodent Reuber hepatoma that has been carried in monolayer culture and is referred to as the H35 cell line (Pitot *et al.*, 1964; Galivan, 1979).

The general format for the drug combination was the use of both the DHFR inhibitor and PDDF at concentrations that caused only modest or no inhibition. Under these conditions with either trimetrexate or metoprin and PDDF, the growth inhibition curve as a function of PDDF was reduced about 10-fold compared with PDDF alone (I_{50} = ~4 μM) (Fig. 3). The effects seen on cell growth were extended to analogous studies with colony formation and *de novo* thymidylate biosynthesis (Tables 1, 2). In both cases the inhibition observed by the combination was greater than predicted when compared with the product of fractional inhibition. A more rigorous analysis was also applied to these data and it was determined by median-effect analysis (Chou, Chapter 2, this volume) that results are consistent with drug synergy (Galivan *et al.*, 1989a).

The combination studies were further extended to DDATHF and this interaction was evaluated by inhibition of growth and colony formation. Results analogous to those observed with PDDF were obtained. That is, the presence of metoprine and trimetrexate caused the inhibition caused by DDATHF to occur at an approximately 10-fold lower concentration (Fig. 4). This was consistent with the cytotoxicity studies (Galivan *et al.*, 1988a). Selected growth inhibition data were also analyzed by the median-effect

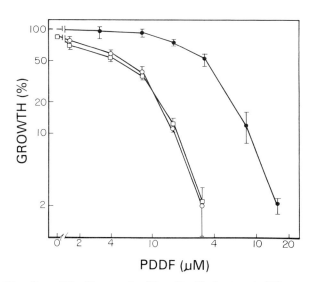

PDDF (μM)

FIGURE 3   The effect of 72 n*M* metoprine (0) and 3 n*M* trimetrexate (□) on growth inhibition by PDDF. The cultures were grown in the presence of the inhibitors for 72 hr and counted after 72 hr (Galivan *et al.*, 1987, 1988b). ●, Control.

TABLE 1
Cytotoxicity of Metroprine and Trimetrexate in Combination with PDDF

Addition			
Metroprine (nM)	Trimetrexate (nM)	PDDF (μM)	Observed survival (%)[a]
			100 ± 7^b
25			86.0 ± 11.25^c
	3.5		91.2 ± 6.4
		2	72.0 ± 1.0
25		2	3.6 ± 1.2
	3.5	2	4.8 ± 2.8

[a] Clonal assays were conducted as described by Galivan *et al.* (1987).
[b] n = 8 ± SD for control. Control cultures had 255 ± 30 colonies.
[c] n = 4 ± SD for drug treated samples.

analysis (Chou, Chapter 2, this volume) and the results were consistent with drug synergy (Galivan *et al.*, 1988b).

The combinations that were used in these studies suggested that the DHFR inhibitors were enhancing the activity of PDDF and DDATHF. However, owing to the complexity of the pathways involved, that was not a certainty. One way to evaluate this possibility is to utilize protection studies with hypoxanthine and thymidine. It is known that protection from toxic levels of the DHFR inhibitors requires both agents whereas protection against PDDF

TABLE 2
Effects of Antifolate Combination on the Release of Tritium from
[5-³H]Deoxyuridine by H35 cells

Addition			
Metroprine (nM)	Trimetrexate (nM)	PDDF (μM)	Tritium released[a] ($10^4 \times$ dpm/h/mg cell protein)
			2.43
72			1.56
	2		2.26
		4	1.85
72		4	0.06
	2	4	0.12

[a] The cells were plated with the indicated concentration of metroprine or trimetrexate at a density of 2×10^5 cells/60 mm plate. After 72 hr PDDF was added and following a 4-hr incubation 20 μM [5-³H]deoxyuridine (3.5×10^4 dpm/nmol) was added. The tritium released was measured. The results are the average of duplicate observations as described by Galivan *et al.* (1985, 1987).

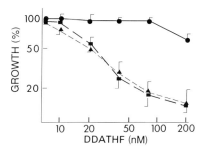

FIGURE 4 The effect of metoprine and trimetrexate on growth inhibition by DDATHF. The experiment was conducted as described in Fig. 3. ● DDATHF alone, ■ plus 2 n*M* trimetrexate, ▲ plus 36 n*M* metoprine.

requires only thymidine (Galivan *et al.*, 1987) and protection against DDATHF requires only hypoxanthine (Galivan *et al.*, 1988b; Beardsley *et al.*, 1989). Thus, if the DHFR inhibitor—PDDF combination is protected by thymidine it would suggest that the combination is acting through PDDF-dependent inhibition of thymidylate synthase. A requirement for both hypoxanthine and thymidine would suggest a more complex mechanism, possibly involving DHFR as the site responsible for cytotoxicity. With DDATHF in combination, the same rationale can be generated and protection by hypoxanthine alone would suggest that the DHFR inhibitor–DDATHF combination is acting primarily by inhibiting GAR formyltransferase. The results shown in Table 3 indicate that the site of action of PDDF in combination is thymidylate synthase and that of DDATHF in combination is GAR formyltransferase since protection is afforded by thymidine and hypoxanthine, respectively. Thus in the hepatoma cell system the DHFR inhibitors appear to be playing an adjunct role, sensitizing the cells to either PDDF and DDATHF.

TABLE 3
Protection against Drug Combinations by Hypoxanthine and Thymidine

	Growth (%)[a]	
Drug combinations	+ Hypoxanthine (50 μ*M*)	+ Thymidine (20 μ*M*)
Trimetrexate (4 n*M*) plus PDDF (4 μ*M*)	10	91
Metoprine (100 n*M*) plus PDDF (4 μ*M*)	12	109
Trimetrexate (2 n*M*) plus DDATHF (200 n*M*)	92	15
Metoprine (100 n*M*) plus DDATHF (200 n*M*)	83	15

[a] Cultures were grown in the presence of the indicated additions for 72 hr and counted as described (Galivan *et al.*, 1987, 1988a). Cell growth was compared with untreated controls and the results are the average of duplicate experiments. In the absence of rescue agents all combinations gave at least 85% inhibition of growth.

Concomitant studies in human lymphoma cells Kisliuk's laboratory (Thorndike *et al.*, 1988; Gaumont *et al.*, 1989; Kisliuk *et al.*, 1989) have provided similar results, suggesting the generality of this phenomenon. As an extension of their earlier work on antifolate combinations against bacteria, the combination of trimetrexate with PDDF or DDATHF was examined in Raji and Manca lymphoma cells. In the Raji cell line the presence of both trimetrexate and PDDF caused a 3- to 8-fold increase in the growth inhibition of the other compared with use of each agent alone (Thorndike *et al.*, 1989). By increasing the folate concentration in the medium from 2.3 to 4.6 μM with Manca cells the sensitivity to drug combinations increased (Gaumont *et al.*, 1989). Similar to the hepatoma cell response, the PDDF–trimetrexate combination was prevented by thymidine and the DDATHF–trimetrexate combination was prevented by the purine percursor, inosine (Kisliuk *et al.*, 1989). These results further substantiated the concept that the DHFR inhibitors were sensitizing the cells to PDDF and DDATHF.

Mammalian cells in culture are routinely grown on folic acid. Since the normal circulating folate is 5-CH_3H_4PteGlu it is possible that certain results of *in vitro* folate metabolism will not be predictive of *in vivo* responses. Gaumont *et al.*, (1989) took the important step of examination these interactions when the cells are grown in 5-CH_3H_4PteGlu and 5-$HCOH_4$PteGlu (folinic acid). They detected synergism between trimetrexate and PDDF or DDATHF when growth was supported by 5-CH_3H_4PteGlu. This was enhanced when the medium concentration was increased from 2.3 μM to 4.6 μM. Analogous results were obtained when cells were grown on 2 nM folinic acid. However when the concentration of folinic acid was increased to 200 nM, antagonism occurred between trimetrexate and PDDF, but synergy with trimetrexate and DDATHF intensified. These important results show the potential for enhanced antimetabolic activity in these combinations not only *in vitro* but also *in vivo*. It should be added that evaluations of possible interactions between PDDF and DDATHF have resulted in either subadditivity or antagonism (Galivan *et al.*, 1988a; Kisliuk *et al.*, 1989).

C. MECHANISM OF ACTION OF ANTIFOLATE COMBINATIONS

The study of the mechanism of action of the antifolate combinations in causing inhibition of cell growth and cytotoxicity is in its early stages. The protection studies described earlier suggest that the DHFR inhibitors are sensitizing the cells (or target enzyme) to inhibition by PDDF and DDATHF. In these studies all inhibitors in combination have been used at relatively low concentrations. Thus it is possible as more cell lines and conditions of inhibition are employed different mechanisms of interaction will be discovered. Certainly this has been the case when fluorouracil and MTX have been combined (Cadman, 1984). Presently evaluation studies on only PDDF have been undertaken, because these drug interactions have been recently described and because a greater knowledge and variety of techniques are available concerning the interaction of PDDF and thymidylate synthase. The combination activity of a DHFR inhibitor and a folate-based thymidylate synthase inhibitor is not restricted to PDDF. Recent studies from the author's laboratory have shown that trimetrexate can enhance the activity of the 2-desamino-2-methyl derivative of PDDF (DMPDDF) (Table 4). Approximately a 10-fold enhancement is produced in the presence of a non-inhibitory level of trimetrexate. DMPDDF is 70-fold more inhibitory than PDDF and there is an additional 10-fold activation with a DHFR inhibitor. DMPDDF enters cells independently of PDDF via the reduced folate/MTX transport system (Patil *et al.*, 1989) as does DDATHF (Boschelli *et al.*, 1988; Galivan *et al.*, 1988a). Thus the lipid soluble DHFR inhibitor can enhance the activity of an antifolate inhibiting thymidylate synthase whether it gains entry via the reduced folate transport system or by another, possibly independent mechanism. It also demonstrates that the effect is not restricted to a specific folate-analog thymidylate synthase inhibitor.

In these studies the DHFR inhibitors are used at nontoxic concentrations which have little ($< 25\%$) inhibition effect on purine and pyrimidine synthesis (Galivan *et al.*, 1987, 1988a). This is consistent with their role as "sensitizers" for PDDF, but evidence is required to show that they are causing some changes that are consistent with greater activity of PDDF.

Examination of thymidylate synthase and DHFR activities from extracts of hepatoma cells treated with trimetrexate and PDDF shows that trimetrexate

TABLE
Effect of Trimetrexate on the
Growth Inhibition of DMPDDF

Trimetrexate (nM)	I_{50}[a] (nM)
0	55[b]
1	12
2	5.6

[a] Trimetrexate and DMPDDF were added concurrently at the outset of growth and cell numbers were measured after 72 hr as previously described (Galivan *et al.*, 1987, 1988a). The concentrations of trimetrexate used caused less than 10% inhibition of growth alone.
[b] Compare with an I_{50} of 3800 for PDDF.

causes a significant inhibition of DHFR and has no effect on thymidylate synthase (Table 5). The inhibition of DHFR is a minimal estimate since the DHFR–trimetrexate complex might well dissociate under the conditions of assay. As expected, PDDF has no effect on DHFR, and its effect on thymidylate synthase cannot be evaluated because of nonlinear reaction rates when extracts are assayed. Further evidence for the inhibition of DHFR by noninhibitory concentrations of trimetrexate comes from the data of Thorndike *et al.* (1989), who showed that a marked inhibition of the reduction of homofolate was caused by 10 nM trimetrexate, a concentration which did not inhibit growth by itself.

Analysis of the effects of metoprine and trimetrexate on cellular folates in hepatoma cells reinforces the DHFR inhibition data (Galivan *et al.*, 1989a,b). Growth of folate-depleted hepatoma cells in 4 μM [^{3}H]folate for 72 hr causes complete labeling of the cellular folates and results in approximately 80 nmol folate/gm cellular protein. Inclusion of noninhibitory metoprine or trimetrexate causes a 75% reduction in total folates (Galivan *et al.*, 1988b, 1989b) (Table 6). When this measurement was extended to the measurement of the substrate for thymidylate synthase 5,10-$CH_2H_4PteGlu_n$, it was also reduced 3- to 4-fold (Galivan *et al.*, 1988b). Moreover the accumulation of PDDF polyglutamates in the DHFR treated cells was nearly doubled (Galivan *et al.*, 1988b, 1989b). The reason for increased PDDF polyglutamates is not known with certainty but it would appear that the lowering in cellular folates causes a reduction in competing substrates/inhibitors for folypolyglutamate synthetase, resulting in more rapid and extensive glutamylation of PDDF. The end result of this is a 6- to 8-fold increase in the ratio of inhibitor to substrate, as a result of treatment of H35 cells with PDDF in the presence of metoprine or trimetrexate. This is quantitatively nearly equivalent to the extent of enhancement of activity of PDDF observed by growth

TABLE 5

Dihydrofolate Reductase and Thymidylate Synthase Activity in Extracts of Hepatoma Cells Exposed to Antifolates[a]

Addition	Dihydrofolate reductase (nmol/min/mg)	Thymidylate synthase (nmol/min/mg)
None	4.1 ± 0.64[b]	0.11 ± 0.01
Trimetrexate (2 nM)	1.03 ± 0.24	0.12 ± 0.02
Trimetrexate (2 nM) plus PDDF (4 μM)	2.84 ± 0.5	NC[c]
PDDF (4 μM)	4.7 + 0.2	NC[c]

[a] DHFR and thymidylate synthase were assayed by published methods (Galivan, 1979).
[b] N = at least 3 ± S.D.
[c] NC, not calculated, since assay was not linear with respect to enzyme concentration or time.

TABLE 6

Alterations in the Parameters Involved in the Inhibition of Thymidylate Synthase as a Result of Exposure of H35 Cells to Low Levels of DHFR Inhibitors

				dUMP	
				No	Plus
			PDDF polyglutamates	addition	PDDF
Inhibitor	Folate	5,10-CH$_2$H$_4$PteGlu$_n$	$(\%)^a$		$(4 \ \mu M)^b$
Metoprine (36 nM)	27	33^c	176	420	2660
Trimetrexate (2 nM)	25	25	192	290	2427

[a] Cultures were incubated for 76 hr in the presence of the indicated DHFR inhibitor and compared with untreated cultures (100%). All measurements were made as described by Galivan *et al.* (1989b).
[b] PDDF was added for the last 4 hr. PDDF alone caused an increase of 871% in dUMP.
[c] The same results occur if 5,10-CH$_2$H$_4$PteGlu$_n$ is measured in the presence or absence of excess formaldehyde. Thus the decrease in H$_4$PteGlu$_n$ plus 5,10-CH$_2$H$_4$PteGlu$_n$ is identical to the decrease in 5,10-CH$_2$H$_4$PteGlu$_n$.

inhibition or cytotoxicity. In addition dUMP levels were markedly elevated by PDDF in the presence of the DHFR inhibitors. Metoprine and trimetrexate cause a 3- to 4-fold increase in dUMP whereas PDDF causes a 10-fold increase. The combination of PDDF and a DHFR inhibitor cause a 30-fold increase in dUMP (Galivan *et al.*, 1989b). Thus it appears that these conditions favor the interaction of PDDF polyglutamate with thymidylate synthase and would result in formation of a stable PDDF polyglutamate–thymidylate synthase–dUMP complex.

IV. Discussion, Conclusions, and Prospects

The results reviewed in this chapter demonstrate that two antifolates acting at different enzyme sites in mammalian cells in culture can exhibit synergistic drug interactions. Thus far this combination has consisted of a lipid-soluble inhibitor of dihydrofolate reductase and a folate analog inhibiting purine biosynthesis or *de novo* thymidylate biosynthesis. It appears that a combination of the purine inhibitor DDATHF and the thymidylate synthase inhibitor PDDF produces primarily antagonistic results (Galivan *et al.*, 1988a; Kisliuk *et al.*, 1989). Combinations of dihydrofolate reductase inhibitors are considered theoretically (Jackson and Harrap, 1973; Harvey, 1978, 1982) and experimentally (Roos and Schimcke, 1988) to be no more than additive in their activity.

The combinations which do show synergy have, as expected, quite complex interactions. Evidence for this comes from the fact that at higher levels of the DHFR inhibitors (in this case, MTX or trimetrexate) with PDDF the combined activity of the agents is less than the product of the fractional inhibition of each component (Galivan et al., 1987). In human lymphoma cells grown in 200 nM folinic acid, antagonism was observed between trimetrexate and PDDF but synergy was observed when PDDF was replaced with DDATHF. A complete accounting for the conditions that cause antagonism to become apparent is not available. The data on hand suggest that the timing of drug administration, the relative concentrations of the inhibitors, and the folate status of the growth medium play an important role in determining whether synergistic or antagonistic interactions predominate.

The most thoroughly established drug interactions involve the combination of low concentration of a DHFR inhibitor with a folate analog inhibitor of either thymidylate synthase or GAR formyltransferase. To the extent that a mechanism has been established the results with metoprine or trimetrexate with PDDF are the most detailed. It appears that the DHFR inhibitors restrict the cellular folates resulting in a decrease in total folates and 5,10-CH$_2$H$_4$PteGlu$_n$ (Galivan et al., 1988b, 1989a,b). This allows an enhancement in the polyglutamates of PDDF and appears to facilitate the inhibition of thymidylate synthase. The way in which the cellular folate concentration is reduced is as yet unknown. Since the cultures are supported by folic acid, two reductive steps are required before conversion to H$_4$PteGlu, which must then be glutamylated before becoming functionally effective (McGuire and Coward, 1984). Thus in the hepatoma culture system the DHFR inhibitors can block the first reductive step (PteGlu $\rightarrow$ H$_2$PteGlu) as well as the reduction of H$_2$PteGlu$_n$ to H$_4$PteGlu$_n$. The latter reaction can be generated by the incoming folate or from the thymidylate synthase cycle (Fig. 2). Since folic acid (PteGlu) is very weak substrate for DHFR relative to H$_2$PteGlu (Blakley, 1984) there is a distinct possibility that the folate restriction caused by low levels of reductase inhibitors is occurring by the inhibition of the reduction of PteGlu and not H$_2$PteGlu.

There is no evidence yet that substrate depletion allows more facile inhibition of GAR formyltransferase by DDATHF. If such a mechanism does exist one must be reminded of a significant difference between DDATHF and PDDF in combination with DHFR inhibitors. If inhibition of H$_2$PteGlu regeneration from thymidylate synthase catalysis is an important site of the DHFR inhibitors, this pathway may be unaffected by relatively high doses of DDATHF. However high levels of PDDF will block the formation of H$_2$PteGlu$_n$ and cause the activity of the DHFR inhibitors to be negated (Fig. 2). Thus high concentrations of PDDF given at the same time as the DHFR inhibitors would prevent the formation of H$_2$-folate. Under these conditions the DHFR inhibitors would not be able to alter folate pools by blocking the regenerative phase (DHFR reduction of H$_2$PteGlu$_n$) of the thymidylate synthase cycle.

Thus far there has not been a stringent evaluation of drug interactions under conditions of high levels on inhibition by both agents. This would seem to be important since ultimately those may be the conditions that would be required to exhibit an *in vivo* effect. Based upon these comments, drug timing and order of addition may become important when higher concentrations are used. It has already been shown that no more than additive effects are seen when PDDF precedes metoprine or trimetrexate (Galivan *et al.*, 1988b, 1989b). It must be remembered that, as the conditions of drug dose and scheduling are changed, additional factors may come into play which will alter not only the objective response but also the way in which it occurs.

Thus far little information is available regarding the possible *in vivo* response to these combinations. However, one study (Ferguson *et al.*, 1990) showed superadditivity with MTX and DDATHF when tested in L1210 leukemia in mice and analyzed by the method of Steel and Peckham (1979). Since cultures are for the most part grown on folic acid and the circulating folate in mammals is 5-CH$_3$H$_4$PteGlu, little precedent is available for *in vivo* results except for the report by Gaumont *et al.* (1989). It should be remembered that use of the combination of DHFR inhibitors with PDDF or DDATHF could enhance the host toxicity of these agents, thus nullifying therapeutic benefits. Further investigations are required to elaborate greater mechanistic detail and the therapeutic potential of these antifolate combinations.

Acknowledgments

This work was generously supported by NIH Grants CA25933, CA34314, and CA46216 from the National Cancer Institute, USPHS/Department of Health and Human Services. The author wishes to thank Dr. Roy L. Kisliuk of the Tufts University School of Medicine and Dr. David Priest of the Medical University of South Carolina for providing information on unpublished studies and for suggestions in the preparation of the manuscript.

References

Allegra, C. J., Chabner, B.A., Drake, J. C., Lutz, R., Rodbard, D., and Jolivet, J. (1985). Enhanced inhibition of thymidylate synthase by methotrexate polyglutamates. *J. Biol. Chem.* **260**, 9720–9726.

Allegra, C. J., Fine, R. L., Drake, J. C., and Chabner, B. A (1986). The effect of methotrexate on intracellular folate pools in human MCF-7 breast cancer cells. *J. Biol. Chem.* **261**, 6478–6485.

Allegra, C. J., Huang, K., Yeh, G. C., Drake, J. C., and Baram, J. (1987). Evidence for direct inhibition of *de novo* purine synthesis in human MCF-7 breast cells as a principal mode of metabolic inhibition by methotrexate. *J. Biol. Chem.* **262**, 13520–13526.

Baram, J., Allegra, C. J., Fine, R. L., and Chabner, B. A. (1987). Effect of methotrexate on intracellular folate pools in purified myeloid precursor cells from normal human bone marrow. *J. Clin. Invest.* **79**, 692–697.

Baram, J., Chabner, B. A., Drake, J. C., Fitzhugh, A. L., Sholar, P. W., and Allegra, C. J. (1988). Identification of biochemical properties of 10-formyldihydrofolate, a novel folate found in methotrexate-treated cells. *J. Biol. Chem.* **263**, 7105–7111.

Beardsley, G. P., Moroson, B. A., Taylor, E. C., and Moran, R. G. (1989). A new folate antimetabolite 5,10-dideaza-5,6,7,8-tetrahydrofolate is a potent inhibitor of *de novo* purine biosynthesis. *J. Biol. Chem.* **264**, 328–333.

Berger, S. H., and Hakala, M. T. (1984). Relationship of dUMP and free FdUMP pools to inhibition of thymidylate synthase by 5-fluorouracil. *Mol. Pharmacol.* **25**, 303–309.

Bertino, J. R. (1979). Toward improved selectivity in cancer chemotherapy. *Cancer Res.* **39**, 293–304.

Bertino, J. R., Sawicki, W. L., Lindquist, C. A., and Gupta, V. S. (1977). Schedule-dependent antitumor effects of methotrexate and fluorouracil. *Cancer Res.* **37**, 327–328.

Blakley, R. L. (1984). Dihydrofolate reductase. *In* "Folates and Pterins" (R. L. Blakley and S. J. Benkovic eds.), Vol. 1, pp. 191–254. Wiley, New York.

Borsa, J., and Whitmore, G. F. (1969). Cell killing studies on the mode of action of methotrexate in L-cells in vitro. *Cancer Res.* **29**, 737–744.

Boschelli, D. H., Webber, S., Whiteley, J. M., Oronsky, A. L., and Kerwar, S. S. (1988). Synthesis and biological properties of 5,10-dideaza-5,6,7,8-tetrahydrofolic acid. *Arch. Biochem. Biophys.* **265**, 43–49.

Browman, G. P. (1984). Clinical application of the concept of methotrexate plus 5-FU sequence dependent "synergy": How good is the evidence? *Cancer Treat. Rep.* **68**, 465–469.

Bunni, M., Doig, M. T., Donato, H., Kesevan, V., and Priest, D. G. (1988). Role of methylenetetrahydrofolate depletion in methotrexate-mediated intracellular thymidylate synthase inhibitors in cultured L1210 cells. *Cancer Res.* **48**, 3398–3404.

Cadman, E. (1984). Synergistic drug interactions involving methotrexate. *In* "Folate Antagonists as Therapeutic Agents" (F. M. Sirotnak, J. J. Burchall, W. B. Ensminger, and J. A. Montgomery eds.), Vol. 2, pp. 23–42, Academic Press, Orlando, Florida.

Cadman, E., Heimer, R., and Benz, C. (1981). The influence of methotrexate pretreatment on 5-fluorouracil metabolism in L1210 cells. *J. Biol. Chem.* **256**, 1695–1704.

Case, G. L., and Steele, R. D. (1989). Mitochondrial and extramitochondrial folate pools in rat liver. *FASEB J.* **3**, 4836.

Cheng, Y-C., Dutschman, G. E., Starnes, M. C., Fisher, M. H., Nanavathi, N. T., and Nair, M. G. (1985). Activity of the new antifolate N^{10} propargyl-5,8-dideazafolate and its polyglutamates against human dihydrofolate reductase, human thymidylate synthase and KB cells containing different levels of dihydrofolate reductase. *Cancer Res.* **45**, 598–600.

Cook, K. D., Cichowicz, D. J, George, S., Lawlor, A., and Shane, B. (1987). Mammalian folylpolyglutamate synthetase. 4. *In vitro* and *in vivo* metabolism of folates and analogues and regulation of folate homeostasis. *Biochemistry* **26**, 530–539.

Danenberg, P. V. (1977). Thymidylate synthetase— A target enzyme for cancer chemotherapy. *Biochim. Biophys. Acta* **73**, 73–92.

Diddens, H., Niethammer, D., and Jackson, R. C. (1983). Patterns of cross-resistance to the antifolate drugs trimetrexate, metoprine, homofolate and CB3717 in human lymphoma and osteosarcoma cells resistant to methotrexate. *Cancer Res.* **43**, 5286–5292.

Duch, D. S., Bowers, S. W., and Nichol, C.A. (1983). Analysis of folate cofactor levels in tissues using high-performance liquid chromatography. *Anal. Biochem.* **130**, 385–392.

Evans, R. M., Laskin, J. D., and Hakala, M. T. (1981). Effect of excess folates and deoxyinosine on the activity and site of acton of 5-fluorouracil. *Cancer Res.* **41**, 3288–3295.

Ferguson, K., Boschelli, D., Hoffman, P., Oronsky, A., Whiteley, J., Webber, S., Galivan, J., Freisheim, J., Hynes, J., and Kerwar, S. S. (1990). Synergy between 5,10-dideaza 5,6,7,8-tetrahydrofolic acid and methotrexate in mice bearing L1210 tumors. *Cancer Chemother. Pharmacol.* **25**, 173–176.

Frei, E., Jaffe, N., Tattersall, M. H. N., Pitman, S., and Parker, L. (1975). New approaches to cancer chemotherapy with methotrexate. *N. Eng. J. Med.* **292**, 846–851.

Fry, D. W., Yalowich, J. C., and Goldman, I. D. (1982). Rapid formation of poly-γ-glutamyl derivatives of methotrexate and their association with dihydrofolate reductase as assessed by high pressure liquid chromatography in Ehrlich ascites tumor cells *in vitro J. Biol. Chem.* **257**, 1890–1896.

Galivan, J. (1979). Transport and metabolism of methotrexate in normal and resistant rat hepatoma cells. *Cancer Res.* **39**, 735–743.

Galivan, J., Maley, G. F., and Maley, R. (1976). Factors effecting substrate binding in *Lactobacillus casei* thymidylate synthase as studied by equilibrium dialysis. *Biochemistry* **15**, 356–362.

Galivan, J., McGuire, J. J., Inglese, J., and Coward, J. K. (1985). γ-Fluoromethotrexate, synthesis and biological activity of a potent inhibitor of dihydrofolate reductase with greatly diminished ability to form poly-γ-glutamates. *Proc. Natl. Acad. Sci. U.S.A.* **82**, 2598–2603.

Galivan, J., Nimec, Z., and Rhee, M. (1987). Synergistic growth inhibition of rat hepatoma cells exposed *in vitro* to N^{10}-propargyl-5,8-dideazafolate with methotrexate with the lipophilic antifolates trimetrexate or metoprine. *Cancer Res.* **47**, 5256–5260.

Galivan, J., Nimec, Z., Rhee, M., Boschelli, D., Oronsky, A. L., and Kerwar, S. S. (1988a). Antifolate drug interactions. Enhancement of growth inhibition due to the antipurine 5,10-dideazatetrahydrofolate acid by the lipophilic dihydrofolate reductase inhibitors metoprine and trimetrexate. *Cancer Res.* **48**, 2421–2425.

Galivan, J., Rhee, M. S., Johnson T. B., Nair, M. G., and Priest, D. (1988b). Antifolate drug interactions. *Proc. Am. Assoc. Cancer Res.* **29**, 1129.

Galivan, J., Rhee, M. S., Johnson, T. B., Chou, T.-C., Nair, M G., Bunni, M., and Priest, D. (1989a). The enhancement in the activity of 10-propargyl-5,8-dideazafolate and 5,10-dideazatetrahydrofolate by inhibitors of dihydrofolate reductase. *Adv. Enzyme Regul.* **28**, 13–21.

Galivan, J., Rhee, M. S., Johnson, T. B., Dilwith, R., Nair, M. G., Bunni, M., and Priest, D. G. (1989b). The role of cellular folates in the enhancement of activity of the thymidylate synthase inhibitor 10-propargyl-5,8-dideazafolate against hepatoma cells *in vitro* by inhibitors of dihydrofolate reductase. *J. Biol. Chem.* **264**, 10685–10692.

Gaumont, Y., Kisliuk, R. L., Emkey, R., Piper, J. R., and Nair, M. G. (1989). Folate enhancement of antifolate synergism in human lymphoma cells. *Proc. Am. Assoc. Cancer Res.* **30**, 1886.

Goldman, I. D. (1977). Effects of methotrexate on cellular metabolism: Some critical elements in drug–cell interaction. *Cancer Treat. Rep.* **61**, 549–558.

Greco, W. R., and Hakala, M. T. (1989). Biochemical pharmacology of lipophilic diaminopyrimidine antifolates in mouse and human cells *in vitro. Mol. Pharmacol.* **18**, 521–528.

Harvey, R. J. (1978). Interaction of two inhibitors which act on different enzymes of metabolic pathway. *J. Theor. Biol.* **74**, 441–437.

Harvey, R. J. (1982). Synergism in the folate pathway. *Rev. Infect. Dis.* **4**, 255–260.

Hitchings, G. H., and Baccanari, D. P. (1984). Folate antagonists are antimicrobial agents. *In* "Folate Antagonists as Therapeutic Agents" (F. M. Sirotnak, J. J. Burchall, W. B. Ensminger, and J. A., Montgomery, eds.) Vol. 1. pp. 151–172, Academic Press, Orlando, Florida.

Houghton, J. A., Maroda, S. J., Phillips, J. O., and Houghton, P. J. (1981). Biochemical determinants of responsiveness to 5-fluorouracil and its derivatives in human colorectal adenocarcinoma xenografts. *Cancer Res.* **41**, 144–149.

Houghton, J. A., Williams, L.G., Radparvar, S., and Houghton, P. J. (1988). Characteristics of the pools of 5,10-methylenetetrahydrofolates in xenografts of human colon adenocarcinoma. *Cancer Res.* **48**, 3062–3069.

Horne, D. W., Cook, R. J., and Wagner, C. (1989). Effects of nitrous oxide on the subcellular distribution of hepatic folate coenzymes, *FASEB J.* **3**, 4837.

Hryniuk, W. M. (1972). Purineless death as a link between growth rate and cytotoxicity by methotrexate. *Cancer Res.* **32**, 1506–1511.

Hughes, L. R., Marsham, P. R., Oldfield, J., Jones, T. T., O'Connor, B. M., Bishop, J. A. M.,

Calvert, A.H., and Jackman, A. L. (1988). Thymidylate synthase (TS) inhibitory and cyto-toxic activity of a series of 2-substituted-5,8-dideazafolates. *Proc. Am. Assoc. Cancer Res.* **29,** 1138 (abstr.).

Jackman, A. L., Taylor, G. A., Moran, R., Bishop, J. A. M., Bisset, J., Pawelezak, K., Balmanno, K. R., and Hughes, A. H. (1988) Biological properties of 2-desamino-2-substituted 5,8-deazofolates that inhibit thymidylate synthase. *Proc. Am. Assoc. Cancer Res.* **29,** 1139 (abstr.)

Jackson, R. C., and Grindey, G. B. (1984). The biochemical basis for methotrexate cytotoxicity. *In* "Folate Antagonists as Therapeutic Agents" (F. M. Sirotnak, J. J. Burchall, W. B. Ensminger, and J. A. Montgomery, eds.), Vol. 1, pp. 290–311, Academic Press, Orlando, Florida.

Jackson, R. C., and Harrap, K. R. (1973). Studies with a mathematical model of folate metabo-lism. *Arch. Biochem. Biophys.* **158,** 827–841.

Jackson, R. C., Jackman, A. L., and Calvert, A. H. (1983). Biochemical effects of a quinazoline inhibitor of thymidylate synthetase, N-(4(N((2-amino-4-hydroxy-6-quinazolinyl)methyl)prop-2-ynylamino)benzyol)-L-glutamic acid (CB3717), on human lymphoblastoid cells. *Biochem. Pharmacol.* **32,** 3783–3790.

Jackson, R. C., Fry, D. W., Boritzki, T. J., Besserer, K., Leopold, N. R., Sloan, B. J., and Elslager, E. F. (1984). Biochemical pharmacology of the lipophilic antifolate, trimetrexate. *Adv. Enzyme Regul.* **22,** 187–206.

Johnson, T. B., Nair, M. G., and Galivan, J. H. (1988). Role of folylpolyglutamate synthetase in the regulation of methotrexate polyglutamate formation in H35 hepatoma cells. *Cancer Res.* **48,** 2426–2431.

Jolivet, J., Cowan, K. H., Curt, G. A., Clendennin, N. J., and Chabner, B. A. (1983). The pharmacology and clinical use of methotrexate. *N. Engl. J. Med.* **292,** 1094–1104.

Jones, T. R., Calvert, A. H., Jackman, A. L., Brown, S. J., Jones, M., and Harrap, K. R. (1981). A potent quinazoline inhibitor of thymidylate synthase: Synthesis, biological evalua-tion and therapeutic results in mice. *Eur. J. Cancer* **17,** 11–32.

Keyomarsi, K., and Moran, R. G. (1988). Mechanism of the cytotoxic synergism of fluoropyrim-idines and folinic acid in mouse leukemic cells. *J. Biol. Chem.* **263,** 14402–14408.

Kisliuk, R. L. (1986). The metabolism of pteroylpolglutamates. *In* "Chemistry and Biology of Pteridines" (B. A. Cooper and V. M. Whitehead, eds.), pp. 743–756. de Gruyter, Berlin.

Kisliuk, R. L., Gaumont, Y., Baugh, C. M., Galivan, J. H., Maley, G. F., and Maley, F. (1979). Inhibition of thymidylate synthetase by poly-γ-glutamyl derivatives of folate and methotrexate. *In* "Chemistry and Biology of Pteridines" (R. L. Kisliuk and G. M. Brown, eds.), pp. 431–435. Elsevier/North-Holland, New York.

Kisliuk, R. L., Gaumont, Y., Kumar, P., Coutts, M., Nair, M. G., Nanavati, N., and Kalman, T. I. (1985a). The effect of polyglutamation on the inhibitor activity of folate analogs. *In* "Folyl and Antifolyl Polyglutamates" (I. D. Goldman, ed.), pp. 319–328. Praeger, New York.

Kisliuk, R. L., Papadapoulos, P., Coutts, M., Koobatian, T., Gaumont, Y., and Nair, M. G. (1985b). Synergistic inhibition of the growth of *Lactobacillus casei* by combination of trimethoprim and 5,8-dideaza-10-propargyl PteGlu. *Pharmacologist* **27,** 51.

Kisliuk, R. L., Gaumont, Y., Powers, J. F., Thorndike, J., Nair, M. G., and Piper, J. R. (1990). Synergistic growth inhibition by combinations of antifolate. *In* "Folic Acid Metabolism in Health and Disease" (M. F. Picciano, E. L. R. Stokstad, and J. F. Gregory, eds.), pp. 79–89. Liss, New York.

Kris, M. G., Kinahan, J. J., Grella, R. J., Fanucci, M. P., Werheim, M. S., O'Connel, J. P., Marks, L. P., Williams, M., Fareg, F., Young, C. N., and Sirotnak, F. M. (1988). Phase I trial and clinical evaluation of 10-ethyl-10-deazaaminopterin in adult patients. *Cancer Res.* **48,** 5573–5579.

Kruger-McDermott, C., Johnson, T. B., Rej, R., van der Hoeven, T., Nair, M. G., and

Galivan, J. (1987). The absence of γ-glutamyl transferase activity in transport-dependent methotrexate-resistant hepatoma cells. *Int. J. Cancer* **40**, 835–839.

Kufe, D. W., and Egan, E. M. (1981). Enhancement of 5-fluorouracil incorporation into human lymphoblastic ribonucleic acid. *Biochem. Pharmacol.* **30**, 129–133.

Matherly, L. H., Barlow, D. K., Phillips, V. M., and Goldman, I. D. (1987). The effects of 4-aminofolates on 5-formyltetrahydrofolate metabolism in L1210 cells. *J. Biol. Chem.* **262**, 710–717.

McGuire, J. J., and Coward, J. K. (1984). Pteroylpolyglutamates: Biosynthesis, degradation and function. *In* "Folates and Pterins" (R. L. Blakley and S. J. Benkovic, eds.), Vol. 1 pp. 135–190. Wiley, New York.

Moran, R. G., Dornin, B. A., and Zabrewski, S. F. (1975). On the accumulation of poly-glutamyldihydrofolate in methotrexate (MTX) inhibited cells. *Proc. Am. Assoc. Cancer Res.* **16**, 49.

Moran. R. G., Mulkins, M., and Heidelberger, C. C. (1979). Role of thymidylate synthetase activity in development of methotrexate cytotoxicity. *Proc. Natl. Acad. Sci. U.S.A.* **76**, 5924–5928.

Moran, R. G., Colman, P. D., Rosowsky, A., Forsch, R. A., and Chen, K. K. (1985). Structural features of 4-amino antifolates required for substrate activity with mammalian folylpolyglutamate synthetase. *Mol. Pharmacol.* **27**, 156–166.

Nair, M. G., Nanavathi, N. T., Nair, I. G., Kisliuk, R. L., Gaumont, Y., Hsio, M. C., and Kalman, T. I. (1985). Folate analogues 26. Synthesis and antifolate activity of 10 substituted derivatives of 5,8-dideazafolic acid and of the poly-γ-glutamylmetabolites of N^{10}-propargyl-5,8-dideazafolic acid (PDDF). *J. Med. Chem.* **29**, 1754–1760.

Nixon, P. F., Slutsky, G., Nahas, A., and Bertino, J. R. (1973). The turnover of folate coenzymes in murine lymphoma cells. *J. Biol. Chem.* **248**, 5932–5936.

Patil, S. D., Jones, C., Nair, M. G., Galivan, J., Maley, F., Kisliuk, R. L., Gaumont, Y., Thorndike, J., Duch, D., and Ferrone, R. (1989). Synthesis and biological evaluation of 2-desamino-2-methyl-N^{10}-propargyl-5,8-dideazafolic acid. *J. Med. Chem.* **32**, 1277–1283.

Pawelczak, K., Jones, T. R., Kempny, M., Jackman, A. L., Newell, D. R., Krzyzanowski, L., and Rzeszotarsko, B. (1988). Quinazoline antifolates inhibiting thymidylate synthase: Synthesis of four oligo (L-γ-glutamyl) conjugates of N^{10}-propargyl-5,8-dideazafolic acid and their enzyme inhibition. *J. Med. Chem.* **32**, 160–165.

Piper, A. A., Nott, S. E., Mackinon, and Tattersall, M. H. N. (1983). Critical modulation by thymidine and hypoxanthine of sequential methotrexate-5-fluorouracil synergism in murine L1210 cells. *Cancer Res.* **43**, 5701–5705.

Piper, J. R., McCaled, G. S., Montgomery, J. R., Kisliuk, R. L., Gaumont, Y., Thorndike, J., and Sirotnak, F. M. (1988). Synthesis and antifolate activity of 5-methyl-5,10-dideaza analogous of aminopterin and folic acid and an alternate synthesis of 5,10-dideazatetrahydrofolic acid, a potent inhibitor of glycinamide ribonucleotide formyl transferase. *J. Med. Chem.* **31**, 2164–2169.

Pitot, H., Periano, C., Morse, P., and Potter, V. R. (1964). Hepatomas in tissue culture compared with adapting liver *in vivo*. Metabolic control. *Natl. Cancer Inst. Monogr.* **13**, 229–245.

Pogolotti, A., Danenberg, P. V., and Santi, D. V. (1986). Kinetics and mechanism of interaction of 10-propargyl-5,8-dideazafolate with thymidylate synthase. *J. Med. Chem.* **29**, 478–482.

Priest, D. G., Veronee, C. D., Mangum, M., Bernards, J. M., and Doig, M. T. (1982). Comparison of folylpolyglutamate hydrolases of mouse liver, kidney, muscle, and brain. *Mol. Cell Biochem.* **43**, 81–87.

Roos, D. S., and Schimke, R. T. (1988). Toxicity of folic acid analogs in cultured human cells. A microtiter assay for the analysis of drug competition. *Proc. Natal. Acad. Sci. U.S.A.* **84**, 4860–4864.

Schimke, R. T. (1986). Methotrexate resistance and gene amplification. *Cancer* **10**, 1912–1917.

Schornagel, J., and McVie, J. G. (1983). The clinical pharmacology of methotrexate. *Cancer Treat. Rev.* **10**, 53–75.

Sikora, E., Jackman, A. L., Newell, D. R., and Calvert, A. H. (1988). Formation and retention and biological activity of N^{10}-propargyl-5,8-dideazafolic acid (CB3717) polyglutamates in L1210 cells *in vitro*. *Biochem. Pharmacol.* **37**, 4047–4054.

Sirotnak, F. M., and DeGraw, J. I. (1984). Selective antitumor activity of folate analogs. *In* "Folate Antagonists as Therapeutic Agents" (F. M. Sirotnak, J. J. Burchall, W. B. Ensminger, and J. A. Montgomery, eds.), Vol. 2, pp. 43–97. Academic Press, Orlando, Florida.

Sirotnak, F. M., DeGraw, J. I., Schmid, F. A., Goutas, L. J., and Moccio, A. M. (1984). Novel folate analogs of the 10-deazaaminopterin series. *Cancer Chemother. Pharmacol.* **12**, 26–30.

Steel, G., and Peckham, M. J. (1979). Exploitable mechanisms in combined radiotherapy–chemotherapy, the concept of additivity. *Int. J. Radiat. Oncol., Biol. Phys.* **5**, 85–91.

Sur, P., Doig, M. T., and Priest, D. G. (1983). Response of methylene tetrahydrofolate levels to methotrexate in Krebs ascites cells. *Biochem. J.* **216**, 295–298.

Tattersall, M. H. N., Jackson, R. C., Jackson, S. T. M., and Harrap, K. R. (1974). Factors determining cell sensitivity to methotrexate: Studies of folate and deoxyribonucleoside triphosphate pools in five mammalian cell lines. *Eur. J. Cancer* **10**, 819–826.

Taylor, E. C., Harrington, B. J., Fletcher, S. R., Beardsley, G. P., and Moran, R. G. (1985). Synthesis of antileukemic agents 5,10-dideazaaminopterin and 5,10-dideaza-5,6,7,8-tetrahydroaminopterin. *J. Med. Chem.* **28**, 914–921.

Thorndike, J., Gaumont, Y., Powers, J., Kisliuk, R. L., and Piper, J. R. (1988). Synergistic growth inhibition of human lymphoma cells by combination of trimetrexate with 5,10-dideazatetrahydrofolate. *Proc. Am. Assoc. Cancer Res.* **29**, 1134 (abstr.).

Thorndike, J., Gaumont, Y., Kisliuk, R. L., Sirotnak, F. M., Murthy, B. R., Nair, M. G., and Piper, J. R. (1989). Inhibition of glycinamide ribonucleotide formyltransferase and other folate enzymes by homofolate polyglutamates in human lymphoma and murine leukemia cell extracts. *Cancer Res.* **49**, 158–163.

Ullman, B., Lee, M., Martin, D. W., Jr., and Santi, D. V. (1978). Cytotoxicity of 5-fluoro-2'-deoxyuridine: Requirement for reduced folate cofactors and antagonism by methotrexate, *Proc. Natl. Acad. Sci. USA* **75**, 980–983.

Washtien, W. R. (1982). Thymidylate synthetase as a factor in 5-fluorodeoxyuridine and methotrexate cytotoxicity in gastrointestinal tumor cells. *Mol. Pharmacol.* **21**, 723–728.

Werbel, L. (1984). Design and synthesis of lipophilic antifols as cancer agents. *In* "Folate Antagonists as Therapeutic Agents" (F. M. Sirotnak, J. J. Burchall, W. B. Ensminger, and J. A. Montgomery, eds.), Vol. 1, pp. 261–290. Academic Press, Orlando, Florida.

White, J. C., and Goldman, I. D. (1976). Mechanism of methotrexate. *Mol. Pharmacol.* **12**, 711–719.

CHAPTER 11

Synergistic and Antagonistic Drug Interactions Resulting from Multiple Inhibition of Metabolic Pathways

Robert C. Jackson
Parke-Davis Pharmaceutical Research Division
Warner-Lambert Company
Ann Arbor, Michigan 48105

SYNERGISM AND ANTAGONISM IN CHEMOTHERAPY
Copyright © 1991 by Academic Press, Inc.
All rights of reproduction in any form reserved.

I. Introduction: Simultaneous Inhibition Effects as a Determinant of the Efficacy of Drug Combinations

A. DRUG–DRUG INTERACTIONS AT DIFFERENT LEVELS OF ORGANIZATION

When a metabolic pathway is inhibited at more than one site, the combined effects of the inhibitors may be additive, or they may show synergism, so that the two agents combined give a greater effect than the sum of the individual effects (sometimes much greater). Alternatively they may be antagonistic, in which case one drug wholly or partly negates the effect of the other. This particular kind of drug–drug interaction is mediated by perturbations in the cellular pool sizes of intermediary metabolites in the pathway. For example, if inhibitor A is a competitive inhibitor of a particular enzyme, and inhibitor B acts in a way that elevates the pool of the substrate that competes with inhibitor A, then it will tend to antagonize A. It is often rather difficult to predict the qualitative nature of the drug–drug interation simply by inspecting the layout of the target pathway, though some generalizations can be drawn. These are considered in Section IV. The case of sequential inhibition, where two inhibitors block different steps of a linear pathway, is of particular theoretical and practical importance. This situation is considered in detail in Section II. The basic approach used to study these questions is that of steady-state enzyme kinetics, extended to treat systems of many enzymes. Section III considers the extent to which the approaches and assumptions of steady-state kinetics are a valid approach to the therapeutic application of cytotoxic drugs. Combined drug effects are highly dependent upon the regulatory properties of the target pathway. Section V examines the extent to which positive and negative feedback influence the response of a pathway to multiple inhibition. While at present it is not possible to predict the nature or extent of multiple-drug effects from a few simple rules, some progress has been made in this direction. This is discussed in Section VI. The synergistic combination of inhibitors tends to maximize inhibitory potency, but the aim of therapeutic combinations is to optimize drug selectivity, i.e., to give the greatest possible safety margin between drug effects on the pathogen or tumor and those on the host normal tissues. This aspect of combination protocol design is discussed in Section VII. Finally, a simple BASIC computer program is described that may be used to predict inhibitor interactions in some simple multienzyme systems.

Drug interactions that arise as a consequence of one drug modifying the clearance rate of another have been studied since the early days of pharmacology. For example, the use of a cholinergic agent in combination with an acetylcholine esterase inhibitor provided one of the classical examples of drug potentiation (Webb, 1963). However, the concept of drug interactions arising from multiple perturbations of a metabolic pathway grew from the use of antimetabolites in cancer chemotherapy, and it is from this field that I shall draw most of my examples. At an early stage, the concept was extended to antiparasitic and antibacterial chemotherapy; the pyrimethamine plus sulfadiazine and trimethoprim plus sulfamethoxazole combinations provide classical examples (Hitchings and Baccanari, 1984). At the present time, the study of antimetabolites in combination is an active area of research in the chemotherapy of serious viral diseases such as HIV[1] (see also Chapters 4 and 8).

The early successes with antifolate combinations against infectious disease, which coincided roughly in time with Potter's concept of "sequential blockade" (which will be discussed in detail in Section IIB) resulted in the notion that multiple simultaneous effects on metabolic pathways were the primary explanation of synergism and antagonism in combination chemotherapy. With the passage of time it became clear that the final outcome of combination chemotherapy would be influenced, even dominated, by interactions at several levels of organization; not only metabolic pathway effects, but also (at the level of whole cells and tissues) cytokinetic interactions exert a major influence on the efficacy of drug combinations. For example, if drug A kills cells only in S phase of the cell cycle, and drug B blocks progression of cells from G_1 phase into S phase, then drug B will probably antagonize the cytotoxicity of drug A. At the level of the whole animal, the kind of drug–drug interactions where one agent modifies the absorption, distribution, or elimination of a second agent are undoubtedly important in certain cases

---

[1] Abbreviations used are a, fractional activity, i.e., activity in presence of inhibitor as fraction of uninhibited activity; ADP, ATP, adenosine 5'-di- and triphosphates; CDP, cytidine 5'-diphosphate; CP, carbamoyl phosphate; CTI, combination toxicity index; dADP, dATP, 2'-deoxyadenosine 5'-di- and triphosphates; dCMP, dCDP, dCTP, 2'-deoxycytidine 5'-mono-, di-, and triphosphates; dGDP, dGTP, 2'-deoxyguanosine 5'-di- and triphosphates; DHFR, dihydrofolate reductase; DHO, dihydroorotic acid; dTTP, thymidine 5'-triphosphate; dUMP, deoxyuridylate; GDP, guanosine 5'-diphosphate; HIV, human immunodeficiency virus; i, fractional inhibition, i.e., 1−a; IC_{50}, concentration of inhibitor that reduces pathway flux to 50% of the control rate; K_i, inhibitor dissociation constant; K_m (or KM), Michaelis constant; MMPR, 6-methylmercaptopurine ribonucleoside; OA, orotic acid; OMP, orotidine 5'-phosphate; PF-MP, pyrazofurin 5'-phosphate; PRPP, 5-phosphoribosyl-1-pyrophosphate; UMP, UDP, UTP, uridine 5'-mono-, di-, and triphosphates; Urd, uridine; V_{max} (or VM) maximal velocity of an enzyme.

(Morselli *et al.*, 1974; Melmon and Gilman, 1980). Combinations of a β-lactam with a β-lactamase inhibitor (e.g., Augmentin®) owe their superior antibacterial spectrum to this kind of pharmacokinetic interaction. Chapters 4, 5, 7, 15, and 20 of this volume discuss various aspects of pharmacokinetic interactions.

In recent years it has become apparent that one of the major constraints on the effectiveness of chemotherapy of large tumors is the emergence of acquired drug resistance. The theoretical basis of this concept rests largely upon the quantitative analyses by Goldie and Coldman (1979), and it has obvious implications for the use of drugs in combination, e.g., where possible combinations should include pairs of non-cross-resistant agents. These ideas have been so influential that they have caused a "paradigm shift" For example, Skipper (1986) has commented that, whereas in the early days of combination chemotherapy the sequential or concurrent relationship of drug target sites in the metabolic pathway was considered to be the primary determinant of the efficacy of the combination, we now know that drug combinations work best when each agent can eliminate tumor cell clones that have acquired resistance to the other agent (see also Chapters 3 and 19). I agree with this argument, but believe that the pendulum may have swung too far in the direction of drug resistance as the major factor determining efficacy of drug combinations. The evidence that other factors are still important is reviewed here.

B. TIME DEPENDENCE OF *IN VITRO* AND *IN VIVO* INTERACTIONS

Perhaps the first clear indication that factors other than prevention of emergence of resistance play a part in drug–drug interactions is that synergism and antagonism may be seen at times too early to be explained by a process of selection. While synergism may sometimes result from elimination by drug B of cells resistant to drug A, and antagonism could result from a drug being mutagenic and thereby increasing the mutation rate for resistance to a second agent, these effects would necessarily take several cell doubling times to become apparent. Interactions observed over the time scale of one cell cycle time or less are likely to be the result of cytokinetic or metabolic effects rather than changes in the relative populations of sensitive and resistant cells.

C. THE COMBINATION TOXICITY INDEX

Although the emergence of tumor cells resistant to anticancer agents is an invariable finding, occurring for every known antitumor drug, there are no documented examples of the normal tissues of the body becoming drug–resistant after repeated exposure. Nevertheless, there are many examples of drug combinations causing synergistic toxicity to normal tissues (Leopold *et al.*, 1987). Since these interactions cannot be caused by effects on drug-resistant normal cells, they are probably cytokinetic or metabolic interactions. Combined drug effects on normal tissues *in vivo* are quantitatively expressed by the combination toxicity index (CTI) (Skipper, 1974). This is defined as the sum of the fractional LD_{10} of the single agents used in the maximum tolerated combination regimen. The CTI is discussed in detail by Leopold *et al.* (1987). For example, if the LD_{10} of compounds A and B are 10 and 20, respectively, and the doses of A and B employed in the optimal combination regimen are 6 and 10, respectively, the CTI would be calculated as:

$$CTI = (6/10) + (10/20) = 1.1$$
$$\text{Drug A} \quad \text{Drug B}$$

If toxicities of the agents in the combination for the target organs for dose-limiting toxicity are additive, the CTI should be close to 1.0. In this case, the individual agents may be considered to have overlapping toxicity. This will often be the case when two closely related agents are used in combination (e.g., methotrexate plus aminopterin). CTI values significantly less than 1 indicate synergistic host toxicity; such combinations, even if they have an adequate therapeutic index, are likely to give unpredictable and dangerous combined toxicity. Examples of such combinations are trimetrexate plus 6-thioguanine and trimetrexate plus methotrexate (Leopold *et al.*, 1987). Values of CTI between 1 and 2 indicate that the two drugs have partially non overlapping toxicity. A CTI of 2.0 indicates that the toxicities are totally non overlapping. For example, if one drug is toxic to intestinal epithelium, but not toxic to bone marrow, and other agent is toxic to bone marrow but not to intestine, then the CTI should be about 2. The significance of this is that both drugs may be administered at full doses in the combination, with a correspondingly greater therapeutic effect. The combinations of trimetrexate plus doxorubicin and trimetrexate plus cytoxan (in mice) both give a CTI of 1.8, and thus approach this situation of non overlapping toxicity. Finally, a CTI of > 2.0 would indicate that the drugs are antagonistic with respect to host toxicity. Note that the definition of CTI is formally identical with the isobol definitions of additivity, antagonism, or synergism, where the fixed-effect end point of the isobol is represented in this instance by the LD_{10}. In summary, the existence of nonunity CTI values demonstrates that factors other than drug resistance may exert a decisive influence on the outcome of combination chemotherapy protocols.

D. SEQUENCE DEPENDENCE OF DRUG INTERACTIONS

As mentioned earlier, drug interactions that are mediated through changes in the subpopulations of drug-resistant cells in a tumor take place over a time scale of several cell doubling times. However, numerous examples are known of combinations where the combined effect is highly dependent on the order of administration of the drugs, e.g., the combination of 5-fluorouracil plus methotrexate is likely to be antagonistic when 5-fluorouracil is administered first, but may be synergistic when the methotrexate is administered first. The time differences in this case are on the order of a few hours, and therefore insignificant from the time frame of cell population shifts. Schedule-dependent effects are discussed in detail in Chapter 11. Such effects provide additional strong grounds for the belief that drug interactions at the level of the biochemical pathway may determine the outcome of combination chemotherapy.

The various lines of evidence outlined above suggest that to design optimal drug combinations it is not sufficient to consider only the elimination of drug-resistant cells, though this is probably the single most important factor. The various considerations (time dependence, nonadditive effects on normal tissues, sequence dependence) establish that other mechanisms of drug interaction are important, but these other mechanisms may include not only multiple inhibition of the target metabolic pathway, but also cytokinetic or pharmacokinetic interactions. The most convincing evidence that multiple inhibition effects are important in determining therapeutic efficacy comes from those studies where the therapeutic outcome can be correlated with measured effects on metabolite pools. A number of good experimental examples will be discussed elsewhere in this volume (see especially Chapter 10).

II. Sequential Inhibition: A Historical Perspective

A. THE KINETICS OF LINKED REACTION SYSTEMS

The science of enzyme kinetics deals with the relationship between the rate of an enzyme-catalyzed reaction and the concentrations of the various reactants, including substrates, cofactors , activators, inhibitors, and the enzyme itself. Many enzyme-catalyzed reactions include complex chains or networks of individual rate processes that contribute to the overall forward reaction, or back reaction, or side reactions leading to dead-end complex formation. Most kinetic approaches handle these complexities by making various simplifying assumptions, the most common of which is the assumption that the enzyme-reactant complexes are at or near a steady state. In principle there is no kinetic difference between a series of partial reactions, catalyzed by a single protein and contributing to the overall enzyme-catalyzed reaction, and a series of linked reactions each catalyzed by a different enzyme that together constitute a metabolic pathway. A number of attempts have thus been made to describe, with varying degrees of rigor, the kinetics of whole pathways.

These, though relevant to the present discussion, are beyond its scope, and the reader is referred to articles by Webb (1963), Reiner (1969), and Grindey *et al.* (1975). The use of such treatments for our present purposes is very limited; the problem is that to draw up manageable rate equations for whole pathways, it is necessary to make so many assumptions and simplifications that the equations are likely to have validity that is too limited for most situations of interest. In my opinion, there are two ways forward: one is to discard the goal of drawing up explicit rate equations for highly complex systems, and to concentrate on the most interesting feature of such systems— their regulation. This approach has led to various theories of metabolic control, such as those of Kacser and Porteous (1987). The other approach is to abandon the requirement for explicit general solutions, and to concentrate on describing particular systems of interest by computer simulation, as has been done by Jackson (1980, 1989) and by White (1986).

B. THE CONCEPT OF "SEQUENTIAL BLOCKADE"

Despite the complexity of describing the kinetics of multienzyme systems, and their inhibition (let alone multiple inhibition), some qualitative conclusions have become apparent. Some of these conclusions can be instructively illustrated by a consideration of the multiple inhibition of sequential mono-linear pathways. Consider the schematic pathway illustrated in Fig. 1A. This is a unidirectional, unbranched, linear pathway. The starting material, A, is assumed to be maintained at a constant concentration. We will consider the validity of these various assumptions for real systems later, and for now consider them as axiomatic. The concept of sequential blockade as an advantageous method of combining inhibitors was proposed by Potter (1951). The rationale for this proposal was stated by Albert (1956) as follows:

> Current biochemical studies make us visualize a growth factor (e.g., folinic acid) being gradually built up from components as it moves along a "production line", each stage of assembly being carried out by a different enzyme. What, then, is the arithmatic of sequential blocking? If we block the first enzyme to the extent of 90%, then only 10% of the partly completed factor reaches the second enzyme. If we are lucky enough to have discovered how to block the second enzyme also by 90%, then only 1% of the (still incomplete) factor emerges, and that is little enough.

This argument, so intuitively reasonable, is completely erroneous. Under certain conditions (including the presence of at least one divergent branch point), a pathway may approach this kind of behavior; an unbranched pathway can never behave this way, as pointed out by Webb (1963). What actually happens? Let us assume that in the uninhibited steady state, rates v1, v2, and v3 all equal 100 units/min, and that none of the enzymes is saturated with its

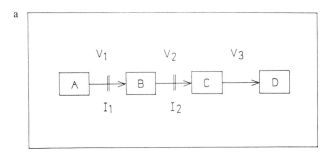

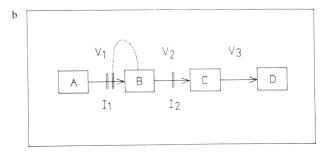

FIGURE 1   An unbranched monolinear pathway subject to sequential inhibition. V_1, V_2, and V_3 indicate reaction velocities for the first, second, and third reactions, respectively, and I_1 and I_2 are inhibitors of reactions 1 and 2, respectively. Figure 1(a) shows an unregulated pathway, and Figure 1(b) shows a similar pathway in which reaction 1 is subject to feedback inhibition by its product, B.

substrate. Let us now assume, with Albert, that we inhibit Reaction 1 by 90%, so the rate of v1 is now 10 units/min. The activity of v2 is unchanged, so substrate B is being consumed faster than it is produced. Its concentration therefore declines and, as it does so, rate v2 falls. When v2 has fallen to 10 units/min, substrate B is once again being produced and consumed at the same rate, and its concentration remains constant. Similarly, the concentration of substrate C will decline until rate v3 is also 10 units/min. At this point, the system has achieved a new, inhibited steady state. Let us now imagine that we are, in the words of Albert, lucky enough to have discovered how to block the second enzyme also by 90%. Rate v2 is decreased to 1 unit/min. This means that B is now being being produced faster than it is consumed, so its concentration starts to rise. This will increase the degree of saturation of the second enzyme with its substrate, so v2 will increase. If the second inhibitor, I2, is a competitive inhibitor (i.e., competitive with B) the accumulation of B will tend to displace I2 from the enzyme. The concentration of B will rise until v2 = v1, at which point the system is again in a steady state, with the total flux through the pathway equal to 10 units/min. This steady state will differ from the first inhibited steady state (where I1 is present but not I2) because the steady-state concentration of B will be higher. However, the presence of I2 has made no difference to the steady state flux through the pathway. Similarly, if a competitive inhibitor of v3 (I3) is added to the system, the concentration of C will increase until v3 again equals 10 units/min. For a sequential monolinear pathway, the first enzyme is always the rate-limiting enzyme, and inhibitors of subsequent reactions can have no effect on the steady-state flux through the pathway. Note that inhibitors of v2 and v3 can, however, affect transient rates, i.e., rates during the period while concentrations of B and C are changing, before a new steady state is achieved. If I2 or I3 is competitive, the system will always achieve a new steady state. If I1 or I3 is noncompetitive, two outcomes are possible, depending on the amount of inhibitor added. Consider the system in which v1 has been reduced by I1 to 10 units/min. When a noncompetitive I2 is added, substrate B will again tend to increase; this will raise the degree of saturation of the second enzyme, but will not displace I2 from its binding site. If the increase in substrate saturation is sufficient to raise rate v2 to 10 units/min, then the system reaches a new, inhibited steady state. However, if the amount of I2 present is such that, even when completely saturated with B, rate v2 will be less than 10 units/min, then the system will not reach a new steady state, and the concentration of B will simply continue to rise. Once again, it may be concluded that inhibitors of enzymes other than the first one in the pathway have no effect on steady-state flux through the pathway. Of course, there may be situations in which the desired therapeutic objective is to perturb a pathway to such a degree that no steady state is possible.

This theoretical consideration of sequential unbranched pathways as described by Webb (1963) was subjected to experimental verification by Rubin *et al.* (1964). Using leukocyte suspensions, they followed the conversion of orotate to uridylate plus carbon dioxide. This process is catalyzed by two enzymes acting sequentially. The first enzyme, orotate phosphoribosyltransferase, is competitively inhibited by 5-azaorotic acid. The second enzyme, orotidylate decarboxylase, is competitively inhibited by 6-azauridine. Rubin *et al.* (1964) showed extensive inhibition in this system using either inhibitor individually, but the two inhibitors combined gave no more inhibition than the separate inhibitors. Similar results were obtained using a cell-free supernatant fraction of rat liver. Handschumacher (1965), discussing these same data, concluded that these results supported the theoretical conclusions of Webb (1963), and that sequential inhibition of enzymes in simple monolinear chains could not explain potentiation of growth inhibition by such inhibitors.

Despite both theoretical and experimental clarification of the nature of sequential inhibition, the literature has been confused by unsupported claims that sequential inhibition necessarily results in synergistic interaction. A frequently cited example is the paper by Black (1963) that attempted to prove that synergism was an intrinsic property resulting from dual inhibition of a monolinear chain of enzymatic reactions. His treatment made many restrictive and unrealistic assumptions, which are listed by Grindey *et al.* (1975); in addition, readers who carefully follow Black's argument will discover mathematical errors in his derivation.

Grindey et al. (1975) considered a variation of the sequential monolinear system in which substrate B acted a feedback inhibitor of the first reaction (Figure 1B). The existence of feedback causes a fundamental change in the properties of a sequential monolinear pathway. In the unregulated system, inhibition of v2 cannot affect the overall steady-state pathway flux (which is always equal to v1). In contrast, when substrate B is a feedback inhibitor of v1, the presence of I2 will increase the pool of B, and thus inhibit v1; I2 thus can decrease overall pathway flux.

The nature of the interaction between I1 and I2 was found to depend on several factors: the competitive or noncompetitive nature of I2, whether the allosteric effector (B) was competitive or noncompetitive with respect to A, and whether B and I1 acted independently (i.e., bound to different sites) or were mutually exclusive (binding to the same site) or gave cooperative binding. Depending upon the combination of these factors, a range of inhibitor interactions was predicted. Discussing this situation of sequential inhibition of a simple monolinear chain constrained by feedback regulation, Grindey and Cheng (1979) concluded that (1) *two inhibitors of sequential enzymes can result in any pattern of interaction (i.e., synergistic, additive, or antagonistic)*; (2) there is no simple correlation between the type of inhibition of each enzyme and the interaction predicted; (3) in the majority of cases the affinity of substrate, inhibitor, or feedback effector, the maximal velocities of the inhibited enzymes, or the substrate concentrations did not affect the type of interaction predicted (i.e., most cases were parameter independent); and (4) the type of interaction depended on the mechanism of inhibition by each agent and the characteristics of feedback regulation.

From the historical perspective, then, the chemotherapist's view of sequential inhibition has progressed from a naive expectation of invariable synergism, through the realization that for unregulated systems in a steady state only inhibition of the rate-determining step can alter the steady-state flux through the pathway, to our present position: that real-life systems are seldom truly unbranched, may not always be at a steady state, and are likely to be highly regulated. All these factors will contribute to the nature of the interaction that will be observed in a sequentially inhibited pathway. Our present view of sequential inhibition will be discussed further in Section IVD.

C. PAST AND PRESENT VIEWS OF THE TRIMETHOPRIM/SULFA COMBINATION

The synergistic interaction between sulfa drugs (which inhibit the biosynthesis of dihydrofolate in microorganisms) and dihydrofolate reductase inhibitors (which inhibit reduction of dihydrofolate to the active tetrahydrofolate form) merits individual consideration, if only because by now several generations of biochemists, microbiologists, physicians and pharmacists have been taught that these classes of agents are synergistic because they form a sequential blockade of tetrahydrofolate production and, conversely, that sequential blockade is a desirable way to combine inhibitors because it has been shown to be so experimentally with these classes of agents. The experimental and clinical value of sulfa drugs + DHFR inhibitors has been shown for two types of chemotherapeutics: antiparasitics, exemplified by the combination of sulfadiazine plus pyrimethamine, and antibacterials, exemplified by sulfamethoxazole plus trimethoprim. Hitchings and Baccanari (1984) have reviewed early explanations of the reason for this synergism, and their own interpretation emphasizes the sequential nature of the blockade, though reference is made to Harvey's view of the interaction as a cycle, with a reaction feeding into the cycle, to be discussed later. Harrap and Jackson (1975) pointed out that this combination could be considered an example of a sequential convergent system; in such systems (unlike the sequential monolinear case), combinations of inhibitors generally give greater inhibition of total flux than either inhibitor acting alone, though the combined effect falls short of additivity. Grindey *et al.* (1975) first proposed that the synergism arose from simultaneous inhibition of a cycle and of a reaction feeding into a cycle. Harvey (1978) also showed that synergism could arise from inhibition of a cyclic process plus inhibition of input to the cycle; as pointed out by Harvey (1978), these treatments abandon the concept of the cell as an open system in a steady state. Instead, the system is assumed to be an open system in a state of exponential growth. It may be that for a bacterial culture undergoing log phase growth this is the most appropriate treatment. The assumption by Grindey *et al.* (1975) and by Harvey (1978) that the pathway of dihydrofolate biosynthesis and oxidoreduction be treated as an open, exponentially growing system, though reasonable, is not essential. A model described by Jackson (1987) treated this system as an open steady state, and also predicted synergism between sulfamethoxazole and trimethoprim. Grindey and Cheng (1979) point out the fact that the basic structure of a metabolic pathway affects the qualitative nature of drug interactions to be expected; since all simple models (such as "sequential convergent pathway" or "cyclic pathway with input reaction") involve rather drastic simplifications, more detailed models are required for a better understanding of the complexities of particular systems, and to make reliable predictions. In the case of folate metabolism in *Escherichia coli*, a detailed kinetic simulation has been published by Harvey and Dev (1975).

In summary, the experimental evidence clearly shows that trimethoprim and sulfamethoxazole cause a synergistic inhibition of bacterial growth (Bushby and Hitchings, 1986; Hitchings, 1969). It is equally clear that the synergistic nature of the interaction cannot be attributed necessarily to sequential relationship of the target sites, since sequential blockade may produce synergistic, additive, or antagonistic interactions, depending upon the detailed kinetics of the system. However, three independent kinetic analyses have concluded that combined inhibition of a cycle reaction with inhibition of a cycle input may be expected to give a synergistic interaction (Grindey et al., 1975; Harvey, 1978; Jackson, 1987). Since the combination of sulfamethoxazole plus trimethoprim is active against organisms that are insensitive to either agent alone, this appears to be a genuine example of therapeutic synergism. However, the therapeutic value of this combination probably also owes something to the ability of each of the separate agents to eliminate organisms with acquired resistance to the other (Bushby, 1969).

III. The Cell as an Open System in a Steady State

A. AN EVALUATION OF THE STEADY-STATE ASSUMPTION

Since the days of the nineteenth century physiologists it has been understood that homeostasis is a fundamental attribute of life, and this concept is epitomized by Claude Bernard's emphasis on "la fixité du milieu intérieur" or the rather more recent definition of life as "a dynamic equilibrium in a polyphasic system". Nowadays we would qualify these concepts in two ways: homeostasis does not mean absolute constancy of the internal milieu, but the ability to maintain it within tolerable limits, and the "dynamic equilibrium" (or "steady state," in present-day terminology) is not a thermodynamic equilibrium. For one thing, equilibrium is a property of closed systems, i.e., systems which matter does not enter or leave. Living organisms are open systems: they take up matter in the form of food and oxygen (sources or inputs to the system) and they tranfer matter to their surroundings (the "sink") in the form of waste products (outputs from the system). In the case of replicating bacteria, or tumor cells in tissue culture, the daughter cells may be considered a form of output. Another difference of living systems from an equilibrium state is that, whereas a system in equilibrium is at its state of maximum entropy (so that the change in entropy is zero), a steady state that is far from equilibrium has a nonzero rate of entropy change, which is, however, the minimal value consistent with the boundary conditions for the system. An open system in a steady state is maintained far from equilibrium by consumption of energy. An interesting discussion of the thermodynamics of nonequilibrium systems is given in the book *Order out of Chaos*, by Prigogine and Stengers (1984).

Traditional chemical kinetics and enzyme kinetics have usually been studied in closed systems. Open systems may sometimes be mimicked in the test tube (e.g., in cell-free extracts or reconstituted multienzyme systems) by maintaining a high concentration of the source material, which will allow the system to sustain a nonequilibrium steady state until the source material becomes significantly depleted. Cultures of microorganisms or tissue culture cells function as open systems, dissipating energy which is liberated from nutrients in the growth medium. Clearly, living systems are open systems; how important is the steady state assumption? In the word of Grindey *et al.* (1975)

> Physiological or pharmacological stress will, of course, cause transient departures from the steady state during the course of adaptation to the new characteristics of the modified steady-state system. As the stress wears off, the metabolism will then gradually and continuously revert to the original state. It must, however, be evident that no sustained departures from the steady state can be tolerated by a living cell. The unrestrained accumulation of a metabolite would eventually be lethal.

In other words, the steady-states characteristic of metabolic pathways are stable to small perturbations, but the system may be unable to recover from large perturbations. The latter may result in death of the cell or organism. Only after death would the system approach thermodynamic equilibrium.

When an open system in a steady state in perturbed by the addition of inhibitors, it will usually respond in one of three ways. (1) The system may adjust the concentrations of the metabolic intermediates to achieve a new steady state in which the overall flux is unchanged. (2) It may settle down to a new, inhibited, steady state. (3) The perturbed system may be unable to reach a steady state. Other, less probable, responses are sometimes seen e.g., the perturbed system may enter a new steady state with increased total flux, or the system may go into stable oscillations. The object of therapy with most pharmacological agents is to restore a system, perturbed by some pathological cause, to a new and more desirable steady state. For example, the object may be to return blood pressure, muscle tone, or brain activity to normal. In contrast, with chemotherapeutic agents, the objective is usually to kill a pathogenic microorganism or tumor cell. This, by definition, means that we are not attempting to achieve a new steady state in the inhibited cell, but to perturb it to a degree where a viable steady state cannot be sustained. Nevertheless, most theoretical treatments of combined effects of inhibitors on metabolic pathways have investigated inhibitor effects *on the steady state of the system.* The main justification for this approach is convenience: it is generally easier to study small perturbations, and it can be argued that the effects of large perturbations will generally be in the same direction as small ones, and that the kinds of drug–drug interactions predicted for lesser degrees of inhibition will be qualitatively similar for greater degrees of inhibition. For this reason, many of the published combined inhibitor studies have measured growth inhibition (which is probably approximately a steady state condition) rather than cell killing.

B. SOURCES AND SINKS FOR OPEN SYSTEMS

Strictly speaking a cell is a single, extremely complex interactive system (indeed, the same could be said, with even greater justification, of the whole organism) and any attempt to study a portion of it in isolation involves making arbitrary assumptions, that will not always be valid, about the inputs and outputs. Grindey *et al.* (1975) defined the origin of a steady-state chain of reactions as "that reaction whose substrate will not undergo a change in concentration as a consequence of any perturbations we may impose upon the system. Application of any treatment which changes the concentration of this substrate will require expansion of the system and redefinition of the origin of the chain." Harrap and Jackson (1975) suggested a number of ways in which the source concentration or input rate might be held stable in experimental subsystems. If microorganisms or tissue culture cells are maintained in growth medium containing high concentrations of glucose or other nutrients that enter the cells at a slow, constant rate, this will constitute a constant, unregulated input. This situation will also be approximated when the source molecule is an intracellular metabolite, but the subsystem under study accounts for only a small fraction of the total turnover of this source material, e.g., the amount of glycine consumed in the purine biosynthetic pathway is much less than the amount used for protein synthesis, and (as a first approximation) it may be assumed that changes in the rate of purine synthesis will have negligible impact on the total glycine pool. A constant input rate may sometimes be achieved because of tight feedback controls regulating the rate of the first reaction, or by the initial metabolite being in rapid reversible equilibrium with a large pool of some other compound. All these situations represent a "buffering" of the initial compound of the system, and make it possible for us to consider the behaviour of the system in isolation (Grindey *et al.*, 1975; Harrap and Jackson, 1975).

The issue of what constitutes the "sink" for systems under consideration is generally less controversial but depends upon the nature of the system being modeled. For modeling of sites of action of antimetabolite drugs used in antimicrobial or anticancer chemotherapy the end product of the pathway is usually a new DNA chain (or some DNA or RNA precursor). In this case the "output" of the system ultimately can be considered to be a daughter cell. However, in the studies of Harvey (1978) on trimethoprim/sulfa interactions, he considered the system to be in exponential growth, rather than in a steady state, and this meant that the exponentially growing system had no outputs.

C. THE CELL CYCLE AS A SLOW TRANSIENT

It is interesting that the seemingly rather radical departure of Harvey (1978) in dispensing with a system output, and indeed dispensing with the whole steady state assumption, should have had so little impact on the properties of the model, and on the nature of the drug interactions predicted. In fact, the assumption of a constantly growing system, rather than a steady state, is probably the most appropriate model for bacterial cells (or tumor cells) in the logarithmic phase of growth. This treatment, instead of assuming that pool sizes of metabolites are constant with time, assumes that their first derivatives with respect to time are constant.

A recurring problem in modeling drug effects is that the events of interest take place on a very wide range of time scales. Pools of the metabolic intermediates in the DNA synthetic pathways typically have turnover times of a few minutes, e.g., the dGTP pool of L1210 cells is sufficient to sustain DNA synthesis for 0.6 min (Jackson, 1989). The plasma clearance half-lives for typical anticancer drugs in mice range from about 20 min to a few hours, and the doubling times of transplantable mouse tumors range from about 9 hr to a few days. Since it would usually be impractical to model pool changes of metabolites with rapid turnover times for period of days, when modeling the slower processes (drug elimination, cell division) it is conventional to assume that the variables with rapid turnover remain at or close to their steady-state values. Again, to quote Grindey *et al.* (1975),

> ... these changes are gradual and will cause the system to move continuously though an infinite succession of steady states, without invalidating the steady-state assumption. Again, the synthesis of DNA, which does not leave the cell, might appear to contradict the steady state; however, the normal cell grows and divides, and the average content of DNA per cell remains constant, and at no time does accumulation occur.

This assumption is a convenient one, and a reasonable one if we are considering a whole population of cells. However, from the point of view of a single cell during S phase of the cell cycle, the amount of DNA is continually changing. Then during mitosis a rather abrupt transition occurs, during which one cell suddenly becomes two cells with half the amount of DNA per cell. Interestingly, when we attempt to model the reactions of DNA precursor production in detail,

> even though the dNTP pools turned over rapidly, with half-lives of a minute or two, the relaxation times following certain perturbations could be extremely long, in some cases approaching the order of a cell cycle time. Perhaps a true steady state is too simplistic a model for understanding the more complex metabolic systems, and the cell should more appropriately be considered as an oscillator, that may exist in a "near-steady" state with respect to the more rapidly turning-over components. (Jackson, 1989)

It is interesting that Bray and Brent (1972) showed that the dATP pool reached a peak in early G_2 phase of the cell cycle. Perhaps the entire S phase is a slow transient phase ending in a state of "shutdown" induced by high dATP.

For our present purposes we will continue to consider the cell as an open system in a steady state, and in considering inhibitor interactions we shall consider the ability of inhibitors to switch the cell to new, inhibited steady states, and sometimes to cause perturbations such that no new steady state is possible. However, it is necessary to bear in mind that this concept is an approximation that holds most closely for those cellular components with rapid turnover times. For some cellular components, including some reactions leading to DNA synthesis which may be important targets for antimetabolite drugs, the steady-state assumption, though convenient, may be questionable.

IV. The Main Patterns of Multiple Inhibitor Use

A. DEFINITIONS AND TERMINOLOGY

In general, the present treatment follows that of Webb (1963). Fractional inhibition (i) is the loss of activity of the inhibited enzyme or pathway as a fraction of activity in the absence of inhibitor; thus $i \times 100 =$ percentage inhibition and $i = 0$ for the uninhibited enzyme. Fractional activity $(a) = 1 - i$. Subscript numbers are used to indicate different inhibitors; thus i_2 is used to indicate the fractional inhibition caused by inhibitor I2, acting alone. The term "additivity" is used by Webb to indicate that

$$i_{1,2} = i_1 + i_2 \tag{1}$$

The situation where a particular inhibitor gives the same degree of fractional inhibition in the presence of a second inhibitor as it does in its absence, so that

$$a_{1,2} = a_1 \times a_2 \tag{2}$$

is termed "summation" by Webb. We will follow the terminology of Harvey (1978) in this instance, and refer to this situation as "independence". This is an intuitively reasonable term, since the degree of inhibition given by I1 is independent of whether I2 is present or not (and vice versa). "Synergism" occurs when two inhibitors together produce an inhibition greater than that expected from independence (Harvey, 1978):

$$a_{1,2} < a_1 \times a_2 \tag{3}$$

Conversely, "antagonism" is when combined effects are less than expected for independence:

$$a_{1,2} > a_1 \times a_2 \tag{4}$$

Though for our present purposes these definitions, based upon the "fractional product" concept, are the most convenient ones, it has been pointed out that in some instances they may lead to apparently paradoxical conclusions. The question of how best to define "additivity" or "synergism" is discussed by Berenbaum (1989).

B. MULTIPLE INHIBITION OF A SINGLE ENZYME

For an enzyme inhibited by more than one inhibitor, the combined effect, at least on the immediate product, can be determined explicitly from the rate equation. This situation has been dealt with in detail by Webb (1963), Segal (1975), Grindey *et al.* (1975), and by Grindey and Cheng (1979). Only the major conclusions will be outlined here. If two inhibitors compete for the same binding site (i.e., are mutually exclusive) the interaction will be additive, regardless of the competitive, noncompetitive, or uncompetitive nature of their binding. If the inhibitors bind at different sites on the enzyme, so that they are mutually nonexclusive, the interaction will be dependent on the kinetic mechanism of the enzyme. In some cases, binding of a ligand at one site may influence the binding affinity of the other site for substrates and inhibitors. If this is not the case, e.g., for a rapid-equilibrium random-order mechanism, then the inhibitor effects will be independent; this is the situation described by Harrap and Jackson (1975) as "bicentric inhibition". Grindey and Cheng (1975) state that this situation leads to synergism; this reflects the fact that their definition of synergism differs from the one used here. Consider an enzyme with two substrates (A and B), with K_m values of K_a and K_b, respectively, and a rapid-equilibrium random-order mechanism. Inhibitor I1 binds at the A site with K_i value of K_{i1} and inhibitor I2 binds at the B site with K_i value K_{i2}. The rate equation is given by:

$$v = V_{max}\ A \cdot B\ /(K_a \cdot K_b)/(1 + A/K_a + I1/K_{i1})/(1 + B/K_b + I2/K_{i2}) \tag{5}$$

Now consider for the sake of example that $V_{max} = 400$, A = 1, B = 1, and all K_m and K_i values = 1. Then the uninhibited rate = 100. Now let I1 = 2. Substituting into Eq. 5 shows that this reduces the rate to 50 (i.e., $a = .5$). Similarly, setting I2 = 2 also gives 50% inhibition. If I1 and I2 are both equal to 2, then $v = 25$ ($a = .25$). This meets Harvey's criterion of independence, since this concentration of I2 gives 50% inhibition regardless of whether the reaction is already inhibited by I1 or not. However, if we take half quantities of both inhibitors (I1 = 1 and I2 = 1), then the reaction rate, v, will be 44.44; this is less than the rate ($v = 50$) given by the full dose of either inhibitor alone, and by Grindey and Cheng's definition qualifies as synergism. This kind of interaction would result in an isobologram plot that was slightly concave upwards.

Grindey and Cheng (1979) also describe another example of bicentric inhibition that results in strong synergism (by any definition). This is the situation where binding of inhibitor at one site potentiates binding of a second inhibitor at the other site. They discuss the example of PRPP amidotransferase, which has separate regulatory sites for adenine and guanine nucleotides. The enzyme was inhibited by methylmercaptopurine riboside (MMPR), which binds at the adenine nucleotide site; if guanosine was present, the inhibitory effect of MMPR was potentiated up to 35-fold. For a more detailed discussion of synergism involving simultaneous binding to the same macromolecular target, see Chapter 12 and the article by Chou and Talalay (1981).

C. CONCURRENT INHIBITION

The term "concurrent inhibition" was first used in print by Skipper *et al.* (1954)—though he attributes the expression to G. B. Elion—to describe the system, illustrated in Fig. 2, where two converging reactions that lead to a single product are simultaneously inhibited. This pattern of multiple inhibition was analyzed in detail by Webb (1963), who commented: "It is interesting that this is one of the few situations where individual inhibitions are additive, i.e., $i_1 + i_3 = i_{1,3}$." This type of effect, constant absolute inhibition, is greater than the degree of inhibition predicted for independence. This additivity is seen both in the transient phase and in the final steady state, and occurs whether the inhibitory effects are competitive or noncompetitive, and whether the system is open or closed.

The conclusion that concurrent inhibition effects are always absolutely additive does not hold, however, when either of the concurrent reactions is subject to feedback inhibition by the common product. This may be readily demonstrated using the NETWORK program presented in the appendix. If the feedback inhibition feature is enabled, so that reaction 1 is subject to feedback inhibition by B, this has two effects: it takes more of either inhibitor to get the same degree of inhibition (because the buffering power of the system is increased by the presence of feedback) and the combined inhibitor effects are now synergistic. An experimental example of concurrently acting agents that give synergism is the combination of rhodamine 123 and deoxyglucose, both of which inhibit ATP synthesis, the former by blocking oxidative phosphorylation, and the latter, glycolysis (Modica-Napolitano *et al.*, 1989).

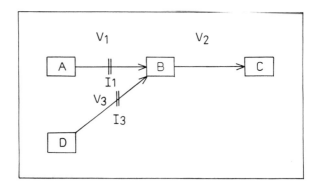

FIGURE 2   Concurrent inhibition of a convergent linear pathway. A and D, Alternative source materials for the intermediate B; C, final reaction product.

D. SEQUENTIAL INHIBITION

1. Sequential Monolinear Systems

The sequential monolinear pathway, such as the system illustrated in Fig. 1, was discussed from a historical perspective in Section II. The conclusions by Webb (1963) remain valid: for an unbranched, unregulated system, only inhibition of the first, rate-limiting enzyme can affect the flux through the pathway. Competitive inhibitors of subsequent enzymes in the chain may cause transient decreases in overall flux, but cannot influence the steady-state flux. Whether these transient effects can be considered therapeutically significant will depend upon the time constants for the system; the consensus view is that in most cases the transient effects will not contribute to therapeutic efficacy. Noncompetitive inhibitors of enzymes other than the first will, similarly, have no effect on steady-state flux unless the degree of inhibition is sufficient to make that enzyme rate-limiting for the system. If that occurs, than the flux through the pathway will be decreased, but the substrate before this step will tend to increase without limit, and the system will not achieve a steady state.

The situation where the input to a sequential monolinear system is subject to feedback inhibition, as studied by Grindey *et al.* (1975) differs in some important respects from an unregulated monolinear pathway. In a system with feedback, inhibitors of an enzymes other than the first may still decrease the overall steady-state flux through the system. Combinations of sequentially acting inhibitors may produce combined effects that are synergistic, additive, or antagonistic, depending upon the mechanism of inhibition and the nature of the feedback regulation. It most cases the qualitative nature of the interaction was parameter independent.

While the concept of a totally unbranched sequential pathway may appear to be something of a theoretical abstraction, there are probably systems within cells that approach this situation as an approximation. The careful consideration of the kinetics of sequential monolinear pathways has provided useful insights into the optimal design of combination drug regimens.

2. Sequential Convergent Systems

These system, exemplified by Fig. 3, represent the deployment of two sequentially acting inhibitors that straddle a convergent metabolic branch point. Note that the enzyme system is the same as that of Fig. 2, but the pattern of inhibitor use is different. Dual inhibition of this pathway was discussed by Harrap and Jackson (1975). The configuration has in common with the sequential monolinear pathway the fact that, because there is no alternative fate for B, competitive inhibitors of $v2$ cannot decrease the steady-state flux through the pathway, and noncompetitive inhibitors can decrease the overall flux only at the cost of making $v2$ rate-limiting for the pathway so

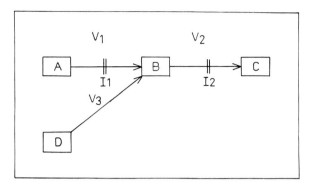

FIGURE 3   Sequential inhibition of a convergent linear pathway.

that a steady state is impossible. In the transient phase, inhibition in the presence of both inhibitors will in general be greater than in presence of one inhibitor alone, but the effects will never reach independence or synergism.

If there is feedback inhibition of reaction 1, it becomes possible for I2 to decrease the steady state flux through the pathway. The combined effects of I1 and I2 on this pathway may give antagonism, independence, or synergism, depending upon the configuration of the system.. The configuration may include such factors as the nature of the feedback inhibition, and the relative capacities of the various branches of the pathway. For the pathway of Fig. 3, let VM1, VM2, and VM3 equal V_{max} values for enzymes 1, 2, and 3, respectively. Clearly, as the ratio of VM3:VM1 becomes low, the system behavior approaches that of the sequential monolinear situation.

Another important determinant is the "sink capacity" of the system. This is defined as the ratio of total capacity for removing substrate B (VM2 in Fig. 3 or VM2 + VM4 in Fig. 12) to the total capacity for producing B (VM1 + VM3). When the sink capacity is high, the steady state concentration of B will be low, and enzyme 2 will be operating low on its saturation curve for substrate B. The presence of moderate amounts of I2 will result in an increase in the steady state level of B, but if the concentration of B in the inhibited steady state is still small compared with its K_m value then the kinetic order of reaction 2 with respect to B will still be close to 1. Thus addition of I1, which will decrease B, will not greatly affect this reaction order. When the sink capacity of the system is low, the steady state concentration of B will be high, and enzyme 2 will be almost saturated with B; in other words, the kinetic order of reaction 2 with respect to B is·close to zero. In this situation, the addition of I1 to the system, with resulting depletion of B, can cause a marked increase of the apparent kinetic order of reaction 2. As pointed out by Harvey (1978), other things being equal, synergism between I1 and I2 will occur if the presence of I1 increases the apparent kinetic order of the reaction catalyzed by enzyme 2. Thus the lower the sink capacity of the system, the greater the tendency for I1 and I2 to have a synergistic interaction. The other factor affecting the nature of the interaction is the proportion of B produced by reaction 1; the higher the ratio of v1 to v3 the greater the degree of synergism.

3. Sequential Divergent Systems

The addition of a divergent branch point to the sequential monolinear system, as in Fig. 4, makes an important difference to the behavior of the system: in the presence of an inhibitor I2, the system can now achieve a steady state, even when reaction 1 is not subject to feedback inhibition. In systems of this kind, I1 and I2 may be less than additive, independent, or synergistic. While the "less than additive" combined drug effect is often termed "antagonistic" (as it generally is in the present treatment, since our terminology follows that of Harvey, 1978) the interaction between I1 and I2 is never antagonistc in the sense that the combination of I1 and I2 gives less inhibition than either inhibitor alone. The interaction seen will depend upon three variables: the relative "sink capacity" for the system, the ratio of VM2 : VM4, and the ratio of $K_m2 : K_m4$ (where K_m2 and K_m4 are K_m values for enzymes 2 and 4, respectively, of Fig. 4). Adding a competitive inhibitor of enzyme 2 or enzyme 4 is equivalent to increasing K_m2 or K_m4, respectively. A noncompetitive inhibitor of enzyme 2 or enzyme 4 is equivalent to decreasing the V_{max} of these enzymes, and thus changing the VM2 : VM4 ratio. An inhibitor of enzyme 1, in effect, changes the relative sink capacity of the system. Any combination of I1, I2, and I4 may be understood in terms of these three effects. In the cases of I1, it is immaterial whether the mechanism of inhibition is assumed to be competitive or noncompetitive, since substrate A is a source material, and thus, by definition, is considered to be constant in concentration. In the case of I2, the interaction seen will depend upon the

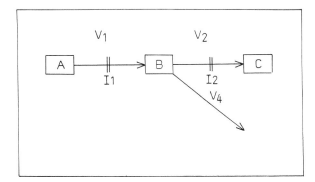

FIGURE 4   Sequential inhibition of a divergent linear pathway.

mechanism of inhibition. For noncompetitive inhibitors, synergism will be most pronounced when $K_m2 > K_m4$ and when $V_m2 < V_m4$, i.e., when the diverging branch accounts for most of the utilization of substrate B. When the two branches are exactly balanced, i.e., if $K_m2 = K_m4$ and $V_m2 = V_m4$, then I1 and I2 will show independence. The effect of decreasing the sink capacity of the system will be to increase the degree of synergism or antagonism given by noncompetitive I2 inhibitors in presence of I1. When I2 is competitive, the effect of the relative capacities of the diverging branches on interaction with I1 is difficult to predict; interactions will often be synergistic, but may be antagonistic. One generalization that can be made is that decreasing the sink capacity of the system will always tend to make the interaction more synergistic when competitive I2 inhibitors are used.

The effect of feedback inhibition of enzyme 1 (by substrate B) on this system is likewise impossible to predict simply by inspecting the kinetic parameters. When the sequential divergent pathway was studied using the NETWORK simulation (described in the appendix), for some parameter values, switching on the negative feedback feature made essentially no difference in the degree of inhibitor interaction. Where differences were seen, they were in the direction of lesser degrees of synergism. This is probably attributable to the fact that synergism tends to be greatest when I2 causes large perturbations in the pool of B, and feedback had the effect of minimizing such perturbations.

A rather different configuration of inhibitors in the sequential divergent pathway is that shown in Fig. 5. This system was discussed by Webb (1963), who pointed out that it has two interesting features: inhibition of enzyme 4 causes an increase in the overall flux through the main pathway (measured by production of C) and inhibition of enzyme 4 will always antagonize an inhibitor acting on enzyme 1. This is anatagonism in the strict sense, i.e., v2 in presence of I1 + I4 will be greater than v2 in presence of I1 alone.

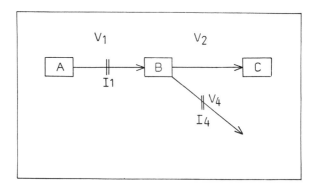

FIGURE 5   Inhibition of a branched linear pathway at the source reaction and at the branch reaction.

E. CYCLIC SYSTEMS

Because of the therapeutic value of combinations of sulfa drugs with dihydrofolate reductase inhibitors, discussed in Section IIC, and because the best description of that pathway is as a cyclic of oxidation and reduction of dihydrofolate, with new dihydrofolate synthesis feeding into the cycle, the kinetics of cyclic systems have received a good deal of attention. The first detailed treatment was by Webb (1963); he considered particularly inhibition of cyclic processes where the cycle intermediates were conserved, i.e., have no inputs or outputs. This will be approximately the case for many cofactor systems. In this case the intermediates have only a limited capacity to accumulate behind a metabolic block, so that inhibitors of two enzymes in a cycle will generally give more inhibition than either inhibitor acting alone, though in general additivity is not obtained. Grindey *et al.* (1975) modeled the system consisting of a cycle with a reaction feeding into it; this system had no outputs, so could not achieve a steady state. They thus considered this model to represent a system in exponential growth, and showed that an inhibitor of the cycle would be synergistic with an inhibitor of the reaction feeding into the cycle, a conclusion confirmed and extended by Harvey (1978). Jackson (1987) showed that this inhibitor combination (inhibition of a cycle enzyme and of a "feed-in" enzyme) could be synergistic even in a system with outputs, that could thus achieve a steady state.

The NETWORK model (Appendix) will simulate the system shown in Fig. 6; this has two outputs from the cycle, reactions 2 and 5. If we regard this as a model of the dihydrofolate cycle, then substrate B represents dihydrofolate, and substrate D is methylenetetrahydrofolate; I1 is then a sulfa drug, I4 a dihydrofolate reductase inhibitor, and I3 a thymidylate synthase

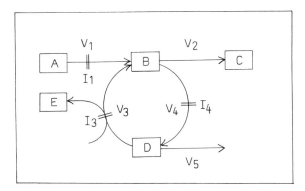

FIGURE 6   A cyclic pathway. V_1 is a reaction feeding into the cycle, and V_2 and V_5 are outputs from the cycle. E and B are both products of reaction 3, but B feeds back into the cycle and E does not.

inhibitor. Product E (thymidylate) may be considered to be the final output of the system, and its rate of production is given by v3. Substrate A is dihydropteroate, which acts as a source of newly-synthesized dihydrofolate in the dihydrofolate synthetase reaction, 1, and reaction 2 represents dihydrofolate breakdown, leading to degradation products (C). These steady state behavior of this system may be demonstrated by running the NETWORK program, with the "Enable cycle" command (no. 8 in the control table) set ON. Now enter the PARAMETER EDITOR (command 3 of the main menu) and set VM1 to 0.2, VM2 to 0, and VM3 to 1; otherwise use the default parameter values. This system achieves a steady state with v3 = 0.50. The IC_{50} value for I1 (the value that decreases v3 to 0.25) is 1, and the IC_{50} for I3 (when I3 is noncompetitive) is also 1. When I1 and a noncompetitive I3 are now used in combination, the steady state rate of v3 is 0.125 (25% control); thus I1 and I3 are independent. For a competitive I3, the IC_{50} value is 2, and I1 used in combination with a competitive I2 gives steady state v3 of 0.10 (20% control). Thus when I3 is competitive, it is synergistic with I1. The degree of synergism is affected by the ratio of K_m3 to K_m5; when this ratio is low, the synergism is greatest. When I3 is noncompetitive, however, I1 and I3 are independent regardless of the K_m ratio. With the parameters set to the same values, the system cannot achieve an inhibited steady state in the presence of an inhibitor of reaction 4 (unless we assume feedback inhibition of reaction 1). However, if we add a divergent branch point at B (reaction 2), the system will show analogous interactions between I1 and I4 to those found between I1 and I3. That is, if I4 is noncompetitive, the combination of I1 and I4 shows independence, and if I4 is competitive, the combination shows synergism. Cyclic systems, in summary, show complex responses to multiple inhibition, and the cofactors involved in redox reactions or one-carbon group transfers undergo multiple overlapping cycles. Probably the best way to predict drug interactions in such systems is to write detailed kinetic models of them.

F. SIMULTANEOUS INHIBITION AND SUBSTRATE DEPLETION OF MULTISUBSTRATE REACTIONS

1. Bicentric Inhibition

The situation where an enzyme with two active centers is simultaneously inhibited at each binding site was discussed in Section IVB. A number of interesting examples have been described of this "multiple hit" approach to metabolic blockade, such as the use of selenazofurin (whose active metabolite blocks IMP dehydrogenase at its NADH product site) plus ribavirin (whose 5'-phosphate inhibits the same enzyme at its IMP site) to give synergistic antiviral activity (Huggins et al., 1983). For many enzymes, binding an inhibitor at one site appears to potentiate binding at the other site; examples are the inhibition of PRPP amidotransferase by methylmercaptopurine ribotide and guanosine 5'-phosphate (Section IVB) or the inhibition of thymidylate synthase by 5-fluorodeoxyuridylate (binding at the deoxyuridylate site) and 5,8-dideaza-10-propargylfolate (which binds at the methylenetetrahydrofolate site). Such inhibitor interactions should be predictable from the enzyme rate equation, though for many of the most interesting cases, accurate rate equations have not been derived. Other combinations of this type are discussed in Chapters 12 and 13.

The pathway shown in Fig. 7 depicts an enzyme (enzyme 2) that catalyzes the reaction of two substrates (B and E) with 1:1 stoichiometry. Substrate B is produced from A in reaction 1, and E is produced from D in reaction 3. The situation where the two required substrates of enzyme 2 are both depleted by inhibition of the input reactions with I1 and I3 was described as "bicentric depletion" (Harrap and Jackson, 1975; Grindey and Cheng, 1979). In systems of this kind, if reactions 1 and 2 are irreversible, and B and E have no other possible routes of removal, a steady state is only possible if v1 = v3, and in this case these input rates will determine the steady-state flux of the system. Otherwise the flux will depend, in the long run, on whichever of v1, v2 or v3 is rate limiting. On inhibition of enzyme 1 or enzyme 3, the level of B or E, respectively, will fall. At first the reduction in v2 will be delayed by a compensatory rise in the other substrate. This rise will be retarded if the second inhibitor is present, and during this transient period the effects of the two inhibitors approach independence. However, v2 continues to decline until it becomes equal to whichever of v1 and v3 is slowest, and at this stage the system is indifferent to the inhibition of the non-rate-limiting process (Harrap and Jackson, 1975).

If the pathway of Fig. 7 is inhibited by I1 and I2 (where I2 inhibits enzyme 2 at the E-binding site), the system presents an example of "bicentric

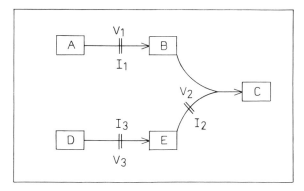

FIGURE 7   A pathway containing a two-substrate enzyme. I_2 is an inhibitor of enzyme 2 that blocks at the binding site for substrate E. The combination of I_1 and I_3 represents a bicentric depletion, and the combination of I_1 and I_2 gives bicentric inhibition and depletion.

inhibition and depletion." In this system inhibition of enzyme 2 will be compensated in the early stages by accumulation of both B and E; this build-up of B is slowed down by I1, so in the early part of the transient phase the effects of I1 and I2 are almost independent; as time goes on the overall flux approaches that of the rate-limiting step. If the system is made more complex by the addition of feedback regulation of enzyme 1 or enzyme 3, or by divergent branch reactions for B or E, or both, the system may be able to attain a steady state in the presence of I2, and combined inhibitor effects for I1 and I2 (or I1 plus I3) may be greater than the effect of either inhibitor separately.

2. Complementary Inhibition

Many experimental observations have demonstrated that DNA-damaging drugs, such as alkylating agents, often show therapeutic synergism with antimetabolites. Sartorelli (1969) suggested the term "complementary inhibition" for this combination of damage to an essential macromolecule with depletion of the precursors of that macromolecule. This is a complex situation to analyze. The potentiating effect of antimetabolites on DNA damage may be caused by inhibition of DNA repair, but many other effects appear to contribute to the therapeutic effect, e.g., antimetabolites are usually most active against proliferating cells, whereas alkylating agents are very active against quiescent tumor cells. From the viewpoint of enzyme kinetics, complementary inhibition may be considered a special case of bicentric inhibition and depletion, if DNA is regarded as one substrate of DNA polymerase, and antimetabolites act by depletion of the other substrates of the DNA polymerase reaction, the deoxyribonucleoside triphosphates. For further discussion of synergism at the level of DNA, see Chapter 15.

V. Multiple Inhibition in Highly Regulated Systems

A. THE EFFECTS OF NEGATIVE FEEDBACK

As we discussed in Section IIIA, a fundamental attribute of living systems is their ability to maintain homeostasis; they are open systems that exist at or near steady states which are locally stable. This stability is largely achieved through negative feedback, typically feedback inhibition of the key rate-determining enzyme of a metabolic pathway either by an immediate reaction product, or more often by a metabolically distant end product of that reaction pathway. Negative feedback has important implications for the regulatory properties of pathways. As pointed out by Grindey *et al.* (1975), feedback inhibition in a sequential monolinear pathway makes it possible for the system to achieve an inhibited steady state after some enzyme in the sequence other than the first, rate-limiting enzyme has been inhibited. Depending upon the mechanism of the feedback effect, the presence of feedback also influences the interaction between multiple inhibitors in a monolinear pathway. For example, the concurrent inhibition of the two arms of a divergent pathway is normally additive, but in the presence of feedback inhibition of either arm the interaction may become synergistic (Section IVC). Conversely, in the sequential divergent pathway the effect of switching on feedback inhibition of input reaction is usually to make the inhibitor interaction less synergistic. In both cases, the difference is attributable to the fact that feedback inhibition tends to oppose the effect of an inhibitor, by minimizing the metabolite pool perturbations caused by it.

B. THE POWER OF POSITIVE FEEDBACK

While the biological function of negative feedback is to maintain stability, the role of positive feedback is to make it possible for the organism to switch rapidly from one state to another. There is probably no living system, however humble, that spends all its life in a single steady state; even the simplest prokaryote must be able to switch between proliferating and quiescent states, or adapt to changing nutrient availability. Biological switch mechanisms, like many electronic switches, exploit the properties of positive feedback. As the name implies, positive feedback is the situation (also termed "autocatalysis") in which a metabolite directly or indirectly stimulates its own production. Examples of this are fairly common in biology. For example, the digestive enzyme trypsin is secreted from the pancreas into the gut as a larger, inert, precursor, trypsinogen. Trypsinogen is activated into trypsin by proteolytic activity in the gut, and once a few active trypsin molecules are formed they catalyze the conversion of the remaining trypsinogen. A system like this thus has two stable states: either all trypsinogen and no trypsin (which is presumably how it is stored in the pancreas) or all active trypsin. The presence of positive feedback in a biological system generally indicates that it may act as a switching mechanism between two (or more) stable states that are generally completely off or completely on. This may be demonstrated with the NETWORK model by turning on the positive feedback feature (command 5 of the Control Menu). This will model the system of Fig. 8. In this system the product of the first reaction (B) stimulates its own production at low concentrations, but inhibits at higher concentrations. This bimodal effect of a reaction product on an enzyme has been described for a number of important regulatory systems; for example, ribonucleotide reductase's GDP-reductase reaction is both stimulated (at low concentrations) and inhibited (at high concentrations) by dTTP. Adenosine kinase is both stimulated and inhibited by its indirect product, ATP. Reaction 2 of the Fig. 8 system must

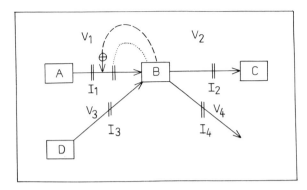

FIGURE 8 A branched linear pathway in which the primary source reaction is subject to feedback activation by low concentrations of its product, B, and to feedback inhibition by high concentrations of B (the dashed line indicates an activation, and the dotted line indicates inhibition).

have higher-order kinetics with respect to substrate B; in the NETWORK program it is assumed to have a Hill coefficient of 2. The switching properties of this system may be demonstrated by running a simulation in the absence of inhibitors, using the default parameter values (some of which will be automatically reset by the model); the system achieves a steady state with B = 0.4 (approximately) and v2 = 1.3. If the effect of inhibitors is now studied, it is found that I1 at very low concentrations has no effect on total flux (v2), but beyond a certain threshold level it will abruptly turn off the pathway almost completely; e.g., setting I1 = 1 gives >94% inhibition of total flux. If I1 is now removed, and the simulation continued, the rate of v2 does not return to its uninhibited value, but increases only to 16% of it, so that the system is still essentially switched off. The implications of this are that, for a given set of parameter values, the system may have more than one stable steady state (an "on'" state and an "off" state), and that the system displays hysteresis, in other words, such kinetic systems "remember" that they have been inhibited, even after the inhibitor has gone away. In fact the concept of these systems having a "memory" is not a bad analogy, because the mathematical properties of these biochemical networks are rather similar to those of the electronic flip–flop circuits used in computer memory chips. A detailed discussion of the kinetics of multistable systems is beyond the scope of this chapter. The most appropriate treatment is that of catastrophe theory, which deals with the study of abrupt transitions in systems of continuous variables (Zeeman, 1976); a biochemical example from the area of purine metabolism is discussed by Jackson (1987).

Not surprisingly, multistable systems, with their very steep inhibitor dose–response curves, can show very marked drug–drug interactions. In the system of Fig. 8, studies with the NETWORK program show that inhibitors of enzyme 1 may show strong synergism with inhibitors of enzyme 3, and strong antagonism with inhibitors of enzymes 2 and 4. These biological switching systems present a very appealing possibility for the development of highly selective anticancer drug combinations. A tumor cell is a cell whose switching mechanisms have become insensitive or unbalanced; if inhibitor combinations could be devised that could switch off DNA replication in the tumor, but (because of small, quantitative biochemical differences) leave it unchanged in normal tissues, this would provide unprecedented selectivity. At present, we are a long way from being able to do this.

C. MULTIPLE INHIBITION OF RIBONUCLEOTIDE REDUCTASE

The enzyme ribonucleoside diphosphate reductase (ribonucleotide reductase) is considered by many to be the single most important rate-determining control point for DNA synthesis in eukaryotic cells (Cory and Cory, 1989). This enzyme is the subject of an unusually elaborate variety of positive and negative feedback regulation, summarized in Fig. 9. This pathway has been modeled for one of the best studied experimental systems, rat hepatoma cells, in a computer model called HEPATOMA (Jackson, 1986). This program is available on request from the author. Not surprisingly, the effect of inhibitors on so complex a system can sometimes be difficult to predict intuitively. Noncompetitive inhibitors of ribonucleotide reductase, such as hydroxyurea,

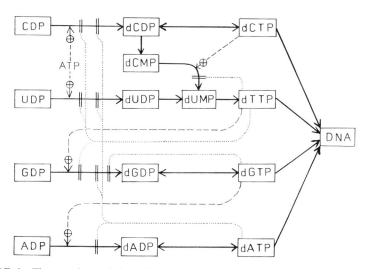

FIGURE 9   The reactions of deoxyribonucleoside triphosphate production and utilization. Dashed lines indicate activation effects and dotted lines show inhibitory effects.

result in inhibited steady states, in which the balance of the four deoxyribonucleoside triphosphate pools is altered, and in which the rate of DNA synthesis is decreased. However, the predicted effect of ribonucleotide reductase inhibitors that bind at a regulatory site is far stranger. An example of such an inhibitor is the 5'-triphosphate of 2-chloro-2'-deoxyadenosine, which is believed to bind at the dATP regulatory site. The HEPATOMA model predicts that this compound gives an "all or nothing" dose–response relationship. This highly nonclassical behavior may seem surprising to enzymologists (though not, perhaps, to neurophysiologists) but it is explicable in terms of the switching properties of a positive feedback system. Consider the highly simplified version of DNA synthesis outlined in Fig. 10. In this simplified pathway we ignore the contribution of dCTP and dATP to DNA synthesis (reaction 3), and consider it to be dependent upon the concentrations of dTTP and dGTP. Reaction 1 represents the rate of dTTP production, which is feedback inhibited by dTTP. Reaction 2 is the GDP reductase reaction of ribonucleotide reductase, which is activated by dTTP at low concentrations, but inhibited by dTTP at high concentrations. If, as is often the case, DNA contains equal amounts of thymidine and deoxyguanosine, the system of Fig. 10 will reach a steady state when v1 = v2 = v3. The pool of dGTP will automatically tend to reach a level such that v2 = v3, since if dGTP is produced faster than it is utilized the pool will increase and accelerate the rate v3. The solid line of Fig. 11 thus represents both v2 and v3 as a function of the dTTP concentration. The broken line in Fig. 11 represents v1 as a function of concentration of dTTP, showing the feedback inhibition by dTTP. Points where the broken line (v1) and the solid line (v2 = v3) intersect represent steady-state points for the system where v1 = v2 = v3; at these points dTTP and dGTP are being produced and consumed at equal rates, and thus remain constant in concentration. Two such points may be seen in Fig. 11

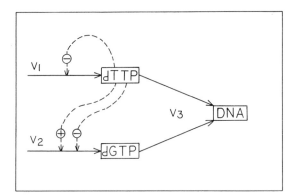

FIGURE 10   A simplified model of some reactions of dTTP and dGTP. Solid lines show reaction pathways, dashed lines indicate positive or negative regulatory effects.

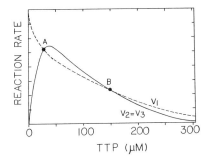

FIGURE 11   Reaction rates for the model of Figure 10. The rate equations were:

$V_1 = VM1/(1 + dTTP/K_{i1})$

$V_2 = VM2 \cdot dTTP/K_a/(1 + dTTP/K_a)/(1 + [dTTP]^2/K_{i2})$

$V_3 = VM3 \cdot dGTP/K_{mG}/(1 + dGTP/K_{mG}) \cdot dTTP/K_{mT}/(1 + dTTP/K_{mT})$

The system was analyzed for the following parameter values:

VM1, 8×10^{-5}; K_{i1}, 5×10^{-5}; VM2, 1×10^{-4}; K_a, 2×10^{-5};

K_{i2}, 6×10^{-9}; VM3, 9.8×10^{-5}; K_{mG}, 3×10^{-6}; K_{mT}, 1×10^{-5}

(marked A and B). In addition, a third steady state exists, in principle, where dTTP tends to infinity, such that v1 = v2 = v3 = 0; we shall refer to this as the OFF state. Now consider the behavior of the system in the vicinity of steady point A. If [dTTP] is perturbed downwards, v1 tends to increase, and v3 decreases. Thus dTTP is produced faster and used more slowly, resulting in restoration of the system to point A. Conversely, if [dTTP] is increased slightly, v1 will decrease, but as long as the system remains to the left of point B, v3 is still greater than v1. Thus dTTP is used faster than it is made, and the system drops back to point A. Point A is stable to local perturbations. Now consider point B: v1 = v2 = v3, so that system is in a steady state. Following a downward perturbation in dTTP, both v1 and v3 will increase, as a result of release from dTTP inhibition, but v3 will increase faster than v1, causing a further drop in dTTP. Thus the system will not revert to point B, but will move down to point A. Conversely, an upward perturbation in dTTP when the system is at point B will result in ever-increasing dTTP as the system moves toward the OFF state. This system is analyzed in detail by Jackson (1989). Point B, though a steady state, is not a stable steady state; it is what is known as a "saddle point" that acts as a watershed between the ON state at point A, and the OFF state. Now consider this system in terms of an inhibitor that perturbs the pool of dTTP: any perturbation that moves the system anywhere to the left of point B will result in the system adjusting back to point A, so there will be no inhibition at the steady state. However, if the system is perturbed so that the dTTP now falls to the right of point B, the system will switch off DNA synthesis. There is no stable intermediate point. Of course, comparing Fig. 11 with the more realistic model of Fig. 9, it is clear that Fig. 11 is a grossly oversimplified model of DNA precursor turnover; however, the detailed model of Fig. 9 (HEPATOMA) shows qualitatively similar behavior. As a result, this model predicts that inhibitors binding to the ribonucleotide reductase dATP site may give this "all or nothing" dose–response relationship.

In the face of these curious dose–response curves, equally striking drug–drug interactions are possible. Thus the HEPATOMA model predicts that reducing the total activity of ribonucleotide reductase with hydroxyurea could shift the "saddle point" of the system and change a cell that does not respond to 2-chloro-2'-deoxyadenosine 5'-triphosphate into a cell that responds by switching off DNA synthesis, a case of striking synergism. Most of these predictions still await experimental verification. However, it may be safely concluded that biochemical systems that exhibit positive feedback are involved in switching behavior, and that such systems are likely to demonstrate exaggerated drug–drug interactions. From examination of Fig. 9, it is clear that the pool of DNA precursors is an excellent example of a cellular switch, and a potentially exploitable site for the design of highly selective therapeutic drug combinations.

VI. Approaches to a General Theory of Multiple Inhibition

A. FLUX CONTROL COEFFICIENTS

The various simplified pathways discussed in Section IV made it possible to draw certain generalizations concerning how inhibitors of different sites are likely to interact, e.g., in the concurrent pattern of inhibition the combined effects are additive (Section IVC). However there were numerous situations where, depending upon the detailed configuration of the system, interactions could be synergistic, additive, or antagonistic, and the situation tends to be further complicated by the presence of feedback effects. Prediction of the effects of inhibitors on metabolic pathways is a branch of biochemical control theory, a subject that is still actively developing. One of the most versatile treatments is the use of control coefficients, as described by Kacser and Porteous (1987), to express the dependence of the flux through a pathway upon the activity of a particular enzyme in the pathway. In a linear pathway of n enzymes, the flux control coefficient for enzyme i is defined as:

$$C_i = dF/dE_i/(dF/dE_1 + dF/dE_2 + \cdots dF/dE_n) \tag{6}$$

where F is the steady-state flux through the pathway, and E_1, E_2, and E_n are concentrations of the first, second, and n^{th} enzymes in the pathway. The control coefficients thus indicate the relative extent to which activity of each enzyme influences steady state flux through the pathway, regardless of the presence of convergent or divergent branches, or of feedback effects. Note that the sum of the control coefficients C_i (for $i = 1$ to $i = n$) = 1. For an unbranched linear pathway in which steady state flux is totally determined by activity of the first enzyme, $C_1 = 1$, and coefficients for subsequent enzymes are zero. We can use the NETWORK model to determine control coefficients by making small changes in the V_{max} values, and noting the effect on v2. Consider the pathway of Fig. 12. If we consider the main pathway to be that leading from A through B to C, then

$$C_1 = dv2/dVM1/(dv2/dVM1 + dv2/dVM2) \qquad (7)$$

and

$$C_2 = dv2/dVM2/(dv2/dVM1 + dv2/dVM2) \qquad (8)$$

where VM1 and VM2 are the V_{max} values for enzymes 1 and 2, respectively. We can use small finite differences to approximate the differentials, say increase the V_{max} values by 5%, and note the effect on v2. Consider the pathway of Fig. 12 in which I1 is present at its IC_{50} value (i.e., the concentration that decreases the steady state v2 value by 50%). We now ask: will inhibiting reaction 2 in addition (by adding I2) make any difference to the steady state flux? If C_2 is zero (as it would be for the unbranched monolinear system) this would suggest that adding I2 could not cause any further inhibition of the steady state value of v2; on the other hand, if C_2 is large then the activity of enzyme 2 has an effect on the steady state flux, even in presence of I1, so adding I2 should cause further inhibition. This approach can be illustrated by considering three configurations of the NETWORK model of Fig. 12. In all of them we let VM3 = 0, so that the system is a sequential divergent one. Consider first a pathway in which the two diverging branches are

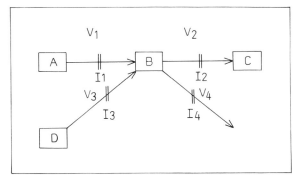

FIGURE 12   Reactions of the NETWORK model. V_1–V_4 indicate reaction rates, and I1 and I4 are inhibitors of reactions 1–4.

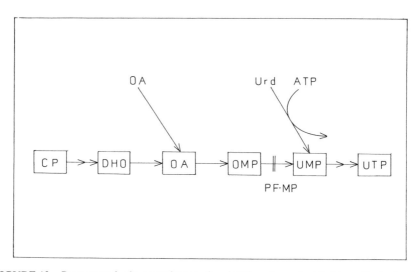

FIGURE 13   *De novo* and salvage pathways of pyrimidine ribonucleotide biosynthesis showing sites of competitive and noncompetitive antagonism of pyrazofurine 5′-phosphate by exogenous orotic acid and uridine, respectively.

equally balanced, with VM2 = 5 and VM4 = 5; unless otherwise indicated, all other parameters are left at their default values. Set I1 at its IC_{50} concentration of 1. This reaction network gives a flux (in the presence of I1) of 0.25 and the control coefficient for enzyme 2 is 0.316, suggesting that inhibition of enzyme 2 should give further reduction in the steady flux. If we now set I2 = 1, competitive (keeping I1 at 1), this concentration of I2 (which alone reduces v2 to 69% of control) gives a v2 value of 34.5%. This represents independence, since $0.5 \times 0.69 = 0.345$. Now set VM2 = 0.3, VM4 = 0.9 and $K_m4 = 10$; the IC_{50} for I1 in this system is 3.46, and the inhibited flux is 0.146. The enzyme 2 control coefficient is 0.40, greater than for the first example, suggesting that the effect of adding I2 should be more than additive; in fact when we add I2 = 1 (competitive) which on its own decreases the flux to 97.6% control, the flux is decreased to 40.5% control. This is a case of synergism, since the activity predicted for independence would be $0.5 \times 0.976 = 0.488$. If we now study the system with VM2 = 0.9, $K_m2 = 10$, VM4 = 0.3, $K_m4 = 1$, this pathway is approaching the unbranched sequential situation, since the capacity of the divergent branch is much less than the capacity of the main (enzyme 2) pathway. When we set I1 to its IC_{50} concentration of 0.629 (giving an inhibited flux of 0.708) the control coefficient for reaction 2 is 0.069, showing that the activity of enzyme 2 has very little influence on steady-state flux. Now adding I2 = 1 (competitive) does, in fact, cause very little additional inhibition. Thus, in the presence of I1, the control coefficients for reaction 2 appear to have some predictive value for additional inhibition caused by small amounts of I2.

This approach, interesting as it is, has two major limitations. First, if we happen to know, for a particular system, what control coefficient corresponds to an independent effect of I2, then we can predict for related systems whether we will get synergism or antagonism for small amounts of I2 from whether the enzyme 2 control coefficient is greater or less than that of the "independent" system; however, we cannot predict, in the first instance, what control coefficient does correspond to independence. Second, control coefficients predict the effect of small perturbations only; strictly speaking, they predict the effect of infinitesimal perturbations. Thus this approach does not help us predict the effect of adding, say, an IC_{50} concentration of I2 to an IC_{50} concentration of I1. Unfortunately, in a therapeutic situation the IC_{50} level of an inhibitor is usually the least amount that is likely to be of interest.

B. HARVEY'S RULE

Undoubtedly the single most important advance in predicting combined inhibitor effects is the rule described by Harvey (1978). Harvey considered systems in which one inhibitor acts directly on an enzyme (which would correspond to enzyme 2 in our diagrams) and the other inhibitor (which would correspond to I1 or I3 in Fig. 12) affects the rate of reaction 2 (v2) by decreasing the concentration of substrate B. Harvey stated that whether the interaction of the two inhibitors was synergistic, independent, or antagonistic depended on whether the effect of the direct inhibitor (I2) upon the apparent kinetic order of its target enzyme is, respectively, less than, equal to, or greater than I2's effect on the fractional substrate decrement produced by the remote inhibitor (I1 or I3). These terms are defined as follows. The apparent kinetic order of reaction 2 in presence of I1 is

$$N_1 = dv_1/dB_1 \cdot B_1/v_1 \qquad (9)$$

and the apparent kinetic order of reaction 2 in presence of both I1 and I2 is given by:

$$N_{1,2} = dv_{1,2}/dB_{1,2} \cdot B_{1,2}/v_{1,2} \qquad (10)$$

where v_1 is the rate of reaction 2 in presence of inhibitor 1, and $v_{1,2}$ is the rate of reaction 2 in presence of both I1 and I2. B_1 is the concentration of substrate B in presence of I1, and $B_{1,2}$ is its concentration in presence of both I1 and I2 (Savageau, 1976; Harvey, 1978). (Note that the present numbering differs from Harvey's: my I1 is Harvey's I2, and vice versa.) For an enzyme that follows a Michaelis–Menten equation, the apparent kinetic order will be close to 1 when B is much less than its K_m value, and will approach zero as B nears saturation. When an enzyme has more than one binding site for a substrate, the apparent kinetic order may be greater than 1, as for an enzyme that follows a Hill equation with Hill coefficient > 1.

In the presence of I1, the concentration of substrate B will be depleted. Harvey defines the specific substrate decrement due to I1 as:

$$F_1 = 1/B_1 \cdot dB_1/dI1 \tag{11}$$

and in the presence of I1 and I2:

$$F_{1,2} = 1/B_{1,2} \cdot dB_{1,2}/dI1 \tag{12}$$

Harvey's Rule then states that the interaction of I1 and I2 will be synergistic if

$$F_{1,2}/F_1 > N_1/N_{1,2} \tag{13}$$

independent if

$$F_{1,2}/F_1 = N_1/N_{1,2} \tag{14}$$

and antagonistic if

$$F_{1,2}/F_1 < N_1/N_{1,2} \tag{15}$$

Of course, determining estimates of apparent kinetic order and specific substrate decrement experimentally may well be more trouble than studying the drug–drug interaction experimentally. If the kinetics of the system under study are well understood, then these parameters may be determined by modeling, using a program such as NETWORK. Of course, NETWORK may also be used to predict drug–drug interactions without having recourse to Harvey's Rule. The real importance of Harvey's Rule is not its predictive power, though it certainly possesses that, as much as its ability to provide a very general theoretical foundation for the otherwise bewildering variety of interactions that can emerge from multiple inhibition of even a very simple metabolic network.

VII. Conclusions: Combining Inhibitors to Optimize Selectivity

A. THE BIOCHEMICAL BASIS OF ANTICANCER DRUG SELECTIVITY

Anticancer drug selectivity arises from differential effects at three levels of organization, namely the pharmacokinetic level (differences in drug distribution, accumulation, retention or elimination between the tumor and the normal tissues), the cytokinetic level (differences in cell cycle time, or cell loss factor, or the proportion of quiescent cells), and the biochemical level. Since this chapter has been concerned with drug interactions at this last level, it is necessary to inquire in some detail about the causes of intrinsic differences (where they exist) between drug sensitivities of malignant and normal cells.

Unlike the situation in antimicrobial chemotherapy, where inhibitors can be designed that inhibit an enzyme in the pathogen without inhibiting a corresponding enzyme of the host organism, this approach to selectivity is not usually possible in cancer chemotherapy. The very high antibacterial selectivity of penicillins, for example, is due to the fact that these compounds inhibit reactions of bacterial cell wall biosynthesis that have no counterpart in mammalian cells. The antibacterial drug, trimethoprim, on the other hand, inhibits bacterial dihydrofolate reductase without strongly inhibiting the human enzyme. This kind of selective inhibition is possible because the dihydrofolate reductases of bacteria and humans have enough structural dissimilarity to enable the design of selective inhibitors. In general tumor enzymes do not show kinetic differences from the corresponding enzymes of normal cells. Occasional reports exist of tumor enzymes that have an altered dissociation constant for a particular inhibitor, or show loss of sensitivity to a particular feedback effector. In some cases such effects may be the result of a somatic mutation; in others, they may represent expression of a fetal isozyme. In general, however, cancer chemotherapy must exploit purely quantitative differences in amounts of enzymes (or other drug target molecules) between the normal and malignant cells.

The various kinds of quantitative differences that can be exploited have been reviewed by Jackson (1986). The major factors are described subsequently.

1. Activity of the Target Enzyme

Elevated activity of a target enzyme decreases sensitivity to inhibitors, whether the inhibitor is competitive or noncompetitive, and whether the inhibition follows classical kinetics or is tightly binding. This conclusion holds whether the enzyme is normally rate limiting for its pathway or not. Simulations with the NETWORK model show that quite small differences in enzyme activity, as may occur between normal and tumor cells, can make an appreciable difference in the IC_{50} value.

2. Drug Transport Differences

Some drugs, including classical antifolates and many nucleoside analogs, require active transport into cells. The activity of the membrane transport carriers involved in mediating this uptake varies widely between different cell types. Fetal cells, which are proliferating rapidly, and require a rich supply of nutrients, tend to have particularly high activity of these transport systems, and so do some tumor cells. Such cells are likely to be particularly sensitive to those drugs that are transported by the carrier systems concerned.

3. Rates of Competing Enzymes at Metabolic Branch Points

In general, a reaction that competes for a particular substrate at a divergent branch point will increase the sensitivity of the system to inhibition of the main enzyme after the branch point; e.g., in Fig. 12, increasing the V_{max} of enzyme 4 increases the sensitivity of the system to I2. In fact, increasing the V_{max} of enzyme 4 has a generally similar effect to decreasing the V_{max} of enzyme 2; the main qualitative difference is that the former will tend to decrease the pool of B, whereas the latter will increase it. This in turn may influence the kind of interaction that is seen with other inhibitors.

4. Rates of Drug Activation or Inactivation

Examples of rates of drug activation or inactivation contributing to drug selectivity are well known, and include differences in conversion of purine bases to active nucleotides, and differences in breakdown of the active nucleotides by nucleotidases. Activity of bleomycin hydrolase is believed to contribute to the relative insensitivity of many normal tissues to that agent. A recent example of selective drug activation by tumor cells is the polyglutamylation of methotrexate, which appears to be more extensive in certain tumors than in normal tissues. Clearly the various reactions of drug activation and inactivation provide points where drug activity may be potentiated by modifying agents. However, in such cases there is a risk that the potentiation will be at the price of selectivity; an inhibitor of bleomycin hydrolase might increase the potency of bleomycin slightly in a tumor cell where the rate of inactivation is low, at the price of greatly increasing bleomycin toxicity to normal tissues.

5. Pool Sizes of Competing Metabolites

In discussing the effects of competing enzymes at metabolic branch points, it was noted that increasing the rate of a competing enzyme depleted the pool of the metabolite at the branch point, particularly in the presence of an inhibitor of the enzyme immediately after the branch point on the main pathway. This pool depletion increases the efficacy of a competitive inhibitor, because there will be less competition at the active site; however, it can also increase the effectiveness of noncompetitive inhibitor by decreasing the degree of saturation of the enzyme with its substrate. Other reactions that modulate metabolite pool sizes will also influence the effectiveness of inhibitors of enzymes that utilize that metabolite. One well studied example is AraCTP, the active metabolite of cytarabine, whose efficacy is decreased in the presence of a large pool of the natural DNA precursor, dCTP (of which AraCTP is an analog).

6. The Activity of Alternative Pathways

The previous example showed that alternative pathways may influence drug selectivity by perturbing precursor pools. The essence of this situation is that the alternative pathway feeds into the main pathway before the site of action of the inhibitor. The pioneer of antimetabolite therapy, D. D. Woods, distinguished between the competitive antagonism give by metabolites that feed in before an antimetabolite site and the noncompetitive antagonism given by metabolites that feed in after the antimetabolite site. In other words, the latter will by-pass the inhibitor effect completely. A well-known example is provided by the salvage pathways of nucleotide biosynthesis: whereas the activity of an orotidylate decarboxylase inhibitor (such as pyrazofurin 5′-phosphate) is competitively antagonized by adding orotate, it is circumvented concompetitively and completely by adding uridine, which feeds into the pyrimidine nucleotide pools through the uridine kinase salvage reaction (Fig. 13). Thus the selectivity of purine and pyrimidine antagonists is extensively influenced by the ratio of *de novo* to salvage biosynthetic activity in different tissues.

B. SELECTIVITY VERSUS POTENCY

Most of the preceding discussion was concerned with the ways in which multiple inhibitors of metabolic pathways can be combined to produce various kinds of drug–drug interactions: antagonism, independence, or synergism. Implicit in this review has been the attitude that synergism is normally a desirable outcome; a synergistic drug interaction tends to be unthinkingly assumed to represent greater therapeutic activity. However, a consideration of the factors involved in anticancer drug selectivity suggests that, for anticancer agents at least, this is not necessarily the case. Drug–drug interactions of anticancer agents are often studied *in vitro*, which provides precise information on drug potency, but none on drug selectivity. When the more complex *in vivo* situation is considered, three general patterns of interaction are observed. Sometimes a synergistic interaction that was found *in vitro* is also seen *in vivo*. This may be expected to be the case when the dose-limiting toxicity of a drug is unrelated to its therapeutic mechanism of action. In the case of cisplatin, as normally used, the dose-limiting toxicity is nephrotoxicity, which appears to be mechanistically unrelated to the DNA cross-linking activity that causes its antitumor effect. In that case, a drug combination that enhances cisplatin's antitumor effect may also increase toxicity to normal tissues, but would not be expected to potentiate the nephrotoxicity, so a therapeutic synergism may be obtained. Sometimes two drugs give an *in vivo* synergistic interaction even when their dose-limiting toxicity is mechanistically related to their antitumor effect. This is often the case when, for cytokinetic or pharmacokinetic reasons, the toxicity profiles of the drugs are non overlapping, e.g., one agent is most toxic to bone marrow, while the dose-limiting toxicity for the other may be gastrointestinal. For such drugs, the combination toxicity index (defined in Section IC) will be greater than 1. Perhaps a more common situation is that in which a particular combination of agents is synergistic *in vitro* but the *in vivo* therapeutic index is unchanged; in this case the drug–drug interaction has increased the dose potency of the primary agent, but not its selectivity. This is a likely outcome where the mechanism of the dose-limiting toxicity is identical with the mechanism of the therapeutic effect. Thus the drug's potency is increased by the interaction, but its toxicity is increased by a similar factor, so that there is no therapeutic advantage.

Synergistic drug interactions may also be observed where the therapeutic index of the combination is actually less than for the best of the single agents; in such cases the therapeutic potency of the primary agent has been enhanced, but its toxicity to normal tissues has been increased even more. The reasons for interactions of this kind are probably usually a complex mixture of biochemical, cytokinetic and pharmacokinetic factors that are almost impossible to predict intuitively, difficult to model, and one of the most compelling reasons for even the most enthusiastic of computer modelers to still recognize the need for continued animal research. However, sometimes careful study of the kinetics of a metabolic pathway may suggest that a particular drug interaction, though synergistic, is unlikely to contribute to *in vivo* selectivity. An example that may be studied with the NETWORK model is the sequential divergent combination of drugs in a pathway with a low sink capacity, and high relative capacity but low substrate affinity of the branch pathway (Fig. 4). In this case, inhibition with I2 will cause a very large rise in the pool of substrate B at the steady state, and addition of I1, by opposing this large rise in [B], will be highly synergistic with I2. However, this particular combination of pathway characteristics, while undoubtedly maximizing the synergistic interaction with I1, also minimizes the potency of I2 as a single agent. Thus, in a mixed population, it is those cells in which the potency of I2 is lowest to begin with that will be potentiated the most. A synergistic effect like this will tend to make naturally resistant cells more sensitive (which may be a desirable effect *in vivo*) but it will not necessarily enhance the therapeutic effect against cells that are sensitive to begin with.

For some pathway configurations (e.g., a sequential divergent system) we have noted that changes in the relative activities of two competing enzymes at a branch point may change the nature or extent of the intraction between two inhibitors. Since tumor cells often differ from normal cells in the ratios of competing enzymes, it should be possible, in principle, to design drug combinations that will be antagonistic in normal tissues, but synergistic in a tumor cell, or in a virus-infected cell.

Of the various pathway configurations that have been discussed, those that seem to suggest the greatest potential for developing drug combinations with radically improved selectivity are the pathways with positive feedback that seem to act as "switching" systems. These pathways present some highly nonclassical responses to inhibition, such as the property of remaining switched off after inhibitor is removed, or the "all or none" dose–response relationship. Probably the future of combination chemotherapy lies in elucidation of the switches involved in normal cell proliferation, and of the abnormalities in these switches in malignancy, and devising innovative drug regimens that can selectively reset the broken switches of tumor cells.

Appendix: A BASIC Program to Illustrate the Conclusions

INTRODUCTION TO THE NETWORK MODEL

Many of the conclusions about the predicted drug interactions in simple model pathways were drawn from studies with a short BASIC program, called NETWORK. This program models the system shown in Fig. 12. In this minimal metabolic network, substrate B is produced by two enzymatic reactions, 1 and 3. Reaction 2 produces the product that is considered the output of the system, and reaction 4 is a divergent branch reaction. Reactions 1–4 have rates v1–v4, respectively, and are inhibited by inhibitors I1-I4, respectively. Inhibitors I1 and I3 are assumed to be noncompetitive, and I2 and I4 may be either competitive or noncompetitive. This program makes it possible to investigate the effects of changing the ratios of input and output reactions, and V_{max} and K_m values, on drug interactions. By setting appropriate V_{max} values to zero, various special case situations can be studied. For example, if both $V_{max}3$ and $V_{max}4$ are set to zero, the network reduces to the sequential monolinear pathway; by setting $V_{max}4$ to zero, and adding inhibitors I1 and I3, the system of concurrent inhibition may be studied, etc. The program allows for an optional feedback inhibition of reaction 1 by substrate B, and reaction 1 can also be stimulated by substrate B (positive feedback). The program listing is given in Section VIIID. Users who do not wish to type out the program into their systems may write to the author, including a floppy disk (3.5 or 5.25 inch) formatted in MS-DOS, and a return addressed envelope, and a copy of the program will be put on their disk.

DESCRIPTION OF THE NETWORK PROGRAM

This brief BASIC program is in six modules. The first is the *initialization routine* (lines 100–290). This routine dimensions arrays, initializes variables and strings, and also contains the output subroutine. This module passes control to the *Command Table* (lines 600–699) which functions as the main menu. This menu allows for entry of inhibitors I1–I4, allows for transfer of control to the program's two other menu tables (the *Parameter Editor* and the *Control Menu*), allows for output to be directed to the printer or just to the screen, and two commands (1 and 2) begin simulation. The difference between these two commands is that 1 will initialize time to zero and B to its default value, whereas command 2, used to continue a current simulation, does not reset these values. If ⟨ENTER⟩ is pressed from the COMMAND TABLE without entering a command number, command 1 is selected as the default command (the asterisk in the command line acts as a reminder of this.) Note: because NETWORK is a BASIC program, all input must be followed by pressing the ⟨ENTER⟩ key.

The *simulation loop* is contained in lines 300–390. These lines integrate the rate equations for reactions 1–4 with respect to time, allowing of course for inhibition and feedback effects. These calculations make calls to the *rate equation integration subroutines*, which are contained in lines 900–990. Lines 900–940 integrate the Michaelis–Menten–Henri equation, using the 4th-order Runga–Kutta algorithm. Lines 970–990 integrate the Hill equation, which is used for reaction 2 when the positive feedback option is selected. As the simulation proceeds, a table of values of time, A, B, D, and rate v1–v4 is displayed (and also printed, if the printer has been selected with command 5). After the requested number of lines of output (15 by default), the system will pause and display the prompt "Continue?". At this point, entering the number 2 will continue the current simulation, just pressing ⟨ENTER⟩ will display the COMMAND TABLE, and entering any other number between 1 and 9 will cause execution of that command (e.g., pressing 6 would elicit a request for a value of I1 to be entered).

The *Parameter Editor* (entered from the COMMAND TABLE by selecting 3) makes it possible for the V_{max} and K_m values for the four enzymes to be set to any required value. The parameters are numbered from 1 to 16. The parameter to be edited is selected by entering its number, then (following the prompt) its new value. The parameter list will then be displayed again, now including the corrected value. If at this point a zero is entered (or just ⟨ENTER⟩) control will return to the COMMAND TABLE. The final module of NETWORK, the CONTROL MENU (lines 800–890) provides for selection of various options. Commands 1 and 2 will decrease or increase the time increment, dt, used by the integration algorithm. Using a larger value for dt will speed up the simulation. If too large a value is used, the program will decrease it automatically, but it does not increase dt automatically. Command 3, by setting a index counter, changes the time interval between successive lines of output. The default value of 4 gives an interval of 0.4 min, and setting the inner loop index to 10, for example will give an output interval of 1.0 min. This command also makes it possible to request any required number of lines of output. One option is to request a very large number of line of output (e.g., 300); then when the system has reached a steady state, pressing any key will interrupt the simulation and return control to the COMMAND TABLE. Simulation can be resumed after an interrupt by command 2. Commands 4 and 5 of the CONTROL MENU are used to switch ON or OFF the negative and positive feedback options, and command 9 turns off the color display for the benefit of users with a monochrome monitor. Command 7 allows new values to be entered for substrates A, B, and D. Command 8 makes reaction 4 into a source of substrate D, thus closing the loop and transforming the system into the cyclic pathway of Fig. 6 (Some studies with this cyclic model are described in Section IVD.) Selecting command 9 of the CONTROL MENU (or just pressing ⟨ENTER⟩) will return control to the COMMAND TABLE.

A DEMONSTRATION RUN WITH THE MODEL

To run NETWORK your BASIC interpreter will have to be loaded, which you can do from DOS by entering the command BASIC. When the BASIC prompt (OK) is displayed, type LOAD"NETWORK (or LOAD" A:NETWORK if you have more than one disk drive and the disk containing NETWORK is in the A: drive). Now type RUN (or press the F2 function key) and the NETWORK Command Table will be displayed. The various commands are selected by typing their number (followed by ⟨ENTER⟩, as all of this program's input must be). Select command 3 (edit parameter values). The Parameter Editor menu is now displayed, showing current values of 16 kinetic parameters; choose 7, and then enter a new value of 1.5 for VM3. The Parameter Editor display will change to show the updated value. Also select parameter 10 (VM4) and enter a new value of 2. Now enter 0 (or just ⟨ENTER⟩) which returns control to the Command Table. Press ⟨ENTER⟩ again to begin the simulation; at 6 min the system has still not reached a steady state, so when the "Continue?" prompt is displayed, press 2, to continue the current simulation, and do this again after 12 min. By 12.8 min the system has reached a steady state, with $B = 1.4$, and the system output ($v2$) = 0.583. Note that in the steady state $v1 + v3$ (the inputs) = $v2 + v4$ (the outputs).

When the program pauses at 18 min, press ⟨ENTER⟩ and the Command Table is displayed. Now select option 6 (change concentration of I1), and enter a value of 7. Press ⟨ENTER⟩ to start a new simulation, and the system will reach a new, inhibited, steady state with $B = 0.412$ and $v2 = 0.292$. This is half the control rate of $v2$, so 7 is an IC_{50} concentration of I1. Reenter the Command Table, set I1 back to zero, and then set I2 to 5.5 (command 7). In this case the program will inquire whether I2 is competitive or noncompetitive. Enter a C (upper or lower case) and begin a new simulation. This system approaches its steady state rather slowly, and it will take about 36 min of simulated time to settle down at a B value of 2.687 and $v2 = 0.292$. Thus 5.5 is the IC_{50} concentration of I2. Return again to the Command Table, reset I1 to 7, and run the simulation again with both inhibitors present. By 7.2 min the system has reached a steady state, with $v2 = 0.090$. Thus the combination of I1 plus I2, each at concentrations that singly inhibit by 50%, reduces the fractional activity to 15.4%, a synergistic interaction.

NETWORK PROGRAM LISTING

```
100 REM Program NETWORK (R.Jackson,1 Oct 89)
110 KEY OFF:Q$="Enter number of selected command:":U$=STRING$(16,45)
120 X$=STRING$(73,45):Y$="  Time      A       B       D       v1      v2      v3
    v4      v5"
140 DIM P(15),P$(15):DT=.1:HD=DT/2:F3=0:IL=4:NL=15:F9=1:FR$(0)="OFF":FR$(1)="ON"

150 FOR I=0 TO 15:READ P(I):NEXT:FOR I=0 TO 15:READ P$(I):NEXT:GOSUB 890:A=1:D=1
    :GOSUB 160:GOTO 601
160 IF F4 THEN D=1
161 B=1:V1=1:V2=.5:V3=0:V4=.5:TI=0:IF F3 THEN B=.394:RETURN ELSE RETURN
200 TX=INT(TI*1000+.5)/1000
210 PRINT USING"###.### ";TX,A,B,D,V1,V2,V3,V4,V5
220 IF F1 THEN LPRINT USING"###.### ";TX,A,B,D,V1,V2,V3,V4,V5:RETURN ELSE RETURN

280 DATA 2,1,1,1,1,1,0,1,1,1,1,1,.2,1,.69,2
290 DATA Vm1,Km1,Ki1,Vm2,Km2,Ki2,Vm3,Km3,Ki3,Vm4,Km4,Ki4,Vm5,Km5
291 DATA Ki for feedback inhib.,Hill coeff. for feedback
300 IF F9 THEN COLOR 2,0
305 IL=IL*DT/.1:DT=.1:GOSUB 160
310 IF F1 THEN INPUT"Enter description: ";A$:C$ = "PROGRAM NETWORK: "+A$:LPRINT:
    LPRINT:LPRINT:LPRINT C$:LPRINT X$:LPRINT Y$:LPRINT X$
315 CLS:PRINT Y$,X$:GOSUB 200
320 V1=VM(1)*A/(A+KM(1))/(1+I1/KI(1)):V3=VM(3)*D/(D+KM(3))/(1+I3/KI(3))
325 FOR I=1 TO NL:J=0
330 IF F2 THEN V1=VM(1)*A/(A+KM(1))/(1+I1/KI(1))/(1+(B/KI(0))^H1)
335 IF F3 THEN V1=VM(1)*A/(A+KM(1))/(1+I1/KI(1))/(1+(B/KI(0))^H1)*(.01+B/(B+KM(1
    ))):S0=B:K=KM(2):V=VM(2)*(1+I2N/KI(2)):IC=I2C/KI(2):H=3:VS=V1+V3-V4:GOSUB
    970:V2=W:GOTO 345
340 S0=B:K=KM(2):V=VM(2):IC=I2C/KI(2):IN=I2N/KI(2):VS=V1+V3-V4:GOSUB 900:V2=W
345 S0=B:K=KM(4):V=VM(4):IC=I4C/KI(4):IN=I4N/KI(4):VS=V1+V3-V2:GOSUB 900:V4=W
346 IF F4 THEN S0=D:K=KM(3):V=VM(3):IC=I3C/KI(3):IN=I3N/KI(3):VS=V4-V3:GOSUB 900
    :V3=W
347 IF F4 THEN S0=D:K=KM(5):V=VM(5):VS=V4-V3:GOSUB 910:V5=W
350 B0=B:B=B+(V1+V3-V2-V4)*DT:IF B<0 THEN B=B0:GOSUB 695
352 IF F4 THEN D0=D:D=D+(V4-V3-V5)*DT:IF D<0 THEN D=D0:GOSUB 695
355 TI=TI+DT:IF INKEY$<>""GOTO 600
360 J=J+1:IF J<IL GOTO 330
365 IF FS THEN IL=OL*2:FS=0
370 GOSUB 200:NEXT I
380 PRINT X$:IF F1 THEN LPRINT X$
390 GOTO 600
600 PRINT:INPUT"Continue";A$:IF A$<>""GOTO 660
601 IF F9 THEN COLOR 1,7
602 CLS:PRINT:PRINT:PRINT:PRINT TAB(25)"NETWORK COMMAND TABLE"
605 PRINT TAB(25)STRING$(21,45):PRINT
610 PRINT TAB(20)"1. *Begin new simulation"
```

```
615 PRINT TAB(20)"2.  Continue present simulation"
620 PRINT TAB(20)"3.  Edit parameter values"
625 PRINT TAB(20)"4.  Display CONTROL SWITCH menu"
630 PRINT TAB(20)"5.  Turn printer ON/OFF (now ";FR$(F1);")"
635 PRINT TAB(20)"6.  Change concn. of I1 (now ";I1;" nM)"
640 PRINT TAB(20)"7.  Change concn. of I2 (now ";I2;" uM)"
645 PRINT TAB(20)"8.  Change concn. of I3 (now ";I3;" uM)"
650 PRINT TAB(20)"9.  Change concn. of I4 (now ";I4;" uM)"
655 IF F9 THEN COLOR 4,7
657 PRINT:PRINT:PRINT Q$;:INPUT " ",A$
660 N=VAL(A$):IF N=0 THEN N=1
665 ON N GOTO 300,690,700,800,675,670,680,685,697,679
666 GOTO 601
670 PRINT:INPUT"Enter new I1 concentration: ";I1:GOTO 601
675 F1=F1*-1+1:GOTO 601
676 PRINT:INPUT"Enter new inner loop index: ";A$:IF A$<>""THEN IL=VAL(A$)
677 PRINT"Number of lines of output? (currently ";NL;") : ";:INPUT A$:IF A$<>""
    THEN NL=VAL(A$)
678 GOTO 800
679 COLOR 7,0:CLS:END
680 PRINT:INPUT"Enter new I2 concentration";I2
681 IF I2>0 THEN INPUT"Is I2 competitive or noncompetitive (C/N)";A$
682 I2C=I2:I2N=0:IF A$="N" OR A$="n" THEN I2C=0:I2N=I2:GOTO 601 ELSE GOTO 601
685 PRINT:INPUT"Enter new I3 concentration";I3
686 IF I3>0 THEN INPUT"Is I3 competitive or noncompetitive (C/N)";A$
687 I3C=I3:I3N=0:IF A$="N" OR A$="n" THEN I3C=0:I3N=I3:GOTO 601 ELSE GOTO 601
690 IF F9 THEN COLOR 2,0
691 GOTO 315
695 DT=DT/2:HD=DT/2:OL=IL:IL=J+2*(IL-J):FS=1:RETURN
697 PRINT:INPUT"Enter new I4 concentration";I4
698 IF I4>0 THEN INPUT"Is I4 competitive or noncompetitive (C/N)";A$
699 I4C=I4:I4N=0:IF A$="N" OR A$="n" THEN I4C=0:I4N=I4:GOTO 601 ELSE GOTO 601
700 IF F9 THEN COLOR 4,7
710 CLS:PRINT:PRINT:PRINT TAB(25)"PARAMETER EDITOR":PRINT TAB(25)U$:PRINT
720 FOR I=1 TO 16:PRINT TAB(24)I;TAB(28)".";TAB(32)P$(I-1);" = ";P(I-1):NEXT
730 PRINT:INPUT"Enter parameter to be edited (0 to return): ";A$:N=VAL(A$):IF N=
    0 GOTO 601
740 PRINT"Enter new value for parameter no.";N;": ":INPUT P(N-1):GOSUB 890:GOTO
    700
800 IF F9 THEN COLOR 14,0
801 CLS:PRINT:PRINT:PRINT:PRINT TAB(26)"CONTROL MENU":PRINT TAB(26)STRING$(12,4
    ):PRINT
805 PRINT TAB(20)"1.  Halve DT value (now ";DT;")"
810 PRINT TAB(20)"2.  Double DT value"
815 PRINT TAB(20)"3.  Change inner loop index  (now ";IL;")"
820 PRINT TAB(20)"4.  Negative feedback ON/OFF (now ";FR$(F2);")"
825 PRINT TAB(20)"5.  Positive feedback ON/OFF (now ";FR$(F3);")"
830 PRINT TAB(20)"6.  Colour ON/OFF            (now ";FR$(F9);")"
832 PRINT TAB(20)"7.  Change substrate concns."
833 PRINT TAB(20)"8.  Enable closed cycle     (now ";FR$(F4);")"
835 PRINT TAB(20)"9.* Return to COMMAND TABLE"
840 PRINT:PRINT:PRINT Q$;:INPUT "",A$:N=VAL(A$):IF N=0 THEN N=9
845 ON N GOTO 850,855,676,865,870,860,880,886,601
850 DT=DT/2:HD=HD/2:IL=IL*2:GOTO 800
855 DT=DT*2:HD=HD/2:IL=IL/2:GOTO 800
860 F9=F9*-1+1:GOTO 800
865 F2=F2*-1+1:F3=0:GOTO 800
870 F3=F3*-1+1:F2=0:P(0)=5.172:P(1)=.345:P(3)=1.293:P(4)=.0318:H1=2:KI(0)=.69:
    GOSUB 890:GOTO 800
880 PRINT:INPUT"Which substrate (A, B or D)";A$:IF A$="a" OR A$="A" THEN INPUT"E
    nter new concn. of A: ";A:GOTO 800
882 IF A$="b" OR A$="B" THEN INPUT "Enter new concn. of B: ";B:GOTO 800
885 INPUT"Enter new concn. of D: ";D:GOTO 800
886 F4=F4*-1+1:GOTO 800
890 VM(1)=P(0):KM(1)=P(1):KI(1)=P(2):VM(2)=P(3):KM(2)=P(4):KI(2)=P(5):VM(3)=P(6
    :KM(3)=P(7):KI(3)=P(8):VM(4)=P(9):KM(4)=P(10):KI(4)=P(11):VM(5)=P(12):KM(5)
    P(13):KI(0)=P(14):H1=P(15):RETURN
900 V=V/(1+IN):K=K*(1+IC)
910 W0=V*S0/(S0+K):S1=S0+(VS-W0)*HD
920 W1=V*S1/(S1+K):S2=S0+(VS-W1)*HD
930 W2=V*S2/(S2+K):S3=S0+(VS-W2)*HD
```

```
940 W3=V*S3/(S3+K):W=(W0+W1+W1+W2+W2+W3)/6:RETURN
970 K=K*(1+IC)
975 W0=V*(S0/K)^H/(1+(S0/K)^H):S1=S0+(VS-W0)*HD
980 W1=V*(S1/K)^H/(1+(S1/K)^H):S2=S0+(VS-W1)*HD
985 W2=V*(S2/K)^H/(1+(S2/K)^H):S3=S0+(VS-W2)*HD
990 W3=V*(S3/K)^H/(1+(S3/K)^H):W=(W0+W1+W1+W2+W2+W3)/6:RETURN
```

Acknowledgments

I am indebted to Drs. Darryl Rideout and William Greco for their helpful comments on the manuscript.

References

Albert, A. (1956). Discussion paper. *Proc. R. Soc. Med.* **49,** 881.

Berenbaum, M. C. (1989). What is synergy? *Pharmacol. Rev.* **41,** 93–141.

Black, M. L. (1963). Sequential blockage as a theoretical basis for drug synergism. *J. Med. Chem.* **6,** 145–153.

Bray, G., and Brent, T. P. (1972). Deoxyribonucleoside 5'-triphosphate pool fluctuations during the mammalian cell cycle. *Biochim. Biophys. Acta* **269,** 184–191.

Bushby, S. R. M. (1969). Combined antibacterial action in vitro of trimethoprim and sulfonamides: The in vitro nature of synergy. *Postgrad. Med. J.* **45,** 10–18.

Bushby, S. R. M., and Hitchings, G. H. (1968). Trimethoprim, a sulphonamide potentiator. *Br. J. Pharmacol.* **33,** 72–90.Chou, T.-C., and Talalay, P. (1981). Generalized equations for the analysis of inhibitions of Michaelis–Menten and higher-order kinetic systems with two or more mutually exclusive and nonexclusive inhibitors. *Eur. J. Biochem.* **115,** 207–216.

Chou, T.-C., and Talalay, P. (1981). Generalized equations for the analysis of inhibitions of Michaelis–Menten and higher-order kinetic systems with two or more mutually exclusive and nonexclusive inhibitors. *Eur. J. Biochem.* **115,** 207–216.

Cory, J. G., and Cory, A. H., eds. (1989). "Inhibitors of Ribonucleoside Diphosphate Reductase Activity". Pergamon, Oxford, England.

Goldie, J. H., and Coldman, A. J. (1979). A mathematic model for relating the drug sensitivity of tumors to their spontaneous mutation rate. *Cancer Treat. Rep.* **63,** 1727–1733.

Grindey, G. B., and Cheng, Y. C. (1979). Biochemical and kinetic approaches to inhibition of multiple pathways. *Pharmacol. Ther.* **4,** 307–327.

Grindey, G. B., Moran, R. G., and Werkheiser, W. C. (1975). Approaches to the rational combination of antimetabolites for cancer chemotherapy. *In* "Drug Design" (E. J. Ariens, ed.), Vol. 5, pp. 169–249. Academic Press, New York.

Handschumacher, R. E. (1965). Some enzymatic considerations in chemotherapy. *Cancer Res.* **25,** 1541–1543.

Harrap, K. R., and Jackson, R. C. (1975). Enzyme kinetics and combination chemotherapy: An appraisal of current concepts. *Adv. Enzyme Regul.* **13,** 77–96.

Harvey, R. J. (1978). Interaction of two inhibitors which act on different enzymes of a metabolic pathway. *J. Theor. Biol.* **74,** 411–437.

Harvey, R. J., and Dev, I. K. (1975). Regulation in the folate pathway of *Escherichia coli. Adv. Enzyme Regul.* **13,** 99–124.

Hitchings, G. H. (1969). Chemotherapy and comparative biochemistry: G.H.A. Clowes Memorial Lecture. *Cancer Res.* **29,** 1895–1903.

Hitchings, G. H., and Baccanari, D. P. (1984). Design and synthesis of folate antagonists as antimicrobial agents. *In* "Folate Antagonists as Therapeutic Agents" (F. M. Sirotnak, J. J.

Burchall, W. D. Ensminger, and J. A. Montgomery, eds.), Vol. 1, pp. 151–172. Academic Press, Orlando, Florida.

Huggins, J. W., Spears C. T., Stiefel, J., Robins, R. K., and Canonico, P. G. (1983). Synergistic antiviral effects of ribavirin and the C-nucleoside analogues tiazofurin and selenazole against togaviruses and arenaviruses. *Abstr. Intersci. Conf. Antibiot. Chemother.* p. 108.

Jackson, R. C. (1980). Kinetic simulation of anticancer drug interactions. *Int. J. Bio-Med. Comput.* **11**, 197–224.

Jackson, R. C. (1986). Kinetic simulation of anticancer drug effects on metabolic pathway fluxes: Two case studies. *Bull. Math Biol.* **48**, 337–351.

Jackson, R. C. (1987). Computer simulation of the effects of antimetabolites on metabolic pathways. *In* "New Avenues in Developmental Cancer Chemotherapy" (K. R. Harrap and T. A. Connors, eds.), pp. 3–35. Academic Press, Orlando, Florida.

Jackson, R. C. (1989). The role of ribonucleotide reductase in regulation of the deoxyribonucleoside triphosphate pool composition: Studies with a kinetic model. *In* "Inhibitors of Ribonucleoside Diphosphate Reductase Activity" (J. G. Cory and A. H. Cory, eds.), pp. 127–150. Pergamon, Oxford, England.

Kacser, H., and Porteous, J. W. (1987). Control of metabolism: What do we have to measure? *Trends Biochem. Sci. Pers. Ed.* **12**, 5–14.

Leopold, W. R., Dykes, D. J., and Griswold, D. P. (1987). Therapeutic synergy of trimetrexate in combination with doxorubicin, vincristine, cytoxan, 6-thioguanine, cisplatin, or 5-fluorouracil against intraperitoneally implanted P388 leukemia. *Natl. Cancer Inst. Monogr.* **5**, 99–104.

Melmon, K. L., and Gilman, A. G. (1980). Drug interactions. *In* "The Pharmacological Basis of Therapeutics" (L. S. Goodman and A. Gilman, eds.), 6th Ed. pp. 1738–1751. Macmillan, New York.

Modica-Napolitano, J. S., Steele, G. D., and Chen, L. B. (1989). Aberrant mitochondria in two human colon carcinoma cell lines. *Cancer Res.* **49**, 3369–3373.

Morselli, P. L., Garattini, S., and Cohen, S. N. (1974). "Drug Interactions." Raven, New York.

Potter, V. R. (1951). Sequential blocking of metabolic pathways in vivo. *Proc. Soc. Exp. Biol. Med.* **76**, 41–46.

Prigogine, I., and Stengers, I. (1984). "Order out of Chaos." Bantam, New York.

Reiner, J. M. (1969). "Behavior of Enzyme Systems," 2nd Ed. Van Nostrand, New York.

Rubin, R. J., Reynard, A., and Handschumacher, R. E. (1964). An analysis of the lack of drug synergism during sequential blockage of de novo pyrimidine biosynthesis. *Cancer Res.* **24**, 1002–1007.

Sartorelli. A. C. (1969). Some approaches to the therapeutic exploitation of metabolic sites of vulnerability of neoplastic cells. *Cancer Res.* **29**, 2292–2299.

Savageau, M. A. (1976). "Biochemical Systems Analysis: A Study of Function and Design in Molecular Biology". Addison-Wesley, Reading, Massachusetts.

Segal, I. H. (1975). Multiple inhibition analysis. *In* "Enzyme Kinetics", pp. 465–504. Wiley (Interscience). new York.

Skipper, H. E. (1974).Combination chemotherapy: Some concepts and results. *Cancer Chemother. Rep.* **4**, (Part 2), 137–146.

Skipper, H. E. (1986). On mathematical modelling of critical variables in cancer treatment. *Bull Math. Biol.* **48**, 253–278.

Skipper, H. E., Thomson, J. R., and Bell, M. (1954). Attempts at dual blocking of biochemical events in cancer chemotherapy. *Cancer Res.* **14**, 503–507.

Webb, J. L. (1963). General principles of inhibition. *In* "Enzyme and Metabolic Inhibitors", Vol. I, pp. 1–949. Academic Press, New York.

White, J. C. (1986). Use of the circuit simulation program SPICE2 for analysis of the metabolism of anticancer drugs. *Bull. Math. Biol.* **48**, 353–380.

Zeeman, E. C. (1976). Catastrophe theory. *Sci. Am.* **234**, 65–83.

CHAPTER 12

Enhanced Effects of Drugs That Bind Simultaneously to the Same Macromolecular Target

Lucjan Strekowski
W. David Wilson
Department of Chemistry
Georgia State University
Atlanta, Georgia 30303

I. Introduction

This review discusses ternary interactions between two drugs and a macromolecular target, from selected papers which have been abstracted by Chemical Abstracts through January 1990. The literature on the subject has been exploding during recent years, and a number of other short reviews have already appeared. In order to avoid unnecessary repetitions and, at the same time, to keep this chapter to a reasonable size, whenever possible these previous summaries are referenced without elaborating on the previously covered primary publications. The reader will notice, thus, that a vast majority of the cited papers has been published within the last five years.

We have attempted to discuss drug interactions with a macromolecular target in a broad perspective of the appropriate background, the comparison to other types of interactions that may produce a similar effect, and the design

SYNERGISM AND ANTAGONISM IN CHEMOTHERAPY
Copyright © 1991 by Academic Press, Inc.
All rights of reproduction in any form reserved.

of new drug combinations. With this in mind, not only synergistic and antagonistic interactions are discussed, but the potentiation effects of nonchemotherapeutic agents on the activity of drugs are also included. Such nonchemotherapeutic compounds are often called amplifiers. Drug amplifiers have no intrinsic activity but enhance the chemotherapeutic effects of the active drug when given in combination. The physical basis for amplification is clearly more complicated than for drug synergism or antagonism, and many features of amplification have yet to be elucidated.

In principle, any drug which induces some chemical event at a target cell can have its activity amplified if appropriate amplifiers can be discovered. There are numerous advantages in working with established drugs—they are known to be at least partly selective for the target cells and their toxic effects in humans are defined. The search for drug amplifiers may be as valuable as the search for new drugs.

II. Interactions with Nucleic Acids

A. NUCLEIC ACID INTERACTION MODES

There are four primary modes for reversible interactions of molecules with nucleic acids: (1) electrostatic interaction with the charged sugar–phosphate backbone of the nucleic acid, (2) interactions in the nucleic acid major groove, (3) interactions with the nucleic acid minor groove, and (4) intercalation between base pairs (Wilson *et al.*, 1990). Some molecules remain reversibly bound and in dynamic equilibrium with the nucleic acid. Other bound ligands undergo chemical events which lead to covalent reaction or ligand induced cleavage of the nucleic acid after the initial reversible binding. In all known cases, however, the ligand first binds reversibly by the modes described earlier prior to subsequent chemical events.

It is now well established that the stability of folded nucleic acid conformations requires the association of metal cations (Na^+, Mg^{2+}, etc.) from solution in a process called counterion condensation (Manning, 1978). Specific interactions of other cationic ligands with nucleic acids neutralize phosphate charges and result in release of some condensed counterion. This ion release provides an entropic contribution to the binding free energy (Manning, 1978; Record *et al.*, 1978). Such charge interactions are quite dependent on the solution's salt concentration and are generally weak at physiological conditions for ligands with one or two charges. This type of interaction does, however, make a significant contribution to the binding fee energy of highly charged molecules including many proteins which interact with DNA (Lavery and Pullman, 1985).

Molecules which bind in the nucleic acid grooves or by intercalation can have an electrostatic component to their binding free energy. However, their complexes are also stabilized by van der Waals, hydrogen-bonding, hydrophobic, and/or other similar interactions. Many protein molecules specifically recognize DNA through major groove contacts, but there are few well-known examples of small molecules which bind primarily to DNA in a major-groove complex.

There are a number of small molecules (Fig. 1), such as netropsin (**1**), distamycin A (**2**), Hoechst 33258 (**3**), 4′,6-diamidino-2-phenylindole (DAPI) (**4**), and related unfused aromatic heterocycles which bind to DNA in a minor-groove complex. As is typical of groove-binding compounds, the molecules listed here contain small unfused aromatic rings linked by bonds with torsional freedom which allows the molecules to assume appropriate twist to fit the helical curvature of the groove. A view of the crystal structure of netropsin bound into the minor groove of d(CGCGAATTCGCG), obtained by Dickerson and co-workers (Kopka *et al.*, 1985) is shown in Fig. 2 as an example of this type of binding (Liquier *et al.*, 1989). The nucleic acid grooves differ in width, chemical composition, hydration, electrostatic potential, hydrogen-bonding capacity, and other similar factors (Kopka *et al.*, 1985; Lavery and Pullman, 1985). In A·T sequences the minor groove in DNA is not as wide as in G·C regions, has a more negative electrostatic potential, and does not have a group such as G-NH$_2$ to prevent molecules from sliding deeply into the groove (Kopka *et al.*, 1985; Zakrzewska and Pullman, 1986). Cations are thus attracted to A·T rich regions where they can form van der Waals contacts with the walls of the groove and hydrogen bonds with the A-N3 and T-O2 groups on the floor of the minor groove.

In addition to the minor groove interactions, intercalation provides the only other mode for strong binding of small molecules with nucleic acids. To create an intercalation site, two base pairs must be separated by ∼ 3.4 Å but must be maintained in an approximately parallel relative orientation, perpendicular to the helix axis. This can be accomplished by torsional angle changes in the nucleic acid sugar–phosphate backbone (Neidle and Abraham, 1984). The double helical structure is thus lengthened and unwound at the intercalation binding site. The normal base-pair twist angle, for example, is 36° in B-form DNA (10 base pairs per turn of 360°). Opening an intercalating site decreases this twist angle and thus causes local unwinding of the double helix (Wilson, 1990). The amount of helix unwinding varies with the type of intercalator from an approximately 10° decrease in base-pair twist for anthracyclines to 26° for ethidium (**5**) (resulting in base-pair twist angles of 26° to 10°, respectively, at the intercalation site). An intercalator stacks between two base pairs and makes van der Waals contact with them. As an example of an intercalator–DNA complex, a crystal structure of the intercalator with the dinucleoside duplex from iodo–UpA is shown in Fig. 3. This type of interaction is less specific than the extensive hydrogen bonds and steric restrictions

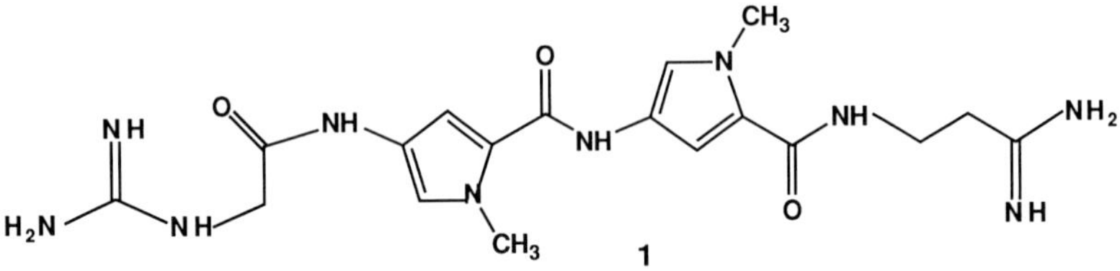

1

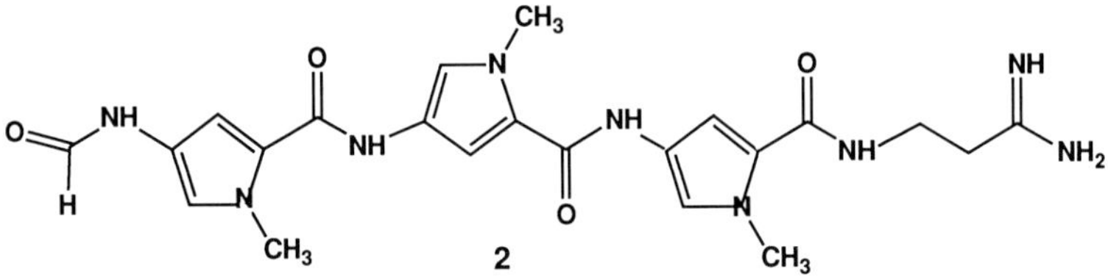

2

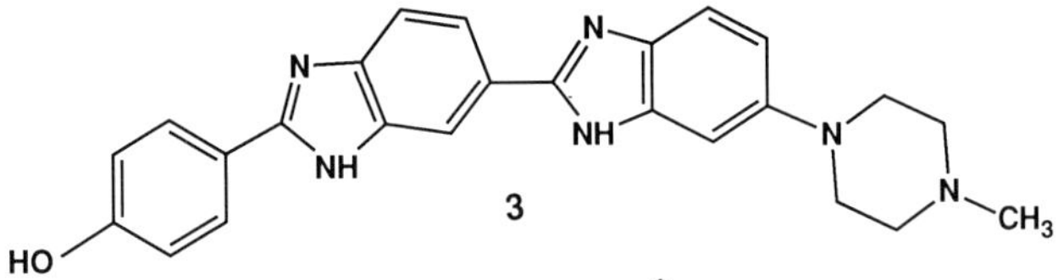

3

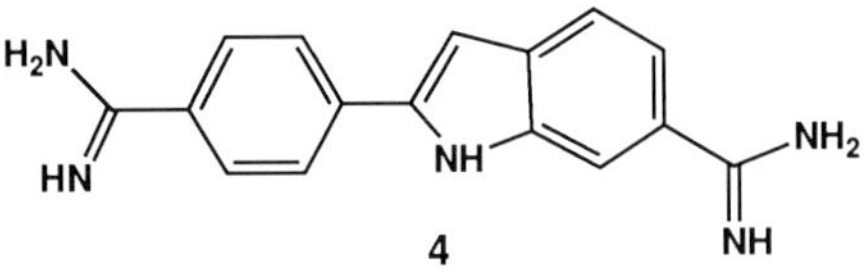

4

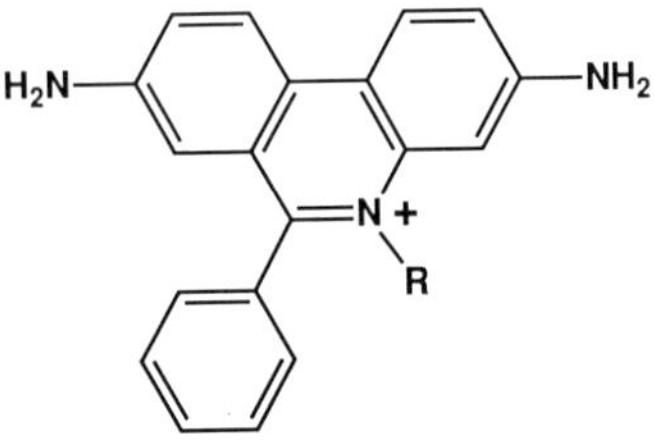

5: $R{=}C_2H_5$, Br^-

6: $R{=}(CH_2)_3\overset{+}{N}(C_2H_5)_2CH_3$, $2I^-$

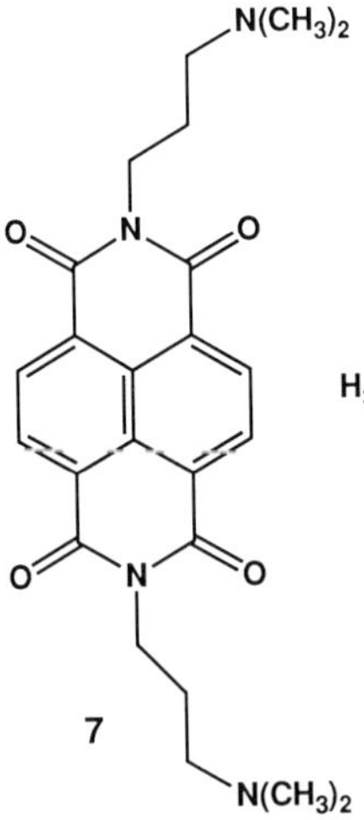

7

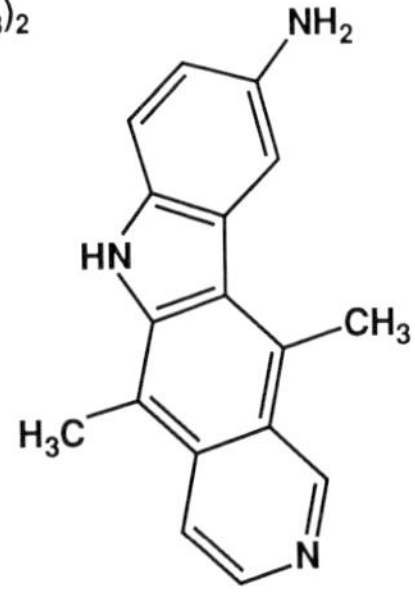

8

FIGURE 1 Structures of classical DNA minor groove-binding compounds, netropsin (**1**), distamycin A (**2**), Hoechst 33258 (**3**), DAPI (**4**), and classical DNA intercalators, ethidium bromide (**5**), propidium iodide (**6**), bis-imide (**7**), and 9-aminoellipticine (**8**).

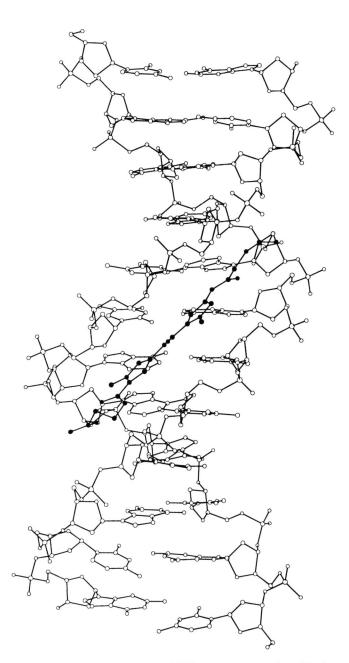

FIGURE 2 A view of the crystal structure of Dickerson and co-workers (Kopka *et al.*, 1985) of netropsin (**1**), bound into the minor groove AATT region of the DNA dodecamer d(CGCGAATTCGCG). Netropsin atoms are black for ease of visualization. The netropsin amide NH groups point to the floor of the minor groove for hydrogen bond formation with the acceptors C2=O of T and N3 of A. The extended pyrrole-amide system of netropsin forms good van der Waals contacts with the walls of the minor groove. Distamycin, DAPI, and other AT selective minor groove agents bind to DNA in a similar mode.

To prepare this figure we used coordinates 6BNA for the netropsin–DNA complex which were deposited by Dickerson and co-workers in the Protein Data Bank (Bernstein *et al.*, 1977; Abola *et al.*, 1987) at Brookhaven National Laboratory in August 1984. The coordinates were read into molecular modeling program MACROMODEL from Dr. Clark Still of Columbia University for visualization.

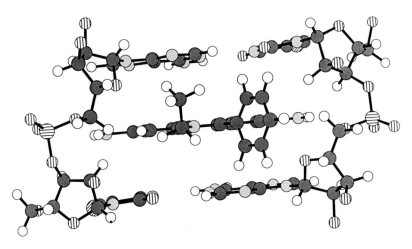

FIGURE 3   A view of the crystal structure of Sobell and co-workers (Tsai *et al.*, 1977) of ethidium (**5**) intercalated into the dinucleoside phosphate duplex formed by iodo-UpA. The view is into the minor groove and the phenyl and *N*-ethyl substituents of ethidium can be seen projecting into this groove. The approximately planar stacking of the ethidium phenanthridinium ring with the base pairs can also be seen. The atom coding is as follows: carbon, dark grey; nitrogen, light grey; hydrogen, white; oxygen, vertical lines; and phosphorus, horizontal lines. We thank Dr. Robert Jones of Emory University for preparation of this figure from the Cambridge Crystallographic Data Base.

involved in groove binding, and intercalators generally show much less base-pair binding specificity than groove-binding molecules (Wilson and Jones, 1981; Wilson, 1987).

Intercalation and groove-binding modes should not be viewed as mutually exclusive processes (W. D. Wilson, *et al.*, 1989a,b). Both A·T and G·C base pairs can form intercalator binding sites. A·T regions, however, also have a very favorable minor groove site for binding aromatic cations. The minor groove in the G·C regions is wider than in the A·T regions, has additional steric constraints on binding due to the G-NH$_2$ groups and thus, in the absence of specific interactions with the G-NH$_2$ groups, does not present as favorable a binding site as in A·T sequences. We have found, for example, that DAPI (**4**) binds very strongly to three or more consecutive A·T base pairs in a minor groove complex. With fewer consecutive A·T base pairs, and particularly in pure G·C regions, DAPI binds by intercalation. The DAPI binding constant for intercalation at G·C sites is similar to the binding constant for known strong intercalators. In A·T regions, however, it makes specific contacts with more base pairs and binds more strongly as long as the site contains the required minimum number of consecutive A·T base pairs (Larsen *et al.*, 1989; W. D. Wilson, *et al.*, 1989a).

Intercalation and groove binding should thus be viewed as a continuum. The mode with the most favorable free energy for a particular ligand will depend on the DNA sequence and conformation as well as on the specific molecular features of the bound molecules.

B. COMPLEXES OF SHORT OLIGONUCLEOTIDES AND ANALOGS WITH NUCLEIC ACIDS

1. Interaction of Antisense Oligonucleotides and Analogs with mRNA

Antisense single-stranded nucleic acids contain sequences that are complementary to specific sequences of messenger RNA. They can bind to the complementary RNA to form an intermolecular double-stranded region. This sequence-specific interaction provides a potent mechanism by which specific transcripts can be translationally inactivated. It has been suggested that this process, called hybridization arrest (Paterson *et al.*, 1977), can be employed for sequence-specific control of gene expression (van der Krol *et al.*, 1988) and in various chemotherapeutic applications (Ts'o *et al.*, 1987; Zon, 1988) including development of a new class of anticancer agents (Miller and Ts'o, 1987; Paoletti, 1988; Stein and Cohen, 1988).

Hybridization to the 5' end coding region of mRNA is necessary for arrest of translation. This hybridization arrest *in vitro* is efficient for long antisense sequences and relatively inefficient for short sequences of synthetic oligonucleotides. The latter result is due to low stability of the small hybrid. One way to increase the stability of the short oligomer–RNA duplex is to attach a compound which binds to nucleic acid duplexes to the oligomer. The presence of the intercalator at the end of the oligonucleotide chain, for example, greatly increases the interaction with RNA and does not interfere with specific base pairing (Asseline *et al.*, 1984; Toulmé *et al.*, 1986; Helen and Thuong, 1987).

Another way to increase efficiency of hybridization arrest is to employ the combination of short oligonucleotides which are more easily transported into cells and whose binding sites are contiguous or separated by one or two nucleotides. Such combinations are synergistic in their ability to specifically inhibit mRNA translation. For example, two 14-mer oligodeoxyribonucleotides selectively inhibited translation of the mRNA coding for human dihydrofolate reductase to about the same extent as a 20-mer binding to the same region at a concentration eight times greater (Maher and Dolnick, 1987). When a third 14-mer was added to the pair of contiguous antisense oligodeoxyribonucleotides, translation of the target RNA was selectively abolished. This effect could not be achieved by two 14-mers even at 10-fold higher oligonucleotide concentration (Maher and Dolnick, 1988). This synergistic effect is apparently a result of binding cooperativity of the short oligomers with the target RNA (Springgate and Poland, 1973; Asseline *et al.*, 1984).

The combination approach may be useful for hybrid arrest by synthetic oligonucleotide analogs *in vivo* where oligomer chain length is likely to be a serious constraint due to permeability limits. Likely candidates for future development are α-anomeric oligodeoxyribonucleotides (Paoletti, 1988), methylphosphonates (**9**), and alkylphosphotriesters (**10**) which are resistant to enzymatic hydrolysis (Fig. 4). Moreover, the removal of the negative charge of the phosphate group allows the analogs **9** and **10** to form complementary duplexes with the natural mRNA with the potential for a higher degree of stability (Lin et al., 1989). The absolute stereochemistry at the phosphorus atoms in antisense oligodeoxyribonucleoside methylphosphonates (**9**) strongly affects the RNA affinity of these oligomers. While a weak oligomer–RNA interaction is observed for methylphosphonate oligomers with a random phosphorus stereochemistry (Maher and Dolnick, 1988), an increased complex stability can be achieved by using stereoregular oligomers with the R absolute configuration at the phosphorus atoms. It has also been shown that the complexes of the R stereoisomers of methylphosphonates with the methyl groups at pseudo-equatorial positions are inherently more stable than those of the corresponding S isomers having the methyl groups at pseudo-axial positions (Ts'o et al., 1983; Bower et al., 1987). The same must be true for the two stereoregular isomers of **10**. It would be of interest to study hybridization arrest with two short oligomer methylphosphonates or alkylphosphotriesters having an intercalator linked covalently to the end of the R-oligomers and binding to the same region of mRNA. Their increased binding with mRNA, as a result of the optimized structures and the

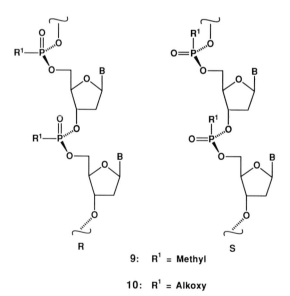

9: R¹ = Methyl

10: R¹ = Alkoxy

FIGURE 4   Absolute stereochemistry (R,S) at the phosphorus atoms of stereoregular deoxyribonucleotide oligomer methylphosphonates (**9**) and alkylphosphotriesters (**10**).

binding cooperativity, is also expected to produce a strong synergistic effect in their biological activity.

Phosphorothioate oligodeoxyribonucleotides ($R^1 = S^-$ in Fig. 4) have also shown very promising chemotherapeutic potential (Zon, 1988). Although ionic, these oligomers are more readily taken up into cells than simple oligo-deoxyribonucleotides. The phosphorothioates are available in stereoregular forms through chemical synthesis and DNA polymerase-mediated synthesis (LaPlanche *et al.*, 1986; Eckstein, 1983).

2. Interactions with DNA

While short oligodeoxyribonucleotides bind more strongly than the corresponding oligonucleoside methylphosphonates to complementary regions of RNA, the reverse is true for the interaction with single-stranded DNA. In comparison to the presence of one methylphosphonate oligomer, the binding affinity increases considerably for two oligomers that bind at adjacent sites on the target DNA (Lin *et al.*, 1989). In both the RNA and DNA systems, this cooperativity may arise from two different interactions: base stacking of the terminal oligomer bases and/or an induced conformational change at contiguous polymer sites.

Strobel and Dervan (1989) have reported cooperativity in specific binding of two different pyrimidine oligonucleotides at adjacent sites of duplex DNA. Such oligomers bind duplex DNA sequence specifically at homopurine sites in the major groove to form a triple helix structure. The binding specificity is imparted by Hoogsteen base pairing between the pyrimidine oligonucleotide and the purine strand of the Watson–Crick duplex DNA (Fig. 5). This triple helix formation is a powerful chemical approach for the sequence-specific recognition of double helical DNA. It may provide a new method for artificial repression of gene expression and treatment of viral diseases. A way to increase stability of the Hoogsteen base pairing is by using the oligomers with methylated and/or brominated pyrimidine bases. The triple helix with such oligomers is stable at physiological pH (Povsic and Dervan, 1989).

C. POTENTIATION OF DNA DAMAGE BY AGENTS THAT BIND COVALENTLY WITH DNA

1. Hypoxia-Mediated Chemosensitization of Alkylating Agents with Nitroaromatics

Oxygen-deficient (hypoxic) cells develop in tumors as a result of a fast growth that outstrips the supply of essential nutrients including oxygen. Such tumors are mostly resistant to radiation and to treatment with chemotherapeutic agents that require oxygen for maximum expression of cytotoxic

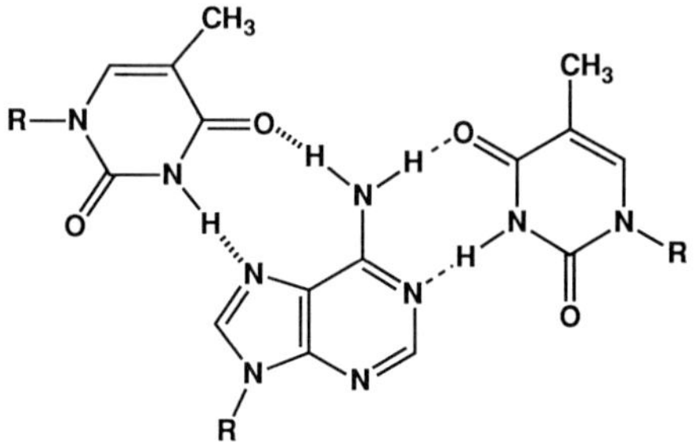

TAT-base triplet

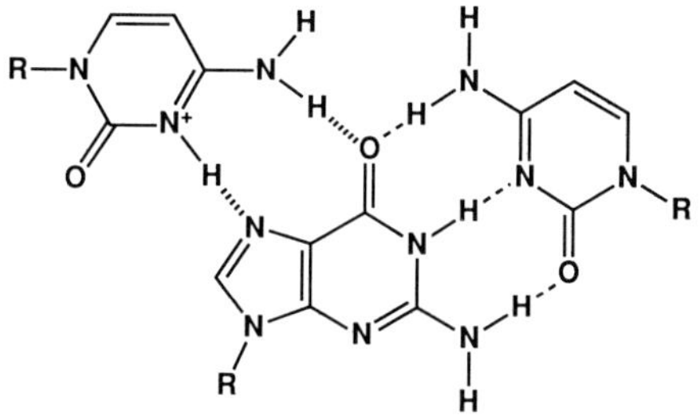

C⁺GC-base triplet

FIGURE 5   Base triplets of TAT and C⁺GC. The pyrimidine oligonucleotide is bound by Hoogsteen hydrogen bonds (|||||) in the major groove to the purine strand in the Watson–Crick duplex.

activity. Unfavorable pharmacodynamic and other factors may also play a role in the resistance to common chemotherapeutic agents (Kennedy *et al.*, 1980). On the other hand, the difference in oxygen concentration between hypoxic and normal cells has provided a unique basis for the development of hypoxia-selective agents (HSA) such as quinones (Lin *et al.*, 1984) and nitroaromatics (Denny and Wilson, 1986; W. R. Wilson, *et al.*, 1989a,b). In general, these HSAs are prodrugs the reduction of which produces transient, highly reactive intermediates that react with cellular macromolecules, principally nucleic acids (Edwards, 1986; Declerck and De Ranter, 1986; Lin *et al.*, 1984). Since oxygen is a scavenger of these reactive intermediates, their toxicity is greatly reduced in oxygenated cells.

A great deal of work has been devoted to understanding the nature of the activated nitro group. It is known that the activation can be achieved by high energy radiation or bioreduction (Adams and Stratford, 1986). The two mechanisms may be quite different, however. The radiation sensitization occurs mainly via a very fast free-radical process while more than one mode of action is apparently involved in the time-dependent bioreduction (Edwards, 1986). There is considerable evidence that certain enzymes in mammalian cells can behave as nitroreductases in an hypoxic environment, as reviewed by Millar (1982). It is interesting to note that ascorbic acid, a known electron donor, increases hypoxic cytotoxicity of nitroaromatics but has little effect on their radiation sensitization (Josephy *et al.*, 1978). Bulky DNA monoadducts are apparently the major cytotoxic lesions induced by bioreduced nitroaromatics (Declerck and De Ranter, 1986; W. R. Wilson, *et al.*, 1989b).

The early attempts of radiation-assisted treatment of hypoxic tumors with nitro compounds has been a limited success. Studies conducted mostly with 3-methoxy-1-(2-nitro-1-imidazolyl)-2-propanol (MISO, **11**, Fig. 6) have shown substantial benefit in the radiotherapy of several tumors (Hall and Roizin-Towle, 1975), but the neurological toxicity of MISO is the major limitation in its application (Brown and Workman, 1980).

Another way to improve cancer chemotherapy with nitro derivatives is through their synergistic action with alkylating agents (Siemann and Mulcahy, 1986). Although many cell constituents including DNA, RNA, proteins, and cell membranes are alkylated, the present working hypothesis for antitumor activity is that most alkylating agents produce their lethal effects by reacting with cellular DNA (Remers, 1984). It has been shown that the alkylation involves primarily DNA bases. The phosphate groups are also alkylated easily, but apparently the phosphate ester thus produced is involved in transalkylation of the bases. Widely tested, common alkylating agents (Fig. 6) are derivatives of aziridine (**12**) and 1-(2-chloroethyl)-1-nitrosourea (**13**). Another class of compounds is the nitrogen mustards (2-chloroethylamines) (**14**) which are activated in aqueous solution through intramolecular cyclization to give an aziridinium cation. The aziridinium cations, obtained in the latter reaction or by protonation of aziridine, are electrophilic species responsible for the alkylation of DNA (Ross, 1962). The activation of the

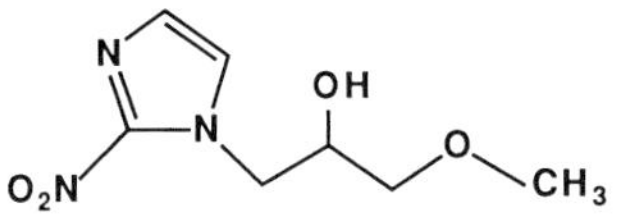

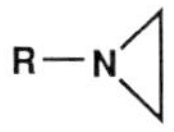

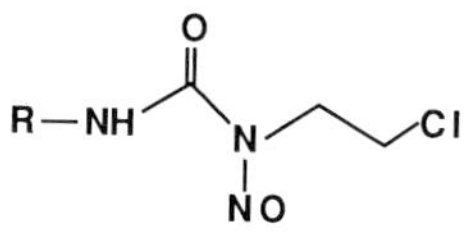

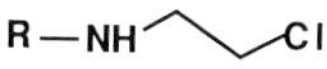

FIGURE 6 Structures of MISO (**11**) and derivatives of aziridine (**12**), 1-(2-chloroethyl)-1-nitrosourea (**13**), and 2-chloroethylamine (**14**) (R=alkyl).

nitrosourea derivatives is more complicated. Depending on their structure and external factors, these compounds may decompose to give electrophilic species responsible for the alkylation of DNA or isocyanates. The isocyanates are believed to react principally with proteins to produce carbamoylated derivatives (Lown et al., 1978, 1985).

The early attempts of the chemosensitization of the alkylating agents by nitroaromatics were reviewed by Millar (1982). The enhancement of the activity has been observed for several combinations both in vitro and in vivo at nontoxic concentrations of the two components. The molecular basis for this synergistic action appears to be complex. In particular, the combinations of nitroaromatics and chloroethylnitrosoureas appear to damage both DNA and proteins. These results can be understood in terms of high reactivity of the transient reactive species resulting from both the reduction of the nitro group and the activation of the alkylating agent. Such reactive species do not discriminate between the many available targets. Changes in pharmacokinetics have also been suggested to account for the observed enhancement in activity.

Several bifunctional compounds composed of a MISO system and an alkylating moiety, such as (12) or (13), have been prepared recently. As expected, the nitro group in these new prodrugs is a potent intramolecular chemosensitizer of the alkylating moiety, especially under hypoxic conditions (Carminati et al., 1988; Mulcahy et al., 1988, 1989; Stratford et al., 1986).

2. Synergism Involving Intercalation and Alkylation

Normally the intercalation of an aromatic compound with DNA does not result in chemical damage of the DNA. A notable exception is the interaction of certain nucleophilic intercalators with the double helix already destabilized by the presence of apurinic and apyrimidinic sites (Malvy et al., 1988). These sites, mostly the apurinic ones, can be generated under the influence of heating, ionizing radiation, alkylating agents, N-glycosylases, and anticancer agents such as bleomycin or methotrexate (Lindahl, 1982). The intercalators such as 9-aminoellipticine (8, Fig. 1), 9-hydroxyellipticine, quinacrine, or daunorubicin enhance the effect of these base-removing agents and conditions. The net result is DNA strand breakage. The exact chemistry involved is not known. The position of the attachment of the nucleophilic group, such as an amine or ring nitrogen, at the intercalator molecule is of primary importance for the strand scission activity. For example, ethidium bromide (5, Fig. 1) is a strong intercalator but completely devoid of strand-breakage ability (Malvy et al., 1986). On the other hand, it has been shown that both types of intercalators, that is, those inducing DNA breakage and those devoid of this activity, can enhance the toxicity of the alkylating agent dimethyl sulfate in bacteria. It has been demonstrated that apurinic sites are involved in this stimulation of toxicity, and the overall effect is synergistic (Malvy et al.,

1988). The synergism may be a result of either the inhibition of DNA repair or a modification of the toxicity at each apurinic site. (Synergism involving inhibition of DNA repair is discussed in more detail in Chapter 15 by Teicher *et al.*,) It appears, however, that regardless of the mechanism involved the intercalating agents bind preferentially in the vicinity of the structurally relaxed apurinic sites, and this binding is involved in the observed biological effect (Woodson and Crothers, 1988). A closely similar effect was observed by Williams and Goldberg (1988) who showed a selective intercalation and strand scission by intercalating drugs at DNA bulges. The bulge is an unpaired nucleotide on one strand of the double helix. The normal helix is highly relaxed and destabilized in the immediate vicinity of this unusual structral feature of the duplex.

Of interest is a synergistic effect between many antitumor, DNA-interacting drugs and *cis*-DDP (**15**, Fig. 7) observed in P388 leukemia in mice (Uchida *et al.*, 1989). The interaction of cis-DDP with DNA is discussed in the next section.

3. DNA-Mediated Reaction of *cis*-Diamminedichloroplatinum(II) (*cis*-DDP) with the Amino Group of DNA Intercalators

The antitumor drug *cis*-DDP (**15**, Fig. 7) forms a covalent complex with DNA by cross-linking two intrastrand bases, preferentially adenines or guanines, at N7 with the loss of two chloride ions. The reaction takes place in the major groove and results in significant distortion of the helix (Marrot and Leng, 1989; Rice *et al.*, 1988). After initial reaction with DNA, the drug can also react with ethidium bromide (**5**) to form a covalent bond between one of the amino groups of ethidium and platinum of *cis*-DDP with the loss of one chloride ion. A ternary covalent complex is thus formed between *cis*-DDP, DNA, and ethidium bromide. The ternary complex consists of platinum coordinated by two *cis* ammines, an amino group of ethidium, and a DNA ligand such as a purine N7. Interestingly, this platination of ethidium by *cis*-DDP in the presence of DNA is much faster than the analogous reaction in the absence of DNA even at higher concentrations of the two reactants (Sundquist *et al.*, 1988). The reaction of *cis*-DDP with proflavin (Malinge and Leng, 1986; Malinge *et al.*, 1987) and several other nucleophilic intercalators

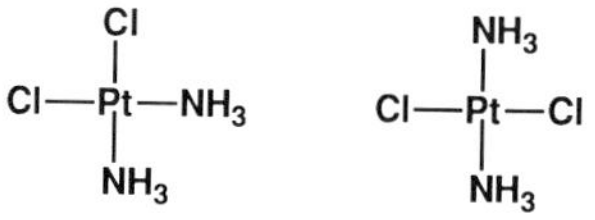

FIGURE 7   Structures of *cis*-DDP (**15**) and *trans*-DDP (**16**).

(M. Leng, personal communication) is also accelerated by DNA. Experimental and computer modeling studies are consistent with the molecular basis for this enhancement effect in which DNA acts as a template to favorably orient the two reactants (Sundquist et al., 1988). In this model the initial monofunctional adduct of cis-DDP with DNA in the major groove and the adjacent intercalation complex are favorably oriented for the subsequent nucleophilic displacement of the remaining chlorine atom at the DNA-bound platinum atom by the amino group of the intercalator. This model is strongly supported by the known facts that although typical intercalation complexes form and dissociate quickly, they usually have well-defined stereochemistry concerning the orientation of the intercalator molecule within the framework of the adjacent base pairs (Neidle and Abraham, 1984; Strekowski et al., 1988b; Wilson et al., 1988). Consistent with this explanation is the observation that 9-aminoacridine, which intercalates preferentially with the amino group projecting into the minor groove [e.g., of the six 9-aminoacridines in a dinucleotide crystal, five have the 9-amino group in a minor groove orientation (Sakore et al., 1979)], does not react with cis-DDP in the presence of DNA. In addition, for stereochemical reasons trans-DDP (16) does not react with any intercalator in the presence of DNA.

D. AMPLIFICATION AND INHIBITION OF
BLEOMYCIN-INDUCED DEGRADATION OF DNA

1. Bleomycin: An Overview

Bleomycin (BLM) is a generic name for a family of metal-complexing glycopeptide antitumor antibiotics differing in the structure of the C-terminal cationic side chain (Fig. 8). The bleomycins were first isolated from a strain of *Streptomyces verticillus* by Umezawa in 1963, as reviewed by Umezawa (1979). Chronologically, the discovery of bleomycins was preceded by isolation of phleomycins (Maeda et al., 1956; Earhart, 1979), relatively less stable molecules due to the presence of a hydrogenated thiazole ring (see Fig. 8). As expected, dehydrogenation of phleomycins gives the respective bleomycins. More recently, closely related talisomycins were isolated. All these antibiotics constitute a family of bleomycin-type drugs (Louie et al., 1985; Remers, 1979; Strong and Crooke, 1978). Their structures have been elucidated by chemical degradation and partial synthesis of the degradation products, as reviewed by Remers (1979). Total synthesis of BLM-A2 was accomplished by two independent groups headed by Umezawa (Takita et al., 1982) and by Hecht (Aoyagi et al., 1982) virtually at the same time but using different synthetic approaches.

The strong interest in these drugs is being generated by their exceptionally broad spectrum of biological activity. They are active against a variety of

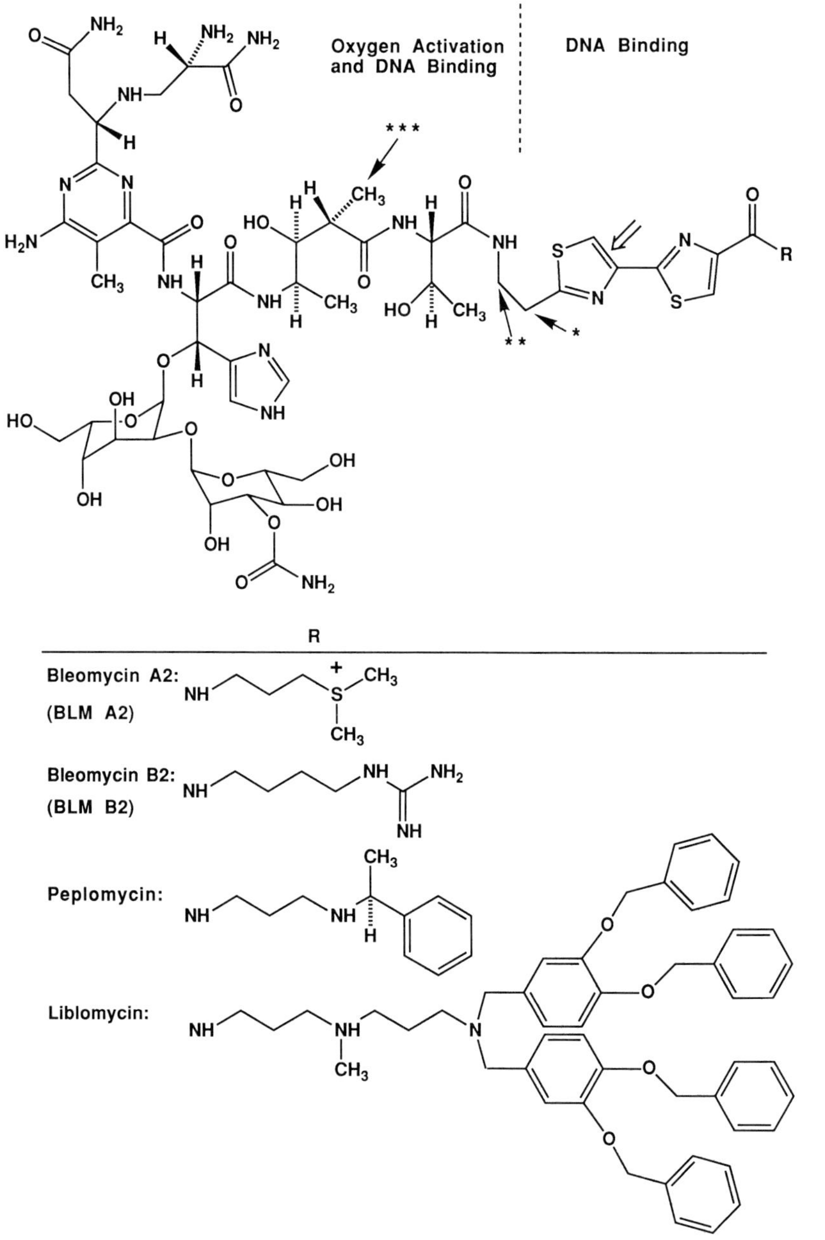

FIGURE 8 Structures of selected bleomycins. Other bleomycins differ in the C-terminal (R) substituent. A double arrow and single arrows indicate sites of structural differences in phleomycins and talisomycins, respectively. The indicated double bond of a thiazole is saturated in phleomycins. The talisomycins contain two additional substituents, namely a hydroxy group (*) and a 4-amino-4,6-dideoxy-L-talose moiety (**) at the indicated positions. In addition, one methyl group of the bleomycin core (***) is replaced by hydrogen in talisomycins.

gram positive and gram negative bacteria, fungi, and tumors at low concentrations (Remers, 1979). The fascinating properties of these drugs are that they virtually do not suppress either antibody- or cell-mediated immunity in humans, and are preferentially accumulated in tumor cells (Lehane and Lane, 1978). Unfortunately, their therapeutic index is low, which is mainly due to a severe pulmonary toxicity (Catravas, 1988). Phleomycin is strongly nephrotoxic. A great deal of work has been devoted to improving the therapeutic properties of bleomycin. As a result, a clinical preparation Blenoxane®, consisting mainly of BLM-A2 and BLM-B2 (Fig. 8), has been developed. Blenoxane is still highly toxic. The drug is used with limited success in combination chemotherapy with other anticancer drugs, such as cisplatin (*cis*-DDP, Fig. 7) and vinblastine, against a large number of tumor types (Carter, 1985; Carlson and Sikic, 1985; Wittes, 1985). Novel bleomycin analogs, peplomycin and liblomycin (Fig. 8), with improved therapeutic properties, have been prepared and tested (Takita and Ogino, 1987; Takahashi *et al.*, 1987; Kuramochi *et al.*, 1988; Lazo *et al.*, 1989). These new bleomycins contain lipophilic terminal amines. The increased lipophilicity may, in part, facilitate membrane permeability of the drugs and thus contribute to their increased activity inside the cell. A further improvement of bleomycin chemotherapy can be achieved through understanding of the molecular basis for the bleomycin activity, as discussed in the next section.

2. Molecular Basis for Bleomycin Activity

Many studies, reviewed by Hecht (1986a,b), Stubbe and Kozarich (1987), Fox (1987), Goldberg (1986), and Yukio (1987), have shown that DNA is the primary target for bleomycin *in vitro*. It is believed that the bleomycin-mediated degradation of DNA is also the major mechanism by which bleomycin exerts its influence *in vivo*, as reviewed by Povirk and Goldberg (1987), Solaiman (1988), Bauch (1989), Lambert and Le Pecq (1987), and Niitani (1987). Although several metal complexes of bleomycin are known to bind with and cause degradation of DNA, it appears that the ferrous chelate-induced chemistry is of the greatest importance *in vivo*. Accordingly, the chemistry of the BLM·Fe(II) chelate has received the most attention. It has been demonstrated that the BLM·Fe(II) chelate reversibly forms a ternary complex with molecular oxygen (Fig. 8), and the ternary complex is then activated through one-electron reduction. In general, this activation can be achieved by disproportionation of two BLM·Fe(II) complexes in the presence of oxygen (Hecht, 1986a,b; Stubbe and Kozarich, 1987), in a bimolecular collision of BLM·Fe(II)·O_2 with free ferrous ion (Strekowski *et al.*, 1988a), or by reduction of the ternary complex with an organic reductant such as thiol or ascorbic acid. The activated bleomycin complex thus obtained reacts with deoxyribose portions of DNA with the release of small DNA fragments (Sugiyama *et al.*, 1988) and the inactive BLM·Fe(III) complex.

The latter complex can be reduced to BLM·Fe(II). Thus, bleomycin acts catalytically, and each turnover requires two electrons and one molecule of oxygen. It has been shown that the BLM·Fe(III) complex can also undergo redox cycling by the microsomal NADH-dependent cytochrome b_5 reductase–cytochrome b_5 system (Mahmutoglu and Kappus, 1988).

The reaction of bleomycin complexes with DNA indicates that bleomycin binding is minor groove specific. Thus, the bleomycin-mediated chemistry which leads to DNA strand scission originates with abstraction of the deoxyribose C4' hydrogen located in the minor groove of the DNA double helix (Hecht, 1986a,b; Wu *et al.*, 1985a,b). In other studies the binding of bleomycin to poly(dA–dT) has been investigated by NMR methods (Booth *et al.*, 1983). The lack of perturbation of the dT-H6 and dT-CH$_3$ located in the major groove is also consistent with minor groove interaction for the BLM–poly(dA–dT) complex. In addition, total cleavage of T4 DNA, which contains bulky 5-(glucosyloxy)methyl groups in the major groove, is little changed from cleavage of T4 DNA lacking these groups (Hertzberg *et al.*, 1988). The same study has shown that Z-DNA is not cleaved efficiently by bleomycin.

Interaction of the cationic bithiazole portion of bleomycin with DNA strongly contributes to the overall DNA binding strength of the bleomycin molecule. Spectroscopic investigations (Lin and Grollman, 1981; Kross *et al.*, 1982a; Sakai *et al.*, 1982) and computer modeling studies (Miller *et al.*, 1985) are consistent with the model of binding in which this bleomycin fragment interacts with the minor groove of B form DNA. This interaction may involve intercalation of one thiazole ring only or partial intercalation of the two rings of the bithiazole system with DNA base pairs (Henichart *et al.*, 1985; Houssin *et al.*, 1986). Available experimental data strongly argue against a complete insertion of the bithiazole system with DNA base pairs to form an intercalation complex in a classical sense.

Although this nonclassical intercalation and electrostatic interaction between the cationic C-terminal substituent and the anionic DNA backbone govern DNA affinity, apparently they do not determine the precise geometry of binding of the active bleomycin complex. The binding followed by DNA strand scission is remarkably sequence specific. For example, BLM-A2 and BLM-B2, which contain different C-substituents, both cleave DNA predominantly at 5'-GT-3' and 5'-GC-3' sequences (Shipley and Hecht, 1988; D'Andrea and Haseltine, 1978; Takeshita *et al.*, 1978; Mirabelli *et al.*, 1982; Kross *et al.*, 1982b). These and additional data obtained with other bleomycin congeners support the interaction model in which the precise geometry of interaction and chemical reactivity are determined by the structural domain of bleomycin containing the chelated metal ion cofactor (Levy and Hecht, 1988; Shipley and Hecht, 1988).

The sequence specificity of DNA strand scission is perturbed by agents that bind to DNA in the minor groove (Sugiura and Suzuki, 1982). In contrast, it has been demonstrated that several agents that bind in the major groove do not alter the DNA cleavage specificity of bleomycin (Suzuki *et al.*, 1983). Again, all these results are consistent with the previous conclusion that the interaction of the active bleomycin complex with DNA is minor groove specific.

One interesting possibility which arises from our analysis of intercalation and groove-binding modes with diamidines such as DAPI (see Section IID5) is that bleomycin diffuses along DNA with the bithiazole system in a minor groove, nonintercalated complex. At certain sites the bithiazole intercalates, as does DAPI. At cleavage sites two conditions must be met: intercalation of the bithiazole and selective fit of the iron containing moiety into the correct region of the minor groove to give attack at the 4′ position of deoxyribose. This two-fold recognition mechanism could lead to the observed high specificity for bleomycin cleavage.

It should be cautioned, however, although bleomycin binding with DNA followed by DNA degradation is the major mechanism, it is apparently not the only mechanism for the well-established antitumor activity of the drug. Recent studies have suggested that bleomycin may (1) induce chromatin aggregation (Woynarowski *et al.*, 1987), (2) eliminate the tumor-specific T-suppressor activity without having any influence on responder T-lymphocytes in the cell-mediated antitumor immune reactivity (Xu *et al.*, 1988; Yamaki *et al.*, 1988; Morikawa, 1988), (3) enhance the production of interleukin-2 (Turk *et al.*, 1989; Ahmed *et al.*, 1988; Hamied and Turk, 1987), (4) inhibit thymidine phosphorylation by decreasing the activity of thymidine kinase (Sharif and Goussault, 1988), and (5) inhibit polymerases and ligase from leukemic cells while being less effective on the enzyme activity of normal thymocytes and lymphocytes (Saulier *et al.*, 1988). All these novel activities of bleomycin may contribute to the antiproliferative properties of the drug, in addition to the well-known DNA degradation.

3. Biological Evidence for Amplification

The biological activity of bleomycin group drugs can be strongly potentiated by many agents. The known potentiation effects include synergistic interactions (Agostino *et al.*, 1984; Scheid and Traut, 1985; Pavelic *et al.*, 1985) and amplification (Stewart and Evans, 1989). The latter term was adopted by Grigg (1970) to describe the increased activity of a combination of two agents of which only one is active alone. However, the term "potentiation" is also used as a synonym (Strekowski *et al.*, 1989). In general the activity is enhanced by agents that (1) inhibit bleomycin hydrolase (Nishimura *et al.*, 1987, 1989; Sebti and Lazo, 1988), (2) increase efficiency with which ferrous ion is recruited into the drug (Burger *et al.*, 1985), (3) increase the intracellular level of the drug (Iwao *et al.*, 1988; Yokoyama *et al.*, 1988), (4) inhibit poly(ADP-ribose) polymerase (Kato *et al.*, 1988), (5) interfere with DNA repair and/or synthesis (Ackland *et al.*, 1988; Tsuruo, 1989; Rosenthal and Hait, 1988; Scheid and Traut, 1988; Pavelic, 1987; Lazo *et al.*, 1985; Leyhausen *et al.*, 1985), and (6) act directly on the DNA. The last mode of action will be analyzed in detail. Other, less understood potentiation mechanisms are discussed in our recent paper (Strekowski *et al.*, 1988b).

Phleomycin was the first drug for which the amplification phenomenon was observed (Grigg, 1970). It was demonstrated that caffeine enhanced the degradation of DNA in cultures of *Escherichia coli* exposed to a low concentration of phleomycin. Caffeine alone had no effect on such a culture, but the effect of caffeine with phleomycin was similar to that of increasing the concentration of phleomycin 30-fold. Following this discovery a large number of potential phleomycin amplifiers was synthesized by Brown and screened for amplification using the simple *E. coli* assay, as reviewed by Brown and Grigg (1982) and by Brown (1987). Many amplifiers were identified that were active *in vitro* not only in combination with phleomycin, but also with bleomycin and talisomycin (Grigg *et al.*, 1977, 1985). The most promising compounds were screened subsequently in combination with phleomycin or bleomycin in experimental animal tumors (Allen *et al.*, 1984, 1985; Brown and Grigg, 1982). Since the enhancement of the activity *in vivo* was obtained, clinical trials have been undertaken in Australia (Brown, 1987). Other independent groups have confirmed the results of the Australian group that bleomycin can be amplified *in vivo* in animal models (Kato *et al.*, 1988; Kaya *et al.*, 1988; Jin *et al.*, 1989).

In addition to ethidium bromide (**5**, in Fig. 1), the structures of other representative amplifiers that enhance the activity of the bleomycin group in biological systems are given in Fig. 9. Their modes of action apparently depend on their structures and may involve several mechanisms. There seem to be two major mechanisms responsible for the increased cytotoxicity: the inhibition of certain DNA repair functions (Grigg *et al.*, 1984) by caffeine (**17**), chlorpromazine (**18**), and ethidium bromide (**5**), and the direct action on the DNA double helix (Agostino *et al.*, 1984; Pavelic *et al.*, 1985; Strekowski *et al.*, 1986a) by DNA-intercalating compounds such as ethidium bromide (**5**), a thiazolylpyridine (**19**), and a thienylpyrimidine (**20**). Thus, the intercalator **5** may be involved in two different modes of action. It is interesting to note that a simple amplification effect was observed in the cleaving action of bleomycin on purified DNA in the presence of ethidium bromide (Strekowski *et al.*, 1987a), but a synergistic effect was found upon the treatment of mouse L1210 leukemia cells with bleomycin and ethidium bromide (Agostino *et al.*, 1984).

4. Methods to Monitor the Effect of DNA-Interacting Compounds on Bleomycin-Mediated Degradation of DNA

Evidence was presented in the preceding sections that the major known biological activity of bleomycin involves DNA degradation and that this

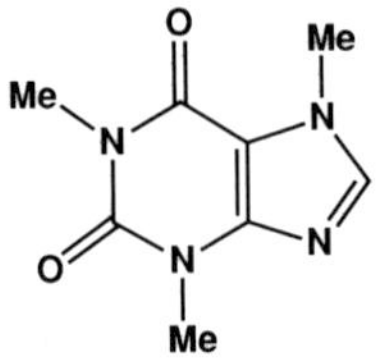

17

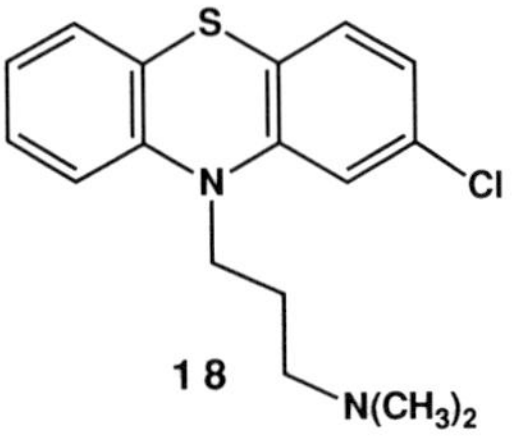

18

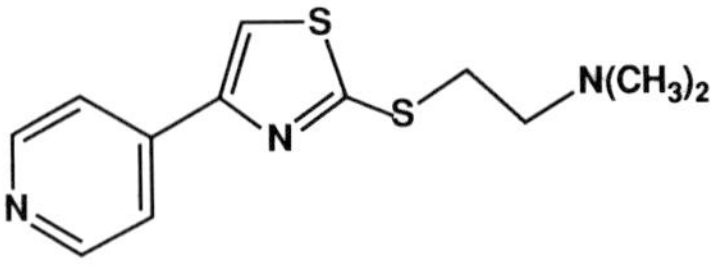

19

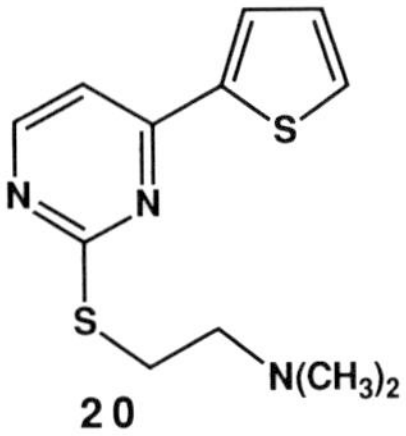

20

FIGURE 9   Selected amplifiers developed by Brown and Grigg (Allen *et al.*, 1984): caffeine (17), chlorpromazine (18), BC55 (19), and LS34 (20).

activity can be potentiated by certain DNA interacting compounds. In order to develop better amplifiers, it is extremely important to understand their structure–activity relationship and the molecular basis for bleomycin amplification. Such work requires the availability of simple but sensitive and reliable assays to monitor the kinetics of bleomycin-mediated degradation of DNA in the presence and absence of potential amplifiers. These must be *in vitro* tests so that the effects of amplifiers on DNA can be distinguished from secondary effects of transport and metabolism. Three such methods are described here. The methods mimic the biological conditions of low concentration of iron and high molecular ratio of DNA nucleotides to bleomycin, where the drug acts as a catalyst with many turnovers. The oxygen-dependent distribution of DNA degradation products (Wu *et al.*, 1985a,b) is minimized by the use of a buffer preequilibrated with air. Dithiothreitol (DTT) is used as an iron reducing agent. In general, stock solutions are added to a buffer of pH 7 in the order given here to reach the approximate final concentration: DNA, 1×10^{-4} M (concentration of base pairs); DNA-intercalating compound, 1×10^{-6}–1×10^{-3} M; $\mathrm{Fe(NH_4)_2(SO_4)_2 \cdot 6H_2O}$, 1×10^{-6} M; DTT, 1×10^{-4} M; BLM, 1×10^{-6} M. The reaction mixture is then incubated at 37°C and promptly analyzed as a function of time in a uv spectrophotometer (Strekowski *et al.*, 1987a), by HPLC (Strekowski *et al.*, 1987a, 1988a), or using a viscometric assay (Strekowski *et al.*, 1987a, 1988a,b,d). It should be noted that with increasing amounts of DNA-interacting compound added to the solution the DNA may collapse, aggregate, or even precipitate. Under such conditions the bleomycin-mediated degradation of DNA is highly inefficient and meaningless data are obtained. A sensitive test for the physical state of the DNA in solution is based on laser light scattering. Thus, similar apparent diffusion constants and similar amounts of light scattered by the solutions of DNA in the absence and presence of the DNA-binding compound are indicative of the absence of the complications mentioned above (Strekowski *et al.*, 1988d).

uv Analysis The method is based on bleomycin-induced hyperchromicity of DNA. Activated bleomycin initially produces widely separated breaks in the DNA strands that have little effect on the DNA uv absorption spectrum. As the number of breaks increase and become more closely spaced, single-stranded regions and short low-melting double-helical pieces of DNA are produced, which result in a cooperative increase in absorption until the double helix is completely separated.

In this simple qualitative assay the reaction solution is prepared in a uv cell thermostated at 37°C and placed in a spectrophotometer set at a constant wavelength, usually 260 nm. Either sonicated or high molecular weight natural or synthetic DNA can be used. The progress of the DNA degradation with time is monitored by changes in the uv absorption.

Compounds that enhance the bleomycin chemistry also cause faster increases in uv absorption of the solution in comparison to the absorption changes of the control solution without the additive. On the other hand, inhibition of the DNA degradation results in slower increases of the absorption in comparison to that of the same control solution. For the reasons already mentioned the method is relatively insensitive for monitoring the initial stage of the DNA degradation (Strekowski *et al.*, 1987a).

HPLC Analysis This is a quantitative and convenient way to monitor the bleomycin-mediated degradation of DNA. Moreover, for the indicated concentration conditions the method is sensitive enough to measure the degradation at very low conversion levels. Using this method it has been demonstrated that the bleomycin-induced chemistry is biphasic (Strekowski *et al.*, 1988a). A fast initial reaction with DNA is followed by a much slower process. The fast degradation is due to the fast activation of the BLM·Fe(II) complex and the subsequent fast reaction of the activated complex with DNA. In the slow process the BLM·Fe(III) complex is reactivated.

It is not necessary to analyze all small molecular products of the DNA degradation, which include four possible bases, four possible base propenals, modified nucleotides, and oligonucleotides (Stubbe and Kozarich, 1987; Hecht, 1986a,b). It has been shown that under the conditions of relatively high and constant concentrations of molecular oxygen and DTT used in the reactant solution the formation of the bases is proportional to the degradation events. Thus, quantitation of the bases alone can conveniently be used to monitor the progress of the bleomycin-induced chemistry. To characterize the effect of a DNA-interacting compound on the bleomycin-mediated DNA degradation the term amplification activity has been adopted (Strekowski *et al.*, 1987a). It is defined as a ratio of the total amounts of bases released from DNA in the presence and absence of a given compound. Thus, the amplification activities larger than 1.0 and smaller than 1.0 correspond to the enhancement and inhibition, respectively, of the bleomycin chemistry.

Viscometric Method   The initial breaks produced in DNA create regions with significantly enhanced flexibility. This results in large decreases in solution viscosity that are quite dependent on the effective length of the DNA molecules. It has been found that the viscosity changes can serve as a basis for a very sensitive and quantitative test to evaluate the effects of the additives on the reaction of bleomycin with DNA. Under the conditions used the DNA viscosity changes can be described by the following biphasic Eq. 1,

$$\left(\frac{\eta}{\eta_0}\right)_t = a\mathrm{e}^{-k_{\mathrm{f}}t} + b\mathrm{e}^{-k_{\mathrm{s}}t} \tag{1}$$

where η_0 is the initial reduced specific viscosity for DNA before the addition of bleomycin, η is the reduced specific viscosity for DNA at the reaction (degradation) time t, k_{f} is the apparent rate constant for the first (fast) process, k_{s} is the apparent rate constant for the second (slow) process, and a and b are the amplitudes for the fast and slow phases, respectively (Strekowski *et al.*, 1988a,b). Both the fast and slow processes of the DNA degradation are accelerated in the presence of an amplifier. The viscometric amplification activity is defined as a ratio of apparent rate constants k_{s} obtained for the DNA degradations conducted in the presence and absence of a given compound.

5. Molecular Basis for Amplification and Inhibition of Bleomycin-Mediated Digestion of DNA

Structure–Activity Relationship for Amplifiers   Following the discovery that ethidium bromide (**5**), a classical DNA intercalator, strongly potentiates the bleomycin-mediated digestion of DNA, systematic studies have been undertaken to understand this enhancement effect (Strekowski *et al.*, 1986a,b, 1987a,b, 1988a–d; W. D. Wilson, *et al.*, 1988, 1989b, 1990). It has been demonstrated that a large variety of DNA binding compounds (Figs. 1 and 10), such as fused-ring planar intercalators **6** and **7**, unfused-ring nonplanar intercalators, such as **20–27**, groove binding compounds, such as **28**, and even polyamines exemplified by putrescine (**29**), spermidine (**30**), and spermine (**31**), can potentiate the effect of bleomycin on DNA at low ratios of compound to DNA base pairs.

As the ratio of compound/DNA increases, the amplification activity of classical intercalators **5–7** decreases, and strong inhibition is observed at high ratios. In sharp contrast to the behavior of these classical intercalators, no bleomycin inhibition is observed in the presence of unfused-ring systems

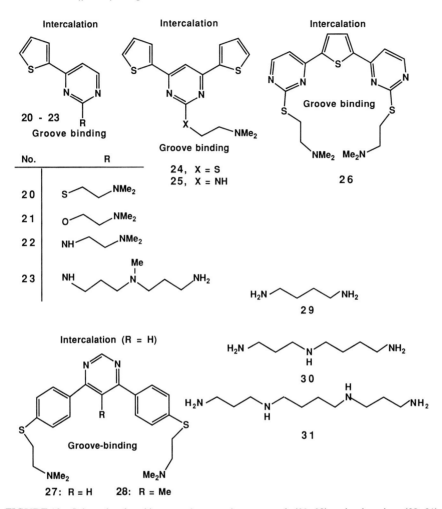

FIGURE 10   Selected unfused heteropolyaromatic compounds (**20–28**) and polyamines (**29–31**) that bind with and amplify the bleomycin-mediated degradation of DNA.

20–28 at any ratio of compound/DNA. For example, compound **26**, one of the best amplifiers, gave a noticeable enhancement at a ratio as low as 0.01 of **26**/DNA base pairs, and its amplification activity reached 11.2 at the ratio of 0.5, as measured by the HPLC method (Strekowski *et al.*, 1987a). The amplification activity depends on the DNA binding mode of an amplifier. Of the two closely related compounds **27** and **28**, the former is an intercalator and is much more active than the groove-binding compound **28** (Wilson *et al.*, 1988). Polyamines are also groove-binding compounds and show weaker activity than unfused-ring intercalators (Strekowski *et al.*, 1988d). On the other hand, a polyamine chain attached to the intercalator system greatly enhances activity. For example, compound **23** is a much better amplifier than compound **22**. Even a simple monoamine attached to the intercalator system increases activity of the intercalator. Thus, 4-(2-thienyl)pyrimidine alone, the biaromatic system of **20–23**, is a weak amplifier, and 2-(dimethylamino)ethanethiol is not active at all. When combined into an integral molecule such as **20**, the activity is strongly enhanced. These effects can be attributed, in part, to the increased binding of the amino-substituted compound with DNA. The amine is protonated under the amplification conditions of pH 7, and the resultant ammonium cation interacts electrostatically with the anionic DNA backbone increasing the stability of the molecule–DNA complex.

The binding strength of an amplifier molecule with DNA is also strongly affected by a heteroatom attached directly to the aromatic system, which constitutes part of a tether linking the aromatic system with the amine. For example, the DNA binding constants for three closely related compounds are in order **20 > 21 > 22**, which corresponds to the heteroatoms S, O, and N, respectively. The thio derivative (**24**) also interacts more strongly with DNA than its imino analog (**25**), and the same order has been observed for other series of *S*- and *N*-substituted amplifiers (L. Strekowski and W. D. Wilson, unpublished observations). This heteroatom-dependent interaction strength of the intercalator molecule with DNA has been explained in terms of the heteroatom-induced polarization and other stereoelectronic effects within the intercalation complexes (Strekowski *et al.*, 1986a, 1988b).

In general, the heteroatom-induced interaction strengths with DNA parallel the amplification activities of the corresponding amplifiers within the same series of compounds using the assays with isolated DNA and the *E. coli* test (Brown and Grigg, 1982). On the other hand, there is no direct correlation between DNA binding constants and bleomycin amplification activities for compounds that differ significantly in structure. The amplifiers bind with DNA but with little and different base sequence specificity (Strekowski *et al.*, 1987b). Some amplifiers alter (Sugiura and Suzuki, 1982) and others do not affect the base sequence-specific mode of DNA cleavage by the bleomycin complex system (Suzuki *et al.*, 1983). uv studies have strongly suggested that the representative amplifiers **20–27** do not complex ferrous ion nor do they interact with bleomycin. In contrast, known ferrous ion chelators, such as (ethylenedinitrilo)tetraacetic acid (EDTA), 2,2′–bipyridine and its derivatives, and *o*-phenanthroline strongly inhibit the BLM-mediated degradation of DNA (Sausville *et al.*, 1978; Strekowski *et al.*, 1988b; Larramendy *et al.*, 1989).

Stereochemical Basis for Amplification and Inhibition   It has been concluded that the amplification phenomenon is due to stereochemical effects in the amplifier–DNA complex (Strekowski *et al.*, 1987a, 1988b,c). More specifically, the distorted helix in the complex is a better target for activated bleomycin than the native form of DNA. Intercalators produce more extensive perturbations of the DNA tertiary structure than groove-binding compounds and, at the same time, are better amplifiers. Recent studies from our laboratories have suggested that widening of the minor groove of the helix upon binding of the amplifier molecule with DNA is directly responsible for the amplification effect. The minor groove widening can result from the interaction with either groove; however, this effect is much stronger for the amplifier–DNA complex in the major groove. In addition, the weak amplification effect for compounds that bind selectively in the minor groove can be seen only at relatively low ratios of compound/DNA. A strong inhibition of the DNA degradation is observed at higher ratios because of the binding competition with bleomycin in the minor groove under such conditions. Examples are the bithiazole portion of bleomycin and closely related analogs which exhibit the same minor-groove affinity as bleomycin (Fisher *et al.*, 1985; Kross *et al.*, 1982a). Another example is distamycin A (**2**) (Sugiura and Suzuki, 1982). As expected, these compounds inhibit bleomycin degradation by competitive binding to the DNA substrate.

The effects of classical planar intercalators **5–7** are consistent with their well-known nonselective interaction with DNA. Relatively strong amplification at low concentrations of **5–7** is a result of the intercalation-induced unwinding of the helix, which is transmitted along the helix many base pairs away from the intercalation site. Such conformational changes of the helix increase the efficiency of the DNA degradation by bleomycin. With increasing concentrations of the intercalators both grooves become progressively blocked, thus making the bleomycin binding increasingly difficult. As a result, an inhibition effect is observed. This inhibition is especially effective for a dicationic intercalator (**7**) (Yen *et al.*, 1982), the two cationic chains of which block the major and minor grooves of DNA simultaneously.

The tricyclic unfused compounds **24–27** and similar derivatives intercalate strongly with DNA, are excellent bleomycin amplifiers, and, in contrast to classical intercalators **5–7**, do not inhibit bleomycin chemistry even when taken in an excess over available DNA binding sites. Since the DNA equilibrium binding constants for **24–27** are several times larger than the binding constant for bleomycin (W. D. Wilson *et al.*, 1988, 1989b; Strekowski *et al.*, 1987b), these rather unusual results show that **24–27** must not compete with bleomycin for binding sites on DNA. It must be concluded, therefore, that these compounds interact with the major DNA groove with high specificity, that is, they show little minor groove affinity. This conclusion is strongly supported by computer modeling studies on the interaction of **27** with DNA (Wilson *et al.*, 1988). It is interesting to note that compounds **24** and **27**, although of different structures, induce similar distortion of DNA upon binding, as measured by ^{31}P-NMR, uv, and hydrodynamic methods. In excellent agreement with the discussed mechanism of the amplification, these compounds show comparable activities as bleomycin amplifiers.

Our recent studies of the bleomycin-mediated degradation of poly(dG–dC)$_2$ and poly(dA–dT)$_2$ in the presence of DAPI are consistent with the major groove-binding requirement for a strong amplification. As discussed in

Section IIA, DAPI binds very strongly to consecutive A·T base pairs in a minor groove complex but intercalates in pure G·C regions. Additional modeling of the DAPI intercalation has indicated that the amidine groups must face the major groove as observed with good amplifiers of bleomycin. Intercalation complexes with the amidines in the minor groove with significant overlap of the phenyl-indole ring system with the G·C base pairs could not be prepared, due to steric clash. These observations suggest that DAPI should be a sequence selective bleomycin amplifier with significant amplification at G·C but not at A·T sites. In excellent agreement with this analysis, at a ratio of 0.8 DAPI per base pair, for example, the bleomycin-catalyzed cleavage of poly(dG–dC)$_2$ is enhanced 6-fold while that of poly(dA–dT)$_2$ is reduced by 60% (Wilson et al., 1990).

Polyamines **29–31** are representative examples of another class of bleomycin potentiators (Strekowski et al., 1988d). To better understand the stereochemical features of amplification, we have recently conducted computer modeling studies of the interaction of **29–31** with a self-complementary B-decamer (dG–dC)$_5$·(dG–dC)$_5$. The results are presented in Figs. 11 and 12. The theoretical studies suggest that **29–31** bind preferentially on the floor of the major groove of the oligomer helix. This interaction results in an enlargement of the minor groove (Fig. 11) and a bend of the helix toward the major groove (Fig. 12), both effects being in the order **29** < **30** < **31** (Strekowski et al., 1989). The polyamine-induced distortion, as obtained from theoretical studies, parallels the experimental values of the amplification activities of

1 2 3 4   5 6 7   8 9 10
5´ G C G C G C G C G C 3´
3´ C G C G C G C G C G 5´
20 19 18 17 16 15 14 13 12 11

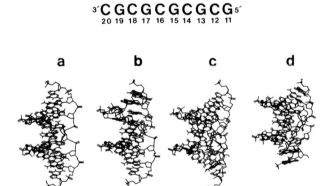

a   b   c   d

FIGURE 11   Base numbering scheme for (dG-dC)$_5$·(dG-dC)$_5$ and polyamine-induced opening of the minor groove of this B decamer duplex upon binding of the polyamine in the middle of the major groove of the helix. The distance between the sugar carbon atoms 4´(·) of the adjacent interstrand deoxyguanylates G7 and G17 and the polyamine complexed are given in order: 8.01Å, no polyamine (**a**); 9.87Å, (**29**), (**b**); 10.64Å, (**30**), (**c**); 10.91Å, (**31**), (**d**). (Reproduced from Strekowski et al., 1989, with permission.)

a   **b**   **c**   **d**

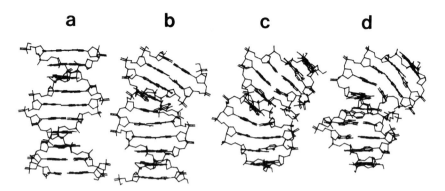

FIGURE 12 Polyamine-induced bend of the B helix of the decamer duplex (dG-dC)$_5$·(dG-dC)$_5$ upon binding of the polyamine in the major groove of the helix. The estimated bend of the helix and the polyamine complexed are given in order: 0°, no polyamine (**a**); 26°, (**29**), (**b**); 30°, (**30**), (**c**); 38°, (**31**), (**d**). The perturbed complexes (**b–d**) are positioned to show the largest bend. (Reproduced from Strekowski *et al.*, 1989, with permission.)

29–31 in the BLM-mediated degradation of poly (dG–dC)·poly (dG–dC). It should be noted that in order to compensate for different binding constants of these polyamines, their concentrations were adjusted so that the same ratio of bound polyamine to DNA-P was obtained in the three cases studied.

These and previously discussed studies on major groove-specific intercalators and nonspecific intercalators strongly support the following amplification mechanism which involves two features: (1) noncompetitive binding of amplifier molecules in the major groove and the active bleomycin complex in the minor groove and (2) the amplifier-induced stereochemical fit of the DNA helix for a favorable interaction with the bleomycin complex. Thus, optimum binding of the active bleomycin complex may depend on the local DNA shape. We strongly suggest that the amplifier-induced increase in affinity of the active bleomycin complex toward the DNA minor groove results in an increased efficiency of the BLM-mediated degradation of the helix.

The proposed mechanism of the amplifier-induced fit for bleomycin is in strong agreement with the results of two studies of binding of bleomycin with DNA under conditions in which the drug does not cut DNA. Thus, it has been demonstrated that the tertiary structure of the double helix is significantly distorted in the complex with bleomycin (Fox and Grigg, 1988). It is also known that binding of bleomycin with DNA increases with increasing temperature, and the maximum interaction occurs at temperatures at which the DNA helix becomes highly relaxed, albeit not melted (Booth *et al.*, 1983).

In summary, it seems to be little doubt that the amplifier-induced conformational changes of the DNA double helix play a role in bleomycin potentiation. It should be cautioned, however, that the stereochemistries of the distorted double helix in complexes with polyamines, aromatic groove-binding compounds, nonclassical unfused intercalators, and planar classical intercalators differ substantially from each other. Clearly, additional systematic studies with these different types of compounds are necessary to delineate differences and fine features in their amplification mechanisms.

III. Interactions with Protein Targets

This section discusses only selected interactions of small molecules with proteins. It is intended only to address this fast developing subject, rather than offer complete coverage. An interested reader is directed to Chapter 13 by E. Mini and J. R. Bertino for additional examples.

Inosine monophosphate (IMP) dehydrogenase has recently been recognized as a sensitive target in anticancer and antiviral chemotherapy. The enzyme is involved in a rate-limiting step in biosynthesis of purine riboside monophosphates, precursors to ATP, GTP, dATP, and dGTP. Thus, by inhibiting the enzyme, many important biological functions including synthesis of nucleic acids and proteins, and polymerization of tubulin can be affected.

Tiazofurin (**32**, Fig. 13) and selenazofurin (**33**) are potent inhibitors of IMP dehydrogenase *in vivo* (Weber *et al.*, 1989). Kinetic and synthetic studies have shown that **32** and **33** are prodrugs acting through their active metabolic derivatives **34** and **35**, respectively, containing a phosphodiester linkage between O5′ of the respective drug and adenosine diphosphate (Gebeyehu *et al.*, 1985). These metabolites bind with the NADH (**38**) site of IMP dehydrogenase (Yamada *et al.*, 1988). Ribavirin 5′-monophosphate (**37**), metabolite of the antiviral drug ribavirin (**36**) is also an inhibitor of the enzyme, which blocks the xanthosine monophosphate (**39**) binding site. It has been shown that combinations of either **32** or **33** with **36** are strongly synergistic in their ability to inhibit the conversion of IMP to guanylates. When studied in hepatoma cells, a synergistic cytotoxicity was observed (Natsumeda *et al.*, 1988).

As mentioned earlier, the inhibition of IMP dehydrogenase strongly affects the formation of microtubules, which is a result of lowering the concentration of guanosine 5′-triphosphate (GTP). The tubulin assembly can also be inhibited by agents that bind with tubulin, such as Bis-ANS (**40**) (Prasad *et al.*, 1986; Horowitz *et al.*, 1984) (Fig. 14), the antimiotic drugs colchicine (**41**) and podophyllotoxin (**42**) (Kumar, 1981; Liu *et al.*, 1989), and a novel antitumor agent amphethinile. It has been shown that amphethinile displaces colchicine bound to tubulin and stimulates the activity of GTP hydrolase *in vitro* (McGown and Fox, 1989).

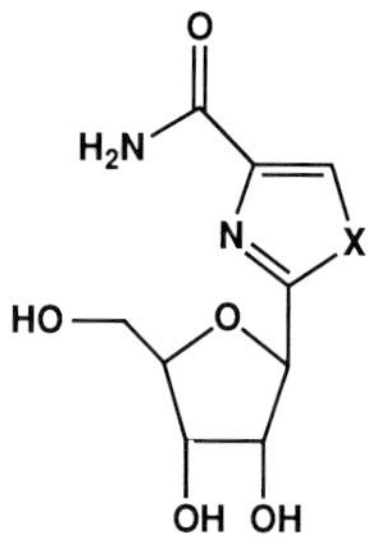

32: X = S
33: X = Se

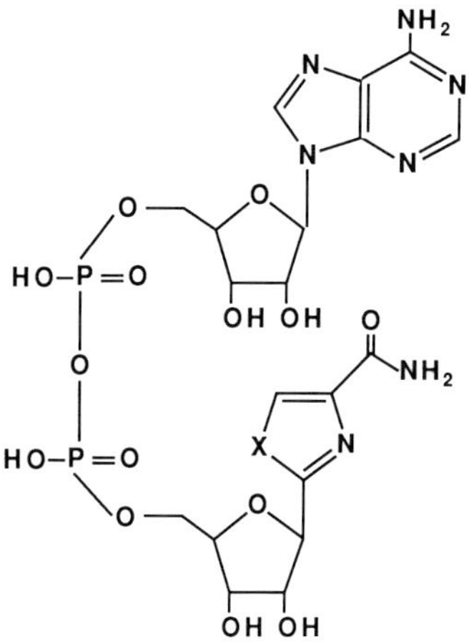

34: X = S
35: X = Se

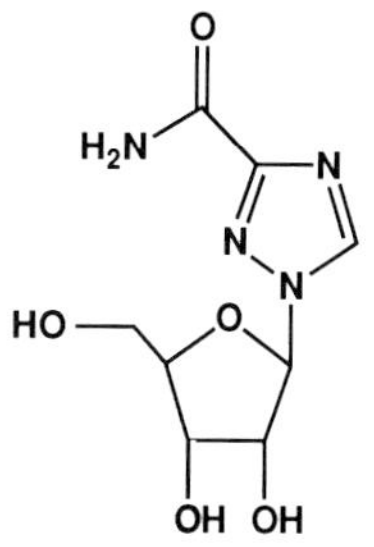

36

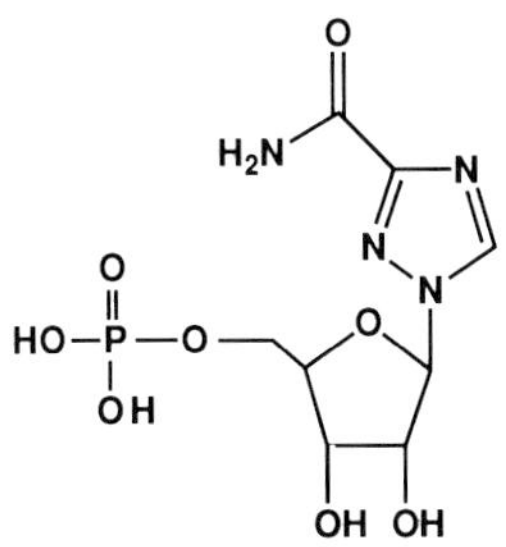

37

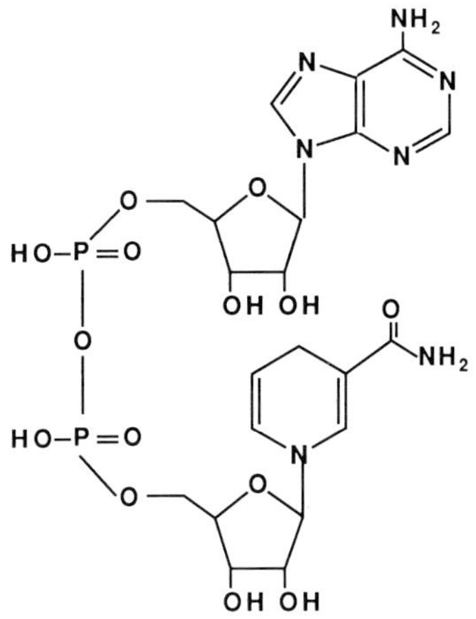

38 (NADH)

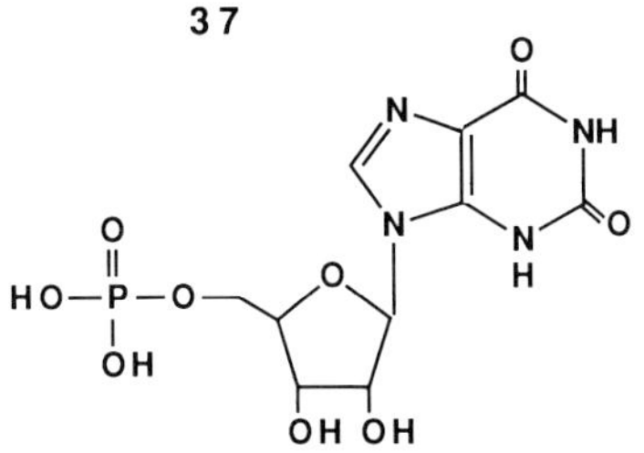

39 (XMP)

FIGURE 13 Structures of tiazofurin (**32**), selenazofurin (**33**), the active form of tiazofurin (**34**), the active form of selenazofurin (**35**), ribavirin (**36**), the active form of ribavirin (**37**), NADH (**38**), and XMP (**39**).

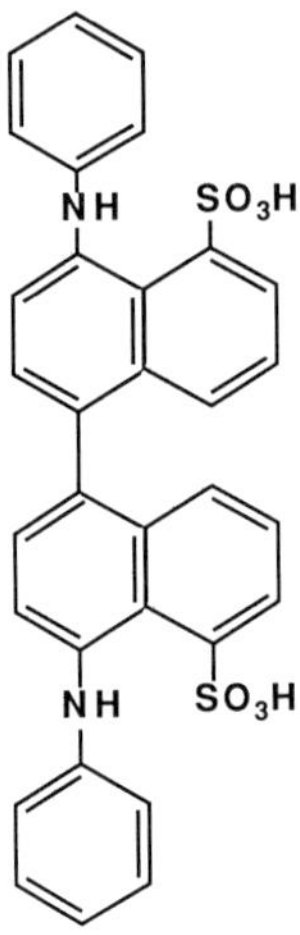

FIGURE 14   Structures of Bis-ANS (**40**), colchicine (**41**), and podophyllotoxin (**42**).

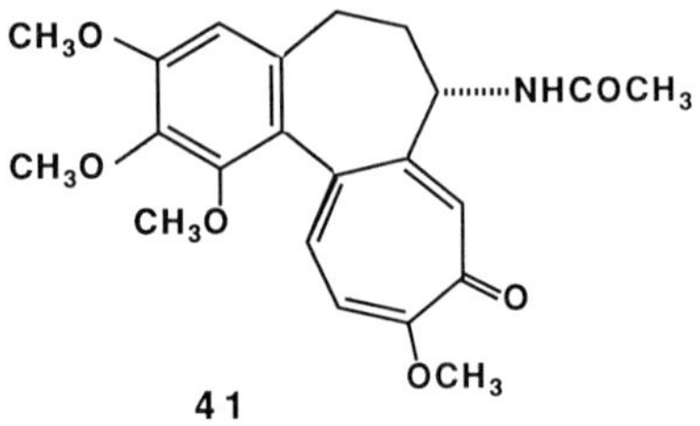

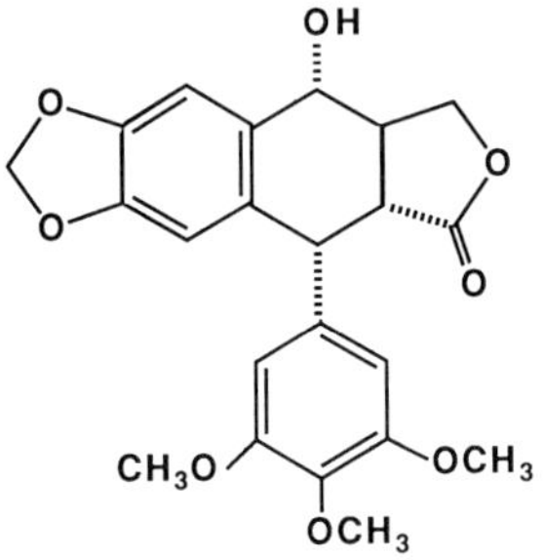

Acknowledgments

Research from our laboratories presented in this review was generously supported by grants from the American Cancer Society (CH-383) and the National Institutes of Health (AI27196).

References

Abola, E. E., Bernstein, F. C., Bryant, S. H., Koetzle, T. F., and Weng, J. (1987). *In* "Crystallographic Databases—Information Content, Software Systems, Scientific Applications" (F. H. Allen, G. Bergerhoff, and R. Sievers, eds.), pp. 107–132. Data Comm., Int. Union Crystallogr., Bonn, Federal Republic of Germany.

Ackland, S. P., Schilsky, R. L., Beckett, M. A., and Weichselbaum, R. R. (1988). Synergistic cytotoxicity and DNA strand break formation by bromodeoxyuridine and bleomycin in human tumor cells. *Cancer Res.* **48**, 4244–4249.

Adams, G. E., and Stratford, I. J. (1986). Hypoxia-mediated nitroheterocyclic drugs in the radio- and chemotherapy of cancer. *Biochem. Pharmacol.* **35**, 71–76.

Agostino, M. J., Bernacki, R. J., and Beerman, T. A. (1984). Synergistic interactions of ethidium bromide and bleomycin on cellular DNA and growth inhibition. *Biochem. Biophys. Res. Commun.* **120**, 156–163.

Ahmed, K., Hamied, T. A., and Turk, J. L. (1988). Effect of anticancer drugs on cytokine release. *Int. J. Immunother.* **4**, 137–143.

Allen, T. E., Brown, D. J., Cowden, W. B., Grigg, G. W., Hart, N. K., Lamberton, J. A., and Lane, A. (1984). Amplification of the antitumor activity of phleomycins in rats and mice by heterocyclic analogues of purines. *J. Antibiot.* **37,** 376–383.

Allen, T. E., Aliano, N. A., Cowan, R. J., Grigg, G. W., Hart, N. K., Lamberton, J. A., and Lane, A. (1985). Amplification of the antitumor activity of phleomycins and bleomycins in rats and mice by caffeine. *Cancer Res.* **45,** 2516–2521.

Aoyagi, Y., Katano, K., Suguna, H., Primeau, J., Chang, L.-H., and Hecht, S. M. (1982). Total synthesis of bleomycin. *J. Am. Chem. Soc.* **104,** 5537–5538.

Asseline, U., Delarue, M., Lancelot, G., Toulmé, F., Thuong, N. T., Montenay-Garestier, T., and Hélene, C. (1984). Nucleic acid-binding molecules with high affinity and base sequence specificity: Intercalating agents covalently linked to oligodeoxynucleotides. *Proc. Natl. Acad. Sci. U.S.A.* **81,** 3297–3301.

Bauch, H. J. (1989). Cytostatics: Part 2. Bleomycins, antimetabolites, mitosis spindle toxins, L-asparaginase, and antihormones. *Med. Monatsschr. Pharm.* **12,** 34–45.

Bernstein, F. C., Koetzle, T. F., Williams, G. J. B., Meyer, E. F., Jr., Brice, M. D., Rodgers, J. R., Kennard, O., Schimanouchi, T., and Tasumi, M. (1977). The protein data bank: A computer-based archival file for macromolecular structures. *J. Mol. Biol.* **112,** 535–542.

Booth, T. E., Sakai, T. T., and Glickson, J. D. (1983). Interaction of bleomycin A2 with poly(deoxyadenylythymidylic acid). A proton nuclear magnetic resonance study of the influence of temperature, pH, and ionic strength. *Biochemistry* **22,** 4211–4217.

Bower, M., Summers, M. F., Powell, C., Shinozuka, K., Rega, J. B., Zon, G., and Wilson, W. D. (1987). Oligodeoxyribonucleoside methylphosphonates. NMR and UV spectroscopic studies of R_p–R_p and S_p–S_p methylphosphonate (Me) modified duplexes of {d[GGAATTCC]}$_2$. *Nucleic Acids Res.* **15,** 4915–4930.

Brown, D. J. (1987). Amplification of bleomycin and phleomycin as antitumor agents. *Chem. Aust.* pp. 116–118.

Brown, D. J., and Grigg, G. W. (1982). Heterocyclic amplifiers of phleomycin. *Med. Res. Rev.* **2,** 193–210.

Brown, J. M., and Workman, P. (1980). Partition coefficient as a guide to the development of radiosensitizers which are less toxic than misonidazole. *Radiat. Res.* **82,** 171–190.

Burger, R. M., Horwitz, S. B., and Peisach, J. (1985). Stimulation of iron(II) bleomycin activity by phosphate-containing compounds. *Biochemistry* **24,** 3623–3629.

Carlson, R. W., and Sikic, B. I. (1985). Effective treatment of malignant ovarian germ-cell tumors with cisplatin, vinblastine, and bleomycin. *In* "Bleomycin Chemotherapy" (B. I. Sikic, M. Rozencweig, and S. K. Carter, eds.), pp. 69–76. Academic Press, Orlando, Florida.

Carminati, A., Barascut, J.-L., Chenut, E., Bourut, C., Mathé, G., and Imbach, J.-L. (1988). On a new class of mixed-function drugs associating nitroimidazoles and CENU: The NICE-NU. *Anti-Cancer Drug Des.* **3,** 57–65.

Carter, S. K. (1985). Bleomycin: More than a decade later. *In* "Bleomycin Chemotherapy" (B.I. Sikic, M. Rozencweig, and S. K. Carter, eds.), pp. 3–35. Academic Press, Orlando, Florida.

Catravas, J. D. (1988). Pulmonary toxicity of anticancer drugs: Alterations in endothelial cell function. *Dev. Oncol.* **53,** 118–127.

D'Andrea, A. S., and Haseltine, W. A. (1978). Sequence specific cleavage of DNA by the antitumor antibiotics neocarzinostatin and bleomycin. *Proc. Natl. Acad. Sci. U.S.A.* **75,** 3608–3612.

Declerck, P. J., and De Ranter, C. J. (1986). *In vitro* reductive activation of nitroimidazoles. *Biochem. Pharmacol.* **35,** 59–61.

Denny, W. A., and Wilson, W. R. (1986). Considerations for the design of nitrophenyl mustards as agents with selective toxicity for hypoxic tumor cells. *J. Med. Chem.* **29,** 879–887.

Earhart, C. F., Jr. (1979). Phleomycin. *In* "Antibiotics" (F. E. Hahn, ed.), Vol. 5, Part 2, pp. 229–312. Springer-Verlag, Berlin.

Eckstein, F. (1983). Phosphorothioate analogues of nucleotides—Tools for the investigation of biochemical processes. *Angew. Chem., Int. Ed. Engl.* **22,** 423–439.

Edwards, D. I. (1986). Reduction of nitroimidazoles *in vitro* and DNA damage. *Biochem. Pharmacol.* **35**, 53–58.

Fisher, R., Kuroda, R., and Sakai, T. T. (1985). Interaction of bleomycin A2 with deoxyribonucleic acid: DNA unwinding and inhibition of bleomycin-induced DNA breakage by cationic thiazole amides related to bleomycin A2. *Biochemistry* **24**, 3199–3207.

Fox, K. R. (1987). Interaction of antitumor drugs and DNA. *ISI Atlas Sci.: Pharmacol.* **1**, 289–294.

Fox, K. R., and Grigg, G. W. (1988). Diethyl pyrocarbonate and permanganate provide evidence for an unusual DNA conformation induced by binding of the antitumor antibiotics bleomycin and phleomycin. *Nucleic Acids Res.* **16**, 2063–2075.

Gebeyehu, G., Marquez, V. E., Van Cott, A., Cooney, D. A., Kelley, J. A., Jayaram, H. N., Ahluwalia, G. S., Dion, R. L., Wilson, Y. A., and Johns, D. G. (1985). Ribavirin, tiazofurin, and selenazofurin: Mononucleotides and nicotinamide adenine dinucleotide analogues. Synthesis, structure, and interactions with IMP dehydrogenase. *J. Med. Chem.* **28**, 99–105.

Goldberg, I. H. (1986). Molecular mechanisms of DNA sugar damage by antitumor antibiotics. *Pontif. Acad. Sci. Scr. Varia* **70**, 425–462.

Grigg, G. W. (1970). Amplification of phleomycin-induced death and DNA breakdown by caffeine in *Escherichia coli. Mol. Gen. Genet* **107**, 162–172.

Grigg, G. W., Gero, A. M., Hughes, J. M., Sasse, W. H. F., Bliese, M., Hart, N. K., Johansen, O., and Kissane, P. (1977). Amplifications of phleomycin and bleomycin-induced antibiotic activity in *Escherichia coli* by aromatic cationic compounds. *J. Antibiot.* **30**, 870–878.

Grigg, G. W., Gero, A. V., Sasse, W. H., and Sleigh, M. J. (1984). Inhibition and enhancement of phleomycin-induced DNA breakdown by aromatic tricyclic compounds. *Nucleic Acids Res.* **12**, 9083–9093.

Grigg, G. W., Hall, R. M., Hart, N. K., Kavulak, D. R., Lamberton, J. A., and Lane, A. (1985). Amplification of the antibiotic effects of the bleomycins, phleomycins and talisomycins: Its dependence on the nature of the variable basic groups. *J. Antibiot.* **38**, 99–110.

Hall, E. J., and Roizin-Towle, L. (1975). Hypoxic sensitizers: Radiobiological studies at the cellular level. *Radiology* **117**, 453–457.

Hamied, T. A., and Turk, J. L. (1987). Enhancement of interleukin-2 release in rats by treatment with bleomycin and Adriamycin *in vivo. Cancer Immunol. Immunother.* **25**, 245–249.

Hecht, S. M. (1986a). The chemistry of activated bleomycin. *Acc. Chem. Res.* **19**, 383–391.

Hecht, S. M. (1986b). DNA strand scission by activated bleomycin group antibiotics. *Fed. Proc., Fed. Am. Soc. Exp. Biol.* **45**, 2784–2791.

Helen, C., and Thuong, N. T. (1987). Oligodeoxynucleotides covalently linked to intercalating agents and to nucleic acid-cleaving reagents. New families of gene regulating substances. *In* "Molecular Mechanisms of Carcinogenic and Antitumor Activity" (C. Chagas and B. Pullman, eds.), pp. 205–274. Adenine, Schenectady, New York.

Henichart, J.-P., Bernier, J.-L., Helbecque, N., and Houssin, R. (1985). Is the bithiazole moiety of bleomycin a classical intercalator? *Nucleic Acids Res.* **13**, 6703–6717.

Hertzberg, R. P., Caranfa, M. J., and Hecht, S. M. (1988). Degradation of structurally modified DNAs by bleomycin group antibiotics. *Biochemistry* **27**, 3164–3174.

Horowitz, P., Prasad, V., and Luduena, R. F. (1984). Bis(1,8-anilinonaphthalenesulfonate) a novel and potent inhibitor of microtubule assembly. *J. Biol. Chem.* **259**, 14647–14650.

Houssin, R., Helbecque, N., Bernier, J.-L., and Henichart, J.-P. (1986). A new bithiazole derivative with intercalative properties. *J. Biomol. Struct. Dyn.* **4**, 219–229.

Iwao, T., Yoshikawa, H., Takada, K., and Muranishi, S. (1988). Potentiation of antitumor effect of bleomycin by fusogenic lipid-surfactant mixed micelles. Tumor-neutralizing assay for inherently bleomycin-resistant murine leukemia. *Chem. Pharm. Bull.* **36**, 2270–2273.

Jin, X., Lu, W., Xu, J., and Zhao, T. (1989). Cisplatin combination therapy of murine S180. *Shanghai Yike Daxue Xuebao* **16**, 50–54.

Josephy, P. D., Palcic, B., and Skarsgard, L. D. (1978). Ascorbate-enhanced cytotoxicity of misonidazole. *Nature (London)* **271,** 370–372.

Kato, T., Suzumura, Y., and Fukushima, M. (1988). Enhancement of bleomycin activity by 3-aminobenzamide, a poly(ADP-ribose) synthesis inhibitor, *in vitro* and *in vivo*. *Anticancer Res.* **8,** 239–243.

Kaya, G., Akin, C., Altug, T., and Devrim, S. (1988). The antitumor effect of bleomycin combined with bestatin against Ehrlich ascites carcinoma in mice. *Proc. Soc. Exp. Biol. Med.* **187,** 292–295.

Kennedy, K. A., Teicher, B. A., Rockwell, S., and Sartorelli, A. C. (1980). The hypoxic tumor cell: A target for selective cancer chemotherapy. *Biochem. Pharmacol.* **29,** 1–8.

Kopka, M. L., Yoon, C., Goodsell, D., Pjura, P., and Dickerson, R. D. (1985). The molecular origin of DNA–drug specificity in retropsin and distamycin. *Proc. Natl. Acad. Sci. U.S.A.* **82,** 1376–1382.

Kross. J., Henner, W. D., Haseltine, W. A., Rodriguez, L., Levin, M. D., and Hecht, S. M. (1982a). Structural basis for the deoxyribonucleic acid affinity of bleomycins. *Biochemistry* **21,** 3711–3721.

Kross, J., Henner, W. D., Hecht, S. M., and Haseltine, W. A. (1982b). Specificity of deoxyribonucleic acid cleavage by bleomycin, phleomycin and tallysomycin. *Biochemistry* **21,** 4310–4318.

Kumar, N. (1981). Taxol-induced polymerization of purified tubulin. *J. Biol. Chem.* **256,** 10435–10441.

Kuramochi, H., Motegi, A., Takahashi, K., and Takeuchi, T. (1988). DNA cleavage activity of liblomycin (NK 313), a novel analog of bleomycin. *J. Antibiot.* **41,** 1846–1853.

Lambert, B., and Le Pecq, J.-B. (1987). Pharmacology of DNA binding drugs. *NATO Adv. Study Inst. Ser., Ser. A* **137,** 141–157.

LaPlanche, L. A., James, T. L., Powell, C., Wilson, W. D., Uznanski, B., Stec, W. J., Summers, M. F., and Zon, G. (1986). Phosphorothioate-modified oligodeoxynucleotides. III. NMR and UV spectroscopic studies of the R_p–R_p, S_p–S_p, and R_p–S_p duplexes, [d(GG$_s$AATTCC)]$_2$, derived from diastereomeric *O*-ethyl phosphorothioates. *Nucleic Acids Res.* **14,** 9081–9093.

Larramendy, M. L., Lopez-Larraza, D., Vidal-Rioja, L., and Bianchi, N. O. (1989). Effect of the metal chelating agent *o*-phenanthroline on the DNA and chromosome damage induced by bleomycin in Chinese hamster ovary cells. *Cancer Res.* **49,** 6583–6586.

Larsen, T. A., Goodsell, D. S., Cascio, D., Grzeskowiak, K., and Dickerson, R. D. (1989). The structure of DAPI bound to DNA. *J. Biomol. Struct. Dyn.* **1,** 477–491.

Lavery, R., and Pullman, B. (1985). The dependence of the surface electrostatic potential of B-DNA on environmental factors. *J. Biomol. Struct. Dyn.* **2,** 1021–1032.

Lazo, J. S., Braun, I. D., Meandzija, B., Kennedy, K. A., Pham, E. T., and Smaldone, L. F. (1985). Lidocaine potentiation of bleomycin A2 cytotoxicity and DNA strand breakage in L1210 and human A-253 cells. *Cancer Res.* **45,** 2103–2109.

Lazo, J. S., Braun, I. D., Labaree, D. C., Schisselbauer, J. C., Meandzija, B., Newman, R. A., and Kennedy, K. A. (1989). Characteristics of bleomycin-resistant phenotypes of human cell sublines and circumvention of bleomycin resistance by liblomycin. *Caner Res.* **49,** 185–190.

Lehane, D. E., and Lane, M. (1978). The immunopharmacology of bleomycin in man. *In* "Bleomycin, Current Status and New Developments" (S. K. Carter, S. T. Crooke, and H. Umezawa, eds.), pp. 143–150. Academic Press, New York.

Levy, M. J., and Hecht, S. M. (1988). Copper(II) facilitates bleomycin-mediated unwinding of plasmid DNA *Biochemistry* **27,** 2647–2650.

Leyhausen, G., Schroeder, H. C., Schuster, D. K., Maidhof, A., Umezawa, H., and Mueller, W. E. G. (1985). Potentiation of the bleomycin-, arabinofuranosylcytosine-, and Adriamycin-caused inhibition of DNA synthesis in lymphocytes by bestatin *in vitro*. *Eur. J. Cancer Clin. Oncol.* **21,** 1325–1330.

Lin, S. Y., and Grollman, A. P. (1981). Interactions of a fragment of bleomycin with deoxyribo-dinucleotides: Nuclear magnetic resonance studies. *Biochemistry* **20**, 7589–7598.

Lin, S.-B., Blake, K. R., Miller, P. S., and Ts'o, P. O. P. (1989). Use of EDTA derivatization to characterize interactions between oligodeoxyribonucleoside methylphosphonates and nucleic acids. *Biochemistry* **28**, 1054–1061.

Lin, T.-S., Antonini, I., Cosby, L. A., and Sartorelli, A. C. (1984). 2,3-Dimethyl-1,4-napthoquinone derivatives as bioreductive alkylating agents with cross-linking potential. *J. Med. Chem.* **27**, 813–815.

Lindahl, T. (1982). DNA repair enzymes. *Annu. Rev. Biochem.* **51**, 61–89.

Liquier, J., Mchami, A., and Taillandier, E. (1989). FTIR study of netropsin binding to poly d(A–T) and poly dA·poly dT. *J. Biomol. Struct. Dyn.* **7**, 119–126.

Liu, S. Y., Hwang, B. D., Haruna, M., Imakura, Y., Lee, K. H., and Cheng, Y. C. (1989). Podophyllotoxin analogs: Effects on DNA topoisomerase II, tubulin polymerization, human tumor KB cells, and their VP-16-resistant variants. *Mol. Pharmacol.* **36**, 78–82.

Louie, A. C., Farmen, R. H., Gaver, R. C., Comis, R. L., Rozencweig, M., and Franks, C. R. (1985). Peplomycin and tallysomycin $S_{10}b$: Two bleomycin analogs. *In* "Bleomycin Chemotherapy (B. I. Sikic, M. Rozencweig, and S. K. Carter, eds.), pp. 277–287. Academic Press, Orlando, Florida.

Lown, J. W., McLaughlin, L. W., and Chang, Y. M. (1978). Mechanism of action of 2-haloethylnitrosoureas on DNA and its relation to their antileukemic properties. *Bioorg. Chem.* **7**, 97–110.

Lown, J. W., Koganty, R. R., Tew, K. D., Oiry, J., and Imbach, J.-L. (1985). Mechanism of action of 2-haloethylnitrosoureas on deoxyribonucleic acid. Pathways of aqueous decomposition and pharmacological characteristics of new anticancer disulfide-linked nitrosoureas. *Biochem. Pharmacol.* **34**, 1015–1024.

Maeda, K., Kosaka, H., Yagishita, K., and Umezawa, H. (1956). A new antibiotic, phleomycin. *J. Antibiot., Ser. A* **9A**, 82–85.

Maher, L. J., III, and Dolnick, B. J. (1987). Specific hybridization arrest of dihydrofolate reductase mRNA *in vitro* using anti-sense RNA or anti-sense oligonucleotides. *Arch. Biochem. Biophys.* **253**, 214–220.

Maher, L. J., III, and Dolnick, B. J. (1988). Comparative hybrid arrest by tandem antisense oligodeoxyribonucleotides or oligodeoxyribonucleoside methylphosphonates in a cell free system. *Nucleic Acids Res.* **16**, 3341–3358.

Mahmutoglu, I., and Kappus, H. (1988). Redox cycling of bleomycin–iron(III) and DNA degradation by isolated NADH–cytochrome b_5 reductase: Involvement of cytochrome b_5. *Mol. Pharmacol.* **34**, 578–583.

Malinge, J.-M., and Leng, M. (1986). Reaction of nucleic acids and *cis*-diamminedichloro-platinum(II) in the presence of intercalating agents. *Proc. Natl. Acad. Sci. U.S.A.* **83**, 6317–6321.

Malinge, J.-M., Schwartz, A., and Leng, M. (1987). Characterization of the ternary complexes formed in the reaction of *cis*-diamminedichloroplatinum(II), ethidium bromide and nucleic acids. *Nucleic Acids Res.* **15**, 1779–1797.

Malvy, C., Prevost, P., Gansser, C., Viel, C., and Paoletti, C. (1986). Efficient breakage of DNA apurinic sites by the indoleamine related 9-aminoellipticine. *Chem–Biol. Interact.* **57**, 41–53.

Malvy, C., Safraoui, H., Bloch, E., and Bertrand, J. R. (1988). Involvement of apurinic sites in synergistic action of alkylating and intercalating drugs in *Escherichia coli*. *Anti-Cancer Drug Des.* **2**, 361–370.

Manning, G. S. (1979). The molecular theory of polyelectrolyte solutions with applications to the electrostatic properties of polynucleotides. *Q. Rev. Biophys.* **11**, 179–246.

Marrot, L., and Lang, M. (1989). Chemical probes of the conformation of DNA modified by cis-diamminedichloroplatinum(II). *Biochemistry* **28**, 1454–1461.

McGown, A. T., and Fox, B. W. (1989). Interaction of the novel agent amphethinile with tubulin. *Br. J. Cancer* **59**, 865–868.

Millar, B. C. (1982). Hypoxic cell radiosensitizers as potential adjuvants to conventional chemotherapy for the treatment of cancer. *Biochem. Pharmacol.* **31**, 2439–2445.

Miller, P. S., and Ts'O, P. O. P. (1987). A new approach to chemotherapy based on molecular biology and nucleic acid chemistry: Matagen (masking tape for gene expression). *Anti-Cancer Drug Des.* **2**, 117–128.

Miller, K. J., Laner, M., and Caloccia, W. (1985). Interactions of molecules with nucleic acids. XII. Theoretical model for the interaction of a fragment of bleomycin with DNA. *Biopolymers* **24**, 913–934.

Mirabelli, C. K., Ting, A., Huang, C.-H., Mong, S., and Crooke, S. T. (1982). Bleomycin and talisomycin sequence-specific strand scission of DNA: A mechanism of double-strand cleavage. *Cancer Res.* **42**, 2779–2785.

Morikawa, K. (1988). Enhancement of antitumor immune response by bleomycin. *Hokkaido Igaku Zasshi* **63**, 772–780.

Mulcahy, R. T., Carminati, A., Barascut, J .L., and Imbach, J. L. (1988). Aerobic and hypoxic toxicity of a new class of mixed-function drugs associating nitroimidazoles and chloroethylnitrosourea in nitrosourea-sensitive (Mer−) and resistant (Mer+) human tumor cells. *Cancer Res.* **48**, 798–801.

Mulcahy, R. T., Gipp, J. J., Carminati, A., and Barascut, J. L. (1989). Chemosensitization at reduced nitroimidazole concentrations by mixed-function compounds combining 2-nitroimidazole and chloroethylnitrosourea. *Eur. J. Cancer Clin. Oncol.* **25**, 1099–1104.

Natsumeda, Y., Yamada, Y., Yamaji, Y., and Weber, G. (1988). Synergistic cytotoxic effect of tiazofurin and ribavirin in hepatoma cells. *Biochem. Biophys. Res. Commun.* **153**, 321–321.

Neidle, S., and Abraham, Z. (1984). Structural and sequence-dependent aspects of drug intercalation into nucleic acids. *CRC Crit. Rev. Biochem.* **17**, 73–95.

Niitani, H. (1987). Bleomycin. *Gan to Kagaku Ryoho* **14**, 3173–3179.

Nishimura, C., Nishimura, T., Tanaka, N., and Suzuki, H. (1987). Potentiation of the cytotoxicity of peplomycin against Ehrich ascites carcinoma by bleomycin hydrolase inhibitors. *J. Antibiot.* **40**, 1794–1795.

Nishimura, C., Nishimura, T., Tanaka, N., Yamaguchi, H., and Suzuki, H. (1989). Inhibition of intracellular bleomycin hydrolase activity by E-64 leads to the potentiation of the cytotoxicity of peplomycin against Chinese hamster lung cells. *Gann* **80**, 65–68.

Paoletti, C. (1988). Anti-sense oligonucleotides as potential antitumor agents: Prospective views and preliminary results. *Anti-Cancer Drug Des.* **2**, 325–331.

Paterson, B. M., Roberts, B. E., and Kuff, E. L. (1977). Structural gene identification and mapping by DNA·mRNA hybrid-arrested cell-free translation. *Proc. Natl. Acad. Sci. U.S.A.* **74**, 4370–4374.

Pavelic, K. (1987). Calmodulin antagonist W13 prevents DNA repair after bleomycin treatment of human urological tumor cells growing on extracellular matrix. *Int. J. Biochem.* **19**, 1091–1095.

Pavelic, K., Beerman, T. A., and Bernacki, R. J. (1985). An evaluation of the effects of combination chemotherapy *in vitro* using DNA-reactive agents. *Cancer Drug Deliv.* **2**, 255–270.

Povirk, L. F., and Goldberg, I. M. (1987). A role for oxidative DNA sugar damage in mutagenesis by neocarzinostatin and bleomycin. *Biochimie* **69**, 815–823.

Povsic, T. J., and Dervan, P.B. (1989). Triple helix formation by oligonucleotides on DNA extended to the physiological pH range. *J. Am. Chem. Soc.* **111**, 3059–3061.

Prasad, A. R. S., Luduena, R. F., and Horowitz, P. M. (1986). Detection of energy transfer between tryptophan residues in the tubulin molecule and bound bis(8-anilinonaphthalene-1-sulfonate), an inhibitor of microtubule assembly, that binds to a flexible region on tubulin. *Biochemistry* **25**, 3536–3540.

Record, M. T., Anderson, C. F., and Lohman, T. M., (1978). Thermodynamic analysis of ion effects on the binding and conformational equilibria of proteins and nucleic acids: The roles of ion association or release, screening, and ion effects on water activity. *Q. Rev. Biophys.* **11**, 103–118.

Remers, W. A. (1979). Bleomycins and phleomycins. *In* "The Chemistry of Antitumor Antibiotics", Vol. 1, pp. 176–220. Wiley (Interscience), New York.

Remers, W. A. (1984). Chemistry of antitumors drugs. *In* "Antineoplastic Agents" (W. A. Remers, ed.), pp. 83–261. Wiley (Interscience), New York.

Rice, J. A., Crothers, D. M., Pinto, A. L., and Lippard, S. J. (1988). The major adduct of the antitumor drug cis-diamminedichloroplatinum(II) with DNA bends the duplex by ~40° toward the major groove. *Proc. Natl. Acad. Sci. U.S.A.* **85**, 4158–4161.

Rosenthal, S. A., and Hait, W. N. (1988). Potentiation of DNA damage and cytotoxicity by calmodulin antagonists. *Yale J. Biol. Med.* **61**, 39–49.

Ross, W. C. J. (1962). "Biological Alkylating Agents," pp. 1–10. Butterworths, London.

Sakai, T. T., Riordan, J. M., and Glickson, J. D. (1982). Models of bleomycin interactions with poly(deoxyadenylylthymidylic acid). Fluorescence and proton nuclear magnetic resonance studies of cationic thiazole amides related to bleomycin A2. *Biochemistry* **21**, 805–816.

Sakore, T. D., Reddy, B. S., and Sobell, H. M. (1979). Visualization of drug–nucleic acid interactions at atomic resolution. IV. *J. Mol. Biol.* **135**, 763–785.

Saulier, B., Prigent, C., Boutelier, R., and David, J. C. (1988). Peplomycin DNA breakage and *in vitro* inhibition of DNA polymerases and ligase from human normal and leukemic cells. *Carcinogenesis* **9**, 965–970.

Sausville, E. A., Peisach, J., and Horwitz, S. B., (1978). Effect of chelating agents and metal ions on the degradation of DNA by bleomycin. *Biochemistry* **17**, 2740–2745.

Scheid, W., and Traut, H. (1985). Synergistic enhancement of the bleomycin and peplomycin induced mitotic index reduction by verapamil. *Arzneim.-Forsch.* **35**, 1717–1719.

Scheid, W., and Traut, H. (1988). Enhancement of the cytogenetic efficacy of the antitumor agent bleomycin by the calcium and calmodulin antagonist fendiline. *Experientia* **44**, 228–230.

Sebti, S. M., and Lazo, J. S. (1988). Metabolic inactivation of bleomycin analogs by bleomycin hydrolase. *Pharmacol. Ther.* **38**, 321–329.

Sharif, A., and Goussault, Y. (1988). *In vitro* study of the modification by bleomycin of thymidine phosphorylation activity in lectin-stimulated normal human lymphocytes. *Cancer Biochem. Biophys.* **10**, 17–24.

Shipley, J. B., and Hecht, S. M. (1988). Bleomycin congeners exhibiting altered DNA cleavage specificity. *Chem. Res. Toxicol.* **1**, 25–27.

Siemann, D. W., and Mulcahy, R. T. (1986). Sensitization of cancer chemotherapeutic agents by nitroheterocyclics. *Biochem. Pharmacol.* **35**, 111–115.

Solaiman, D. (1988). Bleomycin and its metal complexes. *In* "Metal-Based Anti-tumor Drugs" (M. F. Gielen, ed.), pp. 235–256. Freund, London.

Springgate, M. W., and Poland, D. (1973). Cooperative and thermodynamic parameters for oligoinosinate–polycytidylate complexes. *Biopolymers* **12**, 2241–2260.

Stein, C. A., and Cohen, J. S. (1988). Oligodeoxynucleotides as inhibitors of gene expression: A review. *Cancer Res.* **48**, 2659–2668.

Stewart, D. J., and Evans, W. K. (1989). Nonchemotherapeutic agents that potentiate chemotherapy efficacy. *Cancer Treat. Rev.* **16**, 1–40.

Stratford, I. J., O'Neill, P., Sheldon, P. W., Silver, A. R.J., Walling, J. M., and Adams, G. E. (1986). RSU 1069, a nitroimidazole containing an aziridine group. *Biochem. Pharmacol.* **35**, 105–109.

Strekowski, L., Chandrasekaran, S., Wang, Y.-H., Edwards, W. D., and Wilson, W. D. (1986a). Molecular basis for anticancer drug amplification: Interaction of phleomycin amplifiers with DNA. *J. Med. Chem.* **29**, 1311–1315.

Strekowski, L., Tanious, F. A., Chandrasekaran, S., Watson, R. A., and Wilson, W. D. (1986b). New approach to conformational analysis of heterobiaryls in solution. *Tetrahedron Lett.* **27**, 6045–6048.

Strekowski, L., Strekowska, A., Watson, R. A., Tanious, F. A., Nguyen, L. T., and Wilson,

W. D. (1987a). Amplification of bleomycin-mediated degradation of DNA. *J. Med. Chem.* **30**, 1415–1420.

Strekowski, L., Watson, R. A., and Wilson, W. D. (1987b). Selective catalysis os A·T base pair proton exchange in DNA complexes: Imino proton NMR analysis. *Nucleic Acids Res.* **15**, 8511–8519.

Strekowski, L., Mokrosz, J. L., and Wilson, W. D. (1988a). A biphasic nature of the bleomycin-mediated degradation of DNA. *FEBS Lett.* **24**, 24–28.

Strekowski, L., Mokrosz, J. L., Tanious, F. A., Watson, R. A., Harden, D., Mokrosz, M., Edwards, W. D., and Wilson, W. D. (1988b). Molecular basis for bleomycin amplification: Conformational and stereoelectronic effects in unfused amplifiers. *J. Med. Chem.* **31**, 1231–1240.

Strekowski, L., Wilson, W. D., Mokrosz, J. L., Strekowska, A., Koziol, A. E., and Palenik, G. J. (1988c). A non-classical intercalation model for a bleomycin amplifier. *Anti-Cancer Drug Des.* **2**, 387–398.

Strekowski, L., Mokrosz, M., Mokrosz, J. L., Strekowska, A., Allison, S. A., and Wilson, W. D. (1988d). Amplification of bleomycin-mediated degradation of DNA by polyamines. *Anti-Cancer Drug Des.* **3**, 79–89.

Strekowski, L., Harden D. B., Wydra, R. L., Stewart, K. D., and Wilson, W. D. (1989). Molecular basis for potentiation of bleomycin-mediated degradation of DNA by polyamines: Experimental and molecular mechanical studies. *J. Mol. Recognition* **2**, 158–166.

Strobel, S. A., and Dervan, P. B. (1989). Cooperative site specific binding of oligonucleotides to duplex DNA. *J. Am. Chem. Soc.* **111**, 7286–7287.

Strong, J. E., and Crooke, S. T. (1978). Mechanism of action of tallysomycin, a third generation bleomycin. *In* "Bleomycin, Current Status and New Developments" (S. K. Carter, S. T. Crooke, and H. Umezawa, eds.), pp. 343–355. Academic Press, New York.

Stubbe, J., and Kozarich, J. W. (1987). Mechanisms of bleomycin-induced DNA degradation. *Chem. Rev.* **87**, 1107–1136.

Sugiura, Y., and Suzuki, T. (1982). Nucleotide sequence specificity of DNA cleavage by iron–bleomycin. *J. Biol. Chem.* **257**, 10544–10546.

Sugiyama, H., Xu, C., Murugesan, N., and Hecht, S. M. (1988). Chemistry of the alkali-labile lesion formed from iron(II) bleomycin and d(CGCTTTAAAGCG). *Biochemistry* **27**, 58–67.

Sundquist W. I., Bancroft, D. P., Chassot, L., and Lippard, S. J. (1988). DNA promotes the reaction of *cis*-diamminedichloroplatinum(II) with the exocyclic amino groups of ethidium bromide. *J. Am. Chem. Soc.* **110**, 8559–8560.

Suzuki, T., Kuwahara, J., and Sugiura, Y. (1983). Nucleotide sequence cleavage of guanine-modified DNA with aflatoxin B1, dimethylsulfate, and mitomycin C by bleomycin and deoxyribonuclease I. *Biochem. Biophys. Res. Commun.* **117**, 916–922.

Takahashi, K., Ekimoto, H., Minamide, S., Nishikawa, N., Kuramochi, H., Motegi, A., Nakatani, T., Takita, T., Takeuchi, T., and Umezawa, H. (1987). Liblomycin, a new analogue of bleomycin. *Cancer Treat. Rev.* **14**, 169–177.

Takeshita, M., Grollman, A. P., Ohtsubo, E., and Ohtsubo, H. (1978). Interaction of bleomycin with DNA. *Proc. Natl. Acad. Sci. U.S.A.* **75**, 5983–5987.

Takita, T., and Ogino, T. (1987). Peplomycin and liblomycin, new analogs of bleomycin. *Biomed. Pharmacother.* **41**, 219–226.

Takita, T., Umezawa, Y., Saito, S., Morishima, H., Naganawa, H., Umezawa, H., Tsuchiya, T., Miyake, T. Kegeyama, S., Umezawa, S., Muraoka, Y., Suzuki, M., Otsuka, M., Narita, M., Kobayashi, S., and Ohno, M. (1982). Total synthesis of bleomycin A2. *Tetrahedron Lett.* **23**, 521–524.

Toulmé, J. J., Krisch, H. M., Loreau, N., Thuong, N T., and Helene, C. (1986). Specific inhibition of mRNA translation by complementary oligonucleotides covalently linked to intercalating agents. *Proc. Natl. Acad. Sci. U.S.A.* **83**, 1227–1231.

Tsai, C.-C., Jain, S. C., and Sobell, H. M. (1977). Visualization of drug-nucleic acid interactions at atomic resolution. *J. Mol. Biol.* **114**, 301–315.

Ts'O, P. O. P., Miller, P. S., and Greene, J. J. (1983). Nucleic acid analogs with targeted delivery as chemotherapeutic agents. *In* "Development of Target-Oriented Anticancer Drugs" (Y.-C. Cheng, ed.), pp. 189–206. Raven, New York.

Ts'O, P. O. P., Miller, P. S., Aurelian, L., Murakami, A., Agris, C., Blake, K. R., Lin, S.-B., Lee, B. L., and Smith, C. C. (1987). An approach to chemotherapy based on base sequence information and nucleic acid chemistry. Matagen (masking tape for gene expression). *Ann. N. Y. Acad. Sci.* **507**, 220–231.

Tsuruo, T. (1989). Potentiation of antitumor agents by calcium antagonists. *Pharma Med.* **7**, 65–68.

Turk, J. L., Hamied, T. A., and Parker, D. (1989). Potentiation of interleukin-2 release by anticancer drugs. *Agents Actions* **26**, 156–157.

Uchida, T., Nishikawa, K., Shibasaki, C., Okamoto, H., and Takahashi, K. (1989). Combined effect of cisplatin with various types of antitumor drugs against P388 leukemia. *Gan to Kagaku Ryoho* **16**, 2275–2282.

Umezawa, H. (1979). Advances in bleomycin studies. *In* "Bleomycin: Chemical, Biochemical and Biological Aspects" (S. M. Hecht, ed.), pp. 24–36, Springer-Verlag, New York.

van der Krol, A. R., Mol, J. M., and Stuitje, A. R., (1988). Modulation of eukaryotic gene expression by complementary RNA or DNA sequences—Antisense gene action and application. *Biotechniques* **6**, 958–969.

Weber, G., Yamaji, Y., Olah, E., Natsumeda, Y., Jayaram, H. N., Lapis, E., Zhen, W., Prajda, N., Hoffman, R., and Tricot, G. J., (1989). Clinical and molecular impact of inhibition of IMP dehydrogenase activity by tiazofurin. *Adv. Enzyme Regul.* **28**, 335–356.

Williams, L. D., and Goldberg, I. H. (1988). Selective strand scission by intercalating drugs at DNA bulges. *Biochemistry* **27**, 3004–3011.

Wilson, W. D. (1987). Cooperative effects in drug–DNA interactions, *Prog. Drug Res.* **31**, 193–221.

Wilson, W. D. (1990). Reversible interactions of small molecules with nucleic acids. *In* "The Chemistry and Biology of Nucleic Acids" (M. Gait and M. Blackburn, eds.), Chapter 8, Oxford Univ. Press, Oxford, England.

Wilson, W. D., and Jones, R. L. (1981). Intercalating drugs: DNA binding and molecular pharmacology. *Adv. Pharmacol. Chemother.* **18**, 177–222.

Wilson, W. D., Strekowski, L., Tanious, F. A., Watson, R. A., Mokrosz, J. L., Strekowska, A., Webster, G. D., and Neidle, S. (1988). Binding of unfused aromatic cations to DNA. The influence of molecular twist on intercalation. *J. Am. Chem. Soc.* **110**, 8292–8299.

Wilson, W. D., Tanious, F. A., Barton, H. J., Strekowski, L., and Boykin, D. W. (1989a). Binding of 4',6-diamidino-2phenylindole (DAPI) to GC and mixed sequences in DNA: Intercalation of a classical groove-binding molecule. *J. Am. Chem. Soc.* **111**, 5008–5010.

Wilson, W. D., Tanious, F. A., Watson, R. A., Barton, H. J., Strekowska, A., Harden, D. B., and Strekowski, L. (1989b). Interaction of unfused tricyclic aromatic cations with DNA: A new class of intercalators. *Biochemistry* **28**, 1984–1992.

Wilson, W. D., Tanious, F. A., Barton, H. J., Wydra, R. L., Jones, R. L., Boykin, D. W., and Strekowski, L. (1990). The interaction of unfused polyaromatic heterocycles with DNA: Intercalation, groove-binding and bleomycin amplification. *Anti-Cancer Drug Des.* **5**, 31–42.

Wilson W. R., Anderson, R. F., and Denny, W. A. (1989a). Hypoxia-selective antitumor agents. 1. Relationships between structure, redox properties and hypoxia-selective cytotoxicity for 4-substituted derivatives of nitracrine. *J. Med. Chem.* **32**, 23–30.

Wilson, W. R., Thompson, L. H., Anderson, R. F., and Denny, W. A. (1989b). Hypoxia-selective antitumor agents. 2. Electronic effects of 4-substituents on the mechanisms of cytotoxicity and metabolic stability of nitracrine derivatives. *J. Med. Chem.* **29**, 31–38.

Wittes, R. E. (1985). Bleomycins: Future prospects. *In* "Bleomycin Chemotherapy" (B. I. Sikic, M. Rozencweig, and S. K. Carter, eds.), pp. 305–311. Academic Press, Orlando, Florida.

Woodson, S. A., and Crothers, D. M. (1988). Binding of 9-aminoacridine to bulged-base DNA oligomers from a frame-shift hot spot. *Biochemistry* **27**, 8904–8914.

Woynarowski, J. M., Gawron, L. S., and Beerman, T. A. (1987). Bleomycin-induced aggregation of presolubilized and nuclear chromatin from L1210 cells. *Biochim. Biophys. Acta.* **910**, 149–156.

Wu, J. C., Kozarich, J. W., and Stubbe, J. (1985a). Mechanism of bleomycin: Evidence for a rate-determining 4′-hydrogen abstraction from poly(dA–dU) associated with the formation of both free base and base propenal. *Biochemistry* **24**, 7562–7568.

Wu, J. C., Stubbe, J., and Kozarich, J. W., (1985b). Mechanism of bleomycin: Evidence for 4′-ketone formation in poly(dA–dU) associated exclusively with free base release. *Biochemistry* **24**, 7569–7573.

Xu, Z., Hosokawa, M., Morikawa, K., Hatakeyama, M., and Kobayashi, H. (1988). Overcoming suppression of antitumor immune reactivity in tumor-bearing rats by treatment with bleomycin. *Cancer Res.* **48**, 6658–6663.

Yamada, Y., Natsumeda, Y., and Weber, G. (1988). Action of active metabolites of tiazofurin and ribavirin on purified IMP dehydrogenase. *Biochemistry* **27**, 2193–2196.

Yamaki, H., Nishimura, T., Matsunaga, K., Suzuki, H., Yamaguchi, H., and Tanaka, N. (1988). Antitumor effect of a new derivative of bleomycin against a cell line of multidrug resistant murine lymphoblastoma. *J. Antibiot.* **41**, 1500–1502.

Yen, S.-F., Gabbay, E. J., and Wilson, W. D. (1982). Interaction of aromatic imides with DNA. Spectrophotometric and viscometric studies. *Biochemistry* **21**, 2070–2076.

Yokoyama, Y., Ohmori, I., Kohda, K., and Kawazoe, Y. (1988). Potentiation of the cytotoxic activity of anticancer drugs against cultured L1210 cells by *Bacillus thuringiensis* subsp. *israelensis* toxin. *Chem. Pharm. Bull.* **36**, 4499–4504.

Yukio, S. (1987). Molecular mechanism for specific recognition and cleavage of DNA nucleotide sequences by anticancer drugs. *Yakugaku Kenkyu no Shinpo* pp. 46–55.

Zakrzewska, K., and Pullman, B. (1986). Spermine–nucleic acid interactions: A theoretical study. *Biopolymers* **25**, 375–392.

Zon, G. (1988). Oligonucleotide analogues as potential chemotherapeutic agents. *Pharm. Res.* **5**, 539–549.

CHAPTER 13

Biochemical Modulation of 5-Fluorouracil by Metabolites and Antimetabolites

Enrico Mini[*]
Department of Preclinical and Clinical
Pharmacology
University of Florence
Florence, 50134 Italy

Joseph R. Bertino[†]
Memorial Sloan-Kettering
Cancer Center
New York, New York 10021

I. Introduction

The approach to modulation of the therapeutic efficacy of an antineoplastic drug (effector) by a so-called modulating agent which can interact with the drug's metabolic pathways, resulting either in selective enhancement of the

---

[*] Awardee of the Lady Tata Memorial Trust. To whom reprint requests should be addressed.
[†] American Cancer Society Professor.

SYNERGISM AND ANTAGONISM IN CHEMOTHERAPY
Copyright © 1991 by Academic Press, Inc.
All rights of reproduction in any form reserved.

449

sensitivity of tumor cells or selective reduction of the sensitivity of host cells to the drug, has emerged in clinical oncology from laboratory studies (Martin, 1987). Biochemical as well as pharmacological and cytokinetic modulation are mechanisms by which synergism occurs between two or more antineoplastic drugs in combination.

Up to the present, most attempts aimed at potentiation of the effectivenes of antineoplastic drugs have concerned cytotoxic antimetabolites, i.e., drugs for which knowledge of the mechanisms of action is more advanced.

A significant example is represented by attempts to potentiate the anti-tumor activity of 5-fluorouracil (FUra) since this fluoropyrimidine, synthe-sized by Heidelberger *et al.* (1957) at the University of Wisconsin about 30 years ago, is still among the most effective drugs in the treatment of adenocar-cinomas of the gastrointestinal tract (i.e., colon, stomach, pancreas), breast cancer, and squamous cell carcinoma of the head and neck (reviewed by Wasserman *et al.*, 1975). Although FUra is inactive against other solid tumors and leukemia–lymphoma (Costanzi *et al.*, 1979; Hedley, 1987), FUra or 5-fluorodeoxyuridine (FdUrd) may be converted by appropriate modulation to an active drug.

Various drugs have been used to interact with FUra through modula-tion of common metabolic pathways (reviewed by Damon and Cadman, 1988) (*Table 1*). Drugs that potentiate antitumor activity of FUra include antimetabolites inhibiting purine biosynthesis (i.e., methotrexate, MTX; dichloromethotrexate, DCMTX; trimetrexate, TMTX; triazinate, TZT; 6-methylmercaptopurine ribonucleoside, MMPR; 6-diazo-5-oxo-L-norleu-cine; L-alanosine, azaserine), inhibitors of nucleoside transport (dipyrida-mol, DIP), pyrimidine antagonists (phosphonacetyl-L-aspartic acid, PALA; pyrazofurin), inhibitors of ribonucleotide reductase (hydroxyurea, HU), halogenated pyrimidines which inhibit FUra catabolism (bromovinyl-deoxyuridine) (Iigo *et al.*, 1988a), heavy metals (cisplatin, CDDP), and folates (5-formyltetrahydrofolate; leucovorin, LV; 5-methyltetrahydrofolate, $5\text{-CH}_3\text{H}_4\text{PteGlu}$) interfering with intracellular folate pools, purine nucleo-sides (inosine, Ino; deoxyinosine, dIno; guanosine, Guo), thymidine (dThd), and combinations of these agents (e.g., PALA–dThd).

Drugs may selectively protect normal cells of the host from fluoropyrimi-dine toxicity. Examples include antimetabolites (allopurinol, HPP) and pyri-midine nucleosides (uridine, Urd; cytidine).

Combinations of such drugs may act with both mechanisms, i.e., PALA, MTX, and FUra with LV and Urd rescues.

On the basis of laboratory studies some of these combinations have been employed with variable success in the clinic while others are in a preliminary phase of clinical evaluation (reviewed by Bertino and Mini, 1987; Damon *et al.*, 1989; Grem *et al.*, 1987a, 1988; Klubes and Leyland-Jones, 1989; Ley-land-Jones and O'Dwyer, 1986; Mini *et al.*, 1990; O'Dwyer *et al.*, 1987; Pinedo and Peters, 1988).

The clinically tested combination regimens based on biochemical modula-tion will be discussed here.

II. Mechanism of Action of FUra

FUra is not active *per se* but requires intracellular conversion to fluororibo-nucleotides and fluorodeoxyribonucleotides to exert cytotoxicity (Heidelber-ger *et al.*, 1983) (*Figure 1*).

FUra can be converted to its ribonucleoside, fluorouridine (FUrd), by Urd phosphorylase and then to the ribonucleotide derivative, fluorouridine monophosphate (FUMP), via Urd kinase or it can be converted directly to FUMP via orotate phosphoribosyl transferase (OPRTase). In this latter cir-cumstance, phosphoribosyl pyrophosphate (PRPP) acts as a phosphoribose donor. FUMP is further converted to either fluorouridine triphosphate (FUTP) or via ribonucleotide reductase to fluorodeoxyuridine monophos-phate (FdUMP) and then to the triphosphate (FdUTP). Also, FUra might be converted directly to FdUMP via dThd phosphorylase and dThd kinase.

Cytotoxicity may depend on (1) misincorporation of FUTP into RNA and consequent interference with the synthesis and function of some or all species of RNA (Glazer and Peale, 1979; Glazer and Lloyd, 1982; Mandel, 1969); (2) fluorodeoxyuridine triphosphate (FdUTP) incorporation into DNA, and the subsequent process of removal of the altered bases from DNA leading to fragmentation of this macromolecule (Kufe *et al.*, 1983; Lönn and Lönn, 1986a); (3) the formation of FdUMP and subsequent inhibition of thymidy-late (dTMP) synthesis by this molecule (Danenberg and Lockshin, 1981; Hartman and Heidelberger, 19961; Houghton *et al.*, 1981; Santi *et al.*, 1974).

FdUMP is in fact a specific inhibitor of dTMP synthase (TS), the enzyme that catalyzes the conversion of uridylate (dUMP) to dTMP necessary for DNA synthesis (Fig. 1). In this reaction the folate coenzyme 5,10-methylenetetrahydrofolate (5,10-CH_2H_4PteGlu), both as a monoglutamate

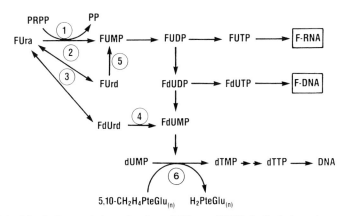

FIGURE 1   Metabolism and sites of action of FUra and FdUrd. Circled numbers refer to the following enzymes: (1) orotate phosphoribosyl-transferase; (2) uridine phosphorylase; (3) thymi-dine phosphorylase; (4) thymidine kinase; (5) uridine kinase; (6) thymidylate synthase. $Glu_{(n)}$, polyglutamate.

(Danenberg *et al.*, 1974; Santi *et al.*, 1974) and with increased efficiency as a polyglutamate (Allegra *et al.*, 1985; Radparvar *et al.*, 1989), acts as the methyl group donor and is oxidized to dihydrofolate (H_2PteGlu) during the course of this reaction.

When FdUMP replaces dUMP as a quasi-substrate in the reaction, the enzyme is inactivated, as demonstrated by studies by Santi *et al.* (1974), Danenberg *et al.* (1974), and Danenberg and Danenberg (1978). In the absence of the folate coenzyme a binary complex is formed by TS and FdUMP which is rapidly dissociable, while in the presence of 5,10-CH_2H_4PteGlu, a ternary covalent complex is formed. dTMP biosynthesis is blocked as is DNA synthesis consequently.

These mechanisms of action cooperate to determine FUra cytotoxicity, but their relative importance to FUra therapeutic effects varies in different tumor cells.

III. Sequential MTX–FUra

A. PRECLINICAL RATIONALE

In *in vitro* and *in vivo* experimental tumor systems, the sequence of administration of combined MTX and FUra has been shown to be a critical determinant of cytotoxicity. Synergism relative to tumor cell kill has been observed only when tumor bearing rodents (Bertino *et al.*, 1977; Brown and Ward, 1978; Heppner and Calabresi, 1977; Herrmann *et al.*, 1985; Lee and Kwaja, 1977; Mulder *et al.*, 1981) or tumor cells in cultures are exposed to MTX prior to FUra (Benz and Cadman, 1981; Benz *et al.*, 1982; Bertino *et al.*, 1983; Cadman *et al.*, 1979, 1981c; Fernandes and Bertino, 1980; Kufe and Egan, 1981; Mini and Bertino, 1982; Mini *et al.*, 1983a,b). Less than additive effects have been noted with the reverse sequence. For a general treatment of schedule-dependent synergism, see Chapter 17 by Chang.

Proposed mechanisms of this synergism are (1) enhancement of FdUMP binding to TS by MTX and H_2PteGlu polyglutamate derivatives (Fernandes and Bertino, 1980) and/or (2) enhancement of FUra nucleotide formation via OPRTase from the increased intracellular PRPP levels accumulated as a consequence of MTX inhibition of purine biosynthesis (Cadman *et al.*, 1981c). PRPP is in fact a co-substrate donor of phosphoribose which is necessary both for the first step in this biosynthetic pathway and for activation of FUra by OPRTase. As a consequence, higher levels of all active metabolites (FdUMP, FUTP, and FdUTP) induce a more elevated cytotoxicity.

In vitro studies have also shown that fluoropyrimidine deoxynucleotides are incorporated into DNA (Kufe *et al.*, 1983) and that misincorporation of uracil into DNA occurs after MTX treatment (Goulian *et al.*, 1980). These biochemical events may contribute to the cytotoxicity of combined MTX–FUra. Enhancement of the incorporation of FdUMP into DNA occurred, in fact, in human leukemia cells pretreated with MTX (Tanaka *et al.*, 1983), but not in human adenocarcinoma cells (Lönn and Lönn, 1986b).

Although a great deal is known about MTX–FUra activity, the interactions between these drugs are complex and may be different for cells of different origins and growth rates.

In vivo studies by Bertino *et al.* (1977) have demonstrated that in mice bearing sarcoma S-180 sequential MTX–FUra had higher therapeutic efficacy in terms of increase of animal life span than simultaneous administration or the reverse sequence: the optimal interval was 1–4 hr. In other animal tumor models the interval was longer (varying from 1 to 24 hr) Brown and Ward, 1978; Heppner and Calabresi, 1977; Herrmann *et al.*, 1985; Lee and Kwaja, 1977; Mulder *et al.*, 1981).

When murine tumor cells were used *in vitro*, short intervals (3–4 hr) between MTX and FUra administration were sufficient for maximum synergism (Cadman *et al.*, 1981c; Fernandes and Bertino, 1980). However, synergistic cell kill of human tumor cells (breast, colon, and T- and B-cell leukemia) was observed when the interval between MTX and FUra was 12–24 hr (Benz and Cadman, 1981; Benz *et al.*, 1982; Mini *et al.*, 1983a,b).

It has also been shown *in vitro* that synergism of sequential MTX–FUra combinations may depend also on concentrations of folate metabolism end products such as hypoxanthine and dThd in the culture medium (Piper *et al.*, 1983). Thus the efficacy of this therapy *in vivo* may depend not only on the individual drug doses used and the timing of administration but also on the regional concentrations of dThd and purines in serum or bone marrow and other tissues (Benz *et al.*, 1985).

The potential importance of using this sequential drug combination in the treatment of leukemias and lymphomas prompted the examination of the interaction of MTX and FUra in a human leukemic lymphoblast cell line, CCRF-CEM.

The effects of dose, duration, and sequence of exposure for MTX, FUra, and their combination on human leukemic CCRF-CEM cell viability were determined (Bertino *et al.*, 1983; Mini and Bertino, 1982; Mini *et al.*, 1983a) (Table 1).

CCRF-CEM cells were exposed for 4 hr to MTX (1, 10, and 100 μM) and FUra (50 μM) either alone or in simultaneous and sequential combination (MTX + FUra, MTX → FUra, FUra → MTX) (Fig. 2). The effects on cell viability were highly dependent upon the sequence of drug administration. Synergistic cell kill was observed when cells were exposed to MTX prior to FUra (MTX → FUra). Maximum synergism was noted with 1 μM MTX. Simultaneous drug administration (MTX + FUra) resulted in either additive or antagonistic cell kill at all MTX concentrations. Marked antagonistic effects were noted with the reverse sequence (FUra → MTX).

TABLE 1
Biochemical Modulation of FUra[a]

Modulating agent	Proposed mechanism	Sequence of administration[b]	Tested tumor systems
Metabolites			
Purines			
Ino	Enhanced FUra anabolism due to increased ribose-1-phosphate levels	→, +	*In vitro*, mice
Guo	Enhanced FUra anabolism due to increased ribose-1-phosphate levels	→, +	*In vitro*, mice
dIno	Enhanced FdUMP synthesis due to increased deoxyribose-1-phosphate levels	→, +	*In vitro*, mice
Pyrimidines			
dThd	Enhanced FUTP synthesis and/or diminished FUra catabolism	→, +	*In vitro*, mice, humans
Urd	Selective rescue of normal tissues due to enhanced UTP synthesis and competition with FUTP	←	*In vitro*, mice, humans

Folate coenzymes			
LV,5-CH$_3$H$_4$PteGlu	Enhanced formation and retention of the ternary complex FdUMP-5,10-CH$_2$H$_4$PteGlu-TS	→, +	*In vitro*, mice, humans
Antimetabolites			
MTX	Enhanced FUra anabolism due to increased PRPP levels and/or enhanced FdUMP binding to TS due to increased H$_2$PteGlu$_n$ levels	→	*In vitro*, rodents, humans
MMPR	Enhanced FUra anabolism due to increased PRPP levels	→	*In vitro*, mice, humans
PALA	Enhanced FUMP synthesis due to decreased competing orotate pool	→	*In vitro*, rodents, humans
HU	Enhanced inhibition of TS by FdUMP due to decreased dUMP levels via ribonucleotide reductase inhibition	←	*In vitro*, mice, humans
DIP	Inhibition of FUra nucleotide efflux (i.e., FUdr and FdUrd)	+	*In vitro*, humans
HPP	Selective protection of normal tissues due to decreased FUMP synthesis via OPRTase by HPP metabolites	+	*In vitro*, mice, humans

[a] Abbreviations: UTP, Urd triphosphate; FUdr, fluorouridine.

[b] →, modulating agent before FUra; +, modulating agent simultaneously with FUra; →, modulating agent after FUra.

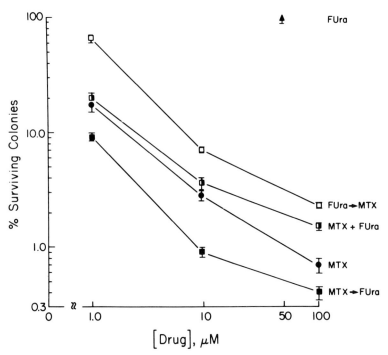

FIGURE 2   Sequence dependent effects of MTX and FUra on CCRF-CEM cell viability. CCRF-CEM cells were exposed to either FUra (50 μM) or MTX (1, 10, or 100 μM). Each drug exposure, either alone or in combination, was for 4 hr. After each 4 hr exposure, cells were washed in drug-free medium, and cell viability was determined by cloning in soft agar. Survival is expressed as percentage of untreated control. Results represent mean values of quadruplicate determinations in a single representative experiment. Bars, S.E.; ▲, FUra; ●, MTX; □, FUra→MTX; ◨, MTX + FUra; ■, MTX→FUra. Drug concentrations, [Drug], refers to MTX concentrations for single-agent exposure to the antifol and for the combinations of MTX and FUra. FUra concentration was 50 μM either alone or in combinations. (From Mini *et al.*, 1983a, with permission.)

When cells were exposed to MTX (1, 10, and 100 μM) for 4 hr followed by FUra (50 μM) in the last 2 hr of MTX exposure, synergistic cell kill was observed. Maximum synergism was observed with 10 μM MTX. Synergistic cell kill was also found when cells were exposed to MTX (0.1, 1, 10, and 100 μM) for a longer time (24 hr) followed by FUra (50 μM) in the last 2 hr of MTX exposure. Maximum synergism was observed with 1 μM MTX (data not shown).

The effects of MTX–FdUrd combinations on CCRF-CEM cell viability were also studied using FdUrd concentrations equimolar to FUra. The results of exposing CCRF-CEM cells for 4 hr to MTX and FdUrd simultaneously (MTX + FdUrd) or in sequence (MTX→FdUrd, FdUrd→MTX) are shown in Fig. 3. Antagonistic cell kill was produced by all combinations tested,

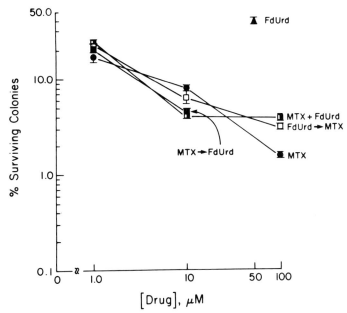

FIGURE 3 Sequence dependent effects of MTX and FdUrd on CCRF-CEM cell viability. CCRF-CEM cells were exposed to either FdUrd (50 μM) or MTX (1, 10, or 100 μM). Each drug exposure, either alone or in combination, was for 4 hr. After each 4 hr exposure cells were washed in drug-free medium, and cell viability was determined by cloning in soft agar. Survival is expressed as percentage of untreated control. Results represent mean values of quadruplicate determinations in a single representative experiment. Bars, S.E.; ▲, FdUrd; ●, MTX; □, FdUrd→MTX; ◧, MTX + FdUrd; ■, MTX→FdUrd. Drug concentrations, [Drug], refers to MTX concentrations for single-agent exposure to the antifol and for the combinations of MTX and FdUrd. FdUrd concentration was 50 μM either alone or in combinations.

independent of sequence of administration and MTX concentrations used. Antagonism was also observed with various sequential MTX–FdUrd combinations (data not shown).

Our results, demonstrating sequence-dependent synergism of MTX–FUra, but not of MTX–FdUrd combinations in a cell line with low levels of dThd phosphorylase (Piper and Fox, 1982), suggested that the mechanism by which MTX enhances FUra cytotoxicity may be the result of increased FUra nucleotide formation. This mechanism would be the consequence of MTX inhibition of purine biosynthesis leading to increased intracellular PRPP.

If enhancement of FdUMP (the active metabolite of both FUra and FdUrd) binding to its target, TS, by MTX polyglutamates or by H_2PteGlu polyglutamates substituting for the natural folate coenzyme 5,10-CH_2H_4PteGlu in the formation of the ternary complex, had occurred in this cell line, synergism would have been observed with both MTX–FUra and MTX–FdUrd sequential combinations.

No effects of FUra on MTX polyglutamate formation and efflux were observed by McGuire *et al.* (1985), regardless of the sequence of administration of these two drugs (i.e., MTX + FUra, FUra→MTX, MTX→FUra). The exact mechanism of sequential MTX–FUra synergism in this cell line must await complete investigation of FUra pharmacokinetics and pharmacodynamics. Data demonstrating enhanced incorporation of FUTP into RNA of CCRF-CEM cells have been presented (Kufe and Egan, 1981).Data from this laboratory (not shown), however, demonstrate that dThd provides partial protection from FUra cytotoxicity in this cell line. This suggests that FUra cytotoxicity is mediated by both RNA- and DNA-directed effects, and that an increase of both FUTP and FdUMP intracellular levels induced by MTX pretreatment might be responsible for the occurrence of synergism.

B. CLINICAL STUDIES

1. Colorectal Cancer

The sequential combination of MTX–FUra has been extensively studied in patients with advanced colorectal cancer (29 phase II trials) (Ajani *et al.*, 1985; Blumenreich *et al.*, 1982; Bruckner and Cohen, 1983; Burnet *et al.*, 1981; Canobbio *et al.*, 1986; Cantrell *et al.*, 1982; Coates *et al.*, 1984; Drapkin *et al.*, 1981, 1983; Glimelius *et al.*, 1986; Hansen *et al.*, 1986; Herrmann *et al.*, 1984b; Kaye *et al.*, 1984; Kemeny *et al.*, 1984; Leone *et al.*, 1986; Mahajan *et al.*, 1983; Mehrotra *et al.*, 1982; Panasci *et al.*, 1985; Rangineni *et al.*, 1983; Solan *et al.*, 1982; Tisman and Wu, 1980; Weinerman *et al.*, 1982, 1984) (Table 2). However, the variability of treatment parameters (dose of both antimetabolites, route and duration of administration, interval between drug administration) employed in these studies has not allowed for an appropriate

TABLE 2

Summary of Phase II Clinical Studies of Sequential MTX/FUra

Cancer	Number of trials	Number of patients	Response (%)		
			CR	PR	CR + PR
Head and neck	4	147	16(11)	56(38)	72(49)
Breast	10[a]	239	18(8)	65(27)	83(35)
Gastric	7[b]	250	19(8)	68(27)	87(35)
Colorectal	29	685	22(3)	130(19)	152(22)
Lung (non-small cell)	1	16	0	0	0

[a] 2 trials: +tamoxifen and Premarin.
[b] 4 trials: +Adriamycin.

comparison between the usually reported overall response rates of single agent FUra (15–20%) and FUra combined sequentially with MTX (22%). This has been done on the basis of results of randomized studies that compared the therapeutic efficacy of sequential MTX→FUra to FUra alone or different intervals of administration for MTX and FUra (Table 3). The study by the Nordic Gastrointestinal Tumor Adjuvant Therapy Group (1989) demonstrated the superiority of the sequential combination with a 3 hr interval as compared to FUra alone (24 vs 3% response rate). A study by Herrmann et al. (1986) reached similar superior therapeutic results using an interval of 7 hr between administration of the two antimetabolites compared with single FUra (28 vs 15% response rate). Poon et al. (1989) also observed higher tumor response using sequential MTX–FUra with intervals of 7 and 24 hr, when compared with single-agent FUra (21 and 26% vs 10% response rate). Petrelli et al. (1987) and Valone et al. (1989) did not however document any significant difference in the response rates of sequential MTX–FUra (intervals 4 and 24 hr, respectively) and single-agent FUra.

The study by Marsh et al. (1989) compared an interval of administration of 1 and 24 hr: using the longer interval, therapeutic results were significantly higher (25 vs 11%). No difference between the sequence MTX→FUra and FUra→MTX with a short interval (1 hr) was observed by MacIntosh et al. (1987).

With the exception of that by Petrelli et al. (1987), these studies utilized doses of MTX adequate for in vivo biochemical modulation of FUra cytotoxicity (200–300 mg/m^2), although administration modalities of the antifol varied as well as doses of the fluoropyrimidine (400–900 mg/m^2/day). These variables may have affected the results obtained.

2. Head and Neck Cancer

The mean response rate observed in four phase II studies in patients with advanced or recurrent squamous cell carcinoma of the head and neck was equal to 49% (72 out of 147 cases) (Jacobs, 1982; Pitman et al., 1983; Ringborg et al., 1983; Stewart et al., 1986) (Table 2). Responding patients survived longer than those not responding with a duration inferior to one year. Because percentage of response obtained with sequential MTX–FUra is about the same in patients that have not responded to chemotherapy (usually including MTX) and in patients not previously treated, it is conceivable that synergism occurs in pretreated patients.

Following these encouraging results, three randomized studies have examined the importance of the sequential use of these drugs (Browman et al., 1983, 1988; Coates et al., 1984) (Table 4).

In the first study by Browman et al. (1983) a 1 hr interval (MTX→FUra) was compared with rapid sequential administration (MTX–FUra), referred to as "simultaneous" administration. The response rate was higher in patients

TABLE 3
Randomized Trials of Sequential MTX/FUra in Advanced Colorectal Cancer

MTX dose (mg/m²)	Interval MTX–FUra (hr)	FUra dose (mg/m²)	LV rescue	Schedule	Number of patients	% Response CR	% Response PR + CR	Reference
250 iv (over 2 hr)	3	500 iv	+	q 2 wk	82	4	24	Nordic Gastrointestinal Tumor Adjuvant Therapy Group (1989)
—	—	600 iv (d 1–2)	—	q 2 wk	91	0	3	
50 iv (over 4 hr)	4	600 iv	+	q 1 wk × 4 then q 2 wk	21	0	5	Petrelli et al. (1987)
—	—	450 iv (d 1–5) then 200 iv q 2 days × 6	—	—	19	0	11	
300 iv (over 4 hr)	7	900 iv	+	q 2 wk	126	NS[a]	28	Herrmann et al. (1986)
—	—	450 iv (d 1–5)	—	q 3 wk		NS	15	

50 × 5 po q 6 hr	24	500 iv	+	q 2 wk	96	6	19	Valone et al. (1989)
—	—	12 mg/kg/day iv (d 1–5) then 15 mg/kg	—	q 1 wk	52	7	17	
200 iv (over 4 hr)	7	900 iv	+	q 3–4 wk	39	NS	21	Poon et al. (1989)
40 iv (d 1, 8)	24	700 iv (d 2, 9)	—	q 4 wk	39	NS	26(p = 0.04)	
—	—	500 iv (d 1–5)	—	q 5 wk	39	NS	10	
250 iv	+	600 iv	+	q 1 wk × 6 then q 2 wk	54	0	28	MacIntosh et al. (1987)
250 iv	−1	600 iv	+	q 1 wk × 6 then q 2 wk		0	28	
200 iv (over 0.5 hr)	1	600 iv	+	NS	87	5.5	11	Marsh et al. (1989)
200 iv (over 0.5 hr)	24	600 iv	+	NS	79	10	25	

[a]NS, not stated.

TABLE 4
Effects of Sequence of MTX and FUra on Response Rate of Patients with Head
and Neck Cancer

Sequence	Interval (hr)	Number of patients	RC + RP (%)	Reference
A. MTX→FUra	1	37[a]	38	Browman et al.
MTX→FUra	simultaneous[b]	42[a]	62	(1983)
B. MTX→FUra	1	26[a]	65	Coates et al.
FUra→MTX	1	23[a]	47	(1984)
C. MTX→FUra	18	58	45	Browman et al.
MTX→FUra	simultaneous[b]	55	47	(1988)

[a] None previously treated with chemotherapy.
[b] MTX was administered first, followed immediately by FUra.

given the drug simultaneously when compared with the sequence (Table 4). There was no difference in survival among the groups (Browman et al., 1983, 1986).

In the study by Coates et al. (1984), the sequence MTX→FUra was compared with the FUra→MTX sequence; the interval in both arms was 1 hr (Table 4). In this investigation the MTX→FUra sequence produced a higher response rate than FUra→MTX although the difference was not significant. Again there were no differences in survival between the two groups. More recently Browman et al., (1988) have also shown that a longer interval between MTX and FUra administration (18 hr) is not superior to simultaneous administration of the two antimetabolites in patients with this malignancy (Table 4).

3. Breast Cancer

Ten phase II studies were carried out in patients with advanced mammary carcinoma (Allegra, 1983; Eisenhower et al., 1984; Ellison et al., 1985; Gewirtz and Cadman, 1981; Herrmann et al., 1984a; Kaye et al., 1984; Panasci and Margolese, 1982; Perrault et al., 1983; Plotkin and Waugh, 1982; Tisman and Wu, 1980) (Table 2). The majority of patients were resistant to previous single agent and/or combination chemotherapy. About one-third of patients were resistant to CMF or other FUra containing regimens. Also in these studies variability of treatment characteristics was marked, thus the mean observed response rate (35%) was not particularly significant. All studies except two (Herrmann et al., 1984a; Kaye et al., 1984) employed an interval of 1 hr between MTX and FUra administration, which is a non-optimal interval based on experimental evidence. Although sequential MTX–FUra demonstrated significant therapeutic efficacy, synergism in treatment of this disease remains to be demonstrated.

Based on cell cycle activity of MTX and FUra, their sequential administration has been shown to be particularly active in the palliative treatment of breast cancer synchronized by hormonal treatment utilizing tamoxifen–Premarin (69% overall response rate) (Allegra, 1983). In a more extensively pretreated patient population, mostly with visceral dominant disease, the results of the same chemo–hormonal therapy were, however, less impressive (10% response rate) (Eisenhower *et al.*, 1984).

Since MTX and FUra are cycle-active drugs and might be expected to be more effective against exponentially growing tumor populations and since both drugs have not been reported to have carcinogenic effects, this combination is attractive both for theoretical as well as toxicological considerations in the adjuvant treatment of breast cancer, as suggested by Bertino (1982).

The National Surgical Adjuvant Breast Program (NSABP) has employed sequential MTX and FUra followed by LV rescue as adjuvant therapy to treat patients with estrogen receptor-negative, node-negative breast cancer (Fisher *et al.*, 1989). This large, randomized trial (vs no treatment) demonstrated that adjuvant therapy with the sequential antimetabolite combination was able to prolong significantly disease-free survival at 4 years (80 vs 71%, $p = 0.003$). This benefit, achieved without the use of an alkylating agent, was associated with tolerable side effects.

4. Gastric Cancer and Other Tumors

Only a few studies in gastric cancer using this sequence have been reported, mostly utilizing high- or moderate-dose MTX (Ayani *et al.*, 1989; Bruckner and Cohen, 1983; Dickinson *et al.*, 1989; Klein *et al.*, 1984; Muro *et al.*, 1986; Sasaki, 1984; Wils *et al.*, 1986) (Table 2). Alternating or combining sequential MTX–FUra with Adriamycin was particularly interesting. With the exception of the study by Muro *et al.* (1986) employing low-dose MTX (9% response rate), the efficacy was satisfactory with response rates varying between 68 and 22% (Ajani *et al.*, 1989; Klein *et al.*, 1984; Wils *et al.*, 1986) and a maximum median survival of responding patients of more than 18 months (Klein *et al.*, 1984).

Non-small-cell carcinoma of the lung appears to be resistant to sequential MTX–FUra (Woods *et al.*, 1985).

5. Sequential Combinations with Other Folate Antagonists

Other folate antagonist inhibitors of DHFR can potentiate FUra antitumor activity when administered prior to the fluoropyrimidine. Data from *in vitro* tumor models are at present available for both "classical" (i.e., DCMTX) (Sobrero and Bertino, 1983) and "nonclassical" antifolates (i.e., TZT and TMTX) (Howard *et al.*, 1987; Mini *et al.*, 1988; Sobrero *et al.*, 1989). Also, in mice bearing P388 leukemia, combined TMTX–FUra showed increased therapeutic efficacy when compared with single agent treatment (Leopold et al., 1987).

Since synergism occurred both for classical antifolates such as MTX and DCMTX substrates of folylpolyglutamate synthetase, and nonclassical ones such as TMTX and TZT, lacking the glutamate moiety and therefore not substrates for this enzyme (McGuire et al., 1980), it is unlikely that polyglutamylation of antifolates plays a significant role in the synergism at the TS level. Other possible mechanisms, such as enhanced FdUMP or FUTP formation due to accumulation of PRPP or elevated levels of $H_2PteGlu$ polyglutamates, may contribute in determining the observed synergism between antifolates and FUra.

The importance of the availability of alternative antifolate–FUra combination regimens for clinical use is related to the different pharmacological properties of the various antifolates. DCMTX is in fact a potentially useful modulating agent of FUra for hepatic artery infusion of patients with metastasis due to gastrointestinal cancer since it undergoes extensive degradation by hepatic aldehyde oxidase (Blakley, 1969). TMTX and TZT are not polyglutamylated (McGuire et al., 1980), do not enter cells via the carrier-mediated transport system utilized by MTX and reduced folates (Kamen et al., 1984; Skeel et al., 1973), and are effective against MTX-resistant tumor cells characterized by impaired transport, low level increase of DHFR or impaired polyglutamylation (Mini et al., 1985; Pizzorno et al., 1988; E. Mini, unpublished observations). Preliminary clinical trials of sequential TMTX–FUra and DCMTX–FUra have been carried out, and deserve further testing on larger patient populations (Hudes et al., 1989; J. R. Bertino, unpublished observations).

The combination of FUra with a folate antagonist inhibitor of TS (N_{10}-propargyl-5,8-dideazafolic acid) did not lead to increased antitumor activity in colorectal cancer patients (Cantwell and Harris, 1988).

IV. Leucovorin–FUra

A. PRECLINICAL RATIONALE

The rationale of combining LV and FUra is based on the idea of enhancing the antiproliferative activity of FUra at the TS level. In tumors from patients, the inhibition of TS by FdUMP appears to be an important mechanism to achieve a clinically relevant response to FUra treatment (Peters et al., 1989; Spears et al., 1984).

It has been demonstrated that the administration of exogenous chemically stable folate precursors of $5,10\text{-}CH_2H_4PteGlu$, such as LV and $5\text{-}CH_3H_4$

PteGlu (Huennekens *et al.*, 1987), can increase the intracellular levels of this cofactor and enhance the formation and stability of ternary complexes between FdUMP and TS, leading to increased FUra-mediated DNA effects (Berger and Hakala, 1984; Houghton *et al.*, 1986a,b; Ullman *et al.*, 1978; Yin *et al.*, 1982).

The synergism of the combination of folates and fluoropyrimidines has been demonstrated *in vitro* against various tumor cell lines including those of human origin (Chang and Bertino, 1989; Evans *et al.*, 1981; Keyomarsi and Moran, 1986; Mini *et al.*, 1987a,b; Park *et al.*, 1988; Ullman *et al.*, 1978; Waxman and Bruckner, 1982). Various concentrations of LV or 5-CH₃H₄PteGlu and both FUra and FdUrd were used. Also, LV was added to the cultures either before, together with, or after FUra or FdUrd. The simultaneous addition of LV to FUra and FdUrd or sequential exposure (LV → fluoropyrimidine) enhanced growth inhibition in all tumor cell lines.

In the human T-lymphoblast leukemic cell line CCRF-CEM, the sequence of administration of FUra in relation to folate (4 hr before, FUra → LV; simultaneous, FUra + LV; 4 hr after, LV → FUra) did not influence the degree of synergism significantly; the inhibition of cell growth was marked under all experimental conditions (Mini *et al.*, 1987b) (Fig. 4). The sequence of administration of FdUrd in relation to the folate was, on the contrary, important in the outcome of the combination: the sequence LV → FdUrd (4 hr before) and the simultaneous combination (LV + FdUrd) were active, while the inverse sequence FdUrd → LV was not (Mini *et al.*, 1987b) (Fig. 4).

In both cases pretreatment with LV or at least simultaneous exposure to these drugs is necessary to allow intracellular uptake of the folate to reach equilibrium with the other folate monoglutamate and polyglutamate pools, including 5,10-CH₂H₄PteGlu, and consequently to maximize the binding of FdUMP to TS. The difference observed in this cell line in the degree and sequence dependence of synergism might be explained by the different rate of conversion of the two fluoropyrimidines to the common active metabolite, FdUMP. In CCRF-CEM cells, in fact, FdUrd will presumably reach higher FdUMP levels in a shorter time than FUra. Since these cells lack Urd and dThd phosphorylases (Piper and Fox, 1982), FdUrd is completely activated to FdUMP via dThd kinase, while FUra is only partially converted to the fluorodeoxynucleotide via a multienzymatic pathway including OPRTase and ribonucleotide reductase, which presumably requires a longer time.

When CCRF-CEM cells were treated with varying concentrations of LV (0.1–100 μM) for 4 hr and with FUra (250 μM) for the last 2 hr of exposure, synergism was dependent on LV concentrations, with a maximum at the higher concentrations (100 μM), but with a significant degree at the intermediate concentration (10 μM) (Mini *et al.*, 1987b). In the clinic it is possible to reach levels between 10 and 100 μM of the physiological diastereoisomer (L) and of its metabolite 5-CH₃H₄PteGlu in the plasma using

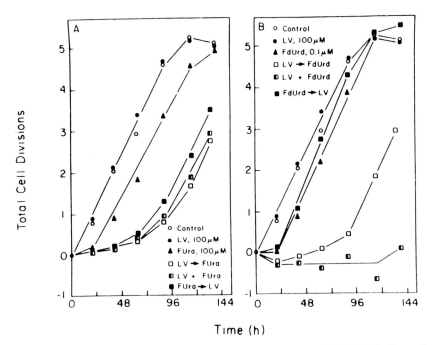

Time (h)

FIGURE 4   Sequence dependent effects of LV and FUra or FdUrd on CCRF-CEM cell growth. Cells were exposed to FUra (100 μM) or FdUrd (0.1 μM) with or without LV (100 μM). Each drug exposure either alone or in simultaneous or sequential combination was for 4 hr. After 4 hr of exposure to the first drug, cells were washed twice and re-exposed to the second drug. After the final wash, cells were resuspended in drug-free medium and their growth was followed. (From Mini et al., 1987b, with permission.)

(DL) LV doses of 200–500 mg/m^2 (Machover et al., 1986; Madajewicz et al., 1984; Newman et al., 1989).

A certain degree of synergism was also observed at the lower concentration (1 μM). This level can be reached in the clinic with LV doses $\geq$ 50 mg (Straw et al., 1984). Synergism did not occur at 0.1 μM LV in this experimental system (Mini et al., 1987b), nor had it been previously observed in others (Keyomarsi and Moran, 1986; Ullman et al., 1978).

Similar results have been obtained using (DL)5-CH$_3$H$_4$PteGlu at equimolar concentrations (Evans et al., 1981; Mini et al., 1987a) (Fig. 5). These in vitro data appear to be particularly important from the clinical point of view since (L)5-CH$_3$H$_4$PteGlu represents the main metabolite of LV, which is formed rapidly in the plasma following oral or parenteral administration, and reaches the level of the physiological diastereoisomer (L) within 2 hr (Mehta et al., 1978; Nixon and Bertino, 1972; Straw et al., 1984).

Polyglutamylation of 5,10-CH$_2$H$_4$PteGlu has been demonstrated to be an important determinant of the formation and stability of the FdUMP–TS–

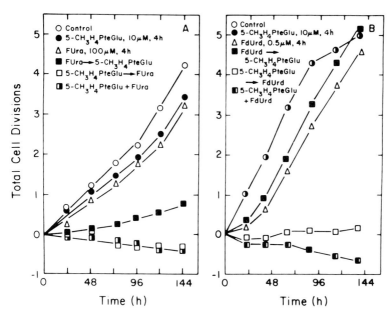

FIGURE 5   Sequence dependent effects of 5-CH$_3$H$_4$PteGlu and FUra or FdUrd on CCRF-CEM cell growth. Cells were exposed to FUra (250 μM) or FdUrd (0.1 μM) with or without 5-CH$_3$H$_4$PteGlu (100 μM). Each drug exposure either alone or in simultaneous or sequential combination was for 4 hr. After 4 hr of exposure to the first drug, cells were washed twice and re-exposed to the second drug. After the final wash, cells were resuspended in drug-free medium and their growth was followed. (From Mini *et al.*, 1987a, with permission.)

folate ternary complex. Radparvar *et al.* (1989) recently demonstrated, using enzyme purified from a human colon carcinoma xenograft (HxVRC$_5$), that long chain polyglutamate species (Glu$_{3-6}$) of (L)-5,10-CH$_2$H$_4$PteGlu at concentrations of 0.9 to 1.6 μM were >200-fold more effective than (L)-5,10-CH$_2$H$_4$PteGlu (335 μM) at stabilizing ternary complexes for a $t_{1/2}$ for dissociation of 100 min.

Polyglutamates of 5,10-CH$_2$H$_4$PteGlu also increased the affinity of [6-^{3}H]FdUMP for TS as demonstrated by a 7- to 10-fold decrease in k_d determined by Scatchard analysis at a folate concentration of 10 μM.

The relevance of polyglutamylation in the potentiation of FdUrd cytotoxicity by LV has also been investigated by Romanini *et al.* (1989) by utilizing CCRF-CEM/P cells resistant to MTX due to impaired polyglutamylation (Pizzorno *et al.*, 1988). Using short-term exposure (4 hr) to LV concentrations up to 100 μM, enhancement of FdUrd cytotoxicity did not occur in CCRF-CEM/P mutants which showed decreased retention of 5,10-CH$_2$H$_4$PteGlu polyglutamate pools and a more rapid recovery of dTMP synthesis following LV exposure, when compared with MTX-sensitive parent CCRF-CEM cells.

The first *in vivo* study of combined folates and FUra by Klubes *et al.* (1981) in mouse L1210 leukemia did not demonstrate enhancement of FUra efficacy by LV. The potentiation of the antitumor activity of FUra has been shown, however, in two different tumor models sensitive or resistant to this fluoropyrimidine (murine colon tumors 26 and 38, respectively) (Nadal *et al.*, 1987). A significantly longer tumor growth delay occurred when LV administration preceded or was simultaneous with FUra, compared with fluoropyrimidine alone, with a similar degree of toxicity. Similar results were obtained in P-388 leukemia bearing mice using folic acid instead of LV administered 12 or 24 hr prior to but not simultaneously with FUra (Parchure *et al.*, 1984). In a study by Martin *et al.* (1988) the combination LV and FUra was tested in mice bearing advanced CD8F1 breast tumor at a maximally tolerated dose (MTD) of FUra with and without LV. Overall, therapy with FUra at the MTD was not improved by LV. Furthermore, although the activity of FUra doses lower than the MTD could be increased by LV, the therapeutic results were comparable to those of single agent FUra at MTD.

Recently it has been shown that high-dose LV can expand 5,10-$CH_2H_4PteGlu$ pools in EMT6 mammary adenocarcinoma tissues of mice, but not in bone marrow (Wright *et al.*, 1989). LV given 2 hr prior to FUra resulted in a modest but significant increase in the delay of tumor growth when compared to FUra alone. The benefit afforded by the use of LV in this regimen was obtained without an increase in host toxicity, suggesting tumor-selective potentiation.

B. CLINICAL STUDIES

In phase I–II clinical studies of the combination of LV and FUra, a variety of dose schedules of both LV and FUra were used. Bolus administration as well as short-term or long-term infusions were used for the two drugs, while time intervals between LV and FUra and dose levels of the two drugs also varied markedly. Recently LV has also been given by oral route (Brenckman *et al.*, 1988; Hines *et al.*, 1989).

Two treatment regimens have primarily been employed. Machover *et al.*, (1982) administered LV by iv bolus and FUra by iv short-term infusion (over 15 min) daily for 5 consecutive days. In the original protocol patients received 200 mg/m² of LV and 370 mg/m² of FUra with adjustment of dosage in subsequent courses of therapy, depending on tolerance. Courses were repeated at 28-day intervals. In further studies utilizing this treatment regimen the two drugs were both administered by iv bolus or iv infusion lasting up to 2 hr. Time intervals between LV and FUra varied between 0 and 2 hr. The LV dose varied in the different studies among "low" and "high" levels (i.e., 60–500 mg/m²/day), which are able to produce plasma concentrations of reduced folates of 1 μM or higher. Dose levels of FUra also varied between 340 and 770 mg/m²/day in the different studies.

The second treatment schedule proposed by Madajewicz *et al.* (1984) utilizes a high dose of LV (500 mg/m^2/day) administered as a 2 hr infusion with escalating bolus doses of FUra given midway through the infusion. This schedule was administered weekly 6 times followed by a 2 week rest period. Although the maximum tolerated dose was 750 mg/m^2, Madajewicz *et al.* (1984) recommended the dose of 600 mg/m^2 which yielded nearly the same response rate as the 750 mg/m^2 dose with less toxicity. Further studies conducted according to this schedule always included high-dose LV but with varied administration modalities (iv bolus or infusion up to 2 hr). Doses of FUra also varied between 300 and 850 mg/m^2, always given by iv bolus. The interval between LV and FUra administration was between -30 min (FUra given before) and $+1$ hr (LV given before).

1. Colorectal Cancer

Of 730 patients with colorectal carcinoma treated in 22 studies, 177(24%) responded to therapy with combined LV-FUra (Barone *et al.*, 1987; Bertrand *et al.*, 1984, 1986, Brenckman *et al.*, 1988; Bruckner *et al.*, 1982, 1983; Byrne *et al.*, 1984; Cunningham *et al.*, 1984; DeGramont *et al.*, 1988; Greene *et al.*, 1986; Hines *et al.*, 1988, 1989; Kaplan and Rivkin, 1984; Laufman *et al.*, 1987; Machover *et al.*, 1986; Madajewicz *et al.*, 1984; Mortimer and Higano, 1988; Van Groeningen *et al.*, 1989; Wilke *et al.*, 1988) (Table 5). Patients who had been treated previously with FUra-containing chemotherapy have been included in these results. In FUra-untreated patients the response rate was higher (30%). Tumor remission was also observed, although in a limited

TABLE 5
Summary of Phase II Clinical Studies of the Combination LV–FUra

| | | Response | | |
| | | Number of FUra | Number of FUra | |
Tumor type	Number of studies	pretreated patients	untreated patients	Total number of patients
Colon and rectum	22	32/250(13%)	145/480(30%)	177/730(24%)
Breast	9	66/221(30%)	18/41(44%)	84/262(32%)
Stomach	4	2/13(15%)	15/54(28%)	17/67(25%)
Head and neck	1	NS[a]	NS	3/16(19%)
Pancreas	1	0/0	2/27(7%)	2/27(7%)
Prostate	1	0/0	0/17	0/17
Liver	1	0/0	0/14	0/14
Kidney	1	0/0	0/14	0/14

[a] NS, not stated.

percentage of cases (13%) in patients resistant to FUra single-agent therapy. These data indicate that with current FUra–LV regimens, the number of patients who will benefit from combination therapy after tumor progression with single-agent FUra is small, and such patients are candidates for other investigational approaches including the addition of other modulating agents to FUra.

Based on phase II data suggesting that FUra combined with LV is active in metastatic colorectal carcinoma, various phase III studies were initiated to compare the efficacy of this combination with standard therapy with FUra alone (Di Costanzo *et al.*, 1989; Doroshow *et al.*, 1987; Erlichman *et al.*, 1988; LaBianca *et al.*, 1989; Nobile *et al.*, 1988; Petrelli *et al.*, 1987, 1989; Poon *et al.*, 1989; Valone *et al.*, 1989). Six trials evaluated 5 days courses of LV–FUra (Table 6) while three studies evaluated a weekly LV–FUra schedule (Table 7).

The results of the majority of these trials (7/9) demonstrate a significant advantage of the combination of LV and FUra over single agent chemotherapy with FUra in terms of response rates (Doroshow *et al.*, 1987; LaBianca *et al.*, 1989; Erlichman *et al.*, 1988, Nobile *et al.*, 1988; Petrelli *et al.*, 1987, 1989; Poon *et al.*, 1989). Only two studies show, however, a significant increase in patient survival (Erlichman *et al.*, 1988; Poon *et al.*, 1989).

In these studies LV was used at low and high doses (between 20 and 500 mg/m^2/day). Other treatment variables included length of LV administration, interval of administration of folate and fluoropyrimidine, and FUra dosage.

The optimal dose levels of LV in regard to the maximal enhancement of the clinical efficacy of FUra have not yet been established. The study by Poon *et al.* (1989) demonstrated that a low-dose (20 mg/m^2/day) was superior to a high dose (500 mg/m^2/day) schedule with regard to both response and survival. Petrelli *et al.* (1989) showed the contrary, that is, high-dose LV (500 mg/m^2/day) had a greater enhancing effect on FUra efficacy than low-dose folate (25 mg/m^2/day). It should be noted, however, that the dose of FUra used in the study by Poon *et al.* with low-dose LV was higher than that with high-dose LV.

The planned dose intensities for single agent FUra ranged from 463 to 760 mg/m^2/week and the actual delivered doses, established in two of the trials (Erlichman *et al.*, 1988; Nobile *et al.*, 1988), were 531 and 533 mg/m^2/week. The planned dose intensity for the FUra–LV combination ranged from 463 to 600 mg/m^2/week compared with the delivered dose intensity of 443 and 463 mg/m^2/week in the same two studies. When used with LV, FUra could not be administered at the single-agent MTD because of unacceptable toxicity. Nevertheless, a lower dose of FUra, when combined with LV, had higher response rates than FUra alone.

No significant differences were observed in response rate and survival as a function of length of administration of FUra (i.e., iv bolus or continuous

TABLE 6
Summary of Phase III Clinical Studies of 5-Day Regimens of LV–FUra in Colorectal Cancer

Number of evaluable patients	Treatment[a]	Interval LV–FUra (hr)	Response (%)	Median survival (mo)	Reference
63	FUra vs HDLV (infusion 144 hr)–FUra (bolus)	– 24	13 44(p = 0.0019)	12.8 14.5(p = 0.25)	Doroshow et al. (1987)
125	FUra vs HDLV (bolus)–FUra (bolus)	– 0	7 33(p = 0.0005)	9.6 12.6(p = 0.05)	Ehrlichman et al. (1988)
163	FUra vs HDLV (bolus)–FUra (bolus)	– 0	17 19	11.5 10.8	Valone et al. (1989)
111	FUra vs HDLV (bolus)–FUra (bolus) vs LDLV (bolus)–FUra (bolus)	– 0 0	10 26(p = 0.04) 43(p = 0.001)	7.7 12.2(p = 0.037) 12.0(p = 0.050)	Poon et al. (1989)
119	FUra vs HDLV (bolus)–FUra (bolus)	– 0	12 19	10.7 10.5	Di Costanzo et al. (1989)
93	FUra vs HDLV (bolus)–FUra (bolus)	– 0	9 23	5.5 5.0	LaBianca et al. (1989)

[a] HD, high dose; 200–500 mg/m^2/day; LD = low dose; 20 mg/m^2/day.

TABLE 7

Summary of Phase III Clinical Studies of Weekly LV–FUra in Colorectal Cancer

Number of evaluable patients	Treatment[a]	Interval LV–FUra (hr)	Response (%)	Median survival[b] (mo)	Reference
44	FUra(5-day regimen) vs HDLV (infusion 2 hr)–FUra (bolus)	– 1	11 48(p = 0.009)	9.6 10.9	Petrelli et al. (1987)
82	FUra vs HDLV (infusion 2 hr)–FUra (bolus)	– 1	5 16(p = 0.05)	NR(NS) NR	Nobile et al. (1988)
328	FUra (5-day regimen) vs HDLV (infusion 2 hr)–FUra (bolus) vs LDLV (infusion 10 min)–FUra (bolus)	– 1 1	12 30(p < 0.01) 19	10.7 12.8(p = 0.08) 10.5	Petrelli et al. (1989)

[a] HD, high dose, 500 mg/m^2/day; LD = low dose, 25 mg/m^2/day.
[b] NR, not reported; NS, no significant difference.

infusion over 96 hr) Budd *et al.*, 1987). All arms received the same LV dose; this study did not include, however, a treatment arm with FUra alone.

On the basis of positive therapeutic results and good tolerance, adjuvant chemotherapy studies using this combination have been initiated by cooperative groups in patients with colorectal cancer, Dukes stages B and C (i.e., NSABP, Wolmark, 1988; NCI of Canada, Erlichman, 1988; NCCTG; an intergroup study of NCCTG, ECOG, CALGB, and SWOG; IST of Genoa; GIVIO of Milan; COG of Florence, Italy). The results of these studies will be of interest in as much as the effectiveness of this combination may be greater and have more impact on survival in the adjuvant situation.

2. Various Other Solid Tumors

A 30% response rate was obtained with FUra and LV in patients with cancer of the breast (Becher *et al.*, 1988; Doroshow *et al.*, 1989; Fine *et al.*, 1988; Hamm *et al.*, 1989; Jabboury *et al.*, 1989; Loprinzi *et al.*, 1988; Margolin *et al.*, 1989; Marini *et al.*, 1987a; Swain *et al.*, 1988) (Table 5). The response rate was higher in previously untreated patients (44%) than in those previously treated with FUra (30%). An overall 25% response rate was also reported in gastric cancer (Arbuck *et al.*, 1987; Becher *et al.*, 1988; Machover *et al.*, 1986; Marini *et al.*, 1987b) (Table 5). Patients not previously treated with FUra responded more frequently than those previously treated (28 vs 15%). A recent preliminary report by Berenberg *et al.* (1989) suggests, however, a lower activity of LV–FUra, both as a 5-day course and as a continuous infusion schedule, in previously untreated gastric cancer patients.

Patients with prostatic, hepatocellular and renal cell carcinomas apparently do not benefit from therapy with the LV–FUra combination (Becher *et al.*, 1988; Zaniboni *et al.*, 1988, 1989). Data from Bruckner *et al.* (1982, 1988) on a limited number of patients (16 and 8, respectively) with squamous cell carcinoma of the head and neck and with pancreatic cancer, indicate that this combination may be active in these diseases. A recent report by DeCaprio *et al.* (1989) demonstrated, however, a low response rate (7%) in 27 evaluable previously untreated patients with advanced pancreatic cancer (Table 5).

Futher clinical studies in other potentially FUra-responsive solid tumors and in leukemia–lymphomas are warranted, based upon experimental tumor results (Keyomarsi and Moran, 1986; Mini *et al.*, 1987b).

3. Toxicity

LV and FUra treatment is usually well tolerated if attention is paid to early signs of gastrointestinal toxicity. Substantial or severe toxicity (grade 3 and 4 of the WHO scale) occurs rarely or not at all when the FUra dose does not exceed 370 mg/m^2/day in treatment schedules utilizing a 5-day administration, or 500 mg/m^2 as a single administration weekly. Thus, in the presence of LV the dose of FUra has to be decreased to approximately 20% of that used when this fluoropyrimidine is used alone. An analysis of type and degree of toxic effects observed has demonstrated that at higher dosages the toxicity of the combination may reach grade 3 and 4, comprising mainly stomatitis, diarrhea, and dehydration. In some studies deaths have been reported (reviewed by Grem et al., 1987b).

C. FUTURE DEVELOPMENTS

The evaluation of the administration of FUra and LV by alternative routes (i.e., oral or regional administration) warrants further investigation in view of the preclinical rationale and recently published preliminary clinical data (Arbuck et al., 1986; Brenckman et al., 1988; Budd et al., 1986; Hines et al., 1989; Smith et al., 1989; Valeri et al., 1987).

Since studies of orally administered LV indicate that absorption of the L-isomer is approximately 5 times greater than that of the D-isomer due to stereoselectivity (Straw et al., 1984), repeated oral administration of LV might be useful in achieving persistently higher levels of the L-isomer compared with the D-isomer. This event would avoid possible competition between the D-and the L-forms for cellular uptake and for formation of stable ternary complexes, which may occur at high D-LV concentrations following iv administration. Results by Bertrand and Jolivet (1989) in an in vitro human model (CCRF-CEM) suggest however that interference with L-LV-FUra synergism would not occur however even at high D-LV concentrations (1 mM). A multicenter double-blind randomized trial comparing FUra with oral LV or placebo is ongoing (Laufman et al., 1989b).

The regional administration of a fluoropyrimidine (FUra and FdUrd) and LV by intraperitoneal and hepatic artery routes, although still investigational, might provide a means of enhancing drug efficacy and decreasing systemic toxicity (Arbuck et al., 1986; Budd et al., 1986; Kemeny et al., 1988; Smith et al., 1989; Valeri et al., 1987).

Clinical studies with combinations of other folates (i.e., DL-5-CH$_3$H$_4$Pte Glu or folic acid) and FUra have also been performed, but the limited number of cases does not allow at present a final evaluation of therapeutic results (Asbury et al., 1987; Nobile et al., 1985; Valeri et al., 1987). Recent data from Houghton et al. (1989) comparing the conversion of DL-LV and DL-5-CH$_3$H$_4$PteGlu to 5, 10-CH$_2$H$_4$PteGlu and H$_4$PteGlu in human colon adenocarcinoma xenografts in mice suggested however that DL-5-CH$_3$H$_4$PteGlu is less effective than DL-LV in expanding 5,10-CH$_2$H$_4$PteGlu and H$_4$PteGlu pools and in modulating polyglutamylation. Possible explanations include competition of D and L isomers of 5-CH$_3$H$_4$PteGlu at the transport level (White et al., 1978), the level of folylpolyglutamate synthetase (McGuire *et al.*, 1980), or inefficient metabolism of 5-CH$_3$H$_4$PteGlu to H$_4$PteGlu via methionine synthetase.

Since preclinical synergism observed with LV–FdUrd has usually been greater than that seen with LV–FUra (Keyomarsi and Moran, 1986; Mini *et al.*, 1987b), preliminary clinical studies of this combination have been performed (Laufman *et al.*, 1989a; Marsh *et al.*, 1988) and warrant further testing.

The integration of LV–FUra treatment with other drugs is also possible and potentially worthwhile, since the toxicity of this combination is mostly gastrointestinal (reviewed by Mini *et al.*, 1990). The activity of LV–FUra in combination with other agents that modulate fluoropyrimidine cytotoxicity, in particular those that enhance inhibition of TS, also deserves further study, e.g., in head and neck squamous cell carcinomas (Loeffler *et al.*, 1988; Vokes *et al.*, 1988; Wendt *et al.*, 1989).

V. Sequential MTX–LV–FUra

Based on the previously illustrated preclinical rationales, it is conceivable that appropriate sequencing of MTX–LV and FUra might result in enhanced cytotoxicity in human tumors, even though the only published preclinical study reports negative results.

In L1210 leukemia cells *in vitro*, in fact, LV was added to the MTX–FUra combination with the intention of augmenting the cytotoxicity: the addition of 10 or 100 μM LV concurrently with or after 10 μM FUra following MTX (1 μM) pretreatment did not increase the inhibition of cell growth or clonogenicity in comparison with MTX given prior to FUra without LV (Danhauser *et al.*, 1985).

Kaizer *et al.* (1986) suggested that low-dose LV administered as a rescue agent in several trials of sequential moderate- to high-dose MTX–FUra might have contributed substantially to the favorable therapeutic outcome observed. Results from studies employing sequential low-dose MTX–FUra without LV rescue were significantly poorer. Preclinical data would suggest use of high-dose LV as a modulating agent for FUra at least 24 hr following MTX administration.

Recent clinical trials have been initiated in breast and in colon cancer using both moderate-dose MTX by iv bolus (125–250 mg/m^2) or repeated administration of low-dose MTX po (30 mg/m^2 q 6 hr × 6) and high-dose LV (200 mg/m^2 iv bolus q 24 hr × 3–4 or 500 mg/m^2 by iv 24 hr infusion) with a long interval between MTX and LV (24 or 36 hr) and standard-dose FUra (300 mg/m^2 q 24 hr × 3–4 or 600 mg/m^2) concurrently or 1 hr after LV (Saidman *et al.*, 1989; Richards *et al.*, 1989, respectively) (Table 8). Encouraging results of sequential MTX–high-dose-LV–FUra have been

TABLE 8

Sequential MTX–HDLV–FUra in Breast and Colon Cancer

MTX dose (mg/m²)	Interval		LV dose (mg/m²)	FUra dose (mg/m²)	Schedule	Number of patients	% Response		Reference
	MTX-LV (hr)	MTX-FUra (hr)					CR	PR + CR	
Breast cancer									
125(−250)iv d 1	24	24	200 iv d 2,3,4,(5) 10 po × 3 d 2	300 iv d 2,3,4,(5)	NS[a]	13	8	15	Saidman et al. (1989)
Colon cancer									
30 po (q 6 hr × 6)	36	37	500 iv (over 24 hr)	600 iv	q 2 wk	70	10	27	Richards et al. (1989)

[a] NS, not stated.

obtained both in colon cancer patients (27% response rate in 70 cases) (Richards *et al.*, 1989) and in heavily pretreated breast cancer patients (up to two prior lines of chemotherapy) (15% response rate in 15 cases) (Saidman *et al.*, 1989).

VI. 6-Methylmercaptopurine Ribonucleoside–FUra

MMPR is an active agent which, when phosphorylated to its active form [6-methylmercaptopurine ribonucleoside 5′-monophosphate (MMPRMP)] by adenosine kinase, exerts antitumor activity *in vitro* by inhibiting amidophosphoribosyl transferase (Henderson, 1963; Henderson *et al.*, 1967), the second committed step in *de novo* purine synthesis whereby 5-phospho-α-D-ribo-1-pyrophosphoric acid is converted to 5-phospho(3-D-ribosylamine) utilizing glutamine as an amino-group donor. Like the indirect inhibition of purine synthesis induced by MTX, this results in intracellular elevation of PRPP (Paterson and Wang, 1970), and enhancement of FUra cytotoxicity through enhancement of intracellular accumulation of FUra. In L1210 leukemia cells *in vitro*, MMPR given prior to FUra results in synergistic cytotoxicity (Cadman *et al.*, 1981a,b). In mice bearing the mammary carcinoma CD8F1, MMPR prior to FUra significantly increased the antitumor effects when compared with FUra alone (Martin *et al.*, 1981). In both models an increase in PRPP levels was noted (16- and 3.5-fold respectively in L1210 cells and CD8F1 tumor). Also MMPR pretreatment of HCT-8 colon adenocarcinoma cells increased intracellular FUra levels by 230% (Benz and Cadman, 1981) and enhanced by 3-fold the incorporation of FUra into RNA in CD8F1 tumor cells (Martin *et al.*, 1981) (Table 1). The effects of FUra prior to MMPR have not been examined. At this time only one phase I trial of combined sequential MMPR–FUra has been done that demonstrated significant increases in bone marrow PRPP levels (W. P. Peters *et al.*, 1984) (Table 9). The effectiveness of this combination remains to be evaluated.

VII. Phosphonacetyl-L-Aspartic Acid and/or Thymidine–FUra

A. PALA–FUra

Phosphonacetyl-L-aspartic acid (PALA) is a transition state analog inhibitor of L-aspartate carbamoyltransferase (ACTase), the enzyme catalyzing the second step of *de novo* pyrimidine synthesis (Collins and Stark, 1971). Thus, PALA results in a reduction in intracellular pyrimidine nucleotide pools

TABLE 9
Summary of Clinical Studies of Other Biochemically Modulated FUra Combinations

Modulating agent	Tumor	Clinical trials		Therapeutic advantage over FUra alone
		Phase	Number	
TMTX	Solid tumors	I	1	—
MMPR	Solid tumors	I	1	—
dThd	Breast	II	1	No
	Colon	II	2	No
	Colon	III	1	No
PALA	Breast	II	2	No
	Melanoma	II	1	No
	Colon	II	6	5 No (HD-PALA, LD-FUra); 1 Yes (LD-PALA, HD-FUra)[a]
	Colon	III	3	1 No, 1 Yes (HD-PALA,LD-FUra); 1 Yes (LD-PALA, HD-FUra)
PALA–dThd	Colon	II	3	1 Yes, 2 No (HD-PALA, HD-dThd, LD-FUra)
PALA–MTX–LV	Solid tumors	I	1	—
	Unknown primary	II	1	No
DIP	Solid tumors	I	1	—
HU	Colon–rectum	II	5	No
HPP	Mostly colon	I–II	8	No
Urd	Solid tumors	I	3	—
PALA–MTX–LV–Urd	Solid tumors	I	1	—

[a] HD, high-dose; LD = low dose.

(Moyer and Handschumacher, 1979; Moyer *et al.*, 1980) and its effects may be reversed by nucleoside salvage. This will serve to reduce the competition between dUMP, UTP, and FUra nucleotides for their metabolic targets (i.e., TS and RNA, respectively) (Anukarahanonta *et al.*, 1980; Ardalan *et al.*, 1981; Kufe and Egan 1981; Liang *et al.*, 1982; Major *et al.*, 1982; Martin *et al.*, 1983a; Spiegelman *et al.*, 1980a,b). In addition, PALA, which blocks a step in pyrimidine synthesis prior to the formation of orotic acid, results in reduced competition between orotate and FUra for the enzyme OPRTase and enhances conversion of FUra to nucleotides (Ardalan *et al.*, 1981; Kufe and Egan, 1981). PALA also blocks the consumption of the intracellular PRPP polls through its effect on the early step in the *de novo* pyrimidine synthesis pathway and consequently increases the activation of FUra (G. J. Peters *et al.*, 1984) (Table 1).

In vivo studies have shown that pretreatment with PALA of mice and rats bearing a variety of tumors, resulted in enhanced antitumor effect of FUra with more tumor responses and more prolonged tumor growth delays than FUra alone (Anukarahanonta *et al.*, 1980; Johnson *et al.*, 1980; Martin *et al.*, 1978a, 1981, 1983a; Sawyer *et al.*, 1984; Spiegelman *et al.*, 1980a,b).

Based on preclinical data, phase I–II clinical trials with sequential PALA–FUra were performed in patients with various solid tumors, mostly colon carcinoma (Ardalan *et al.*, 1984; Bedikian *et al.*, 1981; Erlichman *et al.*, 1982; Meshad *et al.*, 1981; Muggia *et al.*, 1987; O'Connell *et al.*, 1982; Presant *et al.*, 1983; Weiss *et al.*, 1982a,b) (Table 9). In all of these early trials, therapeutic results were disappointing because response rates ranged between 5 and 24% (average 14%), even in untreated patients. In addition, notable toxic effects were also encountered, including primarily either dose-limiting mucositis and diarrhea or incapacitating neurotoxicity. In all these trials high doses of the modulating agent PALA were used (0.4–2.0 g/m^2) which prevented the administration of full doses of FUra, the effector agent, due to the toxic effects encountered. Therapeutic results of the high-dose PALA/low-dose FUra combination were thus disappointing because response rates were not higher than those obtained with single-agent FUra, even in untreated patients (Table 9).

In a randomized multicenter study comparing sequential high-dose PALA–low-dose FUra therapy and treatment with FUra alone in advanced colorectal carcinoma, response rates observed were 12 and 30%, respectively (Buroker *et al.*, 1985). A comparative study of PALA–FUra and single-agent FUra in a limited number of advanced heavily pretreated breast cancer patients demonstrated a higher response rate (28%) for the combination vs the single-agent fluoropyrimidine (14%) (Mann *et al.*, 1985) (Table 9).

More recently Martin *et al.* (1983a) demonstrated that low nontherapeutic doses of PALA which are capable of ACTase inhibition, could still enhance the therapeutic efficacy of FUra without undue toxicity in an experimental murine tumor model. Based on this observation, Casper *et al.* (1983) in a phase I study demonstrated that low dose PALA (250 mg/m^2) is sufficient to inhibit pyrimidine synthesis in humans and can be administered 24 hr before bolus injection of FUra at standard doses (750 mg/m^2) once weekly with minimal toxicity. Using this regimen, one patient responded partially, 3 had minor response, and 7 maintained stable disease.

Other clinical trials used drug administration schedules which closely resembled experimental treatment conditions characterized by synergistic antitumor effects, i.e., low dose of PALA (250 mg/m^2) and protracted exposure time to FUra (24 hr) at MTD (2,600 mg/m^2) by iv infusion and weekly administration (Ardalan et al., 1988; O'Dwyer et al., 1989). In both studies high response rates were obtained in pancreatic and colorectal cancer (46 and 43%, respectively, by Ardalan et al., 1988; O'Dwyer et al., 1989), comparable to results obtained with other means of FUra modulation (Table 9). Further phase III clinical trials are now necessary to define survival benefits with sequential combination.

B. dThd–FUra

When given at high-dose in humans shortly before or concurrently with FUra, dThd is able to inhibit FUra catabolism since it undergoes rapid conversion to thymine which is a competitive inhibitor of dihydrouracil dehydrogenase (Covey and Straw, 1983) and markedly prolongs FUra half-life from 6–38 min to 76–190 min (Au et al., 1982; Kirkwood et al., 1980; Woodcock et al., 1980). Since dThd inhibition of FUra catabolism in vivo would be expected to increase both FUra antitumor activity and host toxicity to levels comparable to those achievable with higher doses of FUra, other factors may be responsible for true enhancement of the therapeutic index of FUra observed in several murine tumor systems (Martin and Stolfi, 1977; Martin et al., 1978b, 1980; Santelli and Valeriote, 1978; Spiegelman et al., 1980a,b). An excess of dThd may enhance FUra incorporation into RNA by producing high levels of dTTP and blocking conversion of FdUrd to FdUMP via dThd kinase and of fluororibonucleotides to FdUMP via ribonucleotide reductase, resulting in the channeling of FUra into FUTP an thereby into RNA (Martin et al., 1978b, 1980) (Table 1).

Although in phase I clinical trials (Kirkwood et al., 1980; Lynch et al., 1985; Ohnuma et al., 1980; Vogel et al., 1979; Woodcock et al., 1980) some objective responses with this combination were observed in patients with advanced colorectal cancer, some of whom were resistant to single agent FUra, phase II clinical trials and one phase III study with this combination have failed to demonstrate an improved response rate in colorectal cancer or in breast carcinoma (Buroker et al., 1985; Presant et al., 1983; Sternberg et al., 1984) (Table 9). On the contrary, dThd appeared to increase the toxicity of FUra because dose-limiting myelosuppression occurred at FUra doses varying from one-half to two-thirds of the standard tolerated dose of FUra alone, and relevant central nervous system toxicity at high dThd doses was observed (Lynch *et al.*, 1985).

C. PALA–dThd–FUra

Improved therapeutic results with the three-drug regimen of PALA–dThd–FUra over those with the combinations of PALA–FUra or dThd–FUra were obtained in the CD8F1 murine mammary tumor (Martin *et al.*, 1983a). Based on these observations pilot clinical trials were carried out to determine the antitumor activity and toxicity of this combination in patients with advanced solid tumors (Casper *et al.*, 1984; Chiuten *et al.*, 1985; O'Connell *et al.*, 1984). O'Connell *et al.* (1984) have reported a promising response rate of 27% in patients with colorectal cancer, with higher responses seen in patients with rapidly growing tumors using high-dose PALA (4 g/m^2, iv 1 hr infusion), high-dose dThd (15 g/m^2, iv 0.5 hr infusion) and low-dose FUra (200 mg/m^2, iv 3 hr infusion) every 4 weeks. Two other studies, however, using a different dosage regimen (PALA 200–2000 mg/m^2, very high-dose dThd, i.e., 30–45 mg/m^2 and low-dose FUra, i.e., 100 or 300 mg/m^2, every 3–4 weeks) did not document any significant activity of this combination in a variety of advanced solid tumors, including colorectal cancer patients, most of them heavily pretreated with fluoropyrimidines (Casper *et al.*, 1984; Chiuten *et al.*, 1985) (Table 9). A high incidence of neurologic toxicity consisting of dizziness, lethargy, and confusion was observed in these two studies that used very high doses of dThd (30 and 45 mg/m^2), limiting possible clinical application of this combination.

D. PALA–MTX–FUra

Based on experimental data from Martin *et al.*, (1983b) demonstrating that concomitant modulation of FUra by PALA and MTX results in improved antitumor activity in murine tumor systems, a phase I study of this combination was conducted (Kemeny *et al.*, 1989). MTX was given at moderate dose (250 mg/m^2) and PALA at low dose (250 mg/m^2) in order to allow escalation of FUra to toxicity. The two drugs were given 24 hr before FUra to enhance maximal incorporation of FUra into RNA. LV (10 mg po every 6 hr for 8 doses) was given starting 24 hr after MTX. Two schedules of administration were used: one every other week and one weekly for 2 weeks. There were no responses in the every other week schedule. There was one partial response and three patients with stable disease in four evaluable patients on the weekly-for-2-weeks schedule. A successive phase II trial of this combination did not show antineoplastic activity superior to that of FUra alone in patients with adenocarcinoma of unknown primary sites (Kelsen *et al.*, 1989) (Table 9).

Attempts are currently underway adding Urd rescue to this combination with the hope of further escalating the FUra dose (Schneider *et al.*, 1989) (also see Section XII).

VIII. Hydroxyurea–FUra

Inhibitors of ribonucleotide reductase such as HU have enhanced the anti-tumor activity of FUra in experimental systems (Frankfurt, 1973; Moran *et al.*, 1982). More recently, Kobayashi and Hoshino (1983) demonstrated potentiation of FUra cytotoxicity on 9L rat brain tumor cells by HU when given after subtoxic doses of FUra (Table 1). Such potentiation occurred with other ribonucleotide reductase inhibitors and was blocked by dThd. These experiments suggest that drugs that lower dUMP levels via inhibition of ribonucleotide reductase, by preventing rescue by normal nucleotides, may potentiate FUra effects and that this enhancement is schedule dependent.

Several clinical trials have been published using different doses and schedules of FUra and HU in colorectal cancer patients (Kao *et al.*, 1984; Krein *et al.*, 1978; Lavin *et al.*, 1980; Lerner, 1974; Lokich *et al.*, 1975). One of these included the sequence which was suggested as optimal in preclinical studies (Kao *et al.*, 1984). Although the regimen did not appear to increase FUra toxicity, response rates for the treatment of bowel cancer did not differ from those observed when FUra was used as a single agent or in analogous combinations (Table 9).

IX. Dipyridamol–FUra

Dipyridamol (DIP) inhibits nucleoside transport without blocking the transport of nucleobases in a variety of cell types (Henderson and Zomber, 1977; Paterson *et al.*, 1980; Plagemann and Wohlbocke, 1980). When it was first combined with FUra, it was thought that this property might enhance the cytotoxicity of the fluoropyrimidine base by preventing nucleoside salvage and the repletion of nucleoside pools (Grem and Fischer, 1985) (Table 1). In a human colon cancer cell line (HCT-116) DIP strongly augmented the growth inhibition and lethality produced by FUra. However, inhibition of dThd transport and the subsequent depletion of dTTP pools did not account for the enhanced toxicity (Grem and Fischer, 1985). It was later suggested that DIP inhibition of nucleoside efflux may have augmented the synthesis of FdUrd from FUra by enhancing the availability of deoxyribose phosphate donors (Grem and Fischer, 1986) (Table 1). Intracellular concentrations of FdUMP were also increased when DIP was given with FdUrd (Schwartz et al., 1987). Preliminary data suggest that DIP enhances FUra toxicity to normal cells and tissues more than to tumors (Lee and Park, 1987) while having the opposite effect on FdUrd toxicity (Schwartz et al., 1987).

The combination of DIP–FUra has not yet received sufficient clinical investigation. In a phase I study of a 72 hr continuous iv infusion of DIP plus FUra, there was no apparent increase in toxicity, although all of the patients experienced headache due to the very high dose of DIP used (Grem and Fischer, 1986) (Table 9). In an attempt to maximize TS inhibition, the combination of oral DIP and iv FUra–LV has been explored in phase I–II studies (Allen et al., 1987; Budd et al., 1989; Schmoll et al., 1988). In a phase II study by Allen et al. (1987) using DIP 75 mg/m^2 po q 8 hr days 1–6, LV 200 mg/m^2 iv days 2–6, and FUra 370 mg/m^2 iv days 2–6, responses were noted in one of seven patients previously treated with chemotherapy and in four out of six previously untreated patients. The three-drug regimen was well tolerated. In 35 patients with advanced colorectal cancer treated with DIP 75 mg po q 12 hr days 1–5, LV 300 mg/m^2 iv and FUra 600 mg/m^2 days 1–3 or 2–4 every 3–4 weeks a 36% response rate was observed (Schmoll et al., 1988). This was not substantially higher than that usually observed with LV–FUra. When given with DIP 50 mg/m^2 po q 6 hr, days 0–6, and LV 200 mg/m^2 iv bolus, days 1–5, the MTD of FUra was the same as the MTD of FUra + LV without DIP (i.e., 375 mg/m^2) (Budd et al., 1989). Because DIP is highly protein-bound, current studies involve dose escalation of DIP with pharmacologic monitoring in order to determine the MTD of oral DIP when given with FUra + LV (at dosages $\geq$ 100 mg/m^2 po q 6 hr days 0–6) (Budd et al., 1989). Oral DIP was also added to FdUrd in the treatment of metastatic colorectal cancer (Schwartz et al., 1987). Three out of five patients achieved an objective response while toxicity was mild.

X. Purines–FUra

Complex interactions might occur between purines and FUra. Either Ino or dIno in combination with FUra have been shown to increase the growth inhibitory potency of FUra against a variety of cell lines in vitro (Evans et al., 1981; Gotto et al., 1969; Peters et al., 1987a) but in one cell line, mouse T-cell lymphosarcoma S49, Ino and dIno protected cells from the effects of FUra (Ullman and Kirsch, 1979). Guo (Iigo et al., 1982) or Guo ribonucleotides (Iigo et al., 1984, 1987) or other purine nucleosides and deoxyribonucleosides (Iigo et al., 1982; Santelli and Valeriote, 1980) markedly enhanced the antitumor activity of FUra without increasing toxicity to the host in various murine tumor systems. The potentiation by Guo may be due to an increase in FUra nucleotides caused by both pyrimidine phosphorylase and uridine–cytidine kinase with increased ribose 1-phosphate derived from Guo (Iigo et al., 1984) (Table 1).

Kessel and Hall (1969) reported that glucose and Ino strongly stimulate the conversion of FUra to ribonucleosides both in vitro and in vivo but this had little effect on the survival time of tumor bearing mice. More recently these observation have been extended to FdUrd (Iigo et al., 1984, 1988b). These findings however have not yet been applied clinically.

XI. Allopurinol–FUra

Preclinical work has shown that 4-hydroxy-pyrazolo-pyrimidine (HPP, allopurinol) can interfere with the biochemical action of FUra, thereby diminishing the toxicity of the latter in both gastrointestinal tissues and bone marrow in rodents (Schwartz et al., 1980; Houghton and Houghton, 1980). The effect is attributed to the inhibition of orotidine 5'-monophosphate decarboxylase (ODCase) by the major metabolite of HPP, oxipurinol (Schwartz et al., 1980). This leads to accumulation of orotic acid within the cell which then interferes with the conversion of FUra to FUMP via a key enzyme, OPRTase (Table 1).

Therefore, different tissues may demonstrate different degrees of modulation of FUra cytotoxicity. If this results in the selective protection of normal tissues, HPP could provide a means of increasing the therapeutic index of FUra in those tumors that may metabolize it predominantly by Urd phosphorylase and kinase.

In vitro investigations using animal tumor cell lines in culture have demonstrated that HPP antagonizes the cytotoxicity of FUra in some lines (Schwartz and Handschumacher, 1979). Studies using three human tumor cell lines in semisolid media have demonstrated however that oxipurinol either did not significantly effect FUra cytotoxicity or increased FUra cytotoxicity (Garewal et al., 1983).

An increase in the therapeutic index of FUra with HPP has been observed in mice bearing the colon tumor 38 but not in P388 leukemia (Schwartz et al., 1980), nor in human colon adenocarcinoma xenografts (Houghton and Houghton, 1982). It is probable that different metabolic pathways of FUra activation exist in different cell lines. Recently Berne et al. (1987) measured the formation of FdUMP and dTMP synthase inhibition in regenerating liver and in tumor tissue in rats bearing colon carcinoma following FUra administration with or without HPP. The highest levels of FdUMP were found in tumor tissue 30 min after administration of FUra. The maximal FdUMP levels in liver were only 4% of those attained in tumor tissue. Rats pretreated with HPP had a maximal FdUMP level in tumor tissue of only 42% of controls at 60 min, while the FdUMP levels in regenerating liver were not affected. These results suggest that activation of FUra to FUMP by OPRTase was quantitatively important in colon carcinoma, but not in rat liver. This finding contradicts earlier impressions that OPRTase was important for FUra activation in normal tissues but not tumors.

Several phase I–II clinical studies of modulation of FUra by HPP have been performed (Ahmann *et al.*, 1986; Campbell *et al.*, 1982; Fine *et al.*, 1982; Fox *et al.*, 1979, 1981; Howell *et al.*, 1981, 1983; Kroener *et al.*, 1982; Woolley *et al.*, 1985) (Table 9). FUra in these studies was given either as an intermittent bolus or as prolonged continuous infusion. Studies of a 5-day continuous infusion of HPP throughout the entire period of FUra administration have shown protection in each case, with mucositis as the dose-limiting toxicity (Ahmann *et al.*, 1986; Fine *et al.*, 1982., Fox *et al.*, 1979, 1981; Howell *et al.*, 1981).

A phase I study has demonstrated that concurrent HPP therapy permitted a doubling of the FUra MTD given by continuous infusion, leading to a 4-fold increase in the tolerated concentration for time exposure to FUra (Howell *et al.*, 1981). These results were also confirmed by later studies (Erlichman *et al.*, 1986).

In the intermittent bolus studies, neurotoxicity was more prominent than mucositis (Campbell *et al.*, 1982; Howell *et al.*, 1983; Woolley *et al.*, 1985). HPP modulated the toxicity of intermittent bolus FUra when it was given every day of the 5-day schedule (Campbell *et al.*, 1982). Shorter courses of HPP (only 1 to 24 hr after FUra bolus) did not protect against the toxicity (i.e., myelosuppression, mucositis, neurotoxicity) of intermittent bolus high-dose FUra (up to 1.9 gm/m^2) (Howell *et al.*, 1983). When HPP administration was maintained for a longer time, presumably continuing inhibition of ODCase and maintaining inhibitory levels of orotic acid during persistence of FUra intracellularly, intermittent bolus FUra-induced myelodepression was diminished, thereby allowing a 50% increase in the MTD of FUra (i.e., 1800 mg/m^2) (Woolley *et al.*, 1985).

Most patients treated with a combination of HPP and FUra had colorectal cancer. In this disease the combination has not yet shown any clinical advantage (Ahmann *et al.*, 1986; Fine *et al.*, 1982; Fox *et al.*, 1981; Kroener *et al.*, 1982). In a few studies, patients with other malignancies were also included and clinical responses were noted (Howell *et al.*, 1983; Kroener *et al.*, 1982). This combination might be proposed for further clinical evaluation when more is known about FUra metabolism in human tumors and healthy tissues.

XII. Uridine–FUra

Biochemical modulation of FUra with high-dose Urd rescue was based on the hypothesis that, while antitumor activity of FUra is mainly caused by inhibition of TS, its toxicity depends primarily upon its incorporation into RNA

(Houghton *et al.*, 1979). It has been demonstrated that when high-dose Urd is given several hours after FUra, the binding of FdUMP to TS will not be affected. However, increased pools of Urd triphosphate (UTP) will replace FUTP in RNA (Martin *et al.*, 1982) (Table 1). Klubes *et al.* (1982), giving a 5-day sc infusion of Urd after a single ip dose of FUra, were able to triple the LD10 in mice. Also, the therapeutic efficacy of FUra against B16 melanoma was increased by using high-dose FUra plus delayed infusion of Urd (Klubes and Cerna, 1983). Treatment of L1210 leukemia with the same combination was not more effective than FUra at MTD.

Martin *et al.* (1982) administered high-dose Urd bolus injections to mice bearing the FUra-resistant colon tumor 26 starting 2 hr after FUra. In a combination with PALA and MMPR, the MTD of FUra was increased by the presence of Urd, resulting in improved therapeutic index.

The hematological toxicity of FUra was diminished when FUra was followed by delayed Urd administration (Martin *et al.*, 1982). Also, in bone marrow the recovery of DNA synthesis was markedly increased by Urd (Martin *et al.*, 1982). This effect might be related to observed increased conversion of Urd to nucleotides under normal physiological nucleotide levels, (2–5 mmol/liter) in bone marrow cells when compared with other tissues (Moyer *et al.*, 1981). In animals and patients treated with high doses, Urd metabolism in bone marrow cells occurs at even higher rates since cells are probably in direct contact with much higher Urd plasma levels (10–20 mmol/liter in mice; 300–1,000 μmol/liter in humans) (Peters *et al.*, 1987b,c, 1988; Van Groeningen *et al.*, 1989). Consequently a high ratio of total Urd nucleotides to fluorouridine nucleotides might be observed in bone marrow cells compared with solid tumors; Urd pools in several tissues are usually higher than in plasma due to a concentrative mechanism for Urd uptake (Darnowsky and Handschumacher, 1986). In colon tumor 38, however, the increase in Urd nucleotides was less than in other tissues, with consequent lack of Urd effect on FUra sensitivity (Peters *et al.*, 1988); this phenomenon might occur in human tumors.

Other observations by Ullman and Kirsch (1979) and by Parker and Klubes (1985) have demonstrated that simultaneous administration of Urd with FUra enhanced the cytotoxicity of FUra against a mouse T-cell lymphoma cell line (S-49). This potentiation was due at least in part to formation of UTP from exogenous Urd, which acted as a feedback inhibitor of *de novo* pyrimidine biosynthesis. The inhibition of pyrimidine biosynthesis presumably decreases orotic acid levels and thereby increases the availability of PRPP and OPRTase for anabolism of FUra to FdUMP.

The clinical application of Urd rescue is currently under investigation (Table 9). A single dose of Urd can be administered safely as a 1-hr infusion up to a dose of 12 gm/m^2 which results in peak plasma levels of 2 mmol/liter (Leyva *et al.*, 1984). Despite achievement of active peak Urd levels (>1 mmol/liter), no protective effects on bone marrow suppression of FUra were observed. Tissue exposure to Urd was probably limited with this schedule due to the rapid elimination half-life of Urd. Subsequently Urd was administered as a continuous infusion, but treatment had to be discontinued due to the occurrence of high fever (Peters *et al.*, 1987c; Van Groeningen *et al.*, 1986).

An intermittent schedule of Urd administration was therefore used to prevent the development of fever (Van Groeningen *et al.*, 1986, 1989). With this schedule, peak Urd levels of 0.5–1.0 mmol/liter were achieved. Urd administration in 3-hr infusion periods alternated with a 3-hr infusion-free period over 72 hr at a dose of 2 gm/m^2/hr induced the reversal of leukopenia but not thrombocytopenia induced by weekly treatments with FUra at a MTD of 720 mg/m^2.

Since intermittent iv administration might be unpractical, oral administration of Urd is currently under investigation for its potency to reverse FUra induced myelosuppression (Van Groeningen *et al.*, 1987). Although bioavailability of Urd after oral administration was only 7% compared with sc administration in mice (Klubes *et al.*, 1986), Urd could provide rescue while cleotides in normal tissues has been recently proposed by Darnowsky and (Martin *et al.*, 1989).

An alternative approach to the selective increase in pools of Urd nucleotides in normal tissues has been recently proposed by Darnowsky and Handschumacher (1985), consisting of the administration of benzylacyclouridine (BAU). This potent inhibitor of Urd phosphorylases can delay the disappearance of Urd from plasma, selectively affect the salvage and use of Urd for nucleic acid biosynthesis in normal tissues, and alone or in combination with Urd can enhance the therapeutic efficacy of FUra without increasing its toxicity in mice bearing colon tumor 38 (Darnowsky and Handschumacher, 1985). In a tumor-bearing murine model, BAU given concomitantly with oral Urd enabled a 50% reduction of the fluoropyrimidine dose without affecting the therapeutic index (Martin *et al.*, 1989). Although the clinical potential of this modality has yet to be investigated, oral Urd plus oral BAU appears to be a promising alternative to parenteral administration of Urd for selective rescue of FUra toxicity.

XIII. Conclusions

Biochemical modulation has recently become a reliable approach to developing more effective drug combinations in cancer chemotherapy. Prediction of synergistic interactions between fluoropyrimidines and various metabolites and antimetabolites based on knowledge of their mechanisms of action and metabolic pathways has been tested *in vitro*, in animal tumor models, and in clinical trials. On the basis of preclinical information, several combination regimens which are characterized by an improved therapeutic index over single-agent FUra in terms of tumor response (i.e., sequential MTX–FUra, LV–FUra, and PALA–FUra) have now been employed in the clinic. The attainment of clear survival benefits in patients with advanced malignancies receiving FUra with modulating agents must await confirmation from further clinical studies. It is conceivable, however, that the optimal setting in which to expect benefit from the synergistic modulation of FUra is the adjuvant treatment of tumors, as demonstrated by the NSABP study for sequential MTX–FUra in node-negative, estrogen receptor-negative breast cancer (Fisher *et al.*, 1989). Research with other promising modulator–effector pairs is underway (e.g., FUra–Urd rescue).

Preclinical data also suggest the possibility of utilizing fluoropyrimidines as modulating agents of other antineoplastic drugs (e.g., cytarabine, iododeoxyuridine).

While improved therapeutic efficacy in FUra sensitive tumors has been demonstrated, the possibility of converting, by appropriate modulation, fluoropyrimidines to active drugs in tumors with little or no inherent sensitivity or acquired resistance during treatment remains to be demonstrated.

Along with a precise laboratory and clinical strategy for the development of modulating combinations, additional information on FUra metabolism and sites of action in human tumor cells of different origin as well as in normal tissues are needed for complete elucidation of these complex drug interactions and for further clinical applications aimed to improve FUra selectivity.

References

Ahmann, F. R., Garewal, H. X., and Greenberg, B. R. (1986). Phase II trial of high-dose continuous infusion 5-fluorouracil with allopurinol modulation in colon cancer. *Oncology* **43**, 83–85.

Ajani, J., Kanojia, M. D., Bedikian, A., Korinek, J., Stein, S. H., Espinoza, E. G., and Bodey, G. P. (1985). Sequential methotrexate and 5-fluorouracil in the primary treatment of metastatic colorectal carcinoma. *Am. J. Clin. Oncol.* **8**, 69–71.

Ajani, J. A., Goudeau, P., Levin, B., Faintuch, J. S., Abbruzzese, J. L., Boman, B. M., and Kanojia, M. D. (1989). Phase II study of Adriamycin with sequential methotrexate and 5-fluorouracil (AMF) in gastric carcinoma. *Cancer Chemother. Pharmacol.* **24**, 41–44.

Allegra, J. C. (1983). Methotrexate and 5-fluorouracil following tamoxifen and Premarin in advanced breast cancer. *Semin. Oncol.* **10** (Suppl. 2), 23–28.

Allegra, C. J., Chabner, B. A., Drake, J. C., Lutz, R., Rodbard, D., and Jolivet, J. A. (1985). Enhanced inhibition of thymidylate synthetase by methotrexate polyglutamates. *J. Biol. Chem.* **260**, 9720–9726.

Allen, S., Fine, S., and Erlichman, C. (1987). A phase II trial of 5-fluorouracil (5FU) and folinic acid (FA) plus dipyridamole (D) in patients with metastatic colorectal cancer (MCC). *Proc. Am. Assoc. Clin. Oncol.* **6**, 95.

Anukarahanonta, T., Holstege, A., and Keppler, D. O. R. (1980). Selective enhancement of

5-fluorouridine uptake and action in rat hepatomas *in vivo* following pretreatment with D-galactosamine and 6-azauridine or N-(phosphonacetyl-L-aspartate. *Eur. J. Cancer* **10**, 1171–1180.

Arbuck, S. G., Trave, F., Douglass, H. O., Jr., Nava, H., Zakrzewski, S., and Rustum, Y. M. (1986). Phase I and pharmacologic studies of intraperitoneal leucovorin and 5-fluorouracil in patients with advanced cancer. *J. Clin. Oncol.* **4**, 1510–1517.

Arbuck, S. G., Douglass, H. O., Jr., Trave, F., Milliron, S., Baroni, M., Nava, H., Emrich, L. J., and Rustum, Y. M. (1987). A phase II trial of 5-fluorouracil and high-dose intravenous leucovorin in gastric carcinoma. *J. Clin. Oncol.* **5**, 1150–1156.

Ardalan, B., Glazer, R. I., Kensler, T. W., Jayaram, H. N., Pham, T. V., MacDonald, J. S., and Cooney, D. A. (1981). Synergistic effects of 5-fluorouracil and N-(phosphonacetyl)-L-aspartate on cell growth and ribonucleic acid synthesis in a human mammary carcinoma. *Biochem. Pharmacol.* **30**, 2045–2049.

Ardalan, B., Jamin, D., Jayaram, H. N., and Presant, C. A. (1984). Phase I study of continuous-infusion PALA and 5-FU. *Cancer Treat. Rep.* **68**, 531–534.

Ardalan, B., Singh, G., and Silberman, H. (1988). A randomized phase I and II study of short-term infusion of high-dose fluorouracil with or without N-(phosphonacetyl)-L-aspartic acid in patients with advanced pancreatic and colorectal cancers. *J. Clin. Oncol.* **6**, 1053–1058.

Asbury, R. F., Boros, L., Brower, M., Woll, J., Chang, A., and Benett, J. (1987). 5-Fluorouracil and high-dose folic acid treatment for metastatic colon cancer. *Am. J. Clin. Oncol.* **10**, 47–49.

Au, J. L.-S., Rustum, Y. M., Ledesma, E. J., Mittelman, A., and Creaven, P. J. (1982). Clinical pharmacological studies of concurrent infusion of 5-fluorouracil and thymidine in treatment of colorectal carcinoma. *Cancer Res.* **42**, 2930–2937.

Barone, C., Astone, A., Garufi, C., Grieco, A., Cavallaro, A., Netri, G., Rossi, S., Cassano, A., Ricevuto, E., Noviello, M. R., and Gambassi, G. (1987). High-dose folinic acid (HDFA) combined with 5-fluorouracil (5-FU) in first line chemotherapy of advanced large bowel cancer. *Eur. J. Cancer Clin. Oncol.* **23**, 1303–1306.

Becher, R., Kurschel, E., Wandl, U., Klohe, O., Scheulen, M., Weinhardt, O., Hoffken, K., Niederle, N., Khan, H., Bergner, S., Sauerwein, W., Ruther, U., and Schmidt, C. G. (1988). 5-Fluorouracil and 5-formyltetrahydrofolate in advanced malignancies. *In* "The Expanding Role of Folates and Fluoropyrimidines in Cancer Chemotherapy" (Y. Rustum and J. J. McGuire, eds.), pp. 225–231. Plenum, New York.

Bedikian, A. Y., Stroelein, J. R., Karlin, D. A., Bennetts, R. W., Bodey, G. P., and Valdivieso, M. (1981). Chemotherapy for colorectal cancer with a combination of PALA and 5-FU. *Cancer Treat. Rep.* **65**, 747–753.

Benz, C., and Cadman, E. (1981). Modulation of 5-fluorouracil metabolism and cytotoxicity by antimetabolite in human colorectal adenocarcinoma, HCT-8. *Cancer Res.* **41**, 994–999.

Benz, C., Tillis, T., Tattelman, E., and Cadman, E. (1982). Optimal scheduling of methotrexate and 5-fluorouracil in human breast cancer. *Cancer Res.* **42**, 2081–2086.

Benz, C., Sambol, N., Yawitz, B., Wilbur, B., and De Gregorio, M. (1985). Serum purines and pyrimidines correlate with methotrexate (MTX) and 5-fluorouracil (5-FU) response and toxicity. *Proc. Am. Soc. Clin. Oncol.* **4**, 43.

Berenberg, J. L., Goodman, P. J., Oishi, N., Fleming, T., Natale, R. B., Hutchins, L. H., Guy, G. T., and MacDonald, J. (1989). 5-Fluorouracil (5FU) and folinic acid (FA): For the treatment of metastatic gastric cancer. *Proc. Am. Soc. Clin. Oncol.* **8**, 101.

Berger, S., and Hakala, M. T. (1984). Relationship of dUMP and free FdUMP pools to inhibition of thymidylate synthase by 5-fluorouracil. *Mol. Pharmacol.* **25**, 303–309.

Berne, M., Gustavsson, B., Almersjo, O., Spears, C. P., and Weldestrom, J. (1987). Concurrent allopurinol and 5-fluorouracil: 5-Fluoro-2'-deoxyuridylate formation and thymidylate synthase inhibition in rat colon carcinoma and in regenerating rat liver. *Cancer Chemother. Pharmacol.* **20**, 193–197.

Bertino, J. R. (1982). Keynote address—Adjuvant chemotherapy and cancer cure. *Int. J. Oncol., Biol. Phys.* **8,** 109–113.

Bertino, J. R., and Mini, E. (1987). Does modulation of 5-fluorouracil by metabolites or antimetabolites work in the clinic? *In* "New Avenues in Developmental Cancer Chemotherapy" (K. R. Harrap and T. A. Connors, eds.), pp. 163–184. Academic Press, Orlando, Florida.

Bertino, J. R., Sawicki, W. L., Lindquist, C. A., and Gupta, V. S. (1977). Schedule-dependent antitumor effects of methotrexate and 5-fluorouracil. *Cancer Res.* **37,** 327–328.

Bertino, J. R., Mini, E., and Fernandes, D. J. (1983). Sequential methotrexate and 5-fluorouracil: Mechanism of synergy. *Semin. Oncol.* **10** (Suppl. 2), 2–5.

Bertrand, R., and Jolivet, J. (1989). Lack of interference by the unnatural isomer of 5-formyltetrahydrofolate with the effects of the natural isomer in leucovorin preparations. *JNCI, J. Natl. Cancer Inst.* **81,** 1175–1178.

Bertrand, M., Doroshow, J. H., Multhauf, P., Newman, E., Blayney, D., Carr, B. I., and Goldberg, D. (1984). Combination chemotherapy with high-dose continuous infusion leucovorin and bolus 5-fluorouracil in patients with advanced colorectal cancer. *In* "Advances in Cancer Chemotherapy. The Current Status of 5-Fluorouracil–Leucovorin Calcium Combination" (H. W. Bruckner and Y. M. Rustum, eds.), pp. 73–76. Wiley, New York.

Bertrand, M., Doroshow, J. H., Multhauf, P., Blayney, D. W., Carr, B. I., Cecchi, G., Goldberg, D., Leong, L., Margolin, K., Metter, G., and Staples, R. (1986). High-dose continuous infusion folinic acid and bolus 5-fluorouracil in patients with advanced colorectal cancer: A phase II study. *J. Clin. Oncol.* **4,** 1058–1061.

Blakley, R. L. (1969). Biochemistry and pharmacology of folate analogues. *In* "The Biochemistry of Folic Acid and Related Pteridines" (R. L. Blakley, ed.), p. 454. Elsevier/North-Holland, Amsterdam.

Blumenreich, M. S., Woodcock, T. M., Allegra, M., Richman, S. P., Kubota, T. T., and Allegra, J. C. (1982). Sequential therapy with methotrexate (MTX) and 5-fluorouracil (5-FU) for adenocarcinoma of the colon. *Proc. Am. Soc. Clin. Oncol.* **1,** 102.

Brenckman, W. D., Jr., Laufman, L. R., Morgan, E. D., Stydnicki, K. K., Collier, M. A., and Knick, V. D. (1988). Pilot trial of 5FU with high doses of oral Wellcovorin (LV) tablets in advanced colorectal carcinoma. *Proc. Am. Soc. Clin. Oncol.* **7,** 98.

Browman, G. P., Archibald, S. D., Young, J. E. M., Hryniuk, W. M., Russell, R., Kiehl, K., and Levine, M. N. (1983). Prospective randomized trial of one hour sequential versus simultaneous methotrexate plus 5-fluorouracil in advanced and recurrent squamous cell head and neck cancer. *J. Clin. Oncol.* **1,** 787–792.

Browman, G. P., Levine, M. N., Russell, R., Young, Y. E., and Archibald, S. D. (1986). Survival results from a phase III study of simultaneous versus 1-hour sequential methotrexate–5-fluorouracil chemotherapy in head and neck cancer. *Head Neck Surg.* **8,** 146–152.

Browman, G. P., Levine, M. N., Goodyear, M. D., Russell, R., Archibald, S. D., Jackson, B. S., Young, J. E. M., Basrur, V., and Johanson, C. (1988). Methotrexate/fluorouracil scheduling influences normal tissue toxicity but not antitumor effects in patients with squamous cell head and neck cancer: Results from a randomized trial. *J. Clin. Oncol.* **6,** 963–968.

Brown, I., and Ward, H. W. C. (1978). Therapeutic consequences of antitumor drug interactions: Methotrexate and 5-fluorouracil in the chemotherapy of C3H mice with transplanted mammary adenocarcinoma. *Cancer Lett.* **5,** 291–297.

Bruckner, H., and Cohen, J. (1983). MTX/5-FU trials in gastrointestinal and other cancers. *Semin. Oncol.* **10** (Suppl. 2), 32–39.

Bruckner, H. W., Ohnuma, T., Hart, R., Jaffrey, I., Spiegelman, M., Ambinder, E., Storch, J. A., Wilfinger, C., Goldberg, J., Biller, H., and Holland, J. F. (1982). Leucovorin (LV) potentiation of 5-fluorouracil (FU) efficiency and potency. *Proc. Am. Assoc. Cancer Res.* **23,** 111.

Bruckner, H. W., Roboz, J., Spiegelman, M., Ambinder, E., Hart, R., and Holland, J. F. (1983). An efficient leucovorin and 5-fluorouracil sequence: Dosage escalation and pharmacological monitoring. *Proc. Am. Assoc. Cancer Res.* **24,** 138.

Bruckner, H. W., Crown, J., McKenna, A., and Hart, R. (1988). Leucovorin and 5-fluorouracil as a treatment for disseminated cancer of the pancreas and unknown primary tumors. *Cancer Res.* **48,** 5570–5572.

Budd, G. T., Schreiber, M. J., Steiger, E., Bukowski, R. M., and Wick, J. K. (1986). Phase I trial of intraperitoneal chemotherapy with 5-fluorouracil and citrovorum factor. *Invest. New Drugs* **4,** 155–158.

Budd, G. T., Fleming, T. R., Bukowski, R. M., McCracken, J. D., Rivkin, S. E., O'Bryan, R. M., Balcerzak, S. P., and MacDonald, J. S. (1987). 5-Fluorouracil and folinic acid in the treatment of metastatic colorectal cancer: A randomized comparison. A Southwest Oncology Group Study. *J. Clin. Oncol.* **5,** 272–277.

Budd, G. T., Bukowski, R. M., Murthy, S., Boyett, J., and Jayaraj, A. (1989). Phase I trial of dipyridamole (DPM) with 5- fluorouracil (5FU) and leukovorin (LV). *Proc. Am. Soc. Clin. Oncol.* **8,** 67.

Burnet, R., Smith, F. P., Hoerni, B., Lagarde, C., and Schein, P. (1981). Sequential methotrexate-5-fluorouracil in advanced measurable colorectal cancer: Lack of appreciable therapeutic synergism. *Proc. Am. Assoc. Cancer Res.* **22,** 370.

Buroker, T. A., Moertel, C. G., Fleming, T. R., Everson, L. K., Cullinan, S. A., Krook, J. E., Mailliard, J. A., Marschke, R. F., Klaasen, D. J., Laurie, J. A., and Moon, M. D. (1985). A controlled evaluation of recent approaches to biochemical modulation or enhancement of 5-fluorouracil therapy in colorectal carcinoma. *J. Clin. Oncol.* **3,** 1624–1631.

Byrne, P. J., Treat, J., McFadden, M., McVie, G., Huiniuk, T. B., Schein, P. S., and Wooley, P. V. (1984). Therapeutic efficacy of the combination of 5-fluorouracil and high-dose leucovorin in patients with advanced colorectal carcinoma: Single daily intravenous dose for five days. *In* "Advances in Cancer Chemotherapy. The Current Status of 5-Fluorouracil–Leucovorin Calcium Combination" (H. W. Bruckner and Y. M. Rustum, eds.), pp. 65–67. Wiley, New York.

Cadman, E., Heimer, R., and Davis, L. (1979). Enhanced 5-fluorouracil nucleotide formation after methotrexate administration: Explanation for drug synergism. *Science* **205,** 1135–1137.

Cadman, E., Benz, C., Heimer, R., and O'Shaughnessy, J. (1981a). The effect of de novo purine synthesis inhibitors on 5-fluorouracil metabolism and cytotoxicity. *Biochem. Pharmacol.* **30,** 2469–2478.

Cadman, E., Benz, C., Heimer, R., and O'Shaughnessy, J. (1981b). The modulation of 5-fluorouracil metabolism by inhibitors of *de novo* purine synthesis. *In* "Nucleosides and Cancer Treatment" (M. H. N. Tattersall and R. M. Fox, eds.), pp. 242–250. Academic Press, New York.

Cadman, E., Heimer, R., and Benz, C. (1981c). The influence of methotrexate pretreatment on 5-fluorouracil metabolism in L1210 cells. *J. Biol. Chem.* **256,** 1695–1704.

Campbell, T. N., Howell, S. B., Pfeifle, C., and House, B. A. (1982). High-dose allopurinol modulation of 5-FU toxicity: Phase I trial of an outpatient dose schedule. *Cancer Treat. Rep.* **66,** 1723–1727.

Canobbio, L., Nobile, M. T., Ardizzoni, A., Tatarek, R., and Rosso, R. (1986). Phase II study of sequential methotrexate and 5-FU combination in the treatment of advanced colorectal cancer. *Cancer Treat. Rep.* **70,** 419–420.

Cantrell, J. R., Jr., Burnet, R., Lagarde, C., Schein, P. S., and Smith, F. P. (1982). Phase II study of sequential methotrexate–5-FU therapy in advanced measurable colorectal cancer. *Cancer Treat. Rep.* **66,** 11563–1565.

Cantwell, B. M. J., and Harris, A. L. (1988). The efficacy of 5-fluorouracil in human colorectal cancer is not enhanced by thymidylate synthetase inhibition with CB3737 (N^{10}-propargyl-5,8 dideazafolic acid). *Br. J. Cancer* **58,** 189–190.

Casper, E. S., Vale, K., Williams, L. J., Martin, D. S., and Young, C. W. (1983). Phase I and clinical pharmacological evaluation of biochemical modulation of 5-fluorouracil with *N*-(phosphonacetyl)-L-aspartic acid. *Cancer Res.* **43**, 2324–2329.

Casper, E. S., Michaelson, R. A., Kemeny, N., Martin, D. S., and Young, C. W. (1984). Phase I evaluation of a biochemically designed combination: PALA, thymidine, and 5-FU. *Cancer Treat. Rep.* **68**, 539–541.

Chang, Y.-M., and Bertino, J. R. (1989). Enhancement of fluoropyrimidine inhibition of cell growth by leucovorin and deoxynucleosides in a human squamous cell carcinoma cell line. *Cancer Invest.* **47**, 557–563.

Chiuten, D. F., Valdivieso, M., Bedikian, A., Bodey, G. P., and Freireich, E. J. (1985). Phase I–II clinical trial of thymidine, 5-FU and PALA given in combination. *Cancer Treat. Rep.* **69**, 611–613.

Coates, A. S., Tattersall, M. H. N., Swanson, C., Hedley, D., Fox, R. M., and Raghavan, D. (1984). Combination therapy with methotrexate and 5-fluorouracil: A prospective randomized clinical trial of order of administration. *J. Clin. Oncol.* **2**, 756–761.

Collins, K. D., and Stark, G. R. (1971). Aspartate transcarbamylase, interaction with the transition state analogue *N*-(phosphonacetyl)-L-aspartate. *J. Biol. Chem.* **246**, 6599–6605.

Costanzi, J. J., Gagliano, R., Coltman, C. A., Jr., and Bickers, J. N. (1979). 5-Fluorouracil in the treatment of acute leukemia: A Southwest Oncology Group Study. *Cancer Treat. Rep.* **63**, 2126–2128.

Covey, J. M., and Straw, J. A. (1983). Nonlinear pharmacokinetics of thymidine, thymine and fluorouracil and their kinetic interactions in normal dogs. *Cancer Res.* **43**, 4587–4595.

Cunningham, J., Bukowski, R. M., Budd, G. T., Weick, J. K., and Purvis, J. (1984). 5-Fluorouracil and folinic acid: A phase I–II trial in gastrointestinal malignancy. *Invest. New Drugs* **2**, 391–395.

Damon, L. E., and Cadman, E. C. (1988). The metabolic basis for combination chemotherapy. *Pharmacol. Ther.* **38**, 73–127.

Damon, L. E., Cadman, E., and Benz, C. (1989). Enhancement of 5-fluorouracil antitumor effects by the prior administration of methotrexate. *Pharmacol. Ther.* **43**, 155–185.

Danenberg, P. V., and Danenberg, K. D. (1978). Effect of 5,10-methylenetetrahydrofolate on the dissociation of 5-fluoro-2'-deoxyuridylate from thymidylate synthase: Evidence for an ordered mechanism. *Biochemistry* **17**, 4018–4024.

Danenberg, P. V., and Lockshin, A. (1981). Fluorinated pyrimidines as tight-binding inhibitors of thymidylate synthetase. *Pharmacol. Ther.* **13**, 69–90.

Danenberg, P. V., Langenbach, J. U., and Heidelberger, C. (1974). Structures of reversible and irreversible complexes of thymidylate synthetase and fluorinated pyrimidine nucleotides. *Biochemistry* **13**, 926–933.

Danhauser, L. L., Heimer, R., and Cadman, E. (1985). Lack of enhanced cytotoxicity of cultured L1210 cells using folinic acid in combination with sequential methotrexate and fluorouracil. *Cancer Chemother. Pharmacol.* **15**, 214–219.

Darnowsky, J. W., and Handschumacherl R. E. (1985). Tissue-specific enhancement of uridine uridine utilization and 5-fluorouracil therapy in mice by benzylacyclouridine. *Cancer Res.* **45**, 5364–5368.

Darnowsky, J. W., and Handschumacher, R. E. (1986). Tissue uridine pools: Evidence *in vivo* of a concentrative mechanism for uridine uptake. *Cancer Res.* **46**, 3490–3494.

DeCaprio. J. A., Arbuck, S. G., and Mayer, R. J. (1989). Phase II study of weekly 5-fluorouracil (5FU) and folinic acid (FA) in previously untreated patients with unresectable, measurable pancreatic adenocarcinoma. *Proc. Am. Soc. Oncol.* **8**, 100.

De Gramont, A., Krulik, M., Cady, J., Lagadec, B., Maisani, J.-E., Loiseau, J.-P., Grange, J.-D., Gonzales-Canali, G., Demuynck, B., Louvet, C., Seroka, J., Dray, C., and Debray, J. (1988). High-dose folinic acid and 5-fluorouracil bolus and continuous infusion in advanced colorectal cancer. *Eur. J. Cancer Clin. Oncol.* **24**, 1499–1503.

Dickinson, R., Presgrave, P., Levi, J., Milliken, S., and Woods, R. (1989). Sequential moderate-dose methotrexate and 5-fluorouracil in advanced gastric adenocarcinoma. *Cancer Che-mother. Pharmacol.* **24**, 67–68.

Di Costanzo, F., Bartolucci, R., Sofra, M., Calabresi, F., Malacarne, P., Belsanti, V., Boni, C., and Bacchi, M. (1989). 5-Fluorouracil (5-FU) alone vs high dose folinic acid (FA) and 5-FU in advanced colorectal cancer (CA): A randomized trial of the Italian Oncology Group for Clinical Research (GOIRC). *Proc. Am. Soc. Clin. Oncol.* **8**, 106.

Doroshow, J. H., Bertrand, M., Multhauf, P., Leong, L., Goldberg, D., Margolin, K., Carr, B., Akman, S., and Hill, R. (1987). Prospective randomized trial comparing 5-FU versus (VS) 5-FU and high-dose folinic acid (HDFA) for treatment of advanced colorectal cancer. *Proc. Am. Soc. Clin. Oncol.* **6**, 96.

Doroshow, J., Leong, L., Margolin, K., Flanagan, B., Goldberg, D., Bertand, M., Akman, S., Carr, B., Odujinrin, O., Newman, E., and Lichtfield, T. (1989). Refractory metastatic breast cancer: Salvage therapy with fluorouracil and high-dose continuous infusion leucovorin calcium. *J. Clin. Oncol.* **7**, 439–444.

Drapkin, R., Griffiths, E., McAloon, E., Paladine, W., Sokol, G., and Lyman, G. (1981). Sequential methotrexate (MTX) and 5-fluorouracil (5-FU) in adenocarcinoma of the colon and rectum. *Proc. Am. Assoc. Cancer Res.* **22**, 453.

Drapkin, R., McAloon, E., and Lyman, G. (1983). Sequential methotrexate (MTX) and 5-fluorouracil (5-FU) in advanced measurable colorectal cancer. *Proc. Am. Soc. Clin. Oncol.* **2**, 118.

Eisenhower, E. A., Bowman, D. H., Pritchard, K. I., Paterson, A. H., Ragaz, J., Geggie, P. H. S., and Maxwell, I. (1984). Tamoxifen and conjugated estrogens (Premarin) followed by sequences methotrexate and 5-FU in refractory advanced breast cancer. *Cancer Treat. Rep.* **68**, 1421–1422.

Ellison, J., Bernarth, A. M., Gallagher, J. G., Porter, P. A., Rine, K. T., and Lewis, G. O. (1985). Toxicity without benefit for sequential MTX/5FU. *Proc. Am Soc. Clin. Oncol.* **4**, 26.

Erlichmann, C. (1988). 5-Fluorouracil (FUra) and folinic acid (FA) therapy in patients with colorectal cancer. *In* "The Expanding Role of Folates and Fluoropyrimidines in Cancer Chemotherapy" (Y. Rustum and J. J. McGuire, eds.), pp. 185–189. Plenum, New York.

Erlichman, C., Donehower, R. C., Speyer, J. L., Klecker, R., and Chabner, B. A. (1982). Phase I–phase II trial of N-phosphonacetyl-L-aspartic acid given by intravenous infusion and 5-fluorouracil given by bolus injection. *JNCI, J. Natl. Cancer Inst.* **68**, 227–231.

Erlichman, C., Fine, S., and Elhakim, T. (1986). Plasma pharmacokinetics of 5-FU given by continuous infusion with allopurinol. *Cancer Treat. Rep.* **70**, 903–904.

Erlichman, C., Fine, S., Wong, A., and Elhakim, T. (1988). A randomized trial of fluorouracil and folinic acid in patients with metastatic colorectal carcinoma. *J. Clin. Oncol.* **6**, 469–475.

Evans, R. M., Laskin, J. D., and Hakala, M. T. (1981). Effects of excess folates and deoxyino-sine on the activity and site of action of 5-fluorouracil. *Cancer Res.* **41**, 3288–3295.

Fernandes, D. J., and Bertino, J. R. (1980). 5-Fluorouracil–methotrexate synergy. Enhance-ment of 5-fluorodeoxyuridine binding to thymidylate synthetase by dihydropteroylpolygluta-mates. *Proc. Natl. Acad. Sci. U.S.A.* **77**, 5663–5667.

Fine, S., Erlichman, C., Klecker, R., and Myers, C. E. (1982). Phase I–II trial of high dose 5-FU infusion with allopurinol in metastatic colorectal cancer. *Proc. Am. Assoc. Cancer Res.* **23**, 93.

Fine, S., Erlichman, C., Kaizer, L., Warr, D., and Elhakim, T. (1988). Phase II trial of 5-FU + folinic acid (FA) as first line treatment for metastatic breast cancer. *Proc. Am. Soc. Clin. Oncol.* **7**, 41.

Fisher, B., Redmond, C., Browman, D., Legault-Poisson, S., Wickerham, D. L., Wolmark, N., Fisher, E. R., Margolese, R., Sutherland, C., Glass, A., Foster, R., Caplan, R., and others (1989). A randomized clinical trial evaluating sequential methotrexate and fluorouracil in the treatment of patients with node-negative breast cancer who have estrogen-receptor-negative tumors. *N. Engl. J. Med.* **320**, 473–478.

Fox, R. M., Woods, R. L., Tattersall, M. H. N., and Brodie, G. M. (1979). Allopurinol modulation of high dose fluorouracil toxicity. *Lancet* **1**, 677.

Fox, R. M., Woods, R. L., Tattersall, M. H. N., Piper, A. A., and Sampson, D. (1981). Allopurinol modulation of fluorouracil toxicity. *Cancer Chemother. Pharmacol.* **5**, 151–155.

Frankfurt, O. S. (1973). Enhancement of the antitumor activity of 5-fluorouracil by drug combinations. *Cancer Res.* **33**, 1043–1047.

Garwal, H. S., Ahmann, F. R., and Alberts, D. S. (1983). Lack of inhibition of oxipurinol of 5-fluorouracil toxicity against human tumor cell lines. *Cancer Treat. Rep.* **67**, 495–498.

Gewirtz, A. M., and Cadman, E. (1981). Preliminary report on the efficacy of sequential methotrexate and 5-fluorouracil in advanced breast cancer. *Cancer (Philadelphia)* **47**, 2552–2555.

Glazer, R. I., and Lloyd, L. S. (1982). Association of cell lethality with incorporation of 5-fluorouracil and 5-fluorouridine into nuclear RNA in human colon carcinoma cells in culture. *Mol. Pharmacol.* **21**, 468–473.

Glazer, R. I., and Peale, A. L. (1979). The effects of 5-fluorouracil on the synthesis of nuclear RNA in L1210 in vitro. *Mol. Pharmacol.* **16**, 270–277.

Glimelius, B., Ginman, C., Graffman, S., Pahlman, L., and Stahle, E. (1986). Sequential methotrexate–5-FU–leucovorin (MFL) in advanced colorectal cancer. *Eur. J. Cancer Clin. Oncol.* **22**, 295–300.

Gotto, A. M., Belkhode, M. L., and Touster, O. (1969). Stimulatory effects of inosine and deoxyinosine on the incorporation of uracil-2^{14}C, 5-fluorouracil-2^{14}C, and 5-bromouracil-2^{14}C into nucleic acid by Ehrlich ascites tumor cells *in vitro*. *Cancer Res.* **29**, 807–811.

Goulian, M., Bleile, B., and Tseng, B. Y. (1980). Methotrexate-induced misincorporation of uracil into DNA. *Proc. Natl. Acad. Sci. U.S.A.* **77**, 1956–1960.

Greene, H., Desai, A., Levick, S., and Tester, W. (1986). Combined 5-fluorouracil infusion and high dose folinic acid in the treatment of metastatic gastrointestinal cancer. *Proc. Am. Soc. Clin. Oncol.* **5**, 89.

Grem, J. L., and Fischer, P. H. (1985). Augmentation of 5-fluorouracil cytotoxicity in human colon cancer cells by dipyridamole. *Cancer Res.* **45**, 2967–2972.

Grem, J. L., and Fisher, P. H. (1986). Alteration of fluorouracil metabolism in human colon cancer cells by dipyridamole with a selective increase in fluorodeoxyuridine monophosphate levels. *Cancer Res.* **46**, 6191–6199.

Grem, J. L., Hoth, D. F., Hamilton, J. M., King, S. A., and Leyland-Jones, B. (1987a). Overview of current status and future direction of clinical trials with 5-fluorouracil in combination with folinic acid. *Cancer Treat. Rep.* **71**, 1249–1264.

Grem, J. L., Shoemaker, D. D., Petulli, N. J., and Douglass, H. O., Jr. (1987b). Severe life-threatening toxicities observed in study using leucovorin with 5-fluorouracil. *J. Clin. Oncol.* **5**, 1704.

Grem, J. L., King, S. A., O'Dwyer, P. J., and Leyland-Jones, B. (1988). Biochemistry and clinical activity of *N*-(phosphonacetyl)-L-aspartate: A review. *Cancer Res.* **48**, 4441–4454.

Hamm, J. T., Seeger, J., Joseph, G., Blumenreich, M., Sheth, S. P., Woodcock, T. M., Gentile, P. S., Kellihan, M. J., Sherril, L., Allegra, J. C., and Hendler, F. J. (1989). Treatment of advanced breast cancer with 5-fluorouracil and folinic acid. *Proc. Am. Soc. Clin. Oncol.* **8**, 42.

Hansen, R. M., Ritch, P. S., and Anderson, T. (1986). Sequential methotrexate, 5-fluorouracil, and calcium leucovorin in colorectal carcinoma. *Am. J. Clin. Oncol.* **9**, 352–354.

Hartman, K. U., and Heidelberger, C. (1961). Studies on fluorinated pyrimidines. XIII—Inhibition of thymidylate synthetase. *J. Biol. Chem.* **236**, 3006–3013.

Hedley, D. W. (1987). DNA flow cytometric study of 5-fluorouracil used to treat end stage non-Hodgkin's lymphoma. *Br. J. Cancer* **55**, 107–108.

Heidelberger, C., Chaudhuri, N. K., Danenberg, P., Mooren, D. X., and Griesbach, L. (1957). Fluorinated pyrimidines, a new class of tumor-inhibitory compounds. *Nature (London)* **179**, 663–666.

Heidelberger, C., Danenberg, P. V., and Moran, R. G. (1983). Rluorinated pyrimidines and their nucleosides. *Adv. Enzymol.* **54**, 57–119.

Henderson, J. F. (1963). Feedback inhibition of purine biosynthesis in ascites tumor cells by purine analogues. *Biochem. Pharmacol.* **12**, 551–556.

Henderson, J. F., and Zomber, G. (1977). Effects of dipyridamole on nucleotide synthesis from adenine in murine tumor cells. *Biochem. Pharmacol.* **26**, 2455–2465.

Henderson, J. F., Cadwell, J. C., and Paterson, A. R. P. (1967). Decreased feedback inhibition in a 6-p(methylmercapto)purine ribonucleoside resistant tumor. *Cancer Res.* **27**, 1773–1778.

Heppner, G. H., and Calabresi, P. (1977). Effect of sequence of administration of methotrexate, leucovorin, and 5-fluorouracil on mammary tumor growth and survival in syngeneic C3H mice. *Cancer Res.* **37**, 4580–4583.

Herrmann, R., Manegold, C., Schroeder, M., Tigges, F. J., Bartsch, H., Jungi, F., and Fritze, D. (1984a). Sequential methotrexate and 5-FU in breast cancer resistant to conventional application of these drugs. *Cancer Treat. Rep.* **68**, 1279–1281.

Herrmann, R., Spehn, J., Beyer, J. H., von Franque, U., Schmieder, A., Holzmann, K., and Abel, U. (1984b). Sequential methotrexate and 5-fluorouracil: Improved response rate in metastatic colorectal cancer. *J. Clin. Oncol.* **2**, 591–594.

Herrmann, R., Kunz, W., Osswald, H., Ritter, M., and Port, R. (1985). The effect of methotrexate treatment on 5-fluorouracil kinetics in sarcoma 180 *in vivo. Eur. J. Cancer Clin. Oncol.* **211**, 753–758.

Herrmann, R., Knuth, A., Kleeberg, U., Middecke, R., and Abel, U. (1986). Randomized multicenter trial of sequential methotrexate (MTX) and 5-fluorouracil (FU) vs FU alone in metastatic colorectal carcinoma (CRC). *Proc. Am. Soc. Clin. Oncol.* **5**, 91.

Hines, J. D., Zakem, M. H., Adelstein, D. J., and Rustum, Y. M. (1988). Treatment of advanced stage colorectal adenocarcinoma with fluorouracil and high-dose leucovorin calcium: A pilot study. *J. Clin. Oncol.* **6**, 142–146.

Hines, J. D., Adelstein, D. J., Spiess, J. L., Girowski, P., and Carter, S. G. (1989). Efficacy of high-dose oral leucovorin and 5-fluorouracil in advanced colorectal carcinoma. Plasma and tissue pharmacokinetics. *Cancer* **63**, 1022–1025.

Houghton, J. A., and Houghton, P. J. (1980). 5-Fluorouracil in combination with hypoxanthine and allopurinol: Toxicity and metabolism in xenografts of human colonic carcinomas in mice. *Biochem. Pharmacol.* **29**, 2077–2080.

Houghton, J. A., and Houghton, P. J. (1982). Combination of 5-FU, hypoxanthine, and allopurinol in chemotherapy for human colon adenocarcinoma xenografts. *Cancer Treat. Rep.* **66**, 1201–1206.

Houghton, J. A., Houghton, P. J., and Wooten, R. S. (1979). Mechanism of induction of gastrointestinal toxicity in the mouse by 5-fluorouracil, 5-fluorouridine, and 5-fluoro-2'-deoxyuridine. *Cancer Res.* **39**, 2406–2413.

Houghton, J. A., Maroda, S. J., Jr., Phillips, J. O., and Houghton, P. J. (1981). Biochemical determinants of responsiveness to 5-fluorouracil and its derivatives in xenografts of human colorectal adenocarcinomas in mice. *Cancer Res.* **41**, 144–149.

Houghton, J. A., Torrance, P. M., Radparvar, S., Williams, L. G., and Houghton, P. J. (1986a). Binding of 5-fluorodeoxyuridylate to thymidylate synthase in human colon adenocarcinoma xenografts. *Eur. J. Cancer Clin. Oncol.* **22**, 505–510.

Houghton, J. A., Weiss, K. D., Williams, L. G., Torrance, P. M., and Houghton, P. J. (1986b). Relationship between 5-fluoro-2'-deoxyuridylate, 2'-deoxyuridylate, and thymidylate synthase activity subsequent to 5-fluorouracil administration, in xenografts of human colon adenocarcinomas. *Biochem. Pharmacol.* **35**, 1351–1358.

Houghton, J. A., Williams, L. G., de Graf, S. S. N., Cheshire, P. J., Wainer, I. W., Jadaud, P., and Houghton, P. J. (1989). Comparison of the conversion of 5-formyltetrahydrofolate and 5-methyltetrahydrofolate to 5,10-methylenetetrahydrofolates and tetrahydrofolates in human colon tumors. *Cancer Commun.* **1**, 167–174.

Howard, C. T., Hook, K. E., and Leopold, W. R. (1987). Sequence and schedule dependent synergy of trimetrexate in combination with 5-fluorouracil or cisplatin. *Proc. Am. Assoc. Cancer Res.* **28,** 417.

Howell, S. B., Wung, W. W., Taetle, R., Hussain, F., and Romine, J. S. (1981). Modulation of 5-fluorouracil toxicity by allopurinol in man. *Cancer (Philadelphia)* **48,** 1281–1289.

Howell, S. B., Pfeifle, C. E., and Wung, W. E. (1983). Effect of allopurinol on the toxicity of high-dose 5-fluorouracil administered by intermittent bolus injection. *Cancer (Philadelphia)* **51,** 220–225.

Hudes, G. R., Lacreta, F., De Lap, R. J., Grillo-Lopez, A. J., Catalano, R., and Comis, R. L. (1989). Phase I clinical and pharmacologic trial of trimetrexate in combination with 5-fluorouracil. *Cancer Chemother. Pharmacol.* **24,** 117–122.

Huennekens, F. M., Duffy, T. H., and Vitols, K. S. (1987). Folic acid metabolism and its disruption by pharmacologic agents. *Natl. Cancer Inst. Monogr.* **5,** 1–8.

Iigo, M., and Hoshi, A. (1984). Effect of guanosine on antitumor activity of fluorinated pyrimidines against P338 leukemia. *Cancer Chemother. Pharmacol.* **13,** 86–90.

Iigo, M., Ando, N., Hoshi, A., and Kuretani, K. (1982). Effect of pyrimidines, purines and their nucleotides on antitumor activity of 5-fluorouracil against L1210 leukemia. *J. Pharm. Dyn.* **5,** 515–520.

Iigo, M., Yamaizumi, Z., Nishimura, S., and Hoshi, A. (1987). Mechanism of potentiation of antitumor activity of 5-fluorouracil by guanine ribonucleotides against adenocarcinoma 755. *Eur. J. Cancer Clin. Oncol.* **23,** 1059–1065.

Iigo, M., Araki, E., Nakajima, Y., Hoshi, A., and DeClerck, E. (1988a). Enhancing effect of bromovinyldeoxyuridine on antitumor activity of 5-fluorouracil against adenocarcinoma 755 in mice. Increased therapeutic index and correlation with increased plasma 5-fluorouracil levels. *Biochem. Pharmacol.* **37,** 1609–1613.

Iigo, M., Yamaizumi, Z., Nakajuma, Y., Nishimura, S., and Hoshi, A. (1988b). Mechanisms of potentiation of antitumor activity of 5-fluoro-2'-deoxyuridine in the adenocarcinoma 755 system by guanosine 5'-monophosphate. *Drugs Exp. Clin. Res.* **13,** 257–263.

Jabboury, K., Holmes, F., Kau, S., and Hortobaygi, G. (1989). Folinic acid modulation of low-dose fluorouracil (FU) infusion in refractory breast cancer: A dose optimization study. *Proc. Am. Soc. Clin. Oncol.* **8,** 40.

Jacobs, C. (1982). Use of methotrexate and 5-FU for recurrent head and neck cancer. *Cancer Treat. Rep.* **66,** 1925–1928.

Johnson, R. K., Clement, J. J., and Howard, W. S. (1980). Treatment of murine tumors with 5-fluorouracil in combination with de novo pyrimidine synthesis inhibitors PALA or pyrazofurine. *Proc. Am. Assoc. Cancer Res.* **21,** 292.

Kaizer, L., Fine, S., and Erlichman, C. (1986). Sequential methotrexate-fluorouracil: The role of leucovorin in improving tumor response. *J. Clin. Oncol.* **4,** 1280–1281.

Kamen, B. A., Eibl, B., Cashmore, A. R., and Bertino, J. R. (1984). Uptake and efficacy of trimetrexate (TMQ, 2,4-diamino-5-methyl-6-[(3,4,5-trimethoxyanilino)methyl]quinazoline), a non-classical antifolate in methotrexate-resistant leukemia cell *in vitro. Biochem. Pharmacol.* **33,** 1697–1699.

Kao, A. F., Muggia, F. M., Dubin, N., Lerner, V. A., Stark, R., Wernz, J. C., Speyer, J. L., and Blum, R. H. (1984). Evolution of a sequential 5-FU and hydroxyurea combination in advanced bowel cancer. *Cancer Treat. Rep.* **68,** 1383–1385.

Kaplan, H., and Rivkin, S. (1984). 5-Fluorouracil (5-FU) and leucovorin (LV) in the treatment of metastatic colon carcinoma. *Proc. Am. Assoc. Cancer Res.* **25,** 144.

Kaye, S. B., Sangster, G., Hutcheon, A., Habeshaw, T., Crossling, F., Ferguson, C., McArdle, C., Smith, D., George, W. D., and Kalman, K. C. (1984). Sequential methotrexate plus 5-FU in advanced breast and colorectal cancers: A phase II study. *Cancer Treat. Rep.* **68,** 547–548.

Kelsen, D., Coit, D., Houston, C., Martin, D., Sawyer, R., and Colofiore, J. (1989). Phase II trial of PALA, methotrexate (MTX), fluorouracil (FU), and leucovorin (L) in adenocarcinoma of unknown primary site (ACUP). *Proc. Am. Soc. Clin. Oncol.* **8,** 293.

Kemeny, N. E., Ahmed, T., Michaelson, R. A., Harper, H. D., and Yip, J. L. C. (1984). Activity of sequential low-dose methotrexate and fluorouracil in advanced colorectal carcinoma: Attempts at correlation with tissue and blood levels of phosphoribosylpyrophosphate. *J. Clin. Oncol.* **2**, 311–315.

Kemeny, N., Cohen, A., Bertino, J. R., Sigurdson, E., and Oderman, P. (1988). A phase I study of continuing intrahepatic infusion of floxuridine (FUDR) and leucovorin (LV) via an Infusaid pump for the treatment of hepatic metastases from colorectal carcinoma. *Proc. Am. Soc. Clin. Oncol.* **7**, 99.

Kemeny, N., Schneider, A., Martin, D. S., Colofiore, J., Sawyer, R. C., Derby, S., and Salvia, B. (1989). Phase I trial of *N*-(phosphonacetyl)-L-aspartate, methotrexate, and 5-fluorouracil with leucovorin rescue in patients with advanced cancer. *Cancer Res.* **49**, 4636–4639.

Kessel, D., and Hall, T. C. (1969). Influence of ribose donors on the action of 5-fluorouracil. *Cancer Res.* **29**, 1749–1754.

Keyomarsi, K., and Moran, R. C. (1986). Folinic acid augmentation of the effects of fluoropyrimidines on murine and human leukemic cells. *Cancer Res.* **46**, 5229–5235.

Kirkwood, J. M., Ensminger, W., Rosowsky, A., Papathanasopoulos, N., and Frei, E., III (1980). Comparison of pharmacokinetics of 5- fluorouracil and 5-fluorouracil with concurrent thymidine infusion in a phase I trial. *Cancer Res.* **40**, 107–113.

Klein, H. O., Wickramanayake, P. D., Schultz, V., Mohr, R., Oerkemann, H., and Farrokh, G. R. (1984). 5-Fluorouracil, Adriamycin and methotrexate: A combination protocol (FAMETH) for the treatment of metastasized stomach cancer. *In* "Fluoropyrimidines in Cancer Therapy" (K. Kimura, S. Fuji, G. P. Ogawa, G. P. Bodey, and P. Alberto, eds.), pp. 280–287. Elsevier, Amsterdam.

Klubes, P., and Cerna, I. (1983). Use of uridine rescue to enhance the antitumor selectivity of 5-fluorouracil. *Cancer Res.* **43**, 3182–3186.

Klubes, P., and Leyland-Jones, B. (1989). Enhancement of the antitumor activity of 5-fluorouracil by uridine rescue. *Pharmacol. Ther.* **41**, 289–302.

Klubes, P., Cerna, I., and Meldon, M. A. (1981). Effect of concurrent calcium leucovorin infusion on 5-fluorouracil cytotoxicity against murine L1210 leukemia. *Cancer Chemother. Pharmacol.* **6**, 121–125.

Klubes, P., Cerna, I., and Meldon, M. A. (1982). Uridine rescue from the lethal toxicity of 5-fluorouracil in mice. *Cancer Chemother. Pharmacol.* **8**, 17–21.

Klubes, P., Geffen, D. B., and Cysyk, R. L. (1986). Comparison of the bioavailability of uridine in mice after oral or parenteral administration. *Cancer Chemother. Pharmacol.* **17**, 236–240.

Kobayashi, S., and Hoshino, T. (1983). Combined effects of low dose 5-fluorouracil and hydroxyurea on 9L cells *in vitro*. *Cancer Res.* **43**, 5309–5313.

Krein, B. N., Conroy, J. F., and Brodsky, I. (1978). Sequential 5-fluorouracil and hydroxyurea therapy for metastatic colorectal adenocarcinoma. *J. Am. Osteopath. Assoc.* **77**, 604–605.

Kroener, J. F., Saleh, F., and Howell, S. B. (1982). 5-FU and allopurinol: Toxicity modulation and phase II results in colon cancer. *Cancer Treat. Rep.* **66**, 1133–1137.

Kufe, D. W., and Egan, E. M. (1981). Enhancement of 5-fluorouracil incorporation into human lymphoblastic ribonucleic acid. *Biochem. Pharmacol.* **30**, 129–133.

Kufe, D. W., Scott, P., Fram, R., and Major, P. (1983). Biologic effect of 5-fluoro-2′-deoxyuridine incorporation into DNA. *Biochem. Pharmacol.* **32**, 1337–1340.

LaBianca, R., Pancera, G., Beretta, G., Aitini, E., Barni, S., Beretta, A., Beretta, G. D., Cesana, B., Comella, G., Cozzaglio, L., Cristoni, M., Deponte, A., Frontini, L., Geccherle, A., Gottardi, O., Martignoni, G., Smeireri, F., Zadro, A., Zaniboni, A., and Luporini, G. (1989). A randomized study of intravenous fluorouracil ± folinic acid (f) in advanced metastatic colorectal carcinoma (AMC). *Proc. Am. Soc. Clin. Oncol.* **8**, 118.

Laufman, L. R., Krzeczowski, K. A., Roach, R., and Segal, M. (1987). Leucovorin plus 5-fluorouracil: An effective treatment for metastatic colon cancer. *J. Clin. Oncol.* **5**, 1394–1400.

Laufman, L., Patterson, D., Gordon, P., Stydnicki, K. A., and Hicks, W, (1989a). Long term continuous infusion (CI) leucovorin (LV)–floxuridine (FUDR): A phase I trial. *Proc. Am. Soc. Clin. Oncol.* **8,** 82.

Laufman, L. R., Brenckman, W. D., Jr., Collier, M. A., Sullivan, B. A., and the Colorectal Multicenter Group (1989b). Interim analysis of a phase III randomized, double-blind, placebo-controlled trial of 5-fluorouracil (FU) with oral leucovorin (LV) vs FU alone in metastatic colorectal cancer. *Proc. Am. Soc. Clin. Oncol.* **8,** 125.

Lavin, P., Mittleman, A., Douglass, H., Jr., Engstrom, P., and Klaassen, D. (1980). Survival and response to chemotherapy for advanced colorectal carcinoma. *Cancer* **46,** 1536–1543.

Lee, M. W., and Park, C. H. (1987). Potentiation of methotrexate (MTX) and 5-FU effect on human cancer and normal bone marrow cells by dypyridamole (DIP). *Proc. Am. Assoc. Cancer Res.* **28,** 410.

Lee, Y. N. and Kwaja, T. A. (1977). Adjuvant postoperative chemotherapy with 5-fluorouracil and methotrexate: Effect of schedule of administration on metastasis of 13762 mammary adenocarcinoma. *J. Surg. Oncol.* **9,** 469–479.

Leone, B. A., Romero, A., Rabinowich, M. G., Perez, J. E., Macchiavelli, M., and Strauss, E. (1986). Sequential therapy with methotrexate and 5-fluorouracil in treatment of advanced colon carcinoma. *J. Clin. Oncol.* **4,** 23–27.

Leopold, W. R., Dykes, D. D., and Griswold, D. P. (1987). Therapeutic synergy of trimetrexate (CI-898) in combination with doxorubicin, vincristine, cytoxan, 6-thioguanine, cisplatin or 5-fluorouracil against intraperitoneally implanted P388 leukemia. *Natl. Cancer Inst. Monogr.* **5,** 99–104.

Lerner, H. J. (1974). Clinical experience using hydroxyurea–5-fluorouracil in hepatic metastatic from adenocarcinoma of the colon. *Proc. Am. Assoc. Cancer Res. Am. Soc. Clin. Oncol.* **15,** 32.

Leyland-Jones, B., and O'Dwyer, P. J. (1986). Biochemical modulation: application of laboratory models to the clinic. *Cancer Treat. Rep.* **70,** 219–229.

Leyva, A., Van Groeningen, L. J., Gall, M., Krall, I., Peters, G. J., Lankelma, J. E., and Pinedo, H. M. (1984). Phase I and pharmacokinetic studies of high-dose uridine intended for rescue from 5-fluorouracil toxicity. *Cancer Res.* **44,** 5928–5933.

Liang, C.-M., Donehower, R. C., and Chabner, B. A. (1982). Biochemical interactions between N-(phosphonacetyl-aspartate and 5-fluorouracil *Mol. Pharmacol.* **21,** 224–230.

Loeffler, T. M., Lindemann, J., Luckhaupt, H., Rose, K. G., and Hausmen, T. U., (1988). Chemotherapy of advanced and relapsed squamous cell cancer of the head and neck with split-dose cis-platinum (DDP), 5-fluorouracil (FUra) and leucovorin (CF). In "The Expanding Role of Folates and Fluoropyrimidines in Cancer Chemotherapy" (Y. Rustum and J. J. McGuire, eds.), pp. 267–283. Plenum, New York.

Lokich, J. J., Pitman, S. W., and Skarin, A. T., (1975). Combined 5-fluorouracil and hydroxyurea therapy for gastrointestinal cancer. *Oncology* **32,** 34–37.

Lönn, U., and Lönn, S. (1986a). DNA lesions in human neoplastic cells and cytotoxicity of 5-fluoropyrimidines. *Cancer Res.* **46,** 3866–3871.

Lönn, U., and Lönn, S. (1986b). The increased cytotoxicity in colon adenocarcinoma of methotrexate–5-fluorouracil is not associated with increased induction of lesions in DNA by 5-fluorouracil. *Biochem. Pharmacol.* **35,** 177–181.

Loprinzi, C. L., Ingle, J. N., Schaid, D. J., Buckner, J. C., and Edmonson, J. H. (1988). Progress report on a phase II trial of 5-fluorouracil plus citrovorum factor in women with metastatic breast cancer. In "The Expanding Role of Folates and Fluoropyrimidines in Cancer Chemotherapy" (Y. Rustum and J. J. McGuire, eds.), pp. 255–259. Plenum, New York.

Lynch, G., Kemeny, N., Chun, H., Martin, D., and Young, C. (1985). Phase I evaluation and pharmacokinetic study of weekly iv thymidine and 5-FU in patients with advanced colorectal cancer. *Cancer Treat. Rep.* **69,** 169–184.

Machover, D., Schwarzenberg, L., Goldschmidt, E., Tourani, J. M., Michalski, B., Hayat, M.

Dorval, T., Misset, J. L., Asmin, C., Maral, R., and Mathé, G. (1982). Treatment of advanced colorectal and gastric adenocarcinomas with 5-FU combined with high-dose folinic acid: A pilot study. *Cancer Treat. Rep.* **66**, 1803–1807.

Machover, D., Goldschmidt, E., Chollet, P., Metzger, G., Zittoun, J., Marquet, J., Vandenbulcke, J.-M., Misset, J.-I., Schwarzenberger, L., Fourtillan, J. B., Gaget, H., and Mathé, G. (1986). Treatment of advanced colorectal and gastric adenocarcinomas with 5-FU and high dose folinic acid. *J. Clin. Oncol.* **4**, 685–690.

MacIntosh, J., Coates, A., Swanson, C., Raghavan, D., and Tattersall, M. H. N. (1987). Chemotherapy of advanced colorectal cancer. A randomized trial of sequential methotrexate and 5-fluorouracil. *Am. J. Clin. Oncol.* **10**, 210–212.

Madajewicz, S., Petrelli, N., Rustum, Y. M., Campbell, J., Herrera, L., Mittelman, A., Perry, A., and Creaven, P. J. (1984). Phase I–II trial of high-dose calcium leucovorin and 5-fluorouracil in advanced colorectal cancer. *Cancer Res.* **44**, 4667–4669.

Mahajan, S. L., Ajani, J. A., Kanojia, M. H., Veilenkoop, L., and Bodey, G. P. (1983). Comparison of two schedules of sequential high-dose methotrexate (MTX) and 5-fluorouracil (5-FU) for metastatic colorectal carcinoma. *Proc. Am. Soc. Clin. Oncol.* **2**, 122.

Major, P. P., Egan, E. M., Sargent, L., and Kufe, D. W., (1982). Modulation of 5-fluorouracil metabolism in human MCF-7 breast carcinoma cells. *Cancer Chemother. Pharmacol.* **8**, 87–91.

Mandel, H. D. (1969). The incorporation of 5-fluorouracil in RNA and its molecular consequences. *Prog. Mol. Subcell. Biol.* **1**, 82–135.

Mann, G. B., Hortobagyi, G. N., Buzdar, K. V., Yapo, H.-U. P., and Valdivieso, M. (1985). A comparative study of PALA, PALA plus 5-FU and 5-FU in advanced breast cancer. *Cancer (Philadelphia)* **56**, 1320–1324.

Margolin, K., Doroshow, J., Leong, L., Akman, S., and Carr, B. (1989). 5-Fluorouracil (5FU) and high dose folinic acid (HDFA) as initial therapy of advanced breast cancer (BrCA). *Proc. Am. Soc. Clin. Oncol.* **8**, 38,

Marini, G., Simoncini, E., Zaniboni, A., Gorni, F., Marpicati, P., and Zaniboni, A. (1987a). 5-Fluorouracil and high-folinic acid as salvage treatment of advanced breast cancer: An update. *Oncology* **44**, 336–340.

Marini, G., Zaniboni, A., Gorni, P., Marpicati, P., Montini, E., and Simoncini, E. (1987b). Clinical experience with 5-fluorouracil (5-FU) and high-dose folinic acid in solid tumors. *Drugs Exp. Clin. Res.* **13**, 373–378.

Marsh, J. C., Davis, C., O'Halloren, K., Durivage, H., Voynick, I., Pasquale, D., Simonich, S., and Bertino, J. R., (1988). Phase II study of 5-fluoro-2′-deoxyuridine (FUdR) and leucovorin (LV) in advanced refractory colorectal cancer. *Proc. Am. Soc. Clin. Oncol.* **7**, 107.

Marsh, J. C., Bertino, J. R., Rome, L. S., Capizzi, R. L., Katz, K. H., Richards, F. H., Koletsky, A. J., Durivage, H. J., Makuch, R. W., O'Hallaren, K., and Davis, C. A. (1989). Sequential methotrexate (MTX), 5-fluorouracil (FU) and leucovorin (LV) in metastatic colorectal cancer: A controlled comparison of two intervals between drug administration. *Proc. Am. Soc. Clin. Oncol.* **8**, 103.

Martin, D. S., (1987). Biochemical modulation: Perspectives and objectives. *In* "New Avenues in Developmental Cancer Chemotherapy" (K. R. Harrap and T. A. Connors, eds.), pp. 163–184. Academic Press, Orlando, Florida.

Martin, D. S., and Stolfi, R. L. (1977). Thymidine (TdR) enhancement of antitumor activity of 5-fluorouracil (FU) against advanced murine (CD8F1) breast carcinoma. *Proc. Am. Assoc. Cancer Res. Am. Soc. Clin. Oncol.* **18**, 126.

Martin, D. S., Nayak, R., Sawyer, R. C., Stolfi, R. L., Young, C. W., Woodcock, T., and Spiegelman, S. (1978a). Enhancement of 5-fluorouracil chemotherapy with emphasis on the use of excess thymidine. *Cancer Bull.* **30**, 219–224.

Martin, D. S., Stolfi, R. L., and Spiegelman, S. (1978b). Striking augmentation of the *in vivo* anti-cancer activity of 5-fluorouracil (FU) by combination with pyrimidine nucleosides: An RNA effect. *Proc. Am. Assoc. Cancer Res.* **19**, 221.

Martin, D. S., Stolfi, R. L., Sawyer, R. C., Nayak, R., Spiegelman, S., Young, C., and Woodcock, T. (1980). An overview of thymidine. *Cancer (Philadelphia)* **45**, 1117–1128.

Martin, D. S., Stolfi, R. L., Sawyer, R. C., Nayak, R., Spiegelman, S., Schmid, F., Heimer, R., and Cadman, E., (1981). Biochemical modulation of 5-fluorouracil and cytosine arabinoside with emphasis on thymidine, PALA, and 6-methylmercaptopurine riboside. *In* "Nucleosides and Cancer Treatment" (M. H. N. Tattersall and R. M. Fox, eds.), pp. 339–382, Academic Press, New York.

Martin, D. S., Stolfi, R. L., Sawyer, R. C., Spiegelman, S., and Young, C. W. (1982). High-dose 5-fluorouracil with delayed uridine "rescue" in mice. *Cancer Res.* **42**, 3964–3970.

Martin, D. S., Stolfi, R. L., Sawyer, R. C., Spiegelman, S., Casper, E. S., and Young, C. W. (1983a). Therapeutic utility of utilizing low doses of *N*-(phosphonoacetyl)-L-aspartic acid in combination with 5-fluorouracil: A murine study with clinical relevance. *Cancer Res.* **43**, 2317–2321.

Martin, D. S., Stolfi, R. L., Sawyer, R. C., Spiegelman, S., and Young, C. W. (1983b). Improved therapeutic index with sequential *N*-phosphonacetyl-L-aspartate plus high-dose methotrexate plus high-dose 5-fluorouracil and appropriate rescue. *Cancer Res.* **43**, 4653–4661.

Martin, D. S., Stolfi, R. L., and Colofiore, J. R., (1988). Failure of high-dose leucovorin to improve therapy with the maximally tolerated dose of 5-fluorouracil: A murine study with clinical relevance? *JNCI, J. Natl. Cancer Inst.* **80**, 496–501.

Martin, D. S., Stolfi, R. L., and Sawyer, R. C., (1989). Use of oral uridine as a substitute for parenteral uridine rescue of 5-fluorouracil therapy, with and without the uridine phosphorylase inhibitor 5-benzylacyclouridine. *Cancer Chemotherapy. Pharmacol.* **24**, 9–14.

McGuire, J. J., Hsieh, P., Coward, J. K., and Bertino, J. R. (1980). Enzymatic synthesis of folylpolyglutamates. Characterization of the reaction and its products. *J. Biol. Chem.* **225**, 5776–5788.

McGuire, J. J., Mini, E., Hsieh, P., and Bertino, J. R., (1985). Role of methotrexate polyglutamylation in methotrexate and sequential methotrexate–5-fluorouracil mediated cell kill. *Cancer Res.* **45**, 6395–6400.

Mehrotra, S., Rosenthal, C. J., and Gardner, B. (1982). Biochemical modulation of antineoplastic response in colorectal carcinoma: 5-fluorouracil (F), high dose methotrexate (M) with calcium leucovorin (L) rescue (FML) in two sequences of administration. *Proc. Am. Soc. Clin. Oncol.* **1**, 100.

Mehta, B.M., Gisolfi, A. L., Hutchinson, D. J., Nirenberg, A., Kellick, M. G., and Rosen, G. (1978). Serum distribution of citrovorum factor and 5-methyltetrahydrofolate following oral and i.m. administration of calcium leucovorin in normal adults. *Cancer Treat. Rep.* **62**, 345–350.

Meshad, M. W., Ervin, T. J., Kufe, D , Johnson, R. K., Blum, R. H., and Frei, E., III (1981). Phase I trial of combination therapy with PALA and 5-FU. *Cancer Treat. Rep.* **65**, 331–334.

Mini, E., and Bertino, J. R., (1982). Time and dose relationships for methotrexate (MTX), fluorouracil (FUra) and combinations of these drugs for maximum cell kill in human CCRF–CEM cells. *Proc. Am. Assoc. Cancer Res.* **23**, 181.

Mini, E., Moroson, B. A., Blair, O. C., McGuire, J. J., and Bertino, J. R., (1983a). Time, dose and sequence relationships for methotrexate, fluorouracil, and their combinations for optimal cell kill in the human T-lymphoid cell line CCRF–CEM. *Chemioterapia* **2** (Suppl. 5), 452–453.

Mini, E., Moroson, B. A., Cashmore, A. R., and Bertino, J. R., (1983b). Synergistic effects of sequential methotrexate and 5-fluorouracil treatment on the growth of human B- and T-lymphoblast cell lines: Modulation by horse and fetal bovine serum. *In* "Proceedings of the 13th International Congress of Chemotherapy" (K. H. Spitzy and K. Karrer, eds.), Part 262, pp. 5–8. Egermann Druckereigesellschaft, Vienna.

Mini, E., Moroson, B. A., Franco, C. T., and Bertino, J. R. (1985). Cytotoxic effects of folate antagonists against methotrexate-resistant human leukemic lymphoblast CCRF–CEM cell lines. *Cancer Res.* **45,** 325–330.

Mini, E., Mazzei, T., Coronnello, M., Criscuoli, L., Gualtieri, M., Periti, P., and Bertino, J. R. (1987a). Effects of 5-methyltetrahydrofolate on the activity of fluoropyrimydines against human leukemia (CCRF–CEM) cells. *Biochem. Pharmacol.* **36,** 2905–2911.

Mini, E., Moroson, B. A., and Bertino, J. R., (1987b). Cytotoxicity of floxuridine and 5-fluorouracil in human T-lymphoblast leukemia cells: Enhancement by leucovorin. *Cancer Treat. Rep.* **71,** 381–389.

Mini, E., Mazzei, T., Coronnello, M., Bertino, J. R., and Periti, P. (1988). Comparative effects of two non-classical antifolates. trimetrexate (TMQ) and triazinate (TZT), on 5-fluorouracil (FUra) cytotoxicity against human CCRF–CEM leukemia cells. *Proc. Am. Assoc. Cancer Res.* **29,** 473.

Mini, E., Trave, F., Rustum, Y. M., and Bertino, J. R., (1990). Enhancement of the antitumor effects of 5-fluorouracil by folic acid. *Pharmacol. Ther.* **47,** 1–19.

Moran, R. G., Danenberg, P. V., and Heidelberg, C. (1982). Therapeutic response of leukemic mice with fluorinated pyrimidines and inhibitors of deoxyuridylate synthesis. *Biochem. Pharmacol.* **31,** 2929–2935.

Mortimer, J. E., and Higano, C. (1988). Continuous infusion 5-fluorouracil and folinic acid in disseminated colorectal cancer. *Cancer Invest.* **6,** 129–132.

Moyer, J. D., and Handschumacher, R. E. (1979). Selective inhibition of pyrimidine synthesis and depletion of nucleotide pools by N-(phosphonacetyl)-L-aspartate. *Cancer Res.* **39,** 3089–3094.

Moyer, J. D., Smith, P. A., and Handschumacher, R. E. (1980). Effects of N-(phosphon-acetyl)-L-aspartate (PALA) on circulating pyrimidine nucleosides, nucleotide pools, and nucleic acid synthesis. *Proc. Am. Assoc. Cancer Res.* **21,** 443.

Moyer, J. D., Oliver, J. D., and Handschumacher, R. E. (1981). Salvage of circulating pyrimidine nucleosides in the rat. *Cancer Res.* **41,** 3010–3017.

Muggia, F. M., Camacho, F. J., Kaplan,, B. H., Green. M. D., Greenwald, E. S., Weinz, J. C., and Engstrom, P. F. (1987). Weekly 5-fluorouracil combined with PALA: Toxic and therapeutic effects in colorectal cancer. *Cancer Treat. Rep.* **71,** 253–256.

Mulder, J., Smink, T., and Van Putten, L. (1981). 5-Fluorouracil and methotrexate combination chemotherapy: The effect of drug scheduling. *Eur. J. Cancer Clin. Oncol.* **17,** 831–837.

Muro, H., Acuna, L. R., Castagnari, A., Blajman, C., Schmilovich, A., Hidalgo, A., Fiori, H., Bade, M., and Marantz, A. (1986). Sequential methotrexate, 5-fluorouracil (high-dose), and doxorubicin for advanced gastric cancer. *Cancer Treat. Rep.* **70,** 1333–1334.

Nadal, J. C., Van Groeningen, C. J., Pinedo, H. M., and Peters, G. J. (1987). In vivo synergism of 5-fluorouracil (5FU) and leucovorin (LV) in murine colon carcinoma. *Proc. Eur. Conf. Clin. Oncol.* **4,** 76.

Newman, E., Straw, J. A., and Doroshow, J. (1989). Pharmacokinetics of diastereoisomers of (6R, S)-folinic acid (Leucovorin) in humans during constant high-dose intravenous infusion. *Cancer Res.* **49,** 5755–5760.

Nixon, P. F., and Bertino, J. R., (1972). Effective absorption an utilization of oral formyltet-rahydrofolate in man. *N. Engl. J. Med.* **286,** 175–179.

Nobile, M. T., Sertoli, M. R., Bruzzone, M., Tagarelli, G., Rubagotti, A., and Rosso, R. (1985). Phase II study with high-dose $N^{5,10}$-methyltetrahydrofolate and 5-fluorouracil in advanced colorectal cancer. *Eur, J. Cancer Clin. Oncol.* **21,** 1175–1177.

Nobile, M. T., Vidili, M. G., Sobrero, A., Sertoli, M. R., Canobbio, L., Fassio, T., Rubagotti, A., Gallo, L., Lo Re, G., Galligioni, E., and Rosso, R. (1988). 5-Fluorouracil (FU) alone or combined with high dose folinic acid (FA) in advanced colorectal cancer patients: A random-ized trial. *Proc. Am. Soc. Clin. Oncol.* **7,** 97.

Nordic Gastrointestinal Tumor Adjuvant Therapy Group (1989). Superiority of sequential methotrexate, fluorouracil, and leucovorin to fluorouracil alone in advanced symptomatic colorectal carcinoma: A randomized study. *J. Clin. Oncol.* **7,** 1437–1446.

O'Connell, M. J., Powis, G., Rubin, J., and Moertel, C. G., (1982). Pilot study of PALA and 5-FU in patients with advanced cancer. *Cancer Treat. Rep.* **66**, 77–80.

O'Connell, M. J., Moertel, C. G., Rubin, J., Hahn, R. G., Kvols, K., and Schutt, A. J., (1984). Clinical trial of sequential N-phosphonacetyl-L-aspartate, thymidine, and 5-fluorouracil in advanced colorectal carcinoma. *J. Clin. Oncol.* **10**, 1133–1138.

O'Dwyer, P. J., King, S. A., Hoth, D. F., and Leyland-Jones, B. (1987). Role of thymidine in biochemical modulation: A review. *Cancer Res.* **47**, 3911–3919.

O'Dwyer, R. J., Paul, A. R., Peter, R., Weiner, L. M., and Comis, R. L. (1989). Biochemical modulation of 5-fluorouracil (5FU) by PALA: Phase II study in colorectal cancer. *Proc. Am. Soc. Clin. Oncol.* **8**, 107.

Ohnuma, T., Roboz, J., Waxman, S., Mandel, E., Martin, S. D., and Holland, J. F., (1980). Clinical pharmacological effects of thymidine plus 5-FU. *Cancer Treat Rep.* **64**, 1169–1177.

Panasci, D., and Margolese, P. (1982). Sequential methotrexate (MTX) and 5-fluorouracil (FU) in breast and colorectal cancer. Results of increasing the dose of FU. *Proc. Am. Soc. Clin. Oncol.* **1**, 101.

Panasci, L., Ford, J., and Margolese, R. (1985). A phase I study of sequential methotrexate and fluorouracil in advanced colorectal cancer. *Cancer Chemother. Pharmacol.* **15**, 164–166.

Parchure, M., Ambaye, R. Y., and Gokhale, S. V., (1984). Combination of anticancer agents with folic acid in the treatment of murine leukaemia P388. *Chemotherapy (Basel)* **30**, 119–124.

Park, J.-G., Collins, J., Gazdar, A., Allegra, C. J., Steinberg, S. M., Greene, B. F., and Kramer, B. S. (1988). Enhancement of fluorinated pyrimidine-induced cytotoxicity by leucovorin in human colorectal carcinoma cell lines. *JNCI, J. Natl. Cancer Inst.* **80**, 1560–1564.

Parker, W. B., and Klubes, P. (1985). Enhancement by uridine of the anabolism of 5-fluorouracil in mouse T-lymphoma (S-49) cells. *Cancer Res.* **45**, 4249–4256.

Paterson, A. R. P., and Wang, M. C. (1970). Mechanism of the growth inhibition potentiation arising from combination of 6-mercaptopurine with 6-(methylmercapto)purine ribonucleoside. *Cancer Res.* **30**, 2379–2387.

Paterson, A. R. P., Lau, E. Y., Dahlig, E., and Cass, C. E., (1980). A common basis for inhibition of nucleoside transport by dipyridamole and nitrobenzylthioinosine. *Mol. Pharmacol.* **18**, 40–44.

Perrault, D. J., Erlichman, C., Hasselback, R., Tannock, I., and Boyd, N. (1983). Sequential methotrexate (MTX) and 5-fluorouracil (5FU) in refractory metastatic breast cancer. *Proc. Am. Soc. Clin. Oncol.* **2**, 100.

Peters, G. J., Laurensse, E., Leyva, A., and Pinedo, H. M. (1984). Fluctuations in phosphoribosyl-pyrophosphate levels. Effects of drugs. *FEBS Lett.* **170**, 277–280.

Peters, G. J., Laurensse, E., Leyva, A., and Pinedo, H. M. (1987a). Purine nucleotides as cell-specific modulators of 5-fluorouracil metabolism and cytotoxicity. *Eur. J. Cancer Clin. Oncol.* **23**, 1869–1881.

Peters, G.J., Van Groeningen, C.J., Laurensse, E. J., Lankelma, J., Leyva, A., and Pinedo, H. M. (1987b). Uridine-induced hypothermia in mice and rats in relation to plasma and tissue levels of uridine and its metabolite. *Cancer Chemother. Pharmacol.* **20**, 101–108.

Peters, G. J., Van Groeningen, C. J., and Laurensse, E., (1987c). Effect of pyrimidine nucleosides on body temperatures of man and rabbit in relation to pharmacokinetic data. *Pharm. Res.* **4**, 113–119.

Peters, G. J., van Dijk, J., Laurensse, E., Van Groeningen, C. J., Lankelma, J., Leyva, A., Nadal, J. C., and Pinedo, H. M. (1988). In vitro biochemical and in vivo biological studies of the uridine "rescue" of 5-fluorouracil. *Br. J. Cancer* **57**, 259–265.

Peters, G. J., Van Groeningen, C. J., Laurensse, E. J., Meijer, S., and Pinedo, H. M. (1989). Inhibition of thymidylate synthase (TS) in patients with colorectal cancer treated with 5-fluorouracil. *Proc. Am. Assoc. Cancer. Res.* **30**, 598.

Peters, W. P., Weiss, G., and Kufe, D. W. (1984). Phase I trial of combination therapy with continuous-infusion MMPR and continuous-infusion 5-FU. *Cancer Chemother. Pharmacol.* **13**, 136–138.

Petrelli, N., Herrera, L., Rustum, Y., Burke, P., Creaven, P., Stulc, J., Emrich, L. J., and Mittelman, A. (1987). A prospective randomized trial of 5-fluorouracil versus 5-fluorouracil and high-dose leucovorin versus 5-fluorouracil and methotrexate in previously untreated patients with advanced colorectal carcinoma. *J. Clin. Oncol.* **5**, 1559–1565.

Petrelli, N., Douglass, H. O., Herrera, L., Russell, D., Stablein, D. M., Bruckner, H. W., Mayer, R. T., Schinella, M. R., Green, M. D., Muggia, F. M., Megibow, A., Greenwald, E. S., Bukowski, R. M., Harris, J., Levin, B., Gaynor, E., Loutfi, A., Kalser, M. H., Barkin, J. S., Benedetto, P., Woolley, P. V., Neuta, R., Weaner, D. W., and Leichman, L. P., for the Gastrointestinal Study Group (1989). The modulation of fluorouracil with leucovorin in metastatic colorectal carcinoma: A prospective randomized phase III trial. *J. Clin. Oncol.* **7**, 1419–1426.

Pinedo, H. M., and Peters, G. D. (1988). Fluorouracil: Biochemistry and pharmacology. *J. Clin. Oncol.* **6**, 1653–1664.

Piper, A. A., and Fox, R. M. (1982). Biochemical basis for the differential sensitivity of human T and B lymphocyte lines to 5-fluorouracil. *Cancer Res.* **42**, 3753–3760.

Piper, A. A., Nott, S. E., Machinnon, W. B., and Tattersall, M. H. N. (1983). Critical modulation by thymidine and hypoxanthine of sequential methotrexate–5-fluorouracil synergism in murine L1210 cells. *Cancer Res.* **43**, 5701–5705.

Pitman, S. W., Kowal, C. D., and Bertino, J. R. (1983). Methotrexate and 5-fluorouracil in sequence in squamous head and neck cancer. *Sem. Oncol.* **10**, (Suppl. 2), 15–19.

Pizzorno, G., Mini, E., Coronnello, M., Moroson, B. A., Cashmore, A.R., Dreyer, R. N., Lin, J. T., Mazzei, T., Periti, P., and Bertino, J. R. (1988). Impaired polyglutamylation of methotrexate as a cause of resistance in CCRF–CEM cells after short-term, high-dose treatment with this drug. *Cancer Res.* **48**, 2149–2155.

Plagemann, P., and Wohlbocke, R. M. (1980). Permeation of nucleosides in animal cells. *Curr. Top. Membr. Transp.* **14**, 225–330.

Plotkin, D., and Waugh, W. J., (1982). Sequential methotrexate 5-fluorouracil (M → F) in advanced breast carcinoma. *Proc. Am. Soc. Clin. Oncol.* **1**, 80.

Poon, M. A., O'Connell, M. J., Moertel, C. G., Wieland, H. S., Cullinan, S. A., Everson, L. K., Krook, J. E., Mailliard, J. A., Laurie, J. A., Tshetter, L. K., and Wiesenfeld, M. (1989). Biochemical modulation of fluorouracil: Evidence of significant improvement of survival and quality of life in patients with advanced colorectal carcinoma. *J. Clin. Oncol.* **7**, 1407–1411.

Presant, C. A., Multhauf, P., Klein, L., Chan, C., Chang, F. F., Hum, G., Opfell, J. R., Lemkin, S., Shiftan, T., and Plotkin, D. (1983). Thymidine and 5-FU: A phase II pilot study in colorectal and breast carcinomas. *Cancer Treat. Rep.* **67**, 735–736.

Radparvar, S., Houghton, P. J., and Houghton, J. A. (1989). Effect of polyglutamylation of 5,10-methylenetetrahydrofolate on the binding of 5-fluoro-2'-deoxyuridylate to thymidylate synthase purified from a human colon adenocarcinoma xenograft. *Biochem. Pharmacol.* **38**, 335–342.

Rangineni, R. R., Ajani, J. A., Bedikian, A. T., McKelvey, E. M., and Bodey, G. P. (1983). Sequential conventional dose methotrexate (MTX) and 5-fluorouracil (5-FU) in the primary therapy of metastatic colorectal carcinoma. *Proc. Am. Soc. Clin. Oncol.* **2**, 125.

Richards, F., II, Capizzi, R. L., Muss, H., Cruz, J., Powell, B., White, D., Jackson, D., Atkins, J., Caldwell, L., Pukett, J., Christian, E., Schmell, F., Scharpe, S., Stanley, V., and Brockschmidt, J. (1989). 5-Fluorouracil (5-FU), high dose folinic acid (FA) and methotrexate (MTX) for advanced colorectal cancer. A phase II trial of the Piedmont Oncology Association. *Proc. Am. Soc. Clin. Oncol.* **8**, 101.

Ringborg, U., Ewert, G., Kinnman, J., Lundquist, P. G., and Strander, N. (1983). Methotrexate and 5-fluorouracil in head and neck cancer. *Semin. Oncol.* **10**, (Suppl. 2), 20–22.

Romanini, A., Lin, J. T., Bertino, J. R., Bunni, M., and Priest, D. G., (1989). Role of polyglutamylation in the potentiation of fluoropyrimidine (FP) cytotoxicity by leucovorin (LV). *Proc. Am. Assoc. Cancer Res.* **30**, 595.

Saidman, B., Hait, W., Marsh, J. C., Reiss, M., Durivage, H., Davis, C., Makuch, R. W., Bertino, J. R., and Portlock, C. S. (1989). Phase II study of sequential methotrexate (MTX) high dose leucovorin (HDLV) and 5-fluorouracil (5-FU) in the treatment of metastatic breast cancer. *Proc. Am. Soc. Clin. Oncol.* **8,** 38.

Santelli, G., and Valeriote, F. (1978). *In vivo* enhancement of 5-fluorouracil cytotoxicity to AKR leukemia cells by thymidine in mice. *JNCI, J. Natl. Cancer Inst.* **61,** 843–847.

Santelli, G., and Valeriote, F. (1980). *In vivo* protentiation of 5-fluorouracil-cytotoxicity against AKR leukemia by purines, pyrimidines and their nucleosides and deoxynucleosides. *JNCI, J. Natl. Cancer Inst.* **64,** 69–72.

Santi, D. V., McHenry, C. S., and Sommer, H. (1974). Mechanism of interaction of thymidylate synthetase with 5-fluorodeoxyuridylate. *Biochemistry* **13,** 471–481.

Sasaki, T. (1984). Low dose methotrexate and sequential 5-fluorouracil in advanced gastric cancer. *In* "Fluoropyrimidines in Cancer Therapy" (K. Kimura, S. Fujii, G. P. Ogawa, G. P. Bodey, and P. Alberto, eds.), pp. 269–279. Elsevier, Amsterdam.

Sawyer, R. C., Stolfi, R. L., Spiegelman, S., and Martin, D. S. (1984). Effect of metabolism of 5-fluorouracil in the CD8F1 murine mammary carcinoma system. *Pharm. Res.* **1,** 69–75.

Schmoll, H. J., Wilke, H., Schober, C., and Stahl, M. (1988). High dose folinic acid (HDFA/5-fluorouracil (FU)/dipyridamol (DP): Salvage after HDFA/FU failure in colorectal cancer: A phase I study. *Proc. Am. Soc. Clin. Oncol.* **7,** 114.

Schneider, A. K., Kemeny, N., Martin, D., Sawyer, R., and Williams, L. (1989). A phase I study of *N*-phosphonacetyl-aspartate (PALA), methotrexate (MTX), 5-fluorouracil (FU) with leucovorin and uridine rescues. *Proc. Am. Assoc. Cancer. Res.* **30,** 244.

Schwartz, P. M., and Handschumacher, R. E., (1979). Selective antagonism of 5-fluorouracil cytotoxicity by 4-hydroxypyrazolopyrimidine (allopurinol) *in vitro. Cancer Res.* **39,** 3095–3101.

Schwartz, J., Alberts, D., Einspahr, J., Peng, Y.-M., and Spears, P. (1987). Dipyridamole (D) potentiation of FUdr activity against human colon cancer *in vitro* and in patients. *Proc. Am. Soc. Clin. Oncol.* **6,** 83.

Schwartz, P. M., Dunigan, J. M., Marsh, J. C., and Handschumacher, R. E. (1980). Allopurinol modulation of the toxicity and antitumor activity of 5-fluorouracil. *Cancer Res.* **40,** 1885–1889.

Skeel, R. T., Sawicki, W. L., Cashmore, A. R., and Bertino, J. R., (1973). The basis for the disparate sensitivity of L1210 leukemia and Walker 256 carcinoma to a new triazine folate antagonist. *Cancer Res.* **33,** 2972–2976.

Smith, J. A., Markman, M., Kelsen, D., Reichman, B., Tong, W. P., Duafala, M. E., and Bertino, J. R., (1989). Phase I study of intraperitoneal (IP) FUDR and leucovorin (LV). *Proc. Am. Soc. Clin. Oncol.* **8,** 67.

Sobrero, A. F., and Bertino, J. R. (1983). Sequence-dependent synergism between dichloromethotrexate and 5-fluorouracil in a human colon carcinoma cell line. *Cancer Res.* **43,** 4011–4013.

Sobrero, A., Romanini, A., Russello, O., Nicolin, A., Rosso, R., and Bertino, J. R., (1989). Sequence-dependent enhancement of HCT-8 cell kill by trimetrexate and fluoropyrimidines: Implications for the mechanism of this interaction. *Eur. J. Cancer Clin. Oncol.* **25,** 977–982.

Solan, A., Vogl, S. E., Kaplan, B.H., Berenzweig, M., Richard, J., and Lanham, R. (1982). Sequential chemotherapy of advanced colorectal cancer with standard or high-dose methotrexate followed by 5-fluorouracil. *Med. Pediatr. Oncol.* **10,** 145–149.

Spears, C. P., Gustavsson, B. G., Mitchell, M. S., Spicer, D., Berne, M., Bernstein, L., and Danenberg, P. V. (1984). Thymidylate synthetase inhibition in malignant tumors and normal liver of patients given intravenous 5-fluorouracil. *Cancer Res.* **44,** 4144–4150.

Spiegelman, S., Nayak, R., Sawyer, R., Stolfi, R., and Martin, D. (1980a). Potentiation of the anti-tumor activity of 5-FU by thymidine and its correlation with the formation of (5-FU) RNA. *Cancer (Philadelphia)* **45,** 1129–1134.

Spiegelman, S., Sawyer, R., Nayak, R., Ritzi, E., Stolfi, R., and Martin, D. (1980b). Improving the anti-tumor activity of 5-fluorouracil by increasing its incorporation into RNA via metabolic modulation. *Proc. Natl. Acad. Sci. U.S.A.* **77,** 4966–4970.

Sternberg, A., Petrelli, N. J., Au, J., Rustum, Y., Mittelman, A., and Creaven, P. (1984). A combination of 5-fluorouracil and thymidine in advanced colorectal cancer. *Cancer Chemother. Pharmacol.* **13**, 218–222.

Stewart, D. J., Maroun, J. A., Cripps, C., Young, V., Laframboise, G., and Gerin-Lajoie, J. (1986). Methotrexate and 5-fluorouracil in the treatment of squamous and other carcinomas of the head and neck. *Cancer Chemother. Pharmacol.* **17**, 91–94.

Straw, J. A., Szapary, D., and Wynn, W. T. (1984). Pharmacokinetics of disastereoisomers of leucovorin after intravenous and oral administration to normal subjects. *Cancer Res.* **44**, 3114–3119.

Swain, S. M., Lippman, M. E., Egan, E. F., Drake, J. C., Steinberg, S. M., and Allegra, C. J. (1989). 5-Fluorouracil and high dose leucovorin in previously treated patients with metastatic breast cancer *J. Clin. Oncol.* **7**, 890–899.

Tanaka, W., Kimura, K., and Yoshida, S. (1983). Enhancement of the incorporation of 5-fluorodeoxyuridylate into DNA of HL-60 cells by metabolic modulation. *Cancer Res.* **43**, 5154–5150.

Tisman, G., and Wu, S. J. G., (1980). Effectiveness of intermediate dose methotrexate and high dose 5-fluorouracil as sequential combination chemotherapy in refractory breast cancer and as primary therapy in metastatic adenocarcinoma of the colon. *Cancer Treat. Rep.* **64**, 829–835.

Ullman, B., and Kirsch, J. (1979). Metabolism of 5-fluorouracil in cultured cells. Protection from 5-fluorouracil cytotoxicity by purines. *Mol. Pharmacol.* **15**, 357–366.

Ullman, B., Lee, M., Martin, D. W., Jr., and Santi, D. V. (1978). Cytotoxicity of 5-fluoro-2'-deoxyuridine: Requirement for reduced folate cofactors and antagonism by methotrexate. *Proc. Natl. Acad. Sci. U.S.A.* **75**, 980–983.

Valeri, A., Crescioli, L., and Mini, E. (1987). Trattamento loco-regionale per metastasi epatiche da carcinoma del colon-retto con acido metiltetraidrofolico e 5-fluorouracile. Risultati preliminari in 6 casi. *Farm. Ter.* **11**, 220–232.

Valone, F. H., Friedman, M. A., Wittlinger, P. S., Drakes, T., Eisenberg, P. D., Malec, M., Hanningan, J. F., and Brown, B. W. (1989). Treatment of patients with advanced colorectal carcinoma with fluorouracil alone, high-dose leucovorin plus fluorouracil, or sequential methotrexate, fluorouracil, and leucovorin: A randomized trial of the Northen California Oncology Group. *J. Clin. Oncol.* **7**, 1427–1436.

Van Groeningen, C. J., Leyva, A., Kraal, I., Peters, G. J., and Pinedo, H. M. (1986). Clinical and pharmacokinetic study of prolonged administration of high-dose uridine intended for rescue from 5-fluorouracil toxicity. *Cancer Treat. Rep.* **70**, 745–750.

Van Groeningen, C. J., Peters, G. J., Nadal, J., Leyva, A., Gall, H. L., and Pinedo, H. M. (1987). Phase I clinical and pharmacokinetics study of orally administered uridine. *Proc. Am. Assoc. Cancer Res.* **28**, 195.

Van Groeningen, C. J., Peters, G. J., and Pinedo, H. M. (1989). Lack of effectiveness of combined 5-fluorouracil and leucovorin in patients with 5-fluorouracil-resistant advanced colorectal cancer. *Eur. J. Cancer Clin. Oncol.* **25**, 45–49.

Van Groeningen, C. J., Peters, G. J., Leyva, A., and Pinedo, H. M. (1989). Reversal of 5-fluorouracil-induced myelosuppression by prolonged administration of high-dose uridine. *JNCI, J. Natl. Cancer, Inst.* **81**, 157–162.

Vogel, S. J., Presant, C. A., Ratkin, G. A., and Klahr, C. (1979). Phase I study of thymidine plus 5-fluorouracil infusions in advanced colorectal carcinoma. *Cancer Treat. Rep.* **63**, 1–5.

Vokes, E. E., Choi, K. E., Schilsky, R. L., Moran, W. J., Guarnieri, C. M., Weichselbaum, R. R., and Panje, W. R., (1988). Cisplatin, fluorouracil, and high-dose leucovorin for recurrent or metastatic head and neck cancer. *J. Clin. Oncol.* **6**, 618–626.

Wasserman, T. H., Comis, R. L., Goldsmith, M., Handelsman, H., Penta, J. S., Slavik, M., Lopez, W. T., and Carter, S. K. (1975). A tabular analysis of the clinical chemotherapy of solid tumors. *Cancer Chemother. Rep.* **6**, (Part 3), 399–419.

Waxman, S., and Bruckner, H. (1982). Enhancement of 5-fluorouracil antimetabolic activity by leucovorin, menodione and α-tocopherol. *Eur. J. Cancer Clin. Oncol.* **18**, 685–692.

Weinerman, B., Schacter, B., Schipper, H., Bowman, D., and Levitt, M. (1982). Sequential methotrexate and 5-FU in the treatment of colorectal cancer. *Cancer Treat. Rep.* **66**, 1553–1555.

Weinerman, B., Maroun, J., Stewart, D. J., Bowman, D., Levit, M., and Schacter, B. (1984). Phase II trial of methotrexate (M) and 5-fluorouracil (F) (M-F) in metastatic colorectal cancer (CRC) II. *Proc. Am. Soc. Clin. Oncol.* **3**, 133.

Weiss, G. R., Ervin, T. J., Meshad, M. W., and Kufe, D. W., (1982a). Phase II trial of combination therapy with continuous-infusion PALA and bolus-injection 5-FU. *Cancer Treat. Rep.* **66**, 229–303.

Weiss, G. R., Ervin, T. J., Meshad, M. W., Schade, D., Branfman, A. R., Bruni, R. J., Chadwick, M., and Kufe, D. W. (1982b). A phase I trial of combination therapy with continuous-infusion PALA and continuous-infusion 5-FU. *Cancer Chemother. Pharmacol.* **8**, 301–304.

Wendt, T. G., Hartenstein, R. C., Wustrow, T. P. U., and Lissner, J. (1989). Cisplatin, fluorouracil with leucovorin calcium enhancement and synchronous accelerated radiotherapy in the management of locally advanced head and neck cancer: A phase II study. *J. Clin. Oncol.* **7**, 471–476.

White, J. C., Bailey, D. D., and Goldman, I. D. (1978). Lack of stereo specificity at carbon 6 of methyltetrahydrofolate transport in Ehrlich ascites tumor cells. *J. Biol. Chem.* **253**, 242–245.

Wilke, H., Schmoll, H. J., Preusser, P., Fink, U., Stahl, M., Shober, C., Link, H., Freund, M., Hanauske, A., Meyer, H.-J., Achterrath, W., and Poliwoda, H. (1988). Folinic acid (CF)/5-fluorouracil (FUra) combinations in advanced gastrointestinal carcinomas. *In* "The Expanding Role of Folates and Fluoropyrimidines in Cancer Chemotherapy" (Y. Rustum and J. J. McGuire, eds.), pp. 233–242. Plenum, New York.

Wils, J., Bleiberg, H., Dalesio, O., Blijharu, G., Mulder, N., Planting, A., Splinter, T., and Duer, N. (1986). An EORTC Gastrointestinal Group evaluation of the combination of sequential methotrexate and 5-fluorouracil, combined with Adriamycin in advanced measurable gastric cancer. *J. Clin. Oncol.* **4**, 1799–1803.

Wolmark, N. (1988). Adjuvant therapy for colorectal cancer: The NSABP clinical trials. *In* "The Expanding Role of Folates and Fluoropyrimidines in Cancer Chemotherapy" (Y. Rustum and J. J. McGuire, eds.), pp. 261–264. Plenum, New York.

Woodcock, T. M., Martin, D. S., Damin, L. A. M., Kemeny, N. E., and Young, C. W. (1980). Combination clinical trials with thymidine and fluorouracil: A phase I and clinical pharmacologic evaluation. *Cancer (Philadelphia)* **43**, 1135–1143.

Wood, C. D., Slevin, M. L., Ponder, B. A., and Wrigley. P. F. M. (1985). Sequential methotrexate and 5-fluorouracil in the treatment of non-small cell carcinoma of the lung. *Eur. J. Cancer Clin. Oncol.* **21**, 587–589.

Woolley, B. P. V., Ayoob, M. J., Smith, F. P., Lokey, J. L., De Green, P., Marantz, A., and Schein, P. S. (1985). A controlled trial of the effect of 4-hydroxypyrazolopyrimidine (allopurinol) of the toxicity of a single bolus dose of 5-fluorouracil. *J. Clin. Oncol.* **3**, 103–109.

Wright, J. E., Dreyfuss, A., El-Magharbel, I., Trites, D., Jones, S. M., Holden, S. A., Rosowsky, A., and Frei, E., III (1989). Selective expansion of 5,10-methylenetetrahydrofolate pools and modulation of 5-fluorouracil antitumor activity by leucovorin *in vivo*. *Cancer Res.* **49**, 2592–2596.

Yin, M. B., Zakrzewski, S. F., and Hakala, M. T. (1982). Relationship of cellular folate cofactor pools to the activity of 5-fluorouracil. *Mol. Pharmacol.* **23**, 190–197.

Zaniboni, A., Simoncini, E., Marpicati, P., and Marini, G. (1988). Phase II study of 5-fluorouracil (5-FU) and high dose folinic acid (HDFA) in hepatocellular carcinoma. *Br. J. Cancer* **57**, 319.

Zaniboni, A., Simoncini, E., Marpicati, P., Montini, E., Ferrari, V., and Marini, G. (1989). Phase II trial of 5-fluorouracil and high-dose folinic acid in advanced renal cell cancer, *J. Chemother.* **1**, 350–351.

CHAPTER 14

Synergism and Antagonism through Direct Bond Formation between Two Agents in Situ

Darryl C. Rideout
Department of
Molecular Biology
The Research Institute of
Scripps Clinic
La Jolla, California 92037

Theodora Calogeropoulou
Viochrom Corporation
Athens, Greece

SYNERGISM AND ANTAGONISM IN CHEMOTHERAPY
Copyright © 1991 by Academic Press, Inc.
All rights of reproduction in any form reserved.

I. Introduction

The simplest possible mechanism for synergistic or antagonistic interaction between two agents involves chemical bond formation between the agents *in vivo* to produce an adduct with altered biological properties. This mechanism can be described by the term "covalent modulation" in the sense that agent A modulates the activity of the agent B by forming a covalent bond with agent B. If the adduct has a greater cytotoxic or antimicrobial effect than the precursors (prodrugs) from which it is formed, the combination is known as a "self-assembling drug" (Rideout *et al.*, 1990a). Because of its relative simplicity, covalent modulation has inherent advantages over other mechanisms for the rational design of combinations exhibiting target-selective synergism or host-tissue-selective antagonism. Both of these properties can result in a combination of agents with an enhanced therapeutic index relative to the individual agents or the preformed adduct.

This chapter discusses nonbiochemical, bimolecular reactions that can occur under nearly physiological conditions and *in vivo*, mechanisms for the enhancement or inhibition of bioactivity in bifunctional molecules relative to the corresponding monofunctional molecules, and synergism and antagonism involving direct covalent combinations as a mechanism of interaction. Leading references and illustrative examples will be provided without attempting to comprehensively review the literature.

This chapter does not discuss nutritional synergism involving covalent bond formation *in situ* between drugs and naturally occurring nutrients or catabolites. For a discussion of antimalarial synergism involving combination of naturally occurring metal ions and chelating drugs, see Chapter 5 by Vennerstrom *et al.*

II. Nonbiochemical Condensation Reactions Occurring *in Vivo* or under Near-Physiological Conditions

A. HYDRAZONE FORMATION

Hydrazone formation from aldehydes (or ketones) and hydrazine derivatives (Fig. 1) is generally rapid in water and, in contrast to imine formation, is often irreversible (see Rideout *et al.*, 1990a and references therein). Hydrazone formation has been observed in animal and human subjects. For example, hydralazine reacts with biogenic aldehydes and ketones such as acetone, pyruvic acid, and pyridoxal in human subjects (O'Donnell *et al.*, 1979).

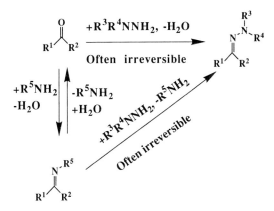

FIGURE 1   Hydrazone formation in water.

Aminoguanidine reacts in animals with the Amidori product, a ketone, which is formed from glucose and proteins. Because of this hydrazone formation, aminoguanidine is useful in preventing protein cross-linking that is responsible for many of the medical problems plaguing diabetics (Ledl and Schleicher, 1990). Aminoguanidine is currently undergoing clinical trials for the treatment of diabetes. Hydrazino-aminovaleric acid forms a hydrazone with pyridoxal bound to ornithine decarboxylase in live mammalian cells (Höltta *et al.*, 1981).

Aldehydes, hydrazine derivatives, hydrazones, and ketones are all biocompatible. The clinical anticancer agent cyclophosphamide is converted to its active form, an aldehyde, in human liver (Kwon and Borch, 1989). Leupeptin, a peptide aldehyde which inhibits proteases only in its aldehyde form, can be administered to mice and exhibit protease inhibition *in vivo* (Salminen, 1984). Benzaldehyde, which occurs naturally in figs and almonds, has been reported to possess anticancer activity against P388 leukemia in mice and (in prodrug form) against a variety of tumors of human subjects (Balazova and Koza, 1988; Kochi *et al.*, 1980). One of the active metabolites of the antineoplastic agent hexamethylene bisacetamide formed in human subjects is an aldehyde, 6-acetamidohexanal (Subramanyam *et al.*, 1989). Clinically useful monofunctional hydrazines include hydralazine, isoniazid, phenelzine, and carbidopa (Gilman *et al.*, 1980). Hydrazine sulfate itself has been tested clinically against cancer cachexia (Chlebowski and Heber, 1986; Chlebowski *et al.*, 1987). It is interesting to note that hydrazide derivatives of oligodeoxynucleotides have been prepared. These have been useful in the preparation of stable enzyme/nucleic acid hydrazones as hybridization probes (Ghosh *et al.*, 1989). Peptide hydrazides and peptide aldehydes have also been described (DeGrado and Kaiser, 1980; Salminen, 1984). For examples of biocompatible hydrazones and ketones, see Huff (1989) and Smith *et al.* (1983).

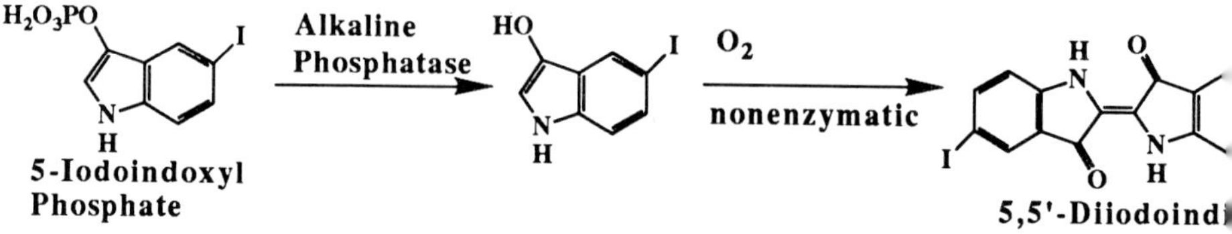

FIGURE 2   Diiodoindigo assembly from an iodoindoxyl ester.

B. INDIGO FORMATION THROUGH OXIDATIVE INDOXYL DIMERIZATION

It is possible to form 5,5'-diiodoindigo inside HeLa human cervical carcinoma cells by administering a monomeric precursor, 5-iodoindoxyl phosphate. The reaction occurs through the mechanism shown in Fig. 2, involving dimerization through nonenzymatic carbon–carbon double bond formation (Tsou, 1968). Cells thus treated become sensitized to treatment with laser light. Similarly, selective formation of 5,5'-diiodoindigo in esterase-rich ciliary body rabbit eyes is achieved by administering 5-iodoindoxyl acetate. It was suggested that 5-iodoindoxyl acetate might by useful for laser eye surgery, since selective destruction of the ciliary body by laser light is possible in eyes treated with this agent (Vucicevic et al., 1969). Indigo is believed to form from indoxyl sulfate in the human bladder through analogous chemistry: sulfate ester cleavage and oxidative dimerization (Dealler et al., 1988).

C. DIELS–ALDER REACTION

Diels–Alder reactions (example in Fig. 3) between lipophilic dienes and lipophilic dienophiles can be substantially more rapid in water and in aqueous surfactant suspensions than in organic solvents because of hydrophobic interactions between the two reactants (Rideout and Breslow, 1980; Breslow et al., 1983). It has been suggested that nonenzymatic Diels–Alder reactions are involved in the synthesis of certain polycyclic natural products (Carruthers, 1978; Joshi et al., 1975). Thus, Diels–Alder reactions that occur in vivo might be useful in terms of achieving synergism or antagonism through direct combination of dienes and dienophiles.

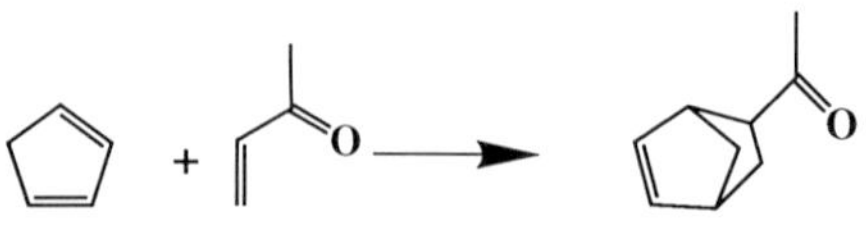

FIGURE 3   A Diels–Alder reaction.

D. METAL–LIGAND BOND FORMATION

The complexation reaction between nickel dication and pyridine-2-aza-*p*-dimethylaniline (Fig. 4) is of interest because it is catalyzed by a number of anionic surfactants. The extent of catalysis varies dramatically with the length and number of lipophilic chains in anionic surfactant micelles, suggesting that reactions of this sort might be catalyzed by biological membranes which contain anionic head groups (Jobe and Reinsborough, 1984; see Section V for example of antagonism involving metal complexation). The intercalators ethidium and proflavine can be linked to DNA using cisplatin: platinum–nitrogen bonding is involved (Malinge and Leng, 1986). When ethidium, cisplatin, and DNA are combined, an ethidium/cisplatin/guanine covalent complex forms (Malinge *et al.*, 1987). It was suggested that this type of self-assembly between cisplatin and amino intercalators might be responsible for a certain type of synergism involving cisplatin in combination with other DNA-binding antineoplastic agents. A structure of an ethidium/cisplatin complex responsible for the lesion has been attained using X-ray crystallography (Sundquist *et al.*, 1988). The reaction is quite specific: although DNA promotes combination between cisplatin and ethidium DNA does not promote combination between transplatin and ethidium nor between acridine and ethidium (Malinge and Leng, 1986; Sundquist *et al.*, 1988). The interactions between cisplatin and aminointercalators may well involve complexation of one or both drugs to the DNA *before* assembly (see Chapter 12 by Strekowski and Wilson).

E. ALKENE PHOTODIMERIZATION

As in the case of Diels–Alder reactions, photodimerization of alkenes can be significantly more rapid in water and aqueous surfactant solutions than in organic solvents because of hydrophobic interaction between the reactants.

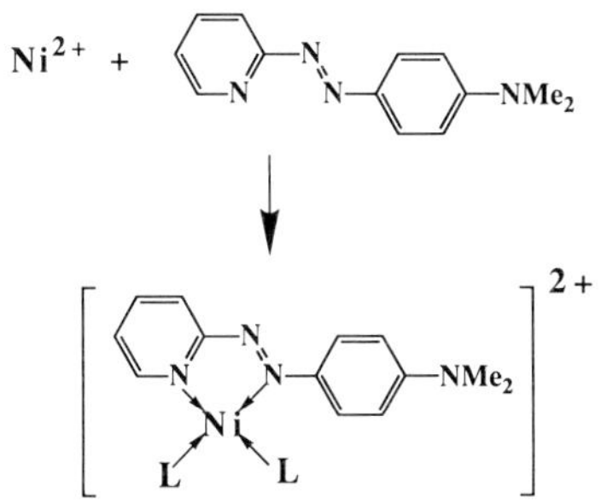

FIGURE 4 Metal chelate formation between nickel dication and pyridine-2-aza-paradimethylaniline.

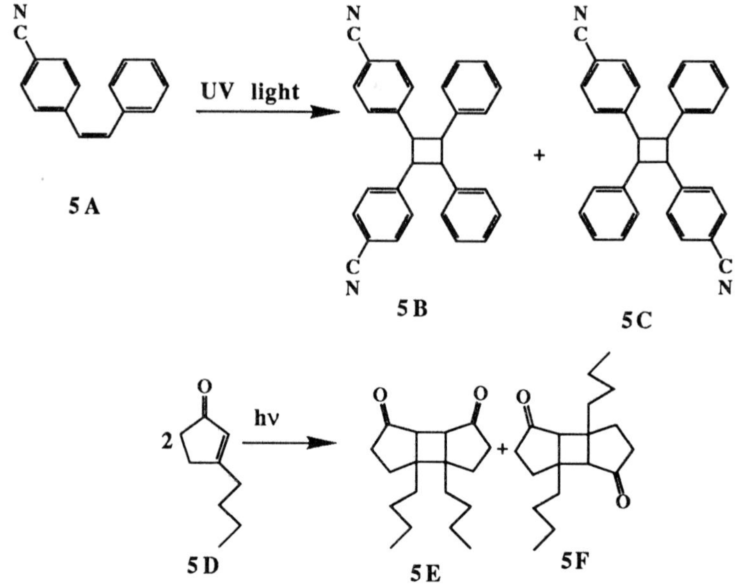

FIGURE 5   Photochemical alkene dimerization to form cyclobutanes.

Photochemical dimerization of the stilbene derivative **5A** at 0.2 mM concentrations, as shown in Fig. 5, is significantly more efficient in water than in organic solvents. The dimerization yield (for **5B** + **5C** combined) is 80% in water, yet only 4% in methanol. In contrast, only traces of the dimerization product are formed in benzene. These differences have been attributed to the hydrophobic interactions between pairs of stilbene molecules in water (Ito *et al.*, 1989). Photodimerization of 5.9 mM **5D** in benzene yields only about 0.01% **5E** and 0.9% **5F**, whereas dimerization of 5.6 mM **5D** in aqueous anionic surfactant micelles yields 98% **5E** and 2% **5F** (Lee and deMayo, 1979). Thymine dimer formation at TT sequences in DNA, a form of alkene photodimerization, can occur in living cells (Kornberg, 1980). Together these results suggest that photochemical cyclobutane formation from pairs of alkene prodrugs should be feasible *in vivo*.

III. Enhanced Cytotoxic and Antimicrobial Bioactivity of Polyfunctional versus Monofunctional Molecules

It is not at all difficult to decrease the bioactivity of a molecule by chemically linking it to a second molecule. If a bifunctional molecule C is formed by linking a monofunctional molecule A to a part of a monofunctional molecule

B that is critical for the binding of B to its receptor, the relative bioactivity of
the bifunctional molecule C will be diminished or eliminated altogether. If A
and B can combine under physiological conditions to form C, and B is more
bioactive than A and C, the result will be antagonism between A and B. (For
examples, see Section V.)

In order to achieve synergism through direct combination, however, the
bifunctional or polyfunctional adduct must be more active than either of the
two monofunctional adducts from which it forms. Although the development
of bifunctional molecules that exhibit enhanced bioactivity relative to their
monofunctional counterparts is not always straightforward, there are still very
many examples of such bifunctional molecules. Several representative examples of such bifunctional molecules will be discussed in this section.

A. DNA BISINTERCALATORS

Molecules formed by linking two monointercalators to one another by a
long tether are frequently much more avid DNA binders than the monointercalators themselves because the dimers are capable of bisintercalation (Wirth
et al., 1988; Jaycox *et al.*, 1987; Kuhlmann and Mosher, 1981; Delbarre
et al., 1987). Often, the dimeric intercalators are much more cytotoxic
against cultured tumor cells and/or have greater anticancer activity in animal
models compared with the corresponding monointercalators. For example,
aminoacridine **6A** shown in Fig. 6 inhibits L1210 leukemia with an ID_{50}
of 2.3 μM whereas the ID_{50} values are 0.4 and 0.081 for bisintercalators

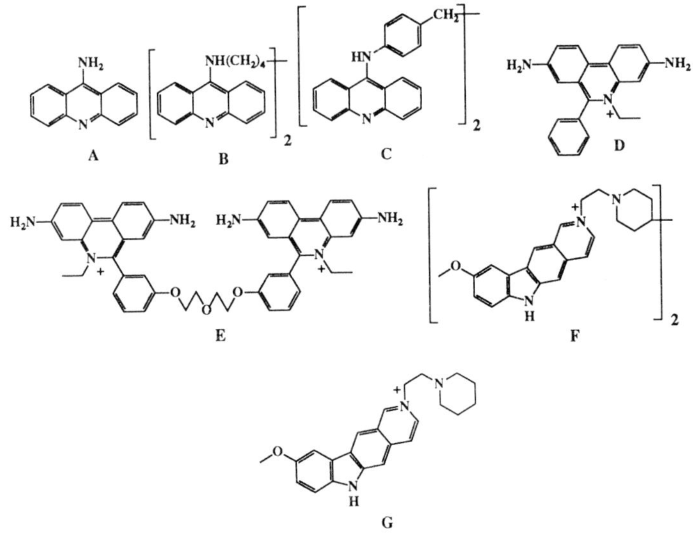

FIGURE 6   Monointercalators and bisintercalators.

6B and **6C**, respectively (Jaycox *et al.*, 1987). A dose of 8.9 μmol/kg of ethidium chloride **6D** is required to extend the life span of L1210 bearing mice by 50% whereas only 1.7 μmol/kg of the related bisintercalator **6E** is sufficient to extend the life span of these mice by 50% (Kuhlmann and Mosher, 1981). The bisintercalator ditercalinium (**6F**) dissociates from DNA 100–1000 times as slowly as the monomer **6G** (Delbarre *et al.*, 1987). In the case of **6F** and **6G** there is also a striking difference in mechanism of action. Ditercalinium (**6F**) causes the complete disappearance of mitochondrial DNA in L1210 leukemia cells without any observable effect on nuclear DNA. In contrast, **6G** only affects the nuclear DNA, does not have an effect on mitochondrial DNA content, and is less cytotoxic than 6F (Segal-Bendirdjian *et al.*, 1988; Fellous *et al.*, 1988). DNA binding constants can increase enormously as the number of intercalating units in the molecule is increased in certain cases. Estimated DNA binding constants for monoacridine (**7A**), bisacridine (**7B**), and trisacridine (**7C**) (Fig. 7) to DNA are 3×10^6, 2×10^{11}, and 10^{14}, respectively (Laugaa *et al.*, 1985). Daunorubicin (**8A**) in Fig. 8 can extend the life span of mice infected with P388 lymphocytic leukemia by only 71% at its optimum dose, whereas the dimeric hydrazones of daunorubicin (**8B** and **8C**) can each extend the life span in the same system by more than 108% at their optimum doses (Henry, 1982).

B. BISALKYLATING AGENTS

Bisalkylating agents often exhibit significantly better antineoplastic activity than monoalkylating agents, by virtue of the ability of bisalkylating agents to form cross-links between DNA strands. A number of bisalkylating agents, including cyclophosphamide and mechlorethamine, are used clinically in cancer treatment (John, 1983; Gilman *et al.*, 1980).

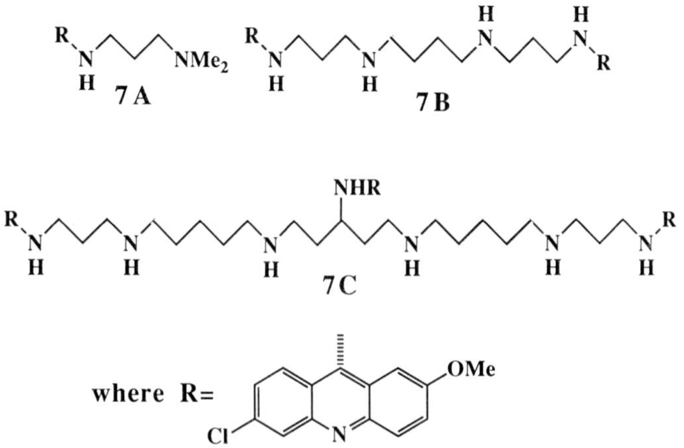

FIGURE 7   Interrelated (A) monointercalator, (B) bisintercalator, and (C) trisintercalator.

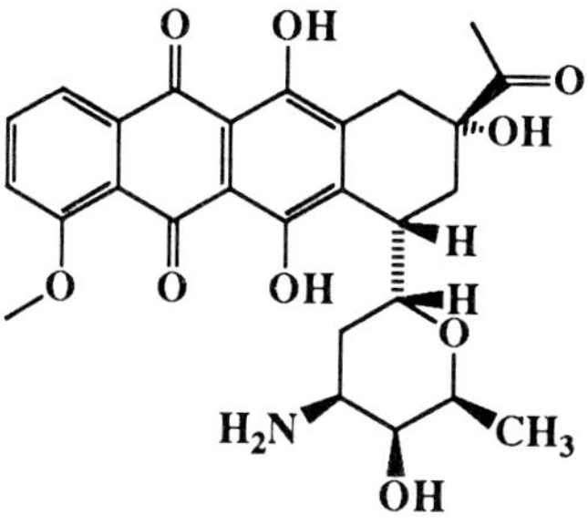

8A

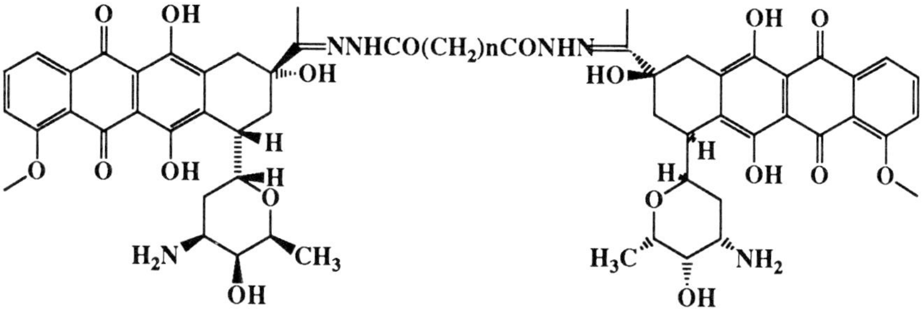

8B: n=2
8C: n=4

FIGURE 8   Daunorubicin and a dimeric daunorubicin bishydrazone.

C. METAL CHELATING AGENTS

Metal chelators formed by linking pyridine units together also show a trend of increasing cytotoxicity with increase in the number of subunits linked together, so that the bipyridine **9A** in Fig. 9 inhibits L1210 leukemic cell growth with an ED_{50} of 48 μM whereas the terpyridine **9B** has an ED_{50} of only 2 μM (McFadyen *et al.*, 1985). For an example of antibacterial synergism through metal chelate formation, see Rideout (1988).

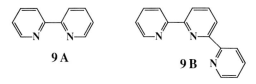

9 A   **9 B**

FIGURE 9   (A) Bipyridine and (B) terpyridine (metal chelators).

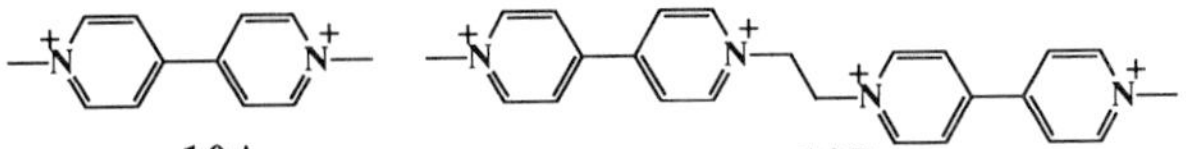

10A   10B

FIGURE 10   (A) Paraquat and (B) a paraquat dimer.

D. POLYCATIONS

Oligomerization of cations often leads to an enhancement in antimicrobial and cytotoxic activity. For example, dication **10A** in Fig. 10 inhibits B16 melanoma cells *in vitro* with an ED_{50} of 40 μM (using a thymidine incorporation assay). In contrast, the tetracation of related structure **10B** has an ED_{50} of only 4 μM in the same system (Minchin *et al.*, 1989). Polylysine can cause the permeabilization of phosphatidylcholine vesicles, a model system for permeabilization of cell membranes. A concentration of polylysine needed to cause 50% release of entrapped dye from phosphatidylcholine vesicles decreases in terms of charge ratio of polylysine to phospholipid as the molecular weight of polylysine increases. Polylysine of molecular size averaging greater than 70 KDa is twice as potent as 1.5–8 KDa polylysine (Gad *et al.*, 1982). The same trend is observed in the *in vitro* inhibition of HeLa cell growth with polylysine. The 70 KDa polylysine is 20 times as potent as 3 KDa polylysine in the inhibition of HeLa cell growth as determined on a per weight basis (Arnold *et al.*, 1979). Furthermore, the ability of 25 mg/kg of polylysine given on 5 consecutive days to extend the immediate life span of leukemia-bearing mice increases steadily with molecular weight. No antileukemic effect is observed for 3 KDa polylysine, compared with an 80% cure rate for the 70 KDa polymer (Arnold *et al.*, 1979). The polycation depicted in Fig. 11 exhibits antibacterial activity which rises dramatically with increasing molecular weight. The polymer with average molecular weight of 80,000 is more than 10,000 times as potent as the polymer of molecular weight 25,000 at a concentration of 2 μM (Ikeda *et al.*, 1986). The dicationic dequalinium (**12A**, Fig. 12) is an antineoplastic agent that selectively inhibits carcinoma cell

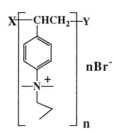

FIGURE 11   Polycations.

Dequalinium
12A

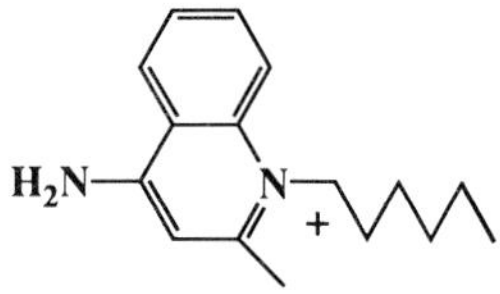

12 B

FIGURE 12   Interrelated (A) monocation and (B) dication.

growth. It is also capable of inhibiting protein kinase C with an IC_{50} less than 20 μM. In contrast, the monomeric analog **12B** fails to significantly inhibit protein kinase C even at 160 μM concentrations (Rotenberg *et al.*, 1990a). Compound **12A** is also significantly more potent than **12B** in its inhibition of MB49 murine bladder carcinoma cell growth *in vitro* (Rideout *et al.*, 1990b).

E. OTHER HOMODIMERS

Compound **13A** in Fig. 13 is an agent that binds in the minor groove of DNA and can also alkylate at that site. The monomeric **13A** has an ID_{50} of 60,000 pM against L1210 leukemia cells *in vitro*. In contrast the dimers **13B**, **13C**, and **13D** are significantly more potent with ID_{50}s of 2, 5, and 3000 pM, respectively, against L1210 under the same conditions (Mitchell *et al.*, 1989).

The dimer *bis*-ANS (**14B** in Fig. 14) binds to tubulin with a dissociation constant of 2 μM whereas the dissociation constant for ANS (**14A**) is 25 μM (Prasad *et al.*, 1986).

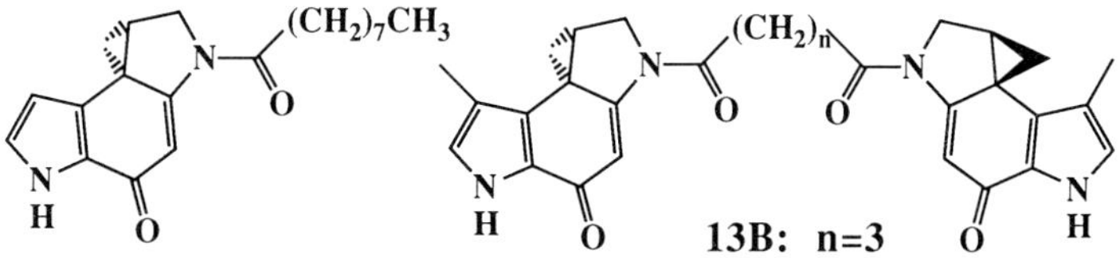

13A

13B: n=3
13C: n=8
13D: n=14

FIGURE 13   DNA alkylating agents that bind in the minor groove.

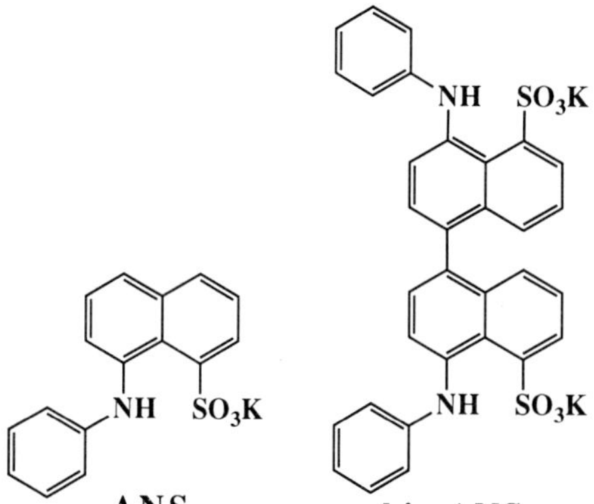

ANS   bis-ANS
14A   14B

FIGURE 14   Tubulin polymerization inhibitors.

The phenomenon of enhanced antineoplastic effects upon linking of two moieties is not limited to cell killing activity. The bisamide hexamethylenebisacetamide is a substantially better inducer of erythroleukemia cell differentiation than the related monoamide *N*-methylacetamide (Reuben et al., 1976).

Digallic acid (**15B** in Fig. 15) is more than 4 times as potent as the monomer gallic acid (**15A**) in terms of inhibiting the mutagenic properties of benzpyrenediolepoxide. For both molecules, anticarcinogenic activity most likely involves direct interaction with the epoxide leading to an acceleration of the rate of hydrolysis (Huang et al., 1985).

F. HETERODIMERS

Linkage of two different kinds of bioactive molecules to form heterodimers can lead to dramatic increases in overall cytotoxic or antimicrobial potency.

The phosphodiester conjugate **16A** in Fig. 16 is significantly more antiviral against HIV in cell culture than either of the precursors (azidodeoxythymi-

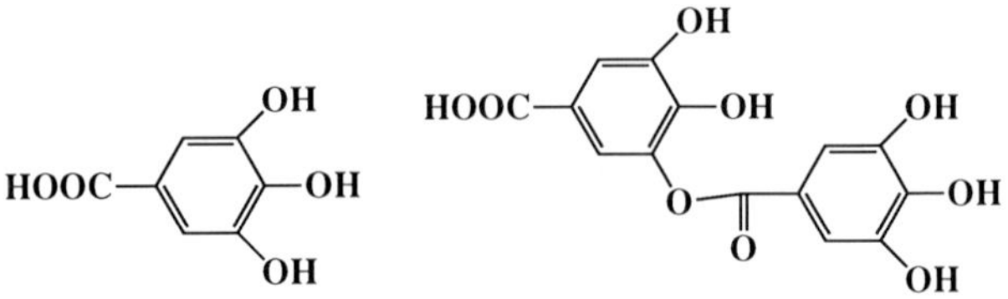

15A   15B

FIGURE 15   Anticarcinogens.

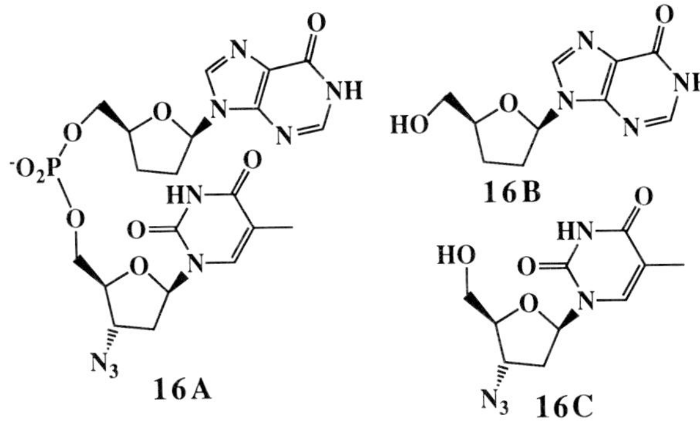

FIGURE 16   Antiretroviral agents.

dine) (**16C**) or (dideoxyinosine) (**16B**) from which it is formed. The greater
activity of **16A** in comparison **16B** and **16C** is believed to be due to intracellular hydrolysis to form a mixture of **16B**, **16C**, and their monophosphates.
Compounds **16B** and **16C** exhibit pronounced synergism (for further discussion, see Chapter 4 by Schinazi).

Linkage of a cyanomorpholine moiety to doxorubicin (**17A**) to form structure **17B** (Fig. 17) leads to a striking 500-fold increase of the cytotoxic potency
in vitro against L1210 cells and a 600-fold decrease in the optimal dose for the
inhibition of P388 leukemia *in vivo*. The cyanomorpholine moiety acts as an
alkylating agent so that compound B is better able than A to form a covalent
link to the DNA duplex after intercalation (Acton *et al.*, 1986).

For a description of synergism involving self-assembly of an amphiphilic
hydrazone from a cationic hydrazine derivative and a lipophilic aldehyde, see
Rideout (1986) and Rideout *et al.*, (1988), and p. 522.

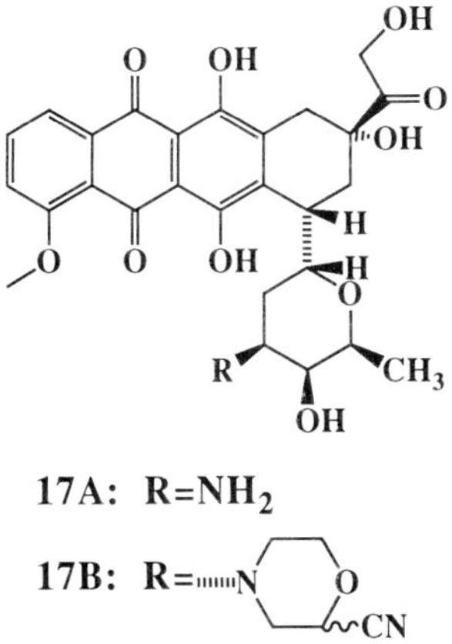

17A: R=NH$_2$

17B: R=

FIGURE 17   (A) Doxorubicin and (B) a derivative capable of alkylating DNA.

IV. Synergism Involving Covalent Self-Assembly of Cytotoxic and Antimicrobial Agents from Less Bioactive Precursors

A self-assembling chemotherapeutic agent is a combination of relatively nontoxic prodrugs which can react with one another irreversibly *in situ* to form a more bioactive product drug (i.e., a more cytotoxic, antimicrobial, or antiviral product). In this chapter, only spontaneous nonbiochemical condensation reactions (such as hydrazone formation from a hydrazine derivative and an aldehyde or ketone) will be considered, although enzyme-catalyzed reactions might also be useful. If the velocity of the reaction between prodrugs is more rapid at the target site than in healthy tissue, then the result will be target-selective synergism (Rideout *et al.*, 1990). The target site could be a tumor cell, a tumor extracellular space, a pathogenic microbe, a virally infected cell, or a subcellular compartment in which the pathogenic microbe resides. The resulting chemotherapeutic effect will be significantly enhanced through selective self-assembly at the target site, while the toxic effects of the drug on healthy tissue (where the velocity of drug formation is lower) will be enhanced only slightly. The combination of prodrugs will be more selective (although less potent) than the preformed product drug because of the unique advantages imparted through bimolecular kinetics. At optimized dose levels, the combination would be capable of providing more therapeutic benefits with fewer side effects than the performed product. The combination will be both more selective and more potent than either prodrug alone.

Figure 18 depicts one way in which the self-assembling drug approach could be used to amplify selectivity. First target and normal cells are exposed to identical concentrations of nontoxic prodrugs X and Y. Then X and Y are taken up by the target cell with slight distribution selectivity. As a result, let us suppose that their concentrations become 3-fold higher in the target than in normal tissues. Finally, X and Y react to form the cytotoxic product X–Y. Because the velocity of X–Y formation is proportional to the product of the concentrations of X and Y, the resulting concentration of cytotoxin X–Y will initially be 9-fold higher in the target. This analysis assumes identical second-order rate constants in both cells. Because of the multiplicative nature of second-order reaction kinetics (i.e., $d[X–Y]/dt = k_2[X][Y]$, where k_2 is the second-order rate constant), the resulting distribution selectivity of X–Y becomes amplified relative to the selectivities of X or Y alone. This 9-fold distribution selectivity will be independent of any selectivity inherent in X–Y as a preformed drug. If X–Y exhibits any selectivity due to distribution properties and/or innate biochemical or cytological differences between target and normal cells, the overall selectivity of target inhibition by combinations of X and Y will ultimately exceed 9. Although the kinetic analysis in this example is simplified, more sophisticated computer simulations which take into

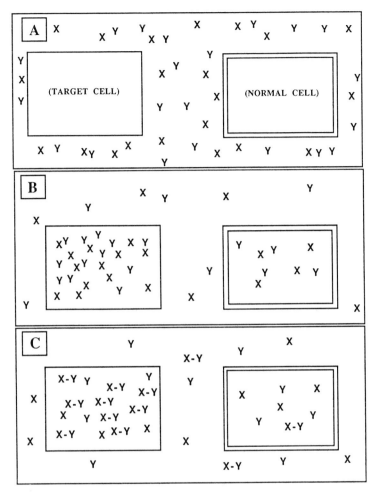

FIGURE 18   The enhancement of distribution selectivity effects using self assembly. Reprinted from Rideout *et al.*, 1990a by permission. Copyright © 1990, John Wiley & Sons, Inc.

(A) Two relatively nontoxic precursors (prodrugs) X and Y are administered.

(B) These accumulate *somewhat* selectively in the target cell.

(C) X and Y combine to form the more cytotoxic (or more antimicrobial) drug X–Y *more* selectively in the target cell because of the second order rate law.

account the simultaneous nature of product formation, prodrug and product efflux, and prodrug and product influx also predict that selectivity amplification should be possible using this approach (Rideout and Rideout, 1991). A possible example of this phenomenon, involving combinations of a triarylalkylphosphonium-derived aldehyde and an acylhydrazide (Rideout *et al.*, 1990), is described on p. 523.

Figure 18 represents a cell-targeting approach. An equally valid approach involves targeting the environment *near* the tumor cell, pathogenic microbe,

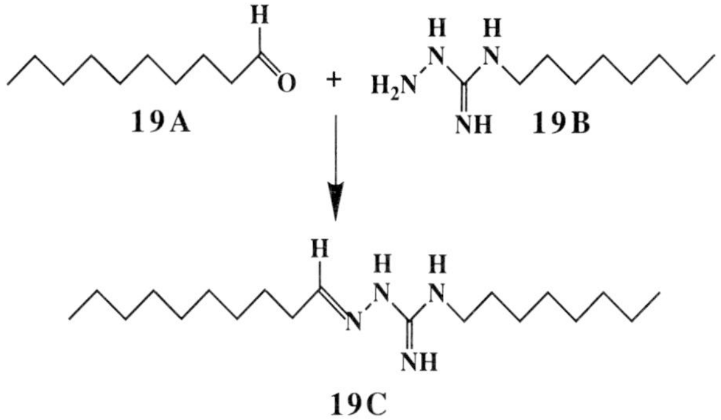

19C

FIGURE 19   A self-assembling amphiphilic cation.

virus, or virally infected cell (Rideout *et al.*, 1988). Hematoporphyrin derivative (fotofrin II) and lipiodol are among the agents which tend to accumulate
in the extracellular environment around tumor cells in solid tumors (Bohmer
and Morstyn, 1985; Ogita *et al.*, 1987). An example of environment-selective
synergism involving an *in vitro* model system has been described: a combination of **19A** and **19B** (Fig. 19) exhibits selective cytotoxic synergism in a
low-serum environment by virtue of more rapid formation of hydrazone **19C**
(Rideout *et al.*, 1988).

   Another approach to target-selective synergism using self-assembling combinations involves the exploitation of differences in the second-order rate
constant k_2 for the reactions in the target compared with normal tissues. If k_2
in the target is larger than k_2 in the normal tissue, then selective formation of
bioactive product from prodrugs would be more rapid in the target even for
prodrugs which could not accumulate selectively in the target. This form of
selectivity, unique to self-assembling chemotherapeutic agents, is called
assembly-rate-constant selectivity (Rideout *et al.*, 1988). Examples are described in Sections IVC and IVD.

A. BIOACTIVITY OF HYDRAZONES

   Hydrazones can exhibit chemotherapeutic activity clinically, in animals,
and/or *in vitro* through a very wide variety of mechanisms, including DNA
intercalation, cell membrane lysis, inhibition of bacterial cell wall synthesis,
reverse transcriptase inhibition, inhibition of tubulin aggregation, and so
forth (reviewed by Rideout *et al.*, 1990a). In addition, viruses can be inhibited
by a hydrazone that inhibits ribonucleotide reductase (see Chapter 8 by
Spector and Fyfe), the antineoplastic hydrazone antibiotic luzopeptin A can
form interduplex DNA cross-links (Searle *et al.*, 1989), the anticoccidial
hydrazone robenzidine inhibits oxidative phosphorylation in mammalian liver
cells (Wong *et al.*, 1972), several thiosemicarbazones of 2-acetylpyridine show

activity *in vivo* against mice bearing the malarial parasite *Plasmodium berghei* (Klayman *et al.*, 1979), and antimitotic hydrazones have been described (Temple, 1990).

B. ANTINEOPLASTIC COMBINATIONS
OF ALDEHYDES AND HYDRAZINE DERIVATIVES

We have developed a self-assembling protein kinase C inhibitor consisting of aldehyde **20A** and acylhydrazide **20B**. Testing of PKC inhibition by this combination has been carried out in collaboration with S. Rotenberg and I. B. Weinstein (Columbia University Cancer Center, New York). The adduct formed from **20A** and **20B** is hydrazone **20C**, a dication that inhibits protein kinase C activity with an IC_{50} value of 20.4 μM. In contrast, the values for **20A** and **20B** are both > 200 mM. Inhibition of α2-PKC by **20A** alone, **20B** alone, or by **20C** alone shows no significant time dependence. However, when α2-PKC is exposed to 200 μM **20A** and 200 μM **20B** in combination, the activity of the enzyme drops from 48% of control at 0 min to 15% of control at 70 min. The percentage of control values are 87 and 40, respectively, for 100 μM + 100 μM. Kinetic studies using HPLC demonstrate the formation of approximately 27 μM **20C** from 200 μM **20A** + 200 μM **20B** under the PKC assay conditions after 60 min. In conclusion, the combination of **20A** and **20B** exhibits synergistic inhibition of PKC in a time-dependent manner through direct assembly to form the more potent inhibitor **20C** (Rotenberg *et al.*, 1990b).

The structures of phosphonium salts **20A** and **20B** resemble tetraphenylphosphonium, a delocalized lipophilic cation that accumulates selectively in many carcinoma cells relative to most untransformed cells (see Rideout *et al.*, 1989, and Chen, 1988). Tetraphenylphosphonium and **20B** all inhibit Ehrlich–Lettre ascites (ELA), PaCa-2 human pancreatic carcinoma, and MB49 murine bladder carcinoma selectively relative to CV-1 untransformed monkey kidney cells, whereas cisplatin and cytarabine are unselective (Rideout *et al.*, 1989, 1990 unpublished observations). We predicted that **20A** and **20B** would accumulate selectively in carcinoma cells, combining selectively therein to form **20C** (as in Fig. 20). Perhaps because it is a better PKC inhibitor, **20C** is more cytotoxic than either **20A** or **20B**. (Rideout *et al.*, 1990). Carcinoma-selective accumulation of **20A** and **20B** should lead to carcinoma-selective formation of the more cytotoxic **20C**, and thus to carcinoma-selective synergism between **20A** and **20B**. As predicted, combinations of **20A** and **20B** exhibit synergism in their inhibition of ELA cells, but not in their inhibition of CV-1 cells, as determined by isobologram and combination index methodologies (Rideout *et al.*, 1990). Combinations related to **20A** + **20B** that cannot form hydrazones (e.g., **20A** + **20D** and **20B** + **20E**) do *not* exhibit synergism in their inhibition of ELA cell growth, consistent with the hypothesis that hydrazone formation is the mechanism of synergism

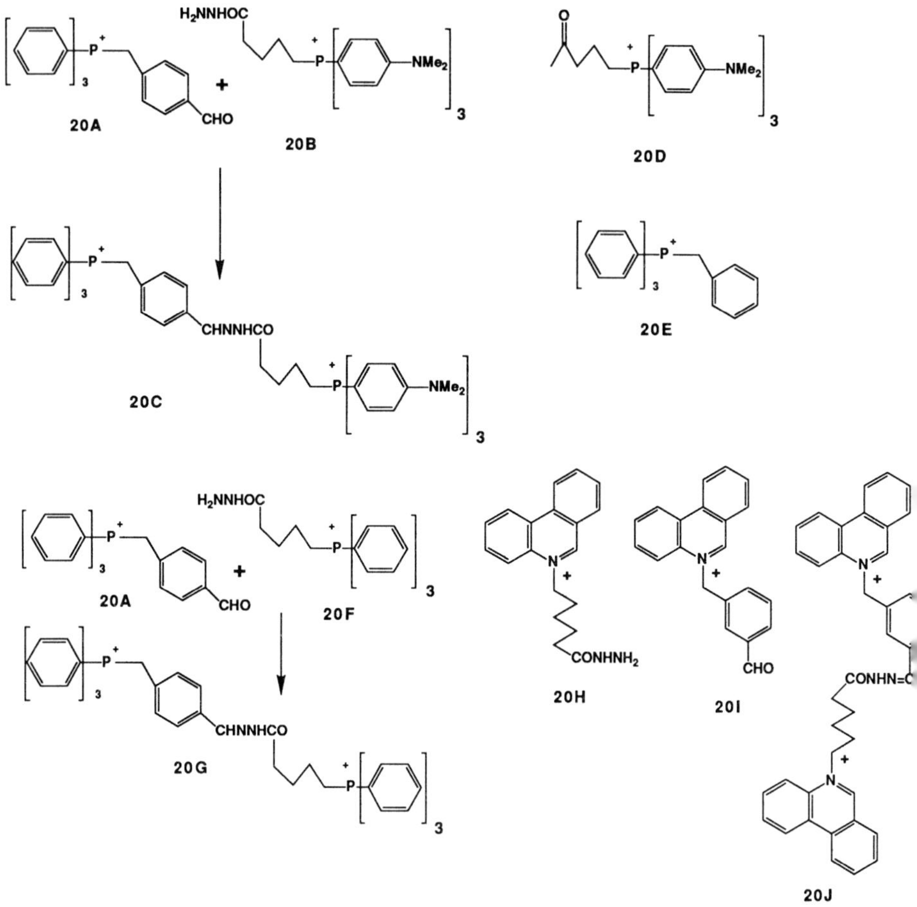

FIGURE 20   Synergistic antineoplastic combinations of delocalized lipophilic cations.

between **20A** and **20B**. In summary, ELA cells probably selectively catalyze hydrazone **20C** formation from **20A** and **20B** by concentrating the two together in the cytoplasm and/or mitochondria. The **20C** thus formed inhibits cell growth more effectively than the precursors (Rideout *et al.*, 1990).

Aldehyde **20A** and acylhydrazine **20F** are both less cytostatic than their hydrazone **20G** in assays using PaCa-2 (2-day exposure) and CV-1 cells (2-day exposure). The cytostatic synergism between **20A** and **20F** against PaCa-2 carcinoma cells (2-day assay, isobologram and combination index analysis) is most likely due to the formation of the more cytostatic hydrazone **20G**. Modest carcinoma-selective synergism against PaCa-2 cells has also been observed for combinations of **20A** and **20F**. In other words, a combination of 212 μg/ml of **20F** with 31 μg/ml of **20A** is more selective than either

424 μg/ml of **20F** alone or 62 μg/ml of **20A** alone (Rideout *et al.*, unpublished observations).

A combination of **20H** and **20I** is quite synergistic against Paca-2 carcinoma cells (2-day exposure). Assuming mutually exclusive mechanism, at a 50% inhibition level the combination indexes (see Chapter 2 by Chou) are less than 0.488 for **20H** + **20I** at a 10.2:1 molar ratio, and less than 0.568 at a 4.4:1 ratio. At a 1:1.62 molar ratio, the combination **20H** + **20I** is also synergistic against MB49 cells, with a combination index less than 0.68 (2-day exposure, 50% effect level). The synergism observed in this combination probably involves the *in situ* formation of the dicationic hydrazone **20J** (Rideout *et al.*, unpublished observations).

C. ANTIMICROBIAL COMBINATIONS OF ALDEHYDES AND HYDRAZINE DERIVATIVES

Mixtures of aldehydes and acylhydrazines that can react *in situ* to form antimicrobial hydrazones demonstrated greater degrees of synergism against the intracellular pathogen *Salmonella typhimurium* at pH 5 than at pH 7.4. These mixtures include 5-nitro-2-furaldehyde (**21D**) plus semicarbazide (**21E**) and 2-hydrazinopyridine (**21A**) plus pyridine-2-carboxaldehyde (**21B**) (Fig. 21). Combinations are more selectively toxic to bacteria at pH 5 (versus pH 7.4) than individual precursors and preformed hydrazone products because acid-catalyzed hydrazone bond formation plays a role only for the combination (Rideout *et al.*, 1988).

The precursors **21D** and **21E** are at least 100 times as soluble in water as their product **21F**. This suggests that self-assembling combinations of soluble prodrugs might be used as a means of administering drugs that are otherwise too insoluble to be used as single agents in preassembled form.

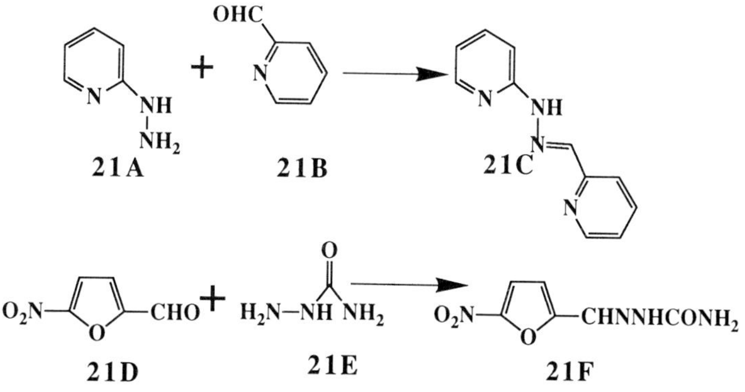

FIGURE 21 Synergistic antibacterial combinations.

Pyridoxal and isoniazid exhibit antimalarial synergism due to the formation of an iron-chelating hydrazone *in situ* (see Chapter 5 by Vennerstron *et al.*).

D. HYDRAZONE FORMATION IN MACROPHAGES: PROGRESS TOWARD SELF-ASSEMBLING ANTIRETROVIRAL AGENTS

Hydrazine **22A** (Fig. 22) accumulates with some selectivity in macrophages through rapid pinocytic accumulation in lysosomes (Swanson *et al.*, 1985). This fact suggests that hydrazones might be found selectively in macrophages from **22A** and various aldehydes. If such a hydrazone possesses antiretroviral activity, the corresponding aldehyde and acylhydrazide could be used to selectively target the hydrazone to macrophages which act as reservoir for the AIDS virus (Fauci, 1988; Pauza, 1988).

Hydrazones **22B**, **22C**, and **22D**, all derived from hydrazine of **22A**, inhibit reverse transcriptase isolated from HIV-1 and/or feline immunodeficiency virus (FIV) (Rideout *et al.*, unpublished observations). FIV is a lentivirus that

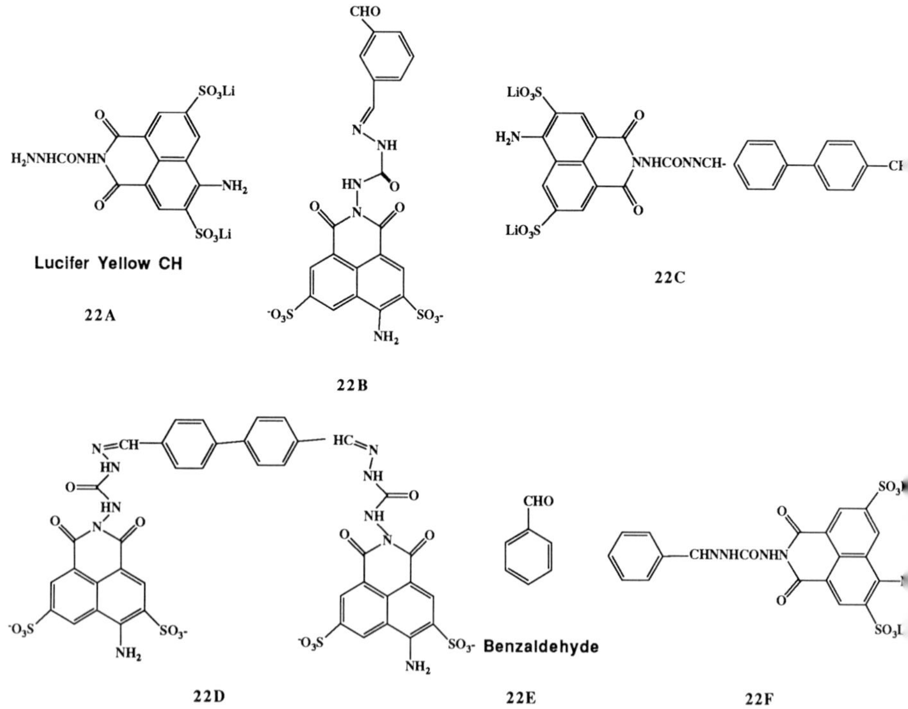

FIGURE 22   (A) Lucifer Yellow CH and (B–D,F) antiretroviral hydrazones.

causes a syndrome resembling AIDS in cats (Pedersen *et al.*, 1987; Talbott *et al.*, 1989). Hydrazone **22C** can inhibit FIV proliferation by >99% at nontoxic doses (Rideout *et al.*, unpublished observations). Hydrazine **22A** has very low cytotoxicity (Swanson *et al.*, 1985) and whole animal toxicity (S. C. Silverstein, unpublished observations) but has no measurable reverse tran-scriptase-inhibiting or antiretroviral activity (Swanson *et al.*, 1985). Together these results suggest that it may be possible to achieve macrophage-selective delivery of antiretroviral hydrazones related to **22C** by administering them as a combination of less toxic aldehyde and acylhydrazide prodrugs (which need not have antiviral activity as single agents in order to be effective as prodrugs).

Hydrazone **22F** forms from benzaldehyde (**22E**) and the hydrazine de-rivative Lucifer Yellow CH (**22A**) at 37°C with a rate constant that is ap-proximately 100 times higher at pH 5.0 than at pH 7.4 (Rideout *et al.*, unpublished observations). The hydrazine **22A** is taken up by macrophages through pinocytosis (Swanson *et al.*, 1985) and concentrated in lysosomes, in which the pH is close to 5 (Roos and Boron, 1981). When TPA-pretreated U937 human leukemic cells are treated for 24 hr with **22A** (2 mg/ml), washed with PBS, and treated with 471 μM benzaldehyde (S3B) in PBS + 8% serum for 20 hr at pH 7.4, half of the **22A** is converted to hydrazone **22F** [TPA-treated U937 cells are a convenient model for human monocyte/macrophages (Harris and Ralph, 1985)]. If the cells are killed by hypoosmotic shock and freezing just prior to **22A** addition, less than 10% of **22A** is converted to **22F**. Addition of chloroquine, which lowers the lysosomal pH, substantially re-duces the rate of hydrazone formation in the cell extracts. If human monocyte/macrophages are treated with **22A** for 24 hr, and **22E** is then added without washing away external **22A**, **22F** forms with a $t_{1/2}$ of 2 hr in the cell extracts. In contrast, there is less than 10% conversion to **22F** in the ex-tracellular medium (pH 7.4) after 20 hr. These data are consistent with accelerated hydrazone formation inside the cell lysosomes through pinocyto-sis of **22A**, followed by diffusion of **22E** into the lysosomes, and acid catalysis of **22F** formation (Rideout *et al.*, unpublished observations).

The reaction between **22A** and **22E** can also occur inside living mice. BALB/c female mice were treated ip with thioglycolate broth to induce macrophages (Campbell, 1984). After 4 days, they were treated with **22A** (120 mg/kg ip). After 24 hr, **22E** (30 mg/kg) was added. After 2 more hr, ip macrophages and extracellular fluids were isolated postsacrifice and analyzed using fluorescence-linked HPLC. The hydrazone **22F** was detected (using HPLC) inside the macrophages and in the extracellular fluid (Rideout *et al.*, unpublished observations). This result is important because it suggests that antiretrovial hydrazones could be delivered to macrophages by aldehyde and acylhydrazide administration *in vivo*. Perhaps more important, the experiment demonstrates that a hydrazone can be formed within a live animal from an aldehyde and a hydrazine derivative, even when *both* of these precursors are administered rather than being endogenous metabolites.

In summary, selective accumulation of one or both prodrug precursors in or near the target and/or an enhanced reactivity between the prodrug precursors in the target could be used to attain target-selective synergism and amplified selective inhibition of the target. The key strength of the approach is flexibility. Useful drug self-assembly could occur inside target cells, in the extracellular environment associated with the target, or in a particular type of subcellular compartment containing a target pathogen. It could occur in solution, in membranes, on proteins, or on nucleic acids. The concept is neither restricted to any one type of bimolecular reaction, nor to any one mechanism of selective accumulation, nor to any type of environmental difference responsible for assembly-rate-constant selectivity, nor to any particular mechanism of target inhibition by the product drug.

V. Antagonism Involving
Covalent Self-Assembly

One strategy for enhancing the therapeutic index of a chemotherapeutic agent involves decreasing the host toxicity through selective chemical combination with a second agent in host tissue, forming a third, less-toxic product. Most of the reported examples of this covalent modulation strategy involve combinations of cisplatinum and ligands. Examples of antagonism involving iron chelation and epoxide hydrolysis will also be discussed here.

Chelators and thiols can protect mammals against the toxicity of metal ions. For example, *N*-acetylcysteine can facilitate the excretion of lead and cadmium from rats. The chelator "tiron" can protect against the toxicity of uranium oxide ions in Swiss mice and the clinically used chelating agent dimercaprol can protect against acute cadmium toxicity in rats (Ottenwälder and Simon, 1987; Domingo *et al.*, 1989; Peele *et al.*, 1988). More interesting from the point of view of chemotherapy is the reduction of the toxicity of platinum-derived antitumor agents such as cisplatin and carboplatin by a variety of nucleophilic ligands, mostly thiols (see next section).

A. CISPLATIN MODULATION WITH
CHLORIDE ANION

Much of the toxicity of the anticancer agent cisplatin to human beings is believed to be due to the reaction of hydrated forms of the drug such as $Pt(NH_3)_2(H_2O)(OH)^+$ with proteins (Dedon and Borch, 1987). One of the key dose-limiting toxicities of cisplatin involves damage to kidneys. The hydrated form of cisplatin inactivates the rat renal proximal tubular enzyme

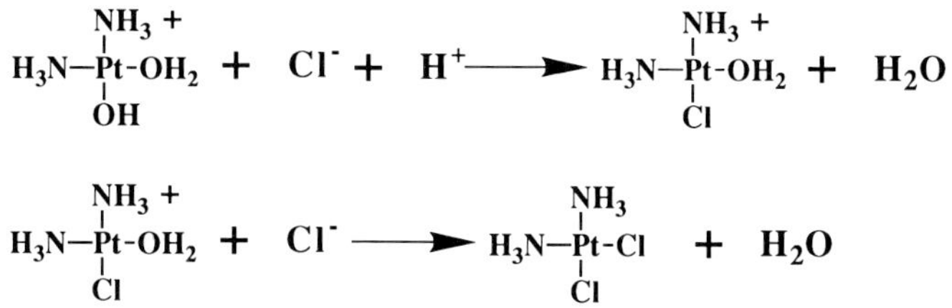

$$\underset{\underset{OH}{|}}{\overset{\overset{NH_3\ +}{|}}{H_3N-Pt-OH_2}} + \ Cl^- + \ H^+ \longrightarrow \underset{\underset{Cl}{|}}{\overset{\overset{NH_3\ +}{|}}{H_3N-Pt-OH_2}} + \ H_2O$$

$$\underset{\underset{Cl}{|}}{\overset{\overset{NH_3\ +}{|}}{H_3N-Pt-OH_2}} + \ Cl^- \longrightarrow \underset{\underset{Cl}{|}}{\overset{\overset{NH_3}{|}}{H_3N-Pt-Cl}} + \ H_2O$$

FIGURE 23   Detoxification of nephrotoxic cisplatin metabolites through self-assembly with chloride ion.

γ-glutamyltranspeptidase (γ-GT) at 2 μM with a half-life of just over 1 hr, wheras cisplatin itself in water requires over 3 hr at the same concentration (Dedon and Borch, 1987). In the presence of 100 mM sodium chloride, cisplatin requires more than 5 hr to inactivate γ-GT by 50% under the same conditions. This is believed to be due to a self-assembly reaction between chloride ion and various toxic hydrated cisplatin species as shown in Fig. 23 (Dedon and Borch, 1987). Chloride ion in the form of hypertonic saline is used to decrease the kidney toxicity of cisplatin in the clinic through this self-assembly reaction (Fuks *et al.*, 1987; Aamdal *et al.*, 1987). Related studies of cisplatin hydration and reformation include those by Corden (1987) and by LeRoy and Thompson (1989).

B. CISPLATIN MODULATION WITH THIOLS AND SELENIUM

A number of sulfur nucleophiles which form complexes with platinum derivatives involving sulfur–platinum bonds have been shown to reverse many of the toxic side effects of cisplatin and related platinum-containing anticancer agents. The chemical structures of several such nucleophiles are shown in Fig. 24. The modulation of cisplatin toxicity by thiols was recently reviewed by Leyland-Jones (1988).

1. Thiosulfate

In a classic study, Howell *et al.* (1983) showed that thiosulfate delivered intravenously can protect human subjects against the renal toxicity of cisplatin administered intraperitoneally. They showed that unreacted cisplatin could be found in high levels in plasma, despite the fact that concentrations of thiosulfate present in the plasma were 280 to 950 times higher than plasma cisplatin concentrations. It was proposed that thiosulfate becomes extensively concentrated in the kidneys, producing a high enough local concentration to

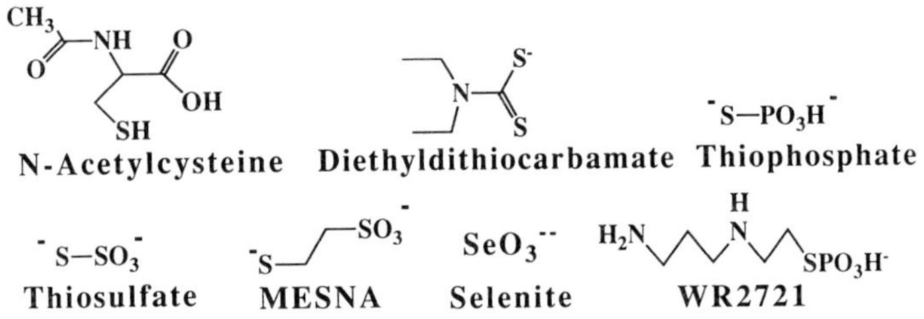

N-Acetylcysteine   Diethyldithiocarbamate   Thiophosphate

Thiosulfate   MESNA   Selenite   WR2721

FIGURE 24 Sulfur nucleophiles for cisplatin and alkylating agent "rescue" (antagonism) involving Pt-S, Pt-Se, or C-S bond formation.

neutralize cisplatin and its metabolites in renal tubules without compromising the antineoplastic activity of cisplatin elsewhere. In other words, kidney-selective detoxification of cisplatin occurs through kidney selective assembly between thiosulfate and cisplatin. In a later study, Goel et al. (1989) showed that sodium thiosulfate can also prevent peripheral neuropathy caused by cisplatin and perhaps even decrease some of the central nervous system effects such as fatigue and vomiting that can be associated with clinical cisplatin therapy. Uozumi and Litterst (1986) showed that although thiosulfate protects against kidney damage by cisplatin, thiosulfate does not induce a significant change in the subcellular distribution of platinum within rat kidney cells. The authors concluded that thiosulfate acts by forming a nontoxic complex with cisplatin in the kidney rather than influencing the total amount of platinum metal that reaches vulnerable targets within kidney cells. Thiosulfate was shown to protect guinea pigs from hearing loss induced by cisplatin. This observation is significant because hearing loss is one of the irreversible side effects which is dose limiting in cisplatin usage (Otto et al., 1988).

Cisplatin has been shown to react with 2 mM solutions of thiosulfate, being completely consumed within a few hours at 37°C and forming convalent adducts in aqueous buffers (Elferink et al., 1986). Carboplatin is considerably less reactive under the same conditions. Electron opaque bodies are found in the lysosomes of kidney tubule cells when cisplatin and thiosulfate are administered together to male Wistar rats, but not when either cisplatin or thiosulfate is administered alone. On microanalysis these opaque bodies have been shown to contain high levels of sulfur and platinum, suggesting that covalent combination between thiosulfate (or its metabolites) and platinum-containing species had occurred in vivo (Berry and Lespinats, 1988). The reaction between thiosulfate and cisplatin has been shown by HPLC to occur at a rate of about 0.06 mol/sec at 37°C (Dedon and Borch, 1987). Thiosulfate is incapable of reacting with cisplatin/thioprotein adducts and thus cannot reverse damage caused by cisplatin to proteins such as those in the renal tubules. As a result, thiosulfate must be given prior to and simultaneous with cisplatin in order to decrease the toxic effects of the drug (Dedon and Borch,

1987; Elferink *et al.*, 1986.) In addition, thiosulfate does reduce the antitumor activity of cisplatin in animals when certain procedures are used (Aamdal *et al.*, 1988).

2. Diethyldithiocarbamate

In contrast to thiosulfate, diethyldithiocarbamate (DDTC) is capable of reversing adducts between cisplatin and cysteine-containing proteins, thereby freeing the protein. As a result, DDTC can be administered after cisplatin and still significantly reduce the host toxicity of the heavy metal complex (Basinger *et al.*, 1989; Dedon and Borch, 1987).

Diethyldithiocarbamate (DDTC) is unusual because it can reverse cisplatin/thiol protein adducts. For example, DDTC reactivates γ-glutamyltranspeptidase (γ-GT) which can become inhibited due to cisplatin derivitization (γ-GT is an important enzyme found at high levels in kidney tissue). On the other hand, DDTC is not able to reverse the cross-links which cisplatin introduces between adjacent guanine residues in DNA. In other words, DDTC can reverse the nephrotoxic lesions of cisplatin while sparing the cell-killing, potentially antineoplastic lesions of cisplatin (Bodenner *et al.*, 1986). Unfortunately, at concentrations which are sufficient to reverse the toxic effects of cisplatin, DDTC itself exhibits quite significant toxicity (Leyland-Jones, 1988).

DDTC is also capable of reversing adducts between cisplatin and human $\alpha2$-macroglobulin (Gonias *et al.*, 1984). It has been shown to protect animals from the kidney and bone marrow toxicity of cisplatin at levels that do not compromise the antineoplastic activity of cisplatin (Gonias *et al.*, 1984; Schmalbach and Borch, 1989). It also suppresses the bone marrow toxicity of the cisplatin analog, carboplatin (Schmalbach and Borch, 1989).

3. Other Sulfur Nucleophiles and Selenium

The radioprotectant S-2-(3-aminopropylamino)-ethylphosphorothioic acid (WR2721) is capable of protecting peripheral nerves from cisplatin toxicity in human subjects (Mollman *et al.*, 1988). Glover *et al.*, (1984) showed in a clinical trial that the organic thiophosphate WR2721 provides some protection against cisplatin-mediated nephrotoxicity in human subjects. Presumably, WR2721 is dephosphonylated to the thiol *in vivo*. More recent clinical trials of WR2721/cisplatin are encouraging (Glover *et al.*, 1989; Mahaney, 1990). The thiol 2-mercaptoethanesulfonate (MESNA) not only protects rats against the nephrotoxicity of cisplatin, but also completely suppresses cisplatin induced benign and malignant neoplasm formation in these animals. On the other hand, it does not seem to have an effect on the antineoplastic effects of cisplatin against L1210 leukemia in mice (Kempf and Ivankovic, 1987). Selenium in the form of sodium selenite dramatically reduces the intestinal and kidney toxicity and associated lethality of cisplatin (Satoh *et al.*, 1989). Electron microprobe and metal distribution studies suggest that direct selenium–platinum bond formation is involved in the detoxification (Satoh *et al.*, 1989; Naganuma *et al.*, 1983; Berry and Lespinats, 1988). Selenium (in the form of sodium selenate) has also been shown to decrease cisplatin toxicity. A direct combination mechanism has been proposed (Ohkawa *et al.*, 1988).

One must be cautious in interpreting many of the results as involving simple direct combination between cisplatin and the antagonizing agent. It is possible in some instances to induce the synthesis of biological thiols such as metallothionein which is induced by bismuth salts (Satoh *et al.*, 1988). It is possible that the administration of selenium or of thiols such as thiosulfate causes the induction of endogenous thiol synthesis or of enhanced levels of enzymes such as glutathione-*S*-transferase which can catalyze the detoxification of agents such as cisplatin. Cisplatin reacts with 5 μM (excess) glutathione to form a product with a sulfur–platinum bond. The reaction half life is 200 min at 40°C (Corden, 1987). There is still only limited direct evidence involving animal studies for significant direct bond information between cisplatinum and thiols or selenium administered exogenously.

C. IRON CHELATION

8-hydroxyquinoline and a one-to-one 8-hydroxyquinoline:Fe(II) complex exhibit antibacterial antagonism by virtue of self-assembly to form 2(8-hydroxyquinoline):Fe(II), which is much less active that the 1:1 complex. At physiological iron concentrations, 8-hydroxyquinoline is active as an antibacterial agent at 10 μM yet much *less* active at 1000 μM concentrations, an unusual phenomenon known as "concentration quenching" (see Chapter 5 by Vennerstrom *et al.*).

D. EPOXIDE HYDROLYSIS

Plant-derived phenols such as tannic and ellagic acid are part of a large and diverse group of compounds that can prevent the mammalian cell mutagenesis and carcinogenesis by certain chemical agents and radiation (Wattenberg, 1985; Mukhtar *et al.*, 1988). Tannic acid is particularly active in preventing the induction of skin tumorigenesis in mice by the carcinogens dimethylbenzanthracene, benzpyrene, and 3-methylcholanthrene. By comparing a large number of plant flavonoids, Huang *et al.* (1983) have shown that phenolic groups are essential for the inhibition of mutagenicity by bay region diolepoxides.

Ellagic acid (**25B** in Fig. 25) is a naturally occurring plant phenol that is normally ingested by humans. A 10 μM concentration of **25B** increases the

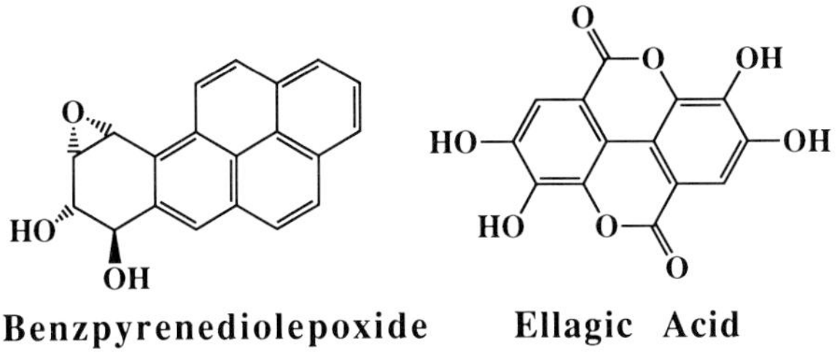

Benzpyrenediolepoxide   Ellagic Acid

FIGURE 25   A carcinogen (benzpyrenediolepoxide) and an anticarcinogen (ellagic acid).

rate of disappearance of the carcinogen benzpyrene-7,8-diol-9,10-epoxide (**25A**) by a factor of 20 at pH7. At 2 μM, **25B** can inhibit the mutagenicity of 0.1 μM **25A** by 50% in the Ames *Salmonella typhimurium* mutagenicity assay (Wood *et al.*, 1982). Despite the simplicity of elagic acid's structure, the rate of the reaction between **25B** and **25A** has a surprisingly high second-order rate constant of 560 molar^{-1} sec^{-1}. The toxic antagonism of **25A**, a metabolite of the ubiquitous carcinogen benzpyrene, by a variety of phenols found in common foods is of possible importance in the prevention of cancer through control of diet. Tannic acid inhibits the mutagenicity of **25A** against *S. typhimurium* by 50% at 1 μM concentrations (Huang *et al.*, 1985).

The rates of hydrolysis of **25A** by a variety of soluble aromatic compounds increase in the order ferullic acid, caffeic acid, chlorogenic acid, flavin mononucleotide, and elagic acid (Wood *et al.*, 1982). The ability of these compounds to protect V79 Chinese hamster cells or *S. typhimurium* cells against mutagenesis increases in the same order, suggesting that in cell culture as well as *in vitro* direct catalysis of **25A** hydrolysis is the mechanism of protection for these compounds. In other words, these aromatic compounds inhibit the mutagenicity of **25A** through a direct interaction (Wood *et al.*, 1982).

E. MODULATION OF ALKYLATING
AGENTS WITH THIOLS

The thiol 2-mercaptoethane sulfonate sodium (MESNA) is known to react with toxic metabolites of cyclophosphamide such as acrolein, forming a carbon–sulfur bond. MESNA greatly diminishes the toxicity of these metabolites, which probably do not contribute the antineoplastic effects of cyclophosphamide. MESNA is now being used clinically in an attempt to decrease the kidney toxicity of alkylating agents such as ifosfamide (Goren *et al.*, 1989). Similarly, glutathione monoethyl ester can selectively protect the liver from high doses of alkylating agents such cyclophosphamide or BCNU without compromising anticancer activity against fibrosarcoma in mice (Teicher *et al.*, 1988). Like MESNA, *N*-acetylcysteine can protect against the toxicity of phosphoramide mustard without compromising its cytotoxic activity. It was shown that although *N*-acetylcysteine can react with acolein, an undesirable toxic metabolite of cyclophosphamide, *N*-acetylcysteine does *not* react with phosphoramide mustard, the active antineoplastic metabolite of cyclophosphomide (Seitz *et al.*, 1989).

In summary, chemotherapeutic synergism and antagonism between two agents can be mediated through direct chemical reaction between the agents, creating a covalent bond and forming a new molecule or molecules. Some of the best characterized examples of this very straight forward mechanism of interaction (covalent modulation) involve carbon–nitrogen double bond or platinum sulfur bond formation. Other types of bimolecular reactions that can be rapid under near-physiological conditions (e.g. Diels–Alder condensations) might someday be applied in novel synergistic or antagonistic combinations. Enhancement of the therapeutic index can be achieved using covalent modulation if reactions leading to a more active product occur more rapidly at the target site or if reactions producing a less toxic product occur more rapidly in normal tissues.

Acknowledgments

This work was supported, in part, by the National Cancer Institute (CA44505, CA02656), the American Foundation for AIDS Research (001163-8-RG) and PPG Industries, Inc.

References

Aamdal, S., Fodstad, O., and Pihl, A. (1987), Some procedures to reduce cis-platinum toxicity reduce antitumor activity. *Cancer Treat. Rev.* **14**, 389–395.

Aamdal, S., Fodstad, O., and Pihl, A. (1988). Sodium thiosulfate fails to increase the therapeutic index of intravenously administered cis-diamminedichloroplatinum (II) in mice bearing murine and human tumors. *Cancer Chemother. Pharmacol,* **21**, 129–133.

Acton, E. M., Tong, G. L., Taylor, D. L., Streeter, D. G., Filppi, J. A., and Wolgemuth, R. L. (1986). N-(Cyanomethyl)- and N-(2-methoxy-1-cynanoethyl)anthracyclines and related carboxyl derivatives. *J. Med. Chem.* **29**, 2074–2079.

Arnold, L. J., Jr., Dagan, A., Gutheil, J., and Kaplan, N. O. (1979), Antineoplastic activity of poly(L-lysine) with some ascites tumor cells. *Proc. Natl. Acad. Sci. U.S.A.* **76**, 3246–3250.

Balazova, E., and Koza, I. (1988). Therapy of P388 leukemia with benzaldehyde. *Neoplasma* **35**, 725–728.

Basinger, M. A., Jones, M. M., Gilbreath, S. G., IV, Walker, E. M., Jr., Fody, E. P., and Mayhue, M. A. (1989). Dithiocarbamate-induced biliary platinum excretion and the control of cis-platinum nephrotoxicity. *Toxicol. Appl. Pharmacol.* **97**, 279–288.

Berry, J. P., and Lespinats, G. (1988). Cis DDP in combination with selenium and sulfur. Subcellular effect in kidney cells. Electron microprobe study. *J. Submicrosc. Cytol. Pathol.* **20**, 59–65.

Bodenner, D. I., Dedon, P. C., Keng, P. C., and Borch, R. F. (1986), Effect of dieth-yldithiocarbamate on cis-diamminedichloroplatinum(II)-induced cytotoxicity, DNA cross-linking, and γ-glutamyl transpeptidase inhibition. *Cancer Res.* **46,** 2745–2750.

Bohmer, R. M., and Morstyn, G. (1985). Uptake of hematoporphyrin derivative by normal and malignant cells: Effect of serum, pH, temperature, and cell size. *Cancer Research* **45,** 5328–5334.

Breslow, R., Maitra, U., and Rideout, D. (1983). Selective Diels–Adler reactions in aqueous solutions and suspensions. *Tetrahedron Lett.* **24,** 1901–1904.

Campbell, A. M. (1984). Animal handling techniques. *In* "Monoclonal Antibody Technology" (R. H. Burdon and P. H. van Knippenberg, eds.), pp. 216–231. Elsevier, New York.

Carruthers, W. (1978). "Some modern methods of organic synthesis." Cambridge University Press, Cambridge.

Chen, L. B. (1988), Mitochondrial membrane potential in living cells. *Annu. Rev. Cell Biol.* **4,** 155–181.

Chlebowski, R. T., and Heber, D. (1986). Metabolic abnormalities in cancer patients: Carbohy-drate metabolism. *Surg. Clin. North Am.* **66,** 957–968.

Chlebowski, R. T., Bulcavage, L., Grosvenor, M., Tsunokai, R., Block, J. B., Heber, D., Scrooc, M., Chlebowski, J. S., Chi, J., Oktay, E., Akman, S., and Ali, I. (1987). Hydrazine sulfate in cancer patients with weight loss. (*Philadelphia*) *Cancer* **59,** 406–410.

Corden, B. J. (1987). Reaction of platinum(II) antitumor agents with sulfhydral compounds and the implications for nephrotoxicity. *Inorg. Chim. Acta* **137,** 125–130.

Dealler, S. F., Hawkey, P. M., and Millar, M. R. (1988). Enzymatic degradation of urinary indoxyl sulfate by *Providencia stuartii* and *Klebsiella pneumoniae* causes the purple urine bag syndrome. *J. Clin. Mircrobiol.* **26,** 2152–2156.

Dedon, P. C., and Borch, R. F. (1987). Characterization of the reactions of platinum antitumor agents with biologic and nonbiologic sulfur-containing nucleophiles. *Biochem. Pharmacol.* **36,** 1955–1964.

DeGrado, W. F., and Kaiser, E. T. (1980). Polymer-bound oxime esters as supports for solid-phase peptide synthesis. Preparation of protected peptide fragments. *J. Org. Chem.* **45,** 1295–1300.

Delbarre, A., Delepierre, M., Garbay, C., Igolen, J., Le Pecq, J.-B., and Roques, B. P. (1987), Geometry of the antitumor drug ditercalinium bisintercalated into d(CpGpCpG)₂ by ¹H NMR. *Proc. Natl. Acad. Sci. U.S.A.,* **84,** 2155–2159.

Domingo, J. L., Ortega, A., Llobet, J. M., Paternain, J. L., and Corbella, J. (1989). The effects of repeated parenteral administration of chelating agents on the distribution and excretion of uranium, *Res. Commun. Chem. Pathol. Pharmacol,* **64,** 161–164.

Elferink, F., van der Vijgh, W. J. F., Klein, I., and Pinedo, H. M. (1986). Interaction of cisplatin and carboplatin with sodium thiosulfate: Reaction rates and protein binding. *Clin. Chem.* **32,** 641–645.

Fauci, A. S. (1988). The human Immunodeficiency virus: Infectivity and mechanisms of pathogenesis. *Science* **239,** 617–672.

Fellous, R., Coulaud, D., Abed, I. E., Roques, B. P., Le Pecq, J. F., Delain, E., and Gouyette, A. (1988). Cytoplasmic accumulation of ditercalinium in rat hepatocytes and induction of mitochondrial damage. *Cancer Res.* **48,** 6542–6549.

Fuks, J. Z., Wadler, S., and Wiernik, P. H. (1987). Phase I and II agents in cancer therapy: Two cisplatin analogues and high-dose cisplatin in hypertonic saline or with thiosulfate protection, *J. Clin. Pharmacol.* **27,** 357–365.

Gad, A. E., Silver, B. L., and Eytan, G. D. (1982). Polycation-induced fusion of negatively-charged vesicles. *Biochim. Biophys. Acta* **690,** 124–132.

Ghosh, S. S., Kao, P. M., and Kwoh, D. Y. (1989). Synthesis of 5'-oligonucleotide hydrazide derivatives and their use in preparation of enzyme–nucleic acid hybridization probes. *Anal. Biochem.* **178,** 43–51.

Gilman, A. G., Goodman, L. S., Gilman, A., Mayer, S. E., and Melmon, K. L., eds, (1980). "The Pharmacological Basis of Therapeutics", 6th Ed.

Glover, D., Glick, J. H., Weiler, C., Yuhas, J., and Kligerman, M. M. (1984). Phase I trials of WR-2721 and cisplatinum. *Int. J. Radiat. Oncol. Biol. Phys.* **10**, 1781–1784.

Glover, D., Grabelsky, S., Fox, K., Weiler, C., Cannon, L., and Glick, J. (1989). Clinical trials of WR-2721 and cis-platinum. *Int. Journ. Rad. Oncol. Biol. Phys.* **16**, 1201–1204.

Goel, R., Cleary, S. M., Horton, C., Kirmani, S., Abramson, I., Kelly, C., and Howell, S. B. (1989). Effect of sodium thiosulfate on the pharmacokinetics and toxicity of cisplatin, *JNCI, J. Natl. Cancer Inst.* **81**, 1552–1560.

Gonias, S. L., Oakley, A. C., Walther, P. J., and Pizzo, S. V. (1984). Effects of diethyldithiocarbamate and nine other nucleophiles on the intersubunit protein cross-linking and inactivation of purified human macroglobulin by cis-diamminedichloroplatinum (II). *Cancer Res.* **44**, 5764–5770.

Goren, M. P., Pratt, C. B., Meyer, W. H., Wright, R. K., Dodge, R. K., and Viar, M. J., (1989). Mesna excretion and ifosfamide nephrotoxicity in children. *Cancer Res.* **49**, 7153–7157.

Harris, P., and Ralph, P. (1985). Human leukemic models of myelomonocytic development: A review of the HL-60 and U937 cell lines, *J. Leukocyte Biol.* **37**, 407–422.

Henry, D. W. (1982). Receptor based drug design. *In* "New approaches to the design of antineoplastic agents" (T. J. Bandos and T. I. Kalman, eds.), Elsevier, Amsterdam.

Hölttä, E., Korpela, H., and Hovi, T. (1981). Several inhibitors of ornithine and adenosylmethionine decarboxylases may also have antiproliferative effects unrelated to polyamine depletion, *Biochim. Biophys. Acta* **677**, 90–102.

Howell, S. B., Pfeifle, C. E., Wung, W. E., and Olshen, R. A. (1983). Intraperitoneal cis-diamminedichloroplatinum with systemic thiosulfate protection. *Cancer Res.* **43**, 1426–1431.

Huang, M-T., Wood, A. W., Newmark, H. L., Sayer, J. M., Yagi, H., Jerina, D. M., and Conney, A. H. (1983). Inhibition of the mutagenicity of bay-region diol-epoxides of polycyclic aromatic hydrocarbons by phenolic plant flavonoids. *Carcinogenesis* **4**, 1631–1637.

Huang, M.-T., Chang, R., L., Wood, A. W., Newmark, H. L., Sayer, J. M., Yagi, H., Jerina, D. M., and Conney, A. H. (1985). Inhibition of the mutagenicity of bay-region diol-epoxides of polycyclic aromatic hydracarbons by tannic acid, hydroxylated anthraquinones and hydroxylated cinnamic acid derivatives, *Carcinogenesis* **6**, 237–242.

Huff, B. B., ed. (1989). "Physicians' desk reference," pp. 1485 and 2392. Medical Economics Company Inc., Oradell, NJ.

Ikeda, T., Hirayama, H., Yamaguchi, H., Tazuke, S., and Watanabe, M. (1986). Polycationic biocides with pendant active groups: Molecular weight dependence of antibacterial activity. *Antimicrob. Agents Chemother.* **30**, 132–136.

Ito, Y., Kajita, T., Kunimoto, K., and Matsuura, T. (1989). Accelerated photodimerization of stilbenes in methanol and water. *J. Org. Chem.* **54**, 587–591.

Jaycox, G. D., Gribble, G. W., and Hacken, M. P. (1987). Potential DNA bis-intercalating agents: Synthesis and antitumor activity of novel, conformationally restricted bis(9-aminoacridines). *J. Heterocyclic Chem.* **24**, 1405–1408.

Jobe, D. J., and Reinsborough, V. C. (1984). Surfactant structure and micellar rate enhancements: Comparison within a related group of one-tailed, two-tailed and two headed anionic surfactants. *Aust. J. Chem.* **37**, 1593–1599.

John, K. E., (1983). Biological aspects of DNA damage by crosslinking agents. *In* "Molecular Aspects of Anti-Cancer Drug Action" (S. Neidle and M. J. Waring, eds.) pp. 315–361. Verlag Chemie, Basel, Switzerland.

Joshi, B. S., Viswanathan, N., Gawad, D. H., Balakrishnan, V., Von Philipsborn, W., and Quick, A. (1975). Piperaceae alkaloids. II. Structure and synthesis of cyclostachine A, a novel alkaloid from Piper trichostachyon. *Experientia* **31**, 880–881.

Kempf, S. R., and Ivankovic, S. (1987) Nephrotoxicity and carcinogenic risk of cis-platin (CDDP) prevented by sodium 2-mercaptoethane-sulfonate (Mesna): Experimental results. *Cancer Treat. Rev.* **14**, 365–372.

Klayman, D. L., Scovill, J. P., Bartosevich, J. F., and Mason, C. J. (1979). N^4,N^4-Disubstituted derivatives as potential antimalarial agents. *J. Med. Chem.* **22,** 1367–1373.

Kochi, M., Takeuchi, S., Mizutani, T., Mochizuki, K., Matsumoto, Y., and Saito. Y. (1980). Antitumor activity of benzaldehyde. *Cancer Treat. Rep.* **64,** 21–23.

Kornberg, A. (1980). "DNA Replication." (A. C. Bartlett, ed.). Freeman, San Francisco, California.

Kuhlmann, K. F., and Mosher, C. W. (1981). New dimeric analogues of ethidium: Synthesis, interaction with DNA, and antitumor activity, *J. Med. Chem.* **24,** 1333–1337.

Kwon, C.-H., and Borch, R. F. (1989). Effects of *N*-substitution on the activation mechanisms of 4-hydroxycyclophosphamide analogues. *J. Med. Chem.* **32,** 1491–1496.

Laugaa, P., Markovits, J., Delbarre, A., Le Pecq, J.-B., and Roques , B. P. (1985). DNA tris-intercalation: First acridine trimer with DNA affinity in the range of DNA regulatory proteins. Kinetic studies. *Biochemistry* **24,** 5567–5575.

Ledl, F., and Schleicher, E. (1990). New aspects of the Maillard reaction in foods and in the human body. *Angewandte Chem. Int. Ed. Eng.* **29,** 565–594.

Lee K.-H., and de Mayo, P. (1979). Biphasic photochemistry: Micellar regioselectivity in enone dimerisation. *J. Chem. Soc., Chem. Commun.* **1979,** 493–495.

LeRoy, A. F., and Thompson, W. C., (1989). Binding kinetics of tetrachloro-1,2-diaminocyclohexaneplatinum (IV) (tetraplatin) and cis-diamminedichloroplatinum(II) at 37°c with human plasma proteins and with bovine serum albumin. Does aquation precede protein binding? *JNCI, J. Natl. Cancer Inst.* **81,** 427–436.

Leyland-Jones, B. (1988). Whither the modulation of platinum? *JNCI, J. Natl. Cancer Inst.* 1432–1433.

Mahaney, F. X., Jr, (1990), Agent designed for war now in doctor's armamentarium, *JNCI, J. Natl. Cancer Inst.* **83,** 255–257.

Malinge, J.-M., and Leng, M., (1986). Reaction of nucleic acids and cis-diamminedichloroplatinum(II) in the presence of intercalating agents. *Proc. Natl. Acad. Sci. U.S.A,* **83,** 6317–6321.

Malinge, J.-M., Schwartz, A., and Leng, M. (1987). Characterization of the ternary complexes formed in the reaction of cis-diamminedichloroplatinum (II), ethidium bromide and nucleic acids. *Nucl. Acids. Res.* **16,** 7663–7672.

McFadyen, W. D., Wakelin, L. P. G., Roos, I. A. G., and Leopold, V. A. (1985). Activity of platinum(II) intercalating agents against murine leukemia L1210. *J. Med. Chem.* **28,** 1113–1116.

Minchin, R. F., Martin, R. L., Summers, L. A., and Ilett, K. F. (1989). Inhibition of putrescine uptake by polypyridinium quaternary salts in B16 melanoma cells treated with difluoromethylornithine *Biochem. J.* **262,** 391–395.

Mitchell, M. A., Johnson, P. D., Williams, M. G., and Aristoff, P. A. (1989). Interstrand DNA cross-linking with dimers of the spirocyclopropyl alkylating moiety of CC-1065. *J. Am. Chem. Soc.* **111,** 6428–6429.

Mollman, J. E., Glover, D. J., Hogan, W. M., and Furman, R. E. (1988). Cisplatin neuropathy. *Cancer* **61,** 2192–2195.

Mukhtar, H., Das, M., Khan, W. A., Wang, Z. Y., Bik, D. P., and Bickers, D. R. (1988). Exceptional activity of tannic acid among naturally occurring plant phenols in protecting against 7,12-dimethylbenz(*a*)anthracene-, benzo(*a*)pyrene-, 3-methylcholanthrene-, and N-methyl-N-nitrosourea-induced skin tumorigenesis in mice. *Cancer Res.* **48,** 2361–2365.

Naganuma, A., Tanaka, T., Maeda, K., Matsua, R., Tabata-Hanyu, J., and Imura, N. (1983). The interaction of selenium with various metals *in vitro* and *in vivo*. *Toxicology,* **29,** 77–86.

O'Donnell, J. P., Proveaux, W. J., and Ma, J. K. H. (1979). High-performance liquid chromatographic studies of reaction of hydralazine with biogenic aldehydes and ketones. *J. Pharm. Sci.* **68,** 1524.

Ogita, S., Tokiwa, K., Taniguchi, H., and Takahashi, T. (1987). Intraarterial chemotherapy with lipid contrast medium for hepatic malignancies in infants. *Cancer* **60,** 2886–2890.

Ohkawa, K., Tsukada, Y., Dohzono, H., Koike, K., and Terashima, Y. (1988). The effects of co-administration of selenium and cis-platin (CDDP) on CDDP-induced toxicity and anti-tumour activity. *Br. J. Cancer* **58**, 38–41.

Ottenwälder, H., and Simon, P. (1987). Differential effect of N-acetylcysteine on excretion of the metals Hg, Cd, Pb and Au. *Arch. Toxicol.* **60**, 401–402.

Otto, W. C., Brown, R. D., Gage-White, L., Kupetz, S., Anniko, M., Penny, J. E., and Henley, C. M. (1988). Effects of cisplatin and thiosulfate upon auditory brainstem responses of guinea pigs. *Hearing Res.* **35**, 79–86.

Pauza, C. D. (1988). HIV persistence in monocytes leads to pathogenesis and AIDS. *Cell. Immunol.* **112**, 414–424.

Pedersen, N. C., Ho, E. W., Brown, M. L., and Yamamoto, J. K. (1987). Isolation of a T-lymphotropic virus from domestic cats with an immunodeficiency-like syndrome. *Science* **235**, 790–793.

Peele, D. B., Farmer, J. D., and MacPhail, R. C. (1988). Behavioral consequences of chelator administration in acute cadmium toxicity. *Fund. Appl. Toxicol.* **11**, 416–428.

Prasad, A. R. S., Luduena, R. F., and Horowitz, P. M. (1986). Bis(8-anilinonaphthalene-1-sulfonate) as a probe for tubulin decay, *Biochemistry* **25**, 739–742.

Reuben, R. C., Wife, R. L., Breslow, R., Rifkind, R. A., and Marks, P. A. (1976). A new group of potent inducers of differentiation in murine erthroleukemia cells. *Proc. Natl. Acad. Sci. U.S.A.* **73**, 862–866.

Rideout, D. (1986). Self-assembling cytotoxins. *Science* **233**, 561–563.

Rideout, D., and Breslow, R. (1980). Hydrophobic Acceleration of Diels–Alder reactions. *J. Am. Chem. Soc.* **102**, 7816–7817.

Rideout, D., Jaworski, J., and Dagnino, R., Jr., (1988). Environment-selective synergism using self-assembling cytotoxic and antimicrobial agents, *Biochem. Pharmacol.* **37**, 4505–4512.

Rideout, D., Calageropoulou, T., Jaworski, J., Dagnino, R., Jr., McCarthy, M. R. (1989), Phosphonium salts exhibiting selective anti-carcinoma activity in vitro. *Anti-Cancer Drug Des.* **4**, 265–280.

Rideout, D., Calageropoulou, T., Jaworski, J., and McCarthy, M. R. (1990a). Synergism through direct covalent bonding between agents: A strategy for rational design of chemotherapeutic combinations *Biopolymers* **29**, 247–262.

Roos, A., and Boron, W. F. (1981). Intracellular pH. *Physiol. Rev.* **61**, 296–434.

Rotenberg, S. A., Smiley, S., Ueffing, M., Krauss, R. S., Chen, L. B., and Weinstein, I. B. (1990a). Inhibition of rodent protein kinase C by the anticarcinoma agent dequalinium. *Cancer Res.* **50**, 677–685.

Rotenberg, S. A., Calogeropoulou, T., Jaworski, J. S., Weinstein, I. B., and Rideout, D. (1990b). A self-assembling protein kinase C inhibitor. *Proc. Natl. Acad. Sci., USA*, in press.

Salminen, A. (1984). Activities and regeneration of mouse skeletal muscles after exercise injuries, *Am. J. Pathol.* **117**, 64–70.

Satoh, M., Naganuma, A., and Imura, N. (1988). Metallothionein induction prevents toxic side effects of cisplatin and adriamycin used in combination. *Cancer Chemother. Pharmacol.* **21**, 176–178.

Satoh, M., Naganuma, A., and Imura, N. (1989). Optimum schedule of selenium administration to reduce lethal and renal toxicities of cis-diamminedichloroplatinum in mice. *J. Pharmacobio-Dyn.* **12**, 246–253.

Schmalbach, T. K., and Borch, R. F. (1989). Diethyldithiocarbamate modulation of marrow toxicity induced by cisdiammine(cyclobutanedicarboxylato)platinum(II). *Cancer Res.* **49**, 6629–6633.

Searle, M. S., Hall, J. G., Denny, W. A., and Wakelin, L. P. G. (1989). Interaction of the antitumour antibiotic luzopeptin with the hexanucleotide duplex d(5′-GCATGC)$_2$. *Biochem. J.* **259**, 433–441.

Segal-Bendirdjian, E., Coulaud, D., Roques, B. P., and Le Pecq, J.-B. (1988). Selective loss of mitochondrial DNA after treatment of cells with ditercalinium (NSC 335153), an antitumor bis-intercalating agent. *Cancer Res.* **48**, 4982–4992.

Seitz, D. E., Katterjohn, C. J., Rinzel, S. M., and Pearce, H. L. (1989). Thermodynamic analysis of the reaction of phosphoramide mustard with protector thiols. *Cancer Res.* **49**, 3525–3528.

Smith, E. L., Hill, R. L., Lehman, I. R., Lefkowitz, R. J., Handler, P., and While, A. (1983). "Principles of biochemistry: Mammalian biochemistry," 7th ed. McGraw-Hill, New York.

Subramanyam, B., Callery, P. S., Egorin, M. J., Synder, S. W., and Conley, B. A. (1989). An active, aldehydic metabolite of the cell-differentiating agent hexamethylene bisacetamide, *Drug Metab. Dispos.* **17**, 398–401.

Sundquist, W. I., Bancroft, D. P., Chassot, L., and Lippard, S. J. (1988). DNA promotes the reaction of cis-diamminedichloroplatinum(II) with the exocyclic amino groups of ethidium bromide. *J. Am. Chem. Soc.* **110**, 8559–8560.

Swanson, J. A., Yirinec, B. D., and Silverstein, S. C. (1985). Phorbol esters and horseradish peroxidase stimulate pinocytosis and redirect the flow of pinocytosed fluid in macrophages. *J. Cell Biol.* **100**, 851–859.

Swanson, J. A., Yirinec, B., Burke, E., Bushnell, A., and Silverstein, S. C. (1986). Effect of alterations in the size of the vacuolar compartment on pinocytosis in J774.2 macrophages. *J. Cell Physiol.* **128**, 195–201.

Talbott, R. L., Sparger, E. E., Lovelace, K. M., Fitch, W. M., Pedersen, N. C., Luciw, P. A., and Elder, J. H. (1989). Nucleotide sequence and genomic organization of feline immunodeficiency virus. *Proc. Natl. Acad. Sci. U.S.A.* **86**, 5743–5747.

Teicher, B. A., Crawford, J. M., Holden, S. A., Lin, Y., Cathcart, K. N. S., Luchette, C. A., and Flatow, J. (1988). Glutathione monoethyl ester can selectively protect liver from high dose BCNU or cyclophosphamide. *Cancer (Philadelphia)* **62**, 1275–1281.

Temple, C., Jr, (1990). Antimitotic agents: Synthesis of imidazo[4,5-c]pyridin-6-ylcarbamates and imidazo]4,5-b]pyridin-5-ylcarbamates. *J. Med. Chem.* **33**, 656–661.

Tsou, K. C. (1968). A cytochemical method for selective intracellular damage with a ruby laser. *Life Sci.* **7**, 785–790.

Uozumi, J. and Litterst, C. L. (1986). The effect of sodium thiosulfate on subcellular localization of platinum in rat kidney after treatment with cisplatin. *Cancer Lett.* **62**, 279–283.

Vucicevic, Z. M., Tsou, K. C., Nazarian, I. H., Scheie, H. G., and Burns, W. P. (1969). A cytochemical approach to the laser coagulation of the ciliary body. *Mod. Probl. Ophthalmol.* **8**, 467–478.

Wattenberg, L. W. (1985). Chemopreve⌐ .ion of cancer. *Cancer Res.* **45**, 1–8.

Wirth, M., Buchardt, O., Koch, T., Niels⌐ P. Γ., and Nordén, B. (1988). Interactions between DNA and mono-, bis-, tris-, tetrakis-, and hexakis(aminoacridines). Linear and circular dichroism, electric orientation relaxation, viscometry, and equilibrium study. *J. Am. Chem. Soc.* **110**, 932–939.

Wong, D. T., Horng, J.-S., and Wilkinson, J. R. (1972). Robenzidene, an inhibitor of oxidative phosphorylation, *Biochem. Biophys. Res. Commun.* **46**, 621–627.

Wood, A. W., Huang, M.-T., Chang, R. L., Newmark, H. L., Lehr, R. E., Yagi, H., Sayer, J. M., Jerina, D. M., and Conney, A. H. (1982). Inhibition of the mutagenicity of bay-region diol epoxides of polycyclic aromatic hydrocarbons by naturally occurring plant phenol is: Exceptional activity of ellagic acid. *Proc. Natl. Acad. Sci. U.S.A.* **79**, 5513–5517.

CHAPTER 15

Chemotherapeutic Potentiation through Interaction at the Level of DNA

Beverly A. Teicher
Terence S. Herman
Sylvia A. Holden
Dana-Farber Cancer Institute
and Harvard Medical School
Boston, Massachusetts 02115

J. Paul Eder
Beth Israel Hospital
Boston, Massachusetts 02215

This chapter discusses several strategies for improving the tumor cell killing ability of chemotherapeutic agents, especially alkylating agents, by enhancing directly or indirectly the action of these drugs at the level of the DNA. Isobologram methodology was used in several examples to determine if combination treatments were additive, greater than additive (synergistic), or less than additive. Combinations of alkylating agents and DNA repair inhibitors focus mainly on topoisomerase II-interactive agents, especially novobiocin, and poly(ADP–ribose) polymerase-interactive agents, especially nicotinamide and 3-aminobenzamide. Lonidamine, which inhibits cellular respiration and indirectly, inhibits DNA repair processes, also produces greater than additive tumor cell killing. The repair of potentially lethal DNA damage by X-rays, bleomycin, and other drugs can be inhibited by a variety of agents including several experimental platinum complexes. Combinations of alkylating agents may be visualized as interacting directly or indirectly (by blocking cytoplasmic inactivation systems) to facilitate bifunctional alkylation of DNA. The combination of 4-hydroperoxycyclophosphamide with thiotepa or melphalan both *in vitro* and *in vivo* is used as an example of this type of

SYNERGISM AND ANTAGONISM IN CHEMOTHERAPY
Copyright © 1991 by Academic Press, Inc.
All rights of reproduction in any form reserved.

541

interaction. Finally, the 2-nitroimidazole radiosensitizers and perfluoro-chemical emulsions with oxygen breathing (PFCE/O_2) have been shown to enhance that antitumor action of alkylating agents in solid tumor models. *In vivo*, both the 2-nitroimidazoles and PFCE/O_2 have been shown to enhance the efficiency of DNA cross-link formation. The exact molecular mechanism of this effect has yet to be elucidated. The clinical implications of several of these treatment combinations are discussed.

I. Isobologram Methodology

In the study of multimodality therapy or combined chemotherapy, it is of interest to determine whether the combined effects of two agents are additive or whether their combination is substantially different than the sum of their parts. Conceptual foundations for this form of analysis were popularized by Steel and Peckham (1979), based on the construction of an envelope of additivity in an isoeffect plot (isobologram). This approach provides a rigorous basis for defining regions of additivity, supraadditivity and subadditivity and protection. This method of analysis is based on a clear conceptual formulation of the way that drugs or agents can be expected to show additivity. The first form of additivity is more conceptually simple and is defined as Mode I by Steel and Peckham (1979). For a selected level of effect (survival in this case) on a log scale, the dose of Agent A to produce this effect for the survival curve is determined. A lower dose of Agent A is then selected, the difference in effect from the isoeffect level is determined, and the dose of Agent B needed to make up this difference is derived from the survival curve for Agent B. For example, 3 mg of Agent A may be needed to produce 0.1% survival (3 logs of kill), the selected isoeffect level. A dose of 2.5 mg of Agent A produces 1.0% survival (2 logs of kill). The Mode I isoeffect point for Agent B would thus be the level of B needed to produce 1 log of kill, to result in the same overall effect of 3 logs of kill. In this instance, we might find that 4 mg of Agent B are needed to produce 1 log of kill.

Mode II additivity is conceptually more complex, but corresponds to the notions of additivity discussed by Berenbaum (1977). For any given level of effect, the dose of Agent A needed to produce this effect is determined from the survival relationship. The isoeffect dose of Agent B is calculated as the amount of Agent B needed to produce the given effect starting at the level of effect produced by Agent A. For example, 3 mg of Agent A may be needed to produce 0.1% survival (3 logs of kill). A dose of 2.5 mg of A produces 1.0% (2 logs of kill). A dose of 6 mg of Agent B is needed to produce 3 logs of kill, and 2 logs of kill are obtained with Agent B at 5 mg. Thus, the Mode II isoeffect point with Agent A at 2.5 mg is equal to the amount of Agent B needed to take Agent B from 2 logs of kill to 3 logs of kill

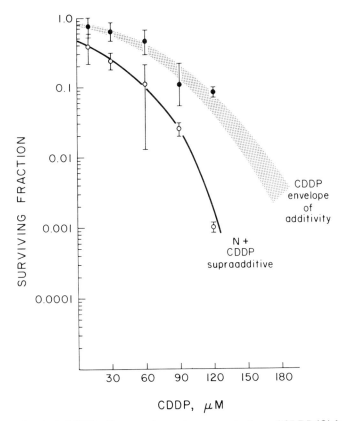

FIGURE 1   Survival of CHO cells exposed to various concentrations of CDDP (●) for 1 hr or to CDDP with exposure to novobiocin (N) (150 μM) for 2 hr prior to, during, and for 24 hr post CDDP exposure (○). *Shaded area*, envelope of additivity for the treatment combinations (Eder *et al.*, 1987).

(6 mg − 5 mg = 1 mg). This can be conceptualized by noting that Agent A should produce 2 logs of kill and is, in this case, equal to 5 mg of Agent B. If Agent A + Agent B are identical in their mode of action, then 1 mg more of Agent B should then be equivalent in effect to 6 mg of B. Graphically, on a linear dose scale, Mode II additivity is defined as the straight line connecting the effective dose of Agent A alone and the effective dose of Agent B alone. This relationship is also described by the equation

$$\frac{\text{Dose of A}}{A_e} + \frac{\text{Dose of B}}{B_e} = 1$$

where A_e and B_e are the doses of Agent A and Agent B, respectively, needed to produce the selected effect.

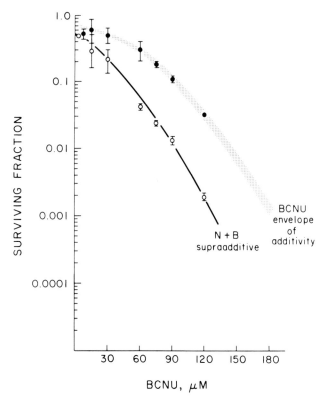

FIGURE 2   Survival of CHO cells exposed to various concentrations of BCNU (●) for 1 hr or to BCNU with exposure to novobiocin (N) (150 μM) for 2 hr prior to, during, and for 24 hr post BCNU exposure (○). *Shaded area*, envelope of additivity for the treatment combinations (Eder *et al.*, 1987).

Overall, combinations that produce the desired effect that are within the boundaries of Mode I and Mode II are considered additive. Those displaced to the left are supraadditive while those displaced to the right are subadditive. Combinations that produce effect outside the rectangle defined by the intersections of A_e and B_e are protective. This type of classical isobologram methodology is cumbersome to use experimentally as each combination must be carefully titrated to produce a constant level of effect. Dewey *et al.* (1977) described an analogous form of analysis for the special case in which the dose of one agent was held constant. Using full survival curves of each agent alone, this method produces envelopes of additive effect for different levels of the variable agent. It is conceptually identical to generating a series of isoeffect curves and then plotting the survivals from a series of these at constant dose of Agent A on a log effect by dose of Agent B coordinate system (Deen and Williams, 1979). This approach can often be applied to the experimental situation in a more direct and efficient manner, and isobolograms can be derived describing the expected effect (Mode I and Mode II) for any level of the variable agent and constant agent combinations. This chapter will concentrate on studies in which isobologram methodology has been used.

II. Combinations of Alkylating Agents and Repair Inhibitors

The alkylating agents are an important class of antitumor agents (Colvin, 1982) and play a major role in the successful treatment of selected human cancers (Frei, 1985). The introduction of *cis*-diamminedichloroplatinum(II) (CDDP), a nonclassical alkylating agent, has further extended the efficacy of alkylating agents in the clinic (Zwelling and Kohn, 1982).

The clinically effective alkylating agents are bifunctional, that is, they have the capacity to form two covalent linkages. Although they may combine with many nucleophilic sites within the cell, there is compelling evidence that their cytotoxicity is the result of DNA monoadduct formation followed by inter- or intrastrand cross-links in DNA (Thomas *et al.*, 1978; Lindahl *et al.*, 1982; Sedgwick and Lindahl, 1982; Meyn *et al.*, 1982). The kinetics of monoadduct and cross-link formation, the nature and sites of binding to nucleic acid bases, and the efficacy and kinetics of DNA repair vary among the alkylating agents. The net quantity of cross-links formed is a function of the extent of initial monoadduct formation and their subsequent progression to DNA–DNA cross-links. An O-6-alkylguanine–DNA alkyl transferase has been demonstrated to remove nitrosourea-induced monoadducts and thereby impede cross-link formation (Erickson *et al.*, 1980). DNA repair mechanisms are believed to be responsible for DNA–DNA cross-link removal (Roberts *et al.*, 1982). After damage by alkylating agents, enhanced sensitivity of cells from individuals with selective deficiencies of DNA repair activity is observed (Setlow, 1978). Furthermore, the disappearance kinetics of DNA–DNA cross-links after exposure of neoplastic cells to bifunctional alkylating agents are consistent with their removal by a repair process (Roberts *et al.*, 1982). The relationship of the cross-link index to cytotoxicity raises the possibility that the effectiveness of alkylating agents may be enhanced by either increasing the number of alkylation adducts in cellular DNA or decreasing their repair (Zwelling *et al.*, 1981).

A wide variety of potential DNA repair inhibitors and agents which may inhibit biological processes involved directly or indirectly in DNA repair have been tested in combination with various alkylating agents. These include topoisomerase II inhibitors, poly(ADP–ribose) polymerase inhibitors, inhibitors of cellular respiration, and potentially lethal damage inhibitors.

A. TOPOISOMERASE II INHIBITORS

DNA topoisomerase type II enzymes are proteins found in both prokary-
otic and eukaryotic cells which control and modify the topological state of
DNA. By transiently breaking a pair of complementary DNA strands and
passing another double-stranded segment through the break, topoisomerase
type II can catalyze many types of interconversions between DNA topological
isomers (Wang, 1987). DNA topoisomerases have been found to affect a
number of vital biological functions, including the replication of DNA, RNA
transcription, and mitosis (Wang, 1987; Holm *et al.*, 1985; Reddy and Pardee,
1983; Ryaji and Worcel, 1984). Although the bacterial type II DNA
topoisomerase, termed DNA gyrase, and the eukaryotic topoisomerase II
show a number of characteristic differences, areas of homology in the amino
acid sequences have been found in the bacterial and yeast enzymes (Wang,
1987). Furthermore, significant amino acid homology exists among eukaryotic type II topoisomerases from yeast, *Drosophila*, and human cells (Goto
and Wang, 1984; Wang, 1985 and 1987; Nolan *et al.*, 1986; Tsai-Pflugfelder
et al., 1988). DNA topoisomerase II is a useful focus for cancer therapy on
two levels: first, as a target itself through the formation of the tertiary complex
of drug–enzyme–DNA which can lead directly to a cytotoxic event and second, formation of the tertiary complex may stabilize the DNA into a conformation which is more susceptible to the actions of antitumor bifunctional
alkylating agents or may prevent repair of monoadducts (Ross, 1985; Mattern
and Scudiero, 1981). Molecules from several families have been shown to
interact with DNA topoisomerase II. These include intercalating agents
such as Adriamycin and 4'-(9-acridinylamino)methanesulfon-*m*-anisidide (*m*-
AMSA), the antitumor epipodophyllotoxins such as etoposide and VM-26,
and the coumarin and quinolone bacterial DNA gyrase inhibitors such as
novobiocin and ciprofloxacin, respectively. The actions of these drugs as well
as others on mammalian cells may be at least in part through this mechanism
(Wang, 1987; Ross, 1985; Glisson *et al.*, 1986a,b; Gellert *et al.*, 1976).

Work from our laboratory first focused on the combination of novobiocin
with CDDP or BCNU because novobiocin can be administered in relatively
high doses with little toxicity (Eder *et al.*, 1987, 1989b). *In vitro* in CHO cells
(Eder *et al.*, 1987) administration of novobiocin prior to, during, and for
several hours after treatment with CDDP or BCNU resulted in greater than
additive cell kill by the combination, as determined by isobologram analysis
(Figs. 1 and 2). *In vitro* this enhanced cytotoxicity correlated with an increased
level of DNA–DNA cross-link formation by CDDP (Eder *et al.*, 1987)
(Fig. 3). A similar effect was obtained by Tan *et al.* (1987) with novobiocin
and nitrogen mustard in a human Raji Burkitts lymphoma cell line.

Recently an epipodophyllotoxin(VM-26)-resistant subline of the CHO-
VPM parental cell line was produced and characterized (Glisson *et al.*,
1986b). This subline CHO-VPMR5 has an altered topoisomerase II (Glisson

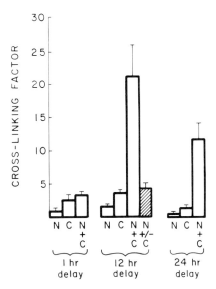

FIGURE 3   The effect of novobiocin treatment on DNA–DNA interstrand crosslinks produced by CDDP. Novobiocin (N) concentration was 150 μM until cell harvest. C, 100 μM CDDP for 1 hr. C + N, 150 μM N for 1 hr preceding and during CDDP exposure and in postexposure media until cell harvest. C +/− N, 150 μM N 1 hr before and during CDDP exposure, but none in postexposure media. The CLF measures the inhibition of DNA elution due to interstrand crosslinks relative to the untreated control. CLF, [log (irradiated control/control)]/[log (irradiated drug/control)] (Eder et al., 1987).

et al., 1986b). There is no change in the CHO-VPMR5 cells' ability to decatenate supercoiled DNA by topoisomerase II, but there were substantial changes in the cytotoxicity and the ability of drug to produce cleavage of DNA in these cells (Glisson et al., 1986b). The major changes in the CHO-VPMR5 line appear to be decreased ability of drugs to cleave DNA, resulting in the cells being 10- to 40-fold more resistant to drug effect (Glisson et al., 1986b). The altered enzyme is more thermally labile and may have decreased affinity for DNA than the enzyme from the parental CHO-VPM cells (Sullivan and Ross, 1987).

The ability of four drugs known to interact with topoisomerase II to enhance the cytotoxicity of CDDP was assessed in the CHO-VPM and CHO-VPMR5 cell lines (Eder et al., 1990). The results of the combination treatments were analyzed by isobologram methodology. Upon 24 hr exposure, there was no significant difference in the cytotoxicity of novobiocin or ciprofloxacin toward CHO-VPM and CHO-VPMR5 cells over a dosage range of each drug (Fig. 4). The CHO-VPMR5 were approximately 9-fold more resistant to m-AMSA and approximately 170-fold resistant to etoposide. The combination of novobiocin and cisplatin produced greater than additive cell

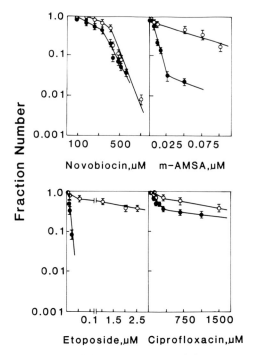

FIGURE 4   Survival of CHO-VPM (●) and CHO-VPMR5 (○) cells exposed to various concentrations of novobiocin, *m*-AMSA, etoposide, or ciprofloxacin for 24 hr. Points represent the mean of three independent experiments; bars represent the SEM (Eder *et al.*, 1989a).

kill over the entire dosage range of cisplatin tested, with 5- to 10-fold greater than additive cell kill being observed in both cell lines (Figs. 5 and 6). All of the combinations of *m*-AMSA and CDDP produced cell kill which fell within the envelope of additivity for this combination in both cell lines. The combination of etoposide and CDDP was slightly greater than additive at the low CDDP concentrations and additive at the highest concentration of CDDP tested in the parental CHO-VPM cell line. In the CHO-VPMR5 cell line, the combination of CDDP and etoposide was slightly greater than additive over the entire concentration range of CDDP tested. Ciprofloxacin in combination with CDDP, like novobiocin, resulted in greater than additive cell kill over the entire range of CDDP concentrations examined in both cell lines, but overall the level of synergy was less than that seen with novobiocin. The enhancement of CDDP cytotoxicity by novobiocin seen in exponentially growing CHO-VPM cells and CHO-VPMR5 cells was lost in the stationary phase cultures of both cell lines (Fig. 7). The cytotoxicity of CDDP can be markedly enhanced by certain drugs which are known to interact with topoisomerase II. In these studies, novobiocin and, to a lesser degree, ciprofloxacin produced greater than additive cell kill in combination with CDDP

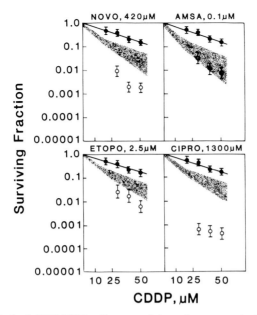

CDDP, µM

FIGURE 5   Survival of CHO-VPM cells exposed to various concentrations of CDDP (●) for 1 hr or to an IC_{90} concentration of novobiocin (NOVO) (350 μM), m-AMSA (AMSA) (0.15 μM), etoposide (ETOPO) (0.02 μM), or ciprofloxacin (CIPRO) (1120 μM) for 1 hr prior to and 1 hr during exposure to various concentrations of CDDP and then for 22 hr post CDDP treatment (○). *Shaded areas*, envelope of additivity for each treatment combination. Points represent the mean of three independent experiments; bars represent the SEM (Eder *et al.*, 1989a).

in the parental CHO-VPM cell line and the epipodophyllotoxin-resistant subline CHO-VPMR5 (Eder *et al.*, 1990).

To study the effect of novobiocin *in vivo*, the drug was administered daily for 3 days prior to alkylating agent treatment, during alkylating agent treatment, and for 2 days after the completion of alkylating agent treatment (Eder *et al.*, 1989b). When combined with CDDP, BCNU, or cyclophosphamide, there was significant enhancement of the growth delay of the FSaIIC fibrosarcoma (Table 1). When single doses of novobiocin were administered to tumor-bearing animals before administration of CDDP and tumor cell survival was measured *in vitro*, there was a 3- to 4-fold increase in tumor cell killing by CDDP (Fig. 8). At the lowest dose of BCNU, there was no significant effect of novobiocin. At a dose of 100 mg/kg of BCNU, there was about a 7-fold increase in tumor cell kill upon addition of novobiocin to the drug treatment; however, at the highest dose of BCNU (200 mg/kg), this effect fell to about a 2-fold increase in tumor cell kill with the combination treatment (Fig. 9). Cyclophosphamide showed increasing effect of novobiocin with increasing dose of the alkylating agent, reaching 13-fold at a dose of 300 mg/kg of cyclophosphamide (Fig. 10). In all cases there was a lesser effect on the

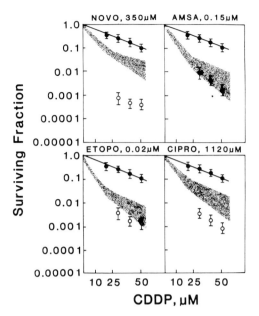

CDDP, µM

FIGURE 6   Survival of CHO-VPMR5 cells exposed to various concentrations of CDDP (●) for 1 hr or to an IC_{90} concentration of novobiocin (NOVO) (420 μM), m-AMSA (AMSA) (0.1 μM), etopside (ETOPO) (2.5 μM), or ciprofloxacin (CIPRO) (1300 μM) for 1 hr prior to an 1 hr during exposure to various concentrations of CDDP and then for 22 hr post CDDP treatment (○). Shaded areas, envelope of additivity for each treatment combination. Points represent the mean of three independent experiments: bars represent the SEM (Eder et al., 1989a).

TABLE 1
Tumor Growth Delay of the FSaIIC Fibrosarcoma Treated with Novobiocin
Alone or in Combination with CDDP, BCNU, or Cyclophosphamide

Treatment group	Tumor growth delay (Days)[a]
Novobiocin, 50 mg/kg[b]	1.0 ± 0.5
Novobiocin, 100 mg/kg	1.0 ± 0.5
CDDP[c]	10.6 ± 0.7
Novobiocin (50 mg/kg) + CDDP	14.0 ± 1.2
Novobiocin (100 mg/kg) + CDDP	29.6 ± 2.5
BCNU[d]	2.9 ± 0.5
Novobiocin (50 mg/kg) + BCNU	4.1 ± 0.6
Novobiocin (100 mg/kg) + BCNU	26.2 ± 2.0
Cyclophosphamide[e]	6.8 ± 1.0
Novobiocin (50 mg/kg) + cyclophosphamide	9.0 ± 1.5
Novobiocin (100 mg/kg) + cyclophosphamide	14.4 ± 1.7

[a] Results are given as the mean ± SE of the days required for treated tumor to reach 500 mm³ compared to untreated controls. Controls reached 500 mm³ in 15.0 ± 0.2 days.
[b] Novobiocin was given once daily on days 6–15 post tumor cell implantation.
[c] CDDP (4.5 mg/kg) was given once daily on days 9–13 post tumor cell implantation.
[d] BCNU (8 mg/kg) was given once daily on days 9–13 post tumor cell implantation.
[e] Cyclophosphamide (100 mg/kg) was given once daily on days 9, 11, and 13 post tumor cell implantation.

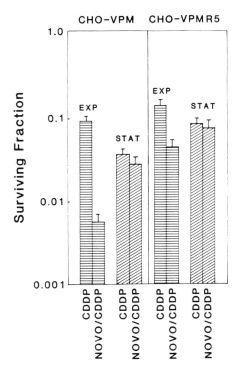

FIGURE 7   Survival of exponentially growing (EXP) or stationary phase (STAT) CHO-VPM and CHO-VPMR5 cells exposed for 1 hr to 50 μM CDDP or to 80 μM novobiocin (NOVO) for 1 hr prior to and during exposure to 50 μM CDDP and then for 22 hr post CDDP treatment. Data presented are the mean of three independent experiments; bars represent the SEM (Eder *et al.*, 1989a).

bone marrow than on the tumor. The combination of novobiocin and alkylating agents may be a clinically useful strategy. Based on these results, a clinical study has been initiated using orally administered novobiocin and cylophosphamide (Schnipper *et al.*, 1988; Eder *et al.*, 1988b).

Etoposide, a podophyllotoxin derivative, has been shown to cause both single-strand and double-strand DNA breaks and DNA protein cross-links in mammalian cells in a process which requires a nuclear protein. Etoposide does not bind directly to DNA, and evidence indicates that the protein-associated DNA breaks resulting from etoposide treatment are mediated through an interaction between the drug and topoisomerase II and perhaps oxygen (Issel *et al.*, 1984; Long and Minocha, 1983; Long *et al.*, 1984; Pommier *et al.*, 1986; Ross, 1985). Flow cytometry in etoposide-treated cells shows a delay in S phase transit before arrest of cells in G_2. Correlating with the S phase delay is a selective inhibition of thymidine incorporation into DNA as well as a concentration-dependent scission of DNA strands (Kalwinsky *et al.*, 1983; Loike and Horwitz, 1976). Inhibition of DNA repair

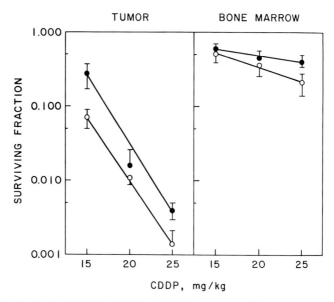

FIGURE 8   Survival of FSaIIC cells from FSaIIC tumors and bone marrow from animals treated with various doses of CDDP (●) or with various doses of CDDP preceded 1 hr earlier by a single dose of novobiocin (100 mg/kg) (○). Points represent the mean of three independent determinations ± SE (Eder *et al.*, 1990).

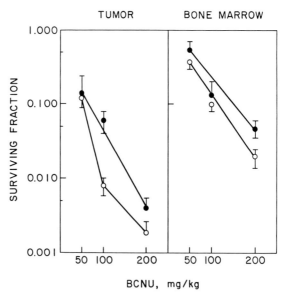

FIGURE 9   Survival of FSaIIC cells from FSaIIC tumors and bone marrow from animals treated with various doses of BCNU (●) or with various doses of BCNU preceded 1 hr earlier by a single dose of novobiocin (100 mg/kg) (○). Points represent the mean of three independent determinations ± SE (Eder *et al.*, 1990).

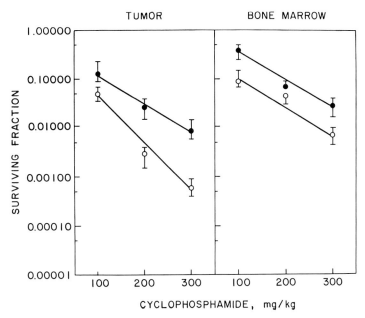

FIGURE 10 Survival of FSaIIC cells from FSaIIC tumors and bone marrow from animals treated with various doses of cyclophosphamide (●) or with various doses of cyclophosphamide preceded 1 hr earlier by a single dose of novobiocin (100 mg/kg) (○). Points represent the mean of three independent determinations ± SE (Eder *et al.*, 1989a).

enhances cytotoxicity (Arnold and Whitehouse, 1982). However, etoposide, when incubated with purified DNA *in vitro*, does not induce such damage (Loike and Horwitz, 1976). Metabolic activation of etoposide to reactive intermediates by rat liver and HeLa microsomal fractions has been reported (Van Maanen *et al.*, 1983), and free radical formation at the pendant phenolic group has been implicated (Wozniak and Ross, 1983; Wozniak *et al.*, 1984). A free hydroxyl group at the 4′ position is essential for DNA breakage activity (Long *et al.*, 1984). Free radical scavengers, dehydrogenase inhibitors, and dehydrogenase substrates prevent the formation of single-strand DNA breaks and inhibit the cytotoxicity of etoposide (Wozniak and Ross, 1983; Wozniak *et al.*, 1984). Verapamil and a number of other calcium channel antagonists potentiate DNA damage by etoposide, perhaps by overcoming resistance to transport or drug action (Yalowich and Ross, 1984). Etoposide shows a selective cytotoxicity to normally oxygenated cells *in vitro* and, when combined with an oxygen-carrying perfluorochemical emulsion, the antitumor activity and therapeutic efficacy of etoposide are enhanced (Teicher *et al.*, 1985d).

Durand and Goldie (1987), using median-effect analysis, examined the cytotoxicity of etoposide and CDDP in CHO-V79 spheroids. They found supraadditive cytotoxicity with these agents in combination at 90% cell kills achievable with clinically relevant drug doses; increasing the toxicity of the combination treatment by increasing the cisplatin level produced a greater relative gain in the noncycling and hypoxic cells than in the aerobic, cycling cells of the spheroids. On the other hand, Tsai *et al.* (1989) used an *in vitro* tetrazolium-based colorimetric assay for cytotoxicity (MTT assay) and an isobologram analysis at 50% cell kill to test combinations of etoposide and CDDP against four human small cell and four human non-small cell lung carcinoma lines. Using a rigorouus test for *in vitro* synergy, they could not establish a greater than additive cytotoxic effect on the four cell lines. Tumor growth delay studies in the Lewis lung carcinoma, FSaIIC fibrosarcoma, and the SW2 human small cell xenograft of etoposide and CDDP or BCNU showed additive effects of the two drugs (Teicher *et al.*, 1988h).

B. POLY(ADP–RIBOSE) POLYMERASE INHIBITORS

The chromosomal enzyme poly(ADP–ribose) polymerase (ADPRP) is involved in the repair of DNA damage caused by X rays and monofunctional alkylating agents (Durkacz *et al.*, 1980). However, the mechanisms are not well understood. Recent evidence suggests that ADPRP is involved in excision repair and that inhibitors of this enzyme retard DNA strand rejoining (Durkacz *et al.*, 1980). The chemical classes of ADPRP inhibitors include thymidine (Preiss *et al.*, 1971), nicotinamides (Shall, 1975), benzamides (Shall *et al.*, 1977), and methylxanthines (Davies *et al.*, 1978).

In 1982 Berger *et al.*, using L1210 cells in culture, found that 6-aminonicotinamide potentiated the cytocidal effect of BCNU; however, it did not significantly potentiate the effects of nitrogen mustard or γ irradiation. *In vivo*, both 6-aminonicotinamide and nicotinamide potentiated the cytocidal effect of BCNU; however, the concentrations of nicotinamide required for this effect were 10- to 20-fold higher than those of 6-aminonicotinamide. None of the analogs significantly potentiated the *in vivo* effect of nitrogen mustard or γ irradiation. Based on the definitions of Schabel *et al.* (1979), treatment of L1210-bearing mice with varying combination of BCNU and 6-aminonicotinamide produced a synergistic increase in life span. In some cases, the combination led to the production of long-term disease-free survivors.

Enhanced cell killing by the bifunctional alkylating agent L-phenylalanine mustard (L-PAM) in the presence of inhibitors of ADPRP in CHO-V79 cells in culture was reported by Brown *et al.* (1984). *In vivo* enhancement of the tumoricidal effects of L-PAM was observed with the ADPRP inhibitor nicotinamide (1000 g/kg), although enhanced myelosuppression was also demonstrated. Nicotinamide also increased the plasma elimination half-life of L-PAM by a factor of at least 2. This alteration of L-PAM pharmacokinetics made it difficult to assess the role that ADPRP inhibition played in the enhancement of L-PAM tumor cell killing *in vivo*.

Nicotinamide and 3-aminobenzamide potentiated the antitumor activity of CDDP on Ehrlich ascites carcinoma in mice (Chen and Pan, 1988). The mean survival times of the mice increased from 21.2–37.0 days in CDDP-treated groups to 47.0–54.6 days in mice treated with CDDP plus nicotinamide or 3-aminobenzamide. These drugs also potentiated CDDP antitumor activity on sarcoma 180, with the inhibition rates increasing from 12.4–20.8% in groups treated daily with CDDP to 29.8–46.4% in those treated with CDDP plus nicotinamide or 3-aminobenzamide. Neither 3-aminobenzamide nor nicotinamide alone showed any antitumor activity. The single-dose lethality and renal toxicity of CDDP on mice were partially reversed by either nicotinamide or 3-aminobenzamide.

C. LONIDAMINE, AN INHIBITOR OF CELLULAR RESPIRATION

Lonidamine, 1-[(2,4–dichlorophenyl)methyl]-1*H*-indazol-3-carboxylic acid, affects the energy metabolism of cells (DeMartino *et al.*, 1984, 1987; Floridi *et al.*, 1981a,b, 1986; Floridi and Lehninger, 1983). In both normal and neoplastic cells, oxygen consumption is strongly inhibited by this drug; furthermore, in tumor cells, aerobic and anaerobic glycolysis are additionally affected (DeMartino *et al.*, 1984, 1987; Floridi *et al.*, 1981a,b, 1986; Floridi and Lehninger, 1983). Based on these data, mitochondria have been considered the primary intracellular targets of the drug. Recent studies by DeMartino *et al.* (1987) indicate that the plasma and mitochondrial membrane of cells are the primary targets of lonidamine. It may be that the inhibition of energy metabolism of cells by lonidamine is a consequence of the structural damage of the inner and outer mitochondrial membranes which leads to inhibition of respiration and glycolysis and finally loss of cell viability (DeMartino *et al.*, 1987).

Lonidamine could be an important component of a combined modality regimen if repair of damage by a cytotoxic treatment is an energy dependent process. Working with Chinese hamster HA-1 cells in culture, Hahn *et al.* (1984) showed that at concentrations achievable *in vivo* lonidamine inhibited the repair of potentially lethal damage caused by X rays, methyl methane sulfonate, bleomycin, and hyperthermia. Kim *et al.* (1984a,b, 1986) showed that lonidamine potentiated the effects of radiation and the effects of hyperthermia (Kim *et al.*, 1984a) in murine tumor models. Lonidamine has also been shown to enhance the cytotoxicity of Adriamycin in culture (Zupi *et al.*, 1986).

The ability of lonidamine to enhance the cytotoxicity of alkylating agents toward MCF-7 human breast carcinoma cells has been examined (Rosbe *et al.*, 1989). A lonidamine concentration of 250 μM was used in the drug

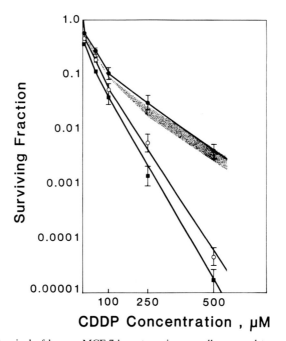

CDDP Concentration , µM

FIGURE 11   Survival of human MCF-7 breast carcinoma cells exposed to various concentrations of CDDP for 1 hr alone (●), with simultaneous exposure to 250 µM lonidamine for 1 hr (○), or with simultaneous exposure to 250 µM lonidamine for 1 hr followed by an additional 12 hrs of exposure to lonidamine (■). *Shaded area*, envelope of additivity for CDDP plus either 1 hr or extended exposure to lonidamine. Points represent the mean of three independent experiments; bars represent the SEM (Rosbe *et al.*, 1989).

combination studies. Lonidamine appeared to have a dose modifying effect on CDDP producing increasingly supraadditive cell kill with increasing CDDP concentration (Rosbe *et al.*, 1989) (Fig. 11). Cell kill was 2 logs greater than additive with 500 µM CDDP when incubated simultaneously with lonidamine for 1 hr. Extending the exposure to lonidamine for 12 hr-post CDDP treatment led to a small additional amount of cell kill of about 2.5-fold over the CDDP concentration range. Lonidamine also appeared to have a dose-modifying effect on melphalan cytotoxicity in the melphalan concentration range of 100–500 µM (Rosbe *et al.*, 1989) (Fig. 12). For 10–100 µM melphalan, the drug combination survivals on a 1-hr exposure fell within the envelope of additivity for the two agents. However, maintaining the presence of lonidamine for an additional 12 hr increased the effect so that the combination was supraadditive over the entire concentration range of melphalan. Simultaneous exposure to 4-HC and lonidamine for 1 hr resulted in greater than additive cell kill; extending the lonidamine exposure period, so that lonidamine was present during and 12 hr after 4-HC treatment, further increased this effect (Rosbe *et al.*, 1989) (Fig. 13). Lonidamine had a moder-

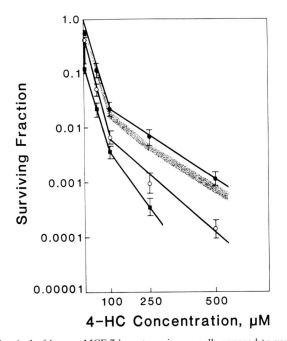

4-HC Concentration, µM

FIGURE 12   Survival of human MCF-7 breast carcinoma cells exposed to various concentrations of melphalan for 1 hr alone (●), with simultaneous exposure to 250 µM lonidamine for 1 hr (○), or with simultaneous exposure to 250 µM lonidamine for 1 hr followed by an additional 12 hrs of exposure to lonidamine (■). *Shaded area*, envelope of additivity for melphalan plus either 1 hr or extended exposure to lonidamine. Points represent the mean of three independent experiments; bars represent the SEM (Rosbe *et al.*, 1989).

ate effect on the cytotoxicity of BCNU with 1-hr simultaneous exposure to the two agents; however, this treatment combination reached greater than additive cytotoxicity only at the highest concentration of BCNU tested (Rosbe *et al.*, 1989) (Fig. 14). Extending the exposure time to lonidamine for an additional 12 hr resulted in supraadditive cell kill over the BCNU concentration range. Therefore, when lonidamine was present during exposure to the alkylating agent and then extended for an additional 12 hr, a synergistic cell kill with all four alkylating agents tested was produced.

D. POTENTIALLY LETHAL DAMAGE INHIBITION

To various extents, mammalian tumor cells have the capacity to repair drug- and radiation-induced damage. The ability of cells to recover from potentially lethal damage has been modeled *in vitro* by maintaining cells under conditions which prevent them from proliferating for various times

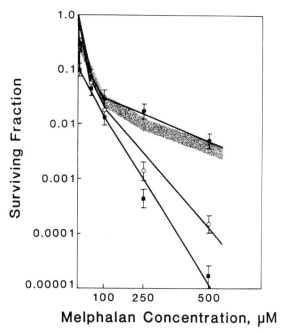

Melphalan Concentration, µM

FIGURE 13   Survival of human MCF-7 breast carcinoma cells exposed to various concentrations of 4-HC for 1 hr alone (●), with simultaneous exposure to 250 µM lonidamine for 1 hr (○), or with simultaneous exposure to 250 µM lonidamine for 1 hr followed by an additional 12 hrs of exposure to lonidamine (■). *Shaded area*, envelope of additivity for 4-HC plus either 1 hr or extended exposure to lonidamine. Points represent the mean of three independent experiments; bars represent the SEM (Rosbe *et al.*, 1989).

and thereby allow time for repair processes to take place (Barranco and Townsend, 1986).

Solid tumors and slow-growing lymphomas are likely to contain large populations of noncycling cells, which may have the capacity to repair potentially lethal damage and contribute to regrowth of the tumor. An *in vitro* system containing cells in stationary phase may be more analogous to the *in vivo* situation than are cells in exponential growth. Such an experimental model can be created by growing monolayer cell cultures to confluency under conditions of constant medium renewal without subculture. These stationary phase cultures contain a large fraction of noncycling, but potentially clonogenic, cells (Hahn and Little, 1972). In such model systems, time-dependent enhancement of cell survival observed with longer presubculture intervals following exposure to cytotoxic agents can be inferred to be caused by potentially lethal damage repair (PLDR) (Ray *et al.*, 1973; Weichselbaum *et al.*, 1982).

The significance of PLDR has remained controversial (Twentyman, 1984; Weichselbaum *et al.*, 1982, 1984); however, it seems reasonable that drug

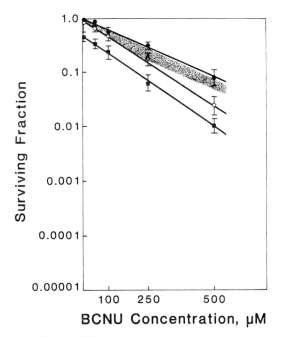

BCNU Concentration, μM

FIGURE 14 Survival of human MCF-7 breast carcinoma cells exposed to various concentrations of BCNU for 1 hr alone (●), with simultaneous exposure to 250 μM lonidamine for 1 hr (○), or with simultaneous exposure to 250 μM lonidamine for 1 hr followed by an additional 12 hrs of exposure to lonidamine (■). *Shaded area*, envelope of additivity for BCNU plus either 1 hr or extended exposure to lonidamine. Points represent the mean of three independent experiments; bars represent the SEM (Rosbe *et al.*, 1989).

combinations which inhibit the ability of tumor cells to repair significant portions of drug-induced damage will lead to improved clinical treatment. The ability of mammalian cells to recover from bleomycin-induced damage has been well documented both *in vitro* and *in vivo* (Barranco and Townsend, 1986). This process can be inhibited with actinomycin D, ethanol, and hyperthermia (Barranco, 1978; Twentyman, 1984) or under hypoxic conditions by misonidazole (Korbelik *et al.*, 1985). More recently it has been shown that some platinum complexes can inhibit the recovery of V79 cells from radiation-induced cell kill (O'Hara *et al.*, 1986).

The SCC-25 cell line is a well-established line derived from a human squamous carcinoma of the head and neck. The capacity of this cell line for recovery from potentially lethal damage following X-ray treatment has been documented. The survival curve of stationary phase SCC-25 cells exposed to various concentrations of bleomycin is biphasic with an initial sensitive phase and a less sensitive second phase as is common for many cell lines (Fig. 15). Stationary phase SCC-25 cells were exposed to 100 mU/ml of bleomycin for 1 hr. The drug was removed, and the cells were allowed various periods to

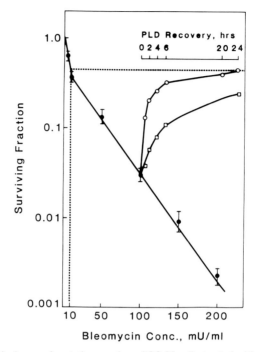

FIGURE 15   Survival curve for stationary phase SCC-25 cells treated with various concentrations of bleomycin for 1 hr (●). PLD recovery (○) showing the loss of effectiveness of bleomycin
(100 mU/ml) due to recovery from potentially lethal damage over 24 hrs and reduced PLD
recovery (□) from the same bleomycin treatment in the presence of 0.5 μM CDDP. Points
represent the mean of three independent experiments; bars represent the SEM (Holden *et al.*,
1987; Wang *et al.*, 1989).

recover from potentially lethal damage (Holden *et al.*, 1987). After 24 hr, the
SCC-25 cells showed a recovery ratio (R/R_0 = surviving fraction (24 hr) surviving fraction [0 hr]) of 7.0 which corresponded to an immediate survival at a
drug level of 27 mU/ml, a dose 3.7-fold less than the exposure concentration
of 100 mU/ml. Over the course of the first 4 hr following bleomycin exposure, 0.5 μM CDDP was a very effective inhibitor of PLDR, giving a R/R_0 of
1.1 or nearly complete inhibition of recovery. Between 2 and 4 hr, the R/R_0
was 1.6–1.8 with CDDP and 4.1–5.3 without CDDP, indicating appreciable
inhibition of recovery. Plant (10 μM) and Plato (10 μM) produced potentially
lethal damage recovery inhibition patterns very similar to that of CDDP
(Teicher *et al.*, 1985c). After 1 hr, the recovery ratios in the presence of Plant
[*trans*-di(2-amino-5-nitrothiazole)dichloroplatinum(II)] and Plato [*cis*-(1,2-
diamino-4-nitrobenzene)dichloroplatinum(II)] were 1.1–1.3. Between 2 and
4 hr, Plato and Plant gave recovery ratios of 1.8–2.3 and 1.6–1.9, respectively. NIPt, *trans*-di(2-nitroimidazole)dichloroplatinum(II), and Pt(terpy),

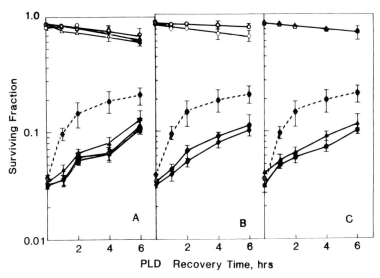

FIGURE 16   Survival of stationary phase SCC-25 cells treated with 100 μM bleomycin which were allowed various periods of time of PLD recovery (---). (A) Survival of these same cells exposed to 0.5 μM CDDP (●), 10 μM Plant (■), or 10 μM Plato (▲) during the PLDR recovery period. Survival of untreated cells exposed to 0.5 μM CDDP (○), 10 μM Plant(□), or 10 μM Plato (△) for the indicated time periods. (B) Survival of these same cells exposed to 10 μM NIPt (●), 25 μM NIPt (◆), 10 μM Pt(terpy) (○), or 25 μM Pt(terpy) (◇) during the PLDR recovery period. (C) Survival of untreated cells exposed to 10 μM NIPt (○), 25 μM NIPt (◇), 10 μM Pt(terpy) (△), 25 μM Pt(terpy) (□) for the indicated time periods. Error bars represent the SEM (Holden *et al.*, 1987).

[Pt(terpyridine)Cl]Cl, were examined at both 10 μM and 25 μM for their ability to inhibit potentially lethal damage recovery after bleomycin treatment. After 1 hr, NIPt gave a recovery ratio of 1.3–1.4, and after 2 to 4 hr the recovery ratio was 1.7–2.6 Pt(terpy) gave recovery ratio of 1.3–1.6 after 1 hr and 1.5–1.8 after 2 to 4 hr (Fig. 16).

The ability of several new complexes of platinum with positively charged dyes (Teicher *et al.*, 1986c, 1987a,d, 1989a,b; Abrams *et al.*, 1986; Teicher and Holden, 1987a) to inhibit bleomycin PLDR was also studied in the SCC-25 squamous cancer cell line. All of the new agents were more cytotoxic against exponentially growing cells than against plateau-phase cell cultures. Exposure of cells to nonlethal drug concentrations for 1–6 hr led to measurable inhibition of bleomycin PLDR in the case of each drug tested. In order of decreasing ability to inhibit bleomycin PLDR, Pt(fast black)$_2$, Pt(thioflavin)$_2$, and Pt(thionin)$_2$ were more effective than CDDP, while Pt(methylene blue)$_2$, Pt(Rh-123)$_2$, and Pt(pyronin Y)$_2$ were less effective. The most directly cytotoxic agents were Pt(thioflavin)$_2$, Pt(pyronin Y)$_2$, and Pt(Rh-123)$_2$, which also proved to be the least selectively toxic drugs toward exponential versus-plateau-phase cells (Fig. 17). These results indicate that several of the new

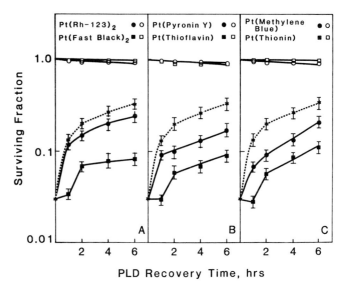

FIGURE 17 Survival of stationary phase SCC-25 cells treated with 100 mU/ml bleomycin which were allowed various periods of time for PLD recovery (---). (A) Survival of these same cells exposed to 5 μM Pt(Rh-123)$_2$ (●) or 5 μM Pt(fast black)$_2$ (■) during the PLD recovery period. Survival of untreated cells exposed to 5 μM Pt(Rh-123)$_2$ (○), or 5 μM Pt(fast black)$_2$ (□) for the indicated time periods. (B) Survival of these same cells exposed to 0.5 μM Pt(pyronin Y)$_2$ (●) or 0.5 μM Pt(thioflavin)$_2$ (■) during the PLD recovery period. Survival of untreated cells exposed to 0.5 μM Pt(pyronin Y)$_2$ (○) or 0.5 μM Pt(thioflavin)$_2$ (□) for the indicated time periods. (C) Survival of these same cells exposed to 5 μM Pt(methylene blue)$_2$ (●), or 5 μM Pt(thionin)$_2$ (■) during the PLD recovery period. Survival of untreated cells exposed to 5 μM Pt(methylene blue)$_2$ (○) or 5 μM Pt(thionin)$_2$ (□) for the indicated time periods. Points represent the mean of three independent experiments; bars represent the SEM (Wang *et al.*, 1989).

platinum complexes may be effective cytotoxic agents as well as effective inhibitors of DNA repair processes following exposure of cells to other DNA interactive modalities (Wang *et al.*, 1989).

III. Combined Alkylating Agent Studies

It has long been recognized that the schedule and sequence of drugs in combination can affect therapeutic outcome. Over the last 15 years the definition of additivity and therapeutic synergism has evolved with increasing stringency. In the work by Schabel *et al.* (1978, 1979, 1983), therapeutic synergism between two drugs was defined to mean that "the effect of the two drugs in combination was significantly greater than that which could be obtained when either drug was used alone under identical conditions of treatment." Using this definition, the combination of CTX and L-PAM administered simultaneously by ip injection every 2 weeks was reported to be therapeutically synergistic in the Ridgeway osteosarcoma growth delay assay (Schabel *et al.*, 1978, 1979, 1983). Similarly, the combination of CTX and L-PAM has been reported to be therapeutically synergistic in L1210 and P388 leukemias (Schabel *et al.*, 1983). CTX plus a nitrosourea (BCNU, CCNU, or MeCCNU) have also been reported to be therapeutically synergistic in increase-in-lifespan and growth delay assays using this definition (Schabel *et al.*, 1983).

In vitro and in vivo studies with the drug combination thiotepa and cyclophosphamide (CPA) were carried out using the MCF-7 human breast carcinoma cell line and the EMT6 mouse mammary carcinoma cell line (Teicher *et al.*, 1988e). *In vitro*, survival curves were essentially linear. The EMT6 cell line was less sensitive to thiotepa than the MCF-7 cell line, with concentrations of 440 μM and 140 μM, respectively, reducing cell survival to 10%. The response of both cell lines to 4-hydroperoxycyclophosphamide was similar. Simultaneous and immediate sequential treatments with these drugs produced supraadditive cell killing of both cell lines, although the magnitude of the supraadditivity was greater in the MCF-7 cell line than in the EMT6 cell line (Frei *et al.*, 1988; Teicher *et al.*, 1988e) (Figs. 18 and 19). Both of these drugs appeared to be as effective as thiol-depleting agents as diethyl maleate. A pattern of increasing DNA cross-linking was seen by DNA alkaline elution techniques that was similar to the increasing levels of cytotoxicity of this drug combination with increasing thiotepa concentrations. In the EMT6 tumor *in vivo*, the maximum tolerated combination therapy (5 mg/kg × 6 thiotepa and 100 mg/kg × 3 CPA) produced about 25 days of tumor growth delay, which was not significantly different than expected for additivity of the individual drugs (Table 2, Fig. 20). The survival of EMT6 tumor cells after treatment of

TABLE 2
Growth Delay of the EMT6 Mouse Mammary Carcinoma
Produced by ThioTEPA, CPA, or the Two-Drug Combination[a]

Drug	Dose (mg/kg)	Tumor growth delay (days)
ThioTEPA[b]	3.3	2.9 ± 0.7
	5.0	3.7 ± 0.6
	6.7	Toxic
CPA[c]	67	10.0 ± 1.1
	100	11.8 ± 1.3
	134	15.0 ± 1.8
	167	15.5 ± 1.7
ThioTEPA and CPA	5.0	25.2 ± 2.1
	100	

[a] The EMT6 tumor was grown sc in BALB/c mice. Treatment began when tumors were 50–100 mm³.
[b] ThioTEPA was administered ip on days 7–12.
[c] CPA was administered ip on days 7, 9, and 11.

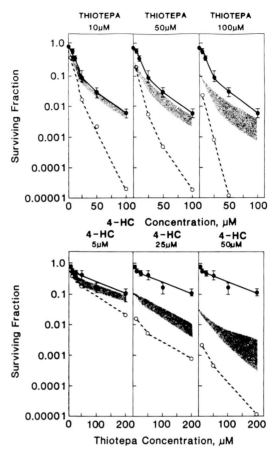

FIGURE 18   *Top*: Isobologram for the simultaneous exposure for 1 hr of MCF-7 cells to thioTEPA at 10, 50, or 200 μM in combination with a range of 4-HC concentrations. Survival curve for 4-HC alone (—). *Shaded area*, envelope of additivity for each combination treatment. Survival for the combination treatment (---). *Bottom*: Isobologram for the simultaneous exposure for 1 hr of MCF-7 cells to 4-HC at 5, 25 or 50 μM in combination with a range of thioTEPA concentrations. Survival curve for thioTEPA alone (—). *Shaded area*, envelope of additivity for each combination treatment. Survival for the combination treatments (---). Points represent the mean of three independent determinations (Teicher *et al.*, 1988e).

the animals with various single doses of thiotepa and CPA was assayed. Tumor cell killing by thiotepa produced a very steep, linear survival curve through 5 logs (Fig. 21). The tumor cell surial curve for CPA out to 500 mg/kg gave linear tumor cell kill through almost 4 logs. In all cases, the combination treatment tumor cell survivals fell well within the envelope of additivity (Fig. 22). Both of these drugs are somewhat less toxic toward bone marrow cells by the granulocyte–macrophage colony-forming unit *in vitro* assay method than to tumor cells (Fig. 23). The combination treatments were

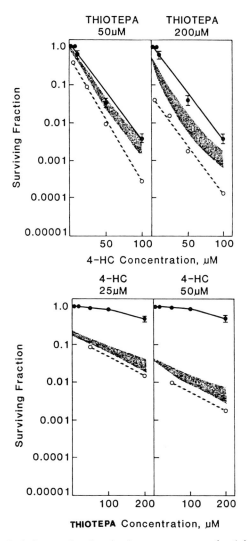

FIGURE 19 *Top*: Isobologram for the simultaneous exposure for 1 hr of EMT6 cells to thioTEPA at 50 or 200 μM in combination with a range of 4-HC concentrations. Survival curve for 4-HC alone (—). *Shaded area*, envelope of additivity for each combination treatment. Survival for the combination treatment (---). *Bottom*: Isobologram for the simultaneous exposure for 1 hr of EMT6 cells to 4-HC at 25 or 50 μM in combination with a range of thioTEPA concentrations. Survival curve for thioTEPA alone (—). *Shaded area*, envelope of additivity for each combination treatment. Survival for the combination treatments (---). Points represent the mean of three independent determinations (Teicher *et al.*, 1988e).

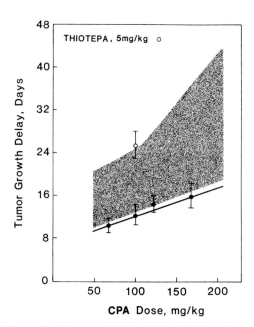

FIGURE 20   Isobologram for the growth delay of the EMT6 murine mammary carcinoma treated with combinations of thioTEPA and CPA. Tumor treatments with CPA alone (—). *Shaded area*, envelope of additivity for treatments with thioTEPA and CPA. Combination treatment of 5 mg/kg thioTEPA × 6 plus 100 mg/kg CPA × 3 (○). See Table 2 for the treatment schedules. Points represent three independent experiments (7 animals/group; 21 animals/point); bars represent the SEM (Teicher *et al.*, 1988e).

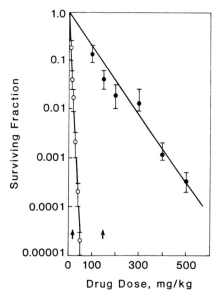

FIGURE 21   Survival of EMT6 tumor cells treated *in vivo* with single doses of thioTEPA (○) or cyclophosphamide (●). *Arrows* indicate 10% lethal dose for each drug. Points represent the mean of three independent determinations; bars represent the SEM (Teicher *et al.*, 1988e).

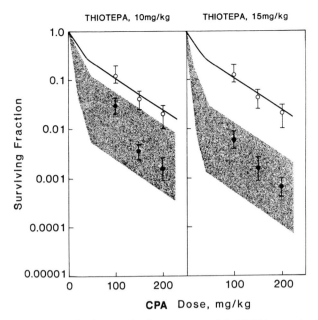

FIGURE 22   Isobolograms for the combination treatment of the EMT6 tumor *in vivo* with 10 or 15 mg/kg thioTEPA and various doses of cyclophosphamide. Survival curve for EMT6 tumors exposed to CPA only (—). *Shaded areas*, envelopes of additivity for the combination treatments. Tumor cell survivals for the combination treatments (●)(Teicher *et al.*, 1988e).

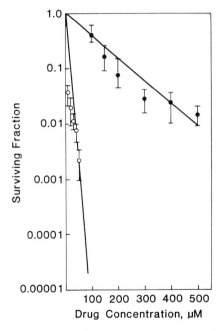

FIGURE 23   Survival of bone marrow (granulocyte–macrophage colony-forming units) from EMT6 tumor-bearing BALB/c mice treated with various doses of thioTEPA (○) or cyclophosphamide (●) (Teicher *et al.*, 1988e).

subadditive or additive in bone marrow granulocyte–macrophage colony-forming unit killing (Fig. 24). When bone marrow is the dose-limiting tissue, there is a therapeutic advantage to the use of this drug combination. The effects of schedule and sequence *in vivo* on the survival of EMT6 tumor cell and bone marrow (CFU-GM) obtained with combinations of cyclophosphamide and thiotepa or melphalan were examined and analyzed using isobologram methodology (Teicher *et al.*, 1989d,e). On a single injection schedule, if cyclophosphamide and thiotepa were administered simultaneously or if thiotepa was administered prior to cyclophosphamide, slightly greater than additive tumor cell kill resulted (Fig. 25). However, if cyclophosphamide preceded thiotepa by 4 hr, there was less than additive cell kill. If the interval between the administration of the two drugs was 8 hr, both sequences of the drugs produced greater than additive tumor cell kill. Simultaneous administration of cyclophosphamide and thiotepa on a multiple-injection schedule resulted in subadditive tumor cell kill (Teicher *et al.*, 1989e). Using the multiple-injection schedule, extending the interval between cyclophos-

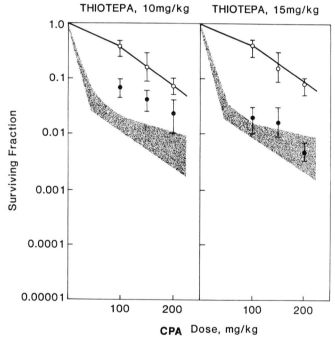

FIGURE 24 Isobolograms for the combination treatment of the bone marrow (granulocyte–macrophage colony-forming units) from EMT6 tumor-bearing BALB/c mice treated with 10 or 15 mg/kg thioTEPA *in vivo* and various doses of cyclophosphamide (CPA). Survival curve for bone marrow exposed to cyclophosphamide only (—). *Shaded areas*, envelopes of additivity for the combination treatments. Bone marrow cell survivals for the combination treatments (●) (Teicher *et al.*, 1988e).

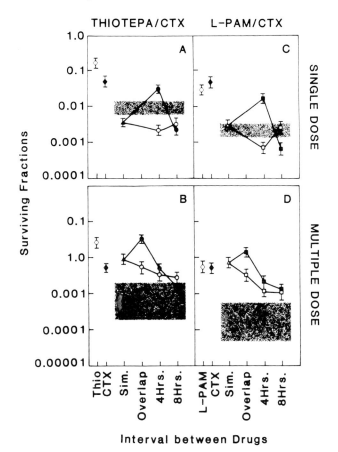

Interval between Drugs

FIGURE 25   Survival of EMT6 tumor cells treated *in vivo* with CTX alone or in combination with thioTEPA (ThioTEPA or L-PAM in various schedules and sequences). CTX was administered as a single dose of 150 mg/kg or as 3 injections of 50 mg/kg at 4.5 hr intervals. ThioTEPA and L-PAM were administered as single doses of 10 mg/kg or as 3 injections of 3.3 mg/kg at 4.5 hr intervals. Survival from exposure to the single agents (◇,◆); survivals from the drug combinations administered simultaneously (Sim.) (▲,△); survival from sequences with CTX administered first (●,■); survivals from sequences with thioTEPA or L-PAM administered first (○,□). Overlap indicates that thioTEPA or L-PAM was given with the first (●,■) or last (○,□) injection of CTX. The 4 and 8 hr intervals were measured from the last drug injection of the first agent on the multiple dose schedules. *Shaded areas*, envelopes of additivity for tumor cell killing of the two drugs as determined by isobologram analysis of the survival curves of each agent given as single or multiple doses. Results are presented as the mean of three independent determinations; bars represent the SEM (Teicher *et al.*, 1989e).

phamide and thiotepa to 4 and then 8 hr resulted in increasing tumor cell kill. With the 4 and 8 hr intervals, there was no significant sequence-dependent difference in tumor cell kill. The results with cyclophosphamide (CTX) and melphalan (L-PAM) combinations paralleled those with cyclophosphamide and thiotepa. Bone marrow (CFU-GM) survival was used as a representative normal

TABLE 3
Ratios of Bone Marrow Survival to EMT6 Tumor Cell Survival from
Mice Treated with Cyclophosphamide and Thiotepa or Melphalan[a]

	Bone marrow/tumor ratio (cfu/gm)		
	Interval between drugs		
Treatment sequence	0 hr	4 hr	8 hr
CTX(sd)/Thio(sd)[b]	10.3	12.8	15.5
Thio(sd)/CTX(sd)	10.3	13.0	6.7
CTX(md)/Thio(sd)	1.4	5.5	11.2
Thio(sd)/CTX(md)	9.6	11.8	10.4
CTX(md)/Thio(md)	10.8	—	—
CTX(sd)/L-PAM(sd)	7.1	10.6	74.0
L-PAM/(sd)/CTX(sd)	7.1	107.0	19.6
CTX(md)/L-PAM(sd)	5.5	46.4	78.3
L-PAM(sd)/CTX(md)	44.4	61.7	39.1
CTX(md)L-PAM(md)	12.3	—	—

[a] The ratios for the single dose agents are CTX(sd), single dose (150 mg/kg) 1.6; CTX(md), multiple dose (3 × 50 mg/kg) 80.0; Thio (sd), single dose (10 mg/kg) 0.21; Thio(md), multiple dose (3 × 3.3 mg/kg) 1.9; L-PAM(sd), single dose (10 mg/kg) 5.7; L-PAM(md), multiple dose (10 mg/kg) 3.1.
[b] Treatment groups correspond to those in Figure 25.

tissue with which to compare tumor cell survival from each treatment in order to obtain a measure of therapeutic effect (Table 3). The trends for the ratios of bone marrow to tumor cell survival are the same for the treatment sequences of cyclophosphamide with thiotepa or melphalan; however, greater magnitudes of differential tumor cell kill were obtained with cyclophosphamide/melphalan combinations. Using this measure, the greatest therapeutic effect was seen with melphalan or thiotepa given as a single dose followed 4 hr later by cyclophosphamide and with cyclophosphamide given as a single or as multiple doses followed 8 hr later by melphalan or thiotepa. These data from tumor model systems may be useful in developing more effective alkylating agent regimens for use in the clinic (Teicher *et al.*, 1989e).

IV. Nitroimidazole and Perfluorochemical Emulsion/O_2 Chemotherapy Combinations

Compounds such as the radiosensitizer misonidazole, which mimic the effect of oxygen in cells, have been shown to enhance the cytotoxicity of several antitumor alkylating agents including melphalan, cyclophosphamide, BCNU,

and CCNU *in vitro* and *in vivo*. This phenomenon has been termed chemosensitization or chemopotentiation (Fowler, 1985). The potentiation of the
antitumor effect of these drugs by misonidazole *in vivo* was seen without a
similar increase in normal tissue toxicity. The presence of hypoxic cells in
solid tumors may account for the preferential effect, since chemosensitization
in vitro occurs only when cells are exposed to misonidazole under hypoxic
conditions, i.e., conditions in which reduction of misonidazole through
formation of oxygen-mimicking free radicals can occur. Increased cross-link
formation has been demonstrated to occur in cells exposed *in vitro* to any of
several alkylating agents and misonidazole. Murray and Meyn (1983) enhanced cross-link formation in tumor cells derived from a murine fibrosarcoma treated with misonidazole and melphalan *in vivo*. Although the
molecular details of misonidazole enhancement of melphalan antitumor activity remain to be elucidated, it has been established that misonidazole
pretreatment results in increased levels of melphalan-induced cross-links in
DNA (Murray and Meyn, 1983; Taylor *et al.*, 1983; Horsman *et al.*, 1984;
Randhawa *et al.*, 1985). Alterations in drug pharmacokinetics and in intracellular nonprotein sulfhydryl levels may also play a role in the chemosensitization effects observed (Horsman *et al.*, 1984; Lee and Workman, 1984;
Begleiter *et al.*, 1983).

Perfluorochemicals have excellent oxygen- and carbon dioxide-carrying
capacity (Geyer, 1975, 1978, 1982). To fully utilize the oxygen-carrying
capacity of these materials, high partial pressures of oxygen are used. Unlike
hemoglobin, in which oxygen is bound to the molecule, the solvent action of
the perfluorochemicals for oxygen does not involve any kind of chemical or
chelating process. The gas molecules situate themselves in the spaces between
the molecules (Geyer, 1975, 1978, 1982). The uptake and release of oxygen
from perfluorochemical emulsions are completely reversible. At least 90% of
the emulsion particles in the preparation which we used are less than 0.2 μm
in diameter, much smaller than RBCs (average diameter, 5 to 10 μm). The
perfluorochemical emulsion, Fluosol-DA, in combination with breathing a
100 or 95% oxygen atmosphere has been shown to enhance the response of
several solid rodent tumors to single-dose and fractionated radiation treatment (Rockwell, 1985; Teicher and Rose, 1984a,b, 1986; Zhang *et al.*, 1984)
with little influence on normal tissues (Mason *et al.*, 1984; Rockwell *et al.*,
1986).

Fluosol-DA with carbogen (95% oxygen and 5% carbon dioxide) breathing can increase the efficacy of melphalan (Teicher *et al.*, 1985b, 1987b,c,e).
Addition of Fluosol-DA to treatment with melphalan leads to a greater
increase in tumor growth delay under conditions of air breathing and carbogen breathing than does the fat emulsion Intralipid; this effect is further
potentiated by hyperbaric oxygen (Teicher *et al.*, 1987c). The ability of
melphalan to kill tumor cells increased with dose over the range of drug
examined. At the lower doses of drug there is some increase in tumor cell
killing seen with the addition of carbogen breathing or Fluosol-DA and air

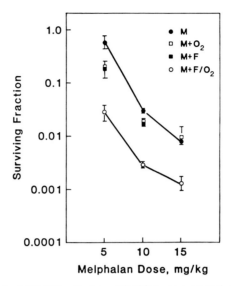

FIGURE 26 Survival of FSaIIC cells from FSaIIC tumors treated with various doses of melphalan and Fluosol-DA. Treatment groups are melphalan (M) and air breathing (●); melphalan followed by carbogen breathing for 1 hr (□); melphalan and Fluosol-DA with air breathing (■); melphalan and Fluosol-DA followed by carbogen breathing for 1 hr (○). Bars represent the SEM (Teicher et al. 1987c).

breathing; however, at the highest dose of the drug this difference disappeared. Thoughout the melphalan dosage range examined there is approximately 1 log greater tumor cell kill observed with the addition of Fluosol-DA and carbogen breathing than with the drug treatment alone (Fig. 26). There was no significant difference in the survival of bone marrow cells under any of the treatment conditions. Fluosol-DA itself with air or carbogen breathing produced no detectable cross-links in DNA from tumors treated in vivo. The cross-linking factors for melphalan with air or carbogen breathing and for melphalan plus Fluosol-DA and air breathing were similar; when carbogen breathing was added to the treatment combination, the cross-linking factor increased almost 3-fold. When melphalan was dissolved in Fluosol-DA, the melphalan moved quickly into the lipophilic fluorochemical particles so that after 1 hr 60% of the drug was in the perfluorochemical layer. At 24 hr, 85–90% of the melphalan was sequestered in the perfluorochemical particles. The pharmacokinetics of [^{14}C]melphalan alone, plus Fluosol-DA, and prepared in Fluosol-DA were studied in several tissues of FSaIIC fibrosarcoma-bearing mice. In general, the tissue absorption and distribution $t_{1/2}$s for melphalan were shortened in the presence of Fluosol-DA (except for kidneys). Shifting the $t_{1/2}$s for absorption and distribution to shorter times produces a much sharper and earlier peak in the drug exposure of the tumor. Fluosol-DA provides a relatively nontoxic means of increasing oxygen delivery to tumors and a therapeutically meaningful way of improving melphalan antitumor activity (Teicher *et al.*, 1987a). Similar effects have been found for a wide variety of antitumor agents including nitrosoureas (Teicher and Holden, 1988b) and bleomycin (Teicher *et al.*, 1986b, 1988c).

The addition of Fluosol-DA followed by carbogen breathing increased the antitumor effect of cyclophosphamide as measured by both tumor growth delay and tumor cell survival assays. Under air breathing conditions, cyclophosphamide (100 mg/kg) administered ip five times on alternate days produced a tumor growth delay in the FSaIIC fibrosarcoma of 8.0 ± 0.8 days (Teicher *et al.*, 1988a). Adding Fluosol-DA (0.3 ml) to treatment with cyclophosphamide followed by carbogen breathing increased tumor growth delay to 11.4 ± 3.6 days, which was not statistically significantly different from that obtained with the drug plus carbogen breathing without Fluosol-DA (Fig. 27). As the dose of Fluosol-DA was increased and administered with drug treatment followed by carbogen breathing for 6 hr, increasing tumor growth delays of 15.0 ± 1.5 days, 18.1 ± 1.7 days, and 29.4 ± 2.2 days were observed with 0.1 ml, 0.2 ml, and 0.3 ml Fluosol-DA, respectively. When 0.1 ml of Fluosol-DA was administered in combination with cyclophosphamide and immediately followed by 1 hr of hyperbaric oxygen (3 atm), a tumor growth delay of 13.7 ± 1.2 days was observed. With 0.2 ml of Fluosol-DA under these conditions, the tumor growth delay increased to 23.2 ± 1.6 days and with 0.3 ml of Fluosol-DA, the tumor growth delay was 35.6 ± 3.2 days. Single doses of cyclophosphamide with and without Fluosol-DA (0.3 ml) and

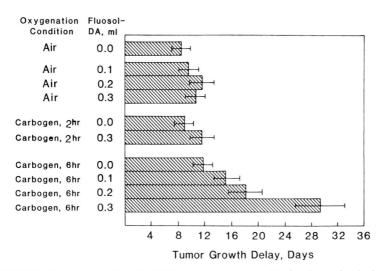

FIGURE 27   Growth delay of the FSaIIC fibrosarcoma produced by five doses of cyclophosphamide (100 mg/kg) on alternate days with and without Fluosol-DA (0.1, 0.2, or 0.3 ml) under various conditions of oxygenation. Results are the mean of three independent experiments with 7 animals per group (21 animals/point); bars represent the SEM (Teicher *et al.*, 1988d).

various conditions of oxygenation were used in a FSaIIC fibrosarcoma tumor cell survival assay. The addition of Fluosol-DA to this single dose protocol produced a 5- to 10-fold increase in tumor cell kill compared to air breathing drug-treated animals. There was no significant difference in the toxicity of any of the treatment conditions toward bone marrow.

Etoposide is much more toxic to normally oxygenated cells (Teicher *et al.*, 1985a). The ratio (hypoxic to oxygenated) of drug concentrations producing 1 log of cell kill was approximately 30:1. Established FSaIIC fibrosarcomas of C3HeB/FeJ mice were treated with 10, 15, or 20 mg etoposide/kg body weight in a 6-day protocol. Fluosol-DA with or without breathing of carbogen (i.e., 95% O_2–5% CO_2) was added to the treatment program on days 1, 3, and 5 (Fig. 28). The combination of etoposide-Fluosol-DA-carbogen markedly enhanced tumor growth delay compared to the result with etoposide alone. The dose-modifying effect observed was 1.9 ± 0.3. With the use of both single-dose and multiple-dose protocols for etoposide and Fluosol-DA with air or carbogen breathing, the survival of bone marrow cells was measured by colony formation *in vitro* (granulocyte–monocyte colony-forming units). Fluosol-DA and carbogen breathing did not increase the toxicity of etoposide to the bone marrow. Thus, the enhancement in antitumor activity produced by the addition of Fluosol-DA and carbogen breathing to etoposide treatment was not accompanied by a concomitant increase in normal tissue toxicity and represents an increase in the therapeutic efficacy of etoposide.

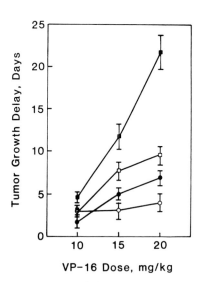

VP–16 Dose, mg/kg

FIGURE 28   Growth delay of the FSaIIC fibrosarcoma produced by 6 daily injections of etoposide (VP-16) at various doses in air (●), followed by 2 hr in a carbogen atmosphere (□), followed by iv administration of Fluosol-DA and air breathing (○), or followed by iv administration of Fluosol-DA and breathing a carbogen atmosphere for 2 hr (■) (Teicher *et al.* 1985a).

Tumor growth delays were obtained in the MX1 human breast carcinoma xenograft treated with various combinations of Adriamycin (doxorubicin), Fluosol-DA, and carbogen breathing (95% oxygen/5% carbon dioxide) (Teicher *et al.*, 1988g). Adding carbogen breathing (6 hr) or Fluosol-DA and air breathing to this drug treatment did not change the tumor growth delay observed (Table 4). When 2 hr of carbogen breathing was delivered immediately after injection with Fluosol-DA and Adriamycin, a tumor growth delay of almost 36 days was observed, which was a significant increase from the tumor growth delay obtained with Adriamycin alone ($p < 0.01$). Increasing the carbogen breathing time to 6 hr immediately following drug administration resulted in a tumor growth delay of about 43 days which was not statistically significantly different from the tumor growth delay seen with the complete treatment and 2 hr of carbogen breathing, but was different from the drug alone ($p < 0.005$). A morphologic evaluation of Adriamycin cardiotoxicity in combination with Fluosol-DA and carbogen breathing was carried out. There was only mild cardiotoxicity in all treatment groups. There was a trend toward increased cardiotoxicity of Adriamycin as the treatment dose was increased from 1 to 4 mg/kg/dose, reaching statistical significance ($p < 0.05$) for the Adriamycin (4 mg/kg/dose) + Fluosol-DA + carbogen breathing group when compared with the three treatment groups at 1 mg/kg/dose Adriamycin. The results of these studies indicate that there may be a therapeutic gain by the use of Fluosol-DA and carbogen breathing in combination with Adriamycin chemotherapy.

TABLE 4
Tumor Growth Delay in MX1 Human Breast Carcinoma Growing in
Outbred Swiss nu/nu Mice Treated with
Adriamycin ± Fluosol-DA ± Carbogen

Treatment group	Tumor growth delay (days)
Adriamycin[a]	8.9 ± 0.3[b] (n = 12)[c]
Adriamycin/carbogen (6 hr)	9.3 ± 1.1 (n = 12)
Adriamycin + Fluosol-DA[d]	8.0 ± 0.7 (n = 12)
Adriamycin in Fluosol-DA	12.9 ± 0.6 (n = 12)
Adriamycin + Fluosol-DA/carbogen (2 hr)	35.7 ± 3.5 (n = 18)
Adriamycin in Fluosol-DA/carbogen (2 hr)	25.3 ± 4.3 (n = 18)
Adriamycin + Fluosol-DA/carbogen (6 hr)	43.5 ± 3.8 (n = 18)

[a] Adriamycin was administered iv at a dose of 1.75 mg/kg daily for 5 days.
[b] The error limits are SEM.
[c] The total number of animals per treatment group. The experiment was done three times.
[d] Fluosol-DA was administered iv at a dose of 0.3 ml (12 mg/kg) daily for 5 days.

Finally, combinations of modifiers can be used to improve the efficacy of current anticancer modalities. For instance, when Fluosol-DA/carbogen plus the hypoxic cell selective cytotoxic drugs misonidazole, etanidazole, or mitomycin C were added to radiation, additive growth delays were produced (Teicher *et al.*, 1989g). Moreover, analysis of environmentally determined tumor subpopulations exposed to these treatments indicated that residual hypoxic cells which escaped reoxygenation by Fluosol-DA/carbogen and were therefore not sensitive to radiation were indeed killed by the hypoxic cell cytotoxin. Ongoing studies of this same approach utilizing alkylating agents in place of radiation demonstrate even more striking degrees of potentiation than were seen with radiation. Especially interesting is the interaction of etanidazole with cisplatin or cyclophosphamide; this radiosensitizing drug was clearly dose-modifying whereas misonidazole was merely additive. The addition of Fluosol-DA/carbogen to cisplatin or cyclophosphamide plus etanidazole resulted in about 2 logs of additional tumor cell kill and far less potentiation in normal bone marrow. Clinical studies testing these approaches at the Dana-Farber Cancer Institute are now being planned.

Acknowledgments

This work was supported by NCI grant 5PO1-CA38493 and grants from the Mathers Foundation and Lederle Inc., Pearl River, NY.

References

Abrams, M. J., Picker, D. H., Frackler, P. H., Lock, C. J. L., Howard-Lock, H. E., Faggiani, R., Teicher, B. A., and Richmond, R. C. (1986). The synthesis and structure of [rhodamine 123]$_2$PtCl$_4$·4H$_2$0: The first tetrachloroplatinate(II) salt with anti-cancer activity. *Inorg. Chem.* **25**, 3980–3983.

Arnold, A. M., and Whitehouse, J. M. (1982). Interaction of VP-16-213 with the DNA repair antagonist chloroquine. *Cancer Chemother. Pharmacol.* **7**, 123–126.

Barranco, S. C. (1978). A review of the survival and cell kinetics effects of bleomycin. *In* "Bleomycin: Current Status and New Developments" (S. K. Carter, S. T. Crooke, and H. Umezawa, eds.), p. 151. Academic Press, New York.

Barranco, S. C., and Townsend, C. M., Jr. (1986). Loss in cell killing effectiveness of anticancer drugs in human gastric cancer clones due to recovery from potentially lethal damage *in vitro*. *Cancer Res.* **46**, 623–628.

Begleiter, A., Grover, J., Froese, E., and Goldenberg, G. J. (1983). Membrane transport, sulfhydryl levels and DNA cross-linking in Chinese hamster ovary cell mutants sensitive and resistant to melphalan. *Biochem. Pharmacol.* **32**, 293–300.

Berenbaum, M. C. (1977). Synergy, additivism and antagonism in immunosuppression. *Clin. Exp. Immunol.* **28**, 1–18.

Berger, N. A., Catino, D. M., and Vietti, T. J. (1982). Synergistic antileukemic effect of 6-aminonicotinamide and 1,3-bis(2-chloroethyl)-1-nitrosourea on L1210 cells *in vitro* and *in vivo*. *Cancer Res.* **42**, 4382–4386.

Brown, D. M., Horsman, M. R., Hirst, D. G., and Brown, J. M. (1984). Enhancement of melphalan cytotoxicity *in vivo* and *in vitro* by inhibitors of poly (ADP-ribose) polymerase. *Int. J. Radiat. Oncol., Biol. Phys.* **10**, 1665–1668.

Chen, G., and Pan, Q. (1988). Potentiation of the antitumor activity of cisplatin in mice by 3-aminobenzamide and nicotinamide. *Cancer Chemother. Pharmacol.* **22**, 303–307.

Colvin, M. (1982). The alkylating agents. *In* "Pharmacologic Principles of Cancer Treatment" (B. Chabner, ed.), pp. 276–306. Saunders, Philadelphia, Pennsylvania.

Davies, M. I., Shall, S., and Skidmore, C. J. (1978). Poly (adenosine disphosphate ribose) polymerase and deoxyribonucleic acid damage. *Biochem. Soc. Trans.* **5**, 949–950.

Deen, D. F., and Williams, M. W. (1979). Isobologram analysis of X-ray–BCNU interactions *in vitro*. *Radiat. Res.* **79**, 483–491.

DeMartino, C., Battelli, T., Paggi, M. G., Nista, A., Marcante, M. L., D'Atri, S., Malorni, W., Gallo, M., and Floridi, A. (1984). Effects of lonidamine on murine and human tumor cells *in vitro*. *Oncology* **41** (Suppl. 1), 15.

DeMartino, C., Malorni, W., Accinni, L., Rosati, F., Nista, A., Formisano, G., Silvestrini, B., and Arancia, G. (1987). Cell membrane changes induced by lonidamine in human erythrocytes and T lymphocytes, and Ehrlich ascites tumor cells. *Exp. Mol. Pathol.* **46**, 15–30.

Dewey, W. C., Stone, L. E., Miller, H. H., and Giblak, R. E. (1971). Radiosensitization with 5-bromodeoxyuridine of Chinese hamster cells X-irradiated during different phases of the cell cycle. *Radiat. Res.* **47**, 672–688.

Durand, R. E., and Goldie, J. H. (1987). Interaction of etoposide and cisplatin in an *in vitro* tumor model. *Cancer Treat. Rep.* **71**, 673–679.

Durkacz, B. W., Omidini, O., Gray, D. A., and Shall, S. (1980). (ADP-ribose)n participates in DNA excision repair. *Nature (London)* **283**, 593–596.

Eder, J. P., Teicher, B. A., Holden, S. A., Cathcart, K. N. S., and Schnipper, L. E. (1987). Novobiocin enhances alkylating agent cytotoxicity and DNA interstrand crosslinks in a murine model. *J. Clin. Invest.* **79**, 1524–1528.

Eder, J. P., Teicher, B. A., Holden, S. A., Senator, L., Cathcart, K. N. S., Frei, E., III, and Schnipper, L. E. (1988a). Novobiocin enhancement of alkylating agent cytotoxicity: Animal model and clinical studies. *Proc. Conf. DNA Topoisomerases Cancer Chemother., 2nd* p. 27 (abstr. 19).

Eder, J. P., Antman, K., Elias, A., Shea, T. C., Teicher, B. A., Henner, D. W., Schryber, S. M., Holden, S. A., Finberg, R., Critchlow, J., Flaherty, M., Mick, R., Schnipper, L. E., and Frei, E., III (1988b). Cyclophosphamide and thiotepa with autologous bone marrow transplantation in patients with solid tumors. *JNCI, J. Natl. Cancer Inst.* **80**, 1221–1226.

Eder, J. P., Teicher, B. A., Holden, S. A., Senator, L., Cathcart, K. N. S., and Schnipper, L. E. (1990). Ability of four topoisomerase II inhibitors to enhance the cytotoxicity of CDDP in Chinese hamster ovary cells and in an epipodophyllotoxin resistant subline. *Cancer Chemother. Pharmacol.* (in press).

Eder, J. P., Teicher, B. A., Holden, S. A., Cathcart, K. N. S., Schnipper, L. E., and Frei, E., III (1989b). Effect of novobiocin on the antitumor activity and tumor cell and bone marrow survivals of three alkylating agents. *Cancer Res.* **49**, 595–598.

Erickson, L. C., Laurent, G., Sharkey, N., and Kohn, K. (1980). DNA crosslinking and monoadduct repair in nitrosourea treated human tumor cells. *Nature (London)* **288**, 727–729.

Floridi, A., and Lehninger, A. L. (1983). Action of the antitumor and antispermatogenic agent lonidamine on electron transport in Ehrlich ascites tumor mitochondria. *Arch. Biochem. Biophys.* **226**, 73–83.

Floridi, A., Bagnato, A., Bianchi, C., Paggi, M. G., Nista, A., Silvestrini, B., and Caputo, A. (1986). Kinetics of inhibition of mitochondrial respiration by antineoplastic agent lonidamine. *J. Exp. Clin. Cancer Res.* **5**, 273–280.

Floridi, A., Paggi, M. G., Marcante, M. L., Silvestrini, B., Caputo, A., and DeMartino, C. (1981a). Lonidamine, a selective inhibitor of aerobic glycolysis of murine tumor cells. *JNCI, J. Natl. Cancer Inst.* **66,** 497–499.

Floridi, A., Paggi, M. G., D'Atri, S., DeMartino, C., Marcante, M. L., Silvestrini, B., and Caputo, A. (1981b). Effect of lonidamine on the energy metabolism of Ehrlich ascites tumor cells. *Cancer Res.* **41,** 4661–4666.

Fowler, J. F. (1985). Chemical modifiers of radiosensitivity—Theory and reality: A review. *Int. J. Radiat. Oncol., Biol. Phys.* **11,** 665–674.

Frei, E., III (1985). Curative cancer chemotherapy. *Cancer Res.* **45,** 6523–6537.

Frei, E., III, Teicher, B. A., Holden, S. A., Cathcart, K. N. S., and Wang, Y. (1988). Preclinical studies and clinical correlation of the effect of alkylating dose. *Cancer Res.* **48,** 6417–6423.

Gellert, M., O'Dea, M. H., Itah, T., and Tomizawa, J. (1976). Novobiocin and coumermycin inhibit DNA supercoiling catalyzed by DNA gyrase. *Proc. Natl. Acad. Sci. U.S.A.* **73,** 4474–4478.

Geyer, R. (1975). Review of perfluorochemical-type blood substitutes. *In* "Proceedings of the Tenth International Congress on Nutrition: Symposium on Perfluorochemical Artificial Blood," pp. 3–19. Elsevier/North-Holland, Amsterdam.

Geyer, R. P. (1978). Substitutes for blood and its components. *Prog. Clin. Biol. Res.* **19,** 1–21.

Geyer, R. P. (1982). Oxygen transport *in vivo* by means of perfluorochemical preparations. *N. Engl. J. Med.* **307,** 304–306.

Glisson, B., Gupta, R., Smallwood-Kentro, S., and Ross, W. (1986a). Characterization of acquired epipodophyllotoxin resistance in a Chinese hamster ovary cell line: Loss of drug-stimulated DNA cleavage activity. *Cancer Res.* **46,** 1934–1938.

Glisson, B., Gupta, R., Hodges, P., and Ross, W. (1986b). Cross resistance to intercalating agents in an epipodophyllotoxin-resistant Chinese hamster ovary cell line: Evidence for a common intracellular target. *Cancer Res.* **46,** 1939–1942.

Goto, T., and Wang, J. C. (1984). Yeast DNA topoisomerase II is encoded by a single copy, essential gene. *Cell* **36,** 1073–1080.

Hahn, G. M., and Little, J. B. (1972). Plateau-phase culture of mammalian cells: An *in vitro* model for human cancer. *Curr. Top. Radiat. Res. Q.* **8,** 39–43.

Hahn, G. M., Van Kersen, I., and Silvestrini, B. (1984). Inhibition of the recovery from potentially lethal damage by lonidamine. *Br. J. Cancer* **50,** 657–660.

Holden, S. A., Teicher, B. A., Boeheim, K., Weichselbaum, R. R., and Ervin, T. J. (1987). Platinum complexes inhibit potentially lethal damage repair following bleomycin treatment. *Br. J. Cancer* **55,** 245–248.

Holm, C. T., Gato, T., Wang, J. C., and Botstein, D. (1985). DNA topoisomerase II is required at the time of mitosis in yeast. *Cell* **41,** 553–563.

Horsman, M. R., Evans, J. W., and Brown, J. M. (1984). Enhancement of melphalan-induced tumor cell killing by misonidazole: An interaction of competing mechanisms. *Br. J. Cancer* **50,** 305–316.

Issel, B. F., Muggia, F. M., and Carter, S. F., eds. (1984). "Etoposide (VP-16): Current Status and New Developments." Academic Press, Orlando, Florida.

Kalwinsky, D. K., Look, A. T., Ducore, J., and Fridland, A. (1983). Effects of the epipodophyllotoxin VP-16-213 on cell cycle traverse, DNA synthesis and DNA strand size in cultures of human leukemic lymphoblasts. *Cancer Res.* **43,** 1592–1597.

Kim, J. H., Kim, S. H., Alfieri, A., Young, C. W., and Silvestrini, B. (1984a). Lonidamine, a hyperthermic sensitizer of HeLa cells in culture and of the Meth-A tumor *in vivo. Oncology* **41** (Suppl. 1), 30–35.

Kim, J. H., Alfieri, A., Kim, S. H., Young, C. W., and Silvestrini, B. (1984b). Radiosensitization of Meth-A fibrosarcoma in mice by lonidamine. *Oncology* **41** (Suppl. 1), 36–38.

Kim, J. H., Alfieri, A. A., Kim, S. H., and Young, C. W. (1986). Potentiation of radiation effects on two murine tumors by lonidamine. *Cancer Res.* **46,** 1120–1123.

Korbelik, M., Palcic, B., and Skarsgard, L. D. (1985). Bleomycin and misonidazole cytotoxicity. *Br. J. Cancer* **51**, 499–504.

Lee, F. Y. F., and Workman, P. (1984). Misonidazole and CCNU: Further evidence for a pharmacokinetic mechanism of chemosensitization and therapeutic gain. *Br. J. Cancer* **49**, 579–585.

Lindahl, T., Karran, P., Demple, B., Sedgwick, B., and Harris, A. (1982). Inducible DNA-repair enzymes involved in the adaptive response to alkylating agents *Biochimie* **64**, 581–583.

Loike, J. D., and Horwitz, S. B. (1976). Effect of VP-16-213 on the intracellular degradation of DNA in HeLa cells. *Biochemistry* **15**, 5443–5448.

Long, B. H., and Minocha, A. (1983). Inhibition of topoisomerase II by VP-16-213 (etoposide), VM-26 (teniposide) and structured congeners as an explanation for *in vivo* DNA breakage and cytotoxicity. *Proc. Am. Assoc. Cancer Res.* **24**, 1271.

Long, B. H., Musial, S. T., and Brattain, M. G. (1984). Comparison of cytotoxic and DNA breakage activity of congeners of podophyllotoxin including VP-16-213 and VM-26: A quantitative structure–activity relationship. *Biochemistry* **23**, 1183–1188.

Mason, K. A., Withers, H. R., and Steckel, R. J. (1984). Acute effects of a perfluorochemical oxygen carrier on normal tissues of the mouse. *Radiat. Res.* **104**, 387–394.

Mattern, M. R., and Scudiero, D. A. (1981). Characterization of the inhibition of replicative and repair type DNA synthesis by novobiocin and nalidixic acid. *Biochim. Biophys. Acta* **653**, 248–258.

Meyn, R. E., Jenkins, S. F., and Thompson, L. H. (1982). Defective removal of DNA crosslink in a repair-deficient mutant of Chinese hamster cells. *Cancer Res.* **42**, 3106–3110.

Murray, D., and Meyn, R. E. (1983). Enhancement of the DNA cross-linking activity of melphalan by misonidazole *in vivo*. *Br. J. Cancer* **47**, 195–203.

Nolan, J. M., Lee, M. P., Wyckoff, E., and Hsieh, T. S. (1986). Isolation and characterization of the gene encoding Drosophila DNA topoisomerase II. *Proc. Natl. Acad. Sci. U.S.A.* **83**, 3664–3668.

O'Hara, J. A., Douple, E. B., and Richmond, R. C. (1986). Enhancement of radiation-induced cell kill by platinum complexes (carboplatin and iproplatin) in V79 cells. *Int. J. Radiat. Oncol., Biol. Phys.* **12**, 1419–1422.

Pommier, Y., Schwartz, R. E., Zwelling, L. A., Kerrigan, D., Mattern, M. R., Charcosset, J. Y., Jacquemin-Sablon, A., and Kohn, K. W. (1986). Reduced formation of protein-associated DNA strand breaks in Chinese hamster cells resistant to topoisomerase II inhibitors. *Cancer Res.* **46**, 611–616.

Preiss, J., Schlaeger, R., and Hilz, H. (1971). Specific inhibition of poly (ADP-ribose) polymerase by thymidine and nicotinamide in HeLa cells. *FEBS Lett.* **19**, 244–246.

Randhawa, V. S., Stewart, F. A., Denekamp, J., and Stratford, M. R. L. (1985). Factors influencing the chemosensitization of melphalan by misonidazole. *Br. J. Cancer* **51**, 219–228.

Ray, G. R., Hahn, G. M., Bagshaw, M. A., and Kurkjian, S. (1973). Cell survival and repair of plateau-phase cultures after chemotherapy—relevance to tumor therapy and to the *in vitro* screening of new agents. *Cancer Chemother. Rep.* **57**, 473–475.

Reddy, G. P., and Pardee, A. (1983). Inhibitor evidence for allosteric interaction in the replicase complex. *Nature (London)* **304**, 86–88.

Roberts, J. J., Pera, M. F., and Rawlings, C. J. (1982). The role of DNA repair in the recovery of mammalian cells from cis-diamminodichloroplatinum (II) (cisplatin)-induced DNA damage. *In* "Progress in Mutation Research" (A. T. Natarajan, G. Obe, and H. Altmann, eds.), pp. 223–246. Elsevier/North-Holland, Amsterdam.

Rockwell, S. (1985). Use of a perfluorochemical emulsion to improve oxygenation in a solid tumor. *Int. J. Radiat. Oncol., Biol. Phys.* **11**, 97–103.

Rockwell, S., Mate, T. P., Irvin, C. G., and Nierenburg, M. (1986). Reactions of tumors and normal tissues in mice to irradiation in the presence and absence of a perfluorochemical emulsion. *Int. J. Radiat. Oncol., Biol. Phys.* **12**, 1315–1318.

Rosbe, K. W., Brann, T. W., Holden, S. A., Teicher, B. A., and Frei, E., III (1989). Effect of

Ionidamine on the cytotoxicity of four alkylating agents *in vitro. Cancer Chemother. Pharmacol.* **25**, 32–36.

Ross, W. E. (1985). DNA topoisomerases as targets for cancer therapy. *Biochem. Pharmacol.* **34**, 4191–4195.

Ryaji, M., and Worcel, A. (1984). Chromatin assembly in Xenopus oocytes: *In vivo* studies. *Cell* **37**, 21–32.

Schabel, F. M., Trader, M. W., Laster, W. R., Wheeler, G. P., and Witt, M. H. (1978). Patterns of resistance and therapeutic synergism among alkylating agents. *Antibiot. Chemother. (Basel).* **23**, 200–215.

Schabel, F. M., Griswold, D. P., Corbett, T. H., Laster, W. R., Mayo, J. G., and Lloyd, H. H. (1979). Testing therapeutic hypotheses in mice and man: Observations on the therapeutic activity against advanced solid tumors of mice treated with anticancer drugs that have demonstrated or potential clinical utility for treatment of advanced solid tumors of man. *Methods Cancer Res.* **17**, 3–51.

Schabel, F. M., Griswold, D. P., Corbett, T. H., and Laster, W. R. (1983). Increasing therapeutic response rates to anticancer drugs applying the basic principles of pharmacology. *Pharmacol. Ther.* **20**, 283–305.

Schnipper, L. E., Eder, J. P., Teicher, B. A., Holden, S. A., Cathcart, K. N. S., and Frei, E., III (1988). Novobiocin enhancement of cisplatin cytotoxicity is mediated by DNA topoisomerase II. *Proc. Conf. DNA Topoisomerases Cancer Chemother., 2nd* p. 16 (abstr. 4S-3).

Sedgwick, B., and Lindahl, T. (1982). A common mechanism for the repair of O-6-methylguanine and O-6-ethylguanine in LDNA. *J. Mol. Biol.* **154**, 169–175.

Setlow, R. B. (1978). Repair deficient human disorders and cancer. *Nature (London)* **271**, 713–717.

Shall, S. (1975). Experimental manipulation of the specific activity of poly (ADP-ribose) polymerase. *J. Biochem. (Tokyo)* **77**, 2p.

Shall, S., Goodwin, P., Halldorsson, H., Khan, H., Skidmore, C., and Tsopanakis, C. (1977). Post-synthetic modification of nuclear macromolecules. *Biochem. Soc. Symp.* **42**, 103–116.

Steel, G. G., and Peckham, M. J. (1979). Exploitable mechanisms in combined radiotherapy–chemotherapy: The concept of additivity. *Int. J. Radiat. Oncol., Biol. Phys.* **5**, 85–91.

Sullivan, D., and Ross, W. R. (1987). Purification and characterization of a topoisomerase II mutant from epipodophyllotoxin-resistant VpmR5 cells. *Proc. Am. Assoc. Cancer Res.* **28**, 293.

Tan, K. B., Mattern, M. R., Boyce, R. A., Hertzberg, R. P., and Schein, P. S. (1987). Elevated topoisomerase II activity and altered chromatin in nitrogen mustard-resistant human cells. *Nat. Cancer Inst.* **4**, 95–98.

Taylor, Y. C., Evans, J. W., and Brown, J. M. (1983). Mechanism of sensitization of Chinese hamster ovary cells to melphalan by hypoxic treatment with misonidazole. *Cancer Res.* **43**, 3175–3181.

Teicher, B. A., and Holden S. A. (1987a). Antitumor and radiosensitizing activity of several platinum-positively charged dye complexes. *Radiat. Res.* **109**, 58–67.

Teicher, B. A., and Holden, S. A. (1987b). A survey of the effect of adding Fluosol-DA 20%/O$_2$ to treatment with various chemotherapeutic agents. *Cancer Treat. Rep.* **71**, 173–177.

Teicher, B. A., and Rose, C. M. (1984a). Oxygen-carrying perfluorochemical emulsion as an adjuvant to radiation therapy in mice. *Cancer Res.* **44**, 4285–4288.

Teicher, B. A., and Rose, C. M. (1984b). Perfluorochemical emulsion can increase tumor radiosensitivity. *Science* **223**, 934–936.

Teicher, B. A., and Rose, C. M. (1986). Effect of dose and scheduling on growth delay of the Lewis lung carcinoma produced by the perfluorochemical emulsion, Fluosol-DA. *Int. J. Radiat. Oncol., Biol. Phys.* **12**, 1311–1313.

Teicher, B. A., Holden, S. A., and Rose, C. M. (1985a). Effect of oxygen on the cytotoxicity and antitumor activity of etoposide. *JNCI, J. Natl. Cancer Inst.* **75**, 1129–1133.

Teicher, B. A., Holden, S. A., and Rose, C. M. (1985b). Differential enhancement of melphalan cytotoxicity in tumor and normal tissue by Fluosol-DA and oxygen breathing. *Int, J. Cancer* **36**, 585–589.

Teicher, B. A., Rockwell, S., and Lee, J. B. (1985c). Radiosensitivity by nitroaromatic Pt(II) complexes. *Int. J. Radiat. Oncol., Biol. Phys.* **11**, 937–940.

Teicher, B. A., Gunner, L. J., and Roach, J. A., (1985d). Chemopotentiation of mitomycin C cytotoxicity *in vitro* by *cis*-diamminedichloroplatinum(II) and four new platinum complexes. *Br. J. Cancer* **52**, 833–839.

Teicher, B. A., Holden, S. A., and Rose, C. M. (1986a). Effect of Fluosol-DA/O_2 on tumor cell and bone marrow cytotoxicity of nitrosoureas in mice bearing FSaII fibrosarcoma. *Int. J. Cancer* **38**, 285–288.

Teicher, B. A., Lazo, J. S., Merrill, W. W., Filderman, A. E., and Rose, C. M. (1986b). Effect of Fluosol-DA/O_2 on the antitumor activity and pulmonary toxicity of bleomycin. *Cancer Chemother. Pharmacol.* **18**, 213–218.

Teicher, B. A., Holden, S. A., Jacobs, J. L., Abrams, M. J., and Jones, A. G. (1986c). Intracellular distribution of a platinum–rhodamine 123 complex in *cis*-platinum sensitive and resistant human squamous carcinoma cell lines. *Biochem. Pharmacol.* **35**, 3365–3369.

Teicher, B. A., Crawford, J. M., Holden, S. A., and Cathcart, K. N. S. (1987a). Effects of various oxygenation conditions on the enhancement by Fluosol-DA of melphalan antitumor activity. *Cancer Res.* **47**, 5036–5041.

Teicher, B. A., Holden, S. A., and Cathcart, K. N. S. (1987b). Efficacy of Pt(Rh-123)$_2$ as a radiosensitizer with fractionated X-rays, *Int. J. Radiat. Oncol., Biol. Phys.* **13**, 1217–1224.

Teicher, B. A., Holden, S. A., and Jacobs, J. L. (1987c). Approaches to defining the mechanism of Fluosol-DA 20%/carbogen enhancement of melphalan antitumor activity. *Cancer Res.* **47**, 513–518.

Teicher, B. A., Herman, T. S., Holden, S. A., and Cathcart, K. N. S. (1988a). The effect of Fluosol^R-DA and oxygenation status on the activity of cyclophosphamide *In vivo*. *Cancer Chemother. Pharmacol.* **21**, 286–291.

Teicher, B. A., Herman, T. S., and Rose, C. M. (1988b). Effect of Fluosol^R-DA on the response of intracranial 9L tumors to X-rays and BCNU. *Int. J. Radiat. Oncol., Biol. Phys.* **15**, 1187–1192.

Teicher, B. A., Holden, S. A., Cathcart, K. N. S., and Herman, T. S. (1988c). Effect of various oxygenation conditions and Fluosol^R-DA on the cytotoxicity and antitumor activity of bleomycin. *JNCI, J. Natl. Cancer Inst.* **80**, 599–603.

Teicher, B. A., Holden, S. A., and Crawford, J. M. (1988d). Effects of Fluosol^R-DA and oxygen breathing on Adriamycin antitumor activity and cardiac toxicity in mice. *Cancer (Philadelphia)* **61**, 2196–2201.

Teicher, B. A., Holden, S. A., Cucchi, C. A., Cathcart, K. N. S., Korbut, T. T., Flatow, J. L., and Frei, E., III (1988e). Combination of N,N′,N″-triethylenethiophosphoramide and cyclophosphamide *in vivo*. *Cancer Res.* **48**, 94–100.

Teicher, B. A., Jacobs., and Kelley, M. J. (1988f). The influence of Fluosol^R-DA on the occurrence of lung metastases in Lewis lung carcinoma and B16 melanoma. *Invasion Metastasis* **8**, 45–56.

Teicher, B. A., McIntosh-Lowe, N. L., and Rose, C. M. (1988g). Effect of various oxygenation conditions and Fluosol-DA on cancer chemotherapeutic agents. *Biomater. Artif. Cells. Artif. Organs* **16**, 533–546.

Teicher, B. A., Bernal, S. D., Holden, S. A., and Cathcart, K. N. S. (1988h). Effect of Fluosol^R-DA/O_2 on etoposide/alkylating agent antitumor activity. *Cancer Chemother. Pharmacol.* **21**, 281–285.

Teicher, B. B., Herman. T. S., and Kaufmann, M. E. (1989a). PtCl$_4$(Nile Blue)$_2$ and PtCl$_4$(Neutral Red)$_2$: DNA interaction, cytotoxicity and radiosensitization. *Radiat. Res.* **120**, 129–139.

Teicher, B. A., Herman, T. S., and Kaufmann, M. E., (1989b). Interaction of PtCl₄(Fast Black)₂ with superhelical DNA and with radiation *in vitro* and *in vivo*. *Radiat. Res.* **119**, 134–144.

Teicher, B. A., Herman, T. S., and Kaufman, M. E. (1989c). Platinum complexes of triamino-triphenylmethanes: Interaction with DNA and radiosensitization. *Cancer Lett.* **47**, 217–228.

Teicher, B. A., Holden, S. A., Eder, P. J., Brann, T. W., Jones, S. M., and Frei, E., III (1989d). Influence of schedule on alkylating agent cytotoxicity *in vitro* and *in vivo*. *Cancer Res.* **49**, 5994–5998.

Teicher, B. A., Holden, S. A., Jones, S. M., Eder, J. P., and Herman, T. S. (1989e). Influence of scheduling on two drug combinations of alkylating agents *in vivo*. *Cancer Chemother. Pharmacol.* **25**, 161–166.

Teicher, B. A., Waxman, D. J., Holden, S. A., Wang, Y., Clarke, L., Alvarez, E., Jones, S. M., and Frei, E., III (1989f). N,N′,N″-Triethylenethiophosphoramide: Evidence for enzymatic activation and oxygen involvement. *Cancer Res.* **49**, 4996–5001.

Teicher, B. A., Herman, T. S., Holden, S. A., and Jones, S. M. (1989g). Addition of misonida-zole, etanidazole or hyperthermia to treatment with Fluosol[R]-DA[R]/carbogen/radiation. *JNCI, J. Natl. Cancer Inst.* **81**, 929–934.

Thomas, C. B., Osieka, R., and Kohn, K. W. (1978). DNA crosslinking by *in vivo* treatment with 1-(2-chloroethyl)-3-(4-methylclohexyl)-1-nitrosourea of sensitive and resistant human colon carcinoma xenographs in nude mice. *Cancer Res.* **38**, 2448–2454.

Tsai, C., Gazdar, A. F., Venzon, D. J., Steinberg, S. M., Dedrick, R. L., Mulshine, J. L., and Kramer, B. S., (1989). Lack of *in vitro* synergy between etoposide and *cis*-diamminedichloroplatinum(II). *Cancer Res.* **49**, 2390–239.

Tsai-Pflugfelder, M., Liu, L. F., Liu, A. A., Tewey, K. M., Whang-Peng, J., Knutson, T., Huebner, K., Croce, C. M., and Wang, J. C. (1988). Cloning and sequencing of CDNA encoding human DNA topoisomerase II and localization of the gene to chromosome region 17q21–22. *Proc. Natl. Acad. Sci. U.S.A.* **85**, 7177–7181.

Twentyman, P. R. (1984). Bleomycin—mode of action with particular reference to the cell cycle. *Pharmacol. Ther.* **23**, 417–441.

Van Maanen, J. M .S., Holthuis, J. J., Gobas, F., de Vries, J., van Oort, W. J., Emmelot, P., and Pinedo, H. M. 1983). Role of bioactivation in covalent binding of VP-16 to rat liver and HeLa cell microsomal proteins. *Proc. Am. Assoc. Cancer Res.* **24**, 319.

Wang, J. C. (1985). DNA topoisomerases. *Annu. Rev. Biochem.* **54**, 665–697.

Wang, J. C. (1987). Recent studies of DNA topoisomerases. *Biochim. Biophys. Acta* **909**, 1–9.

Wang, Y., Herman, T. S., and Teicher, B. A. (1989). Platinum–dye complexes inhibit repair of potentially lethal damage following bleomycin treatment. *Br. J. Cancer* **59**, 722–726.

Weichselbaum, R. R., Schmit, A., and Little, J. B. (1982). Cellular repair factors influencing radiocurability of human malignant tumors. *Br. J. Cancer* **45**, 10–16.

Weichselbaum, R., Dahlberg, W., Little, J. B., Ervin, T. J., Miller, D., Hellman, S., and Rheinwald, J. G., (1984). Cellular X-ray repair parameters of early passage squamous cell carcinoma lines derived from patients with known responses to radiotherapy. *Br. J. Cancer* **49**, 595–601.

Wozniak, A. J., and Ross, W. E. (1983). DNA damage as a basis for 4′-dimethylepipo-dophyllotoxin-9-(4,6-O-ethylidene β-D-glucopyranoside) (etoposide) cytotoxicity. *Cancer Res.* **43**, 120–124.

Wozniak, A. J., Glisson, B. S., Hande, R., and Ross, W. E. (1984). Inhibition of etoposide-induced DNA damage and cytotoxicity in L1210 cells by dehydrogenase inhibitors and other agents. *Cancer Res.* **44**, 626–632.

Yalowich, J. C., and Ross, W. E., (1984). Potentiation of etoposide-induced DNA damage by calcium antagonists in L1210 cells *in vitro*. *Cancer Res.* **44**, 3360–3365.

Zhang, W. L., Pence, D., Patten, M., Levitt, S. H., and Song, C. W. (1984). Enhancement of tumor response to radiation by Fluosol-DA. *Int. J. Radiat. Oncol., Biol. Phys.* **10**, (Suppl. 2). 172–175.

Zupi, G., Greco C., Laudonia, N., Benassi, M., Silverstrini, B., and Caputo, A. (1986). *In vitro* and *in vivo* potentiation by lonidamine of the antitumor effect of Adriamycin. *Anticancer Res.* **6,** 1245–1249.

Zwelling, L. A., and Kohn, K. W. (1982). Platinum complexes. *In* "Pharmacologic Principles of Cancer Treatment" (B. Chabner, ed.), pp. 309–339. Saunders, Philadelphia, Pennsylvania.

Zwelling, L. A., Michaels, S., Schwartz, H., Dobson, P., and Kohn, K. W. (1981). DNA crosslinking as an indicator of sensitivity and resistance of mouse L1210 leukemia to cisplatin and L-phenylalanine mustard. *Cancer Res.* **41,** 640–649.

CHAPTER 16

Drug Synergism, Antagonism, and Collateral Sensitivity Involving Genetic Changes

Vassilios I. Avramis
Sheng-He Huang
John S. Holcenberg
Division of Hematology–Oncology
Childrens Hospital of Los Angeles
and Department of Pediatrics,
Pharmacy, and Biochemistry
University of Southern California
Los Angeles, California 90054

I. Introduction

Drug synergism is defined as a significant improvement in pharmacologic action when two or more drugs are used in combination over the effect of either agent alone. Drug antagonism is then defined as a reduced pharmacologic effect when two or more drugs are administered in combination. Synergism has been observed with combinations of drugs in antibacterial, antiviral, and anticancer therapy. The antidotes for poisons are typical examples of antagonistic drug combinations.

Great advances have been seen in cancer chemotherapy in the last decade. Nevertheless many cancers either do not respond to current drug combinations or recur with drug resistance tumors. Pharmacologic factors that account for these therapeutic failures include a low fraction of the tumor cells in a sensitive part of the cell cycle; suboptimal doses, dosing schedule, or route of administration leading to inadequate circulating or intracellular drug levels; altered drug transport; rapid drug inactivation; reduced drug activation; and rapid DNA repair (Momparler, 1974; Avramis *et al.*, 1987a). Other factors that can diminish drug effects are failure of the drug to reach high

SYNERGISM AND ANTAGONISM IN CHEMOTHERAPY
Copyright © 1991 by Academic Press, Inc.
All rights of reproduction in any form reserved.

concentrations in the tumor cells due to inadequate vascular supply, poor permeability through tight vascular junctions like the blood–brain barrier, and the supply of normal nutrients from necrotic parts of the tumor that compete with antimetabolites (Avramis *et al.*, 1987a; DeVita, 1983; Goldie and Coldman, 1984, 1985). Although all these factors have been studied, and despite protocols developed to circumvent some of the obstacles, clinical tumor drug resistance still is a major problem. Combinations of drugs for clinical protocols are primarily based on the use drugs with different mechanisms of action and nonoverlapping toxicities so that each agent can be used in maximal doses and selection of drug-resistant tumor cells is delayed. More extensive study of the types and mechanisms of drug resistance and synergism is needed to provide a basis for more rational combination therapy.

The action of drugs will be altered when specific gene products that control their transport, activation, degradation, and binding to targets are expressed in abnormal quantities. These genes can be over expressed by amplification of DNA as homogenous staining regions or epigenomic material, by increased transcription of mRNA, or by increased stability of the mRNA. Expression can be decreased by alterations in structural genes leading to transcription and translation of no material or a nonfunctional product. Alternatively, modification of the regulatory components of the gene by mutation or changes in methylation can decrease expression. These genetic alterations often lead to drug resistance and failure of the cancer chemotherapy. As more is known about the function of the key enzymes and targets of chemotherapy, changes in their expression will be exploited to augment response to certain therapeutic agents. Genetic alterations have also been observed that affect antimicrobial and antiviral therapy. For example, some bacteria produce high levels of an episome that encodes penicillinase (Wittman and Wong, 1988), and certain strains of human immunodeficiency virus produce reverse transcriptase enzymes that are resistant to azidothymidine (Larder and Kemp, 1989). Detailed examples of these resistances are beyond the scope of this discussion. This chapter will review major genetic changes and present examples of their effect on the activity of anticancer drugs. It will concentrate on our recent studies of the effects of cytosine arabinoside and 5-azacytidine on methylation of DNA and activity of nucleoside analogs.

II. Over-Expression of Genes

One of the first examples of overexpression of drug related genes was the overproduction of dihydrofolate reductase (DHFR) (Alt *et al.*, 1978; Melera *et al.*, 1980). Overexpression of DHFR can increase the amount of this protein to levels where clinically achievable intracellular concentrations of methotrexate are insufficient to completely inhibit the activity of DHFR or other target enzymes. Antifols can potentiate the action of 6-mercaptopurine, thioguanine, and 5-fluorouracil in cells with active *de novo* nucleotide synthesis by inhibition of purine biosynthesis and elevation of PRPP levels (Bokkerink *et al.*, 1986; Cadman *et al.*, 1979). Amplification of DHFR prevents this synergy. Amplification of other target enzymes has been seen in resistance to the antimetabolites 5-fluorouracil, PALA, and deoxycoformycin. For a discussion of the biochemistry of synergistic interactions of antifolates with other agents, see also Chapters 13 and 19.

Two major theories have been advanced to explain the amplification of DHFR and other genes: Schimke and colleagues (Schimke, 1988; Sharma and Schimke, 1989) have argued that drugs, irradiation, or other events that inhibit DNA synthesis and alter cell cycle progression can cause repeated starts in regions of DNA that contain origins of replications. These lead to multiple DNA strands containing the gene, forming an "onion skin" of replication bubbles which eventually collapse to form expanded chromosomal regions or double minutes. Expanded chromosomal regions (called homogenous staining regions) segregate during mitosis and are generally retained without selective pressure. Double minutes are paired, acentric extrachromosomal structures that replicate and segregate independently of the chromosomes. With time they can be lost unless there is continued selective pressure.

In contrast, Wohl (1989) theorized that disruption of normal replication leads first to the formation of submicroscopic circular precursors of double minutes called episomes, each of which contains an origin of replication. Concatamerization and polymerization of episomes lead to double minutes which can eventually reintegrate into DNA to form homogeneous staining regions. They showed that double minutes are the prominent form of these amplified genes *in vivo* while in tissue culture homogenous staining regions predominate. Thus, the presence of these amplified genes as episomes or double minutes means that some selective pressure is continuing. Analysis of the effects of drug combinations, sequences, and timing may lead to mechanisms to eliminate them or decrease their accumulation. In bacteria, ethidium bromide and·acriflavine treatment can eliminate antibiotic drug resistance that is associated with extrachromosomal elements (Mitsuhashi *et al.*, 1961; Bouanchaud *et al.*, 1969). In other words, ethidium bromide and antibiotics exhibit synergism that involves a genetic alteration. Experiments are needed with models of eukaryotic drug resistance to determine whether similar agents can eliminate the episomes and double minutes.

A more general problem occurs with the overexpression of the surface glycoprotein called P-glycoprotein or gp 170 (Kartner *et al.*, 1985; Fojo *et al.*, 1987; Hamada and Tsuruo, 1986; Ueda *et al.*, 1987; Thiebaut *et al.*, 1987). This protein can actively transport many of the common anticancer agents out of cells. Consequently, its overexpression leads to resistance to agents with a variety of structures and modes of activity including vinca alkaloids, anthracyclines, and podophyllotoxins. Thus, it has been called a multiple drug resistance (MDR) protein. The MDR protein associated with resistance in human cells has been cloned and called MDR1 (Ueda *et al.*, 1987). This protein is expressed in tissues that are involved with excretion (gut epithelium, bile ducts, kidney tubules) and transport (adrenal gland, blood–brain barrier, pancreatic ductules) and tumors derived from these organs (Thiebaut *et al.*, 1987). Expression appears to be heterogeneous in these tumors before treatment. Exposure to anticancer drugs should selectively kill the cells with low MDR1 activity and enrich the population in cells with high MDR1. This may explain the poor results of treatment of these cancers. Expression of MDR1 protein, mRNA, and occasionally DNA has been shown to increase after drug treatment of model cancers and tissue culture lines. Some examples of drug resistance in patients appears to be associated with increased MDR1 expression.

MDR1 activity can be inhibited by some calcium channel blockers, quinidine, and phenothiazines (Croop *et al.*, 1988; Pastan and Gottesman, 1987; Ozols *et al.*, 1987). Clinical trials of combinations of these inhibitors with anticancer drugs that are effluxed by MDR1 are difficult because of the toxicity of the inhibitors. Current research is developing new less-toxic inhibitors of MDR1 activity and analogs of anticancer agents that are not effluxed by MDR1. More detailed treatments of the biochemistry and pharmacology of MDR1 activity reversal in mammalian cells and malarial parasites are described elsewhere in this volume.

Some cells appear to develop resistance to multiple drugs by other mechanisms. These include overexpression of a novel anionic glutathione transferase that protects cells from oxygen radicals (Batist *et al.*, 1986) and altered activity of DNA topoisomerase II (Wolverton *et al.*, 1989).

Overexpression of proteins that are targets of drug action or involved in their transport or activity may lead to collateral sensitivity to other agents. For example, Wallerstein *et al.* (1972) showed that L1210 cells that were resistant to methotrexate regained sensitivity when treated with 6-mercaptopurine and folate. Resistance to 6-mercaptopurine was reversed by treatment with methotrexate and hypoxanthine. Restoration of sensitivity to methotrexate was associated with a *decrease* in dihydrofolate reductase activity. In contrast the cells were sensitive to 6-mercaptopurine when hypoxanthine phosphoribosyltransferase activity increased. Cheng and Brockman (1983) reviewed other examples of collateral sensitivity. A drug that requires metabolism by dihydrofolate reductase would be more effective in cells with overexpresssion of this enzyme. Unfortunately such an agent has not been found yet. Cheng and Brockman did describe a human KB cell line that showed partial resistance to hydroxyurea because of a 10-fold elevation in ribonucleotide reductase. 6-Thioguanine exerts its antitumor activity by incorporation of its deoxyribosyltriphosphate derivative into DNA; ribonucleotide reductase is required for these conversions. Consequently, the cells with elevated ribonucleotide reductase were 10-fold more sensitive to 6-thioguanine than the parent cells.

III. Under-Expression of Genes

The major cause of underexpression of genes associated with drug resistance appears to be mutation within the coding region rather than the promoter or introns. Monnat (1989) recently analyzed the mechanisms of decreased hypoxanthine–guanine phosphoribosyl transferase activity in cloned HL-60 cell lines selected for resistance to thioguanine. The most prevalent class of mutation (47%) had no detectable change in the DNA gene structure by restriction enzyme fragment analysis or in the amount or size of its mRNA. This class may have point mutations, small deletions, or small insertions that cannot be detected by these methods. Other classes had large alterations detected by blotting of DNA and many produced no mRNA. These mutants would produce permanent decreases in the activity of the enzyme.

We have studied a new mechanism for decreased expression: altered control of transcription of genes required for activity of specific drugs (Huang *et al.*, 1989). If this alteration is caused by changes in the methylation pattern of the gene, the expression of the gene may be modified by agents that change methylation of cytidine residues in DNA. The remainder of this chapter will review our experiments with this type of resistance to nucleoside analogs.

The cellular pharmacology and biochemistry of nucleoside analog drugs in mammalian tumor cells has been the subject of intense investigation both *in vitro* and *in vivo* in several laboratories during the last few years and some general features have begun to emerge. Cell resistance to the cytotoxic effects of nucleoside analog compounds such as cytosine arabinoside (araC) and fludarabine phosphate (F-araAMP) can be seen *in vitro* after a single treatment with the drugs (Crane *et al.*, 1989). In addition, heterogeneity has been observed both *in vivo* and *in vitro* in the response of tumor cell populations to these drugs. Such variability in pharmacologic response is due in part to genetic expression of different phenotype(s) and in part to nongenetic factors, such as methylation of DNA.

Resistance to analogs of purine and pyrimidine bases of nucleosides is developed primarily as a consequence of decreased enzymic metabolism to their respective active anabolites, which are inhibitors of DNA polymerase α or other enzymes (Momparler, 1974; Cheng and Brockman, 1983). This mechanism is recognized as an acquired drug resistance in tumor cells. In addition to decreased drug activation, other forms of resistance to these drugs include (1) defective drug transport, (2) increased drug inactivation,

(3) altered cellular pools of nucleotides, (4) altered DNA repair, and (5) altered target proteins (Momparler, 1974; Avramis *et al.*, 1987a). Resistant cells may have altered genes or reduced expression of the genes that code for transport proteins, drug receptors, and enzymes that are necessary for the cellular activation of certain antitumor agents, such as deoxycytidine kinase (dCK).

Several approaches are now available to counteract drug resistance to nucleoside analog drugs. Certain agents have been found to have increased antitumor effect on mutant tumor cells resistant to specific drugs, a phenomenon that is called collateral sensitivity (Cheng and Brockman, 1983). Another approach is to directly alter the expression of the genes by pharmacologically producing changes in DNA methylation patterns favourable to re-expression of the gene (Avramis *et al.*, 1989d; Jones and Taylor, 1980; Antonson *et al.*, 1987). This chapter will examine collateral sensitivity and the modification of DNA methylation as a means by which we may be able to reverse tumor drug resistance to nucleoside analog drugs. Both alteration in DNA methylation patterns and collateral sensitivity have shown promise in counteracting drug resistance to cytosine arabinoside (araC) in leukemia cells (Avramis *et al.*, 1989d; Antonson *et al.*, 1987).

A. ALTERATION IN METHYLATION PATTERNS OF DNA

The degree and pattern of methylation at the 5-position of deoxycytosine bases in DNA appear to play a major role in mammalian gene regulation. Hypermethylation is associated with decreased gene expression, whereas hypomethylation is associated with gene re-expression (Jones and Taylor, 1980; Antonson *et al.*, 1987). Treatment of cells with 5-azacytidine (5-azaC) and related 5-substituted congeners leads to their activation to triphosphates and their incorporation into DNA. This action prevents DNA methyltransferase from duplicating with high fidelity the methylation pattern in the newly synthesized DNA strand during S phase (Jones and Taylor, 1980; Antonson *et al.*, 1987; Schabel *et al.*, 1982; Gandi and Plunkett, 1988; Avramis *et al.*, 1989c; Jones, 1984) (Fig. 1). Once a dC residue is demethylated, further DNA replication will produce cells, which will retain the demethylated pattern, since DNA replication introduces unmethylated C and the methylase methylates the complementary DNA strand of [meCpG]·[CpG] but not [CpG]·[azaCpG] or [CpG]·[CpG]. Thus, even if the 5-azaC anabolites are no longer present, the methylation pattern is inherited with each cell division giving rise to progeny with an altered phenotype. Treatments with 5-azaC and congeners have caused cell differentiation of tumor cells; reactivation of X-linked genes for HGPRTase, glucose-6-phosphate dehydrogenase, and phosphoglucokinase in hybrid cells; restoration of glucocorticoid sensitivity; and activation of latent genes and viruses (Jones, 1984).

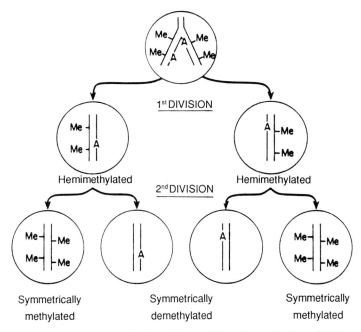

FIGURE 1   Possible mechanism of action of 5-azacytidine. The replication of DNA containing 5-methylcytosine (Me) is shown at the top of the diagram. If 5-azacytidine monophosphate (A) were incorporated into newly synthesized DNA and inhibited DNA methylation, the daughter cells of the first division would contain hemimethylated DNA. Symmetrically demethylated DNA would result after the second division posttreatment because of the specificity of the DNA methyl-transferase. Reproduced with permission from Jones, 1984.

We have shown that this treatment can also increase expression dCK expression as measured by increased araCTP concentrations in the cells and thereby reverse resistance to araC (Avramis *et al.*, 1987a, 1989d; Antonson *et al.*, 1987). We have been partially successful in producing revertants from the drug resistant cell lines both in the murine leukemia model and the human lymphoid cell lines. These efforts to reverse resistance to the nucleoside analog drugs [cytosine arabinoside, araC (Fig. 2)] in murine and human lymphoid cell lines as well as in the patients with acute lymphocytic leukemia (ALL) may identify mechanisms by which tumor cells develop drug resistance to other nucleoside analog drugs.

We are now extending our efforts in producing resistant leukemia lines to araC in the T-cell leukemia cell lines CEM/0 and Jurkat E6-1 *in vitro*. We have undertaken a major effort to analyze whether similar effects occur with antiviral nucleoside analog drugs. We hope to be able to modulate the development of drug resistance by turning on or off specific genes that play a critical role in the development of drug resistance to nucleoside analogs.

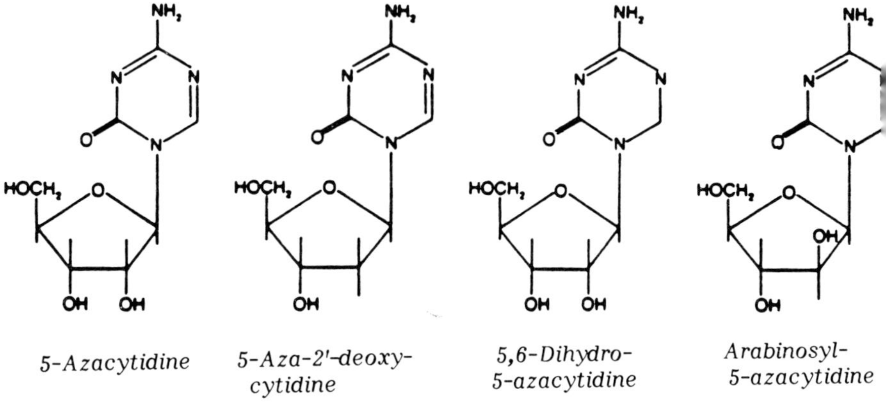

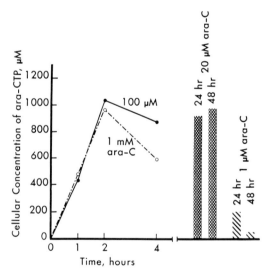

5-Azacytidine   5-Aza-2'-deoxy-   5,6-Dihydro-   Arabinosyl-
                 cytidine   5-azacytidine   5-azacytidine

FIGURE 2   Structures of 5-azacytidine and congener nucleoside analog drugs.

1. Deoxycytidine Kinase "Silencing" Confers Drug-Resistance to Cytosine Arabinoside (araC) in Leukemic Cells *in Vivo* and *in Vitro*

We tested whether methylation patterns are altered as drug resistance develops to araC (Crane *et al.*, 1989). Human lymphoid cells sensitive to araC (CEM/0) were treated with 1, 20, 100, and 1000 μM araC for up to 48 hr; these represent subtherapeutic, therapeutic, and toxic concentrations of

FIGURE 3   Cellular concentrations of araCTP in CEM lymphocytic cells after treatment with 1, 20, and 100 μM and 1 mM araC. AraCTP formation appears to be saturated above 20 μM of the prodrug.

araC. CEM/0 cells accumulated about 1 mM intracellular araCTP by 2 hr after incubations with 0.02, 0.1, and 1 mM araC. Cells incubated with 1 μM araC accumulated an average 181 μM cellular araCTP at 24 hr (Fig. 3). The amount of araC anabolite incorporated into DNA was 15-fold higher after treatment with 20–1000 μM than after treatment with 1 μM araC, reaching at least 0.015 nmol/10^7 cell DNA (Fig. 4). Total genomic DNA methylation levels increased with time and concentration of araC exceeding 210% after treatment at 1 or 20 μM concentrations for 24 or 48 hr. Similar levels of DNA hypermethylation were obtained after treatment with 0.1 and 1.0 mM araC for 4 hr (Table 1). The cells that survived treatment with araC (1 $\mu M \times$ 24 hr) were tested for sensitivity to the drug. A 50% decrease in dCK expression was observed after the 1 μM araC treatment (Table 2). Thus the increase in DNA methylation levels caused by araC may be responsible for the development of tumor drug resistance to araC due to "silencing" of dCK. In contrast we have shown that treatment of the CEM cell line lacking dCK activity, CEM/dCK$^-$, with the DNA hypomethylating agents 5-azaC and dihydro-5-azaC (DHAC) resulted in reexpression of dCK gene and the establishment of partial sensitivity to araC (Antonson *et al.*, 1987). Therefore DNA methylation may control the development of drug resistance to araC in tumor cells through dCK silencing.

2. 5-Azacytidine and Its Congeners Must Be Activated to Triphosphate Anabolites to Induce DNA Hypomethylation and Deoxycytidine Kinase (dCK) Reexpression in Human Leukemic Cell Lines *in Vitro*

5-azaC and its congeners were incubated *in vitro* with human leukemia cell lines CEM/0 and the araC-resistant CEM/dCK$^-$, which lacks the deoxycytidine kinase (dCK) required for activation of this prodrug to its active anabolite, araCTP (Fig. 5). The acid-soluble extracts from these incubations were assayed for mono-, di-, and triphosphate anabolites. 5-azaC and DHAC, which are activated by uridine–cytidine kinase, were activated by both CEM cell lines. 5-azaCdR and araAC, which are activated by dCK, were not activated to the triphosphate anabolites in measurable amounts by the CEM/dCK$^-$ cells. The amount of the incorporated anabolites of these azacytidine analogs into the DNA of these cell lines followed an identical pattern (Avramis *et al.*, 1989a).

5-azaC and its congeners were used to induce DNA hypomethylation in the human lymphoid cell lines CEM/0 and CEM/dCK$^-$. The cell lines were incubated with IC$_{50}$ concentrations of these drugs for 24 hr; then their DNA was purified and hydrolyzed for determination of DNA methylation levels (Antonson *et al.*, 1987; Avramis *et al.*, 1989a; Powell *et al.*, 1987). 5-azaCdR and araAC were not activated by CEM/dCK$^-$ cells, so no DNA hypomethylation occurred.

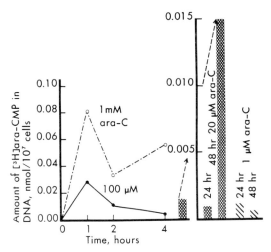

FIGURE 4   Amount of tritiated araC anabolite incorporated into cellular DNA of CEM cells.

The methyl-cytidine levels of DNA in the CEM/0 cells reached a nadir of 38.5% of control 24 hr after the end of treatment with the 100 μM DHAC (Fig. 6). By day 6 after drug treatment the fraction of methyl-cytidine residues had returned to near control levels. The decrease of the total DNA methylation patterns appears to be transient in nature in both murine and human leukemic cells after treatment with DNA hypomethylating agents,

TABLE 1
DNA Methylation Studies in CEM/0 Cells after Treatment with araC[a]

Time (hr)	Concentration of araC (μM)	% Methylation (mean ± SD)[b]	% of Untreated control
1	100	2.88 ± 2.0%	
2	100	3.60 ± 2.1%	
4	100	4.17 ± 0.71%	136%
1	1000	4.45 ± 0.85%	
2	1000	5.65 ± 1.54%	
4	1000	5.92 ± 1.43%	194%
24	1	6.11 ± 1.72%	
48	1	6.99 ± 1.22%	214%
24	20	7.46 ± 1.25%	230%
48	20	6.60 ± 1.94%	

[a] Control DNA methylation in CEM/0 cells is 3.06% ± 0.75%.
[b] n = 5.

TABLE 2
Effect of Subtherapeutic Concentration of araC on araC
Cellular Metabolism

Treatment	[araCTP] ± SD. (μM)	% of Control
Control	56.58 ± 18.24[a]	100.00%
1 μM × 24 hr	22.79 ± 2.63[a]	40.28%
Control	671.45 ± 34.40[b]	100.00%
1 μM ± 24 hr	361.16 ÷ 21.50[b]	53.78%

[a] 1 μM araC, 1 hr, PCA extraction.
[b] 20 μM araC, 1 hr, PCA extraction.

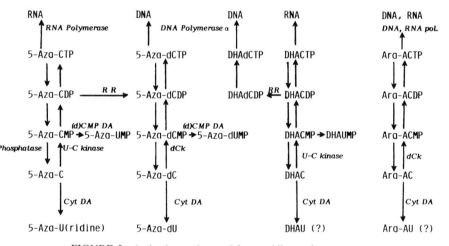

FIGURE 5   Activation pathway of 5-azacytidine and congeners.

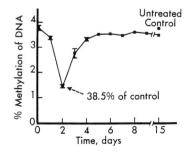

FIGURE 6   Effect of DHAC treatment on DNA methylation levels in CEM cells.

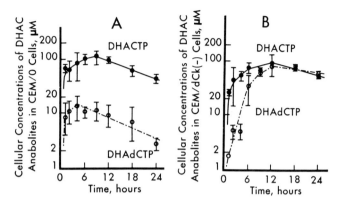

FIGURE 7   (A) Kinetics of the intracellular concentrations of DHAC anabolites in CEM/0 cells. (B) Kinetics of intracellular concentrations of DHAC anabolites in CEM/dCk(−) cells. The cells were treated with their respective IC_{50} concentrations of the drug, extracted with PCA, and assayed by HPLC on an anion exchange column (SAX-10). Symbols represent the means ± SD (n = 3).

reverting to pretreatment levels after a sufficient number of cell replications occurred. However, some genes may still remain hypomethylated after many successive cell replications (Razin *et al.*, 1984). DHACTP accumulated linearly with time for up to 6 hr in both cell lines, to a cellular concentration of approximately 120 μM (Fig. 7). The deoxy-derivative of this anabolite, DHAdCTP, was detected at 10% of DHACTP levels in CEM/0 and at 100% of the DHACTP levels in CEM/dCK⁻ cells. A possible explanation for these differences could be that the CEM/dCK⁻ cells may have increased activity of the ribonucleoside diphosphate reductase, which is responsible for the production of the deoxy-derivatives of nucleotides. The increased activity of ribonucleotide reductase could be an effect secondary to the lack

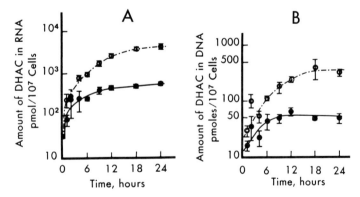

FIGURE 8   Incorporation of DHAC anabolites in RNA (A) and DNA (B) in CEM/0 (●) and CEM/dCk(−) (○) cells. Symbols represent the means ± SD (n = 3).

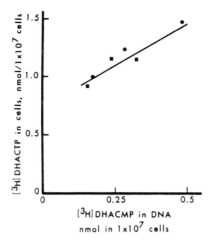

FIGURE 9   Relationship between the intracellular concentrations of DHACTP and amount of DHACMP incorporated into DNA in CEM/0 (●) and CEM/dCk(−) (■) cells.

of dCK, compensating for the lack of activation of deoxycytidine to dCTP. DHAdCTP was incorporated gradually into DNA of both cells lines over time (Fig. 8). A linear relationship was observed between the intracellular concentrations of DHACTP and the amount of DHACdCMP incorporated into DNA (Fig. 9). A curvi-linear relationship was obtained between the amount of DHAC anabolites incorporated into DNA and the DNA methylation levels in both cell lines. This may indicate that DNA hypomethylation cannot continue to zero, or that certain methyl-cytidine residues cannot be eliminated by treatment with these drugs (Fig. 10).

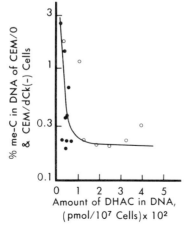

FIGURE 10   Relationship between the amount of DHAC anabolite incorporated into DNA and hypomethylation of DNA in CEM/0 (●) and CEM/dCk(−) (○) cells.

dCK was re-expressed in five revertant monoclonal colonies from CEM/dCK⁻ cells that were treated with 100 μM DHAC (Antonson *et al.*, 1987) (Tables 3 and 4). The relative levels of dCK in the revertants were 113, 93, 49, 97, and 31% in comparison with CEM/0 cells. dCK re-expression in the revertant colonies ranged from 6- to 55-fold higher than in the untreated CEM/dCK⁻ cells. Thus the phosphorylation and incorporation of 5-azaC and its congeners into DNA of cells is responsible for the DNA hypomethylation and subsequent dCK reexpression.

3. *In Vivo* Drug Synergism Studies between 5-azaC and araC

In order to understand the effect of the DNA hypomethylating agent 5-azacytidine (5-azaC) on araC *in vivo* we need to review the cellular pharmacology of araC in leukemic cells.

4. Accumulation and Kinetics of araC in Plasma and araCTP in Leukemic Cells

The pharmacodynamic parameters of araC in pediatric patient plasma and its active anabolite araCTP in circulating and bone marrow blast cells have been studied in 20 evaluable patients with acute leukemia (Avramis *et al.*, 1987b). Multiply relapsed pediatric patients with both acute nonlymphoblas-

TABLE 3
Revertant Frequency of CEM/dCk⁻ Cells
after Treatment with DHAC[a]

Experiment	Treatment DHAC (μM)	Revertant frequency $\times 10^{-5}$
1	0[b]	0
	100	61
	0[b]	160
	100	585
2	0	0
	20	1
	100	81
3	20	290
	100	360

[a] The cells were treated with DHAC for 24 hr, grown in suspension for 72 hr, and plated in soft agar plates selective for dCk expression. After 14 to 21 days of incubation the colonies on the plates were counted and the revertant frequency was calculated.
[b] Two different CEM/dCk⁻ lines were used in the experiment.

TABLE 4
Deoxycytidine Kinase Activity Measured as araCTP Formation in
CEM/0 and CEM/dCk⁻ Clones Isolated from Soft Agar Plates
Selective for Cells Expressing dCk⁻

Cells	Formation of araCTP (nmol of araCTP 1×10^{-7} cells)	% of CEM/0 control
CEM/0		
Control	1.12	100
Clone	1.20	102
Clone	0.85	68
CEM/dCk⁻		
Control	0.02^b	1.8
Control	0.08^b	6.9
Control	0.04^b	3.8
Revertant	1.10	93
Revertant	0.58	49
Revertant	1.27	113
Revertant	1.08	97
Revertant	0.35	31

[a] The clones were removed from the agar plates, grown up in solution, and assayed for araCTP formation.
[b] Quantitated manually.

tic leukemia (ANLL) and acute lymphoblastic leukemia (ALL) achieved complete remission with high dose cytosine arabinoside treatment (HDaraC, 3 gm/m² every 12 hr × 8 doses). The peak plasma concentrations of araC ranged from 0.02 to 5.6 mM after the first dose of araC. The elimination kinetics of araC were biexponential with an average ± SD elimination $t_{1/2}$ of 2.4 ± 1.5 hr in the ANLL and 4.8 ± 4.1 hr in the ALL patients. The plasma concentrations of the catabolite of araC, araU, were 2- to 12-fold higher than araC and it was eliminated monoexponentially with a $t_{1/2}$ in the range of 4 to 5 hr.

The formation and retention of araCTP by circulating and bone marrow blasts in patients was similar among the ANLL patients; the average peak araCTP concentrations was 430 ± 14 μM and the average $t_{1/2}$ of elimination of araCTP was 3.3 ± 0.8 hr (Avramis *et al.*, 1987b). In contrast, considerable variation was seen in the araCTP concentrations among the patients with ALL; the peak concentration ranged from 220 to 1100 μM with a mean of 540 ± 330 μM and the $t_{1/2}$ of araCTP elimination was 6.9 ± 2.8 hr. All pharmacokinetic parameters of araCTP were estimated after the data were fitted in a one-compartment open model (Avramis *et al.*, 1987b).The trough araCTP concentrations in peripheral blasts at 24 hr ranged from 35 to 1010 μM in the patients with ALL and 80 to 470 μM in the patients with ANLL. Bone marrow specimens were obtained from patients prior to (control) and 24 hr after treatment with HDaraC. The trough bone marrow

TABLE 5
Pharmacokinetic and Pharmacodynamic Parameters of araC in Pediatric Patients with Leukemi

Patient	Type	Schedule[a]	Pk araCTP[b]	$T^c_{1/2}$	AUC 12 h[d]	TR 12 h[e]	TR 24 h[f]	TR 3(
1. S. K.	ANNL	8	425.1	2.7	2,372	36.1	94.0	164.
2. A. P.	ANNL	8	420.8	3.0	2,248	58.9	80.3	88.
3. K. J.	ANNL	8	477.8	4.2	2,363	117.0	93.5	
4. M. H.	ANNL	8					1,144.5	
5. J. M.	ANNL	8					465.3	
6. J. R.	ANNL	8					92.8	
7. J. C.	ALL	4	258.5	3.9	1,073	50.2	112.1	137.
8. S. C.	ALL	4	483.7	2.9	2,509	62.1	35.2	25.
9. P. M.	ALL	4	405.3	6.8	1,761	162.9	180.6	
10. S. S.	ALL	4	1,164.0	8.0	14,751	493.4	1,009.8	
11. S. B.	ALL	4					51.1	
12. M. E.	ALL	8	216.4	9.4	2,838	92.5	206.1	
13. B. E.	ALL	8	438.1	10.8	6,679	208.3	178.5	181.
14. R. G.	ALL	8	459.7	4.9	2,658	120.5	170.1	223.
15. S. H.	ALL	8	928.6	8.2	12,907	507.7	885.4	765.
16. S. S.	ALL	8					2.0	
17. M. D.	ALL	8					148.4	
18. H. R.	ALL	8					44.0	
19. D. J.	ALL	8					417.0	
20. J. M.	ALL	4					465.3	

[a] 4 = 4 + 4 doses; 8 = 8 doses.
[b] Peak concentration of araCTP in PBC ($\mu M2$)
[c] Half-line of cellular araCTP in PBC (hr).
[d] Area under the concentration-time curve from 0 to 12 (μMh).
[e] Trough araCTP concentration at 12 h (μM).
[f] Trough araCTP concentration at 24 h (μM).
[g] Trough araCTP concentration at 36 h (μM).
[h] Trough araCTP concentration at 48 h (μM).
[i] DNA synthetic capacity at 24 h (% of control).

cellular araCTP concentrations showed variability similar to the peripheral blasts in the ALL patients, ranging from 2 to 1140 μM (Table 5). We determined araCTP concentrations in both peripheral and bone marrow blasts at the same time in four patients. Three of these had different values in the two leukemic subpopulations suggesting heterogeneity of tumor in the same patients. The exposure of leukemic cells to araCTP, expressed as area under the concentration–time curve (AUC), varied very little in the patients with ANLL averaging 2300 ± 69 $\mu M \cdot$hr. In the ALL patients, the AUC varied much more, averaging 5650 ± 5340 $\mu M \cdot$hr.

R 48 h[h]	DSC 24 h[i]	PL $T^{k}_{1/2}$	PL AUC[l]	PBC $T^{m}_{1/2}$	Response[n]	Log PBC kill[o]	Nadir PBC[p]	Duration of response[q]
207.8	15.3	2.2	7,005	8.5	1	3.6	0	670
83.8	10.2	4.0	26,298	8.1	1	4.3	0	52
	89.7	1.0	13,483	6.1	3	3.9	60	0
							0	126
	24.3							
							0	90
		2.7	3,845	12.6	3	5.0	0	15
		8.1	1,957	15.7	3	4.1	480	0
174.8								
		3.0	200		3	4.5	0	20
47.0	42.2							
		1.3	270	11.2	1	4.6	0	29
	10.7							
					3		0	11
	22.9							
		2.1	19,785	8.1	3	4.9	0	8
		14.4	4,543	7.8	3	4.4	0	23
	40.9							
		4.1	5,154	7.1	3	4.3	0	17
354.0	13.0							
		4.5	3,730		1	4.0	0	50
143.5	18.7							
					2	3.4	0	18
320.4					3	3.1	0	28
					2		0	33
	14.1							
		2.9	302		1	1.3	0	50
	5.2							
					1			

aCTP concentration in BM blast cells.
Ialf-life of araC in plasma (hr).
rea under the plasma concentration HD-araC (μMh).
Ialf-life of PBC from blood (hr).
R = 1, partial remission = 2, no remission = 3.
og peripheral blast cells killed.
Iadir peripheral blast cells.
)uration of response, days.

Other biochemical parameters such as (1) inhibition of thymidine incorporation, (2) disappearance of peripheral blast cells from blood, and (3) achievement of zero nadir blast cells in blood were correlated with response and, to a lesser extent, with peak or trough cellular araCTP concentrations (Avramis *et al.*, 1987b). Similar correlations have been made in larger numbers of adult patients (Plunkett *et al.*, 1987). In the responding patients, there was a direct linear correlation between cellular araCTP concentrations and inhibition of thymidine incorporation. On the other hand, patients who did not respond showed no correlation. This probably indicates the variable effect of araC in a highly heterogeneous tumor population during the treatment period or the existence of other mechanisms of araC resistance.

5. Deoxycytidine Kinase (dCK) and Cytidine Deaminase Activities in Pediatric Patient Leukemic Cells

The effect of cytidine deaminase on araC catabolism was measured both directly and indirectly in patients. The indirect estimation of deaminase was done by measuring the ratio of araU to araC in plasma. In infant patients younger than 3 years of age, the ratio was less than 1; in patients aged 3–7 years, the ratio was 2–4, indicating that significant deamination was taking place; the patients older than 8 years of age had a ratio of 6 or greater, indicating high deamination capacity by the host. This group of pediatric patients behaved more like adults as far as deamination of araC was concerned. This age-dependence of deamination could explain the observed variability in the plasma araC concentrations (Avramis *et al.*, 1987b).

Another study was conducted in mice bearing the araC sensitive murine leukemia, L1210/0, to investigate the efficacy of combination chemotherapy of araC and araU or tetrahydrouridine (THU) (Avramis and Powell, 1987). We showed that araC can be protected from deamination by either araU or THU. Nevertheless, the higher plasma araC concentrations achieved with these deaminase inhibitors had no effect on the cellular araCTP concentrations. This is probably due to saturation of transport and dCK by the araC concentrations without cytidine deaminase inhibitors (Avramis *et al.*, 1989b). Consequently the combination regimens had no additional biological effect in the antitumor efficacy of araC (Avramis and Powell, 1987).

The key enzyme activating araC to araCTP is deoxycytidine kinase (dCK). To investigate the critical parameter(s) of araC activation, we developed a method of purification of dCK from leukemic or lymphoma cells using two HPLC protein columns (Avramis *et al.*, 1989b). The final dCK preparation was purified about 2500-fold and it produced two bands on a nondenaturing polyacrylamide gel. This preparation was assayed for the activation of various nucleoside analogs including araC. Lineweaver–Burk plots showed a bimodal curve with all substrates that were activated by dCK, suggesting negative homotopic cooperativity in the binding of the substrate to the active site of dCK.

Inhibition by other substrates or by the end product, dCTP, was complex, particularly when the substrate (dCyt or araC) varied over a wide range of concentrations (Avramis, 1989). The higher k/m value for araC from 12 patients' malignant cells who responded to HDaraC and from two healthy volunteers ranged from <1 to 9.7 μM with an average of 4.7 ± 2.5 μM. On the other hand, the k/m values for araC from malignant cells from eight patients who did not respond to HDaraC, ranged from 3.1 to 35 μM with an average of 16.6 ± 9.4 μM (Tables 6 and 7). This value was statistically sig-

TABLE 6

Michaelis–Menten Rate Constants of araC and dCyt on Deoxycytidine Kinase from Patients' Tumor and WBC Cells

Patient	Disease	Response	araC		dCVt	
			Km_1 (μM)	Km_2 (μM)	Km_1 (μM)	Km_2 (μM)
B. P.	Lymphoma	C. R.	1.10	9.70	0.16	0.76
S. H.	Lymphoma	P. R.	0.46	4.40	<0.1	1.86
D. W.	All–BM	C. R.	0.83	8.00	0.22	4.60
D. W.	ALL–PBC	C. R.	0.10	5.00	0.41	4.60
L. R.	ALL–PBC	C. R.	0.52	1.94	1.08	4.08
D. P.	ALL–PBC	C. R.	0.14	5.80	0.20	1.30
D. P.	24h$\bar{p}$RxPBC[a]	C. R.	0.46	3.01	0.51	1.83
D. P.	24h$\bar{p}$RxBM	C. R.	0.15	3.34	0.65	1.20
K. W.	WBC	—	0.19	2.70	0.54	1.25
D. F.	WBC	—	0.27	3.53	0.73	1.37
Mean, n = 10 ± SD			0.40 ± 0.3	4.74 ± 2.5	0.46 ± 0.3	2.30 ± 1.5

[a] Rx: HDaraC.

nificantly higher to the value from the responding patients at $p = 0.007$ (Avramis, 1989).

6. Pharmacodynamic Modeling of araC Administration

Theoretical and mathematical considerations led to a mathematical model to predict the antitumor effectiveness of araC. A computer program for this model is currently being developed. This model suggests that the intermittent HDaraC regimen provides sufficient plasma araC levels to saturate dCK only in the small group of patients with low K_ms for araC. The araC K_m for dCK is so high in the blasts from most of the patients that their leukemic cells can optimally phosphorylate this drug for only the first few hours after the araC infusion. Then the araC concentration falls well below the K_m for dCK so activation slows markedly for the rest of the 12 hr time interval between doses (Avramis, 1989). This model relies on the observation that the K_m for araC transport across the cell in lower than the K_m for dCK (White *et al.*, 1987). Very high peak plasma concentrations of araC may lead to a greater chance of central nervous system toxicity, so increasing the dose of araC is not the answer to a decreased rate of araC activation between doses (Weinberg *et al.*, 1989) (Table 8).

TABLE 7

Km Values of araC and dCyt on Deoxycytidine Kinase from Patients' Malignant Cells

Patient	Disease	araC		dCvt	
		Km_1 (μM)	Km_2 (μM)	Km_1 (μM)	Km_2 (μM)
D. V.	ALL-BM	1.01	34.73	1.13	22.84
L. L.	ALL-PBC	2.55	18.70	2.20	38.62
E. H.	ALL-PBC	1.20	22.70	10.75[a]	22.20
J. G.	ALL-PBC	0.36	3.13	0.75	1.33[a]
J. B.	ALL-BM	1.16	15.24	1.38	6.35
J. B.	ALL-PBC	1.32	12.24	2.13	4.20
S. G.	ALL-BM	1.35	6.40	0.21	2.25[a]
U. P.	ALL-PBC	0.67	14.40	0.45	6.57
H. K.	ALL-BM	0.61	21.49	0.36	27.31
Mean ± SD		$\bar{n}_9 = 1.14 \pm 0.6$	$\bar{n}_9 = 16.60 \pm 9.4$	$\bar{n}_8 = 1.08 \pm 0.8$	$\bar{n}_7 = 18.30 \pm 1.$

[a] Statistical outliers not considered for the estimation of the mean.

In order to improve the response rates in patients with ALL and incorpo-rate the just-described biochemical considerations we developed a new regi-men of biochemically directed loading bolus (LB) followed by continuous infusion (CI) of araC for relapsed leukemia (Avramis *et al.*, 1989b). This regimen provides steady-state plasma araC concentrations that are 3 times the K_m2 value of araC on the individual patient's tumor cell dCK (Fig. 11). Thus optimal activation of araC to araCTP occurs for the entire infusion period. The pharmacokinetic data from 10 patients thus far shows that this treatment

TABLE 8

Toxicity of araC in Pediatric Patients after an HDara-C of 3 g/m² (8 doses) and after Continuous Infusion of araC

Route of administration	Mean [araC] in plasma (μM)	Toxicity	
		CNS	Lung
i.v. Bolus	>300	3/3	1/3
	100–299	2/4	—
	<100	0/4	—
Continuous infusion	21.2	0/14	3/14

TABLE 9

Cellular Concentrations of araCTP in Leukemic Cells from Patients Responding and Not Responding to the 5-azaC + HDaraC Regimen

Patient	[araCTP][a] before 5-azaC (μM)	[araCTP][a] after 5-azaC (μM)	Increased expression of dCk	[araCTP][b] peak, 3 hr (μM)	[araCTP][b] trough, 24 hr (μM)	Outcome
Responders						
1. R. G.	84.0	229.0	2.96-fold	—	573.4	CR
2. L. L. (BM)[c]	78.0	208.0	2.67-fold	—	359.4	CR
L. L. (PBC)[d]	42.0	216.0	5.14-fold	—	—	
3. K. W.	90.0	108.0	1.20-fold	—	—	PR
Nonresponders						
1. D. K.	329.6	1187.0	3.60-fold	1259.5	864.0	NR
2. R. G.	97.4	171.0	1.76-fold	460.0	120.5	NR
3. K. K.	86.8	94.3	1.10-fold	191.6	65.6	NR
4. S. S.	223.4	1365.0	6.11-fold	—	1025.0	NR
5. L. M.	92.0	142.0	1.54-fold	—	182.0	NR
6. O. M.	<5.0	30.3	6.00-fold	—	20.0	NR
7. H. K.	70.0	78.0	1.12-fold	—	99.0	NR
8. E. H.	29.0	168.0	5.79-fold	471.4	60.3	NR
9. U. P. (BM)[c]	408.0	573.4	1.41-fold	—	687.0	NR
U. P. (PBC)[d]	13.6	14.0	1.02-fold	—	—	NR
10. L. R.	297.0	—	—	—	—	NE
11. J. R.	168.0	143.0	0.85-fold	—	186.0	NR
12. D. W.	86.0	147.0	1.72-fold	—	389.0	NR
13. D. P.	10.0	465.0	46.50-fold[e]	—	579.3	NR
	$X_{18} = 122.7 \pm 116.8$	$X_{17} = 314.1 \pm 390.0$	$X_{16} = 2.75\text{-fold} \pm 1.95$		$X_{14} = 372.2 \pm 325.3$	

[a] AraCTP concentration data from the sensitivity test of blasts *in vitro*, 1 m M araC for 1 hr.
[b] AraCTP concentration data after *in vivo* HDaraC regimen.
[c] Data in bone marrow blasts.
[d] Data in peripheral blasts cells.
[e] Statistical outlier not used for the estimation of the mean.

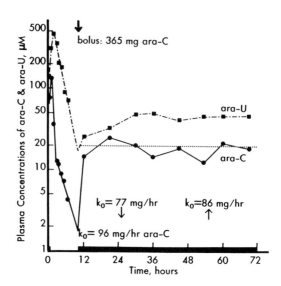

FIGURE 11   Kinetic profile of the test dose of araC followed by a loading bolus plus continuous infusion rate of araC at 12 hr after treatment began. Arrows show adjustment of the constant rate of infusion as suggested by the pharmacokinetic parameters and the plasma steady-state concentrations of araC for this individual patient.

plan can greatly decrease the variability in the AUC of ARAC in plasma and the AUC of araCTP in the leukemic cells (Avramis *et al.*, 1987b). Additional pharmacodynamic studies in patients with leukemia and lymphoma are needed to fully evaluate the relationship between the metabolic changes of araC occurring *in vivo* and clinical response to the drug. Ultimately, the use of pharmacokinetic and biochemical data early in the course of therapy could lead to optimization in dose and schedule of araC designed to increase its clinical efficiency in cancer patients.

7. Determination of *in Vivo* DNA Methylation Levels in Patients' Leukemic Cells and Partial Reversal of Tumor Drug Resistance to HDaraC

The biochemical pharmacology studies of HDaraC before and after 5-azaC treatment in pediatric patients with refractory disease (ALL) have been conducted (Avramis *et al.*, 1989d). Seventeen patients with clinical resistance to HDaraC received a 5-day continuous infusion of 5-azaC 150 mg/m²/day after not responding to (13/17) or relapsing from (4/17) HDaraC regimen. Three days after the end of the 5-azaC infusion, HDaraC was administered again. DNA methylation and dCK activity as araCTP formation were determined in leukemic blasts before and after 5-azaC treatment following incubation of the blasts with araC *in vitro*. After 5-azaC treatment, cellular araCTP concentrations were increased in 15 of 16 patients studied to varying degrees (average, 2.8-fold). This is a much greater effect than that seen with the CEM/dCK(−) cell line (Table 8). DNA methylation was measured in 12 of these patients before and after 5-azaC treatment; the average DNA methylation level in the blasts was 56 ± 16% of control [each patient's pretreatment cells were used as controls (Table 9)]. Thymidine incorporation was measured in 9 of these patients' blasts before and after 5-azaC treatment. There was a profound decline in thymidine incorporation in the 2 patients who achieved a complete remission after the 5-azaC + HDaraC regimen but no decrease was observed in the patients who failed to achieve a remission (Table 10). This data suggests that high cellular araCTP concentrations are necessary but may not be sufficient to achieve remission in all patients. The nonresponding patients may have other mechanisms for their araC resistance or the increased dCK activity may have increased dCTP pools that blocked the antileukemic effects of the elevated araCTP concentrations.

These results suggest that treatment with a cytostatic but DNA-modulatory regimen of 5-azaC causes DNA hypomethylation *in vivo*, which could be associated with dCK re-expression in the patient's leukemic cells (Avramis *et al.*, 1989d). This reversal of drug resistance to araC by 5-azaC yielded two complete remissions in this poor-prognosis multiply relapsed patient population with refractory ALL. The preliminary data indicate that the 5-azaC + HDaraC may be effective treatment for some patients with leukemia resistant to HDaraC alone. These studies continue for higher patient accrual to optimize these effects and further define their mechanisms. *The critical point is that partial resistance may be developing with each course of araC treatment.* Combination with 5-azaC may prevent the observed DNA hypermethylation and thus the development of resistance.

8. Effect of 5-azaC Treatment on Expression of Oncogenes

Leukemic cells from patients who were treated with 5-azaC and HDaraC regimen were harvested before and after the 5-azaC infusion. The DNA and RNA of the cells were extracted and digested with restriction endonucleases specific for the methylation of cytidine residues (*Hpa*II and *Msp*I) and Southern analyses of the DNA RFLPs and Northern analyses of the RNA were run. The nylon membranes were then probed with various probes. Slot blot analyses of the RNA samples were also done. The *in vivo* treatment with 5-azaC did not affect gene specific expression of HGPRTase, EGF, or the protooncogenes c-*sis*, c-*erb*, c-*ets* and c-*myb*. The DNA was hypomethylated but the RNA slot blot analysis showed no increased levels of transcription of these genes in comparison with pediatric leukemic cells isolated before 5-azaC treatment (Avramis *et al.*, 1988) (Figs. 12 and 13). These studies are continuing with other patients' leukemic cells and other oncogene probes.

TABLE 10
DNA Methylation Studies in Patient Leukemic Cells before and after 5-azaC Treatment

Patient	Percentage of 5-mC in DNA before 5-azaC (control methylation)[a]	Percentage of 5-mC in DNA after 5-azaC	DNA (% hypomethylation in leukemic cells)	Outcome
1. R. G.	3.37 ± 0.4	1.20 ± 0.2	61.3	CR
2. L. L.	3.26 ± 0.3	1.29 ± 0.4	60.3	CR
3. K. W.	3.22 ± 0.3	1.16 ± 0.3	63.9	PR
4. U. P.	4.49 ± 0.4	1.17 ± 0.6	73.9	NR
5. J. R.	3.29 ± 0.02	1.57 ± 0.2	52.3	NR
6. D. W. (BM)	3.79 ± 0.2	1.26 ± 0.1	66.8	NR
D. W. (PBC)	3.63 ± 0.2	2.61 ± 0.4	28.1	NR
7. E. H.	3.54 ± 0.2	2.44 ± 0.1	31.1	NR
8. L. M.	4.04 ± 0.3	1.59 ± 0.1	60.4	NR
9. O. M.	3.94 ± 0.6	2.36 ± 0.3	40.0	NR
10. D. P.	3.96 ± 0.2	2.23 ± 0.5	43.2	NR
11. L. R.	3.01 ± 0.8	0.59 ± 0.5	80.4	NE
12. K. K.	2.47	0.97	60.8	NE
			$X_{11} = 55.6\% \pm 15.8\%$	

[a] Mean $\pm$ SD ($n = 4$).

TABLE 11
DNA Synthetic Capacity Studies in Patient Leukemic Cells before and after 5-azaC Followed by HDaraC

Patient	Pretreatment control (%)	Pretreatment +1 mM araC[a] (% of control)	Posttreatment control (%)	Posttreatment +1 mM araC[a] (% of control)	24 hr post HDaraC[b] (% of control)	Outcome
1. R. G.	100	–	57.1	–	6.9	CR
2. L. L.	100	26.4	100	4.8	3.8	CR
3. K. W.	100	72.5	–	–	–	PR
4. U. P.	100	25.8	100	51.0	–	NR
5. J. R.	–	–	100	314.6	–	NR
6. D. W.	100	45.0	–	–	–	NR
7. E. H.	100	28.2	100	25.0	31.6	NR
8. L. M.	100	136.0	100	52.6	112.0	NR
9. D. K.	100	13.7	100	21.6	34.0	NR
10. O. M.	100	21.8	100	28.0	180.7	NR
		$X_8 = 46.2$ $\pm 40.6^c$		$X_6 = 82.1$ $\pm 115.0^d$	$X_4 = 89.6$ $\pm 71.3^d$	

[a] 1 mM araC × 1 hr incubation.
[b] DSC values at 24 hr after the HDaraC regimen was begun and before the third dose of araC, corresponding to trough araCTP cellular concentrations.
[c] Mean ± SD of all patient data.
[d] Mean ± SD of the nonresponding patient 4–10 data.

Hpall restriction

c-myb probed   c-sis probed

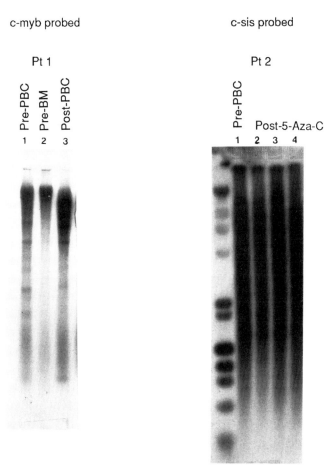

FIGURE 12   RFLP analysis on the DNA from leukemic cells of patients before and after *in vivo* treatment with 5-azacytidine (5-azaC) showing that the DNA was hypomethylated.

9. Evidence of *in Vivo* Development of Tumor Resistance to araC and Correlation with DNA Methylation Patterns in a Pediatric Patient with ALL

A 15-year-old boy with ALL was treated during his second bone marrow relapse with HDaraC regimen. The *in vitro* araC sensitivity test showed that his untreated bone marrow lymhpoblasts accumulated 710 μM araCTP after a 1 hr incubation with 1 mM araC. After the patient was treated with a 3 hr araC infusion of 3 gm/m^2, the peak araCTP concentration was 2000 μM and

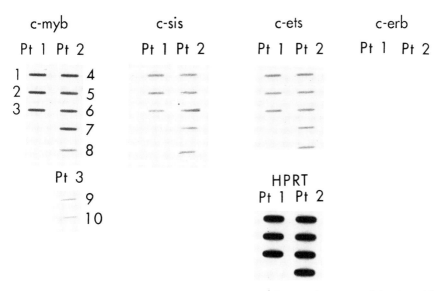

FIGURE 13   RNA slot blot analysis on leukemic cells from patients treated *in vivo* with 5-azacytidine showing no increased levels of transcription of certain oncogenes. Conditions: 2 μg of RNA was used per slot (Bio-Rad). Probes were labeled using the Feinberg method (0.1 μg). Exposures: for oncogene-probed nitrocellulose, 5 days; for HPRT-probed samples, 1 day. Samples 1, 4, and 9 were leukemic blasts before 5-azacytidine treatments; samples 2, 3, 5–8, and 10 were harvested after 5-azacytidine treatment.

the trough was 700 μM. The $t_{1/2}$ of elimination of araCTP was 5.9 hr after the first dose of araC. The tough values of araCTP after the seond and third doses of araC declined to 110 and 84 μM, respectively, indicating that the blast cells 36 hr after HDaraC regimen retained much less araCTP than after the first dose. This change was probably due to selective killing of sensitive tumor cells by araC. The patient achieved a partial remission but relapsed soon afterwards. Then he was treated with 5 days of an infusion of 5-azaC followed by HDaraC again. The *in vitro* araC sensitivity test showed that his lymphoblasts now accumulated only 320 μM araCTP concentration before the 5-azaC treatment (16% of the first test). Three days after 5-azaC they accumulated 1070 μM, a 3.4-fold increase. Forty-eight hours after the HDaraC treatment his lymphoblasts retained a trough araCTP concentration of only 109 μM. Four weeks later, the *in vitro* araC sensitivity test showed that his lymphoblasts accumulated only 17.6 μM araCTP or approximately 2.5% of the results before the first treatment with araC. The patient did not achieve a remission. Thus the low values of araCTP that we obtained in this patient after relapse are indicative of the selection of resistant lymphoblast clones.

The DNA methylation results in this patients' blasts were as follows:

1. Before HDaraC: $4.38 \pm 0.2\%$ mC;
2. After HDaraC and before 5-azaC: $5.22 \pm 0.36\%$ mC or 19.2% hypermethylation;
3. After 5-azaC: $1.37 \pm 0.009\%$ or 73.75% hypomethylation.

After HDaraC and before 5-azaC, there appeared to be a correlation of DNA hypermethylation and reduction in cellular araCTP, probably due to "silencing" of dCK expression in the blast cells. Following the 5-azaC treatment, there was a correlation between DNA hypomethylation and enhancement of araCTP formation, probably due to re-expression of dCK.

These results support the premise that araC treatment facilitates the development of araC resistant subclones of leukemic cells and that treatment with 5-azaC can reverse this resistance. The difference in responding and nonresponding patients may be due to the degree of hypomethylation or the persistence of hypermethylation of dCK after the 5-azaC is stopped. This possibility will be tested with the dCK cDNA probe. However, other factors such as dCTP pools, effect of araCTP on DNA pol α, incorporation rates of araCTP into DNA, or selection of subpopulations with silenced dCK may play an important role in achieving clinical efficacy.

10. Cloning of Human Deoxycytidine Kinase

To analyze the structure, function and control of this clinically important enzyme we raised polyclonal antibodies against human dCK and isolated four different sets of cDNA clones from lambda gt 11 thymus and Molt 4 libraries. The largest cDNA of the first set is 2.9 kb and encodes a 645-amino-acid open reading frame. The fusion protein had low levels of dCK activity so we assumed that it was human dCK (Huang et al., 1989). The DNA and deduced amino-acid sequences of this clone show a high degree of homology with a previously unidentified murine cDNA clone p3.4J (EMBL:MM34j) and murine ERp72, an endoplasmic reticulum protein (Mazzarella et al., 1990). Furthermore, it has cysteine-rich regions that are homologous with thioredoxin, the β subunit of prolyl 4-hydroxylase, phosphoinositide-specific phospholipase C, thyroid hormone-binding protein, protein disulfide isomerase. Using this antibody and the cDNA as probes, no differences were seen in the amount and size of protein, mRNA and genomic restriction fragments between CEM and L1210 leukemic cells lines that express and do not express dCK activity. The results were unexpected since these cell lines re-expressed dCK activity after treatment with 5-azacytidine. These findings suggest that this first set of clones does not represent the major dCK but another enzyme or activator with low araC kinase activity. Consequently, we are now characterizing three other sets of clones isolated with antibody preabsorbed with the fusion protein of the first set clones. These clones will be tested for enzyme activity and differences in DNA and mRNA between the cell lines that express and do not express dCK activity. After cDNA for dCK is definitively identified, this probe will be used to characterize the methylation patterns of the promoter and structure regions of the dCK gene before and after 5-azacytidine and araC treatments.

B. COLLATERAL SENSITIVITY

Shabell and co-workers have shown that L1210 and P388 murine leukemia cell lines resistant to araC frequently have increased sensitive to inhibitors of the *de novo* pyrimidine biosynthesis (Schabel *et al.*, 1982). For example, PALA, acivicin, pyrazofurin, and 5,6-dihydro-5-azacytidine produced 10^3 to 10^5-fold greater tumor cell kill in the deoxyctidine kinase (dCK) mutant than in the sensitive cell lines. Recently, a sequence selective synergism of fludarabine phosphate, F-araA(MP), a purine nucleoside analog and araC has been observed (Gandi and Plunkett, 1988; Avramis *et al.*, 1989c). When F-araA is administered first to leukemic cells either *in vitro* or *in vivo* in a concentration sufficient to produce its active anabolite followed by a pulse araC treatment, cellular araCTP concentrations increase from 2- to 10-fold and the thymidine incorporation is greatly inhibited (Gandi and Plunkett, 1988; Avramis *et al.*, 1989c). When araC is coadministered or is administered before F-araA no drug synergism is observed.

1. Drug Synergism Studies between F-araA(MP) and araC

Human lymphocytic cells (CEM) were treated with F-araA and araC in all possible sequence combinations. The treatment with F-araA first followed by araC was the only synergistic one. The addition of both drugs to the cells had only additive effect and the treatment with araC first followed by F-araA was antagonistic. The median effect principle was applied to determine the synergistic, additive, or antagonistic character of these antileukemic drugs. For this, aliquots of CEM cells were incubated in the presence or absence of F-ara-A or araC and their combination (Chou and Talalay, 1984). F-araA was used because the phosphorylated agent, F-araAMP, is not taken up by the cells. The cells were preincubated with F-araA prior to adding araC in the wells that received both drugs. The incubation with both drugs continued and the thymidine incorporation was determined in all the cells. At least one log of synergism is obtained from the combination in comparison with either drug alone. When the equations were used to calculate the CI_{90}, the resultant value was 0.012, which suggests strong synergism betwen F-araA and araC.

Drug synergism between these two drugs was demonstrated also by the increased cellular concentrations of araCTP and by the greater inhibition

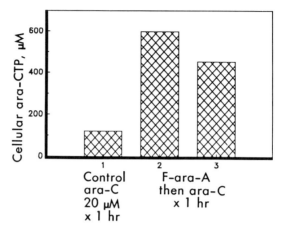

FIGURE 14   Effect of F-araA pretreatment on cellular araCTP concentrations in CEM/0 cells.

of thymidine incorporation caused by combinations relative to individual agents (Fig. 14). Concentrations of 10 to 20 μM araC and 5 to 10 μM F-araA were required for drug synergism to be exhibited (Fig. 15).

2. *In Vivo* Treatment with Fludarabine Phosphate

Leukemic cells isolated from pediatric patients with relapsed ALL who were treated with a 5-day continuous infusion F-ara AMP were treated with araC for 1 hr *in vitro*. These lymphoblasts showed 10- to 12-fold synergism

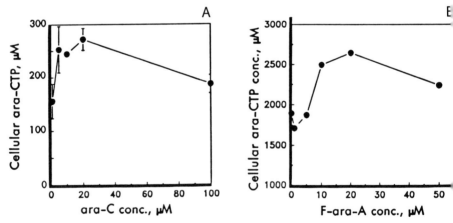

FIGURE 15   (A) Effect of araC concentration on cellular araCTP accumulation in CEM/0 cells. (B) Effect of F-araA concentration in achieving synergism with araC in CEM/0 cells. Treatment with 10 to 20 μM F-araA are necessary for a significant increase in cellular araCTP concentrations (synergism).

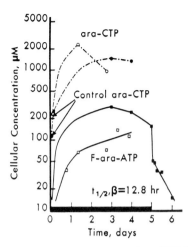

FIGURE 16   Cellular concentrations of F-araATP and araCTP in leukemic cells from two patients who were treated with fludarabine phosphate (F-araAMP) alone. The leukemic cells were isolated and treated *in vitro* with araC (araC sensitivity assay). The cellular concentrations of araCTP increased significantly ($\approx$ 12-fold) after 36 to 96 hr of treatment with F-araA(MP) in comparison with pretreatment control values. F-araATP appears to be disappearing bi-exponentially from the leukemic cells of a patient; the terminal half-life is approximately 13 hr.

increase in araCTP levels over levels before F-araAMP treatment started (Avramis *et al.*, 1990). Preliminary studies show that this sequence of drugs can cause hypomethylation of DNA. Thus this synergism may also be an epigenetic change. F-araAMP treatment for 24-36 hr was needed to achieve the maximal increase in cellular araCTP formation (Avramis *et al.*, 1989c) (Fig. 16.) The time of pretreatment with fludarabine required for the collateral sensitivity to become significant in both the *in vivo* and *in vitro* studies was 1-2 cell division times. A similar time of pretreatment was required for the re-expression of dCK with 5-azaC and congener agents.

As a result of these observations, a clinical protocol has been approved to administer 2 days of F-araAMP followed immediately by a 3-day araC continuous infusion in pediatric patients with ALL. Figure 6 shows the treatment schedule and the times for blood and bone marrow samples. The frequent sampling of peripheral blood will permit us to study both the plasma and the cellular pharmacology of these drugs.

3. Drug Synergism Studies between Anti-HIV Drugs

Drug synergism studies between zidovudine (AZT) and dideoxyinosine (ddI) have been carried out in T-cell lines (CEM and Jurkat E6-1). The combination of these two drugs at 1:1 molar ratio appears to produce a 2-log (100-fold) synergism in as low as 10^{-8} M concentration range. In these studies both the inhibition of DNA synthesis and the inhibition of HIV-produced P24

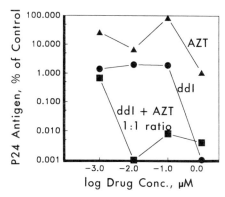

FIGURE 17 Synergistic effect between AZT and ddI on HIV P24 antigen production. The combination of these antiviral drugs (1:1 molar ratio) produces a minimum of 100-fold synergism against HIV.

antigen were monitored in the Jurkat E6-1 T-cell lines (Fig. 17). Results from these studies have supported a new clinical protocol using a combination of AZT and ddI against AIDS.

Similar studies between $2',3'$-dideoxy-$2',3'$-dehydrothymidine (d4T) have shown that these drugs have neither synergistic not additive effect against the T-cells even in as high as $10^{-3} M$ concentrations. However, in the HIV-infected cells the production of P24 antigen was completely suppressed even at $10^{-9} M$ (1:1 molar ratio) up to 7 days of incubation. These results indicate that a 5- to 6-log range of selective drug synergism against the HIV virus proliferation. In these results the median-effect principle equations could not be applied because most of the data on the matrix were zero or near zero, thus not allowing the formation of a line over the 50% affected/unaffected ratio. Preclinical and clinical studies are planned to examine this phenomenon in greater detail.

References

Alt, F. W., Kellems, R. E., Bertino, J. R., and Schimke, R. T. (1978). Selective multiplication of dihydrofolate reductase genes in methotrexate-resistant variants of cultured murine cells. *J. Biol. Chem.* **251**, 1357–1370.

Antonson, B. E., Avramis, V. I., Nyce, J., and Holcenberg, J. S. (1987). Effect of 5-azacytidine and congeners on DNA methylations and expression of deoxycytidine kinase (dCk) in the human lymphoid cell lines CCRF/CEM/0 and CCRF/dCk. *Cancer Res.* **47**, 3672–3678.

Avramis, V. I. (1989). Tumor heterogeneity in deoxycytidine kinase (dCk) levels in patients' leukemia or lymphoma cells. *Proc. Am. Soc. Clin. Oncol.* **8**, 64.

Avramis, V. I., and Powell, W. (1987). Pharmacology of combination chemotherapy of cytosine arabinoside (ara-C) and uracil arabinoside (ara-U) or tetrahydrouridine (THU) against murine leukemia L1210/0 in tumor bearing mice. *Cancer Invest* **5**, 221–299.

Avramis, V. I., Biener, R., and Holcenberg, J. S. New (1987a). New approaches to overcome drug resistance. *In* "The Role of Pharmacology in Pediatric Oncology" (D. Poplack, ed), pp. 97–112. Nijhoff, The Netherlands.

Avramis, V. I., Biener, R., Krailo, M., Finklestein, J., Ettinger, L., Willoughby, M., Siegel, S. E., and Holcenber, J. S. (1987b). Biochemical pharmacology of high-dose cytosine arabinoside (HDara-C) in a childhood acute leukemia. *Cancer Res.* **47**, 6786–6792.

Avramis, V. I., Mecum, R., Nyce, J., Steele, D. A., and Holcenberg, J. S. (1988). DNA methylation and pharmacodynamic studies of HDara-C before and after 5-azacytidine treatment in children with refractory acute lymphocytic leukemia (ALL) to ara-C. *Proc. Am. Assoc. Cancer Res.* **29**, 190.

Avramis, V. I., Powell, W. C., and Mecum, R. (1989a). Cellular metabolism of 5,6-dihydro-5-azacytidine (DHAC) and incorporation into DNA and RNA of human lymphoid cells [CEM/0 and CEM/dCk(−)]. *Cancer Chemother. Pharmacol.* **24**, 155–160.

Avramis, V. I., Wienberg, K. I., Sato, J. K., Lenarsky, C., Willoughby, M. L., Coates, T., Ozkaynak, M. F., and Parkman, R. (1989b). Pharmacology of cytosine arabinoside (ara-C) in pediatric patients with leukemia and lymphoma after a biochemically optimal region of loading bolus plus continuous infusion of ara-C. *Cancer Res.* **49**, 247–251.

Avramis, V. I., Wiersma, S. R., Cheng, A., Sato, J., and Holcenberg, J. S. (1989c). Selective pharmacological synergism between F-araA and ara-C in human leukemic cells *in vitro* and *in vivo*. *Proc. Am. Soc. Clin. Oncol.* **8**, 215.

Avramis, V. I., Mecum, R. A., Nyce, J., Steele, D. A., and Holcenberg, J. S. (1989d). Pharmacodynamics, DNA methylation studies and clinical response of high-dose 1-β-D-arabinofuranosylcytosine before and after *in vivo* 5-azacytidine treatment in pediatric patients with refractory acute leukemia. *Cancer Chemother. Pharmacol.* **24**, 203–210.

Avramis, V. I., Champagne, J., Sato, J., Krailo, M., Ettinger, L., Poplack, D. G., Finklestein, J., Reaman, G., Hammond, D., and Holcenberg, J. S. (1990). Pharmacology and Phase I/II trial of fludarabine as a loading bolus and continuous infusion in pediatric patients. *Cancer Res.* In press.

Batist, G., Tupule, A., Sinha, B. K., Katki, A., Myers, C. E., and Cowan, K.H. (1986). Overexpression of a novel anion glutathione transferase in multidrug-resistant human breast cancer cells. *J. Biol. Chem.* **261**, 15544–15549.

Bokkerink, J., Bakkar, M., Hulscher, T., De Abreu, R., Schretlen, E., Van Laarhoven, J., and De Bruijn, C. (1987). Sequnce-, time- and dose-dependent synergism of methotrexate and 6-mercaptopurine in malignant human T-lymphoblasts. *Biochem. Pharmacol.* **35**, 2549–3555.

Bouanchaud, D. H., Scavizza, M. R., and Chabber, Y. A. (1969). Elimination by ethidium bromide of antibiotic resistance in enterobacteria and staphylococci. *J. Gen. Microbiol.* **54**, 417–425.

Cadman, E., Heimer, R., and Davis, L. (1979). Enhanced 5-fluorouracil nucleotide formation after methotrexate administration: Explanation for drug synergism. *Science* **205**, 1135–1137.

Cheng, Y.-C., and Brockman, R. W. (1983). Mechanisms of drug resistance and collateral sensitivity: Bases for development of chemotherapeutic agents. *In* "Development of Target-Oriented Anticancer Drugs" (Y. C., Cheng, B. Goz, and M. Minkoff, eds.), pp. 107–117. Raven, New York.

Chou, T.-C., and Talalay, P. (1984). Quantitative analyzed at dose–effect relationships: The combined effects of multiple drugs or enzyme inhibitors. *Adv. Enzyme Regul.* **22**, 27–55.

Crane, L. R., Jackson, R., and Avramis, V. I. (1989). DNA hypermethylation studies in CEM/0 cells after treatment with therapeutic and sub-therapeutic concentrations of cytosine arabinoside (ara-C). *Proc. Am. Assoc. Cancer Res.* **30**, 496.

Croop, J. M., Gros, P., and Housman, D. E. (1988). Genetics of multidrug resistance. *J. Clin. Invest.* **81**, 1303–1309.

DeVita, V. T. (1983). The relationship between tumor mass and resistance to chemotherapy. Implications for surgical adjuvant treatment of cancer. *Cancer (Philadelphia)* **51**, 1209–1220.

Fojo, A. T., Ueda, K., Slamon, D. J., Poplack, D. G., Gottesman, M. M., and Pastan, I.

(1987). Expression of a multidrug resistance gene in human tumors and tissues. *Proc. Natl. Acad Sci. U.S.A.* **84**, 265–269.

Gandi, V., and Plunkett. W. (1988). Modulation of arabinosylnucleoside metabolsm by arabinosylnucleotides in human leukemia cells. *Cancer Res.* **48**, 329–334.

Goldie, J. H., and Coldman, A. J. (1984). The genetic origin of drug resistance in neoplasms: Implications for systematic therapy. *Cancer Res.* **44**, 3643–3653.

Goldie, J. H., and Coldman, A. J. (1985). A model of tumor response to chemotherapy: An integration of stem cell and somatic mutation hypothesis. *Cancer Invest.* **3**, 553–564.

Hamada, H., and Tsuruo, T. (1986). Functional role of 170,000–18,000 dalton glycoprotein specific for drug-resistant tumor cells as revealed by monoclonal antibodies. *Proc. Natl. Acad. Sci. U.S.A.* 7785–7789.

Huang, S.-H., Tomich, J. M., Wu, H., Jong, A., and Holcenberg, J. (1989). Human deoxycytidine kinase: Sequence of cDNA clones and analysis of expression in cell lines with and without enzyme activity. *J. Biol. Chem.* **264**, 14762–14768.

Jones, P. A. (1984). Gene activation by 5-azacytidine. *In* "DNA Methylation: Biochemistry and Biological Significance" (A. Razin, H. Cedar, and A. D. Riggs, eds.), pp. 165–187. Springer-Verlag, New York.

Jones, P. A., and Taylor, S. M. (1980). Cellular differentiation, cytidine analogs and DNA methylation. *Cell* **20**, 85–93.

Kartner, N., Evenden-Porelle, D., Bradley, G., and Ling, V. (1985). Detection of P-glycoprotein in multidrug-resistant cell lines by monoclonal antibodies. *Nature (London)* **316**, 820–823.

Larder, B. A., and Kemp, S. D. (1989). Multiple mutations in HIV-1 revose transcriptase confer high-level resistance to Zidovudine (AZT). *Science,* **246**, 1155–1158.

Mazzarella, R. A., Srinivasan, M., Haugejorden, S. H., and Green, M. (1990). ERp72, an abundant luminal endoplasmic reticulum protein, contains three copies of the active site sequences of protein disulfide isomerase. *J. Biol. Chem.* **265**, 1094–1101.

Melera, P. W., Lewis, J. A., Bielder, J. L., and Hession, C. (1980). Antifolate-resistant Chinese hamster cells: Evidence for dihydrofolate reductase gene amplification among independently derived sublines overproducing different dihydrofolate reductases. *J. Biol. Chem.* **255**, 7024–7028.

Mitsuhashi, S., Harada, K., and Kameda, M. (1961). Elimination of transmissible drug-resistance by treatment with acriflavine. *Nature (London).* **139**, 947.

Momparler, R. L. (1974). A model for chemotherapy of acute leukemia with 1-β-D-arabinofuranoside. *Cancer Res.* **34**, 1775–1787.

Monnat, R. J., Jr. (1989). Molecular analysis of spontaneous hypoxanthine phosphoribosyltransferase mutations in phioguanine-resistant HL-60 human leukemia cells. *Cancer Res.* **49**, 81–87.

Ozols, R. F., Cunnion, R. E., Klecker, R. W., Jr., Hamilton, T. C., Ostchega, Y., Parillo, J. E., and Young, R. C. (1987). Verapamil and Adriamycin in the treatment of drug-resistant ovarian cancer patients. *J. Clin. Oncol.* **5**, 641–647.

Pastan, I., and Gottesman, M. (1987). Multiple-drug resistance in human cancer. *N. Engl. J. Med.* **316**, 1388–1393.

Plunkett, W., Liliemark, J. O., Adams, T. M., Nowak, B., Estey, E., Kantragian, H., and Keating, M. J. (1987). Saturation of 1-β-D-arabinofuranosylcytosine 5'-triphosphate accumulation in leukemia cells during high-dose 1-β-D-arabinofuranosylcytosine therapy.*Cancer Res.* **47**, 3005–3011.

Powell, W. C., Mecum, R., and Avramis, V. I. (1987). Biochemical pharmacology of dihydro-5-azacytidine (DHAC) in tumor bearing (L1210) mice. *Proc. Am Assoc. Cancer Res.* **28**, 324.

Razin, A., Cedar, H., and Riggs, A. D., eds (1984). "DNA Methylation: Biochemistry and Biological Significance," Chaps. 1–3 and 7–10. Springer-Verlag, New York.

Schabel, F. M., Jr., Skipper, H. H., Trader, M. W., Brockman, R. W., Laster, W. R., Jr.,

Corbett, T. H., and Griswold, D. P. (1982). Drug control of ara-C resistant tumor cells. *Med. Pediatr. Oncol. (Suppl.)* **1**, 125–148.

Schimke, R. T. (1988). Gene amplification in cultured cells. *J. Biol. Chem.* **263**, 5989–5992.

Sharma, R. C., and Schimke, R. T. (1989). Enhancement of the frequency of methotrexate resistance by gamma radiation in Chinese hamster ovary and mouse 3T6 cells. *Cancer Res.* **49**, 3861–3866.

Thiebaut, F., Tsuruo, T., Hamada, H., Gottesman, M. M., Pastan, I., and Willingham, M. C., (1987). Cellular localization of the multidrug-resistance gene product P-glycoprotein in normal human tissues. *Proc. Natl. Acad. Sci. U.S.A.* **84**, 7735–7738.

Ueda, K., Cardarelli, C., Gottesman, M. M., and Pastan, I. (1987). Expression of a full length cDNA for the human "MDR1" gene confers resistance to colchicine, doxorubicin and vinblastine. *Proc. Natl. Acad. Sci. U.S.A.* **84**, 3004–3008.

Wallerstein, H., Lewis, M. S., Eng., B., and Calman, N. (1972). The return of antimetabolite sensitivity in methotrexate- and 6-mercaptopurine-resistant L1210 murine leukemia by the process of adaptive selection. *Cancer Res.* **32**, 2235–2240.

Weinberg, K. I., Avramis, V. I., Lenarsky, C., Falk, P., Mecum, R. and Parkman, R. (1989). Variability in the pharmacokinetic handling of intermittent high dose ara-C in pediatric and adult bone marrow transplant recipients. *J. Clin. Oncol.* (submitted).

White, C. J., Rathmell, J. P., and Capizzi, R. L. (1987). Membrane transport influences the rate of accumulation of cytosine arabinoside in human leukemia cells. *J. Clin. Invest.* **79**, 380–387.

Wittman, V., and Wong, H. C. (1989). Regulation of the penicillinase gene of *bacillus-lichenformis*: Interaction of the pen repressor with its operators. *J. Bacteriol.* **170**, 3106–3112.

Wohl, G. M. (1988). The importance of circular DNA in mammalian gene amplification. *Cancer Res.* **49**, 1333–1340.

Wolverton, J. S., Danks, M. K., Schmidt, C. A., and Beck, W. T. (1989). Genetic characterization of the multidrug-resistant phenotype of VM-26-resistant human leukemia cells. *Cancer Res.* **49**, 2422–2426.

PART III

Condition-Selective Synergism and Antagonism

CHAPTER 17

Schedule-Dependent Effects in Antineoplastic Synergism and Antagonism

Barbara K. Chang
Department of Medicine
Hematology/Oncology Section
Medical College of Georgia
Augusta, Georgia 30912

I. Introduction

The present review will focus upon the schedule-dependent effects of antineoplastic agents used in combination specifically from the standpoint of *spacing*

SYNERGISM AND ANTAGONISM IN CHEMOTHERAPY
Copyright © 1991 by Academic Press, Inc.
All rights of reproduction in any form reserved.

and *sequencing* of agents, rather than the frequency or duration of administration. In the early 1970s, there was considerable interest in deliberate spacing and sequencing which were intended to induce synchronization and recruitment in attempts to improve cytotoxic effects [as reviewed by Van Putten *et al.* (1976) and more critically by Tannock (1978)]. More recently, exploitation of other (i.e., biochemical or pharmacological) mechanisms of interaction by sequencing of agents has been attempted. Finally, some approaches to scheduling agents in combinations have been purely empirical. Thus, an overview of the various attempts at combining antineoplastic agents with deliberate investigation of the spacing or the order of administration of agents will be presented, with an emphasis on the purported mechanism(s) of the interaction as the organizing principle of the review. A summary of the history of schedule-dependent combinations of antineoplastic agents to be covered in this present review is presented in Table 1. Much of the data cited will be based upon experimental models; however, whenever possible, clinical correlation will be included. It should be pointed out that there is a paucity of studies of schedule-dependent interactions examined in a rigorous fashion, i.e., most studies tend to be "intuitive" in their conclusions, rather than attempting a more definitive, mathematical characterization of the interaction. Therefore, one of the goals of this review will be to suggest areas that may require further investigation.

For over twenty years, antineoplastic agents have been used in combinations with curative or palliative intent in the treatment of selected human cancers (DeVita *et al.*, 1970, 1975). Most of the successful applications of combination chemotherapy have involved the administration of three or more agents at *more or less* the same time and without regard to the order or sequence of administration. "Scheduling" with respect to antineoplastic drug combinations refers *conventionally* to how frequently the courses of treatment are repeated. In the clinical setting, such scheduling of antineoplastic agents depends largely upon the ability of the host to recover from the toxicity of the cytotoxic therapy and hence upon bone marrow kinetics, which generally limit the frequency of administration of combination chemotherapy to every 3 to 4 weeks (Marsh, 1976; Silver *et al.*, 1977; Doll and Weiss, 1983; Tannock, 1989). In addition, scheduling refers to considerations of the *duration* of administration; however, questions relating to duration of administration have been most extensively investigated with respect to single agents and not combinations of agents (Ludwig *et al.*, 1984). That is, individual agents may have different effects depending upon the schedule (here referring to the duration) of administration. For example, low-doses of alkylating agents over prolonged periods of time tend to have more immunosuppressive effects, whereas high-dose bolus administration has greater cytotoxic effects (Blum and Frei, 1979); contrarily, other classes of agents, such as etoposide (VP16) and the epipodophyllotoxins, have more pronounced anticancer effects when the same total dose is given over 3 to 5 days rather than on a single day (Issell and Crooke, 1979).

TABLE 1
History of Schedule Dependent Combination Chemotherapy

Dates	Combination [and investigation] types	Cytokinetic or other parameters monitored	Analytic end points[a]
1950s to early 60s	Empirical [animal tumor models]	None	ILS[a]
Late 1960s to 1970s	Kinetics-based [animal models, clinical]	LI, PLM, DLI, mitotic index, PDPLI	ILS, growth or regression of tumors, clinical response
Early 1980s	Kinetics-directed [clinical, cell lines]	Flow cytometry [DNA histogram: cell cycle/phase analysis]	ILS, cell survival, clinical response
1980s	Biochemical and pharmacologic rationales [cell and animal models]	Biosynthetic enzymes, end-product/target analysis, drug transport, or accumulation	Cell growth/survival, ILS, tumor growth/regression

[a] The abbreviations used are LI, labeling index; PLM, percentage of labeled mitoses; DLI, double-LI; PDPLI, primer-dependent DNA polymerase LI; ILS, increase in median survival time or lifespan.

II. Schedule-Dependent Interactions
Based upon Cell Cycle Considerations

A. COMBINATIONS OF AGENTS BASED UPON CYTOKINETIC CONSIDERATIONS

1. Empirical Approaches

Considerable investigation of the effects of drug sequencing and scheduling was carried out in the 1950s and early 1960s using murine leukemia models (largely L1210 and L5187Y) *in vivo*. The research in question has been reviewed by Goldin (1984); although the work cited was important for subsequent research and emphasized the importance of drug scheduling in combinations of chemotherapeutic agents, the approach taken at that time was unfortunately, largely empirical and this body of literature will not be reviewed in detail here. However, two references will be summarized to illustrate the type of work reported during the period cited.

For example, in L1210 leukemia, methotrexate (MTX) and 6-mercaptopurine (6-MP) were said to be "synergistic" if the MTX was administered every 3 days and the 6-MP daily, but daily administration of both MTX and 6-MP was no more effective than MTX alone (Venditti *et al.*, 1956). The authors used the term synergism "in its broad sense as meaning the cooperation or effective combination of agents." Nevertheless, from the data presented and by present definitions of synergism (*cf.* Chapter 2.), it is difficult to say that any synergism had occurred, since the most effective schedule tested was the daily use of MTX as a single agent.

In another example, 1-β-D-arabinofuranosylcytosine hydrochloride (cytosine arabinoside; araC) and 1,3-*bis*(2-chloroethyl)-1-nitrosourea (BCNU) were combined on various schedules and doses; however, BCNU was always given first on day 7 or 8 after inoculation of L1210 leukemia and araC begun one day later and continued daily or every 2 days or every 4 days thereafter; the intermittent araC was slightly more effective than daily araC, i.e., 600%-range increase in median survival time (ILS or increased length of survival or increased lifespan) over controls for 2 to 4 day treatment as opposed to 400%-range for daily araC (Tyrer *et al.*, 1967). Not suprisingly, the responses were highly dose dependent (especially with respect to toxicity and long-term survivors), but, since neither the sequence nor the interval between the BCNU and araC was altered in any of the treatment protocols, no statement can be made regarding these aspects of the drug scheduling.

2. Cytokinetics-Based Combinations

During the late 1960s and early 1970s, there were numerous attempts to combine agents with the intent of synchronizing cells in order to gain greater efficacy of phase-specific agents, i.e., agents which exert their effects [either inhibiting progression to the next phase (*cytostasis*) or actual cell killing (*cytotoxicity*)] only at a certain phase of the cell cycle (as illustrated in Fig 1), or cycle-specific agents, which exert more cytotoxicity in cell populations with a large growth fraction (proportion of cells in S phase; Fig. 1). Synchronization is here taken to imply a perturbation of cell population kinetics whereby larger than usual fractions of cells pass through the various phases of the cell cycle simultaneously. The ideal conditions for this approach would be to apply a synchronizing agent, which would cause a specific cell cycle progression delay, with rapid simultaneous recovery from the block, followed by administration of the effector agent, which, provided the tumor cells displayed minimal variability in cell cycle parameters and a large fraction of cycling cells, would have a greater cytotoxic effect (Van Putten *et al.*, 1976). Again, under ideal conditions, an improved therapeutic index should result if

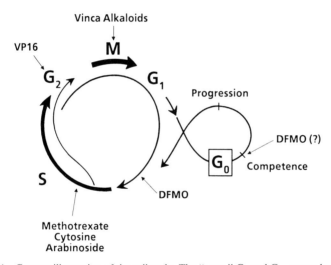

FIGURE 1   Current illustration of the cell cycle. The "gaps," G_1 and G_2, are no longer viewed merely as time intervals—the kinetic view of the cell cycle has been supplemented by a functional view, which permits overlap of phases of the cycle (Pardee, 1989). Thus, as shown, the G_1 events, which are necessary for S (DNA synthesis) phase may begin in the previous cell cycle. Likewise, the G_2 events may begin during S phase. Mitosis (M) is the phase in which duplicate chromosomes condense into a form suitable for transport, and the cell undergoes division with partition and distribution of chromosomes (McIntosh and Koonce, 1989). In order to enter S phase, G_0 or quiescent cells must attain *competence*, which requires stimulation by various growth factors and/or activation of certain genes. [One of these so-called competence genes is the gene for ornithine decarboxylase (Ferrari *et al.*, 1986), an enzyme which is inhibited by α-difluoro-methylornithine or DFMO, which is discussed later in this chapter.] Once competence has been attained, the provision of other growth factors and nutrients (found in plasma) permit *entry* and *progression* of the competent cells into the S phase of the cycle. Several representative phase-specific agents (which are discussed in the text) are indicated on the figure. Examples of vinca alkaloids include vincristine, vinblastine, and vindesine. VP-16, etoposide or epipodophyllotoxin. (Figure adapted from Pardee, 1989, with permission. Copyright 1989 by the AAAS.)

the tumor cell population responds differentially to the synchronizing agent and/or the normal cell populations (i.e., bone marrow and gastrointestinal tract) display cytokinetic parameters sufficiently distinct from those of the tumor cells.

Interestingly, efforts at synchronization were more common when the techiques of studying cell kinetics were more limited, tedious, and less sensitive. The most commonly used techniques were the labeling index (LI, the percentage of cells labeled with [^{3}H]thymidine or another radioactively labeled DNA precursor) (Denekamp and Kallman, 1973; Meyer and Bauer, 1975), which has been used to estimate the proportion of cells in S phase for a specified time of exposure to the label, and percent labeled mitoses (PLM, the percentage of cells in mitosis labeled at various intervals after a single injection or exposure to a DNA precursor, commonly [^{3}H]thymidine) Steel and Hanes, 1971). [A review of these concepts and techniques can be found in the article by Tannock (1989).] Both the LI and PLM require autoradiography and can be carried out on only small numbers of cells, as opposed to the more automated method of flow cytometric determination of DNA content (Krishan, 1975). Flow cytometry is often combined with bromodeoxyuridine labeling and thus permits determination of the percentage of S phase cells in a population and, by pulse-labeling, determination of the cell cycle times (Gratzner, 1982; Dolbeare et al., 1983). However, attempting to synchronize cells has fallen into disfavor not because of difficulty in monitoring the effects of various interventions on cell cycle progression (which remains a problem in the clinical setting despite the advances noted), but because of lack of success of the approach clinically (Tannock, 1978). One of the major reason for this lack of success clinically has to do with the contrast between the more uniform proliferation of cells in animal tumor models as opposed to the wide variation seen in human cancers (Van Putten et al., 1976).

Thus, while there is much data to support the concept that cytotoxic agents affect cell kinetics, few studies are available which have attempted to measure the kinetic disturbances directly and to correlate kinetic effects with cytotoxic effects of particular drug administration schedules. Attempts to translate experimental approaches into clinical treatment protocols have been notoriously difficult to carry out in practice (Simpson-Herren, 1982). Accordingly, in considering the studies for review, the distinction made by Barranco et al. (1982a) between "kinetics-directed" and "kinetics-based" treatment schedules has proved useful, with the former referring to studies in which the cell kinetics parameters are measured directly and used to guide the timing of therapy and the latter referring to studies in which cytokinetic effects are sought and timing of drug administration is based upon historical and empirical data.

Illustrative of the cytokinetics-based approach is the work by Braunschweiger and Schiffer, in which they found that sequential chemotherapy against either 13762 (1980a) or T1699 (1980b) rat mammary tumors responded best when cyclophosphamide (CP) was administered after doxorubicin (DOX) at a time predicted to coincide with the proliferative recovery from DOX (on the basis of LI or other methods). Their data on the rats with T1699 mammary tumors show clear schedule-dependent antagonistic *and* synergistic effects on mean survival. When the drugs are given simultaneously, the survival is no better than CP alone (35 vs 36 days); however, when CP was administered 1, 2, 3, 4, or 5 days after DOX, not only was a maximal synergistic effect seen at 3 days post-DOX, which coincided with the peak LI and recovery of proliferation as shown in Fig. 2, but also an antagonistic effect was seen when CP followed DOX by only 1 day (a time of declining LI and other measures of cellular proliferation).

Another example of a cytokinetics-based chemotherapeutic combination is detailed in the report by Vadlamudi and Goldin (1971) in which agents known to cause metaphase arrest (Fig. 1), demecolcine (Colcemid) and vinblastine (Velban, VLB), were used as synchronizing agents in combination with araC, an S phase specific inhibitor. In this report, mitotic indices were used to

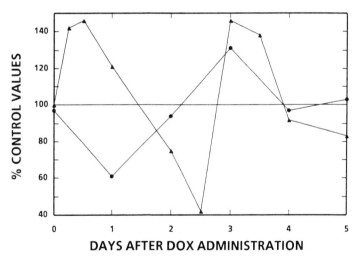

DAYS AFTER DOX ADMINISTRATION

FIGURE 2   Relationship of survival and labeling index (LI) in mice bearing T1699 mammary tumors following doxorubicin (DOX), 5 mg/kg (Braunschweiger and Schiffer, 1980b). The [³H]thymidine LIs were determined on resected tumor specimens at the various times shown following administration of DOX alone and are expressed as percentage of control values for Day 0 of DOX (▲). The mean survivals following DOX and then at various intervals, cyclophosphamide (CP), 100 mg/kg (●) are shown as the percentage survival of animals treated with CP alone as the controls (rather than untreated controls). The two independently derived sets of data are juxtaposed for the purposes of the present review in order to illustrate better the schedule dependent antagonistic and synergistic effects on survival of the combination of DOX followed by CP and the relationship of these effects to changes in the LI as influenced by time interval from the administration of the DOX.

determine the time of maximal mitotic arrest and recovery in mice bearing L1210 leukemia and given *in vivo* therapy with either demecolcine or VLB. AraC was then administered at various intervals (4, 6, 9, 12, or 24 hr) after demecolcine or 8, 16, or 24 hr after VLB. For each combination, the maximal ILS was seen at the time predicted to coincide with the recovery from mitotic arrest, and the ILS was minimal at the time of the maximal mitotic arrest. Demecolcine was totally without antitumor effect and VLB was only marginally effective at the dosage used. Therefore, schedule-dependent synergism and antagonism could be said to have been demonstrated and to correlate with the monitor used to assess the effect on the cell progression.

In contrast, the combination of other vinca alkaloids, i.e., vincristine (VCR) and vindesine (VDS), were shown to be synergistic against L1210 leukemia with another antimetabolite, methotrexate (MTX), when the sequence of administration was MTX followed in 24 to 72 hr by either VCR or VDS (Chello *et al.*, 1979; Chello and Sirotnak, 1981). These authors extended their work to other combinations of S phase specific agents and agents inducing blockade at mitosis or G_2 in L1210 leukemia. Combinations of araC with VDS and MTX with teniposide (VM-26) were synergistic when the antimetabolite was given 24 hr before the second agent, but were additive with simultaneous or reversed order of administration (Sirotnak *et al.*, 1983). While no direct cytokinetic parameters were monitored in these studies, the authors presume that a cytokinetic basis exists for the synergism because of the biochemical and pharmacologic diversity of the agents studied, the schedule-dependency of the response, and their prior studies of the recovery of DNA synthesis in cells treated with MTX (Sirotnak *et al.*, 1976).

Perhaps because VCR does not have myelosuppression as its dose-limiting toxicity and because of its phase-specific activity (metaphase arrest), VCR has been extensively studied in combination with other agents. The work by Jackson *et al.* (1984, 1986) provides another example of sequence-dependent effects of VCR and VP16 in mice bearing P388 murine leukemia. Interestingly, while neither agent was very effective in producing an ILS or probable cures (> 60 day survival) when given alone, both sequences, i.e., VCR followed by VP16 and VP16 followed by VCR, produced effects which were maximal and synergistic at 4-hr and 48-hr intervals between the two agents and which fell off dramatically at a 96-hr interval to less than additive (or antagonistic). The sequence of VP16 followed by VCR also appeared to be somewhat more effective than VCR followed by VP16, although this conclusion must be taken with caution since the two sequences were not compared directly and were studied in separate experiments. Again, no direct measurements of cytokinetic parameters were undertaken; however, it is known that the two agents act on different phases of the cell cycle: VCR causes arrest of cells in the metaphase of mitosis by binding to microtubules and inhibiting spindle formation (Owellen *et al.*, 1976) and VP16 causes an accumulation of cells in G_2 by virtue of causing DNA breaks by interaction with topoisomerase II (Ross *et al.*, 1984).

3. Cytokinetics-Directed Chemotherapy

While the previously mentioned examples illustrate the value of cyto-kinetics-based scheduling of antineoplastic agents in animal tumor models, the work by Barranco *et al.* (1973, 1982a,b) is noteworthy for highlighting the potential utility of a cytokinetics-*directed* approach. The investigators also provide some of the very limited data available on clinical monitoring of drug effects in adult solid tumors. In their initial report (Barranco *et al.*, 1973), they demonstrated in five patients with human malignant melanoma that bleomycin at a dose of 25 units/day for 4 days as a continuous infusion could be used as a synchronizing agent. Multiple tumor biopsies before and after administration bleomycin showed by LI and by pulse-labeling with tritiated thymidine that bleomycin reversibly inhibited cell progression at the S to G_2 phase followed by a partially synchronized recovery that was maximal at about 3 days after stopping bleomycin treatment. Once the recovery parameters were defined, the patients received a second course consisting of bleomycin followed by araC at the time of maximal cells in S phase.

In subsequent work using flow cytometry, Barranco *et al.* (1982a) showed that nontoxic doses of 1,2:5,6-dianhydrogalactitol (DAG) could increase the proportion of CHO (Chinese hamster ovary) cells in G_2–M and thereby increase their *in vitro* sensitivity to bleomycin by proper timing of administration of the bleomycin. In an ambitious sequel to this study, Barranco *et al.* (1982b) studied the cytokinetic effects of low-dose DAG in 17 patients (who were not amenable to further therapy) with tumors accessible for multiple biopsies for flow cytometry analysis and demonstrated that cytokinetic change could be induced in human adult solid tumors. A 30–60% increase in the proportion of S phase cells was seen in squamous cell carcinoma of the head and neck region (3 patients) or skin (3 patients), a 60–100% increase in adenocarcinoma of the rectum (1 patient) or squamous carcinoma of the anus (3 patients), and a 50% increase in 1 patient with breast cancer. S phase increases of 60, 90, 140, and 240% were seen in 4 patients with cervical cancer. One patient each with melanoma and cervical cancer showed no change in distribution of cells by DNA histogram.

4. Summary

In summary, experiments using a number of animal tumor models have demonstrated marked schedule-dependence with respect to sequence and timing of administration of two drugs in combination. Nevertheless, the ultimate test of a cytokinetics-based (or -directed) approach is in its clinical application. Unfortunately, clinical data is more limited and more difficult to interpret. In particular, the clinical situation is more complex due to tumor heterogeneity and generally much lower growth fractions in human tumors than in experimental models (Poste, 1982). The appropriate studies have not been carried out to show that cytokinetic defferences between tumor tissue and normal bone marrow are exploitable in designing schedule-dependent regimens in humans. Likewise, despite promising pilot studies in childhood acute leukemia and neuroblastoma showing high remission and response rates with cytokinetically-timed sequencing of agents (Lampkin *et al.*, 1970; Mauer *et al.*, 1976; Hayes and Mauer, 1976; Hayes *et al.*, 1977), no studies have been conducted that show, in direct, randomized comparisons that sequential, cytokinetic-based or -directed therapy shows any therapeutic advantage over the same drug combinations given on more conventional schedule. Thus, the clinical relevance of the cytokinetic approach to scheduling chemotherapeutic agents in combination remains to be demonstrated.

B. INTERACTIONS OF POLYAMINE INHIBITORS WITH CONVENTIONAL CYTOTOXIC AGENTS

1. Introduction

Unlike the studies reviewed in the preceding section, investigations using polyamine inhibitors in combination with other, more conventional antineoplastic agents have sought mainly to exploit the biochemical effects of these inhibitors, i.e., the polyamine depletion that they induce, and to take advantage of the aberrant polyamine biosynthesis and metabolism in malignantly transformed cells (O'Brien, 1976; Scalabrino *et al.*, 1978; Takano *et al.*, 1981). However, since polyamines are essential for cell growth, proliferation, and DNA synthesis but *not* for cell viability (Pegg and McCann, 1982), successful polyamine depletion generally results in cytostasis rather than cell killing [with the exception of certain small cell lung cancer cell lines (Luk *et al.*, 1981, 1982] and may be accompanied by a block at the G_1 to S phase transition (Alhonen-Hongisto *et al.*, 1984b; Oredsson *et al.*, 1983; Seidenfeld, *et al.*, 1981).

Thus, in addition to having beneficial effects when combined with cytotoxic agents (to be covered in a subsequent section), polyamine inhibitors may have antagonistic effects when administered with certain (especially S phase specific) agents and on certain schedules (for an extensive review of the combination of polyamine inhibitors with cytotoxic agents, see Porter and Jänne, 1987). In particular, if the schedule of administration of polyamine inhibitors is not considered, the induction of a lower rate of cellular proliferation may protect cells from the lethal effects of cytotoxic agents which are most effective against rapidly dividing cells.

2. Combinations of DFMO with Phase-Specific Agents

Evidence for such schedule-dependent effects has been obtained in the well-studied rat 9L brain tumor model *in vitro*. D,L,α-difluoromethylornithine

(DFMO) is a specific inhibitor of ornithine decarboxylase (ODC) (Metcalf *et al.*, 1978; Poso and Pegg, 1982). DFMO has been extensively studied as a prototype polyamine inhibitor which inhibits cell proliferation by inducing depletion of putrescine and spermidine, but generally does not affect spermine (Porter and Sufrin, 1986). Not surprisingly, antagonistic effects were demonstrated in rat 9L brain tumor cells when DFMO as a 48-hr pretreatment was followed by araC; these effects correlated with changes in the DNA histogram distribution by flow cytometry (Oredsson *et al.*, 1983a). Similar findings were observed with DFMO as a 96 hr pretreatment followed by either MTX or VCR (Alhonen-Hongisto *et al.*, 1984b). In this latter study, DFMO was shown to cause an 80% reduction in the mitotic index.

Nevertheless, investigators using different tumor models, i.e., L1210 *in vivo* (Prakash and Sunkara, 1983) and HeLa cells *in vitro* (Sunkara *et al.*, 1980), have demonstrated synergistic effects with the combination of DFMO and araC, when the araC administration was timed to correspond to the maximal accumulation of cells in S phase. Thus, the apparent conflict in the results in the various tumor models is resolved by consideration of the timing of the antimetabolite in conjunction with the cytokinetic parameters monitored. In addition, Sunkara *et al.* (1979, and 1981) have claimed that such combinations may be selective, based upon their data that intracellular polyamine depletion caused normal cells to accumulate in G_1, whereas malignant cells accumulated in S phase. However, in light of the data in the rat 9L cells (cited in the previous paragraph), one must assume that the cytokinetic effects of polyamine depletion may be cell-line specific.

Another report of the combination of DFMO with the antimetabolite 5FU reported *additive* inhibition of the growth of a human colon adenocarcinoma cell line (Colo 205) when the two agents were added simultaneously at 3 days to cells in culture; cells were harvested and counted on days 3, 5, 7, and 10 (Kingsnorth *et al.*, 1983). Notably, beneficial (additive) antiproliferative effects were seen at relatively low DFMO concentrations of 0.2 to 0.5 mM, but cells were not sensitized to nontoxic concentrations of 5FU. Kingsnorth *et al.* may have avoided the previously cited antagonistic effects of DFMO by the *simultaneous* addition of the two agents, thereby allowing 5FU to exert its cytotoxic effects before DFMO could affect the rate of cellular proliferation.

3. DFMO Combined with DNA-Damaging Agents

More perplexing are the antagonistic effects seen in combinations of DFMO with agents that should be cell cycle independent, i.e., alkylating agents and DNA cross-linking agents. Since polyamines have been shown to stabilize DNA at critical times in the cell cyle (Bloomfield and Wilson, 1981; Feuerstein *et al.*, 1986; Basu and Marton, 1987; Bakic *et al.*, 1987), polyamine depletion was predicted to enhance the effectiveness of such agents and indeed has been shown to potentiate the cytotoxic effects of BCNU in several tumor models (Hung *et al.*, 1981; Alhonen-Hongisto *et al.*, 1984c; Cavanaugh *et al.*, 1984; Seidenfeld and Komar, 1985). Nevertheless, Heston *et al.* (1982) were unable to detect any therapeutic advantage in the combination of DFMO and the alkylating agent CP *in vivo* against a model of prostate cancer in the rat.

Interestingly, not all nitrosoureas were found to share the same interaction as BCNU with DFMO. As reported by Oredsson *et al.* (1983b), the structural and activity differences in the nitrosoureas influence their interactions with DFMO pretreatment (for 48 hr) in 9L cells. BCNU and 1-(2-chloroethyl)-3-*trans*-4-methylcyclohexyl-1-nitrosourea (MeCCNU), which alkylate, carbamoylate, and cross-link DNA, and chlorozotocin, which alkylates and cross-links DNA, both showed enhanced cytotoxicity following polyamine depletion. In contrast, 1,3-*bis*(*trans*-4-hydroxycyclohexyl)-1-nitrosourea (BHCNU, which carbamoylates but neither alkylates nor cross-links) and *N*-ethyl-*N*-nitrosourea and *N*-methyl-*N*-nitrosourea (both of which alkylate and carbamoylate but do not cross-link) did not show any potentiation of cytotoxicity. BHCNU actually showed a antagonistic interaction with DFMO. Therefore, the authors felt that the synergistic interaction of DFMO with nitrosoureas required the cross-linking function.

However, work with two additional cross-linking agents, namely, *cis*-diamminedichloroplatinum II (cisplatin) (Oredsson *et al.*, 1982) and aziridinyl-benzoquinone (AZQ) (Alhonen-Hongisto *et al.*, 1984a), showed that DFMO pretreatment (48 to 72 hr) resulted in decreased cytotoxicity in 9L cells. Furthermore, DFMO pretreatment was shown to cause decreased cross-link formation by cisplatin and increased cross-linking by BCNU. The specificity of the observed effects for DFMO-induced polyamine depletion was demonstrated by the reversal of the effects on cross-linking when exogenous putrescine was added to DFMO treated cells 24 hr prior to the addition of BCNU or cisplatin (Tofilon *et al.*, 1983).

Schedule dependence of the antagonistic interaction between DFMO and cisplatin was suggested by studies showing an enhancement of the cytotoxicity of cisplatin when its administration was *followed* by DFMO in pancreatic adenocarcinoma cell lines (Chang *et al.*, 1984). This study was interesting in that two of the four cell lines tested displayed relative drug resistance to cisplatin and other platinum analogs (Chang, 1983) and suggested that cisplatin followed by DFMO might be a mechanism of reversing drug resistance. In addition, unlike previous *in vitro* studies using 1–10 mM DFMO, the studies in the pancreatic cancer cell lines focused on clinically achievable plasma concentrations, i.e., 0.3 to 0.6 mM, as documented in phase I trials (Abeloff *et al.*, 1984; Griffin *et al.*, 1984).

The schedule-dependent nature of the interaction was more clearly demonstrated in subsequent *in vitro* studies showing antagonistic effects when pancreatic cancer cells were exposed to DFMO prior to cisplatin and synergistic effects when DFMO followed the cisplatin (Chang *et al.*, 1987). This study will be presented in some detail since it represents one of the few reports in which schedule-dependent interactions are analyzed quantitatively. The dose–effect relationships were analyzed by the median-effect principle, the multiple drug-effect equation, and the concept of the combination index or CI (see chapter 2 for full definitions). The CI allows for quantitative assessment of synergism (CI <1.0), antagonism (CI >1.0), and additivity (CI = 1.0), and is compatible with the isobologram method (Chou and Talalay, 1984, and 1987; Chou and Chou, 1986). Representative results for the fractional inhibitions and CIs for each agent alone and their combinations in three separate pancreatic cancer lines [PANC-1 of human origin and WD (well-differentiated) PaCa and PD (poorly differentiated) PaCa, both of hamster origin] are shown in Table 2. The colony formation assay was used to assess drug effects; DFMO was administered continuously for the times indicated and cisplatin for 1 hr only. As shown by the CIs, DFMO pretreatment interacted with cisplatin to produce antagonistic effects. In contrast, in most intances in which postcisplatin DFMO was combined with cisplatin, synergistic effects were seen; the only exceptions were in the PD PaCa cell line in which weakly antagonistic or additive effects were occasionally seen. It was hypothesized that, although cisplatin is generally considered a cycle-non-specific agent, the lower rate of cell division induced by DFMO pretreatment may have influenced its cytotoxicity since cycle-nonspecific agents may display a cell cycle age response (i.e., be affected by the overall rate of cellular proliferation). The enhancement of cisplatin's effects with DFMO posttreatment suggests that DFMO may prevent recovery from sublethal damage induced by the cisplatin.

Unfortunately, the *in vivo* testing of the combination of DFMO and cisplatin failed to reveal selectivity of the combination for an athymic (nude) mouse pancreatic adenocarcinoma xenograft. While producing a marked suppression of tumor growth, 3% DFMO in the drinking water combined with cisplatin, 1.5 mg/kg daily for 5 days, was unacceptably toxic at aproximately the LD_{50} (50% lethal dose level), whereas either agent alone was ineffective and nontoxic. The *in vivo* combination of cisplatin and 2% DFMO in drinking water was nontoxic, but produced no enhancement of the antitumor effect (B. K. Chang, unpublished observations).

The report by Allen and Natale (1986) examined the combination of DFMO with either DOX (doxorubicin), cisplatin, or VLB (vinblastine or Velban), using a simultaneous and continuous schedule of *in vitro* exposure against P3J, a Burkitt's lymphoma cell line. Significantly, they used quite low concentrations of DFMO (15 and 25 $\mu g/ml$) and found that all combinations tested potentiated the reduction of colony formation seen with single-agent exposure. In addition, they calculated a performance index, as described by Drewinko *et al.* (1976), to determine when the interactions were additive or synergistic, based on whether the colony reduction was 3-fold greater than the predicted additive effect (synergism) or less than 3-fold greater than

TABLE 2

Fractional Inhibitions and CIs of Cisplatin and DFMO in Three Pancreatic Cancer Cell Lines

Exposure	D1[a] [cisplatin][b]	D2 [DFMO]	PANC-1		WD PaCa		PD PaCa	
			fa	CI	fa	CI	fa	CI
2-Day								
precisplatin	1.0	50	0.09	37.4	0.08	12.8	0.74	1.19
DFMO	5.0	50	0.91	3.2	0.77	1.7	0.97	1.88
	1.0	100	0.56	7.0	0.07	22.0	0.80	2.71
	5.0	100	0.98	—	0.58	4.6	0.98	1.73
5-6-Day[c]								
precisplatin	1.0	50	0.11	32.7	0.29	3.9	0.81	2.24
DFMO	5.0	50	0.91	3.2	0.75	1.7	0.97	1.77
2-Day pre- and								
postcisplatin	0.1	50	0.95	0.25	0.93	0.50	0.77	0.65
DFMO	1.0	50	0.95	0.58	0.93	0.52	0.95	1.13
	5.0	50	0.99	0.63	0.97	0.37	0.99	0.80

Post-cisplatin DFMO		fa	CI	fa	CI	fa	CI
0.1	50	0.92	0.38	0.86	0.71	0.44	1.04
1.0	50	0.96	0.48	0.88	0.70	0.92	1.39
5.0	50	0.99	0.49	0.95	0.48	0.99	0.70
Cisplatin alone							
0.1	—	0.13	—	0.09	—	0.20	—
1.0	—	0.63	—	0.19	—	0.86	—
5.0	—	0.99	—	0.64	—	0.99	—
DFMO alone							
—	25	0.27	—	0.46	—	0.0002	—
—	50	0.88	—	0.73	—	0.18	—
—	100	0.96	—	0.92	—	0.56	—
—	500	0.97	—	0.99	—	1.00	—

[a] $D1$ = dose 1 (concentration of the first drug in the combination); $D2$ = dose 2 (of the second drug); fa = fraction affected, or [1-(treated/control)]; $CI = (D1/Dx1)+(D2/Dx2)$, where $Dx = Dm[fa/(1 - fa)]^{1/m}$ or the dose required to affect $x\%$; Dm = the median effect dose or ID_{50}, and m = the slope of the median-effect plot (Chou and Talalay, 1984) which signifies the sigmoidicity of the dose-effect curve. [$CI = 1.0$, additive; <1.0, synergism; >1.0, antagonism.] For a conservative calculation of CI, a third term $(D1/Dx1)(D2/Dx2)$ should be added for mutually nonexclusive drugs. This will result in slightly higher CI values. [Adapted from Table 2, Chang et al., 1987, with permission.]

[b] For PD PaCa (the most sensitive cell line), the highest [cisplatin] was 2.5 μg/ml (instead of 5.0 μg/ml). All drug concentrations are given in μg/ml. [DFMO] 50 = 0.21mM and 100 μg/ml = 0.42mM. [Cisplatin] 0.1 to 5 μ/ml = 0.33 to 16.7 μM.

[c] Pre-cisplatin incubation with DFMO was 6 days for PANC-1 and 5 days for WD PaCa and PD PaCa.

the predicted additive effect (supraadditive, in their terminology). By this method, they considered all DFMO–DOX combinations synergistic, DFMO–cisplatin combinations synergistic at 25 μg/ml DFMO and supraadditive at 15 μg/ml DFMO, and DFMO–VLB combinations additive and possibly synergistic at the higher concentrations of both VLB and DFMO. Reanalysis (for the purpose of the present review) of Allen and Natale's (1986) published data using the concept of CI and the software developed by Chou and Chou (1989) largely confirms their interpretation of their results: (1) all CIs for the combinations of DFMO–DOX and DFMO–cisplatin were less than 1.0 (i.e., synergistic) and in the case of DFMO–DOX were less than 0.4 (highly synergistic); (2) the CIs for the combination of DFMO–VLB were about 1.0—1.04 for 15 μg/ml DFMO with 10^{-7} μg/ml VLB and 1.19 for 15 μg/ml DFMO with 10^{-6} μg/ml VLB (roughly additive to weakly antagonistic)— or less than 1.0 (synergistic) for all other dosages of the combination tested.

Another study showing potentiation of DOX cytotoxicity by DFMO is that by Chang *et al.* (1986), who reported roughly additive inhibitory effects in four pancreatic adenocarcinoma cell lines *in vitro*. Bartholeyns and Koch-Weser (1981) also demonstrated an improved ILS with DFMO plus DOX, but not VDS (compared with DFMO or DOX or VDS alone) in mice with L1210 leukemia and an enhanced inhibition of the growth of EMT6 mouse mammary tumors with DFMO plus either DOX or VDS. Notably, in all of these studies, the DFMO was given continuously *after* the administration of the conventional cytotoxic agent, which confirms the need to consider the schedule of administration and suggests that polyamine depletion may be more effectively used to augment conventional chemotherapeutic agents if induced *following* their administration.

III. Schedule-Dependent Interactions Based Primarily upon Biochemical Considerations: Combinations (Largely of Antimetabolites) Affecting the Enzymes Involved in DNA Synthesis

A. METHOTREXATE AND 5-FLUOROURACIL

Exploitation of metabolic differences between malignantly-transformed and normal cell populations has been a long sought after and often elusive goal (Sartorelli, 1969). One of the combinations of antimetabolites for which a clear schedule dependence has been demonstrated is MTX and 5FU. Importantly, the rational use of the two agents in combination in experimental models was preceded by the development of a greater knowledge of the biochemical effects of each, whereby 5FU was shown to have as its active metabolite FdUMP (5-fluorodeoxyuridine monophosphate) which effectively inhibits dTMP (thymidylate) synthetase only in the presence of N^5, N^{10}-methylenetetrahydrofolate (5,10-CH$_2$-FH$_4$) (Santi *et al.*, 1974).

Thus, Bertino *et al.* (1977) predicted and were able to show in mouse Sarcoma 180 *in vivo* that pretreatment with MTX (a weak inhibitor of the dTMP synthetase which had been shown to increase the binding of FdUMP to the enzyme) would augment the antitumor activity of 5FU by increasing the binding of FdUMP to dTMP synthetase. Later, Cadman *et al.* (1981) reported that the modulating effect of MTX pretreatment was due to inhibition of *de novo* purine synthesis and documented increased amounts of phosphorylated derivatives of 5FU in L1210 cells pretreated with MTX.

Clinical studies, however, with sequential MTX followed by 5FU have given mixed results. Phase I–II trials of the sequential therapy have suggested enhanced response rates in breast cancer (Gerwitz and Cadman, 1981), head and neck cancer (Ringborg *et al.*, 1983), and colorectal cancer (Hermann *et al.*, 1984). Benz *et al.* (1985), in a pharmacologically guided study, felt that the regimen was well-tolerated and worthy for further study, while Harding *et al.* (1988) felt that no therapeutic benefit could be demonstrated due to excess toxicity. In two propectively randomized clinical trials in squamous cell carcinoma, sequential MTX–5FU was not superior to simultaneous or reversed-order administration of the two agents (Browman *et al.*, 1983; Coates *et al.*, 1984). Because of the results of these trials, current clinical interest in the use of 5FU in combinations with other agents has focused upon its sequential combination with leucovorin (see the next section).

B. 5-FLUOROURACIL AND LEUCOVORIN

As discussed earlier, the antitumor activity of 5FU depends upon its inhibition of dTMP synthetase, for which reduced folates and in particular 5,10-CH$_2$FH$_4$ act as cofactors which stabilize the binding of FdUMP to the ternary complex with dTMP synthetase. Leucovorin (folinic acid, or DL-5-formyl-tetrahydrofolate, 5-formyl-FH$_4$) is a synthetic reduced folate analog that is metabolized to 5-methyl-FH$_4$ and has been shown in the laboratory of Hakala to potentiate the cytotoxicity of 5FU against mouse sarcoma 180 and a 5FU-resistant human cancer cell line (Hep-2) (Evans *et al.*, 1980, 1981; Berger and Hakala, 1984). Based upon the work by Mini *et al.* (1987) in a human T-lymphoblast cell line *in vitro*, the synergism of the combination of leucovorin and 5FU or FdUrd (floxuridine or fluorodeoxyuridine) was found to be highly dependent upon the *dose* of leucovorin ($100 > 10 > 1 \ \mu M$ and not occurring at all at $0.1 \ \mu M$) and upon the *sequence* of administration only for FdUrd but not 5FU.

Nevertheless, in clinical trials, which were begun quite early for the combination, leucovorin has been administered *prior* to 5FU in order to achieve high steady-state levels ($> 10 \mu M$) of reduced folates that are felt to be necessary to potentiate FU's cytotoxicity. The trend in trial design has been away from lower doses of leucovorin (Laufman *et al.*, 1987) toward use of higher doses of leucovorin (Hines *et al.*, 1988). Several doses and schedules of administration have been used and many of these have been compared in Phase III randomized trials to 5FU alone, with a distinct advantage demonstrated for the combination, as summarized by Gem (1988).

Only one of the randomized trials (from Roswell Park Memorial Institute; Petrelli *et al.*, 1987) will be summarized in detail here since it illustrates the antitumor effects of the combination of 5FU and high-dose leucovorin and bears on the preceding section (sequential MTX and 5FU). In this trial, previously untreated patients with advanced colorectal cancer were prospectively randomized to one of three regimens or treatment arms. The respective treatment and the results of therapy in this trial are summarized in Table 3. Based upon this representative study and others referred to above (see Gem, 1988), the combination of leucovorin and 5FU is one for which a good correlation between the pre-clinical and clinical efficacy exists.

C. DFMO AND 4'-(9-ACRIDINYLAMINO) METHANESULFON-*M*-ANISIDIDE OR OTHER TOPOISOMERASE II INHIBITORS

Another example of a rational and schedule-dependent antineoplastic drug combination based upon a knowledge of the biochemical mechanisms of action is that of the polyamine inhibitor DFMO and *m*-AMSA (also known as acridinyl anisidide or amsacrine), which has been shown to be a DNA intercalating agent that also has topoisomerase II as its target (Liu *et al.*, 1983; Nelson *et al.*, 1984; Zwelling *et al.*, 1981). DNA topoisomerase II is an enzyme involved in the breakage–reunion cycle that occurs during DNA replication and has recently been identified as an intracellular target for a number of antineoplastic agents which act by stabilizing an enzyme–DNA intermediate called the "cleavable complex" (for reviews, see Chen and Liu, 1986; Glisson and Ross, 1987). Since spermidine had been shown to alter topoisomerase-II-mediated reactions (Brown and Cozzarelli, 1981; Kreuzer and Cozzarelli, 1980), Zwelling *et al.* (1985) reasoned that polyamine depletion should enhance the effects of *m*-AMSA. Indeed, they were able to show that pretreatment of L1210 cells with 1 m*M* DFMO effectively depleted putrescine and spermidine and was associated with a 40% increase in *m*-AMSA-induced DNA single-strand breaks at 24 hr and an almost 2-fold increase in breaks at 48 hr. Putrescine was able to reverse DFMO's potention

TABLE 3

Randomized Trial of 5FU versus 5FU plus High-Dose Leucovorin versus Sequential MTX and 5FU in Advanced Colorectal Carcinoma[a]

Treatment arm	Dose of drug			Response rate (Complete and partial responses, %)
	5FU[b]	MTX	Leucovorin	
1	450 mg/m^2 × 5D, then 200 mg/m^2 QOD × 6	—	—	11
2	600 mg/m^2 following MTX	50 mg/m^2 over a 4-hr infusion	—	5
3	600 mg/m^2 1 hr after leucovorin begun	—	500 mg/m^2 over a 2-hr infusion	48

[a] From Petrelli et al., 1987. The doses given are for each *course* of treatment, which for Arm 1 was every 4 weeks; Arm 2 was given weekly for 4 weeks, then every 2 weeks; and Arm 3 was given every week for 6 weeks followed by a 2 week rest period. 5FU was administered as a bolus iv injection on all treatment arms. Fairly mild toxicity was reported in the high-dose leucovorin plus 5FU arm (mainly diarrhea). Significant differences in response rates were seen when Arm 3 was compared to the other 2 arms, although there was no significant effect on overall survival.

[b] The abbreviations used are 5FU, 5-fluorouracil; MTX, methotrexate; D, day; and QOD, every other day.

of *m*-AMSA's induction of strand breaks and the effect observed was not associated with any changes in intracellular levels of *m*-AMSA.

In a subsequent report, Zwelling and co-workers (Bakic *et al.*, 1987) found that the effects of polyamine depletion on the DNA cleavage of *m*-AMSA and VP16 (a non-DNA-binding topoisomerase-II-reactive drug) were variable in two L1210 murine leukemia and two human HL-60 leukemia cell lines, one of which was resistant to *m*-AMSA (HL-60/AMSA). In this study, cells were pretreated with 1 m*M* DFMO for 5 days, which depleted intracellular putrescine and spermidine and greatly inhibited cell growth in all cell lines; however, DNA cleavage was *decreased* in all cells with the addition of VP16 and in two of the four cell lines with *m*-AMSA (one L1210 line and the *m*-AMSA sensitive HL-60 line). DNA cleavage by *m*-AMSA was unaffected by DFMO pretreatment in an HL-60 cell line that was resistant to *m*-AMSA and was increased in the other L1210 cell line. Cytotoxicity as assessed by colony formation paralleled the changes in DNA cleavage, i.e., antagonistic effects were observed with the combination in three of the cell lines (all except that of DFMO–*m*-AMSA in the one L1210 cell line in which DNA cleavage was enhanced by DFMO). A putrescine analog, (2R,5R)-6-heptyne-2,5-diamine, was tested in the HL-60 cells and found to be more potent, but to have essentially similar effects to DFMO.

How dose one reconcile these results with the earlier, more encouraging findings? One major difference between the two reports is the schedule in terms of the *duration* of the exposure to DFMO, which was 24–48 hr in the first report (Zwelling *et al.*, 1985) and 5 days in the second (Bakic *et al.*, 1987). As the investigators point out, the reduction in DNA cleavage with the 5 day exposure could be a result of secondary effects of polyamine depletion, such as decreased cellular proliferation which has been reported to reduce topoisomerase II and to render cells less sensitive to the cytotoxic effects of modulators of this enzyme (Hsiang *et al.*, 1988; Nelson *et al.*, 1987; Sullivan *et al.*, 1986; Zwelling *et al.*, 1987). While data were provided to show the effect of the length of exposure (1, 3, and 5 days) on the polyamine levels, cellular proliferation (by ^{3}H-thymidine incorporation), and DNA single-strand breaks induced by VP16 in HL-60 cells, no data showing the actual topoisomerase II levels or enzyme activity were presented. Therefore, it is difficult to assess whether the effects of polyamine depletion are direct (i.e., affecting the drug–topoisomerase II interaction directly) or indirect (e.g., by effects on the rate of cellular proliferation, which did, of course, decline).

Some support for the latter hypothesis comes from work with DOX, another intercalating agent that also has some of its cytotoxicity mediated by topoisomerase II interaction. As summarized earlier (see Section II), the reports by Allen and Natale (1986), Bartholeyns and Koch-Weser (1981), and Chang *et al.* (1986) found the antitumor effects of DOX to be enhanced by DFMO when the DFMO was administered continuously *after* DOX. In contrast, Seidenfeld *et al.* (1986) found that a 48 hr pretreatment with 5 m*M* DFMO against four human adenocarcinoma cell lines "significantly protected . . . cells against the lethal effects of Adriamycin." The investigators ruled out any effect on the intracellular accumulation of DOX and considered proliferation-dependent effects to be the most likely explanation of the observed antagonistic interaction. Thus, the reports summarized further emphasize the need to consider the schedule in the use of polyamine inhibitors in combination with other agents, including topoisomerase II-interacting agents, and in addition, that the interactions are likely to be complex and should take into account rates of cell proliferation.

IV. Interactions Based upon Altered Drug Transport

A. INCREASED UPTAKE OF ONE AGENT INDUCED BY ANOTHER AGENT

Perhaps one of the best studied examples of drug synergy or enhancement for schedule-dependent, pharmacologic reasons is the combination of polyamine analog methylglyoxal bis(guanylhdrazone) (MGBG) and DFMO. As a polyamine analog, MGBG utilizes a spermidine/spermine transport system (Dave and Cabelles, 1973) and intracellular polyamine depletion increases the uptake of MGBG (Sëppänen *et al.*, 1981). Several reports document the potentiation of the effects of MGBG by pretreatment with DFMO in cell lines and tumor models (Herr *et al.*, 1984a,b; Sëppänen *et al.*, 1983). Indeed, the sequential combination has led to encouraging results in small numbers of patients with leukemia (Jänne *et al.*, 1981; Siimes *et al.*, 1981), although another study failed to confirm any benefit for the combination and substantially greater toxicity than was seen with DFMO alone (Warrell *et al.*, 1983). Alhonen-Hongisto *et al.* (1985) examined the response of two cell lines with markedly different sensitivities to MGBG alone, 9L rat brain tumor cells (resistant) and hamster V79 cells (sensitive), and found that both sequential and simultaneous administration of DFMO and MGBG increased MGBG uptake and inhibited growth. However, only in 9L cells did intracellular polyamine levels correlate with the observed antiproliferative effects.

With respect to the question of selective toxicity of the sequential combination, three recent studies are of interest. First, Heston *et al.* (1984) showed that, in the Dunning rat prostate cancer model *in vivo*, DFMO pretreatment did not affect the intracellular accumulation of MGBG in tumor tissue.

Kramer *et al.* (1985) showed that DFMO pretreatment increased MGBG levels in L1210 cells but did so to the same extent as seen in normal tissues and thus failed to improve the selectivity of MGBG. Finally, in a detailed pharmacologic study of patients with refractory acute (7 patients) or chronic (1 patient) myeloblastic leukemia or multiple myeloma (1 patient), Maddox *et al.* (1988) found that (1) no patient achieved a clinical remission, although 6 patients had transient (1 week) decreases in peripheral blood blast cells; (2) the DFMO–MGBG combination was more toxic than DFMO alone; (3) DFMO administration took a long time (average 9 days) to deplete peripheral mononuclear cell polyamine levels; and (4) in 3 of 4 patients who received multiple doses of MGBG, an increased MGBG mononuclear content after DFMO was obtained, but failed to correlate with the degree of polyamine depletion or clinical response. The authors also noted that after an initial dose of MGBG by itself (in order to determine baseline MGBG levels) that levels of all natural polyamines increased in peripheral blood and bone marrow mononuclear cells. This finding is in contrast with the results in experimental models in which MGBG increased putrescine, but not spermidine or spermine (Porter *et al.*, 1980; Käpyaho *et al.*, 1984). The authors concluded that the schedule of administration of DFMO with MGBG was crucial and suggested that potentiation was most likely to result if DFMO was given for 7 to 10 days prior to MGBG. Consequently, despite a strong pharmacologic rationale for the combination of DFMO preceding MGBG, its ultimate clinical utility remains to be established.

B. INCREASED NET DRUG ACCUMULATION (BY DECREASED EFFLUX) OF ONE ANTINEOPLASTIC AGENT INDUCED BY ANOTHER (NONNEOPLASTIC) AGENT

By the time of this writing, the literature on multidrug and other types of cytotoxic drug resistance and ways to attempt to overcome it is extensive (for reviews, see Curt *et al.*, 1984; Dexter and Leith, 1986; Deuchars and Ling, 1989). Although the question of the schedule dependence of the combination of calcium antagonists with natural product chemotherapeutic agents has not been addressed systematically, the clinical studies to date have generally (in a manner analogous to the leucovorin–5FU combination) administered the calcium antagonist first and continued it for some time after the other agent in an attempt to emulate as closely as possible the *in vitro* levels of verapamil necessary to demonstrate the modulation of cytotoxicity. However, in practice it has not been possible to attain the levels of verapamil that are most effective *in vitro* due to its acute cardiac toxicity.

In particular, calcium-channel blockers, which have been used to treat cardiovascular disorders (Chauffert *et al.*, 1984; Ramu *et al.*, 1983, 1984; Rogan *et al.*, 1984; Slater *et al.*, 1982; Tsuruo *et al.*, 1981, 1982a,b, 1983, 1985;

Yalowich and Ross, 1984, 1985), and calmodulin inhibitors (Tsuruo *et al.*, 1982a; Ganapathi and Grabowski, 1983) have been shown to enhance the responsiveness to anthracyclines and other natural product agents in resistant, but generally not in sensitive, cell lines. Reports examining the cellular pharmacology of the interactions have generally suggested that an inhibition of efflux with a net increase in accumulation or retention mediates the enhancement of cytotoxicity observed with calcium antagonists. Nevertheless, when critically examined, the accumulation differences do not always account for the differences in cytotoxicity (Kessel and Wilberding, 1985a). For example, Tsuruo *et al.* (1982a) report a verapamil-induced 13-fold enhancement of sensitivity to DOX but only a 2.4-fold enhancement of cellular accumulation. Second, the concentrations of verapamil used in most of the *in vitro* studies (i.e., 6.6 μM or higher) are beyond those clinically obtainable, which have been reported as peak levels of 0.55 to 2.2 μM and steady-state levels of 0.35 to 1.3 μM (Benson *et al.*, 1985; Chew *et al.*, 1981; Dalton *et al.*, 1989; Frishman *et al.*, 1982). Likewise, data obtained in cells or cell lines that display *primary* or intrinsic drug resistance or clinically acquired resistance (as opposed to cells with resistance acquired by long-term *in vitro* exposure) suggest that drug resistance is neither mediated by differences in net accumulation nor is accumulation enhanced by calcium antagonists (Chang *et al.*, 1989; Chang and Gregory, 1985; Chervinsky and Wang, 1976; Donelli *et al.*, 1979; Louie *et al.*, 1986; Silvestrini *et al.*, 1973). Furthermore, the work by Harker *et al.* (1986) is noteworthy, since it suggests that topoisomerase II may be involved in the mediation of the calcium-antagonist enhancement of drug activity. However, regardless of the mechanism of interaction, studies on cells from clinically resistant patients (Goodman *et al.*, 1987) and *in vivo* studies with verapamil added to chemotherapy (Dalton *et al.*, 1989) suggest that calcium antagonists may have a role in modulation of drug resistance. Accordingly, while the ultimate utility of combination is still in question and may depend upon the development of less toxic calcium antagonists (Kessel and Wilberding, 1985b), the results of studies to date suggest that clinically useful modulation of DOX and other natural products may be achievable.

Other novel combinations designed to affect cytotoxic drug accumulation have been sought. In an effort to reverse resistance by altering drug transport, Inaba and Maruyama (1988) tested 23 lipophilic drugs with polycyclic structures for their effects on net accumulation (by prevention of outward transport or efflux) of VCR in VCR-resistant P388 mouse leukemia cells. While 14 of the 23 were found to potentiate the net accumulation of VCR, quinacrine (an antimalarial agent) had the greatest effect on accumulation and cytotoxicity *in vitro* and was shown to be well tolerated and synergistic *in vivo* with VCR at the highest dosage tested of quinarcine. Schedule dependence was not specifically evaluated, but the quinacrine was administered concurrently with the VCR since continuous exposure (to the quinacrine) was necessary for the effect in cultured cells.

V. Miscellaneous

A. COMBINATIONS OF BIOLOGICAL
RESPONSE MODIFIERS

A recent area of investigation has been the study of combinations of biological response modifiers of different types, with the mechanisms of interaction varying with the types of agents used. The area of investigation is still in its infancy. Only two examples are cited, for illustrative purposes and because the second one emphasizes the potential for schedule-dependent interactions of biological response modifiers. Chouaib *et al.* (1988) demonstrated *in vitro* synergy between interleukin 2 (IL-2) and tumor necrosis factor (TNF) which, they provide data to suggest, is mediated by increased specific binding of TNF to large granular lymphocytes. The schedule dependence of the combination of IL-2 and TNF was systematically investigated by Zimmerman *et al.* (1989) in mouse models of melanoma and fibrosarcoma. TNF followed by IL-2 had extraordinary antitumor efficacy, which was much greater than IL-2 followed by TNF. The sequential TNF–IL-2 causes complete tumor regression, whereas simultaneous administration only slowed tumor growth. Other investigators have also reported synergy for the combination of IL-2 and TNF (Winkelhake *et al.*, 1987), the combination of IL-2 and α-interferon (Brunda *et al.*, 1987), and the combination of γ-interferon and TNF (Brouckaert *et al.*, 1986). Thus, combinations of biological response modifiers which take sequencing into account may improve on their single-agent efficacy which is somewhat limited at the present time (Rosenberg, 1988).

B. COMBINATIONS OF HEMATOPOIETIC
STIMULATING FACTORS WITH
ANTINEOPLASTIC AGENTS

Although this area will not be covered in detail, the use of hematopoietic stimulating factors needs to be mentioned as an area that is currently under investigation that should ultimately have considerable clinical applications (Clark and Kamen, 1987). While these agents or factors do not enhance the spectrum of antitumor activity of antineoplastic agents, the rationale for their administration is to promote recovery of the normal hematopoietic stem cells and precursors, especially of the myeloid series, in order to permit more intensive chemotherapy and/or to maintain certain schedules of administration. Although the colony-stimulating factors (CSF), especially granulocyte–macrophage CSF (GM-CSF), have been studied for only a short period of time (Gasson *et al.*, 1984), they are already being tested in a variety of clinical settings, i.e., acute leukemia and other hematologic malignancies, breast cancer, lung cancer, and sarcoma, in an effort to improve the therapeutic index of cancer chemotherapy by increasing the host tolerance to myelosuppression (Bhalla *et al.*, 1988; Bücher *et al.*, 1989; Theriault *et al.*, 1989; Morstyn *et al.*, 1989; Vadhan-Raj *et al.*, 1989).

Virtually every protocol has used a slightly different schedule of administration and, therefore, the optimal schedule of administration of GM-CSF in combination with chemotherapeutic agents has yet to be worked out. However, Morstyn *et al.* (1989) investigated three different schedules of administration (with respect to duration) and suggests that administration of GM-CSF for 7 days beginning on day 4 following cytotoxic chemotherapy is adequate to prevent myelosuppression in patients treated for small cell carcinoma of the lung. Although the exact role of the CSFs in cancer chemotherapy remains to be defined, they are virtually certain of a place in the oncologic pharmacopoeia for amelioration of host toxicity, thereby permitting the administration of more intensive chemotherapy and possibly its administration to patient populations that currently do not receive optimal doses of chemotherapy (e.g., the elderly).

VI. Summary

The schedule dependence of antineoplastic agents used in various combinations has been reviewed with emphasis on the mechanisms of interaction, as summarized in Table 1. Beginning with purely empirical studies which did demonstrate the importance of scheduling in animal tumor models, research then focused upon attempts at synchronization and recruitment, i.e., cytokinetics-based or -directed protocols. Despite demonstration of the feasibility of a cell kinetics-oriented approach in experimental models, the clinical utility of the approach has not been demonstrated, in large part because of the greater heterogeneity of human tumors and the practical problems associated with monitoring cell kinetics parameters in a clinical study.

The more recent and more successful approaches to sequencing of agents has directed towards achieving a biochemical or pharmacologic advantage through rationally designed combinations. In all such combinations, the major question is whether there is a *selective* advantage with the combination which will result in an improvement in the therapeutic index. Optimal combination of antineoplastic agents must take into consideration host toxicity, drug mechanism of action, dosage and pharmacokinetics, tumor heterogeneity (especially from the standpoint of drug resistance), cell kinetics, and tumor size (Day, 1986). Thus, designing schedules of administration is extremely difficult when all the relevant parameters are taken into account.

The ultimate test of a combination is in its successful clinical application. By this criterion, the only combination of those reviewed that can be said to have demonstrated improved therapeutic efficacy is leucovorin and 5FU. Nevertheless, a number of promising combinations are in the preclinical and early clinical stages of testing. In particular, the combinations of polyamine inhibitors with other agents may provide enhanced antitumor effects (Porter and Sufrin, 1986). However, as the present review has emphasized, attention to the schedule of administration of polyamine inhibitors with other agents is critical and available data suggest that pretreatment with such agents, i.e., whenever polyamine depletion is induced and maintained for a sufficient time to affect the rate of proliferation, is detrimental and may result in antagonistic interactions with most agents, apart from BCNU and possibly other cross-linking agents.

In addition, attention to the doses of the agents used in sequential combinations, is important, since the synergistic effects observed in cell culture may translate into formidable toxicity when combinations are used *in vivo*. For this reason, preclinical animal studies and Phase I trials of the combinations are necessary, even when the maximal tolerated doses for the individual agents in a combination are well known. In fact, toxicity remans the major limiting factor in applying the combinations developed experimentally, as exemplified by the combinations of sequential MTX–5FU and DFMO–MGBG. Thus, it may be that future trials of modulating agents may focus on agents which protect the host tissues from toxicity rather than increasing the antineoplastic potency of a given agent (Howell and Goel, 1988) and, as illustrated by the studies with GM-CSF, to enhance recovery from the myelosuppresive side effects. Accordingly, in addition to using relatively non-toxic drugs like leucovorin, the biological response modifiers, including hematopoietic stimulating factors, are likely to be extensively investigated.

References

Abeloff, M. D., Slavik, M., Luk, G., Griffin, C. A., Hermann, J., Blanc, O., Sjoerdsma, A., and Baylin, S. B. (1984). Phase I trial and pharmacokinetic studies of α-difluoromethylornithine—An inhibitor of polyamine biosynthesis. *J. Clin. Oncol.* **2**, 124–130.

Alhonen-Hongisto, L., Deen, D. F., and Marton, L. J. (1984a). Decreased cytotoxicity of aziridinylbenzoquinone caused by polyamine depletion in 9L rat brain tumor cells *in vitro*. *Cancer Res.* **44**, 39–42.

Alhonen-Hongisto, L., Hung, D. T., Deen, D. F., and Marton, L. J. (1984b). Decreased cell kill of vincristine and methotrexate against 9L brain tumor cells *in vitro* caused by α-difluoromethylornithine-induced polyamine depletion. *Cancer Res.* **44**, 4440–4442.

Alhonen-Hongisto, L., Deen, D. F., and Marton, L. J. (1984c). Time dependence of the potentiation of 1.3-bis(2-chloroethyl)-1-nitrosourea cytotoxicity caused by α-difluoromethylornithine-induced polyamine depletion in 9L rat brain tumor cells. *Cancer Res.* **44**, 1819–1822.

Alhonen-Hongisto, L., Levin, V.A., and Marton, L. J. (1985). Modification of uptake and antiproliferative effect of methylglyoxal bis(guanylhydrazone) by treatment with α-difluoromethylornithine in rodent cell lines with different sensitivities to methylglyoxal bis(guanylhydrazone). *Cancer Res.* **45**, 509–514.

Allen, E. D., and Natale, R. B. (1986). Effect of α-difluoromethylornithine alone and in combination with doxorubicin hydrochloride, cis-diamminedichloroplatinum (II), and vinblasine sulfate on the growth of P3J cells *in vitro. Cancer Res.* **46**, 3550–3555.

Bakic, M., Chan, D., Freireich, E. J., Marton, L. J., and Zwelling, L. A. (1987). Effect of polyamine depletion by α-difluoromethylornithine or (2R,5R)-6-heptyne-2,5-diamine on drug-induced topoisomerase II-mediated DNA cleavage and cytotoxicity in human and murine leukemia cells. *Cancer Res.* **47**, 6437–6443.

Barranco, S. C., Luce, J. K., Romsdah, M. M., and Humphrey, R. M. (1973). Bleomycin as a possible synchronizing agent for human tumor cells *in vivo. Cancer Res.* **33**, 882–887.

Barranco, S. C., May, J. T., Boerwinkle, W., Nichols, S., Hakanson, K. M., Schumann, J., Göhde, W., Bryant, J., and Guseman, L. F. (1982a). Enhanced cell killing through the use of cell kinetics-directed treatment schedules for two-drug combinations *in vitro. Cancer Res.* **42**, 2894–2898.

Barranco, S. C., Townsend, C. M., Costanzi, J.J., May, J. T., Baltz, R., O'Quinn, A. G., Leipzig, B., Hokanson, K. M., Guseman, L. F., and Boerwinkle, W. R. (1982b). Use of 1,2:5,6-dianhydrogalactitol in studies on cell kinetics-directed chemotherapy schedules in human tumors *in vitro. Cancer Res.* **42**, 2899–2905.

Bartholeyns, J., and Koch-Weser, J. (1981). Effects of α-difluoromethylornithine alone and combined with Adriamycin or vindesine on L1210 leukemia in mice, EMT6 solid tumors induced by injection of hepatoma tissue culture cells in rats. *Cancer Res.* **41**, 5158–5161.

Basu, H. S., and Marton, L. J. (1987). The interaction of spermine and pentamines with DNA. *Biochem. J.* **244**, 243–246.

Benson, A. B., Trump, D. L., Koeller, J. M., Egorin, M. I., Olman, E. A., Witte, R. S., Davis, T. E., and Tormey, D. C. (1985). Phase I study of vinblastine and verapamil given by concurrent iv infusion. *Cancer Treat. Rep.* **49**, 795–799.

Benz, C., DeGregorio, M., Saks, S., Sambol, N., Hollerman, W., Ignofo, R., Lewis, G., and Cadman, E. (1985). Sequential infusions of methotrexate and 5-fluorouracil in advanced cancer: Pharmacology, toxicity, and response. *Cancer Res.* **45**, 3354–3358.

Berger, S. H., and Hakala, M. T. (1984). Relationship of dUMP and free FdUMP pools to inhibition of thymidylate synthetase by 5-fluorouracil. *Mol. Pharmacol.* **25**, 190–197.

Bertino, J. R., Sawicki, W. L., Lindquist, C. A., and Gupta, V. S. (1977). Schedule-dependent antitumor effects of methotrexate and 5-fluorouracil. *Cancer Res.* **37**, 327–328.

Bhalla, K., Birkhofer, M., Arlin, Z., Grant, S., Lutzky, J., and Graham, G. (1988). Effect of recombinant GM-CSF on the metabolism of cytosine arabinoside in normal and leukemic human bone marrow cells.

Bloomfield, V. A., and Wilson, R. W. (1981). Interaction of polyamines with polynucleotides. *In* "Polyamines in Biology and Medicine" (D. R. Morris and L. J. Marton, eds.), pp. 183–206. Dekker, New York.

Blum, R. H., and Frei, E., III (1979). Combination chemotherapy. *Methods Cancer Res.* **17**, 215–257.

Braunschweiger, P. G., and Schiffer, L. M. (1980a). Effect of Adriamycin on the cell kinetics of 13762 rat mammary tumors and implications for therapy. *Cancer Treat. Rep.* **64**, 293–300.

Braunschweiger, P. G., and Schiffer, L. M. (1980b). Cell kinetic-directed sequential chemotherapy with cyclophosphamide and Adriamycin in T1699 mammary tumors. *Cancer Res.* **40**, 737–743.

Brouckaert, P. G. G., Lereoux-Roels, G. G., Guisez, Y., Tavernier, J., and Fiers, W. (1986). *In vivo* anti-tumor activity of recombinant human and murine TNF, alone and in combination with murine IFN, on a syngeneic murine melanoma. *Int. J. Cancer* **38**, 763–769.

Browman, G. P., Archibald, S. D., Young, J. E. M., Hyrniuk, M. N., Russell, R., Kiehl, K., and Levine, M. N. (1983). Prospective randomized trial of one-hour sequential versus simultaneous methotrexate plus 5-fluorouracil in advanced and recurrent squamous cell head and neck cancer. *J. Clin. Oncol.* **1**, 787–792.

Brown, P. O., and Cozzarelli, N. R. (1981). Catenation and knotting of duplex DNA by type 1 topoisomerases: A mechanistic parallel with type 2 topoisomerases. *Proc. Natl. Acad. Sci. U.S.A.* **77**, 2936–2940.

Brunda, M. J., Bellantoni, D., and Sulich, V. (1987). *In vivo* antitumor activity of combinations of interferon alpha and interleukin-2 in a murine model. Correlation of efficacy with the induction of cytotoxic cells resembling natural killer cells. *Int. J. Cancer* **40**, 365–371.

Bücher, T., Hiddemann, W., Koenigsmann, M., Žühlsdorf, M., Wörmann, B., Boeckmann, A., Maschmeyes, G., Ludwig, W., and Schulz, G. (1989). Chemotherapy followed by recombinant human granulocyte macrophage colony stimulating factor (GM-CSF) for acute leukemias at higher age or after relapse. *Proc. Am. Soc. Clin. Oncol.* **8**, 198.

Cadman, E., Heimer, R., and Benz, C. (1981). The influence of methotrexate pretreatment on 5-fluorouracil metabolism in L1210 cells. *J. Biol Chem.* **256**, 1695–1704.

Cavanaugh, P. F., Jr., Pavelic, Z. P., and Porter. C. W. (1984). Enhancement of 1,3-bis(2-chloroethyl)-1-nitrosourea-induced cytotoxicity and DNA damage by α-difluoromethylornithine in L1210 leukemia cells. *Cancer Res.* **44**, 3856–3861.

Chang, B. K. (1983). Differential sensitivity of pancreatic adenocarcinoma cell lines to chemotherapeutic agents in culture. *Cancer Treat. Rep.* **67**, 355–361.

Chang, B. K., and Gregory, J. A. (1985). Comparison of the cellular pharmacology of doxorubicin in resistant and sensitive models of pancreatic cancer. *Cancer Chemother. Pharmacol.* **14**, 132–134.

Chang, B. K., Black, O., Jr., and Gutman, R. (1984). Inhibition of growth of human or hamster pancreatic cancer cell line by α-difluoromethylornithine alone and combined with cis-diamminedichloroplatinum (II). *Cancer Res.* **44**, 5100–5104.

Chang, B. K., Gutman, R., and Black, O., Jr., (1986). Combined effects of α-difluoromethylornithine and doxorubicin against pancreatic cancer cell lines in culture. *Pancreas* **1**, 49–54.

Chang, B. K., Gutman, R., and Chou, T.-C. (1987). Schedule-dependent interaction of α-difluoromethylornithine and cis-diamminedichloroplatinum (II) against human and hamster pancreatic cancer cell lines. *Cancer Res.* **46**, 2247–2250.

Chang, B. K., Brenner, D. E., and Gutman, R. (1989). Dissociation of the verapamil-induced enhancement of doxorubicin's cytotoxicity from changes in cellular accumulation or retention of doxorubicin in pancreatic cancer cell lines. *Anticancer Res.* **9**, 341–346.

Chauffert, B., Martin, F., Caignard, A., Jeannin, J.-F., and Leclerc, A. (1984). Cytofluorescence localization of Adriamycin in resistant colon cancer cells. *Cancer Chemother. Pharmacol.* **13**, 14–18.

Chello, P. L., and Sirotnak, F. M. (1981). Icreased schedule-dependent synergism of vindesine versus vincristine in combination with methotrexate against L1210 leukemia. *Cancer Treat. Rep.* **65**, 1049–1053.

Chello, P. L., Sirotnak, F. M., Dorrick, D. M., and Moccio, D. N., (1979). Schedule-dependent synergism of methotrexate and vincristine against murine L1210 leukemia. *Cancer Treat. Rep.* **63**, 1889–1894.

Chen, G. L., and Liu, L. F. (1986). DNA topoisomerases as therapeutic targets in cancer chemotherapy. *Annu. Rep. Med. Chem.* **21**, 257–262.

Chervinsky, D. S., and Wang, J. J. (1976). Uptake of Adriamycin and daunomycin in L1210 and human leukemia cells: A comparative study. *J. Med.* **7**, 63–79.

Chew, C. Y. C., Hecht, H. S., Collett, J. T., McAllister, R. G., and Singh, B. N. (1981). Influence of severity of ventricular dysfunction on hemodynamic responses to intravenously administered verapamil in ischemic heart disease. *Am. J. Cardiol.* **47**, 917–922.

Chou, J., Chou, T.-C. (1986). "Dose–effect Analysis with Microcomputers," computer software for Apple II and IBM-PC series. Elsevier–Biosoft, Cambridge, England.

Chou, J., and Chou, T.-C. (1989). "Dose–effect Analysis with Microcomputers," computer software for IBM-PC DOS 3.3. Elsevier–Biosoft, Cambridge England.

Chou, T.-C., and Talalay, P. (1984). Quantitative analysis of dose–effect relationships: The combined effects of multiple drugs or enzyme inhibitors. *Adv. Enzyme Regul.* **22**, 27–55.

Chou, T.-C., and Talalay, P. (1987). Applications of the median–effect principle for the assessment of low-dose risk of carcinogens and for the quantitation of synergism and antagonism of chemotherapeutic agents. *In* "New Avenues in Development Chemotherapy" (K. R. Harrap and T. A. Connor, eds.), Bristol-Myers Symp. Ser., pp. 37–64. Academic Press, Orlando, Florida.

Chouaib, S., Bertoglio, J., Blay, J.-Y., Marchiol-Fournigault, C., and Fradelizi, D. (1988). Generation of lymphokine-activated killer cells: Synergy between tumor necrosis factor and interleukin 2. *Proc. Natl. Acad. Sci. U.S.A.* **85**, 6875–6879.

Clark, S. C., and Kamen, R. (1987). The human hematopoietic colony-stimulating factors. *Science* **236**, 1229–1237.

Coates, A. S., Tattersall. M. H. N., Swanson, C., Hedley, D., Fox, R. M., Raghavan, A., and the staff of the Combined Head and Neck Clinic, Royal Prince Alfred Hospital (1984). Combination therapy with methotrexate and 5-fluorouracil: A prospective randomized clinical trial of order of administration. *J. Clin. Oncol.* **2**, 756–761.

Curt, G. A., Clendeninn, N. J., and Chabner, B. A. (1984). Drug resistance in cancer. *Cancer Treat. Rep.* **68**, 87–99.

Dalton, W. S., Grogan, T. M., Meltzer, P. S., Scheper, R. J., Durie, B. G. M., Taylor, C. W., Miller, T. P., and Salmon, S. E. (1989). Drug-resistance in multiple myeloma and non-Hodgkin's lymphoma: Detection of P-glycoprotein and potential circumvention by addition of verapamil to chemotherapy. *J. Clin. Oncol.* **7**, 415–424.

Dave, C., and Cabelles, L. (1973). Studies on the uptake of methylglyoxal bis(guanylhydrazone) (CH_3-G) and spermidine in mouse leukemia sensitive and resistant to CH_3-G. *Fed. Proc., Fed. Am. Soc. Exp. Biol.* **32**, 736.

Day, R. S. (1986). Treatment sequencing, asymmetry, and uncertainty: Protocol strategies for combination chemotherapy. *Cancer Res.* **46**, 3876–3885.

Denekamp, J., and Kallman, R. F. (1973). *In vitro* and *in vivo* labelling of animal tumours with tritiated thymidine. *Cell Tissue Kinet.* **6**, 217–227.

Deuchars, K. L., and Ling, V. (1989). P-Glycoprotein and multidrug resistance in cancer chemotherapy. *Semin. Oncol.* **16**, 156–165.

DeVita, V. T., Serpick, A. A., and Carbone, P. P. (1970). Combination chemotherapy in the treatment of advanced Hodgkin's disease. *Ann. Intern. Med.* **73**, 891–895.

DeVita, V. T., Jr., Young, R. C., and Canellos, G. P. (1975). Combination *versus* single agent chemotherapy: A review of the basis for selection of drug treatment of cancer. *Cancer (Philadelphia)* **35**, 98–110.

Dexter, D. L., and Leith, J. T. (1986). Tumor heterogeneity and drug resistance *J. Clin. Oncol.* **4**, 244–257.

Dolbeare, F., Gratzner, H., Pallavicini, M. G., and Gray, J. W. (1983). Flow cytometric measurement of total DNA content and incorporated bromodeoxyuridine. *Proc. Natl. Acad. Sci. U.S.A.* **80**, 5573–5577.

Doll, D. C., and Weiss, R. B. (1983). Chemotherapeutic agents and the erythron. *Cancer Treat. Rev.* **10**, 185–200.

Donelli, M. G., Barbieri, B., Erba, E., Pacciarini, M. A., Salmona, A., Garattini, S., and Morasca, L. (1979). *In vitro* uptake and cytotoxicity of Adriamycin in primary and metastatic Lewis lung carcinoma. *Eur. J. Cancer* **15**, 1121–1129.

Drewinko, B., Loo, T. L., Brown, B., Gottlieb, J. A., and Freireich , E. J. (1976). Combination chemotherapy *in vitro* with Adriamycin. Observations of additive, antagonistic, and synergistic effects when used in two-drug combinations on cultured human lymphoma cells. *Cancer Biochem. Biophys.* **1**, 187–195.

Evans, M., Laskin, J. D., and Hakala, M. T. (1980). Assessment of growth limiting events caused by 5-fluorouracil in mouse cells and in human cells. *Cancer Res.* **40**, 4113–4122.

Evans, R. M., Laskin, J. D., and Hakala, M. T. (1981). Effect of excess folates and deoxyinosine on the activity and site of action of 5-fluorouracil. *Cancer Res.* **41**, 3288–3295.

Ferrari, S., Narni, F., Mars, W., Kaczmarek, L., Venturelli, D., Anderson, B., and Calabretta, B. (1986). Expression of growth-regulated genes in human acute leukemia. *Cancer Res.* 5162–5166.

Fuerstein, B. G., Pattabiraman, N. and Marton, L. J. (1986). Spermine-DNA interactions: A theoretical study. *Proc. Natl. Acad. Sci. U.S.A.* **83**, 5948–5952.

Frishman, W., Kirsten E., Klein, M., Pine, M., Johnson, S. M., Hillis, L. D., Packer, M., and Kates, R. (1982). Clinical relevance of verapamil plasma levels in stable angina pectoris. *Am. J. Cardiol.* **50**, 1180–1184.

Gasson, J. C., Weisbart, R. H., Kaufman, S. E., Clark, S. C., Hewick, R. M., Wong, G. G., and Golde, D. W. (1984). Purified human granulocyte-macrophage colony-stimulating factor: Direct action on neutrophils. *Science* **226**, 1339–1342.

Ganapathi, R., and Grabowski, D. (1983). Enhancement of sensitivity to Adriamycin in resistant P388 leukemia by the calmodulin inhibitor trifluoperazine. *Cancer Res.* **43**, 3696–3699.

Gem J. L. (1988). 5-Fluorouracil plus leucovorin in cancer therapy. *Updates: Cancer: Princ. Pract. Oncol.* **2**, 1–12.

Gerwitz, A. M., and Cadman, E. (1981). Preliminary report on the efficacy of sequential methotrexate and 5-fluorouracil in advanced breast cancer. *Cancer* **47**, 2552–2555.

Glisson, B. S., and Ross, W. E. (1987). DNA topoisomerase II: A primer on the enzyme and its unique role as a multidrug target in cancer chemotherapy. *Pharmacol. Ther.* **32**, 89–106.

Goldin, A. (1984). Dosing and sequencing for antineoplastic synergism in combination chemotherapy. *Cancer (Philadelphia)* **54**, 1155–1159.

Goodman, G. E., Yen, Y. P., Cox, T. C., and Crowley, J. (1987). Effect of verapamil on *in vitro* cytotoxicity of Adriamycin and vinblastine in human tumor cells. *Cancer Res.* **218**, 474–2304.

Gratzner, H. G. (1982). Monoclonal antibody of 5-bromo- and 5-iododeocyuridine: A new reagent for detection of DNA replication. *Science* **218**, 474–475.

Griffin, C., Abeloff, M., Slavik, M., Thompson, G., Luk, G., Hermann, J., Blanc, O., Sjoerdsma, A., and Baylin, S. (1984). Phase I trial and pharmacokinetic study of iv and high dose oral α-difluoromethylornithine (DFMO). *Proc. Am. Soc. Clin. Oncol.* **3**, 34.

Harding, M. J., Kayer, S. B., Soukop, M., and Ferguson, J. C. (1988). A Phase II study of sequential methotrexate and 5-fluorouracil in colorectal cancer. *Med. Oncol. Tumor Pharmacother.* **5**, 239–241.

Harker, W. G., Bauer, D., Etiz, B. B., Newman, R. A., and Sikic, B. I. (1986). Verapamil-mediated sensitization of doxorubicin-selected pleiotropic resistance in human sarcoma cells. Selectivity for drugs which produce DNA scission. *Cancer Res.* **46**, 2369–2373.

Hayes, F. A., and Mauer, A. M. (1976). Cell kinetics and chemotherapy in neuroblastoma. *JNCI, J. Natl. Cancer Inst.* **57**, 697–699.

Hayers, F. A., Green, A. A., and Mauer, A. M. (1988). Correlation of cell kinetic and clinical response to chemotherapy in disseminated neuroblastoma. *Cancer Res.* **37**, 3766–3770.

Hermann, R., Spehn, J., Beyer, J. H., von Franqué, U., Schmieder, A., Holzmann, K., and Abel, U., for the Arbeitsgemeinschaft Internistische Onkologie (1984). Sequential methotrexate and 5-fluorouracil: Improved response rate in metastatic colorectal cancer. *J. Clin. Oncol.* **2**, 591–594.

Herr, H. W., Kleinert, E. L., Conti, P. S., Burchenal, J. H., and Whitmore, W. F., Jr. (1984a). Effect of α-difluoromethylornithine and methylglyoxal bis(guanylhydrazone) on the growth of experimental renal adenocarcinoma in mice. *Cancer Res.* **44**, 4382–4385.

Herr, H. W., Kleinert, E. L., Relyea, N. M., and Whitmore W. F., Jr. (1984b). Potentiation of methylglyoxal bis(guanylhydrazone) by α-difluoromethylornithine in rate prostate cancer. *Cancer (Philadelphia)* **53**, 1294–1298.

Heston, W. D. W., Kadmon, D., and Fair, W. R. (1982). Growth inhibition of a prostate tumor by α-difluoromethylornithine and cyclophosphamide. *Cancer Lett.* **16**, 71–79.

Heston, W. D. W., Kadmon, D., Covey, D. F., and Fair, W. R. (1984). Differential effect of α-difluoromethylornithine on the *in vivo* uptake of ^{14}C-labeled polyamines and methylglyoxal bis(guanylhydrazone) by a rat prostate-derived tumor. *Cancer Res.* **44**, 1034–1040.

Hines, J. D., Zakem, M. H., Adelstein, D. J., and Rustum, Y. M. (1988). Treatment of advanced-stage colorectal adenocarcinoma with fluorouracil and high-dose leucovorin calcium: A pilot study. *J. Clin. Oncol.* **6**, 142–146.

Howell, S. B., and Goel, R. (1988). Cytostatic modifiers: Current status and considerations for the future. *Acta Chir. Scand., Suppl.* **541**, 22–31.

Hsiang, Y.-H., Wu, H-Y., and Liu, L. F. (1988). Proliferation-dependent regulation of DNA topoisomerase II in cultured human cells. *Cancer Res.* **48**, 3230–3235.

Hung, D. T., Deen, D. F., Seidenfeld, J., and Marton, L. J. (1981). Sensitization of 9L rat brain gliosarcoma cell to 1,3-bis(2-chloroethyl)-1-citrosourea by α-difluoromethylornithine, an ornithine decarboxylase inhibitor. *Cancer Res.* **41**, 2783–2785.

Inaba, M., and Maruyama. E. (1988). Reversal of resistance to vincristine in P388 leukemia by various polycyclic clinical drugs, with a special emphasis on quinacrine. *Cancer Res.* **48**, 2064–2067.

Issell, B., and Crooke, S. T. (1979). Etoposide (VP-16-213). *Cancer Treat. Rev.* **6**, 107–124.

Jackson, D. V., Jr., Long, T. R., Trahey, T. F., and Morgan, T. M (1984). Synergistic antitumor activity of vincristine and VP-16-213. *Cancer Chemother. Pharmacol.* **13**, 176–180.

Jackson, D. V., Jr., Long, T. R., Rice, D. G., and Morgan, T. M. (1986). Combination vincristine and VP-16-213: Evaluation of drug sequence. *Cancer Biochem. Biophys.* **8**, 265–275.

Jänne, J., Alhonen-Hongisto, L., Sëppänen, P., and Siimes, M. (1981). Use of polyamine antimetabolites in experimental tumors and in human leukemia. *Med. Biol.* **59**, 448–457.

Käpyaho, K., Kallio, A., and Jänne, J. (1984). Differential effects of α-difluoromethylornithine and methylglyoxal bis(guanylhydrazone) on the testosterone-induced growth of the ventral prostate and seminal vesicles of castrated rats. *Biochem. J.* **219**, 811–817.

Kessel, D., and Wilberding, C. (1985a). Anthracycline resistance in P388 murine leukemia and its circumvention calcium antagonists. *Cancer Res.* **45**, 3360–3365.

Kessel, D., and Wilberding, C. (1985b). Promotion of daunorubicin uptake and toxicity by the calcium antagonist tiapamil and its analogs. *Cancer Treat. Rep.* **69**, 673–676.

Kingsnorth, A. N., Russell, W. E., McCann, P. P., Diekema, K. A., and Malt, R. A. (1983). Effects of α-difluoromethylornithine and 5-fluorouracil on the proliferation of a human colon adenocarcinoma cell line. *Cancer Res.* **43**, 4035–4038.

Kramer, D. L., Paul, B., and Porter, C. W. (1985). Effect of pretreatment with α-difluoromethylornithine on the selectivity of methylgyloxal bis(guanylhydrazone) for tumor tissue in L1210 leukemic mice. *Cancer Res.* **45**, 2512–2515.

Kreuzer, K. N., and Cozzarelli, N. R. (1980). Formation and resolution of DNA catenanes by DNA gyrase. *Cell* **20**, 245–254.

Krishan, A. (1975). Rapid flow cytofluorometric analysis of mammalian cell cycle by propidium iodide staining. *J. Cell Biol.* **66**, 188–193.

Lampkin, B. C., Nagao, T., and Mauer, A. M. (1970). Synchronization recruitment in acute leukemia. *J. Clin. Invest.* **50**, 2204–2214.

Laufman, L. R., Krzeczowski, K. A., Roach, R., and Segal, M. (1987). Leucovorin plus 5-fluorouracil: An effective treatment for metastatic colon cancer. *J. Clin. Oncol.* **5**, 1394–1400.

Liu, L. F., Rowe, T. C., Yang, L., Tewey, K. M., and Chen, G. L. (1983). Cleavage of DNA of mammalian DNA topoisomerase II. *J. Biol. Chem.* **258**, 15365–15370.

Louie, K. G., Hamilton, T. C., Winker, M. A., Behrens, B. C., Tsuruo, T., Klecker, R. W., Jr., McKoy, W. M., Grotzinger, K. R., Myers, C. E., Young, R. C., and Ozols, R. F. (1986). Adriamycin accumulation and metabolism in Adriamycin-sensitive and -resistant human ovarian cancer cell lines. *Biochem. Pharmacol.* **35**, 467–472.

Ludwig, R., Alberts, D. S., Miller, T. P., and Salmon, S. E. (1984). Evaluation of anticancer drug schedule dependency using an *in vitro* human tumor clonogenic assay. *Cancer Chemother. Pharmacol.* **12**, 135–141.

Luk, G. D., Goodwin, G., Marton, L. J., and Baylin, S. B. (1981). Polyamines are necessary for the survival of human small cell lung carcinoma in culture *Proc. Natl. Acad. Sci. U.S.A.* **78** 2355–2358.

Luk, G. D., Goodwin, G., Gazdar, A. F., and Baylin, S. B., (1982). Growth-inhibitory effects of DL-α-difluoromethylornithine in the spectrum of human lung carcinoma cells in culture. *Cancer Res.* **42**, 3070–3073.

Maddox, A.-M., Freireich, E. J., Keating, M. J., and Haddox, M. K. (1988). Alterations in bone marrow and blood mononuclear cell polyamine and methylglyoxal bis(guanylhydrazone) levels: Phase I evaluation of α-difluoromethylornithine and methylglyoxal bis(guanylhydrazone) treatment of human hematological malignancies. *Cancer Res.* **48**, 1367–1373.

Marsh, J. C. (1976). The effects of cancer chemotherapeutic agents in normal hematopoietic precursor cells: A review. *Cancer Res.* **36**, 1853–1882.

Mauer, A. M., Murphy, S. B., and Hayes, F. A. (1976). Evidence for recruitment and synchronization in leukemia and solid tumors. *Cancer Treat. Rep.* **60**, 1841–1844.

McIntosh, J. R., and Koonce, M. P. (1989). Mitosis. *Science* **246**, 622–628.

Metcalf, B. W., Bey, P., Danzin, C., Jung, M. J., Casara, P., and Vevert, J. P. (1978). Catalytic irreversible inhibition of mammalian ornithine decarboxylase (E. C. 4.1.1.17) by substrate and product analogues. *J. Am. Chem. Soc.* **100**, 2511–2553.

Meyer, J. S., and Bauer, W. C. (1975). *In vitro* determination of tritiated thymidine labeling index (LI). Evaluation of a method utilizing hyperbaric oxygen and observations on the LI of human mammary carcinoma. *Cancer (Philadelphia)* **36**, 1374–1380.

Mini, E., Moroson, B. A., and Bertino, J. R. (1987). Cytotoxicity of floxuridine and 5-fluorouracil in human T-lymphoblast leukemia cells: Enhancement by leucovorin. *Cancer Treat. Rep.* **71**, 381–389.

Morstyn, G., Stuart-Harris, R., Bishop, J., Raghavan, D., Kefford, R., Lieschke, G., Bonnem, E., Olver, I., Green, M., Rallings, M., and Fox, R. (1989). Optimal scheduling of granulocyte macrophage colony-stimulating factor (GM-CSF) for the abrogation of chemotherapy induced neutropenia in small cell lung cancer (SCLC). *Proc. Am. Soc. Clin. Oncol.* **8**, 219.

Nelson, E. M., Tewey, K. M., and Liu, L. F. (1984). Mechanism of antitumor drug action: Poisoning of mammalian DNA topoisomerase II on DNA by 4'-(9-acridinylamino)-methanesulfon-*m*-anisidide. *Proc. Natl. Acad. Sci. U.S.A.* **81** 1361–1365.

Nelson, W. G., Cho, K. R., Hsiang, Y.-H., Liu, L. F., and Coffey, D. S. (1987). Growth-related elevations of DNA topoisomerase II Levels found in Dunning R3327 rat prostatic adenocarcinomas. *Cancer Res.* **47**, 3246–3250.

O'Brien, T. G. (1976). The induction of ornithine decarboxylase as an early, possible obligatory, event in mouse skin carcinogenesis. *Cancer Res.* **36**, 2644–2653.

Oredsson, S. M., Deen, D. F., and Marton, L. J. (1982). Decreased cytotoxicity of cis-diamminedichloroplatinum (II) by α-difluoromethylornithine depletion of polyamines in 9L rat brain tumor cells *in vitro. Cancer Res.* **42**, 1296–1299.

Oredsson, S. M., Gray, J W., Deen, D. F., and Marton, L. J. (1983). Decreased cytotoxicity of 1-β-D-arabinofuranosylcytosine in 9L rat brain tumor cells pretreated with α-difluoromethylornithine *in vitro. Cancer Res.* **43**, 2541–2544.

Oredsson, S. M., Deen, D. F., and Marton, L. J. (1984). Influence of polyamine depletion caused by α-difluoromethylornithine, an enzyme-activated irreversible inhibitor of ornithine decarboxylase, on alkylation- and carbamolylation-induced cytotoxicity in 9L rat brain tumor cells *in vitro. Cancer Res.* **43**, 4606–4609.

Owellen, R. J., Hartke, C. A., Dickerson, R. M., and Hains, F. O. (1976). Inhibition of tubulin-microtubule polymerization by drugs of the vinca alkaloid class. *Cancer Res.* **36**, 1499–1502.

Pardee, A. B. (1989). G_1 events and regulation of cell proliferation. *Science* **246**, 603–608.

Pegg, A. E., and McCann, P. P. (1982). Polyamine metabolism and function. *Am. J. Physiol.* **243**, C212–C221.

Petrelli, N., Herrera, L., Rustum, Y., Burke, P., Creaven, P., Stulc, J., Emrich, L. J., and Mittelman, A. (1987). A prospective randomized trial of 5-fluorouracil and methotrexate in previously untreated patients with advance colorectal carcinoma. *J. Clin. Oncol.* **5,** 1559–1565.

Porter, C. W., and Jänne, J. (1987). Modulation of antineoplastic drug action by inhibitors of polyamine biosynthesis. *In* "Inhibition of Polyamine Metabolism: Biological Significance and Basis for New Therapies" (P. P. McCann, A. E. Pegg, and A. Sjoerdsma, eds.), pp. 203–248. Academic Press, Orlando, Florida.

Porter, C. W., and Sufrin, J. R. (1986). Interference with polyamine biosynthesis and/or function by analogs of polyamines or methionine as a potential anticancer chemotherapeutic strategy. *Anticancer Res.* **6,** 525–542.

Porter, C. W., Dworaczyk, D., Gains, B., and Weiser, M. M. (1980). Polyamines and biosynthetic enzymes in the rat intestinal mucosa and the influence of methylglyoxal bis(guanylhydrazone). *Cancer Res.* **40,** 2330–2335.

Poso, H., and Pegg, A. E. (1982). Effect of α-difluoromethylornithine on polyamine and DNA synthesis in regenerating rat liver. Reversal of inhibition of DNA synthesis by putrescine. *Biochim. Biophys. Acta* **696,** 179–186.

Poste, G. (1982). Cellular heterogeneity in malignant neoplasms and the therapy of metastases. *Ann. N.Y. Acad. Sci.* **397,** 34–48.

Prakash, N. J., and Sunkara, P. S. (1983). Combination chemotherapy involving α-difluoromethylornithine and 1-β-D-arabinofuranosylcytosine in murine L1210 leukemia. *Cancer Res.* **43,** 3192–3196.

Ramu, A., Shan, T.-C., and Glaubiger, D. (1983). Enhancement of doxorubicin and vinblastine sensitivity in anthracycline-resistant P388 cells. *Cancer Treat. Rep.* **67,** 895–899.

Ramu, A., Spanier, R., Rahamimoff, H., and Fuks, Z. (1984). Restoration of doxorubicin responsiveness in doxorubicin-resistant P388 murine leukemia cells. *Br. J. Cancer* **50,** 501–507.

Ringborg, U., Ewert, G., Kinnman, J., Lundquist, P., and Strander, H. (1983). Sequential methotrexate–5-fluorouracil treatment of squamous cell carcinoma of the head and neck. *Cancer (Philadelphia)* **52,** 971–973.

Rogan, A. M., Hamilton, T. C., Young, R. C., Klecker, R. W., and Ozols, R. F. (1984). Reversal of Adriamycin resistance by verapamil in human ovarian cancer. *Science* **224,** 994–996.

Rosenberg, S. A. (1988). Cancer therapy with interleukin-2: Immunologic manipulations can mediate regression of cancer in humans. *J. Clin. Oncol.* **6,** 403–406.

Ross, W., Rowe, T., Glisson, B., Yalowich, J., and Liu, L. (1984). Role of topoisomerase II in mediating epipodophyllotoxin-induced DNA cleavage. *Cancer Res.* **44,** 5857–5860.

Santi, D. V., McHenry, C. S., and Sommer, A. (1974). Mechanism of interaction of thymidylate synthetase with 5-fluorodeoxyuridylate. *Biochemistry* **13,** 471–480.

Sartorelli, A. A. (1969). Some approaches to the therapeutic exploitation of metabolic sites of vulnerability of neoplastic cells. *Cancer Res.* **29,** 2292–2299.

Scalabrino, G., Poso, H., Hölttä, E., Hannonen, P., Kallio, A., and Jänne, J. (1978). Synthesis and accumulation of polyamines in rat liver during chemical carcinogenesis. *Int. J. Cancer* **21,** 239–245.

Seidenfeld, J., and Komar, K. A. (1985). Chemosensitization of cultured human carcinoma cells to 1,3-bis(2-chloroethyl)-1-nitrosourea by difluoromethylornithine-induced polyamine depletion. *Cancer Res.* **45,** 2132–2138.

Seidenfeld, J., Gray, J. W., and Marton, L. J. (1981). Depletion of 9L rat brain tumor cell polyamine content by treatment with D,L-α-difluoromethylornithine inhibits proliferation and the G_1 to S transition. *Exp. Cell Res.* **131,** 209–216.

Seidenfeld, J., Komar, K. A., Naujokas, M. F., and Block, A. l. (1986). Reduced cytocidal efficacy for Adriamycin in cultured human carcinoma cells depleted of polyamines by difluoromethylornithine treatment. *Cancer Res.* **46,** 1155–1159.

Seppänen, P., Alhonen-Hongisto, L., and Jänne, J. (1981). Polyamine deprivation-induced enhanced uptake of methyglyoxal bis(guanylhydrazone) by tumor cells. *Biochim. Biophys. Acta* **674,** 169–177.

Sëppänen, P., Alhonen-Hongisto, L., and Jänne, J. (1983). Combined use of α-difluoromethylornithine and methylglyoxal bis(guanylhydrazone) in normal and leukemia-bearing mice. *Cancer Lett.* **18**, 1–10.

Siimes, M., Sëppänen, P., Alhonen-Hongisto, L., and Jänne, J. (1981). Synergistic action of two polyamine antimetabolites leads to a rapid therapeutic success in childhood leukemia. *Int. J. Cancer* **28**, 567–570.

Silver, R. T., Young, R. C., and Holland, J. F. (1977). Some new aspects of modern cancer chemotherapy. *Am. J. Med.* **63**, 772–787.

Silvestrini, R., Lenaz, L., DiFronzo, G. and Sanfillippo, O. (1973). Correlations between cytotoxicity, biochemical effects, drug levels and therapeutic effectiveness of daunomycin and Adriamycin on sarcoma 180 ascites in mice. *Cancer Res.* **33**, 2954–2958.

Simpson-Herren, L. (1982). Kinetic perturbations during cancer therapy. *Ann. N.Y. Acad. Sci.* **397**, 88–100.

Sirotnak, F. M., Donsbach, R. C., Dorick, D. M., and Moccio, D. M. (1976). Tissue pharmacokinetics, inhibition of DNA synthesis and tumor cell kill after high-dose methotrexate in murine tumor models. *Cancer Res.* **36**, 4672–4678.

Sirotnak, F. M., Schmid, F. A., Temple, C., Jr., and Montgomery, J. A. (1983). Optimum scheduling during combination chemotherapy of murine leukemia: Additional examples of schedule-dependent synergism between S-phase-specific antimetabolites and agents inducing mitotic or pre-mitotic (G_2) arrest. *Cancer Chemother. Pharmacol.* **11**, 205–207.

Slater, L. M., Murray, S. L., Wetzel, M. W., Wisdom, R. M., and DuVall, E. M. (1982). Verapamil restoration of daunorubicin responsiveness in daunorubicin-resistant Ehrlich ascites carcinoma. *J. Clin. Invest.* **70**, 1131–1134.

Steel, G. G., and Hanes, S. (1971). The technique of labelled mitoses: Analysis by automatic curve-fitting. *Cell Tissue Kinet.* **4**, 505–518.

Sullivan, D. M., Glisson, B. S., Hodges, P. K., Smallwood-Dentro, S., and Ross, W. E. (1986). Proliferative dependence of topoisomerase II mediated drug action. *Biochemistry* **25**, 2248–2256.

Sunkara, P. S., Rao, P. N., Nishioka, K., and Brinkley, B. R. (1979). Role of polyamines in cytokinesis of mammalian cells. *Exp. Cell Res.* **119**, 63–68.

Sunkara, P. S., Fowler, S. K., Nishioka, K., and Rao, P. N. (1980). Inhibition of polyamine biosynthesis by α-difluoromethylornithine potentiates the cytotoxic effects of arabinosyl cytosine in HeLa cells. *Biochem. Biophys. Res. Commun.* **95**, 423–430.

Sunkara, P. S., Fowler, S. K., and Nishioka, K. (1981). Selective killing of transformed cells in combination with inhibitors of polyamine biosynthesis and S-phase specific drugs. *Cell Biol. Int. Rep.* **10**, 991–997.

Takano, S., Matsushima, M., Erturk, E., and Bryan, G. T. (1981). Early induction of rat colonic epithelial ornithine and S-adenosyl-L-methionine decarboxylase activities by N-methyl-N'-nitrosoguanidine or bile salts. *Cancer Res.* **41**, 624–628.

Tannock, I. (1978). Cell kinetics and chemotherapy: A critical review. *Cancer Treat. Rep.* **62**, 1117–1133.

Tannock, I. F. (1989). Principles of cell proliferation: Cell kinetics. *In* "Cancer: Principles and Practice of Oncology" (V. T. DeVita, Jr., S. Hellman, and S. A. Rosenberg, eds.), 3rd Ed., pp. 3–13. Lippincott, Philadelphia, Pennsylvania.

Theriault, R. L., Frye, D., Fraschini, G., Vadhan, S., (1989). Continuous infusion (CI) vinblastine (VLB) and recombinant human granulocyte macrophage colony-stimulating factor (GM-CSF) in metastatic breast cancer patients. *Proc. Am. Soc. Clin. Oncol.* **8**, 44.

Tofilon, P. J., Deen, D. F., and Marton, L. J. (1983). α-Difluoromethylornithine-induced polyamine depletion of 9L tumor cells modifies drug-induced DNA cross-link formation. *Science* **222**, 1132–1135.

Tsuruo, T., Iida, H., Tsukagoshi, S., and Sakurai, Y. (1981). Overcoming vincristine resistance in P388 leukemia *in vitro* and *in vivo* by calcium influx blockers. *Cancer Res.* **41**, 1967–1972.

Tsuruo, T., Idia, H., Yamashiro, M., Tsukagoshi, S., and Sakurai, Y. (1982a). Enhancement of vincristine- and Adriamycin-induced cytotoxicity by verapamil in P388 leukemia and its sublines resistant to vincristine and Adriamycin. *Biochem. Pharmacol.* **31**, 3138–3140.

Tsuruo, T., Iida, H., Tsukagoshi, S., and Sakurai, Y. (1982b). Increased accumulation of vincristine and Adriamycin in drug-resistant P388 tumor cells following incubation with calcium antagonists and calmodulin inhibitors. *Cancer Res.* **42**, 4730–4733.

Tsuruo, T., Iida, H., Norjiri, M., Tsukagoshi, S., and Sakurai, Y. (1983). Circumvention of vincristine and Adriamycin resistance *in vitro* and *in vivo* by calcium influx blockers. *Cancer Res.* **43**, 2905–2910.

Tsuruo, T., Iida, H., Tsukagoshi, S., and Sakurai, Y. (1985). Cure of mice bearing P388 leukemia by vincristine in combination with a calcium channel blocker. *Cancer Treat. Rep.* **69**, 523–525.

Tyrer, D. D., Kline, I., Venditti, J. M., and Goldin, A. (1967). Separate and sequential chemotherapy of mouse leukemia L1210 with 1-β-D-arabinofuranosylcytosine hydrochloride and 1,3-bis(2-chloroethyl)-1-nitrosourea. *Cancer Res.* **27**, 873–878.

Vadhan-Raj, S., Fenoglio, C., Chawla, S. P., Broxmeyer, H. E., *et al.* (1989). Effects of recombinant human granulocyte macrophage colony-stimulating factor (GM-CSF) on chemotherapy-induced myelosuppression in patients with sarcoma. *Proc. Am. Soc. Clin. Oncol.* **8**, 322.

Vadlamudi, S., and Goldin, A. (1971). Influence of mitotic cycle inhibitors on the antileukemic activity of cytosine arabinoside (NSC-63878) in mice bearing leukemia L1210. *Cancer Chemother. Rep.* **55**, 547–555.

Van Putten, L. M., Keizer, H. J., and Mulder, J. H. (1976). Perspectives in cancer research: Synchronization in tumour chemotherapy. *Eur. J. Cancer* **12**, 79–85.

Venditti, J. M., Humphreys, S. R., Mantel, N., and Goldin, A. (1956). Combined treatment of advanced leukemia (L1210) in mice with amethopterin and 6-mercaptopurine. *JNCI, J. Natl. Cancer Inst.* **17**, 631–638.

Warrell, R. P., Jr., Coonley, C. J., and Burchenal, J. H. (1983). Sequential inhibition of polyamine synthesis: A phase I trial of DFMO (α-difluoromethylornithine) and methyl-GAG [methylglyoxal bis(guanylhydrazone)]. *Cancer Chemother. Pharmacol.* **11**, 134–136.

Winkelhake, J. L., Stampfl, S., and Zimmerman, R. J. (1987). Synergistic effects of combination therapy with human recombinant interleukin-2 and tumor necrosis factor in murine tumor models. *Cancer Res.* **47**, 3948–3953.

Yalowich, J. C., and Ross, W. E. (1984). Potentiation of etoposide-induced DNA damage by calcium antagonists in L1210 cells *in vitro*. *Cancer Res.* **44**, 3360–3365.

Yalowich, J. C., and Ross, W. E. (1985). Verapamil-induced augmentation of etoposide accumulation in L1210 cells *in vitro*. *Cancer Res.* **45**, 1651–1656.

Zimmerman, R. J., Gauny, S., Chan, A., Landre, P., and Winkelhake, J. L. (1989). Sequence dependence of administration of human recombinant tumor necrosis factor and interleukin-2 in murine tumor therapy. *JNCI, J. Natl. Cancer Inst.* **81**, 227–231.

Zwelling, L. A., Michaels, S., Erickson, L. C., Ungerleider, R. S., Nichols, M., and Kohn, K. W. (1981). Protein-associated deoxyribonucleic acid strand breaks in L1210 cells treated with the deoxyribonucleic acid intercalating agents 4′-(9-acridinylamino)methanesulfon-*m*-anisidide and Adriamycin. *Biochemistry* **20**, 6553–6563.

Zwelling, L. A., Kerrigan, D., and Marton, L. J. (1985). Effect of difluoromethylornithine, an inhibitor of polyamine biosynthesis, on the topoisomerase II-mediated DNA scission produced by 4′-(9-acridinylamino)methanesulfon-*m*-anisidide in L1210 murine leukemia cells. *Cancer Res.* **45**, 1122–1126.

Zwelling, L. A., Estey, E., Silberman, L., Doyle, S., and Hittelman, W. (1987). Effect of cell proliferation and chromatin conformation on intercalator-induced, protein-associated DNA cleavage in human brain tumor cells and human fibroblasts. *Cancer Res.* **47**, 251–257.

CHAPTER 18

Effects of Drug Distribution and Cellular Microenvironment on the Interaction of Cancer Chemotherapeutic Agents

Ralph E. Durand

Medical Biophysics Unit
British Columbia Cancer Research Centre
Vancouver, British Columbia
Canada V5Z 1L3

I. Overview

Choosing drug combinations and schedules which produce synergy and avoid antagonism would seem to be obvious goals for chemotherapy of solid

SYNERGISM AND ANTAGONISM IN CHEMOTHERAPY
Copyright © 1991 by Academic Press, Inc.
All rights of reproduction in any form reserved.
659

tumors, yet little current research directly addresses these issues. Instead, emphasis is generally placed on the more immediate problems of adequacy of drug delivery, amelioration of systemic toxicity, and overcoming the emergence of drug resistance. Since each of these questions ultimately impacts on drug efficacy, this chapter begins with a review of the unique features of solid tumor architecture and cellular heterogeneity which complicate drug delivery and assessment of response to single or multiple agents. We then describe techniques which permit quantitative evaluation of drug action and interaction in cell subpopulations of solid tumor models. Finally, examples of synergy between common cancer chemotherapeutic agents in a tumor model system are presented, and the potential relevance of these to human cancer chemotherapy is critically examined.

II. Introduction

A. CANCER CHEMOTHERAPY— A UNIQUE CHALLENGE

Modern cancer chemotherapy has evolved to the stage where most types of human malignancies initially will respond to treatment, where response is defined as a reduction in tumor growth rate and subsequently in tumor volume. Unfortunately, however, chemotherapy is rarely curative. In many cases, tumors which regrow after (or during) the initial therapy are capable of continued growth even during additional chemotherapy treatments. This phenomenon has been given the operational label "drug resistance" and unquestionably is the major obstacle facing cancer chemotherapy today.

By definition, tumor cells most resistant to treatment preferentially survive that treatment. Consequently, understanding mechanisms of action (and, for the purpose of this chapter, *interaction*) of antitumor drugs in turn requires some knowledge of the basis of cellular resistance. Nowell (1976) first drew attention to the genetic diversity of tumor cell subpopulations (tumor heterogeneity), and the impact of resistant cells on tumor chemotherapy, first enunciated by Goldie and Coldman (1983), now dominates protocol design. Numerous other mechanisms of resistance may also occur (Curt *et al.*, 1984), including inadequate drug delivery (Sobrero and Bertino, 1986), cell kinetic changes (Drewinko and Barlogie, 1984), gene amplification (Schimke, 1984; Stark and Wahl, 1984) leading to drug inactivation (Alt *et al.*, 1976) or enhanced drug efflux (Juliano and Ling, 1976), and alterations in cellular sensitivity to, or repair of, drug-induced damage (Kaye and Merry, 1985; de Graeff *et al.*, 1988). Recent studies have focused attention on the role of the tumor cell microenvironment in contributing to (Rice *et al.*, 1986, 1987;

Sutherland, 1988; Young *et al.*, 1988) and possibly overshadowing (Durand, 1986a, 1989a) other mechanisms of drug resistance.

Irrespective of the reasons for drug resistance in human tumors, effective chemotherapy must deal with the problem. Similarly, therapeutic synergism is of little practical significance in sensitive cells. Drug resistance is not, however, an easy problem to study: data showing production or emergence of drug resistant clones in human tumors *in situ* are virtually nonexistent. This is in part due to the complexity of the clinical situation and, additionally, the ethical considerations involved. Consequently, detailed studies can be effectively conducted only in experimental systems.

B. PRACTICAL CONSEQUENCES OF TUMOR HETEROGENEITY

Leaving, for the moment, the question of how heterogeneity arises in a solid tumor, it seems appropriate to first discuss the practical question of how heterogeneity in a tissue will be reflected when cells of that tissue are exposed to a given cytotoxic agent. Two important features of the net response of a system containing heterogeneous cell subpopulations are shown schematically in Fig. 1. For the purposes of this figure, we postulate that a tumor is composed of two pure cell subpopulations, cell types A and B. Suppose as

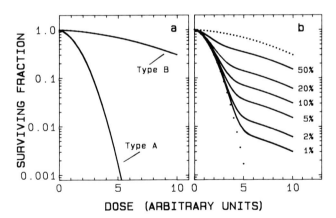

FIGURE 1   A schematic representation of cellular survival as a function of dose of an arbitrary chemotherapeutic agent, for two populations of cells (types A and B) where type A is 5-fold more sensitive to the therapeutic agent. Dose-response curves for the 2 cell types are shown in panel a; the response of a composite population, comprised primarily of sensitive type A cells with the indicated percentages of resistant type B cells is shown in panel b. The dotted curves reproduce the expected survival levels for homogenous populations of types B or A; note that the response of the heterogeneous population always falls between these limits.

well that the response of each to a particular chemotherapeutic agent is as shown in Fig. 1a: 5 times as much drug is required to produce the same amount of cell kill in subpopulation B as in subpopulation A. While noting that a factor of 5 really represents only minimal resistance, it is nonetheless adequate to make an important point when considering heterogeneous tumors.

In Fig. 1b, we illustrate the consequences of changing the proportions of cell types A and B in the tumor on the net response to therapy. We have assumed that the tumor is composed predominantly of sensitive (type A) cells, and the curves show the expected response for increasing fractions of resistant cells also being present. These survival curves for the mixed population (i.e., the net response of the tumor to therapy) should not be a surprise, but two key points may be less obvious. First, the measured response of the heterogeneous tumor, even with only two cell subpopulations, is always intermediate between that of either pure subpopulation (i.e., the full extent of the sensitivity or resistance of the cell populations at those extremes will *never* be fully appreciated if the tumor is assayed as a whole). Second, and even more important, the apparent resistance of the heterogeneous tumor is entirely a function of the dosage of cytotoxic agent used. Stated differently, at low doses (high survival levels), a resistant subpopulation will not be detected unless it is a significant fraction of the overall tumor. Thus, one is always faced with an inherent limitation when assaying a heterogeneous system; at low doses of a toxic agent, the response is primarily determined by the most *sensitive* cells in the population. Only at very high doses is the response determined by the most *resistant* cell subpopulation.

C. TUMOR HETEROGENEITY AND COMBINATION THERAPY

The situation becomes even more complex for multiagent exposures. Consider the tumor that was just described, now treated with two drugs: drug 1 (as just described) and drug 2, which has, for convenience, exactly the opposite effects on the two cell subpopulations (subpopulation B is 5 times more sensitive to drug 2 than is subpopulation A). Functionally, this would be described as a limited degree of non-cross-resistance for subpopulations A and B to the two chemotherapeutic agents. The response of a tumor composed of 95% type A and 5% type B cells to each of the drugs would then be as shown in Fig. 2a. Based on *those* survival curves, one can easily calculate the expected survival for combination treatments with both drugs 1 and 2, using either independent action or additive action models [isobologram analysis, using the terminology defined by Steel (1979)]. These curves are shown in Fig. 2b; the actual response that would be observed, *without invoking any interaction between the drugs,* clearly falls in the supraadditive region at high

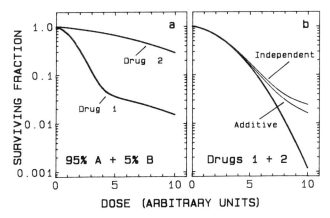

FIGURE 2 A schematic dose response curve for a cell population composed of 95% type A and 5% type B cells (see Figure 1), exposed to drug 1 (as in Figure1) or drug 2, to which the cell subpopulations have an exactly inverse response to that shown in Figure 1 (i.e., type B cells are 5 times more sensitive to drug 2 than are type A cells). The response of the composite population to each drug is shown in panel a; panel b shows the expected response for simultaneous treatment of the mixed population with equal quantities of drugs 1 and 2, without invoking any interaction between the drugs. For reference, the expected curves for independent action and for an additive response (based on the information available in panel a) are also shown.

dose levels. Thus, even a rigorous test for synergy would incorrectly classify drugs 1 and 2 as being strongly synergistic, despite the fact that the greater efficacy shown is entirely due to their specificity of action against the two cell subpopulations.

D. THERAPEUTIC IMPLICATIONS OF SOLID TUMOR ARCHITECTURE

It should be emphasized that the preceding analysis of the effects of multiple drugs in nonhomogeneous cell subpopulations was conducted under the most favourable conditions possible—all cells were presumed to receive identical drug exposures. That assumption may not always hold for cells in tumors treated by systemic drug exposure.

At least in the case of solid tumors, the physical structure (or the 3-dimensional architecture) of the tumor has long been recognized as one of the more useful diagnostic and prognostic features, yet has received little attention in the general field of cancer chemotherapy, or the specific areas of drug resistance and drug interaction. The basic features of tumor induction, neovascularization, and eventual outstripping of vascular expansion are generally well understood. Unlike normal tissues, a significant fraction of solid tumor cells will be displaced from the blood supply by more than 1 or 2 cell diameters. If these are less easily reached by chemotherapeutic agents, a

physiological barrier to effective therapy can be envisaged (Jain, 1989).
Additionally, the changes in nutrient, oxygenation, proliferation, and other
environmental factors which rapidly change as a function of distance from the
blood vessel may impact markedly on the efficacy of chemotherapeutic agents
once they reach their cellular target (Durand, 1986a, 1989a; Sutherland,
1988). A further consideration of obvious importance is the 3-dimensional
cellular interaction within tumors. Similar degrees and consequences of that
intercellular contact can never be duplicated in conventional cell culture
systems (Vitkauskas and Canellakis, 1985).

E. THE INFORMATION REQUIREMENT

Untangling complexities such as those just described has occupied a substantial fraction of our research interests for some years. It is not difficult to
believe that microenvironmental fractors, cell cycle effects, regional differences in drug activation and inactivation, drug delivery problems, intercellular communication, and numerous additional complexities will separately,
and in combination, influence tumor cell responsiveness to systemically administered drugs. Unfortunately, conventional assay techniques are generally
capable of quantifying only net effects, rather than providing data from which
mechanistic information can be deduced. As described in the subsequent
section, this provided the impetus for us to invest a major effort in developing
techniques for recovering and analyzing individual cell subpopulations from
solid tumors and multicell tumor models.

III. Quantitative Techniques for Heterogeneous Systems

A. CHOICE OF MODEL SYSTEMS

Choosing a model system to study the many factors complicating the
response of human tumors to cancer chemotherapeutic agents first requires
that a number of fundamental questions be addressed. Primary among these
is the recognition that a particular model may be adequate for an individual
question, but the only general model, incorporating all of the complexities of
the *in situ* human tumor, is in fact that human tumor. Thus, to study the
inherent responsiveness of human tumor cells to a particular type of drug, one
is constrained to use human tumor cell lines. If, however, one wishes to infer
general principles about a drug's activity, or to study the influence of cellular
microenvironment on the activity of a drug or combination, more flexibility is
available in the choice of model systems.

B. THE *IN VITRO* SPHEROID SYSTEM

Our favourite model system for mechanistic studies is the *in vitro* multicell spheroid. We primarily use the V79-171b Chinese hamster lung fibroblast cell line, which spontaneously forms spheroids, has been extensively characterized (Sutherland and Durand, 1976; Durand, 1976, 1983, 1986a), and grows well in stirred suspension cultures where good reproducibility of nutrient and other growth conditions can be maintained (Sutherland and Durand, 1976; Acker *et al.*, 1984). Mechanistic studies can thus be carried out under rigorously controlled conditions which cannot be achieved in any *in vivo* system (Acker *et al.*, 1984; Nederman and Twentyman, 1984; Durand, 1986a). Spheroids can be grown from virtually all cultured cell lines; some grow spontaneously while others require a much more heroic effort (Acker *et al.*, 1984). While we have yet to find a qualitative difference between the response of V79 and human spheroids to clinically used chemotherapeutic agents, we nonetheless accept the necessity of verifying fundamental observations in both rodent and human tumor spheroids, and have found the WiDR human colon carcinoma cell line to be convenient for the latter studies.

C. DEALING WITH HETEROGENEITY

As should be evident from Sections IIB—IID of this chapter, even the availability of the ultimate model system will be of little consequence for mechanistic studies unless one can deal with heterogeneity in the system. For these reasons, it is of key importance to be able to identify, collect, and separately study cells from different regions of the spheroid (or of solid tumors). This has been approached historically in spheroids by techniques ranging from differential trypsinization (Sutherland and Durand, 1976; Freyer and Sutherland, 1980; Kwok and Twentyman, 1985a,b) to sedimentation on velocity or density gradients (Durand, 1975). More recently, we have introduced a technique of separating cells based on their access to nontoxic fluorescent perfusion probes, using a fluorescence activated cell sorter for the final cell selection step (Durand, 1982, 1986b).

We routinely use the commercially available DNA binding agent, Hoechst 33342, or selected carbocyanine dyes as fluorescent perfusion probes to stain cells differentially on the basis of their distance from the surface of the spheroid, or blood supply of the tumor (Chaplin *et al.*, 1985; Olive and Durand, 1987). Use of these probes relies on the rapid binding of the agent as it diffuses from the stain reservoir (the medium in culture systems or the systemic blood supply in experimental animals) through the cells of interest. The stain must, of course, remain localized in undamaged target cells after the multicell structure is disaggregated into a single cell population (Durand and Olive, 1982; Durand, 1982; Olive *et al.*, 1985). It then follows that sorting cells of given fluorescent intensities permits recovery of subpopulations of cells from equivalent microenvironmental regions (i.e., from regions of the tumor or spheroid which are geometrically and architecturally similar).

Figure 3 schematically summarizes the procedure we have adopted for spheroids. The intact spheroid is stained, washed several times to remove free drug, and rapidly disaggregated to a single cell suspension by trypsinization. As the cells flow through the cytometer, the forward scatter signal is used to identify cells and partially eliminate debris. Additionally, as shown in the bivariate plot in Fig. 3, propidium iodide will stain cells with nonintact

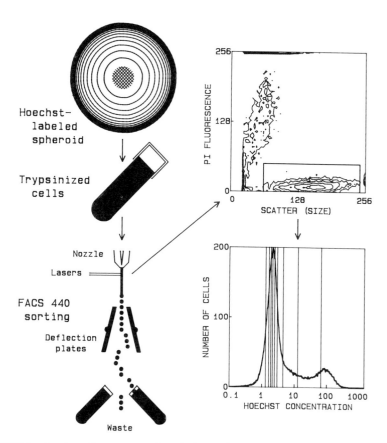

FIGURE 3   A schematic representation of the sorting techniques used with the spheroid system. The intact spheroid is stained with a slowly penetrating, nontoxic fluorescent stain (Hoechst 33342), then reduced to a single cell suspension by trypsinization. When run through the fluorescence activated cell sorter, the cell population is gated on the basis of cell size and, when applicable, a viability stain (as in *top right* panel), and signals from the Hoechst and peripheral light scatter channels are ratioed to provide a Hoechst "concentration" profile, which is then integrated to provide 10 sorting windows such that 10% of the cell population is recovered in each sort.

membranes, so that viable cells can be identified (those within the box in the lower right corner of the distribution). Only those viable cells are sorted. Typically, the Hoechst distribution of those cells is then integrated and fractionated into 10 cell subpopulations as shown in the lower right panel of Fig. 3 (Durand, 1986a,b). Our software completely monitors and controls the actual sorting procedure (Durand, 1986b); sorting is automatically terminated and the operator alerted when the desired number of cells has been collected in each sort fraction.

D. MICROENVIRONMENTAL INFLUENCES

A major advantage of using the *in vitro* spheroid system for mechanistic studies is the option of performing the drug exposure and sorting procedures in either sequence; cell sorting can either precede or follow drug exposure (Durand, 1986a). Thus, cellular response *in situ* (in spheroids) can be compared with that obtained for identical cells dispersed into known conditions of drug concentration and microenvironment. Our studies to date have indicated that, with the exception of the nitrosoureas, the activity of every cancer chemotherapeutic drug evaluated varies significantly as a function of position within the spheroid (Durand and Olive, 1981a; Durand, 1981b, 1986a, 1989a). For example, one of the more complex responses we see is the cytotoxicity profile for cisplatin shown in Fig. 4. When the intact spheroid is exposed to the drug, external cells are killed more effectively than inner cell subpopulations. However, when spheroid disaggregation and cell sorting precedes drug exposure, the same concentration of drug is *less* effective in all subpopulations of the spheroid, and the differential response between the inner and outer cell populations is lost (Durand, 1986a).

Since trypsinizing monolayer cultures prior to cisplatin exposure does not produce similar survival perturbations, we conclude that the differences seen in Fig. 4 are largely a result of the intraspheroid microenvironment increasing the efficiency of the administered cisplatin. While the relative increase in resistance of the inner cells would be consistent with lack of drug delivery (poor penetration), data using radioactive analogs (Durand, 1989a) and data addressing the radiosensitizing capacity of cisplatin (Durand and Vanderbyl, 1988) suggest that the drug diffuses adequately to the central regions, but is relatively less active there due to the unfavorable microenvironmental conditions.

The experimental protocol in which disaggregated cells are subsequently exposed to cisplatin is identical to "*in vitro* chemosensitivity testing" (Hamburger and Salmon, 1977), in which the relative capacity of tumor cells to form colonies in soft agar is evaluated in the presence of selected drugs. As is evident in Fig. 4, the response of an intact spheroid to cisplatin (or most other drugs) cannot be predicted from the average response of the disaggregated

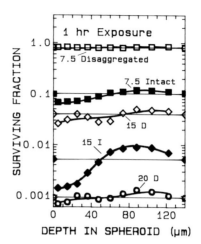

FIGURE 4   Clonogenicity of cells recovered from V79 spheroids after a 1-hr exposure to cisplatin, expressed as a function of cellular location in the spheroid. Two experimental protocols were employed. The open symbols indicate when the sorting step was performed first and the monodispersed single cells were then exposed to the drug for the 1-hr period (D). Closed symbols show the response of cells exposed to cisplatin while in the intact spheroid, and then subsequently separated and sorted (I). Numbers indicate cisplatin concentrations in μg/ml. Note that, under the disaggregated conditions, the cells were neither as responsive to cisplatin nor was a differential survival based on cell location evident. (Redrawn from Durand, 1986a, with permission.)

cells, or even from the individual responses of each cell subpopulation when exposed under disaggregated conditions. It seems exceedingly optimistic to hope that the situation will be more favorable for tumors.

E. DRUG DELIVERY—A PROBLEM?

The capacity of common cancer chemotherapeutic agents to diffuse through the spheroid mass is an obvious question of clinical importance (Sutherland et al., 1979; Deen et al., 1980; West et al., 1980; Wilson et al., 1981; Nederman and Twentyman, 1984; Kwok and Twentyman, 1985a,b). Much of our data addressing this issue has previously been published (Durand and Olive, 1981a, 1987; Durand, 1981b, 1986a,b, 1989a; Durand and Chaplin, 1987; Durand and Vanderbyl, 1988). With the exception of the anthracyclines, we have not observed significant penetration problems with clinically effective cancer chemotherapeutic agents. In view of its activity against a broad spectrum of solid and hematological human tumors, Adriamycin is, however, a compound of particular interest (Kaye and Merry, 1985). The fluorescent nature of that drug also makes it particularly advantageous for simultaneous study of drug distribution and cytotoxicity (Durand,

1981a,b, 1989a; Durand and Olive, 1981b; Luk and Tannock, 1989; Ross *et al.*, 1989).

To illustrate the significance of drug delivery problems, Fig. 5a shows cytotoxicity profiles for Adriamycin when intact spheroids were exposed to the drug, and Fig 5b indicates dose–response curves for selected cell subpopulations. The innermost cells show an apparent drug resistance *at least as large as for MDR-positive cells* (Durand, 1986a, 1989a). Adriamycin cytotoxicity is time, dose, and position dependent in the spheroid (Durand, 1989a). The interrelation among these variables can be seen more clearly in isometric plots like those in Fig. 6, where the top panel shows Adriamycin uptake (intracellular fluorescence) as a function of administered dose, and of cell position in the spheroid. In the lower panel, the biological effect (fraction of cells killed) is similarly displayed. Good correspondence between drug uptake and drug cytotoxicity, like that shown in Fig. 6, leads to the conclusion that drug delivery is a major problem for Adriamycin. Nontheless, other factors (including cell cycle and growth status) also impact markedly on Adriamycin cytotoxicity (Sutherland *et al.*, 1979; Durand, 1981; Durand and Olive, 1981b).

In the context of drug delivery, yet another factor which must be considered is the timing of the uptake or cytotoxicity assay relative to that of delivery of the agent. For the data just reviewed, both the fluorescence intensity profiles and the cytotoxicity profiles were generated *immediately*

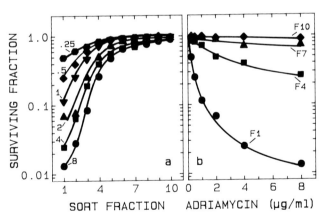

FIGURE 5   Clonogenicity of cells from V79 spheroids following a 2-hr exposure to the indicated concentrations (μg/ml) of Adriamycin. Panel a shows the sensitivity profile of cells to Adriamycin as a function of position in the spheroid (sort fraction 1 is the outermost and sort fraction 10 is the innermost cell subpopulation). Panel b shows the same data points as in panel a, grouped according to selected fraction numbers and thus showing response as a function of dose for the different subpopulations.

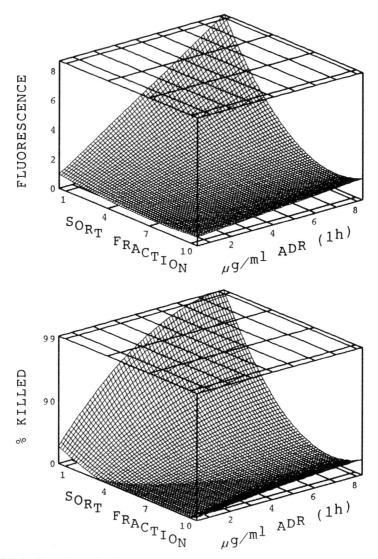

FIGURE 6   Retention and activity of Adriamycin ADR in V79 spheroids, determined with cell sorting techniques to isolate cells as a function of position in the spheroid. Sort fraction 1 represents the outermost 10% of the cells and fraction 10 the innermost 10%. Top: Drug retention based on intracellular fluorescence of Adriamycin. Bottom: Cytotoxicity of Adriamycin as measured by the clonogenicity of the sorted cells. Note the similarity of the panels, in as much as Adriamycin retention and cytotoxicity preferentially occurred in the outermost cells of the spheroids. (Redrawn from Durand, 1989a, with permission.)

after terminating a 2-hr adriamycin exposure. Inherent in this is the assumption that liberating cells from the spheroid has no impact on their ultimate survival. The validity of that assumption should be questioned for a number of reasons (Acker *et al.*, 1984; Olive *et al.*, 1987). Disaggregating damaged cells may inflict further damage. Conversely, repair processes (and particularly, repair of potentially lethal drug damage) may be very different in quiescent cells *in situ* than in those recruited into the proliferating compartment by reversal of the growth constraints inherent in the spheroid (see Acker *et al.*, 1984; Olive *et al.*, 1987).

With Adriamycin, not only does the drug diffuse slowly *into* the spheroid, but it also diffuses *out of* the cells of the intact spheroid very slowly. Thus, if a cell is isolated from the spheroid immediately after exposure and placed in drug-free medium, the intracellular to extracellular drug gradient is clearly much higher than that expected within the intact spheroid (where other drug-containing cells are adjacent to the cell of interest). Consequently, increased cytotoxicity is expected if cells remain in the intact spheroid. Figure 7 confirms this prediction, and also shows that, with increasing time, drug

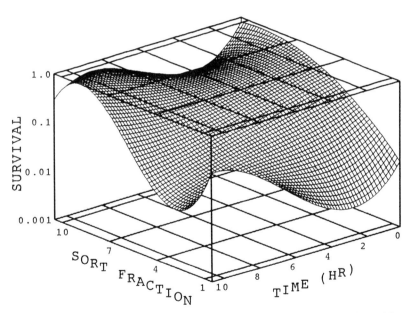

FIGURE 7 Clonogenicity of cells recovered from spheroids exposed to 6 μg/ml Adriamycin for 1 hr, with clonogenicity determined at the indicated times following the termination of the drug exposure. Sort fraction 1 again represents the outermost 10% of the spheroid cells. Note that all cell subpopulations showed increased cytotoxicity if the clonogenicity assay was delayed 2–3 hr following exposure. Subsequent to this, the outermost cells exhibited recovery, presumably due to extrusion of Adriamycin back into the culture medium. Inner cells were not similarly protected.

from the outermost cell subpopulations was presumably released into the medium. Note, however, that cells somewhat deeper into the spheroid (e.g., fractions 3–6) showed a progressive decrease in survival over the 12 hr time period observed; this suggests that cells deeper within the spheroid were unable to eliminate the drug as effectively.

While complex, the time-dependent response of the spheroids to Adriamycin shown in Fig. 7 serves to illustrate that multiple processes and responses are expected when a dynamic multicellular system is perturbed by treatment with cytotoxic agents. In addition to raising questions concerning the relevance of the MDR phenotype and P-glycoprotein expression when cells are surrounded by other drug-containing cells (as opposed to free cells in contact only with the large reservoir of culture medium), these results emphasize the difficulties in assessing drug interactions in complex systems.

F. GENERALIZING THE RESULTS

The methodology and results that we have just described are not specific to the V79 spheroid system. In addition to using other spheroids, including those of human cells, the same techniques can be used in experimental tumors growing in rodents (Chaplin *et al.*, 1985, 1986, 1987; Olive *et al.*, 1985; Loeffler *et al.*, 1987; Siemann and Keng, 1988; Smith *et al.*, 1988; Teicher *et al.*, 1988). Hoechst 33342 (Reinhold and Visser, 1983; Smith *et al.*, 1988) or the carbocyanine dyes (Trotter *et al.*, 1989; Zwi *et al.*, 1989) can also be administered intravenously for analytical purposes; the half-time of these stains in the murine circulation is on the order of only a few minutes (Olive *et al.*, 1985; Trotter *et al.*, 1989), and the stains thus provide a picture of blood flow (tumor perfusion) over a short postinjection time period. For tumor cells which will grow *in vitro*, it is a fairly simple procedure to disaggregate the tumor, run the resulting cell suspension through the fluorescence activated cell sorter, and assess the clonogenicity of the tumor cell subpopulations in a totally analogous way to the spheroids (Chaplin *et al.*, 1985, 1986, 1987; Durand and Chaplin, 1987; Siemann and Keng, 1988).

IV. Interaction of Chemotherapeutic Agents

A. ANALYTICAL TECHNIQUES

Given the technology that we have just described, it is of obvious interest to determine not only the activity of single cancer chemotherapeutic agents, but also their ability to interact with other agents of interest. Quantitative techniques for evaluating drug interactions in homogeneous systems have been thoroughly discussed in the initial chapters of this book; a number of alternatives are available (see also Berenbaum, 1981). Of these, we initially published data using the envelope-isobologram analysis (Steel and Peckham, 1979) in 1980 (Durand and Brown, 1980), and the isobologram equation–combination index analysis (Chou and Talalay, 1984) in 1987 (Durand and Goldie, 1987; Durand and Chaplin, 1987; Durand and Olive, 1987). Clearly, somewhat different answers can be forthcoming from these two types of analysis. In our view, the envelope-isobologram analysis offers two major advantages: it distinguishes among subadditive, additive, and supraadditive effects, and permits the investigator a choice of dose–effect mathematical relationships. A major detraction, however, of the envelope-isobologram analysis is the cumbersome nature of the data display; each isobologram plot can address only one isoeffect level, and provides only a *qualitative* impression as to whether a combination treatment lies in the supraadditive, additive, or subadditive regions of the plot.

In contrast, the combination index model provides a numerical index of interaction for a given set of data if the dose–effect relationship for the data is adequately described by the median-effect principle. A major factor of relevance to complex systems (recall Fig. 2) is the fact that the combination index can easily be expressed as a function of dose or treatment intensity, and therefore is much more amenable to demonstrating dose-dependent changes in interactions than are (multiple) isobologram plots. Further, for data sets of the type just described, where cell location in the tumor is clearly a determinant of response, the combination index can be much more easily generalized into a third dimension for display than can an isobologram with its associated envelope of additivity.

B. CISPLATIN AND ETOPOSIDE

Despite the fact that the authors of numerous clinical papers have speculated on the possible synergistic interactions of anticancer drugs, rigorous documentation of true synergistic effects in the clinic is rare. Further, it is even more unusual to have data which can directly address the role of drug distribution, cellular microenvironment, or other tumor-dependent variables on the interaction of chemotherapeutic agents. The spheroid system and cell sorting techniques just described are, however, amenable to rigorously evaluating both the degree and mechanisms of such interactions.

Cisplatin and etoposide appear to exhibit synergy in clinical combinations (Einhorn, 1986; Evans *et al.*, 1984; Heyden *et al.*, 1987; Klastersky *et al.*, 1982; Lopez *et al.*, 1985; Murray *et al.*, 1986). As shown in Fig. 8, high intensity treatments of the spheroid system produced a true synergistic interaction (combination index < 1). The observed interaction was dependent upon overall treatment cytotoxicity, cellular position within the spheroid, and

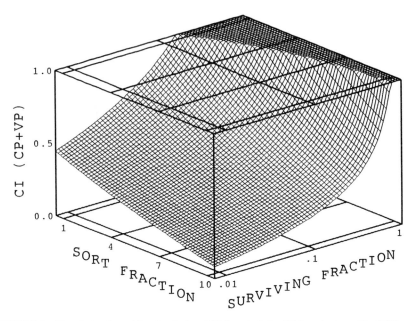

FIGURE 8   The derived combination index (CI) for cisplatin (CP) and etoposide (VP) exposures at a 1:1 ratio, plotted as a function of effect level (surviving fraction) and depth in the spheroid (sort fraction, where fraction 1 is the outermost cells). Values of CI < 1 indicate synergism between the agents. Note that the interaction was more marked in the innermost cells, which were most resistant to each drug when used as a single agent (see Durand and Goldie, 1987).

choice of drug dosages (ratio of cisplatin to etoposide) in the combination treatment (Durand and Goldie, 1987). This laboratory result was of particular interest since both cisplatin and etoposide are preferentially toxic to the external cell subpopulations of the spheroid; the innermost (quiescent) cells are as resistant to etoposide as disaggregated cells expressing the MDR phenotype (Durand and Goldie, 1987). As previously discussed (Section IIC), resistant cells potentially pose a problem for analysis of drug interactions. Although the problem is minimized in the present case where the cytotoxicity profiles of cisplatin and etoposide are qualitatively similar throughout the spheroids, it is still necessary (and considerably more informative) to separately evaluate the interaction between the agents in all resolvable cell subpopulations. Indeed, the potential for developing clinical treatment cycles based on maximizing interactions in chosen cell subpopulations is an exciting extension suggested by these results.

Although the question of whether therapeutic synergy is maximal at a particular dose ratio is important, a combination that is optimal for a given target subpopulation must also show adequate activity against other sub-

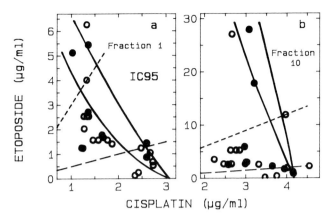

FIGURE 9   Isobologram analysis showing the envelope of additivity of combination treatments with cisplatin and etoposide, for an isoeffect level of 95% cell kill (IC95). Cells from the outermost 10% of the spheroid are shown in panel a, and from the innermost 10% in panel b. Solid circles represent experiments in which complete dose response curves were available for each drug as a single agent, as well as the combination; open circles are from less comprehensive experiments, where a complete dose response curve was generated for the combination treatments, but only 1 or 2 survival points were determined for the single agents to verify that the single agent response was as expected. The broken lines indicate cisplatin to etoposide ratios of 3:1 and 1:2 respectively; note that all points falling below the envelope of additivity were within these ranges.

populations of the tumor. A method of determining maximal synergy is illustrated in Fig. 9, using the envelope-isobologram analysis at the IC_{95} level, rather than the combination index analysis. This choice is due to the practical consideration that the IC_{95} concentration can be estimated from a 2 or 3 point survival curve for the isobologram analysis whereas, for the combination index, a series of dose–response curves at each drug dose ratio is required. The data in Fig. 9 essentially confirm those of Fig. 8, and add another interesting feature. The true synergistic interaction between the drugs, as judged by this somewhat more rigorous analysis, was limited to a fairly small range of effective cisplatin to etoposide dose ratios (1:0.5 to 1:3). It should be noted that this is not unexpected since the cisplatin/etoposide interaction has been suspected in the clinic at similar dose ratios.

A factor of particular interest which emerged in our cisplatin/etoposide studies was the observation that changing the dose ratio of the two active agents could preferentially target the interaction toward different cell subpopulations of the spheroids (Durand and Goldie, 1987). This has the practical consequence of allowing the role of quiescent cells to be clinically evaluated by comparing protocols that should preferentially kill quiescent cells to treatment that should preferentially kill rapidly growing subpopulations. Our observation also suggests the need for a more rigorous evaluation of chemotherapy regimens that selectively target different cell subpopulations, and for the development of treatment combinations with tailored patterns of cytotoxicity and interaction.

C. HYPOXIA-TARGETED INTERACTIONS

An obvious way to attempt to produce tumor-specific interactions (and cytotoxicity) is to use agents which are known to be influenced by location-specific microenvironmental changes. Internal regions of spheroids, like poorly vascularized areas of solid tumors, contain cells at reduced oxygen tensions (Sutherland and Durand, 1976). Use of a drug which is preferentially activated (becomes cytotoxic) under hypoxic conditions should thus lead to a therapeutic gain in solid tumors.

A number of nitroheterocyclics (first studied as radiosensitizers of hypoxic cells) are preferentially toxic to cells under hypoxic conditions (Sutherland, 1974; Olive and McCalla, 1975; Moore et al., 1976; Durand and Olive, 1981a). Unlike the radiosensitizing capacity of those compounds, cytotoxicity

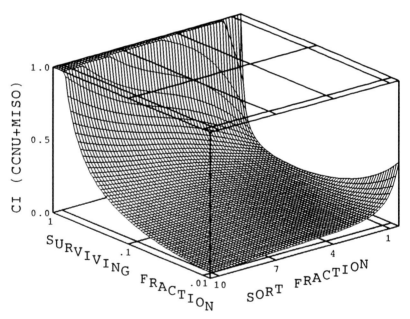

FIGURE 10   The combination index for interaction of misonidazole (MISO) and CCNU in V79 spheroids, plotted as a function of depth in the spheroid (sort fraction 1 represents the outermost 10% of the cells) and survival level produced by the combination treatment. Increasing potentiation is indicated by decreasing values of the combination index. Note the constancy of the interaction with respect to cell position, despite the known variation in oxygen tension throughout the spheroid (see Durand and Chaplin, 1987).

is a result of enzymatic reduction of the drugs, which under aerobic conditions is rapidly reversed by autooxidation, but under hypoxic conditions results in reactive nucleophiles which are cytotoxic. An additional feature of this class of compounds is their ability to increase or "chemopotentiate" the cytotoxicity of a number of common anticancer agents (Kelly *et al.*, 1979; Siemann, 1984; Durand and Olive, 1986, 1987; Durand and Chaplin, 1987). We have had a particular interest in studying chemosensitization in spheroids, because it often seems to be more marked than one might expect based on the size of the hypoxic fraction of cells (Siemann, 1984; Durand and Olive, 1986, 1987; Durand and Chaplin, 1987).

Both dose- and position-dependent synergism between misonidazole and CCNU were seen in V79 spheroids (Fig. 10) and murine tumors (Durand and Chaplin, 1987). Surprisingly, synergism was observed in *both* the oxic and hypoxic subpopulations of the spheroids and tumors. In fact, with relatively nontoxic combinations, greater potentiation was produced in the better-oxygenated cells. From a mechanistic point of view, it is well known from other studies (as well as our own) that hypoxia is required for any interaction between these drugs to occur (reviewed by Siemann, 1984). Our results thus indicate that regional, as opposed to local, hypoxia is adequate to promote the synergistic interaction. Unfortunately, misonidazole (and even some of the newer nitroimidazole radiosensitizers) have rather severe toxicities and are therefore limited in terms of dose that can be administered to a patient.

D. CISPLATIN AND MITOMYCIN C

Mitomycin C is a hypoxia-activated compound which can be used clinically at proven therapeutic doses, unlike the nitroimidazoles. This bioreductive agent produces an active metabolite when reduced under hypoxic conditions, and is consequently about 3-fold more toxic to anaerobic cells than to aerobic cells (Rockwell and Kennedy, 1979; Kennedy *et al.*, 1980, 1981). Based on this observation, which we confirmed in V79 cells growing as monolayers, we characterized the cytotoxicity pattern of mitomycin C in spheroids, and investigated its use in conjunction with cisplatin (Fig. 11). The cytotoxicity profile of mitomycin C through the spheroids was suggestive of a slightly increased cytotoxicity toward the inner (hypoxic) cells (Fig. 11a), but less so than might be predicted on the basis of the cytotoxicity of mitomycin C to anoxic cells in monolayer systems. Recently (R. Durand, unpublished data), we have found that the combination of hypoxia and high cell density may in fact be a severe disadvantage with mitomycin C; the drug is metabolized so rapidly under anaerobic conditions that appreciable quantities of the parent compound apparently cannot penetrate into the innermost regions of the spheroid. Nonetheless, the fact that mitomycin C had an inverse cytotoxicity profile to that of cisplatin suggested that even if the drugs interacted only

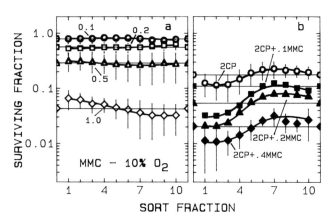

FIGURE 11   Clonogenicity of cells sorted from V79 spheroids following a 30-min equilibration in 10% O_2 + 5% CO_2 + 85% N_2, then a 2-hr exposure to mitomycin C (MMC), cisplatin (CP), or the combination (solid symbols) under the same atmosphere. Sort fraction 1 represents the outermost 10% of the spheroid cells. Panel a shows survival for 10 subpopulations collected from increasing depths into the spheroid after the indicated (μg/ml) mitomycin C exposures; the horizontal lines indicate the average survival for each unsorted population. Panel b shows the "reference" survival level following a 2 μg/ml cisplatin exposure alone, or cisplatin plus increasing concentrations of mitomycin C. Uncertainties indicate 95% confidence limits; in the right hand panel, for clarity, confidence intervals are shown only on the upper and lower curves.

through non-cross-resistance, the combination would still be useful. Further, and of considerable practical interest, the cytotoxicity produced in combination treatments was greater than that expected for independent activity of the agents (Durand, 1989b).

We therefore conducted an extensive series of experiments with different administered dose ratios of cisplatin and mitomycin C, aimed at determining both the degree and location of the apparent synergistic interaction between the drugs. Figure 12 shows that, at a 5:1 dose ratio (which is clinically relevant), mitomycin C significantly potentiated the cisplatin cytotoxicity. Further, the potentiation was greatest in the innermost cell subpopulations which were most resistant to cisplatin as a single agent. With a lower ratio of mitomycin C in the combination less potentiation was observed, and synergism, particularly in the innermost, hypoxic regions of the spheroids, was diminished (Durand, 1989b). These results seem to have two important implications for clinical therapy. First, the ability to evaluate effects of cellular microenvironment on the cytotoxicity of single chemotherapeutic agents adds an additional dimension to the standard clinical definition of non-cross-resistance. Second, the identification of mitomycin C as a drug which can potentiate the activity of cisplatin (a widely used anticancer agent), particularly against cells resistant to cisplatin itself, seems of considerable relevance (Durand, 1989b).

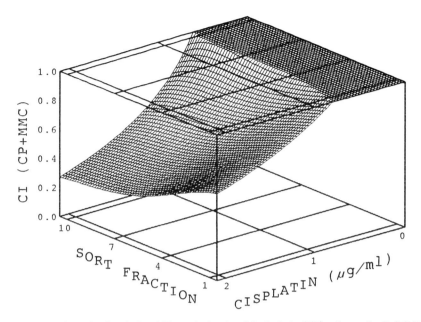

FIGURE 12   Combination index (CI) analysis of a 5:1 cisplatin (CP) mitomycin C (MMC) combination as a function of position in the spheroid and exposure level. Sort fraction 1 is again the outermost 10% of the cells of the spheroid. For ease of comparison with other figures, only the cisplatin dose is shown on the panel scale; 2 μg/ml of cisplatin thus corresponds to 2 μg/ml cisplatin plus 0.4 μg/ml mitomycin C. Interpolation between the sort fractions used a polynomial fitting routine. The region with CI < 1.0 indicates synergism between cisplatin and mitomycin C. For clarity, CI values exceeding unity (antagonism) were truncated to CI = 1.0.

E. CISPLATIN AND CCNU

Another drug combination, also suggested for evaluation based on the complementary cytotoxicity profiles of the two agents, was CCNU and cisplatin. As illustrated in Fig. 13, CCNU exhibits very good activity against the innermost, hypoxic cell subpopulation of the spheroids. For CCNU, the microenvironment at the time of exposure has no effect on outcome since disaggregated single cells from the spheroids respond identically to cells exposed in intact spheroids (Durand, 1990). The increased sensitivity of the internal cells is consistent with other reports that cellular response to the nitrosoureas is primarily determined by position in the cell cycle (Deen *et al.*, 1980). Recent data from Deen's laboratory (Sarkar *et al.*, 1989) has also suggested that chronic hypoxia may play a sensitizing role. Interestingly, the differential in cytotoxicity between the aerobic and hypoxic cells of the spheroids (Fig. 13a) is as great with CCNU as with any of the putatively hypoxic-cell-specific agents we have studied (Durand and Olive, 1981a, 1986,

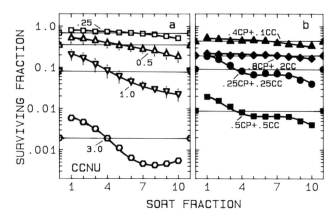

FIGURE 13 Clonogenicity of cells sorted from V79 spheroids (fraction 1 respresents the outermost cells) following a 2-hr exposure to CCNU, cisplatin, or the combination (closed symbols). Panel a shows survival for 10 subpopulations collected from increasing depth into the spheroid after the indicated (μg/ml) CCNU exposure; the horizontal lines indicate the average survival for each unsorted population. Panel b shows selected combination treatments with cisplatin plus CCNU. Uncertainties indicate 95% confidence limits.

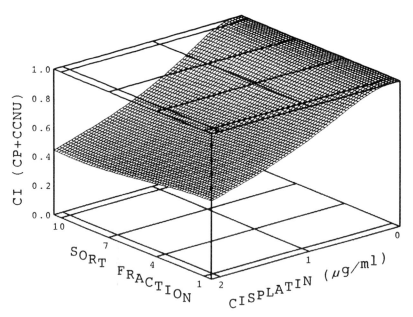

FIGURE 14 Combination index (CI) analysis of a 2:1 cisplatin (CP)/CCNU combination as a function of position in the spheroid (fraction 1 respresents the outermost cells) and exposure level. Again, only the cisplatin dose is shown on the panel scale; 2 μg/ml cisplatin thus corresponds to 2 μg/ml cisplatin plus 1.0 μg/ml CCNU. Interpolation between the sort fractions used a polynomial fitting routine. The region with CI < 1.0 indicates synergism between cisplatin and CCNU. For clarity, CI values exceeding unity (antagonism) were truncated to CI = 1.0.

1987; Durand and Chaplin, 1987; Olive and Durand, 1978; Olive *et al.*, 1987). When CCNU was combined with cisplatin (Fig. 13b) we found both the expected complementary cytotoxicity as a function of depth in the spheroid, and apparent synergy between the drugs.

Using the combination index analysis, Fig. 14 indicates that synergism was indeed seen for CCNU/cisplatin combinations. As with the agents previously described, this synergism was dependent both on total treatment cytotoxicity and on cellular location in the spheroid at the time of treatment. Once again, the cellular microenvironment was clearly influencing both the action and the interaction of these chemotherapeutic agents. As with mitomycin C, the use of CCNU at reduced dose ratios led to a diminished interaction, and a preferential decrease of that interaction in the hypoxic cell region.

V. Additional Considerations

A. SINGLE VERSUS MULTIPLE COURSE THERAPY

The data reviewed in this chapter address the effects of drug distribution and cellular microenvironment on the interaction of cancer chemotherapeutic agents in an *in vitro* test system, Chinese hamster V79 multicell spheroids. All our available results with other *in vitro* spheroid systems and with experimental tumors in rodents qualitatively support those presented here. Mechanistic studies can therefore be undertaken in this simple, user-friendly system. Nonetheless, there are potential dangers in overestimating the relevance and importance of laboratory results for a problem as complex as cancer chemotherapy. We will now discuss two recent experiments to emphasize this point.

It should be quite obvious that the data reviewed here, directed toward understanding mechanisms of drug/drug interaction in complex multicell systems, do not directly relate to the clinical situation of multiple courses of such treatments. The spheroid system and sorting techniques are amenable to such studies, though the "mass culture" aspects of spheroid growth make the system very expensive for multifraction studies. Nonetheless, we have had considerable interest in determining the response of this spheroid system to more clinically relevant multifraction treatments.

With the V79 spheroid system, we are easily able to determine the average number of viable cells present in individual spheroids both before and after each treatment in a multifraction exposure regimen. This allows a separate evaluation of cellular resistance to each treatment, and regrowth (repopulation) between treatments. Figure 15 shows typical multifraction data for cisplatin exposures. We have separated out those two components of the

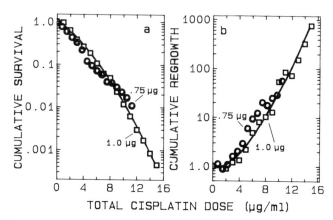

FIGURE 15   Survival and regrowth of cells in V79 spheroids during a multifraction experiment with daily doses of cisplatin (doses in μg/ml for 2 hr/day). The curves are drawn through only the more comprehensive data set, the 1.0 μg/ml data. Drug was administered Monday to Friday, with the weekend off, and the cycle repeated a second and third week. Panel a shows cumulative cytotoxicity, and panel b shows cumulative regrowth (see text for details). Note that following a brief lag, the regrowth was essentially equal and opposite to the cumulative cell kill, producing a quasi-steady-state situation.

response to independently show the cumulative cytotoxicity (Fig. 15a) and cumulative regrowth (Fig. 15b) produced by the multiple treatments.

Two different doses of cisplatin were used: 0.75 μg/ml per exposure or 1.0 μg/ml per exposure. In both cases, the drug was added daily for a 2 hr interval; at the end of the daily exposures, the drug-containing medium was removed by aspiration, the spheroids were washed twice, and fresh growth medium then added. Several assays were performed prior to and immediately following each exposure; of these, the number of cells per spheroid and the fraction of those cells remaining clonogenic were used to construct Fig. 15. The left panel shows the ratio of the post- to pretreatment numbers of viable cells per spheroid, with each point plotted as the product of this ratio and the previous viable fraction of cells. Figure 15b shows the ratio of the number of viable cells per spheroid immediately before treatment to that after the preceding treatment (i.e., the net increase in viable cells per spheroid between exposures), again plotted as a cumulative product. In both cases, drugs were administered daily for 5 days, with the weekend off and the cycle then repeated for a second and third week.

Several conclusions can be deduced from Fig. 15. First, dose rate (or dose intensity) had little effect, since similar responses were seen for the lower and higher doses per cycle. Second, the cumulative cytotoxicity curves (Fig. 15a) are, if anything, concave downward—no resistance to cisplatin emerged during the experiment. Perhaps the most interesting observation is that the induced regrowth (Fig. 15b) essentially compensated for the cell kill, producing a quasi-steady-state situation. The magnitude of the cell turnover during such experiments deserves emphasis. As shown in Fig. 15b, each cell initially present in the spheroid produced as many as 10^3 progeny; with more aggressive treatments in other experiments, as many as 10^6 progeny per starting cell have been followed. Since the spheroids typically contain 10^5 cells at the beginning of such experiments, this means that the fate of as many as 10^{11} cells per spheroid can be followed (not unlike the expected cell numbers in human tumors/metastases under treatment).

B. TIMED SEQUENTIAL TREATMENT

The repopulation capacity seen in cisplatin-treated spheroids suggested that concurrent cisplatin plus etoposide might not be the most efficient treatment in a multifraction regimen. To experimentally evaluate this supposition, we conducted experiments like the one shown in Fig. 16, in which identical total doses of cisplatin and etoposide were administered as twice-weekly doses of the two agents given simultaneously, etoposide followed by cisplatin

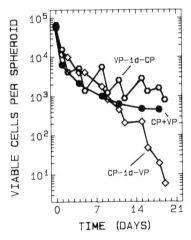

FIGURE 16   The number of viable cells per spheroid as a function of treatment time, for three different treatment protocols using identical doses of cisplatin (CP) and etoposide (VP). For the upper curve, 4 µg/ml etoposide was administered Mondays and Thursdays, with 0.8 µg/ml cisplatin Tuesdays and Fridays (VP-1 day-CP). The middle curve (solid circles) consisted of the same doses administered concurrently Mondays and Thursdays. The lower curve again represents the same drug doses, with cisplatin administered Mondays and Thursdays, and etoposide on Tuesdays and Fridays. Unlike other figures, the substructure of each curve is meaningful within this experiment, and the lines are drawn from point to point to emphasize the progressive changes. Data points show the response immediately following each drug exposure; consequently, for treatments on consecutive days, separate points are plotted with the open symbols, while the solid circles show the results of each simultaneous treatment. Note the marked superiority of cisplatin followed by etoposide.

one day later, or cisplatin followed one day later by etoposide. This protocol thus addressed the question of whether the synergism expected for concurrent administration of the two agents would be more or less effective than giving cisplatin to induce repopulation, and following this a day later by etoposide to kill the cycling cells. The negative control of etoposide followed a day later by cisplatin was also evaluated.

As can be clearly seen in Fig. 16, by far the most effective treatment was twice-weekly administration of cisplatin, with each dose followed 24 hr later by etoposide. Thus, we reached the rather unexpected conclusion that identical total doses of cisplatin and etoposide can produce markedly different results depending upon the dose scheduling used. Further, even dose scheduling designed to exploit proven (molecular) synergism between the agents was not as effective as scheduling based on cell kinetics. These results also emphasize the ambiguity in terminology discussed in Section II: the enhanced response seen for alternating treatments can be shown mathematically to be synergistic, even though no direct (molecular) interaction between the treatment agents occurred.

Results like these suggest that the mechanistic principles learned through single-treatment, single-agent studies may ultimately lead to better results in the clinic. They are, however, unlikely to immediately identify optimal treatments, particularly if clinical responses are partially determined by cell growth kinetics (including recruitment rate) within tumors. While there is still a clear rationale and need for detailed mechanistic information at the cellular and molecular levels, the practical problem of effectively treating human cancers consequently remains one which must also be addressed in the clinic.

Acknowledgments

Supported by USPHS Grant CA-37775, MRC Canada Grant MA-10238, and NCI Canada

References

Acker, H., Carlsson, J., Durand, R., and Sutherland, R. M. (eds.) (1984). Spheroids in Cancer Research. *Recent Results Cancer Res.* **95**.

Alt, R. W., Kellems, R. E., and Schimke, R. T. (1976). Synthesis and degradation of folate reductase in sensitive and methotrexate-resistant lines of S-180 cells. *J. Biol. Chem.* **251**, 3063–3074.

Berenbaum, M. C. (1981). Criteria for analyzing interactions between biologically active agents. *Adv. Cancer Res.* **35**, 269–335.

Chaplin, D. J., Durand, R. E., and Olive, P. L. (1985). Cell selection from a murine tumour using the fluorescent probe Hoechst 33342. *Br. J. Cancer* **51**, 569–572.

Chaplin, D. J., Durand, R. E., and Olive, P. L. (1986). Acute hypoxia in tumors: Implications for modifiers of radiation effects. *Int. J. Radiat. Oncol., Biol. Phys.* **12**, 1279–1282.

Chaplin, D. J., Olive, P. L., and Durand, R. E. (1987). Intermittent blood flow in a murine tumor: Radiobiological effects. *Cancer Res.* **47**, 497–601.

Chou, T.-C., and Talalay, P. (1984). Quantitative analysis of dose–effect relationships: The combined effects of multiple drugs or enzyme inhibitors. *Adv. Enzyme Regul.* **22**, 27–55.

Curt, G. A., Clendennin, N. J., and Chabner, B. A. (1984). Drug resistance in cancer. *Cancer Treat. Rep.* **68**, 87–99.

Deen, D. F., Hoshino, T., Williams, M. E., Muraoka, I., Knebel, K. D., and Barker, M. (1980). Development of a 9L rat brain tumor cell multicellular spheroid system and its response to 1,3-bis(2-chloroethyl)-1-introsourea and radiation. *JNCI, J. Natl. Cancer Inst.* **64**, 1373–1382.

de Graeff, A., Slebos, R. J. C., and Rodenhuis, S. (1988). Resistance to cisplatin and analogues: Mechanisms and potential clinical implications. *Cancer Chemother Pharmacol.* **22**, 325–332.

Drewinko, B., and Barlogie, B. (1984). Cell cycle perturbation effects. *In* "Antitumor Drug Resistance" (B. W. Fox and M. Fox, eds.), pp. 101–141. Springer-Verlag, Berlin.

Durand, R. E. (1975). Isolation of cell subpopulations from *in vitro* tumor models according to sedimentation velocity. *Cancer Res.* **35**, 1295–1300.

Durand, R. E. (1976). Cell cycle kinetics in an *in vitro* tumor model. *Cell Tissue Kinet.* **9**, 403–412.

Durand, R. E. (1981a). Calibration of flow cytometry detector systems. *Cytometry* **2**, 192–193.

Durand, R. E. (1981b). Flow cytometry studies of intracellular Adriamycin in multicell spheroids *in vitro*. *Cancer Res.* **41**, 3495–3498.

Durand, R. E. (1982). Use of Hoechst 33342 for cell selection from multicell systems. *J. Histochem. Cytochem.* **30**, 117–122.

Durand, R. E. (1983). Oxygen enhancement ratio in V79 spheroids. *Radiat. Res.* **96**, 322–334.

Durand, R. E. (1986a). Chemosensitivity testing in V79 spheroids: Role of drug delivery and cellular microenvironment. *JNCI, J. Natl. Cancer Inst.* **7**, 247–252.

Durand, R. E. (1986b). Use of a cell sorter for assays of cell clonogenicity. *Cancer Res.* **46**, 2775–2778.

Durand, R. E. (1989a). Distribution and activity of antineoplastic drugs in a tumor model. *JNCI, J. Natl. Cancer Inst.* **81**, 146–152.

Durand, R. E. (1989b). Synergism of cisplatin and mitomycin C in sensitive and resistant cell subpopulations of a tumor model. *Int. J. Cancer.* **44**, 911–917.

Durand, R. E. (1990). Cisplatin and CCNU synergism in spheroid cell subpopulations. *Br. J. Cancer*, in press.

Durand, R. E., and Brown, S. M. (1980). Effects of lucanthone on Chinese hamster V79 cells. I. Interaction with radiation in monolayers and spheroids. *Int. J. Radiat. Oncol., Biol. Phys.* **6**, 1525–1530.

Durand, R. E., and Chaplin, D. J. (1987). Chemosensitization by misonidazole in CCNU-treated spheroids and tumors. *Br. J. Cancer* **56**, 103–109.

Durand, R. E., and Goldie, J. H. (1987). Interaction of etoposide and cisplatin in an *in vitro* tumor model. *Cancer Treat. Rep.* **71**, 673–679.

Durand, R. E., and Olive, P. L. (1981a). Evaluation of nitroheterocyclic radiosensitizers using spheroids. *Adv. Radiat. Biol.* **9**, 75–104.

Durand, R. E., and Olive, P. L. (1981b). Flow cytometry studies of intracellular Adriamycin in single cells *in vitro*. *Cancer Res.* **41**, 3489–3494.

Durand, R. E., and Olive, P. L. (1982). Cytotoxicity, mutagenicity and DNA damage by Hoechst 33342. *J. Histochem. Cytochem.* **30**, 111–116.

Durand, R. E., and Olive, P. L. (1986). Potentiation of CCNU toxicity by AF-2 in V79 spheroids: Implications for mechanisms of chemosensitization. *Int. J. Radiat. Oncol., Biol. Phys.* **12**, 1375–1378.

Durand, R. E., and Olive, P. L. (1987). Enhancement of toxicity from *N*-(2-chloroethyl)-*N'*-cyclohexyl-N-nitrosourea in V79 spheroids by a nitrofuran. *Cancer Res.* **47**, 5303–5309.

Durand, R. E., and Vanderbyl, S. L. (1988). Response of cell subpopulations in spheroids to radiation–drug combinations. *Natl. Cancer Inst. Monogr.* **6**, 95–100.

Einhorn, L. (1986). Initial therapy with cisplatin plus VP-16 in small-cell lung cancer. *Semin. Oncol.* **13**, 5–9.

Evans, W. K., Feld, R., Osoba, D., Shepherd, F. A., Dill, J., and Deboer, G. (1984). VP-16 alone and in combination with cisplatin in previously treated patients with small cell lung cancer. *Cancer* **53**, 1461–1466.

Freyer, J. P., and Sutherland, R. M. (1980). Selective dissociation and characterization of cells from different regions of multicell tumor spheroids. *Cancer Res.* **40**, 3956–3965.

Goldie, J. H., and Coldman, A. J. (1983). Quantitative model for multiple levels of drug resistance in clinical tumors. *Cancer Treat. Rep.* **67**, 923–931.

Hamburger, A. W., and Salmon, S. E. (1977). Primary bioassay of human tumor stem cells. *Science* **197**, 461–463.

Heyden, V., Scherpe, H. W., and Nagel, G. A. (1987). cis-Dichlorodiamine-platinum(II) (cis-platinum) and etoposide for patients with refractory lymphomas. *Cancer Treat. Rep.* **9**, 45–52.

Jain, R. K. (1989). Delivery of novel therapeutic agents in tumors: Physiological barriers and strategies. *JNCI, J. Natl. Cancer Inst.* **81**, 570–576.

Juliano, R. L., and Ling, V. (1976). A surface glycoprotein modulating drug permeability in Chinese hamster ovary cell mutants. *Biochim. Biophys. Acta* **455**, 152–162.

Kaye, S., and Merry, S. (1985). Tumor cell resistance to anthracyclines—A review. *Cancer Chemother. Pharmacol.* **14**, 96–103.

Kelly, J. P., Hannam, T. W., and Biles, G. R. (1979). The cytocidal action of metronidazole in combination with other antineoplastic agents. *Cancer Treat. Rev.* **6**, 53–61.

Kennedy, K. A., Rockwell, S., and Sartorelli, A. C. (1980). Preferential activation of mitomycin C to cytotoxic metabolites by hypoxic tumor cells. *Cancer Res.* **40**, 2356–2360.

Kennedy, K. A., Teicher, B. A., Rockwell, S., and Sartorelli, A. C. (1981). Chemotherapeutic approaches to cell populations of tumors. *In* "Molecular Actions and Targets for Cancer Chemotherapeutic Agents" (A. C. Sartorelli, J. S. Layo, and J. R. Bertino, eds.), pp. 85–101. Academic Press, New York.

Klastersky, J., Longeval, E., Nicaise, C., and Weerts, D. (1982). Etoposide and cis-platinum in non-small-cell bronchongenic carcinoma. *Cancer Treat. Rev.* **9**, 133–138.

Kwok, T. T., and Twentyman, P. R. (1985a). The relationship between tumor geometry and the response of tumor cells to cytotoxic drugs—An *in vitro* study using EMT6 multicellular spheroids. *Int. J. Cancer* **35**, 675–682.

Kwok, T. T., and Twentyman, P. R. (1985b). The response to cytotoxic drugs of EMT6 cells treated either as intact or disaggregated spheroids. *Br. J. Cancer* **51**, 211–218.

Loeffler, D. A., Keng, P. C., Wilson, K. M., and Lord, E. M. (1987). Comparison of fluorescence intensity of Hoechst 33342-stained EMT6 tumour cells and tumour-infiltrating host cells. *Br. J. Cancer* **56**, 571–576.

Lopez, J. A., Mann, J., Grapski, R. T., Nassif, E., Vanilcola, P., Krikorian, J. G., and Finkel. H. (1985). Etoposide and cisplatin salvage chemotherapy for small cell lung cancer. *Cancer Treat. Rep.* **69**, 369–371.

Luk, C. K., and Tannock, I. F. (1989). Flow cytometric analysis of doxorubicin accumulation in cells from human and rodent cell line. *JNCI, J. Natl. Cancer Inst.* **81**, 55–58.

Moore, B. A., Palcic, B., and Skarsgard, L. D. (1976). Radiosensitizing and toxic effects of the 2-nitroimidazole Ro-07-0582 in hypoxic mammalian cells. *Radiat. Res.* **67**, 459–473.

Murray, N., Shah, A., Brown, E., Kostashuk, E., Laukkanen, E., Goldie, J. H., Band, P., Hoek J. V., Murphy, K., Sparling, T., and Noble, M. (1986). Alternating chemotherapy and thoracic radiotherapy with concurrent cisplatin-etoposide for limited-stage small-cell carcinoma of the lung. *Semin. Oncol.* **13**, 24–30.

Nederman, T., and Twentyman, P. (1984). Spheroids for studies of drug effects. *Recent Results Cancer Res.* **95**, 84–102.

Nowell, P. C. (1976). The clonal evolution of tumor cell populations. *Science* **194**, 23–28.

Olive, P. L., and Durand, R. E. (1978). Activation of radiosensitizers by hypoxic cells. *Br. J. Cancer* **37**, 124–128.

Olive, P. L., and Durand, R. E. (1987). Characterization of a carbocyanine derivative as a fluorescent penetration probe. *Cytometry* **8**, 571–575.

Olive, P. L., and McCalla, D. R. (1975). Damage to mammalian cell DNA by nitrofurans. *Cancer Res.* **35**, 781–784.

Olive, P. L., Chaplin, D. J., and Durand, R. E. (1985). Pharmacokinetics, binding and distribution of Hoechst 33342 in spheroids and murine tumours. *Br. J. Cancer* **52**, 739–746.

Olive, P. L., Durand, R. E., and Chaplin, D. J. (1987). Cytotoxicity of RSU 1069 in spheroids and murine tumors. *Int. J. Radiat. Oncol., Biol. Phys.* **13**, 1361–1366.

Reinhold, H. S., and Visser, J. W. M. (1983). *In vivo* fluorescence of endothelial cell nuclei stained with the bis-benzamide H 33342. *Int. J. Microcirculation Clin. Exp.* **2**, 143–146.

Rice, G. C. Hoy, C., and Schimke, R. T. (1986). Transient hypoxia enhances the frequency of dihydrofolate reductase gene amplification in Chinese hamster ovary cells. *Proc. Natl. Acad. Sci. U.S.A.* **85**, 5978–5982.

Rice, G. C., Ling, V., and Schimke, R. T. (1987). Frequencies of independent and simultaneous selection of Chinese hamster cells for methotrexate and doxorubicin (Adriamycin) resistance. *Proc. Natl. Acad. Sci. U.S.A.* **84**, 9261–9264.

Rockwell, S., and Kennedy, K. A. (1979). Combination therapy with radiation and mitomycin C. Preliminary results with EMT6 tumor cells *in vitro* and *in vivo. Int. J. Radiat. Oncol., Biol. Phys.* **5**, 1673–1676.

Ross, D. D., Thompson, B. W., Ordonex, J. V., and Joneckis, C. C. (1989). Improvement of flow-cytometric detection of multidrug-resistant cells by cell-volume normalization of intracellular daunorubicin content. *Cytometry* **10**, 185–191.

Sarkar, A., Ho, J. T., Marton, L. J., and Deen, D. F. (1989). Comparison of BCNU-induced cell killing and DNA damage in oxic and hypoxic 9L cells. *Proc. Radiat. Res. Soc. Annu. Meet., 37th* p. 119.

Schimke, R. T. (1984). Gene amplification in cultured animal cells. *Cell* **37**, 705–713.

Siemann, D. W. (1984). Modification of chemotherapy by nitroimidazoles. *Int. J. Radiat. Oncol., Biol. Phys.* **10**, 1585–1594.

Siemann, D. W., and Keng, P. C. (1988). Characterization of radiation resistant hypoxic cell subpopulations in KHT sarcomas. (ii) Cell sorting. *Br. J. Cancer* **58**, 296–300.

Smith, K. A., Hill, S. A., Begg, A. C., and Denekamp, J. (1988). Validation of the fluorescent dye Hoechst 33342 as a vascular space marker in tumours. *Br. J. Cancer* **57**, 247–253.

Sobrero, A., and Bertino, J. R. (1986). Clinical, aspects of drug resistance. *Cancer Surv.* **5**, 93–107.

Stark, G. R., and Wahl, G. M. (1984). Gene amplification. *Annu. Rev. Biochem.* **53**, 447–491.

Steel, G. G. (1979). Terminology in the description of drug–radiation interactions. *Int. J. Radiat. Oncol., Biol. Phys.* **5**, 1145–1150.

Steel, G. G., and Peckham, M. J. (1979). Exploitable mechanisms in combined radiotherapy–chemotherapy: The concept of additivity. *Int. J. Radiat. Oncol., Biol. Phys.* **5**, 85–91.

Sutherland, R. M. (1974). Selective chemotherapy of noncycling cells in an *in vitro* tumor model. *Cancer Res.* **34**, 3501–3503.

Sutherland, R. M. (1988). Cell and environment interactions in tumor microregions: The multicell spheroid model. *Science* **240**, 177–184.

Sutherland, R. M., and Durand, R. E. (1976). Radiation response of multicell spheroids—An *in vitro* tumour model. *Curr. Top. Radiat. Res. Q.* **11**, 87–139.

Sutherland, R. M., Eddy, H. A., Bareham, B., Reich, K., and Vanantwerp, D. (1979). Resistance to Adriamycin in multicellular spheroids. *Int. J. Radiat. Oncol., Biol. Phys.* **5**, 1225–1230.

Teicher, B. A., Herman, T. S., and Holden, S. A. (1988). Combined modality therapy with bleomycin, hyperthermia and radiation. *Cancer Res.* **48**, 6291–6297.

Trotter, M. J., Chaplin, D. J., Durand, R. E., and Olive, P. L. (1989). The use of fluorescent probes to identify regions of transient perfusion in murine tumors. *Int. J. Radiat. Oncol., Biol. Phys.* **16,** 931–934.

Vitkauskas, G. V., and Canellakis, E. S. (1984). Intercellular communication and cancer chemotherapy. *Biochim. Biophys. Acta* **823,** 19–34.

West, G. W., Weichselbaum, R., and Little, J. B. (1980). Limited penetration of methotrexate into human osteosarcoma spheroids as a proposed model for solid tumor resistance to adjuvant chemotherapy. *Cancer Res* **40,** 3665–3668.

Wilson, W. R., Whitmore, G. F., and Hill, R. P. (1981). Activity of 4'-(9-acridinylamino)-methanesulfon-*m*-anisidide against Chinese hamster cells in multicellular spheroids. *Cancer Res.* **41,** 2817–2822.

Young, S. D., Marshall, R. S., and Hill, R. P. (1988). Hypoxia induces DNA over-replication and enhances metastatic potential of murine tumor cells. *Proc. Natl. Acad. Sci. U.S.A.* **85,** 9533–9537.

Zwl, L. J., Baguley, B. D., Gavin, J. B., and Wilson, W. R. (1989). Blood flow as a major determinant in the antitumor action of flavone acetic acid. *JNCI, J. Natl. Cancer Inst.* **81,** 1005–1013.

CHAPTER 19

Synergistic Interactions at the Solid Tumor Level through Targeting of Therapies against Two or More Different Tumor Cell Subpopulations

Dietmar W. Siemann
Peter C. Keng
Experimental Therapeutics Division
University of Rochester Cancer Center
Rochester, New York 14642

I. Introduction

Conventional studies evaluating treatment synergism or antagonism typically are directed toward assessing the combination of multiple agents against a single cell population. This is particularly the case for *in vitro* investigations of the interaction between anticancer agents. However, tumors are not uniform cell populations. Rather, a single tumor may comprise a variety of cell subpopulations including, for example, infiltrating host cells, nutritionally deprived cells, and cells in various phases of the cell cycle and proliferation

SYNERGISM AND ANTAGONISM IN CHEMOTHERAPY
Copyright © 1991 by Academic Press, Inc.
All rights of reproduction in any form reserved.

states. Although such intratumor heterogeneity has long been implicated as a major determinant in the response of tumors to radiation or chemotherapy, the specific role of these subpopulations in the overall tumor response to *in situ* therapy has been difficult to assess. This is a consequence primarily of an inability to isolate directly from solid tumors cells constituting the various subpopulations (Tannock, 1978; Meistrich *et al.*, 1977a; Grdina *et al.*, 1978, 1979; Howell and Koch, 1980). Yet direct *in vivo* assessment of therapies may be essential since it has already been demonstrated that cells may respond differently to treatment if grown *in vitro* or *in vivo* (Sutherland *et al.*, 1979; Martin and McNally, 1980; Keng *et al.*, 1984; Siemann, 1989). The central theme of this chapter is the study of treatment interactions in terms of overall tumor response through effects on different cell subpopulations. This approach considers the concept of synergism in terms of one or more therapies aimed at multiple cell populations, i.e., a multitarget approach. With this approach, we will explore the possibility of using combinations of therapies directed at different treatment-refractory cell subpopulations in order to improve the overall success of the treatment.

II. Background on Cell Subpopulations

Solid tumors are composed of multiple subpopulations of cells with diverse morphologic, functional, biochemical, and biophysical properties. The cellular heterogeneity of neoplasms was first documented more than a century ago from the morphologic and histochemical studies of different cells within the same tumor. Since then, the study of tumor biology has revealed remarkable levels of phenotypic heterogeneity among subpopulations of cells from the same primary tumors, from primary versus metastatic tumors of the same host, and in different metastases arising from the same primary tumor (Cifone and Fidler, 1981; Fidler and Hart, 1982a; Poste *et al.*, 1981; Tsuruo and Fidler, 1981). Analysis of the cellular heterogeneity and the mechanisms by which cellular diversity is induced and controlled has become a major interest in contemporary cancer research (Fidler and Hart, 1982b; Nicolson and Poste, 1983; Welch *et al.*, 1983; Hill *et al.*, 1984). The awareness of the tumor cell heterogeneity as a major obstacle to effective therapy has also generated a number of new concepts and approaches in cancer research (Welch *et al.*, 1983; Hill *et al.*, 1984; Siemann and Keng, 1988b).

Because tumors are cellularly diverse, architecturally complex, and composed of both normal and malignant cells, a major difficulty encountered in such studies has been the isolation and purification of the various types of cells from the heterogeneous cell mass. Modern cell separation techniques have provided many advantages for studying the cell subpopulations prepared from solid tumors. These subpopulations include cells which differ in cell cycle and growth kinetics, proliferative potential, differentiation status, drug or radiation sensitivities, immunological reactivities, and metastatic propensities. Studies of the biological significance of these cell subpopulations may have important implications for (1) understanding tumor regression or progression, (2) developing treatment regimens, and (3) predicting therapeutic responses.

III. Characterization of Tumor Cell Subpopulations

Several cell separation procedures that developed for studies such as those described earlier have been reviewed in detail (Pretlow and Pretlow, 1982; Keng *et al.*, 1987). Methods employed for this purpose have included (1) isopycnic density sedimentation, (2) unit gravity sedimentation, (3) isokinetic gradient separation, (4) procedures combining multiple methodologies, (5) surface affinity binding, (6) fluorescence-activated cell sorting, and (7) centrifugal elutriation. In this chapter we will review some of the work performed using the last two techniques.

Using the techniques of centrifugal elutriation and flow cytometry, relatively pure populations of infiltrated host cells, neoplastic cells, synchronized cells, and hypoxic cells have been isolated from several animal tumors (Keng *et al.*, 1981b, 1987; Meistrich *et al.*, 1977b; Pretlow and Pretlow, 1980; Siemann and Keng, 1987; Siemann *et al.*, 1981). The morphological properties and functional aspects of these cells as well as their responses to radiation and chemotherapeutic agents have been well characterized (Hill *et al.*, 1984; Siemann and Keng, 1987, 1988b; Siemann *et al.*, 1981). Recent developments in cell separation procedures have now also provided the capability to investigate the tumor heterogeneity and cell subpopulations in human tumor xenograft models (Keng *et al.*, 1987, 1990).

The principles and applications of counterflow centrifugation (centrifugal elutriation) were first described by Lindahl (1948). Briefly, cells suspended in the separation chamber of an elutriator rotor are subjected to two opposite forces; the centrifugal force generated from the rotation of the rotor in an outward direction and the fluid force through the chamber in an inward direction. The special shape of the separation chamber makes it possible to generate a gradient of flow rate, decreasing toward the center of rotation. Cells of a given size are sedimented to a position in the separation chamber where the two opposite forces are equal, i.e., where the sedimentation rate is balanced by the flow rate of the fluid. The cells remain in the chamber as long as the two opposite forces are in balance. By incremental increase in the flow rate of the fluid or by decrease in the centrifugal force, distinct populations of cells with relatively homogeneous sizes can be eluted out of the rotor sequentially. Since the sizes of a given cell population are closely related to their morphologies (host vs neoplastic), cell cycle position (G_1, S, G_2M), proliferation state, and oxygen content, centrifugal elutriation has become the most effective cell separation method for obtaining bulk tumor cell subpopulations. Typically 10^8 tumor cells can readily be separated into sort fractions containing on average 10^6 cells.

In an early study with the "counterstreaming" centrifuge designed by Lindahl, a partial purification of host and tumor cells was established from mouse Ehrlich ascites tumor cells (Lindahl and Klein, 1955). However, this approach to cell separation was not explored further until a commercially available centrifugal elutriation system was developed (McEwen *et al.*, 1969). In addition, subsequent improvements in the design of the separation chamber coupled with modifications in the separation procedures have made possible isolations of cell subpopulations with purities as high as 95% directly from many animal and human tumors (Sanderson *et al.*, 1976; Keng *et al.*, 1981a, 1990).

Since the basic principle of operation was described by Coulter in 1957, flow cytometric (FCM) analysis has become one of the most powerful tools for studying tumor heterogeneity. The flow cytometer has the advantage of multiparameter analysis by recording several measurements on each cell simultaneously. Typical measurements include Coulter volume, fluorescence at one or more wavelengths, forward angle light scatter (FALS), and 90° light scatter (90° LS). As many as six simultaneous measurements, including cell size (FALS), granularity (90° LS), and fluorescence intensity (DNA content, FITC-conjugated antibody content, etc.) can be made on the same cell. Using multiparameter analysis, one can identify a particular cell subpopulation within heterogeneous tumor cell suspensions. Subsequent isolation of these subpopulations of cells is made possible by the electronic cell sorting function of the flow cytometer.

One example of the use of the flow cytometer for cell separation is the recent development of this sorting technology in conjunction with Hoechst 33342 dye staining to isolate oxic and hypoxic cells from solid tumors (Chaplin *et al.*, 1985, 1986; Olive *et al.*, 1985). The fluorescent dye Hoechst 33342, when administered intravenously to tumor-bearing animals, diffuses outward from tumor blood vessels. A gradient of fluorescence intensity is established, with cells close to blood vessels staining more brightly then cells further away. The fluorescence intensity of individual cells is therefore a function of their location, i.e., distance from blood vessels supplying the tumor. Consequently this use of flow cytometry to sort tumor cells with different Hoechst 33342 intensities, has allowed the effective separation of hypoxic (dim fluorescence) from oxygenated (bright fluorescence) cells in both multicell spheroids and solid tumors (Chaplin *et al.*, 1985, 1986; Olive *et al.*, 1985; Siemann and Keng, 1988a).

IV. Cell Subpopulations Influencing
the Overall Tumor Response
to Therapy

In this chapter we will review approaches to isolating critical cell subpopulations from solid tumors and discuss experiments aimed at evaluating their response to anticancer agents. We will then evaluate the overall tumor response to combination therapies based on the knowledge of the individual therapy's effects against the various cell subpopulations. Thus we will discuss the concept of therapy synergism, not in terms of multiple agents acting against a single cell population, but rather in terms of one or more therapies directed against multiple cell populations. In such an approach it is essential to have characterized the individual cell subpopulation's inherent treatment responses in order to understand how combinations of specific agents might lead to a synergistic interaction in a tumor. Studies with four potentially critical cell subpopulations are described in the following sections.

A. CELLS IN VARIOUS PHASES OF THE CELL CYCLE

Tumor cell subpopulations resistant to therapy due to their position in the cell cycle can have significant impact on the outcome of chemotherapy or radiotherapy (Tannock, 1978; Keng *et al.*, 1987). Although this has been well established in tissue culture evaluations, the impact of cell cycle position on the *in vivo* outcome of therapies has been more difficult to ascertain. However, with the development of effective cell isolation techniques, such as those described in the previous section, it now has become possible to determine the *in situ* drug or radiation response across the cell cycle in a number of solid tumor models. An important consideration in such investigations is the well established fact that treatment with anticancer agents can lead not only to cell cycle specific cell killing but also to significant perturbations in the distribution of cells within the cell cycle as determined by FCM analysis (Siemann and Keng, 1988b). From a therapeutic standpoint, FCM data alone do not provide information concerning the viability of the treated cells. Yet it is clearly important that the cell cycle position of the tumor cells which survive the treatment be determined. To overcome this difficulty we have used centrifugal elutriation in conjunction with FCM analysis. This approach allows the isolation, from treated tumors, of cells in the various phases of the cell cycle with the same degree of purity as from untreated tumors (Siemann and Keng, 1988b). Consequently, it is possible to determine, with respect to cell cycle phase, the *in situ* survival characteristics of tumor cells treated with drugs or radiation. For example, Fig. 1 shows the survival values of cells in various

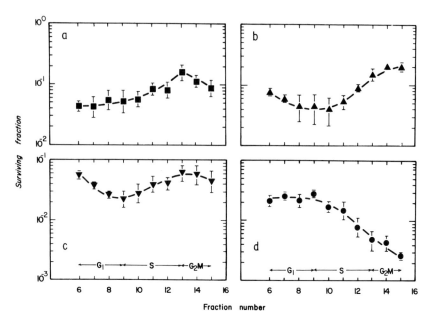

FIGURE 1   Clonogenic cell survival as a function of cell cycle phase for KHT sarcomas treated *in vivo* and separated by centrifugal elutriation immediately (radiation) or 24 hr later (chemotherapeutic agents). Results shown are (a) 6 mg MIT C/kg, (b) 7.5 mg CCNU/kg. (c) 50 mg cyclophosphamide/kg, and (d) 15 Gy radiation. Data shown are the mean ± SE of 3–5 experiments. (From Siemann and Keng, 1988b.)

phases of the cell cycle derived from KHT sarcomas after the *in vivo* treatment with mitomycin C (MIT C), 1-(2-chloroethyl)-3-cyclohexyl-1-nitrosourea (CCNU), cyclophosphamide (CTX), or radiation. These results illustrate the treatment dependent survival in the isolated cell subpopulations. Such information can be used to predict which treatment strategies might be used successfully in combination to yield a synergistic therapeutic response in the tumor (Section V).

B. HYPOXIC CELLS

Oxygen-deficient or hypoxic cells have been shown to not only severely hamper the success of radiation treatment but have also been found to be refractory to certain clinically active anticancer drugs (Siemann, 1984; Sartorelli, 1986). It is clear that most rodent tumors contain hypoxic cells (Moulder and Rockwell, 1984) and the evidence is strongly suggestive that such cells may also influence the clinical therapeutic outcome in at least some human cancers (Bush *et al.*, 1978). Yet it has been difficult to isolate or characterize this relatively small subpopulation (typically 10 to 20%) of radiobiologically hypoxic cells in the presence of an overwhelming majority of well-oxygenated cells. To approach this problem we have utilized centrifugal elutriation in conjunction with FCM analysis and cell sorting and applied these techniques directly to solid tumors.

Initial studies focused on whether hypoxic cells in KHT sarcomas were equally distributed throughout the cell cycle phases or located predominantly in one cell cycle stage (Siemann and Keng , 1987). These investigations used centrifugal elutriation plus cell survival curve analysis to show that hypoxic cells were found predominantly in the G_1 phase of the cell cycle. This finding was confirmed (Fig. 2) in subsequent experiments which utilized the Hoechst 33342 sorting technique (Chaplin *et al.*, 1985, 1986; Olive *et al.*, 1985; Siemann and Keng, 1988a). In addition, when the tumors were irradiated prior to cell sorting, the dimmest cells survived preferentially. Complete dose–response curves for the brightest and dimmest 20% of cells, sorted on the basis of fluorescence intensity then were determined (Fig. 3). The data showed that the brightest fluorescent tumor cells (oxic) were 2- to 3-fold more radiosensitive than cells in the dimmest sort fraction (hypoxic). Also, the slopes of these survival curves were comparable to those of KHT cells irradiated under fully oxic or anoxic conditions (Siemann and Keng, 1988a). Together these results indicate that it is possible to separate treatment-refractory hypoxic cells directly from solid tumors. Thus these cell isolation techniques provide the possibility of evaluating directly the *in vivo* efficacy of therapeutics aimed against oxic and hypoxic tumor cell subpopulations (Section V).

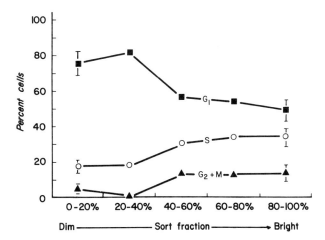

FIGURE 2   Cell cycle distribution of the neoplastic cells in the sort fractions obtained following separation on the basis of Hoechst 33342 fluorescence intensity. Data shown for 0–20% and 80–100% fractions are the mean ± SD of 5 individual experiments. (From Siemann and Keng, 1988a.)

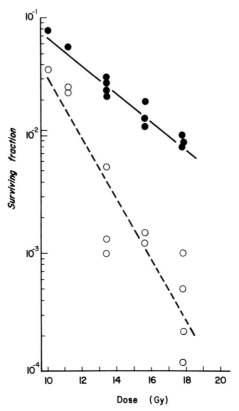

FIGURE 3   Clonogenic cell survival in the sort fractions containing the 20% most dim (●) or 20% most bright (○) cells following *in situ* irradiation. Mice were irradiated with a range of doses while breathing air. (Modified from Siemann and Keng, 1988a.)

C. QUIESCENT CELLS

Evaluations of tumor responsiveness to therapeutic regimens require consideration of the clonogenic or "stem cell" populations of tumors which retain the capacity for tumor regrowth. This population may be subdivided into proliferating (P) and quiescent (Q) or nonproliferating subpopulations (Kallman *et al.*, 1980). Although Q cells have been known to exist for 20 years and usually constitute the majority of tumor cells in solid tumors Kallman *et al.*, 1980; Dethlefsen, 1980), little is known about the nature of these cells. Consequently the relevance of Q stem cells to cancer therapy has not been well documented.

Q cells appear to have relevance to radiation sensitivity by virttue of previous data which suggest that Q cells from *in vitro* plateau phase monolayer cultures have an increased capacity for the repair of potentially lethal radiation damage (PLD) relative to P cells (for representative historical perspectives and review, see Phillips and Tolmach, 1966; Little, 1973). Other data suggest that the propensity of Q cells to repair PLD may occur in some (Hahn *et al.*, 1974) but not all (Siemann, 1989) *in vivo* solid tumors. An additional potential relevance of Q cells to the repair of PLD comes from findings indicating that factors such as low temperature (Elkind *et al.*, 1965), low pH (Varnes *et al.*, 1986), low glucose content (Koch *et al.*, 1977), incubation in balanced salt solutions (Raaphorst and Dewey, 1979), and the addition of antimetabolic agents (Iliakis and Nusse, 1982), all of which decrease cell metabolic activity, thereby favoring a quiescent cell state, have been reported to potentiate the capacity for repair of PLD. However, despite the apparent relevance of Q cells, no definitive study exists directly relating Q cells to PLD repair.

The cellular proliferation state also plays an important role in determining the sensitivity of mammalian cells to a variety of cancer chemotherapy agents. Many of the clinically useful anticancer drugs exhibit proliferation-dependent cytotoxicity (Bhuyan *et al.*, 1977; Hill and Whelan, 1981) and often are more efficacious towards S phase cells because of their direct or indirect involvement in DNA damage and repair (Tannock, 1978; Bhuyan *et al.*, 1973). Thus tumor cells grown as plateau phase cultures (quiescent) are generally more resistant to these drugs than those in the exponential phase of growth (proliferative). Although the mechanisms responsible for this resistance have not been clearly identified, it has been shown that drug–enzyme interactions such as the DNA–topoisomerase II complex, as well as intracellular thiols, may be involved (Sullivan *et al.*, 1987; Allalunis-Turner *et al.*, 1988).

The major reason for the difficulty of Q cell related radio- and chemobiological investigations to date results from the fact that evidence for the existence of Q cells has come only from [^{3}H]thymidine labeling studies which are tedious and difficult to interpret. In addition, this method cannot distinguish tumors composed of uniformly proliferating and nonproliferating subpopulations from tumors which show wide ranges of cell proliferation (Dethlefsen, 1980; Dethlefsen *et al.*, 1980). Perhaps most critically, this technique provides no information regarding the viability of the P and Q cell subpopulations.

One approach which we have pursued in our laboratories to identify and to enrich Q cells from tumor cell cultures involves centrifugal elutriation and flow cytometry with acridine orange (AO) staining (Bauer *et al.*, 1982). Our previous data have shown the enrichment of Q cells from fed plateau cultures, and multicell spheroids could be achieved by this approach (Luk *et al.*, 1982, 1986; Ng *et al.*, 1987). Also, we have observed that isolated Q cells, which were subsequently irradiated, appear to be more radiosensitive than proliferating cells (Luk *et al.*, 1982, 1986; Ng *et al.*, 1987). In addition, Q cells isolated from fed plateau cultures of human tumor cells exhibited 3–5 times more

PLD repair than did P cells (Wilson and Keng, 1989). Although our observation of greater Q cell radiosensitivity is in contrast to the general belief that Q cells *in situ* are more resistant to radiation than are P cells, the influence of hypoxia, cell-to-cell contact, and cell age could change the inherent radiation sensitivities of Q cells *in vivo*. Unfortunately little information concerning the *in situ* impact of this cell subpopulation on tumor response to treatment is currently available.

The presence of resistant, clonogenic, quiescent cells may also pose one of the major problems in the management of tumors by drugs. This is because there are only a limited number of anticancer agents available that are effective against such cells (Tannock, 1978). In order to develop more effective drugs against quiescent tumor cells, a better understanding of the drug activities in these cells is required. Since *in vivo* quiescent tumor cells are often located far away from blood vessels (Siemann and Keng, 1988a), these cells may also be relatively hypoxic or otherwise differ metabolically or morphologically for cycling tumor cells. Consequently, the use of centrifugal elutriation and/or FCM with Hoechst 33342 sorting techniques may make feasible the isolation of quiescent cells directly from solid tumors. Such an approach may allow the direct study of the relationship between drug sensitivity and proliferative activity and provide insight into the *in situ* interaction between anticancer agents.

D. CELLS WITH VARYING THIOL CONTENTS

It is now generally accepted that glutathione (GSH) can confer tumor cells resistance to many anticancer agents (Arrick and Nathan, 1989; Biaglow *et al.*, 1983). Consequently there has been considerable interest in the combined use of conventional chemotherapy and thiol depleting agents that manipulate cellular GSH levels. Attention has focused in particular on one agent, buthionine sulfoximine (BSO), which specifically inhibits GSH synthesis and enhances the response to many cytotoxic drugs both *in vitro* and *in vivo* (Hamilton *et al.*, 1985; Ozols, 1985; Lee *et al.*, 1988a).

Although effective GSH depletion by BSO can lead to increased cancer cell killing, as is illustrated in Fig. 4, many questions remain. For instance, it is not as clear whether steady state GSH levels *per se* are sufficiently different in various cell populations to cause differential treatment sensitivity. In our studies we therefore sought to establish whether or not differences in constitutive GSH levels, without prior manipulation or treatment, were sufficient by themselves to impact on the cellular sensitivity to Adriamycin (ADR). Using the GSH-specific fluorescent probe monochlorobimane in conjunction with flow cytometry allowed the isolation of cell subpopulations with heterogeneous GSH contents from the human ovarian tumor cell line MLS (Lee and Siemann, 1989). Of particular interest, subpopulations isolated at a fixed

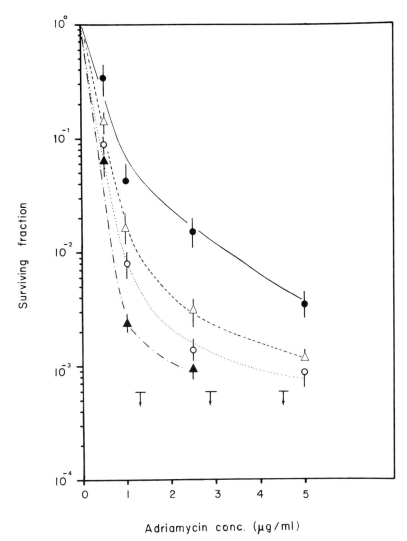

Adriamycin conc. (µg/ml)

FIGURE 4   Effects of pretreatment with 1 m*M* BSO for different times on the clonogenic survival of HEp3 cells following a 1 hr incubation with 1.0 µg ADR/ml. ●, no BSO pretreatment; △, 1 hr; ○, 2 hr; and ▲, 3 hr BSO pretreatment. Error bars represent ± 1 SD. (From Lee *et al.*, 1988a.)

size (cell cycle position) were shown to have differential sensitivity to ADR; cells which had the highest GSH content (increased by 2.1-fold) were also 1.6 times more resistant to this chemotherapeutic agent (Fig. 5).

Since many anticancer agents exhibit cell cycle phase specific cytotoxicity, we also investigated whether such differences for the agent ADR were the

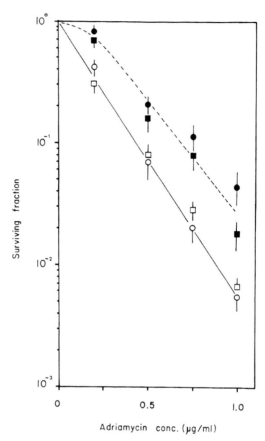

FIGURE 5   Survival of cells sorted on the basis of GSH content following a 2 hr exposure to ADR. ●, ■, MLS/bright; ○, □, MLS/dim. (From Lee and Siemann, 1989).

consequence of variations in the cellular GSH levels (Lee *et al.*, 1988b). Synchronization of cells derived from either tissue culture or solid tumors was achieved by centrifugal elutriation. For cell lines of both rodent and human origin, no significant variation in GSH content was seen for cells in the various phases of the cell cycle when cell volume changes were taken into account (Fig. 6). Consequently the cell cycle dependent cytotoxicity of ADR was not related to cell cycle dependent variations in GSH content.

Irrespective of the origin, i.e., human or rodent, tumor cells grown *in vivo* invariably have lower GSH levels than do those growing exponentially *in vitro* (Allalunis-Turner *et al.*, 1988). While the mechanism(s) responsible for these differences remain unknown, endogenous factors which may influence the growth environment in tumors may play a significant part. *In vivo*, where the nutritional supply may vary regionally throughout a tumor, heterogeneous

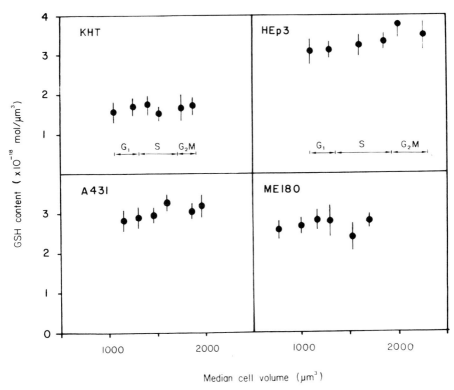

FIGURE 6   Relationship between GSH content and cell volume or cell cycle phases of cells derived from four solid tumors grown in mice. KHT, a mouse sarcoma; HEp3, A431, and ME180, human squamous cell carcinomas. Data are from a representative experiment. Each experiment was repeated at least twice. Single cells obtained from excised tumors were separated according to cell volume by centrifugal elutriation. GSH values were determined by HPLC and were normalized according to cell volume. Bars represent ± 1 SD. (From Lee *et al.*, 1988b.)

cell subpopulations with varying GSH contents may result (Allalunis-Turner *et al.*, 1988). Since GSH is vital in protecting cells from the cytotoxic effects of a variety of agents including radiation and anticancer drugs, variations in GSH in critical tumor cell subpopulations could have significant impact on the overall tumor response to single and combined modality therapies. In particular, cell subpopulations containing high GSH levels or having the ability to rapidly resynthesize GSH (Lee *et al.*, 1989), may have a detrimental effect on the therapeutic outcome. Through the use of cell separation and sorting techniques such as those described in this chapter, it should be possible to study the importance of thiols and their metabolism in such critical subpopulations. Once obtained, such information would aid our understanding and our ability to possibly exploit therapies directed against the protective thiol pool.

V. Improving the Therapeutic Interaction through Therapy Combinations Directed against Specific Tumor Cell Subpopulations

In the previous sections we described how particular tumor cell subpopulations can impact the resultant response to therapies. We also emphasized how cell separation techniques could be used to successfully isolate and characterize some of these subpopulations in an effort to elucidate mechanisms of action. In the present section we will illustrate how, through the choice of specific agents directed at eliminating treatment-refractory cell subpopulations, it is possible to achieve not only a therapeutic synergism in a tumor but also a therapeutic benefit.

A. ATTACKING CELL CYCLE PHASE SPECIFIC RESISTANCE

Figure 1 illustrates the cell cycle phase dependent survival characteristics of KHT sarcoma cells treated *in situ* with a variety of anticancer agents. In this tumor, following irradiation, late S and G_2M cells were found to be sensitive but G_1 cells were preferentially spared. The radiation resistance seen in the G_1 cells was in part due to the observation that in KHT tumors, hypoxic cells are found primarily in this phase of the cell cycle (Siemann and Keng, 1987, 1988a). For all chemotherapeutic agents evaluated, survival of cells at the G_1/S boundary 24 hr after drug exposure was at a nadir, whereas cells in late S or G_2M at this time were always drug resistant (Siemann and Keng, 1988b). However, KHT tumors treated with MIT C showed sensitivity in the G_1 phase, probably reflecting the reported preferential toxicity of this agent in hypoxic cells (Sartorelli, 1986).

Because of the survival characteristics of the tumor cell subpopulations following treatment with MIT C and radiation (Fig. 1), the combination of these two agents was evaluated in detail. MIT C was administered 24 hr prior to radiation therapy in order to allow for full expression of damage and to avoid the complication of MIT C sensitizing the hypoxic cells to radiation. The resultant cell survival was found to be uniform throughout all cell subpopulations (Siemann and Keng, 1988b), indicating that this combination reduced the importance of both hypoxia and cell cycle specificity in the overall treatment response.

To study the nature of the interaction between these two agents, isoeffect plot analysis (Steel and Peckham, 1979) was applied to the MIT C–radiation protocol. Complete *in vivo* dose–response curves for the individual agents were established and used to construct envelopes of additivity at isoeffect survival levels of 10^{-3} and 10^{-2} (Fig. 7). When MIT C preceded tumor

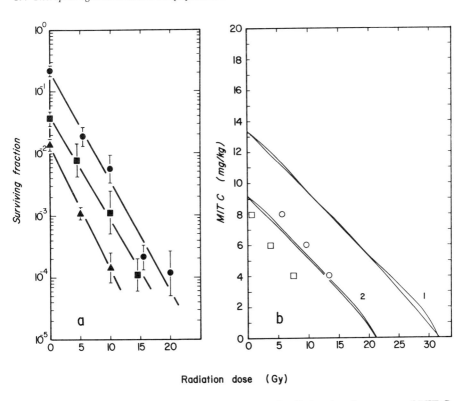

Radiation dose (Gy)

FIGURE 7 (a) Clonogenic cell survival as a function of radiation dose for a range of MIT C doses (●, 4 mg/kg; ■, 6 mg/kg; ▲, 8 mg/kg) given 24 hr before radiation. Points are the mean ± SE of 3–4 experiments. (b) Envelopes of additivity (curves 1 and 2) constructed from the dose–response curves shown in Fig. 4 at isoeffects of 10^{-3} (curve 1) and 10^{-2} (curve 2) tumor cell survival according to the methods of Steel and Peckham (1979). Data points shown in (b) were determined from (a) at isoeffect survival values of 10^{-3} (○) and 10^{-2} (□). (From Siemann and Keng, 1988b.)

irradiation by 24 hr, the combination data points required to achieve these isoeffects (as determined from Fig. 7a) fell well below the envelopes of additivity shown in Fig. 7b. Based on the location of the combination data points within the isobologram, it can be concluded that the combination of radiation and MIT C resulted in a synergistic interaction in the tumor.

B. REDUCING TREATMENT-REFRACTORY HYPOXIC CELLS

The combination of FCM analysis and cells sorting techniques to study directly the *in vivo* therapeutic response of oxic and hypoxic tumor cell subpopulations (described in Section IV) is applicable to a variety of anticancer agents. It may be of particular importance in (1) mechanistic studies of drug actions and (2) the development of new agents with modes of action directed specifically against hypoxic tumor cells.

This approach has been particularly useful in the evaluation of agents targeted specifically against treatment resistant hypoxic cells. Over the last decade there has been considerable interest in the use of nitroimidazole compounds in radiation therapy because of their *in vitro* ability to raise the sensitivity of hypoxic cells to radiation by as much as 2.5 times (Adams, 1977). In addition, nitroheterocyclics such as misonidazole (MISO) have been shown to increase the efficacy of certain chemotherapeutic agents, most notably the alkylating agents melphalan, cyclophosphamide, and CCNU (McNally, 1982; Siemann, 1982, 1984). *In vitro*, both phenomenan (radiosensitization and chemopotentiation) could readily be demonstrated to be hypoxia-mediated. In contrast evidence for the need for hypoxia for these processes *in situ* was by and large indirect. To address this problem, we administered a 2.5 mmol/kg dose of the hypoxic cell sensitizer MISO prior to tumor irradiation, dissociation, and cell sorting on the basis of Hoechst 33342 staining (see Section IV). If hypoxic tumor cells were being preferentially sensitized, then the dimmest sort fraction (hypoxic cells) would demonstrate enhanced cell killing in the presence of MISO whereas the brightest sort fraction (oxic cells) would not. This was in fact observed (Fig. 8). In the presence of MISO, the two subpopulations become equally sensitive to radiation. Similarly, when CCNU and MISO were combined, the sensitizer enhanced CCNU cytotoxicity primarily in the G_1 cell fraction (Hill *et al.*, 1984), i.e., the cell cycle phase containing the majority of hypoxic cells in the KHT sarcoma (Siemann and Keng, 1987, 1988a) (Fig. 2). These studies offer direct *in situ* evidence that MISO sensitizes hypoxic cells preferentially.

C. ACHIEVING A THERAPEUTIC BENEFIT

Most chemopotentiation studies to date have reported increases in the antitumor activity resulting from such drug–sensitizer combinations that outweigh increases in concomitant normal tissue toxicities, i.e., a therapeutic benefit can be realized (McNally, 1982; Siemann, 1982, 1984). Since chemotherapeutic agents are often combined with radiotherapy in the treatment of human cancer, we initiated studies aimed at determining whether the inclusion of a radiosensitizer in such a combination therapy could improve the resultant tumor response. The KHT sarcoma, which has been used extensively in chemopotentiation investigations, was chosen for study. When tumors are treated at macroscopic size with radiation, cures cannot be achieved in this rodent model because of rapid metastatic spread of the tumor. Alternatively, aggressive chemotherapy including the combination of CCNU plus

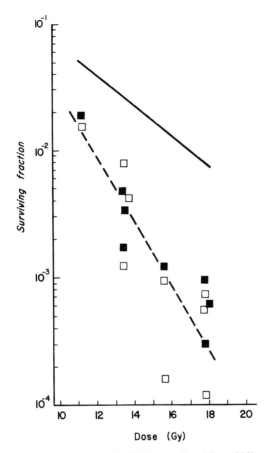

FIGURE 8   Clonogenic cell survival in the 20% most dim (■) or 20% most bright (□) sort fractions following *in situ* irradiation. Mice were irradiated while breathing air in the presence of a 2.5 mmol/kg dose of MISO. The lines are redrawn from Figure 3 and represent the survival of the 20% most dim (————) or 20% most bright (— — —) cells in the absence of MISO. (Modified from Siemann and Keng, 1988a.)

MISO, while effective against the metastases, fails, to cure the animal because of primary tumor failures. In view of this, studies were carried out which combined chemotherapy, sensitizers, and radiotherapy (Siemann *et al.*, 1985).

Previous investigations had already established some of the basic properties of cells in the viable subpopulations derived from solid KHT sarcomas by centrifugal elutriation (see previous sections). This included the determination that in this tumor model radiobiologically hypoxic cells were found primarily in the G_1 phase of the cell cycle (Siemann and Keng, 1987, 1988a) (Fig. 2). Unlike the response to radiation, when CCNU and MISO are combined, cell killing in the hypoxic G_1 cell population is increased and survivors are found preferentially in the late S and G_2M phases (Hill *et al.*, 1984).

When CCNU and MISO were administered 24 hr prior to irradiation, the resultant survival was found to be uniform throughout the various cell fractions (Fig 9). This probably occurred because the CCNU–MISO combination significantly reduced the radiation-resistant cell subpopulations, while the radiation treatment effectively reduced the cell populations preferentially surviving the chemotherapeutic agent–sensitizer treatment.

It was also of interest to evaluate in detail the nature of the interaction between the three agents involved. Isoeffect plot analysis was therefore applied to the CCNU–MISO–radiation treatment regimen (Siemann *et al.*, 1985). The results showed (Fig. 10) that the combination data points from this

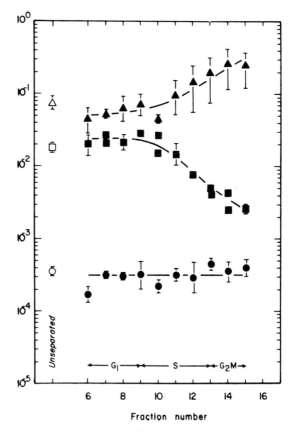

FIGURE 9 *In situ* cell cycle response of KHT sarcomas treated with 3.0 mg/kg CCNU plus 5.0 mmol/kg MISO (▲), 15 Gy radiation (■), or the combination (●) of CCNU (3.0 mg/kg) plus MISO (5.0 mmol/kg) followed 24 hr later by radiation (15 Gy). (Redrawn from Siemann *et al.*, 1985.)

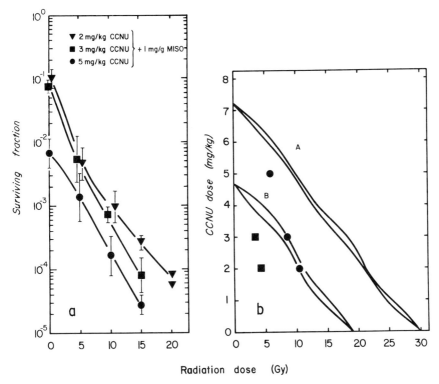

FIGURE 10 (a) Clonogenic cell survival as a function of radiation dose for a range of CCNU doses combined with MISO, 1.0 mg/g, 24 hr prior to radiation. Points, mean of 3 to 4 separate experiments; bars, SE (b) Envelopes of additivity for cell survival isoeffects of 10^{-3} (Envelope A) and 10^{-2} (Envelope B) calculated for a treatment combination in which KHT sarcomas were irradiated 24 hr after treatment with a range of CCNU doses administered in simultaneous combination with MISO, 1.0 mg/g. Points for combinations were determined from at isoeffect survival values of (a) 10^{-3} (●) and (b) 10^{-2} (■). (Modified from Siemann *et al.*, 1985.)

multimodality therapy fell well below the envelopes of additivity indicating that a synergistic interaction had taken place in the tumor (Siemann *et al.*, 1985).

To evaluate whether such a synergistic interaction could be translated into a therapeutic benefit, the potentiation of CCNU by MISO in combination with localized tumor irradiation was evaluated in terms of both tumor cure and normal tissue toxicity (Siemann and Hill, 1986). In those investigations, two forms of MISO administration were studied: single-dose treatment or chronic exposure. The chronic MISO dosing schedule led to plasma MISO levels which can be achieved in humans (Siemann, 1984; Siemann and Hill, 1986). The results showed (Table 1 and Fig. 11) that the inclusion of MISO in the treatment schedule enhanced the animal tumor-free survival between 1.5-

TABLE 1
Tumor Control Probabilities in KHT Sarcoma-Bearing Mice Treated with CCNU, MISO, and Radiation

Experiment	Treatment	$TCD_{50/100}$ $(Gy)^a$	ER^b
A	CCNU 24 hr before radiation	$35.8 \ (31.9–46.5)^c$	1.9
	CCNU + 2.5 mmol MISO/kg 24 hr before radiation	19.0 (16.3–22.8)	
B	CCNU 24 hr before radiation	35.4 (32.0–43.7)	1.5
	CCNU + chronic MISO 24 hr before radiation	23.9 (21.4–27.1)	
C	CCNU 24 hr before radiation	32.5 (28.9-55.1)	1.8
	CCNU + chronic MISO 24 hr before radiation	18.3 (12.4–23.4)	

[a] $TCD_{50/100}$: Radiation dose to cure 50% of the mice of their tumors in 100 days after treatment.
[b] Dose of radiation in the absence of the sensitizer divided by the dose of radiation in the presence of the sensitizer to cure 50% of the mice of their tumors.
[c] Numbers in parentheses, 95% confidence intervals.

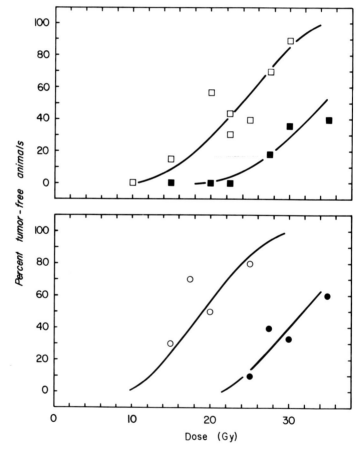

FIGURE 11   Percentage of tumor cure as a function of radiation dose assessed 100 days after treatment. CCNU (20 mg/kg) was administered either alone (●, ■) or 3 hr into a chronic MISO-dosing schedule 24 hr prior to localized irradiation of the tumor (○, □). Two separate experiments are shown. Points represent 10–17 mice. (From Siemann and Hill, 1986.)

and 1.9-fold. Chronic MISO exposures also failed to enhance normal tissue reactions (Siemann and Hill, 1986) but still increased the tumor control probability by a factor of 1.5 to 1.8 (Table 1). In addition, when MISO was included in the combination protocol it was possible to cure 80–100% of the mice of their tumors (Fig. 11). In the absence of the sensitizer, such as high tumor control rate could not be achieved (maximally 40–60%) because of dose limitations necessary to avoid severe normal tissue complications in the treatment field (Siemann and Hill, 1986). These experiments demonstrate how the choice of therapies aimed at select cell subpopulations can yield a synergistic interaction in the tumor and an increased therapeutic gain.

VI. Conclusions

When discussing how an interaction between agents may lead to antagonistic or synergistic effects in cancer therapy, it is usually implicitly assumed that one is dealing with two or more agents acting against a single cell population. Consequently this is a *multihit approach* with particular emphasis on *the agents being combined*. In this case the interaction is typically the result of one agent influencing the activity of the other. Many such interactions are discussed in other chapters of this book. An alternative view is to consider a tumor as an entity composed of a *variety of cell subpopulations. Each subpopulation may have its own inherent properties and intrinsic sensitivities* to therapies. These different sensitivities can arise from environmental or genetic factors or a combination of both. The overall tumor response then is a reflection of the responses of the individual subpopulations. This is a *multitarget approach* with particular emphasis on *the cell subpopulations comprising tumors*. With such an approach it is possible to view the nature of the interaction between agents in terms of the overall tumor response.

In this chapter we have focused on approaches aimed at identifying, characterizing, isolating, and eliminating treatment resistant tumor cell subpopulations. Using techniques of centrifugal elutriation coupled with FCM analysis and sorting, it has been possible to develop *in vivo* mechanisms of action for single–agent or combination therapies. In addition, the targeting of selected agents against particular treatment-refractory cell subpopulations can be shown to lead to synergistic interactions in the tumor and improved overall treatment outcome.

Acknowledgments

These studies were supported by USPHS Grants CA-36858 and CA-38637.

References

Adams, G. E. (1977). Hypoxic cell sensitizers for radio-therapy. *In* "Cancer. A Comprehensive Treatise" (F. F. Becker, ed., Vol. 6, pp. 181–224. Plenum, New York.

Allalunis-Turner, M. J., Lee, F. Y. F., and Siemann, D. W. (1988). Comparison of glutathione levels in rodent and human tumor cells grown *in vitro* and *in vivo*. *Cancer Res.* **48**, 3657–3660.

Arrick, B. A., and Nathan, C. F. (1984). Glutathione metabolism as a determinant of therapeutic efficacy: A review. *Cancer Res.* **44**, 4224–4232.

Bauer, K. D., Keng, P. C., and Sutherland, R. M. (1982). Quiescent cell isolation in multicellular tumor spheroids using centrifugal elutriation. *Cancer Res.* **42**, 72–78.

Bhuyan, B. K., Fraser, T. J., Gray, L. G., Kuentzel, S. L., and Neil, G. L. (1973). Cell-kill kinetics of several S-phase-specific drugs. *Cancer Res.* **33**, 888–894.

Bhuyan, B. K., Fraser, T. J., and Day, K. J. (1977). Cell proliferation kinetics and drug sensitivity of exponential and stationary populations of cultured L1210 cells. *Cancer Res.* **37**, 1057–1063.

Biaglow, J. E., Clark, E. P., Epp, E. R., Morse-Guardio, M., Varnes, M. E., and Mitchell, J. B. (1983). Non-protein thiols and the radiation response of A549 human lung carcinoma cells. *Int. J. Radiat. Biol.* **44**, 489–495.

Bush, R. S., Jenkin, R. D. T., Allt, W. E. C., Beale, F. A., Bean, H., Dembo, A. J., and Pringle, J. F. (1978). Definitive evidence for hypoxic cells influencing cure in cancer therapy. *Br. J. Cancer* **37**, 302–306.

Chaplin, D. J., Durand, R. E., and Olive, P. L. (1985). Cell selection from a murine tumour using the fluorescent probe Hoechst 33342. *Br. J. Cancer* **51**, 569–572.

Chaplin, D. J., Durand, R. E., and olive, P. L. (1986). Acute hypoxia in tumors: Implications for modifiers of radiation effects. *Int. J. Radiat. Oncol., Biol. Phys.* **12**, 1279–1282.

Cifone, M. A., and Fidler, I. J. (1981). Increasing metastatic potential is associated with increasing genetic instability of clones isolated from murine neoplasms. *Proc. Natl. Acad. Sci. U.S.A.* **781**, 6949–6952.

Dethlefsen, L. A. (1980). In quest of the quaint quiescent cells. *In* "Radiation Biology in Cancer Research" (R. E. Meyn and H. R. Withers, eds.), pp. 415–435. Raven, New York.

Dethlefsen, L. A., Bauer, K. D., and Riley, R. M. (1980). Heterogeneous cells in solid tumors measured by acidic acridine orange staining. *Cytometry* **1**, 89–97.

Elkind, M. M., Sutton-Gilbert, H., Moses, W. B., Alescio, T., and Swain, R. W. (1965). Radiation response of mammalian cells grown in culture. V. Temperature dependence of the repair of X-ray damage in surviving cells. *Radiat. Res.* **25**, 359–376.

Fidler, I. J., and Hart, I. R. (1982a). Biological diversity in metastatic neoplasms: Origins and implications. *Science* **217**, 998–1003.

Fidler, I. J., and Hart, I. R. (1982b). Principles of cancer biology: Biology of cancer metastasis. *In* "Cancer—Principles and Practice of Oncology" (V. T. DeVita, S. Hellman, and S. A. Rosenberg, eds.), pp. 1926–1948. Lippincott, Philadelphia, Pennsylvania.

Grdina, D. J., Peters, L. J., Jones, S., and Chan, E. (1978). Separation of cells from a murine fibrosarcoma on the basis of size. 1. Relationship between cell size and age as modified by growth *in vivo* or *in vitro*. *JNCI, J. Natl. Cancer Inst.* **61**, 209–214.

Grdina, D. J., Sigdestad, C. P., and Peters, L. J. (1979). Phase specific cytoxicity *in vivo* of hydroxyurea on murine fibrosarcoma cells synchronized by centrifugal elutriation. *Br. J. Cancer* **39**, 152–158.

Hahn, G. M., Rockwell, S., Kallman, R. F., Gordon, L. F., and Frindel, E. (1974). Repair of potentially lethal damage *in vivo* in solid tumor cells after X-irradiation. *Cancer Res.* **34**, 351–354.

Hamilton, T. C., Winker, M. A., Lowe, K. H., Batist, G., Behrens, B. C., Tsuruo, T., Grotzinger, K. R., McKoy, W. M., Young, R. C., and Ozols, R. F. (1985). Augmentation of Adriamycin, melphalan, and cisplatin cytotoxicity in drug-resistant and sensitive human ovarian carcinoma cell lines by buthionine sulfoximine mediated glutathione depletion *Biochem. Pharmacol.* **34**, 2533–2536.

Hill, B. T., and Whelan, R. D. (1981). Comparative cell killing and kinetic effects of vincristine or vindesine in mammalian cell lines. *JNCI, J. Natl. Cancer Inst.* **67**, 437–443.

Hill, S. A., Keng, P. C., and Siemann, D. W. (1984). Survival in subpopulations of cells derived form solid KHT sarcoma by centrifugal elutriation following treatment with CCNU and MISO. *Int. J. Radiat. Oncol., Biol. Phys.* **10**, 1615–1618.

Howell, R. L., and Koch, C. J. (1980). The disaggregation, separation and identification of cells from irradiated and unirradiated EMT6 mouse tumors. *Int. J. Radiat. Oncol., Biol. Phys.* **6**, 311–318.

Iliakis, G., and Nusse, M. (1982). Aphidicolin promotes repair of potentially lethal damage in irradiated mammalian cells synchronized in S-phase. *Biochem. Biophys. Res. Commun.* **104**, 1209–1214.

Kallman, R. F., Combs, C. A., Franko, A. J., and Furlong, B. M. (1980). Evidence for the recruitment of noncycling clonogenic tumor cells. *In* "Radiation Biology in Cancer Research" (R. E. Meyn and H. R. Withers, eds.), pp. 397–414. Raven, New York.

Keng, P. C., Li, C. K., and Wheeler, K. T. (1981a). Characterization of the separation properties of the Beckman elutriation system. *Cell Biophys.* **3**, 41–56.

Keng, P. C., Wheeler, K. T., Siemann, D. W., and Lord, E. M. (1981b). Direct synchronization of cells from solid tumors by centrifugal elutriation. *Exp. Cell Res.* **134**, 15–22.

Keng, P. C., Siemann, D. W., and Wheeler, K. T. (1984). Comparison of tumor age response to radiation for cells derived from tissue culture or solid tumors. *Br. J. Cancer* **50**, 519–526.

Keng, P. C., Siemann, D. W., and Lord, E. M. (1987). Separation of malignant cells from host cells using centrifugal elutriation. *In* "Cell Separation: Methods and Selected Applications" (T. P. Pretlow and T. G. Pretlow, eds.), Vol. 5, pp. 51–74. Academic Press, Orlando, Florida.

Keng, P. C., Allalunis-Turner, J., and Siemann, D. W. (1990). Evaluation of cell subpopulations isolated from human tumor xenografts by centrifugal elutriation. *Int. J. Radiat. Oncol., Biol. Phys.* **18**, 1061–1067.

Koch, C. J., Meneses, J. J., and Harris, J. W. (1977). The effect of extreme hypoxia and glucose on the repair of potentially lethal and sublethal radiation damage by mammalian cells. *Radiat. Res.* **70**, 542–551.

Lee, F. Y. F., and Siemann, D. W. (1989). Isolation by flow cytometry of a human ovarian tumor cell subpopulation exhibiting a high glutathione content phenotype and increased resistance to Adriamycin. *Int. J. Radiat. Oncol., Biol. Phys.* **16**, 1315–1319.

Lee, F. Y. F., Vessey, A. R., and Siemann, D. W. (1988a). Glutathione as a determinant of cellular response to doxorubicin. *Natl. Cancer Inst. Monogr.* **6**, 211–215.

Lee, F. Y. F., Siemann, D. W., Allalunis-Turner, J., and Keng, P. (1988b). Glutathione contents in human and rodent tumor cells in various phases of the cell cycle. *Cancer Res.* **48**, 3661–3665.

Lee, F. Y. F., Siemann, D. W., and Sutherland, R. M. (1989). Changes in cellular glutathione content during Adriamycin treatment in human ovarian cancer—a possible indicator of chemosensitivity. *Br. J. Cancer* **60**, 291–298.

Lindahl, P. E. (1948). Principle of a counter-streaming centrifuge for the separation of particles of different sizes. *Nature (London)* **161**, 648–649.

Lindahl, P. E., and Klein, G. (1955). Separation of Ehrlich ascites tumor cells from other cellular elements. *Nature (London)* **176**, 401–402.

Little, J. B. (1973). Factors influencing the repair of potentially lethal radiation damage in growth-inhibited human cells. *Radiat. Res.* **56**, 320–333.

Luk, C. K., Keng, P. C., and Sutherland, R. M. (1982). Quiescent cell isolation in multicellular tumor spheroids using centrifugal elutriation. *Cancer Res.* **42**, 72–78.

Luk, C. K., Keng, P. C., and Sutherland, R. M. (1986). Radiation response of poliferating and quiescent subpopulations isolated from ulticellular spheroids. *Br. J. Cancer* **54**, 25–32.

Madoc-Jones, H. S., and Mauro, F. J. (1970). Age response to X-rays, vinca alkaloids, and hydroxyurea of murine lymphoma cells synchronized *in vivo*. *JNCI, J. Natl. Cancer Inst.* **45**, 1131–1143.

Martin, W. M. C., and McNally, N. J. (1980). The cytotoxic action of Adriamycin on tumor cells *in vitro* and *in vivo*. *Br. J. Cancer* **41**, 306.

McEwen, C. R., Stallard, R. W., and Juhos, E. T. (1968). Separation of biological particles by centrifugal elutriation. *Anal. Biochem.* **23**, 369–377.

McNally, N. (1982). Enhancement of chemotherapy. *Int. J. Radiat. Oncol., Biol. Phys.* **8**, 593–598.

Meistrich, M. L., Meyn, R. E., and Barlogie, B. (1977a). Synchronization of mouse L-P59 cells by centrifugal elutriation separation. *Exp. Cell Res.* **105**, 169–177.

Meistrich, M. I., Grdina, D. J., Meyn, R. E. S., and Barlogie, B. (1977b). Separation of cells from mouse solid tumors by centrifugal elutriation. *Cancer Res.* **37**, 4291–4296.

Moulder, J. E., and Rockwell, S. (1984). Hypoxic fractions of solid tumors; experimental techniques, methods of analysis, and a survey of existing data. *Int. J. Radiat. Oncol., Biol. Phys.* **10**, 695–712.

Ng, C. E., Keng, P. C., and Sutherland, R. M. (1987). Characterization of radiation sensitivity of human squamous carcinoma A431 cells. *Br. J. Cancer* **56**, 301–307.

Nicolson, G. L. and Poste, G. (1983). Tumor cell diversity and host response in cancer. *Curr. Probl. Cancer* **7**, 1–42.

Olive, P. L., Chaplin, D. J., and Durand, R. E. (1985). Pharmacokinetics, binding and distribution of Hoechst 33342 in spheroids and murine tumors. *Br. J. Cancer* **52**, 739–746.

Ozols, R. F. (1985). Pharmacological reversal of drug resistance in ovarian cancer. *Semin. Oncol.* **4**, 7–11.

Phillips, R. A., and Tolmach, L. J. (1966). Repair of potentially lethal damage in X-irradiated HeLa cells. *Radiat. Res.* **29**, 413–432.

Poste, G., Doll, J., and Fidler, I. J. (1981). Interaction among clonal subpopulations affect stability of the metastatic phenotype in polyclonal populations of B16 melanoma cells. *Proc. Natl. Acad. Sci. U.S.A.* **781**, 6226–6231.

Pretlow, T. G., and Pretlow, T. P. (1980). Purification of malignant cells and lymphocytes from a rat transplantable mucinous adenocarcinoma. *Contemp. Top. Immunobiol.* **10**, 21–29.

Pretlow, T. G., II, and Pretlow, T. P. (1982). Purification of specific kinds of cells from cancer. *In* "Tumor Immunity in Progrosis: The Role of Mononuclear Cell Infiltration" (J. S. Haskill, eds), pp. 245–277. Dekker, New York.

Raaphorst, G. P., and Dewey, W. C. (1979). A study of the repair of potentially lethal and sublethal radiation damage in Chinese hamster cells exposed to extremely hypo- or hypertonic NaCL solutions. *Radiat. Res.* **77**, 325–340.

Sanderson, R. J., Bird, K. E., Palmer, N. F., and Brenman, J. (1976). Design principles for a counterflow centrifugation cell separation chamber. *Anal. Biochem.* **71**, 615–622.

Sartorelli, A. C. (1986). The role of mitomycin antibiotics in the chemotherapy of solid tumors. *Biochem. Pharmacol.* **35**, 67–69.

Siemann, D. W. (1982). Response of murine tumors to combinations of CCNU with misonidazole and other radiation sensitizers. *Br. J. Cancer* **45**, 272–281.

Siemann, D. W. (1984). Modification of chemotherapy by nitroimidazoles. *Int. J. Radiat. Oncol., Biol. Phys.* **10**, 1585–1594.

Siemann, D. W. (1989). Do *in vitro* studies of potential lethal damage repair predict for *in situ* results? *Int. J. Radiat. Biol.* **56**, 567–571.

Siemann, D. W., and Hill, S. A. (1986). Increased therapeutic benefit through the addition of misonidazole to a nitrosourea–radiation combination. *Cancer Res.* **46**, 629–632.

Siemann, D. W., and Keng, P. C. (1987). Characterization of radiation resistant hypoxic cell subpopulations in KHT sarcomas. (I) Centrifugal elutriation. *Cancer* **55**, 33–36.

Siemann, D. W., and Keng, P. C. (1988a). Characterization of radiation resistant hypoxic cell subpopulations in KHT sarcomas. (ii) cell sorting. *Br. J. Cancer* **58**, 296–300.

Siemann, D. W., and Keng, P. C. (1988b). Tumor cell subpopulation response to single and combined modality therapy. *Natl. Cancer Inst. Monogr.* **6**, 101–105.

Siemann, D. W., Lord, E. M., Keng, P. C., and Wheeler, K. T. (1981). Cell subpopulations dispersed from solid tumors and separated by centrifugal elutriation. *Br. J. Cancer* **44**, 100–107.

Siemann, D. W., Maddison, K., Wolf, K., Hill, S. A., and Keng, P. C. (1985). *In vivo* interaction between radiation and 1-(2-cloroethyl)-3-cyclohexyl-1-nitrosourea in the absence or presence of misonidazole in mice. *Cancer Res.* **45**, 198–202.

Steel, G. G., and Peckham, M. J. (1979). Exploitable mechanisms in combined radiotherapy–chemotherapy: The concept of additivity. *Int. J. Radiat. Oncol., Biol. Phys.* **5**, 85–91.

Sullivan, D. M., Chow, K.-C., Gilsson, B. S., and Ross, W. E. (1987). Role of proliferation in determining sensitivity to topoisomerase II active chemotherapy agents. *Natl. Cancer Inst. Monogr.* **4**, 73–78.

Sutherland, R. M., Eddy, H. A., Bareham, B., Reich, K., and VanAntwerp, D. (1979). Resistance to Adriamycin in multicellular spheroids. *Int. J. Radiat. Oncol., Biol. Phys.* **5**, 1225–1230.

Tannock, I. (1978). Cell kinetics and chemotherapy: A critical review. *Cancer Treat. Rep.* **62**, 117–133.

Tsuruo, T., and Fidler, I. J. (1981). Differences in drug sensitivity among tumor cells from parental tumors, selective variants, and spontaneous metastases. *Cancer Res.* **41**, 3058–3064.

Varnes, M. E., Dethlefsen, L. A., and Biaglow, J. E. (1986). The effect of pH on potentially lethal damage recovery in A549 cells. *Radiat. Res.* **108**, 80–90.

Welch, D. R., Milas, L., Tomasovic, S. P., and Nicolson, G. L. (1983). Heterogeneous response and clonal drift of sensitivities of metastatic 1376NF mammary adenocarcinomas clones to gamma-radiation *in vitro*. *Cancer Res.* **43**, 6–10.

Wilson, K. M., and Keng, P. C. (1989). Radiation-induced DNA damage and repair in quiescent and proliferating human tumor cells *in vitro*. *Int. J. Radiat. Biol.* **55**, 385–395.

CHAPTER 20

Selective Synergism against the Target versus Host Bone Marrow Progenitor Cells

Ellin Berman
Leukemia–Lymphoma Service
Department of Medicine
Memorial Sloan-Kettering Cancer Center
New York, New York 10021

Tai-Tsung Chang
Department of Pediatrics
Kaohsiung Medical College
Kaohsiung, Taiwan, Republic of China

I. Introduction

Clinical drug trials designed to assess synergy *in vivo* are difficult to interpret given such variables as drug absorption, protein binding, and metabolic degradation pathways. Unlike *in vitro* systems that may utilize homogeneous target cell populations, drugs combined *in vivo* must demonstrate some

SYNERGISM AND ANTAGONISM IN CHEMOTHERAPY
Copyright © 1991 by Academic Press, Inc.
All rights of reproduction in any form reserved.

degree of target selectivity if the host is to survive. Compounding these difficulties are the particular sensitivities of the separate organ systems and often it is not possible to predict with reasonable certainty what cumulative toxicity of two (or more) drugs in combination may be.

Fortunately, bone marrow function (which can often prove to be the dose-limiting variable in clinical drug trials) can be assayed reproducibly _in vitro_. Assays of myeloid and erythroid progenitor cells can therefore provide a model against which drugs can be tested, singly and in combination, and results can be analyzed for synergism, additive effect, or antagonism. This chapter will discuss examples of drug synergism studies performed in two separate fields of clinical medicine, infectious disease and oncology, and demonstrate that important information can be gained by using a quantitative approach to data interpretation which may prove useful to the clinician designing combination drug trials.

II. Azidothymidine and Recombinant Interferon Alpha: Synergistic Effects against the Human Immunodeficiency Virus versus Normal Human Bone Marrow Progenitor Cells

A. BACKGROUND

Azidothymidine (AZT) and recombinant interferon alpha (rIFN-α) are among the drugs showing strong _in vitro_ activity against the human immunodeficiency virus type-1 (HIV-1). Patients with the acquired immune deficiency syndrome (AIDS) and those with AIDS-related complex (ARC) who have received AZT have shown prolongation of survival, a decrease in the incidence of opportunistic infections, and improvement in skin reactivity and CD4 cell numbers compared with randomized placebo-treated control subjects (Fischl _et al._, 1987). Recombinant IFN-α has shown significant antitumor activity in patients with AIDS-associated Kaposi's sarcoma (Real _et al._, 1986; Groopman _et al._, 1984), although few beneficial effects have been observed in patients with ARC or with nonneoplastic manifestations of AIDS. A rationale for combined treatment with these agents, which affect distinct stages of retrovirus replication, is provided by studies demonstrating a marked and synergistic suppression of HIV-1 _in vitro_ when the latter was assayed by reverse transcriptase activity, p24 antigen production, or virus yield (Hartshorn _et al._, 1987). Additional support comes from _in vivo_ studies by Ruprecht _et al._, (1988), who showed synergistic suppression of the pathogenic Rauscher murine leukemia retrovirus after treating mice with AZT and rIFN-α. Recent clinical trials have now begun to explore the safety and efficacy of combined AZT and rIFN-α in the treatment of patients with AIDS and ARC.

The potential for significant hematologic toxicity was a major consideration for the clinical investigators planning this trial as both drugs are, independently, toxic to the bone marrow. Richman *et al.*, (1987) have reported that 45% of patients with AIDS or ARC treated with AZT at a dose of 250 mg orally every 4 hr had evidence of grade 3 marrow suppression. Approximately one-third of their patients required dosage adjustments, often multiple times, and three patients were reported as having persistently hypoplastic marrows 5, 6, and 9 months after cessation of treatment. Similarly, Real *et al.*, (1986) have reported leukopenia and thrombocytopenia in patients with Kaposi's sarcoma treated with rIFN-α2a (Hoffman–La Roche, Inc., Nutley, New Jersey). In this study which utilized two dose levels, 25% of patients who received the low dose preparation [3 million units (U) daily] and 50% of patients who received the high dose (36–54 million U daily) developed neutrophil counts < 1,000/μl.

Few guidelines exist, however, for the clinical investigator faced with combining these clinically active but marrow toxic drugs. One approach is to examine the *in vitro* effects of each drug alone and in combination in a standard marrow clonogenic assay and to compare the results with those obtained from *in vitro* anti-HIV-1 assays. The doses required for the different degrees of effect in these test systems can then be compared with the plasma level of each drug clinically achievable in patients. From these data one can approximate the optimal dose of drug that could be used in initial clinical trials. An example of the methodology is now given.

B. DATA ANALYSIS USING THE MULTIPLE-DRUG-EFFECT EQUATION AND MEDIAN-EFFECT PLOT

The multiple-drug-effect equation and median-effect plot described by Chou and Talalay (1984, 1987) were used to determine whether AZT and rIFN-α2a were synergistic, antagonistic, or produced an additive effect against normal human bone marrow progenitor cells, CFU-GM (colony forming units—granulocytic/monocytic) and BFU-E (burst forming units—erythroid). The multiple-drug-effect equation was chosen because, unlike more traditional methods of analyzing drug interactions, it is not limited in its application. The classical isobologram method (Loewe, 1953), for example, is valid only for drugs whose effects are mutually exclusive (drugs with similar modes of action) and the fractional product method (Webb, 1963) is valid only for mutually nonexclusive drugs (drugs with different modes of action) which have hyperbolic dose–effect curves. Further, the fractional product method tends to give more indication of synergism since it underestimates the summation effect of mutually nonexclusive drugs that have sigmoidal dose–effect curves. The median effect equation is devoid of these limitations since it is not influenced by (1) whether the dose–effect relationships are hyperbolic or sigmoidal, (2) whether the effects of the drugs are mutually exclusive or nonexclusive, (3) whether the ligand interactions are competitive or noncompetitive, and (4) whether the drugs are agonists or antagonists (Chou and Talalay, 1984, 1987).

The median-effect method involves plotting the dose effect data for each drug and their mixtures at different combination ratios by using the median-effect equation:

$$f_a/f_u = (D/D_m)^m$$

In this equation, D is dose, D_m is the dose required for 50% effect (e.g., 50% inhibition of growth), f_a is the fraction affected by dose D (e.g., 0.9 if growth activity is inhibited by 90%), f_u is the unaffected fraction, and m is a coefficient of the sigmoidicity of the dose–effect curve. The dose–effect curve is then plotted by using a logarithmic conversion of this equation which determines the m and D_m values. The dose (D_x) required for a given degree of effect (f_a) is then calculated by

$$D_x = D_m[f_a/(1 - f_a)]^{1/m}$$

A combination index (CI) for drugs with different modes of action is then determined with the conservative isobologram equation

$$CI = \frac{(D)_1}{(D_x)_1} + \frac{(D)_2}{(D_x)_2} + \alpha \frac{(D)_1(D)_2}{(D_x)_1(D_x)_2}$$

where $(D_x)_1$ is the dose of drug 1 required to produce $x\%$ effect alone, and $(D)_1$ is the dose of drug 1 required to produce the same $x\%$ in combination with $(D)_2$. Similarly, $(D_x)_2$ is the dose of agent 2 required to produce the same effect alone. Different values of CI are obtained by solving the equation for different values of f_a (e.g., different degrees of inhibition of growth). CI values of < 1 indicate synergy, values > 1 indicate antagonism, and values equal to 1 indicate additive effects. Computer programs based on the median effect plot and the CI equation can be used for data analysis (Chou and Talalay, 1984; Chou and Chou, 1987).

The same data for combination studies can be analyzed using the classical isobologram equation for drugs assumed to have similar or mutually exclusive modes of action in which the CI equation just shown will be the sum of the first two terms ($\alpha = 0$; the third term is dropped). The CI values obtained from both the classical ($\alpha = 0$) and conservative ($\alpha = 1$) isobologram equations are presented in the following example.

C. EFFECT OF AZT AND rIFN-α2a AS SINGLE AGENTS IN CFU-GM AND BFU-E ASSAYS

The CFU-GM and BFU-E assay have been described in detail elsewhere (Strife *et al.*, 1987). Briefly, bone marrow aspirates obtained from normal volunteers were separated using Ficoll–Hypaque gradient centrifugation. Mononuclear cells were then incubated with the selected drug or drug combinations and plated in cell culture systems containing 1 ml of Iscove's modified Dulbecco's medium (IMDM) containing 24% deionized bovine serum albumin, 10^{-4} mol/liter 2-mercaptoethanol, 10% conditioned medium [prepared, in this instance, from the Mo T lymphoblast cell line (Golde *et al.*, 1980)], 0.2 μmol/liter hemin (ferric chloride protoporphyrin IX), and methylcellulose at a final concentration of 1.3% in 35-mm tissue culture plates. One unit of partially purified human urinary erythropoietin was added at the initiation of culture. Culture plates were incubated in a humidified atmosphere of 5% CO_2 in air for 14 days and scored using an inverted microscope. Aggregated of more than 40 cells were considered colonies; those containing hemoglobinized cells were counted as BFU-E.

Concentrations of AZT were chosen that correspond to plasma levels achievable *in vivo*, which have been reported to be in the range of 0.5–4 μmol/liter (Yarchoan *et al.*, 1986; Fanucchi *et al.*, 1988). We therefore tested 0.5, 1.0, 2.0, and 4.0 μmol/liter *in vitro*. Serum interferon levels are more difficult to interpret given the wide variability noted among patients treated at the same dose level (Bornemann *et al.*, 1985; Gutterman *et al.*, 1982). Test concentrations of 100, 200, 500, and 2000 U/ml were chosen; these are within a broad range of peak serum interferon levels reported with doses of 3–36 million U (Bornemann *et al.*, 1985; Gutterman *et al.*, 1982).

Table 1 (Berman *et al.*, 1989) shows the percentage inhibition of BFU-E and CFU-GM from two (BFU-E) or three (CFU-GM) different experiments in which marrow was incubated with different concentrations of AZT, rIFN-α2a, and the two drugs in combination. AZT had approximately equal inhibitory activity against BFU-E and CFU-GM. At higher concentrations of rIFN-α2a, however, BFU-E were inhibited to a greater degree than CFU-GM.

AZT and rIFN-α2a were then combined at an ED_{50} equipotency ratio of 1 μmol/liter:50 U and serially diluted. At the lowest concentration tested (0.2 μmol/liter AZT plus 10 U/ml rIFN-α2a), approximately 50% of BFU-E colony growth was inhibited. Increasing the concentration of both drugs at a constant ratio demonstrated a dose-dependent supression of both BFU-E and CFU-GM.

TABLE 1
Dose–Effect of AZT, rIFN-α2a, and Combinations on Normal Marrow CFU-GM and BFU-E Expressed as Percentage of Inhibition

Drug	BFU-E Experiment 1 Number of colonies[a]	Experiment 1 % Inhibition	Experiment 2 Number of colonies	Experiment 2 % Inhibition	Average % inhibition	CFU-GM Experiment 1 Number of colonies	Experiment 1 % Inhibition	Experiment 2 Number of colonies	Experiment 2 % Inhibition	Experiment 3 Number of colonies	Experiment 3 % Inhibition	Average % inhibition
—	108 ± 2		63 ± 3			35 + 3		94 + 5		60 + 1		
AZT (μM)												
.5	86 + 4	20	47 ± 1	25	22.5	22 ± 1	37	72 ± 4	23	40 + 2	33	31
1.0	72 ± 2	33	43 ± 3	32	32.5	20 ± 2	43	73 ± 1	22	41 + 3	32	32
2.0	50 ± 4	54	37 ± 3	41	47.5	20 ± 2	43	58 ± 5	38	42 + 1	30	37
4.0	56 ± 3	48	37 ± 3	41	44.5	15 ± 1	57	59 ± 3	37	38 ± 2	37	44
rIFN-α2a(U)												
100	71 ± 4	34	48 + 2	24	29	27 ± 2	23	77 ± 3	18	38 ± 2	37	26
200	54 ± 4	50	40 + 4	37	43.5	30 ± 2	14	72 ± 4	23	46 ± 1	23	20
500	32 ± 2	70	28 ± 4	56	63	26 ± 2	26	58 ± 3	38	29 ± 1	52	39
2000	18 ± 1	83	16 ± 2	75	79	13 ± 1	63	60 ± 3	36	27 ± 2	55	51
AZT (μM) rIFN-α2a(U)												
0.2 + 10	52 + 3	52	33 ± 4	48	50	33 ± 2	6	62 ± 4	34	36 ± 2	40	27
0.5 + 25	43 + 4	60	30 ± 2	52	56	27 ± 1	23	63 ± 3	33	35 ± 3	42	33
1 + 50	31 + 2	71	25 ± 3	60	65.5	23 ± 2	34	54 ± 2	43	33 ± 5	45	41
2 + 100	24 + 4	78	28 ± 3	56	67	20 ± 2	43	53 ± 2	44	35 ± 2	42	43
4 + 100	16 + 1	85	27 ± 3	57	71	16 ± 2	54	40 ± 1	57	25 ± 4	58	56

[a] Data expressed as mean ± SEM of the triplicate plates for each experiment. Low density bone marrow cells were plated at a concentration of 2×10^4 cells/ml and scored on day 14. Reproduced by permission from the editorial board of *Blood*.

CI values were then determined using the median-effect equation as described earlier; these are given in Table 2 (Berman *et al.*, 1989). CI values < 1 (drug synergism) were noted for BFU-E at 50% and 75% levels of inhibition (CI values of 0.083 and 0.290, respectively). At the ED_{90} dose level, moderate antagonism was indicated (CI value, 1.175). CI values were < 1 (0.146–0.308) for all levels of CFU-GM inhibition calculated.

D. COMPARISON OF PLASMA DRUG LEVELS, MARROW INHIBITION, AND INHIBITION OF HIV-1 REPLICATION

The plasma levels of each drug achievable in patients, the concentration of each drug alone and in combination required for 50, 75, and 90% inhibition of HIV-1 replication, and the concentration of each drug in combination required to suppress 50, 75, and 90% of normal marrow BFU-E and CFU-GM are compared in Table 3 (Berman *et al.*, 1989). In our study, the peak plasma levels of AZT occurred at 1 hr and were in the range of 1.68 μmol/liter (Fanucchi *et al.*, 1988); higher levels up to 4 to 6 μmol/liter have been reported by others (Yarchoan *et al.*, 1986). Peak plasma levels of rIFN-α2a after an intramuscular injection of 9 million U occurred at about 4 hr and were in the range of 20 to 283 U/ml (median, 167 U/ml) (S.E. Krown, unpublished data).

Plasma levels of AZT were clearly in excess of the concentration required to achieve 95% HIV-1 inhibition *in vitro* using AZT/rIFN-α2a combinations (0.0102 to 0.012 μmol/liter). Similarly, the maximum plasma levels of rIFN-μ2a were much higher than the *in vitro* ED_{95} for the drug when it was combined with AZT (8.15 to 19 U/ml). Plasma levels of each drug were higher than the ED_{50} for BFU-E (BFU-E $ED_{50} = 0.215$ μmol/liter AZT + 10.7 μ/ml rIFN-α2a) and approximated the ED_{50} for CFU-GM (CFU-GM $Ed_{50} = 2.90$ μmol/liter AZT plus 145 U/ml rIFN-α2a) (Berman *et al.*, 1989).

The strong degree of synergism of these drugs in combination against HIV-1 replication is also outlined in Table 3 (Berman *et al.*, 1989). In the day-14 HIV-1 replication experiment, 0.08 μmol/liter AZT was required to inhibit 90% of viral growth (ED_{90}). Similarly, 517 U of rIFN-α2a was required for 90% viral growth inhibition. When the two drugs were combined, only 0.0045 μmol/liter AZT and 7.27 U/ml rIFN-α2a were necessary to inhibit 90% of viral replication. This represents an approximate 18-fold reduction in AZT concentration and 70-fold reduction in rIFN-α2a concentration, levels that should have minimal effect on myeloid and erythroid precursors.

Similarly, the same experiment (day 14) indicated that AZT 0.012 μmol/liter plus rIFN-α2a 19 U/ml inhibited HIV-1 replication by 95%, which repre-

TABLE 2

Synergistic Effects of AZT and rIFN-α2a on Normal Human Bone Marrow Progenitor Cells

Drugs	Median-effect plot parameters			Combination index[a] at inhibition of:		
	D_m	m	r	50%	75%	90%
BFU-E						
AZT	4.638 uM	0.489	0.968			
rIFN-α2a	292 U/ml	0.740	0.994			
AZT + rIFN-α2a (1:50)	0.215 uM + 10.7 U/ml	0.356	0.978	0.083 (0.085)	0.290 (0.310)	1.157 (1.384)
CFU-GM						
AZT	12.620 uM	0.273	0.965			
rIFN-α2a	1,870 U/ml	0.453	0.917			
AZT + rIFN-α2a (1:50)	2.903 uM + 145 U/ml	0.409	0.979	0.308 (0.326)	0.161 (0.167)	0.146 (0.148)

[a] Combination index CI <1, = 1, and >1 indicate synergism, additive effect, and antagonism, respectively. CI values given are from the classical isobologram equation and those in parentheses are from the conservative isobologram equation. Reproduced by permission from the editorial board of *Blood*.

sents a 9%-fold and 56-fold dose reduction, respectively, when compared with the percentage inhibition produced by each drug alone. The peak plasma and serum levels, for both AZT (1.68 μmol/liter) and rIFN-α2a (167 U/ml) were 140-fold and 9-fold, respectively, higher than those required for inhibiting replication by 95% *in vitro* in combination after a 14-day exposure (Berman *et al.*, 1989).

The selectivity indices for AZT inhibition of HIV-1 versus bone marrow progenitor cells at different effect levels were quite large, in the range of 500 to > 10^4 (Berman *et al.*, 1989). In experiments with rIFN-α2a, however, the selectivity indices were generally lower (Berman *et al.*, 1989); nonetheless, for both drugs, BFU-E demonstrated higher sensitivity at a given dose effect, i.e., more inhibition, than CFU-GM.

E. RECOMMENDATIONS FOR CLINICAL TESTING OF AZT AND rIFN-α2a

We recognize that significant difficulty may exist in applying these results directly to the clinical setting given such variables as individual viral load and recent evidence demonstrating emergence of HIV-1 isolates with reduced sensitivity to AZT. Nonetheless, preliminary studies by Ruprecht *et al.*, (1986, 1987, 1988) using a retrovirus-infected murine model suggest that a significant reduction in the dose of interferon is possible, without toxicity or loss of antiviral activity, when AZT is added at suboptimal doses. In these experiments, rIFN-αA/D administered at a dose of 500,000 U intraperitoneally twice daily to mice infected with Rauscher murine leukemia virus inhibited virus-induced splenomegaly by 94% (Ruprecht *et al.*, 1986). Similarly, when AZT was administered at a dose of 1 mg/ml in drinking water, splenomegaly was inhibited by 98% and viremia was completely suppressed. However, all mice developed severe drug related toxicity. At an AZT dose of 0.1 mg/ml there was no toxicity and survival was prolonged but splenomegaly was not prevented, suggesting that this was a suboptimal dose of AZT. When combined with 0.1 mg/ml AZT, a greatly reduced dose of rIFN-αA/D (to 10,000 U) completely prevented clinical and laboratory signs of infection without toxicity (Ruprecht *et al.*, 1988).

Based on the data presented here, we suggest that it may be possible to adjust AZT and rIFN-α2a to levels at which > 90% of HIV-1 replication would be inhibited while < 50% of marrow progenitor growth would be suppressed (Berman *et al.*, 1989). Such dosage reductions may permit fewer treatment interruptions, more continuous drug exposure, and allow more successful treatment of HIV-1-related complications without compromise of potential antiviral synergy.

TABLE 3
Comparison of Doses of AZT, rIFN-α2a, and Their Combinations Required for a Given Degree of Effect in Various Test Systems

Test system	Dose for a given effect[a]	Compound AZT (μM)	rFN-α2a (U/ml)	AZT + rIFN-α2a (μM) (U/ml)	Dose reduction index (fold)[b] AZT	rIFN-α2a
Plasma levels of AZT after a 100 mg dose (n = 12 pts) and serum IFN levels measured by bioassay after a single dose of rIFN-α2a 9 million units (n = 8 pts)[c]	1 hr	1.68 ± 0.26	39 (11–257)			
	2 hr	1.42 ± 0.29	79 (23–342)			
	3 hr	0.92 ± 0.25	105 (23–514)			
	4 hr	0.45 ± 0.12	167 (20–283)			
	5 hr	—	143 (11–378)			
	6 hr	—	134 (38–378)			
	24 hr	—	29 (<10–72)			
Inhibition of bone marrow progenitor cells, day 14						
BFU-E	ED$_{50}$	4.64	292.3	0.215 + 10.7	21.6	27.2
	ED$_{75}$	43.8	1,290	4.71 + 235	9.3	5.5
	ED$_{90}$	413.6	5,692	103.3 + 5,160	4.0	1.1
CFU-GM	ED$_{50}$	12.6	1,870	2.90 + 145	4.35	12.9
	ED$_{75}$	707.7	21,152	42.6 + 2,130	16.62	9.9
	ED$_{90}$	39,690	239,220	624.5 + 31,220	63.55	7.7

Inhibition of HIV-1 replication, PBL, RT[d]

Experiment A (Day 13)

ED$_{50}$	0.0081	4.2	0.0009 + 0.75	9.0	5.6
ED$_{75}$	0.017	13.0	0.0023 + 1.82	7.4	7.1
ED$_{90}$	0.037	2860	0.0056 + 4.45	6.6	643
ED$_{95}$	0.062	>5000	0.0102 + 8.15	6.1	—

Experiment B (Day 14)

ED$_{50}$	0.029	62.1	0.0003 + 0.44	96.7	141
ED$_{75}$	0.048	179	0.001 + 1.18	48.0	101
ED$_{90}$	0.080	517	0.0045 + 7.27	17.8	71.1
ED$_{95}$	0.113	1060	0.012 + 19.0	9.4	56

[a] Dose for a given effect is calculated by $D_x = D_m [f_a/(1-f_a)]^{1/m}$ (Eq. 2) where D_m and m are dose effect curve parameters.

[b] Dose reduction index (DRI) indicates how many *fold* this dose must be reduced to achieve a given degree of inhibition in combination as compared with each drug alone, e.g., for BFU-E, ED$_{50}$, the DRI for AZT is calculated as $4.64/0.215 = 21.6$.

[c] Krown *et al.* Unpublished data. Serum levels of rIFN-α2a were measured using a standard cytopathic effect inhibition bicassay which measures the ability of serum to protect MDBK cells from infection with vesicular stomatitis virus. rIFN-α2a values are expressed as median (range) values. AZT values are mean ± SEM. AZT in plasma was measured with HPLC using an Alltech Adsorbosphere C18 column, mobile phase, isocratic 25 mM ammonium phosphate and 15% ACN, pH 2.2. The retention time for AZT was 12 min as monitored at 267 nm.

[d] For method of assay, see Hartshorn *et al.* (1987). PBL refers to stimulated peripheral blood lymphocytes and RT to reverse transciptase assay. The CI values at $f_a = 0.5, 0.75,$ and 0.9 were 0.02, 0.07, and 0.12, respectively (classical isobologram equation, $\alpha = 0$). These CI values, all <1, indicate the strong synergistic effect of the combination of AZT and rIFN-α2a against HIV replication. Reproduced by permission from the editorial board of *Blood.*

III. Cancer Cell Purging in Autologous Bone Marrow Transplantation

Several recent studies have suggested that intensive chemotherapy followed by reinfusion of autologous bone marrow (autologous bone marrow transplantation, autoBMT) may be curative in various forms of malignancy, including lymphomas, solid tumors, and leukemia (Appelbaum, 1988; Barbasch et al., 1983; Wolff et al., 1983; Yeager et al., 1986). This form of treatment is of particular interest in patients with leukemia, since it can be used in patients who do not have a genetically identical sibling donor; moreover, it avoids the risk of graft-vs-host disease and graft rejection. One major problem, however, is that of eliminating residual tumor cells obtained at time of autologous harvest. Theoretically, the development of suitable methods to purge tumor cells from bone marrow should improve the overall survival rate, and a number of studies, especially in the leukemia literature, suggest that some form of marrow manipulation is important (Gorin, 1987).

Various purging strategies are currently undergoing investigation; these include physical techniques such as lectin agglutination (Morecki et al., 1988), density gradient centrifugation (Dicke et al., 1984), photoradiation (Sieber, 1987; Atzpodien et al., 1987), immunologic methods such as monoclonal antibodies targeted against specific antigens on tumor cells (Ball et al., 1986; Ferrero et al., 1987; Gee and Boyle, 1988; Takvorian et al., 1987; Kaizer et al., 1985), or pharmacological approaches using one or more drugs in combination (Gulati et al., 1988; Yeager et al., 1986; Kaizer et al., 1985; Hartmann et al., 1987). Ideally, a regimen that offered a synergistic effect against the target tumor cell but spared residual marrow progenitor cells would be optimal. It follows that an effective analytical method for determining synergism, antagonism, or additive effect would be extremely useful for making these analyses.

COMPARATIVE CYTOTOXICITY OF DIFFERENT DRUG COMBINATIONS AGAINST HUMAN LEUKEMIA CELLS VERSUS NORMAL HEMATOPOIETIC PRECURSORS

The first set of experiments was designed to assess the efficacy of three different drugs and their combinations against the human myeloid leukemia cell line HL-60 (Chang et al., 1987). The drugs tested were 4-hydroperoxy-cyclophosphamide (4-HC), etoposide (VP-16), and Adriamycin (ADR). HL-60 cells were grown in logarithmic growth conditions and incubated at a concentration of 10 million cells/ml with each drug alone or in specific combination ratios for 1 hr at 37°C. The combination ratios were selected to approximate the equal potency ratios and were varied in a way such that one drug effect was augmented and another drug effect diminished. Cells were then washed and resuspended in fresh medium from which cell count and viability studies were performed at 1 hr, 24 hr, 48 hr, and 72 hr intervals (Chang *et al.*, 1987).

The same drugs and their combinations were then tested against normal marrow myeloid precursor CFU-GM. Mononuclear cells were incubated for 1 hr with the test combination, washed, and plated *in vitro* as decribed earlier in this chapter. Percentage inhibition of CFU-GM was measured after 14 days of incubation.

Table 4 (Chang *et al.*, 1985) outlines the degree of cytotoxicity against HL-60 cells after incubation with 4-HC and VP-16 (concentration ratio of

TABLE 4
HL-60 Cell Counts after Exposure to 4-HC and/or VP-16

Drug concentration[a] (μg/ml)		Total number of viable cells ($\times 10^6$) at the following times after exposure to drugs(s)[b]			
4-HC	VP-16	1 hr	24 hr	48 hr	72 hr
0	0	18.90 ± 0.61[c]	28.04 ± 2.28	39.92 ± 1.29	54.42 ± 1.34
2.92		15.90	21.25	23.00	27.00
5.84		17.30	19.00	10.38	10.50
8.76		19.10	15.50	4.75	1.63
11.68		18.20	13.50	4.50	0.50
14.60		17.40	11.75	2.50	0.44
17.52		16.60	11.25	2.38	0.25
20.44		17.10	11.00	2.00	0.19
29.20		15.80	10.00	0.63	0.02
	1	16.60	21.00	32.25	35.50
	2	15.30	20.00	29.13	30.00
	3	12.90	16.25	19.00	18.00
	4	17.30	14.50	18.75	15.75
	5	15.30	13.25	16.63	15.00
	6	15.70	9.50	13.25	13.25
	7	17.70	9.25	13.00	12.50
	10	17.70	8.50	11.75	8.00
2.92	1	16.60	15.50	23.38	20.25
5.84	2	16.30	11.50	6.38	4.25
8.76	3	17.20	7.25	3.63	0.50
11.68	4	18.40	6.75	1.50	0.25
14.60	5	16.90	5.25	1.38	0.13
17.52	6	13.90	6.00	0.75	0.06
20.44	7	14.60	4.00	0.50	0.02
29.20	10	14.30	1.75	0.25	0.00

[a] 4-HC and VP-16 drug concentration was varied at a constant ratio (1:0.342) of the two drugs.
[b] Cells (20×10^6) were exposed to drugs and incubated at 37°C for 1 hr and then washed twice with RPMI 1640 prior to long-term culture.
[c] Mean $\pm$ SE (n = 3) (Chang *et al.*, 1985).

1:0.342) and measured at the various timepoints noted earlier. Table 5 (Chang *et al.*, 1985) analyzes cytotoxicity data after incubation of the two drugs at a concentration ratio of 1:0.856. Similar cytotoxicity analyses were performed with the combinations of ADR + 4-HC (combination ratios of 1:034 and 1:051) and VP-16 + ADR(combination ratios of 1:0.033 and 1:0.06) (data not shown) (Chang *et al.*, 1987).

From this information, dose–response curves, graphed as percentage inhibition, were generated as shown in Fig. 1 (Chang *et al.*, 1987). Also shown is the percentage inhibition of normal marrow CFU-GM. Figure 1A outlines the dose–response curves for 4-HC alone, VP-16 alone, and the combinations at a ratio of 1:0.342 and 1:0.856. Figure 1B provides the dose–response

TABLE 5
HL-60 Cell Counts after Exposure to 4-HC and/or VP-16

Drug concentrations[a] (μg/ml)		Total number of viable cells ($\times 10^6$) at the following times after exposure to drug(s)[b]			
4-HC	VP-16	1 hr	24 hr	48 hr	72 hr
0.00	0.00	18.60 ± 0.26^c	28.46 ± 1.03	46.17 ± 0.93	70.92 ± 2.14
2.92		18.13	18.13	23.13	26.13
5.84		15.50	15.13	8.13	6.75
8.76		16.05	13.38	5.00	2.00
11.68		17.48	11.88	3.75	1.13
14.60		18.53	11.75	2.38	1.00
17.52		17.93	9.25	1.50	0.37
20.44		16.50	9.63	1.50	0.13
26.28		17.18	7.38	1.25	0.07
	2.50	16.73	19.38	26.88	30.13
	5.00	15.23	14.63	16.38	20.25
	7.50	16.20	8.75	14.75	13.63
	10.00	16.65	9.00	9.13	6.75
	12.50	15.90	8.63	9.50	6.25
	15.00	14.10	6.88	5.63	4.38
	17.50	16.13	8.13	6.25	3.88
	22.50	11.25	8.50	5.38	1.75
2.92	2.50	12.00	12.00	9.13	11.38
5.84	5.00	13.88	8.63	4.88	8.50
8.76	7.50	15.30	7.63	3.75	1.56
11.68	10.00	16.58	6.63	2.38	1.00
14.60	12.50	15.38	4.25	1.25	0.94
17.52	15.00	12.45	4.25	0.75	0.31
20.44	17.50	12.10	3.50	0.46	0.11
26.28	22.50	8.96	2.38	0.22	0.07

[a] 4-HC and VP-16 drug concentration was varied at a constant ratio (1:0.856) of the two drugs.
[b] Cells (20 × 10⁶) were exposed to drugs and incubated at 37°C for 1 hr and then washed twice with RPMI 1640.
[c] Mean ± SE (n = 3) (Chang *et al.* 1985).

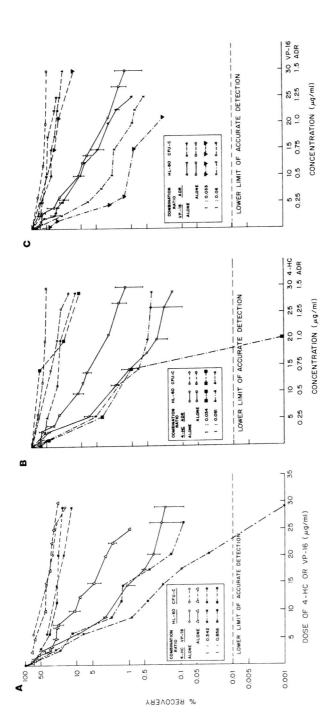

FIGURE 1 Percentage of recovery of HL-60 cells or normal BM CFU-GM. The cells were treated with various drug combinations for 1 hr, washed, and then allowed to grow. Data collected at 72 hr of long term liquid culture for HL-60 cells and the day 10 colony count for treated normal BM CFU-GM are depicted. (A) Effect of 4-HC + VP-16 (Reproduced by permission from the editorial board of *Cancer Research*). (B) Effect of 4-HC + ADR. (C) VP-16 + ADR. CFU-C, colony forming units-cell. (Chang, T. T., *et al.*, *Cancer Res.* **47**, 119–122, 1987.)

curves of ADR, 4-HC, and the two ADR + 4-HC combinations tested and Fig. 1C gives the curves for the VP-16, ADR, and the two VP-16 + ADR combinations tested.

Comparison of these three dose–response curves suggested that the optimal drug combination, i.e., the combination that was most cytotoxic to HL-60 cells but was least toxic to marrow CFU-GM, was 4-HC + VP-16. Median effect plots were then constructed using the data in Tables 4 and 5 along with data in Fig. 1 for HL-60 cells and normal CFU-GM and are given in Figs. 2 and 3, respectively (Chang *et al.*, 1985). The regression line slope for the drug combination at the ratio of 1 : 0.342 was steeper than for either drug alone. Similarly, the two drugs at the same ratio had the least effect on CFU-GM, i.e., had flatter curves, as shown in Fig. 3. These results indicate that the improved cytotoxicity against HL-60 cells and the relative marrow sparing effect can be achieved at high effect levels.

With the *m* and D_m values for each drug and its combination at a known concentration ratio determined from the median-effect plot, the combination index (CI) for each drug combination could be determined. The CIs for each of the drug combinations are shown graphically in Fig. 4, A–C (Chang *et al.*, 1987). The combination of 4-HC + VP-16 at the combination ratio of 1 : 0.342 demonstrated a synergistic effect against HL-60 cells when the fractional cytotoxicity was greater than 73% (Fig. 4A$_1$), yet gave an antagonistic effect

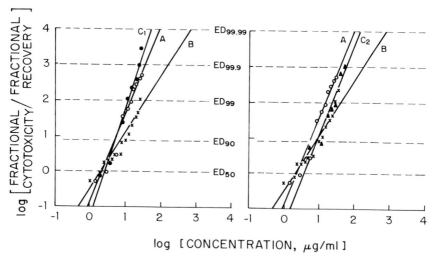

FIGURE 2 Dose–effect relationship of 4-HC and VP-16 alone and in combination against HL-60 cells. Data were analyzed by linear regression. (A) 4-HC alone, $y = 2.4629X - 0.85894$ ($r = 0.994$). (B) VP-16 alone, $y = 1.5163X - 0.4803$ ($r = 0.965$). (C$_1$) 4-HC/VP-16, 1 : 0.342; $y = 3.1215X - 1.2778$ ($r = 0.985$). (C$_2$) 4-HC: VP-16, 1 : 0.856; $y = 2.539X - 1.4483$ ($r = 0.958$). ED, effective dose (subscript, percentage of effectiveness). (Chang, T. T., *et al.*, *Cancer Res.* **45**, 2434–2439, 1985.)

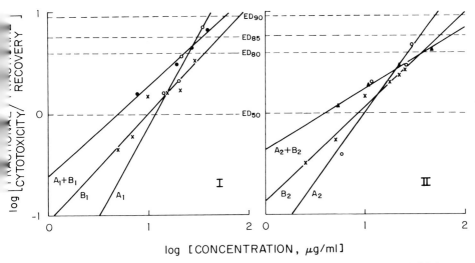

FIGURE 3   Dose–effect relationship of 4-HC and VP-16 alone and in combination of CFU-cs of normal bone marrow. Data were analyzed by linear regression. *Left:* (A₁) 4-HC alone, $y = 1.7718X - 1.8842$ ($r = 0.96$). (B₁) VP-16 alone, $y = 1.0368X - 1.0338$ ($r = 0.954$). (A₁ + B₁) 4-HC:VP-16,   1:0.342; $y = 0.8859X - 0.6092$ ($r = 0.992$). *Right:*   A₂   4-HC   alone, $y = 1.3608X - 1.3426$ ($r = 0.95$). (B₂) VP-16 alone, $y = 0.9375 \times - 0.8659$ ($r = 0.99$). (A₂ + B₂) 4-HC: Vp-16, 1:0.856; $y = 0.5883 \times - 0.3399$ ($r = 0.996$). (Chang, T. T., *et al., Cancer Res.* **45,** 2434–2439, 1985.)

against marrow (i.e., marrow sparing) when the fractional cytotoxicity was greater than 85% (Fig. 4A₂). CIs for the other combinations of 4-HC + VP-16 are given in Figs. 4B₁ and 4B₂ and demonstrate an antagonistic effect on both HL-60 cells and CFU-GM at some level of effect. The CIs for the combination of VP-16 + ADR are given in Figs. 4C₁ and 4C₂ and show that, at the ratio of 1:0.06, the drugs are antagonistic for both HL-60 cells and CFU-GM. At the combination ratio of 1:0.033, a synergistic effect against HL-60 cells was demonstrated but the cytotoxicity was still less than 3 $\log_{10}$ (Fig. 1C).

These curves suggest the following conclusions. (1) The combination of VP-16 + ADR at the ratio of 1:0.033 or 1:0.06 resulted in less than a 3-log decrease in HL-60 cells, which is considered ineffective in terms of marrow purging residual leukemia cells. (2) The combination of 4-HC + ADR at a dose ratio of 1:0.051 exhibited an antagonistic effect on HL-60 and could therefore be considered unsuitable for purging. (3) The 4-HC + ADR combination at a ratio of 1:0.034 produced a synergistic cytotoxic effect on HL-60 cells but also significantly inhibited CFU-GM. (4) The combination of 4-HC + VP-16 is optimal for purging, especially at the dose ratio of 1:0.342, because of the steep cytotoxicity curve against HL-60 and the relative marrow-sparing effect against normal marrow progenitors.

Based on these studies, the combination of 4-HC + VP-16 is undergoing clinical testing at Memorial Sloan–Kettering Cancer Center for the purging of

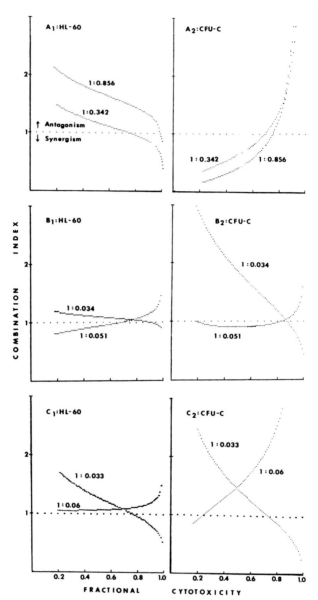

FIGURE 4   Combination effect of various drug combinations on HL-60 cell line (A_1, B_1, and C_1) and normal BM CFU-GM (A_2, B_2, and C_2). (A) 4-HC + VP-16. (B) 4-HC + ADR. (C) VP-16 + ADR. Data points above the horizontal dotted line, antagonism; data points below the horizontal dotted line, synergism. (Chang, T. T., *et al.*, *Cancer Res.* **47**, 119–122, 1987.)

marrow harvested from patients with acute myelogenous leukemia (AML) in remission prior to autoBMT. Preliminary data suggest that this may prove to be a curative procedure in certain patients (Gulati *et al.*, 1989).

IV. Selective Synergism: The Role for Data Analysis Using the Median-Effect Equation in the Design of Clinical Drug Trials

When two or more drugs are active in the treatment of a particular disease, the clinician will frequently combine each at the highest tolerable dose under the impression that "more is better." However, as demonstrated in the first portion of this chapter, high doses of AZT and rIFN-α2a when given together are far in excess of what is necessary to inhibit HIV-1 viral replication (Berman *et al.*, 1989). Moreover, such doses are extremely toxic to the bone marrow, a finding that had already been determined by preliminary clinical trials (Richman *et al.*, 1987). Based upon these data, we can suggest that a dose reduction of each drug would not compromise antiviral activity but would substantially reduce dose-limiting hematologic toxicity.

Analysis of effective drug combinations for the elimination of residual leukemia cells also yielded important information. VP-16 and ADR, each active drugs when tested as single agents, were not particularly synergistic in combination (Chang *et al.*, 1987). The combination of 4-HC + ADR at one ratio was not synergistic in killing HL-60 cells; at a combination ratio that incorporated a higher concentration of ADR, a synergistic effect was elicited but this ratio proved too toxic to normal marrow CFU-GM (Chang *et al.*, 1987). The combination of 4-HC + VP-16 was synergistic in killing HL-60 cells but relatively marrow sparing; these data have formed the basis for an ongoing clinical trial of these drugs in purging residual leukemia cells from patients with AML undergoing autoBMT.

The median-effect equation and combination index have been used in a variety of other *in vitro* model systems which have potential clinical application. Within the field of developmental hematology, for example, Paquette *et al.*, (1988) have demonstrated that the combination of two specific hematopoietic growth factors, recombinant (gibbon) interleukin-3 (IL-3) and granulocyte colony-stimulating factor (G-CSF), was strongly synergistic in a myeloid colony forming assay; the combination of IL-3 and recombinant granulocyte–macrophage CSF (GM-CSF) was less synergistic. This method of analysis also has extensive relevance in the area of oncology. Howell *et al.*, (1989) have used the median-effect equation and combination index to demonstrate that dipyridamole enhanced sensitivity to etoposide (VP-16), doxorubicin, and vinblastine in an ovarian carcinoma cell line. Ratain *et al.*

(1988) have reported that the combination of hydroxyurea and etoposide is synergistic in the murine leukemia model L1210 and have used these data as the basis for performing a phase I trial of these drugs. Kohno *et al.* (1987) have shown that sequential incubation of the PC-10 squamous lung carcinoma cell line with cisplatin followed by doxorubicin was more synergistic in a cytotoxicity assay than simultaneous exposure or sequential exposure with doxorubicin followed by cisplatin.

V. Conclusion

We have outlined two sets of experiments that have utilized the median-effect plot and combination index for data analysis and have demonstrated the model's versatility and flexibility. Additionally, we briefly discuss studies in other fields of clinical research where this model has been used to obtain important information on drug synergy and antagonism. We therefore suggest that this approach be incorporated into the clinician's methodology in the planning of clinical trials.

References

Appelbaum, E. R. (1988). Intensive chemotherapy or chemoradiotherapy with autologous marrow support as treatment for patients with solid tumors. *Hematol./Oncol. Clin. North Am.* **2,** 345–352.

Atzpodien, J., Gulati S. C., Strife, A., and Clarkson, B. D. (1987). Photoradiation models for the clinical ex vivo treatment of autologous bone marrow grafts. *Blood* **70,** 484–489.

Ball, E. D., Mills, L. E., Conghlin, C. T., Beck, J. R., and Cornwell, G. G., III (1986). Autologous bone marrow transplantation in acute myelogenous leukemia: In vitro treatment with myeloid cell-specific monoclonal antibodies. *Blood* **68,** 1311–1315.

Barbasch, A., Higby, D. J., Brass, C., Bakri, K., Karakousis, C., Pontes, J. E., Wajsman, L. Z., Beckley, S., Freeman, A., Killion, K., and Burnett, D. (1983). High dose cytoreductive therapy with autologous bone marrow transplantation in advanced malignancies. *Cancer Treat. Rep.* **67,** 143–148.

Berman, E., Duigou-Osterndorf, R., Krown, S. E., Fanucchi, M. P., Chou, J., Hirsch, M. S., Clarkson, B. D., and Chou, T.-C. (1989). Synergistic cytotoxic effect of azidothymidine and recombinant interferon alpha on normal human bone marrow progenitor cells. *Blood* **74,** 1281.

Bornemann, L. D., Spiegel, H. E., Dziewanowska, Z. E., Krown, S. E., and Colburn, W. A. (1985). Intravenous and intramuscular pharmacokinetics of recombinant leukocyte A interferon. *Eur. J. Clin. Pharmacol.* **28,** 469–471.

Chang, T. T., Gulati, S., Chou, T.-C., Vega, R., Gandola, L., Ezzat-Ibrahim, S. M., Yopp, J., Colvin, M., and Clarkson, B. D. (1985). Synergistic effect of 4-hydroperoxycyclophosphamide and etopodise on a human promyelocytic leukemic cell line (HL-60) demonstrated by computer analysis. *Cancer Res.* **45,** 2434–2439.

Chang, T. T., Gulati, S., Chou, T.-C., Colvin, M., and Clarkson, B. D. (1987). Comparative cytotoxicity of various drug combinations for human leukemic cells and normal hematopoietic precursors. *Cancer Res.* **47**, 119–122.

Chou, J., and Chou, T.-C. (1987). "Dose Effect Analysis with Microcomputers. Quantitation of ED_{50}, LD_{50}, Synergism, Antagonism, Low-Dose Risk, Receptor Ligand Binding and Enzyme Kinetics," IBM-PC series. Elsevier–Biosoft, Cambridge, England.

Chou, T.-C., and Talalay, P. (1984). Quantitative analysis of dose–effect relationships: The combined effects of multiple drugs or enzyme inhibitors. *Adv. Enzyme Regul.* **22**, 27–55.

Chou, T.-C., and Talalay, P. (1987). Applications of the median-effect principle for the assessment of low-dose risk carcinogens and for the quantitations of synergism and antagonism of chemotherapeutic agents. *In* "New Avenues in Developmental Cancer Chemotherapy" (K. R. Harrap and T. A. Connors, eds.), Bristol-Myers Symp. Ser. 8, pp. 37–64. Academic Press, Orlando, Florida.

Dicke, K. A., Jagannath, S., Spitzer, G., Poynton, C., Zander, A., Vellekoop, L., Reading, C. L., Jehn, U. W., and Tindle, S. (1984). The role of autologous bone marrow transplantation in various malignancies. *Semin. Hematol.* **21**, 109–123.

Fanucchi, M., Tong, W., Krown, S. E., Niedzwiecki, D., and Armstrong, D. (1988). Pharmacokinetic interaction of Zidovudine (ZDV) and interferon-α (IFN-α) in patients with AIDS-associated Kaposi's sarcoma. *Proc. Int. Conf. AIDS, 4th* **2**, (abstr.), p. 173.

Ferrero, D., Fabritiis, P., Amadori, S., Felice, L., Gallo, E., Meloni, G., Pregm, P., Pulsoni, A., Simone, F., Tarella, A., Pileri, A., Rovera, G., and Mandelli, F. (1987). Autologous bone marrow transplantation in acute myeloid leukemia after in-vitro purging in an anti-lacto-N-furopentaose III antibody and rabbit complement. *Leuk. Res.* **11**, 265–272.

Fischl, M. A., Richman, D. D., Grieco, M. H., Gottlieb, M. S., Volberding, P. A., Laskins, O. L., Leedom, J. M., Groopman, J. E., Mildvan, D., Schooley, R. T., Jackson, G. G., Durack, D. T., King, D., and the AZT Collaborative Working Group (1987). The efficacy of azidothymidine (AZT) in the treatment of patients with AIDS and AIDS-related complex. *N. Engl. J. Med.* **317**, 185–191.

Gee, A. P., and Boyle, M. D. P. (1988). Purging tumor cells from bone marrow by use of antibody and complement: A critical appraisal. JNCI, *J. Natl. Cancer Inst.* **80**, 153.

Golde, D. W., Bersch, N., Quan, S. G., and Lusis, A. J. (1980). Production of erythroid-potentiating activity by a human T-lymphoblast cell line. *Proc. Natl. Acad. Sci. U.S.A.* **77**, 593–597.

Gorin, N. C. (1987). Autologous bone marrow transplantation: A review of recent advances in acute leukemia. *In* "Progress in Bone Marrow Transplantation" (R. P. Gale and R. Champlin, eds.), p. 723. Liss, New York.

Groopman, J. E., Gottlieb, M. S., Goodman, J., Mitsuya, R. T., Conant, M. A., Jacob, J., Fauci, A., Masur, H., and Longo, D. (1984). Recombinant alpha-2 interferon therapy for Kaposi's sarcoma associated with acquired immune deficiency syndrome. *Ann. Intern. Med.* **100**, 671–676.

Gulati, S., Shank, B., Black, P., Yopp, J., Koziner, B., Straus, D., Filippa, D., Kempir, S., Castro-Malaspina, H., Cunningham, I., Berman, E., Coleman, M., Langleben, A., Colvin, M., Fuks, Z., O'Reilly, R., and Clarkson, B. (1988). Autologous bone marrow transplantation for patients with poor prognosis lymphoma. *J. Clin. Oncol.* **6**, 1303.

Gulati, S., Shank, B., Sarris, A., Berman, E., Gee, T., Cunningham, I., Flomenberg, N., O'Reilly, R. J., and Clarkson, B. (1989). Autologous bone marrow transplantation using 4HC + VP-16 purged bone marrow for acute nonlymphocytic leukemia. *Bone Marrow Transplant.* **4**, 116.

Gutterman, J. U., Fine, S., Quesada, J., Horning, S. J., Levine, J. F., Alexanian, R., Bernhardt, L., Kramer, M., Spiegel, H., Colburn, W., Trown, P., Merigan, T., and Dziewanowska, Z. (1982). Recombinant leukocyte A interferon: Pharmacokinetics, single-dose tolerance, and biologic effects in cancer patients. *Ann. Intern. Med.* **96**, 549–556.

Hartshorn, K. L., Vogt, M. W., Chou, T.-C., Blumberg, R. S., Byington, R., Schooley, R. T., and Hirsch, M. S. (1987). Synergistic inhibition of human immunodeficiency virus *in vitro* by azidothymidine and recombinant alpha A interferon. *Antimicrob. Agents Chemother.* **31**, 168–172.

Howell, S. B., Hom, D., Sanga, R., Vick, J. S., and Abramson, I. S. (1989). Comparison of the synergistic potentiation of etoposide, doxorubicin, and vinblastine cytotoxicity by dipyridamole. *Cancer Res.* **49**, 3178.

Hurd, D. D., LeBien, T. W., and Lasky, L. C. (1988). Autologous bone marrow transplantation in non-Hodgkin's lymphoma: Monoclonal antibodies plus complement for ex vivo marrow treatment. *Am. J. Med.* **85**, 829.

Kaizer, H., Stuart, R. K., Brookmeyer, R., Beschorner, W. E., Braine, H. G., Burns, W. H., Fuller, D. J., Korbling, M., Mangan, K. F., Saral, R., Sensenbrenner, L., Shadduck, R. K., Shende, A. C., Tutschka, P. J., Yeager, A. M., Zinkham, W. H., Colvin, O. M., and Santos, G. W. (1985). Autologous bone marrow transplantation in acute leukemia: A phase I study of in vitro treatment of marrow with 4-hydroperoxycyclophosphamide to purge tumor cells. *Blood* **65**, 1504–1510.

Kohno, N., Ohnuma, T., Holland, J. F., and Greenspan, E. M. (1987). Combination chemotherapy of solid tumor in vitro based on differential drug penetration and inoculum effects. *Am. Assoc. Clin. Res.* p. 7816 (abstr.).

Loewe, S. (1953). The problem of synergism and antagonism of combined drugs. *Arzneim.-Forsch.* **3**, 285.

Morecki, S., Margel, S., and Slavin, S. (1988). Removal of breast cancer cells by soybean agglutinin in an experimental model for purging human marrow. *Cancer Res.* **48**, 4573–4577.

Paquette, R. L., Zhou, J.-Y., Yang, Y.-C., Clark, S. C., and Koeffler, H. P. (1988). Recombinant gibbon interleukin-3 acts synergistically with recombinant human G-CSF and GM-CSF in vitro. *Blood* **71**, 1596.

Ratain, M. J., Schilsky, R. L., Wojack, B. R., Simon, T., Sanekjian, E. K., and Vogelzang, N. J. (1988). Hydroxyurea and etoposide: *In vitro* synergy and Phase I clinical trial. *JNCI, J. Natl. Cancer Inst.* **80**, 1410.

Real, F. X., Oettgen, H. F., and Krown, S. E. (1986). Kaposi's sarcoma and the acquired immunodeficiency syndrome: Treatment with high and low doses of recombinant leukocyte A interferon. *J. Clin. Oncol.* **4**, 544–551.

Richman, D. D., Fischl, M. A., Grieco, M. H., Gottlieb, M. S., Volberding, P. A., Laskin, O. L., Leedom, J. M., Groopman, J. E., Mildvan, D., Hirsch, M. S., Jackson, G. G., Durack, D. T., Nusinoff-Lehrman, S., and the AZT Collaborative Working Group (1987). The toxicity of azidothymidine (AZT) in the treatment of patients with AIDS and AIDS-related complex. *N. Engl. J. Med.* **317**, 192–197.

Ruprecht, R., O'Brien, L. G., Rossoni, L. D., and Nusinoff-Lehrman, S. (1986). Suppression of mouse viraemia and retroviral disease by 3'-azido-3'-deoxythymidine. *Nature (London)* **322**, 467.

Ruprecht, R., O'Brien, L., Rosas, D., and Anderson, J. (1987). Recombinant human interferon A/D (rHuIFN-αA/D) enhances the antiretroviral effect of 3'-azido-3'-deoxythymidine (AZT) in mice. *Proc. Am. Assoc. Cancer Res.* **28**, 456.

Ruprecht, R., Gama-Sosa, M., Mullaney, S., and Rosas, D. (1988). Post-exposure therapy with AZT and interferon prevents retroviral infection. *Proc. Int. Conf. AIDS, 4th* **2**, 174.

Sieber, F. (1987). Elimination of residual tumor cells from autologous bone marrow grafts by dye-mediated photolysis: Preclinical data. *Photochem. Photobiol.* **46**, 71–76.

Strife, A., Lambek, C., Wisniewski, D., Gulatis, Gasson, J. C., Golde, D. W., Weltek, Gabrilove, J. L., and Clarkson, B. (1987). Activities of four purified growth factors on highly enriched human hematopoietic progenitor cells. *Blood* **69**, 1508.

Takvorian, T., Canellos, G. P., Ritz, J., Freedman, A. S., Anderson, K. C., Mauch, P., Tarbell, N., Coral, F., Daley, H., Yeap, B., Schlossman, S. F., and Nadler, L. M. (1987).

Prolonged disease-free survival after autologous bone marrow transplantation in patients with non-Hodgkin's lymphoma with a poor prognosis. *N. Engl. J. Med.* **316**, 1499.

Webb, J. L. (1963). Effect of more than one inhibitor. *In* "Enzyme and Metabolic Inhibitors" (R. M. Hochster, M. Kates, and J. H. Quasteo, eds.), Vol. 1, p. 66. Academic Press, New York.

Wolff, S. N., Fer, M. F., McKay, C. M., Hande, K. R., Hainsworth, J. D., and Greco, F. A. (1983). High dose VP-16-213 and autologous bone marrow transplantation for refractory malignancies: A phase I study. *J. Clin. Oncol.* **1**, 701–705.

Yarchoan, R., Weinhold, K. J., Lyerly, H. K., Gelmann, E., Blum, R. M., Shearer, G. M., Mitsuya, H., Collins, J. M., Myers, C. E., Klecker, R. W., Markham, P. D., Durack, D. T., Lehrman, S. N., Barry, D. W., Fischl, M. A., Gallo, R. C., Bolognesi, D. P., and Broder, S. (1986). Administration of 3'-azido-3'-deoxythimidine, an inhibitor of HTLV-III/LAV replication, to patients with AIDS or AIDS-related complex. *Lancet* **1**, 575–580.

Yeager, A. M., Kaizer, H., Santos, G., Saral, R., Colvin, O. M., Stuart, R. K., Braine, H. G., Burke, P. J., Ambinder, R. F., Burns, W. H., Fuller, D. S., Davis, S. M., Karp, J. E., Stratford, M. W., and Wingard, J. R. (1986). Autologous bone marrow transplantation in patients with acute nonlymphocytic leukemia using ex-vivo marrow treatment with 4-hydroxycyclophosphamide. *N. Engl. J. Med.* **315**, 141.

Index

BERTINO